A Technical History of America's Nuclear Arms
Volume I – Introduction and Weapon Systems Through 1960

Dr. Peter A. Goetz

Cover Credits
A B61 Tactical Thermonuclear Bomb's Components (Credit DOE)

Copyright © by Peter A. Goetz, 2018
Copyright is registered.

All Rights Reserved.

The use of any part of this publication, reproduced, transmitted in any form, or by any means electronic, mechanical, printing, photocopying, recording, or otherwise, is an infringement of the copyright law.

TABLE OF CONTENTS

FOREWORD..6

    CHAPTER 1 NUCLEAR PHYSICS, WEAPON DESIGN, AND TESTING......................12

    CHAPTER 2 PRODUCING AND FINISHING "SPECIAL MATERIALS"......................41

    CHAPTER 3 STOCKPILE LOGISTICS..74

    CHAPTER 4 INTELLIGENCE, TARGETING, AND EARLY WARNING....................102

    CHAPTER 5 THE EFFECTS OF NUCLEAR WEAPONS AND NUCLEAR WAR.........128

    CHAPTER 6 INITIAL DEPLOYMENTS: NUCLEAR FORCES AND STRATEGIC DOCTRINE FROM 1944 – 1968..146

    CHAPTER 7 DEFEATING THE EVIL EMPIRE: NUCLEAR FORCES AND STRATEGIC DOCTRINE FROM 1969 – 1991..170

    CHAPTER 8 PEACE AND WAR 1992 - 2016: AMERICAN ARMS REDUCTION AND REARMAMENT..186

    CHAPTER 9 THE COSTS, BENEFITS, AND CONSEQUENCES OF NUCLEAR DETERRENCE..210

    CHAPTER 10 NON-PROLIFERATION AND THE MANAGEMENT OF THE US NUCLEAR STOCKPILE..226

    CHAPTER 11 BIRTH OF THE ARSENAL: THE CLASSIC FAT MAN ATOMIC BOMB..235

    CHAPTER 12 LIGHTWEIGHT STRATEGIC WARHEADS: ATOMIC WEAPONS IN TRANSITION..271

    CHAPTER 13 THE MK / W7 TACTICAL ATOMIC WARHEAD..............................292

    CHAPTER 14 GUN-ASSEMBLED FISSION BOMBS AND "ATOMIC ANNIE"..........324

    CHAPTER 15 "BOUNCING THE RUBBLE HIGHER" WITH HYDROGEN BOMBS....342

    CHAPTER 16 BOOSTED WARHEADS, SEALED PITS, ELECTRONIC NEUTRON INITIATORS, ADVANCED FIRING-SETS, AND DERIVATIVE WEAPON SYSTEMS...377

    CHAPTER 17 HUNTER-KILLERS WITH NUCLEAR CLAWS..................................402

    CHAPTER 18 BERYLLIUM TAMPERS, LINEAR IMPLOSION, PLASTIC BONDED MINI-NUKES, AND DERIVATIVE WEAPON SYSTEMS..420

- SELECTED BIBLIOGRAPHY……………………………………………………….455
  - NUCLEAR WEAPONS…………………………………………………………...455
  - NUCLEAR DELIVERY SYSTEMS / TECHNOLOGY……………………………477
  - NUCLEAR POLICY / ORGANIZATION………………………………………….495
- MUSEUMS / DISPLAYS……………………………………………………………..508
- ACRONYMS…………………………………...…………………………………….510
- APPENDICES………………………………………………………………………...534
  - APPENDIX 1 ORDER OF BATTLE – UNITED STATES NUCLEAR FORCES………...534
  - APPENDIX 2 US STRATEGIC WAR PLANS……………………………………………..536
  - APPENDIX 3 CURRENT AND HISTORIC DELIVERY SYSTEMS OF THE STRATEGIC TRIAD AIR FORCE STRATEGIC BOMBERS……………………………………….538
  - APPENDIX 4 HISTORICALLY DEPLOYED NUCLEAR BOMBS AND WARHEADS..552
  - APPENDIX 5 US NUCLEAR TEST PROGRAM……………………………………………569
  - APPENDIX 6 BASIC PROPERTIES OF NUCLEAR MATERIALS………………………..571
  - APPENDIX 7 CHEMICAL AND NUCLEAR REACTIONS……………………………….578
  - APPENDIX 8 YIELDS OF NUCLEAR MATERIALS……………………………………..579
  - APPENDIX 9 TABLES OF NUCLEAR EFFECTS…………………………………………580

FOREWORD

Since 1945, the United States Armed Forces have fielded 69 different nuclear and thermonuclear devices on approximately 120 weapon systems. During this period, the manufacturing operations of the Manhattan Project, the Atomic Energy Commission, and the Department of Energy produced 70,300 nuclear bombs and warheads. Most of these are now retired. The Air Force and the Navy currently deploy seven types of thermonuclear device on nine weapon systems and the Department of Defense maintains about 2,500 bombs and warheads (tactical and strategic) on active duty. It keeps approximately the same number in reserve. These numbers are in decline.

The United States developed atomic bombs to counter a threat that Germany might develop them first and then use them to win World War II. Although the United States perfected atomic bombs too late to see service during the war's European Campaign, the US Army Air Force dropped them on two Japanese cities to help win the Pacific Campaign. With the advent of peace, the United States disarmed unilaterally. Personnel from the American Armed Services returned to their civilian occupations and manufacturing concerns once again focused on the production of consumer goods. The Armed Services scrapped a large amount of war materiel and the Manhattan Project put its nuclear programs on hold. In order to restore the war-ravaged economies of Europe and Asia, the United States formulated the Truman Doctrine to shield Eastern Europe from Soviet aggression and initiated the European Recovery Project (Marshall Plan) and an Asian Plan to provide cash and technical assistance. The United States hoped to build a sound world order that had no further need for aggression.

Unfortunately, Russian Dictator Joseph Stalin stood in opposition to American plans for world peace because he wished to draw Europe and Asia into the Soviet sphere of influence. Threatened by American economic intentions and concerned that the West might respond to his designs in a military manner, Stalin instituted a post war expansion of Soviet armed forces. In parallel with this action, he ordered his spy agencies, the KGB and the GRU, to conduct espionage operations that eventually obtained the plans for America's atomic bomb. The collapse of a first minister meeting over German reunification in 1947 marked the end of a less than enthusiastic Soviet cooperation with the West and the start of the Cold War.

The Soviet Union exploded a homemade copy of America's atomic bomb in September 1949, the same month that Mao Tse Tung's Communists Forces declared victory over Chang Kai-Shek's Nationalist Forces in China. The following June, Communist North Korea invaded the partitioned southern half of the Korean Peninsula. Asian nations were seen to be "falling like dominoes" to Communist aggression, a situation that caused the United Nations to send military reinforcements to aid in the defense of South Korea. The dire circumstances in which United Nations and American expeditionary forces soon found themselves brought the realization that atomic weapons might be required to end the conflict. The entry of China into the war in 1951 brought the further realization that the Korean "Police Action" might escalate into a global conflict.

In order to maintain American ascendancy in the world order and offset Soviet atomic developments, President Truman demanded that the Atomic Energy Commission begin work on a thermonuclear or "Super" Bomb in 1950. In both eastern and western hemispheres, proliferation ran unchecked as the armed forces of nuclear capable nations called for advanced weapons and improved delivery systems. Air Forces moved nuclear tipped missiles into hardened underground silos and Navies placed them in submarines to hide them in the ocean's depths. In the United States, these activities resulted in the creation of a "Nuclear Triad" that consisted of strategic bombers, ICBMs, and SLBMs; each weapon system with its own strengths and capabilities. During the 1962 Cuban Missile Crisis, at the high-water mark of the Cold War, American combat forces had more than 30,000 nuclear weapons available for immediate use.

The worldwide buildup of nuclear forces so alarmed President Richard Nixon that he visited Russia and China in 1972 to negotiate détente. There followed the *Strategic Arms Limitation Treaty* (*SALT I*) that restricted the further deployment of strategic launchers. The *Anti-Ballistic Missile* (*ABM*) *Treaty*, also signed at this time, limited defensive weapons. Under a policy of "Mutually Assured Destruction," both sides in the Cold War left themselves vulnerable to counterattacks, reducing the possibility of an initial surprise attack. Nevertheless, the numerous tactical nuclear weapons deployed in Europe remained a destabilizing factor. Because target distances in this region were minimal, political leaders had insufficient time to assess a potential threat in the event of an alert. A very real possibility existed that an all-out nuclear war might arise from a false alarm.

Carl Sagan published the consequences of such a war in a 1983 study entitled: *Global Atmospheric Consequences of Nuclear Winter*. The main thesis of this study stated that dust and smoke particles arising from a major nuclear exchange and subsequent fires would cause much of the earth's surface to drop below freezing – resulting in massive crop failure. The extinction of the human race from such an event was a very real possibility. In the previous year, the World Health Organization had estimated that a 10,000-megaton nuclear exchange would cause two billion casualties. There was great relief for everyone when the *1987 Intermediate Nuclear Forces Treaty* (*INF*) banned the deployment of tactical nuclear weapons in Europe.

The signing of the *Strategic Arms Reduction Treaty* (*START I*) and the collapse of the Soviet Union in 1991 allowed further reductions to the nuclear stockpile. Apart from submarine launched cruise and ballistic missiles, the American Navy relinquished all of its nuclear weapons in 1992. The Army followed suit by retiring its nuclear artillery shells and demolition munitions in 1994, completely divesting itself of any nuclear capability. These developments were part of an ongoing trend. The Air Force had already phased out the majority of its large, strategic bombs during the era of détente and now reduced the number of SAC's ICBMs. Conventional antiaircraft missiles had completely replaced their nuclear predecessors in 1984 and the Air Force had removed the last nuclear air-to-air missile from its inventory in 1986. As a means to prevent the blurring of the line between conventional and nuclear weapons, Congress outlawed the development of weapons with yields smaller than 5 kilotons by passing the *Spratt-Furse Amendment* in 1994.

For most of the last decade, the *Strategic Offensive Reductions Treaty* (*SORT*), signed in Moscow on May 24, 2002, has governed the level of readily available nuclear weapons in the stockpile. The *Moscow Treaty* acknowledged that the *START II Treaty* was still in force and called for further reductions to a level of 1,700 - 2,200 strategic nuclear warheads by December 31, 2012. The treaty allowed each signatory to determine the composition and structure of its strategic offensive arms based on the established aggregate limit for the number of warheads. In the United States, the Bush Administration regarded the nuclear arsenal that became the subject of this treaty as outdated. *The 2002 Nuclear Posture Review* (*NPR*), stated that the existing stockpile, developed to defeat the former Soviet Union by means of the *Single Integrated Operational Plan* (*SIOP*), was not suitable for the future.

As a response to tensions between Pakistan and India, Chinese ambitions for East Asian hegemony; Russian ambitions for Eurasian hegemony; and the actions of "rogue" states like Iran and North Korea, the United States established advanced warhead design teams at its three national weapons laboratories – Los Alamos, Sandia, and Lawrence Livermore. The Bush Administration directed these laboratories to focus on evolving Department of Defense requirements for weapons with improved accuracy and selectable yields to defeat hardened and buried targets. It also directed the development of agent defeat weapons for attacking chemical and biological warfare sites. Congress next lifted the *Spratt-Furse* ban on small warheads in 2003 to permit the development of a bunker buster called the "Robust Nuclear Earth Penetrator" (RNEP). Selected for production in early 2007, a design for the new "Reliable Replacement Nuclear Warhead" (RRW) is currently on hold.

Looking to the future, the *New START Treaty* will govern the composition of the nuclear arsenals of the United States and the Russian Federation (former Soviet Union). President Obama and Premier Medvedev signed this agreement on April 8, 2010, and it came into force on February 5, 2011. Before this, in an April 2009 speech in Prague, Czechoslovakia, President Obama highlighted the dangers of nuclear confrontation, declaring that to overcome this growing threat, the United States had to "seek the peace and security of a world without nuclear weapons." With their ratification of The *New START Treaty*, the United States and Russia must reduce their strategic nuclear arsenals by 2018. As with *SORT*, each signatory has the flexibility to determine the structure of its strategic forces within the limits set by the Treaty. American negotiators based these limits on an analysis conducted by DOD planners in support of the *2010 Nuclear Posture Review*. They are 74 percent lower than the *1991 START Treaty*'s limits and 30 percent lower than the deployed strategic warhead limits of the *2002 SORT Treaty*.

The *New START Treaty* limits each side to no more than 1,550 deployed bombs and warheads. All of the warheads deployed on ICBMs and SLBMs count toward this limit and each nuclear capable deployed heavy bomber counts as one bomb. Each side is also limited to a combined to total of 800 deployed and non-deployed ICBM launchers, SLBM launchers, and nuclear capable heavy bombers. The signatories are further limited to 700 deployed ICBMs, SLBMs, and nuclear capable heavy bombers. Verification measures under the Treaty include onsite inspections, data exchanges, notifications related to the strategic offensive arms and facilities covered by the Treaty, and provisions to facilitate the use of national technical means for monitoring. To increase confidence and transparency, the *New START Treaty* provides for the exchange of telemetry.

The Treaty's duration is ten years unless the signatories agree to a subsequent amendment. They may agree to extend the Treaty for a period of no more than five years. The Treaty includes a withdrawal clause that is standard in arms control agreements. The *2002 Moscow Treaty* terminated when the *New START Treaty* entered force. The *New START Treaty* does not limit the testing, development, or deployment of current or planned US missile defense programs or long-range conventional strike capabilities.

The Author grew up in the "duck and cover" era and has witnessed many changes to the American nuclear deterrent. He wrote this book to satisfy his curiosity about the weapons that formed (and still form) this deterrent. In addition to satisfying their curiosity about these weapons, readers can use the information in this book as a handy tool to better understand the events of the Cold War, assess the capability of new warheads and delivery systems (both domestic and foreign), evaluate worldwide developments in this field, and estimate the probability of their use.

Ordinarily, a prominent working specialist in the field would write a review such as the one presented here. In the field of nuclear weapons, however, knowledgeable military and civilian specialists have all signed non-disclosure agreements with the government as a condition of their employment. This legally prohibits them from revealing their knowledge. A comprehensive discussion of nuclear weapons technology therefore remains in the domain of nuclear historians and non-governmental researchers.

A forbidden topic for many years, the open discussion of nuclear weapons only became possible through the actions of Howard Morland and the Editorial Staff of *The Progressive Magazine*. In 1979, these individuals successfully challenged the United States Government when it attempted to restrict publication of an article detailing the inner workings of the hydrogen bomb. Since fundamental scientific principles govern the processes in nuclear weapons, secrecy has been of little use in preventing the proliferation of weapons technology. University science curriculums teach many weapons related topics such as radiation transport, shock wave dynamics, and plasma physics. Morland's victory set a legal precedent for all subsequent writers, including Richard Rhodes who was awarded the Pulitzer Prize for his book: *The Making of the Atomic Bomb* and John Coster-Mullens who recently released: *Atom Bombs,* a book that contains detailed technical descriptions of the two nuclear bombs exploded over Japan. Chuck Hansen, another prominent Author and nuclear researcher, pioneered the *Freedom of Information Act* in obtaining government documents for an epic 2,900-page digital book: *The Swords of Armageddon,* and Hans Kristensen, the Director of the Nuclear Information Project at the Federation of American Scientists has been tireless in his efforts to bring US nuclear policy and developments to light.

Because of the efforts by these and other individuals, a large amount of public domain material is now available in books, articles, and on the World Wide Web. In particular, the Federation of American Scientists maintains an online library of digital reports produced by Los Alamos National Laboratory. Carey Sublette and Gregory Walker, who had taken the time to download many of the laboratory's reports, provided them to the FAS after Los Alamos blocked public access to their online library in 2002. This act was a response to the Al Qaeda terrorist attacks on New York and Washington. Another site, *The Nuclear Weapons Archive*, contains a wide range of pertinent information, including annotated lists of American nuclear warheads and related tests. It also hosts Carey Sublette's: *Nuclear Weapons Frequently Asked Questions*, an unfinished but very serviceable exposition on basic nuclear principles and engineering. Thomas Cochran, William Arkin, and Milton Hoenig produced a set of *Nuclear Weapons Databooks* covering most of the world's nuclear weapons for the Natural Resources Defense Council during the period 1984 - 1989, and Stephen Schwartz at the Brookings Institution edited a "cost analysis" of the entire nuclear program in 1989.

More declassified information is available at such sites as the National Nuclear Security Agency's Virtual Reading Room, the Defense Threat Reduction Agency, Air Force Historical Studies, and the National Atomic Museum, among others. The information used in the preparation of this book comes from hundreds of primary and secondary (public) sources such as these. It also comes from visits to many military museums where the Author was able to examine and measure nuclear weapon systems first hand. The Author has not used any classified or restricted material in the writing of this book, nor has he worked in the nuclear industry. The

Author sent preprints of this book to the NNSA, DOE, DOD, and the NSC to ensure that no classified information would be revealed.

Nuclear publications prior to the recently published e-book, *A Technical History of America's Nuclear Weapons – Their Design, Operation, Deployment and Delivery,* have concentrated mostly on specific topics such as weapon physics, testing, warheads, delivery systems, deployments, or doctrine. *The e-book* was intended to provide "one stop shopping" with a broad treatment of the subject. In providing a wide scope, some depth was sacrificed in order to produce a volume of manageable size. In republishing the *A Technical History of America's Nuclear Weapons – Their Design, Operation, Deployment and Delivery* e-book as a paperback, it was necessary to divide it into two volumes due to limitations of the printing process. While unfortunate that the book had to be published in two volumes, it allowed the author to lavishly illustrate each chapter of the book.

For the convenience of the Reader, Nuclear publications prior to the e-book, *A Technical History of America's Nuclear Weapons – Their Design, Operation, Deployment, and Delivery,* have concentrated mostly on specific topics such as weapon physics, testing, warheads, delivery systems, deployments, or doctrine. *A Technical History of America's Nuclear Weapons* was intended to provide "one stop shopping" with a broad treatment of the subject. In providing a wide scope, some depth was sacrificed in order to produce a volume of manageable size. In republishing the e-book as a paperback, it was necessary to divide it into two volumes due to limitations of the printing process. While unfortunate that the book had to be published in two volumes, it allowed the author to lavishly illustrate each chapter of the book.

A Technical History of America's Nuclear Weapons: Volume II – Weapon Systems From 1960 to the Present takes up where *Volume I* ends. *Volume II* is written in the same straightforward, easy to understand, manner as *Volume I*. American nuclear weapons and delivery systems are presented in a rough chronological order with some weapons treated individually and others in functional or family groupings. Like *Volume I*, *Volume II* combines development histories with engineering descriptions to illustrate the performance characteristics of each weapon described and the design challenges that faced their developers. Basic data about weapon operation, delivery systems, and deployments are also included.

Volume II:
- Has about 500 technical references, grouped into related categories
- Uses official Military Characteristic (parts) Numbers for components where available, a very useful tool for internet searches
- Discusses second generation multi-megaton hydrogen bombs including the three-stage, 25-megaton MK 41, America's most powerful weapon
- Outlines the evolution of jet-powered, medium and heavy, strategic nuclear bombers
- Discusses the continued development of tactical nuclear bombs and their delivery systems
- Explains the mechanism of Dial-a-Yield, used in the B61 and B83 bombs,
- Follows the evolution of liquid and solid fuel ICBMs, which now form the core of the nuclear triad – based in silos and on-board Fleet Ballistic Missile Submarines (FBMSs).

- Describes the evolution of Multiple Independently-targetable Reentry Vehicles (MIRVs) that allowed Moscow to be targeted with four hundred thermonuclear warheads
- Takes up the development of tactical nuclear missiles where Volume I left off
- Describes the development of Enhanced Radiation or "neutron" weapons to minimize casualties in the European Theater
- Outlines how tactical nuclear missiles and nuclear artillery were combined to provide "Terrain Fire", in which thousands of overlapping nuclear bursts to depths of 50 or 100 kilometers along an adversary's border would annihilate its forward forces leaving millions of dead
- Follows the development air-to-air and surface-to-air missiles, including the specially tailored W61 warhead, which used gold to produce extremely high intensity radiation for use in space

The units of measurement in this book are mixed. Rather than create confusion by converting metric and imperial dimensions and rounding off the results, the book restates measurements in their original (?) units. Specific measurements are frequently in conflict from different sources since many sources are secondary in nature. Where possible, the Author has verified the measurements and specifications of weapon systems from museum displays and declassified documents obtained from military sources and contractors. Even these show a degree of variation in specifications and statistics.

In some cases, the Author has made an educated guess regarding which of several conflicting specifications is correct and in others cases he provides a range of estimates. All measurements and statistics are therefore approximations. To improve the flow of the narrative, the Author has minimized the use of qualifying adverbs and adjectives. For the convenience of the Reader, a number of appendices supply a best estimate of the technical specifications and production statistics for selected weapons. The book also contains a table that outlines the effects of nuclear explosions over a wide range of yields.

Preprints of this paperback were sent to the NNSA, DOE, and DOD for review, thus, the Author can guarantee that the men-in-black will never come to your home if you purchase this book. He cannot guarantee, however, that the ballistic missile submarine that a Reader constructs in a backyard pool will not implode before it reaches its specified collapse depth or that a thermonuclear bomb reverse engineered as the centerpiece display for a day of national celebration will not be a dud.

CHAPTER 1

NUCLEAR PHYSICS, WEAPON DESIGN, AND TESTING

In 1896, Henri Becquerel developed some photographic plates that he had stored in a drawer, along with a container of uranium salt. To his surprise, he found that some sort of invisible "rays," emitted by the salt, had fogged his plates. Marie and Pierre Curie, who went on to make a lifetime study of this phenomenon, subsequently called the phenomenon "radioactivity." They divided radioactivity into alpha, beta, and gamma rays (x-rays) based on its interactions with magnetic fields. As it turned out, two of these rays are actually particles. Alpha rays are helium nuclei composed of two protons and two neutrons, beta rays are electrons produced during the spontaneous disintegration of a neutron, and gamma rays are highly energetic photons of electromagnetic energy.

At the time of radioactivity or radiation's discovery, scientists knew very little about the atom from which it emanated. The realization that this tiny bit of matter was actually a collection of smaller sub-atomic particles did not come until a year later. J. J. Cavendish, while making a study of electrical discharges in gases, discovered the negatively charged "electron" in 1897. To his surprise, Cavendish found that the newly discovered particle possessed only a small fraction of the total mass of a hydrogen atom, which he knew to contain both negative and positive "unit" charges. This suggested that the carrier of the positive charge was quite massive in comparison.

Following World War I, Ernest Rutherford extended the work of Cavendish by conducting experiments in which he fired alpha particles into containers of nitrogen gas. This procedure created scintillations in a detector that looked very much like the response exhibited by hydrogen atoms that had been "ionized" or stripped of their single electron. He correctly concluded that nitrogen atoms were disintegrating after impacts with alpha particles and emitting elementary positive particles that were synonymous with hydrogen ions. Rutherford named these particles "protons."

The last step in the discovery of the three basic atomic building blocks followed a prediction made by Rutherford in 1920. He theorized the existence of a neutral particle or "neutron," in which a proton was bound directly to an electron. Despite ten years of research, Rutherford could find no evidence for this particle. Then, in 1930, German physicists Hans Bothe and Herbert Becker announced that alpha particles fired into a beryllium target produced a neutral radiation able to penetrate 20 centimeters of lead. Rutherford's assistant, James Chadwick, took up the study of this neutral radiation and in 1932 he determined that it was composed of particles with a mass similar to that of the proton. Chadwick had finally confirmed the existence of the long sought-after neutron!

With this discovery, Rutherford was able to complete a model for the basic structure of the atom. In earlier studies, he had directed Hans Geiger and Ernst Madsen to bombard a variety of metal targets with alpha particles. To their astonishment, the assistants found that some of the alpha particles bounced back from a sheet of gold foil. The only way for this to have happened was if the protons in the gold atoms were concentrated in a very small volume, a structure to

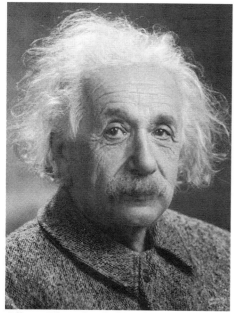 

Albert Einstein (left) who conceived of the equivalence of matter and energy and Leo Szilard (right) who conceived and patented the concept of a nuclear chain reaction – the basis for the atomic bomb. (Left Credit: Orren Jack Turner, Right Credit: Cropped DOE Photo; both via Wikimedia Commons)

The Los Alamos Scientific Laboratory, New Mexico, in the early 1950s. Los Alamos was America's first nuclear weapons laboratory. It created the Little Boy and Fat Man bombs used in WW II. (Credit: LANL)

The "Gadget" – Los Alamos' first nuclear device – wired up and ready to go at the top of its test tower. The x-unit is located in the cone attached on the left. The radar and barometric fuzing units attached on the right, but were not required for the test. (Credit: LANL Photo TR 297)

J. Robert Oppenheimer (left), the Scientific Director of the Manhattan Project and the first Director of the Los Alamos Scientific Laboratory, and General Leslie Groves (right), Military Director of the Manhattan Project, stand at the Trinity Test Site after the 1945 test. (Credit: USACE)

Edward Teller (left) and Stanislaw Ulam (right), two important scientists at LASL during its early years. Together, they developed the concepts needed to build the hydrogen bomb. (Right Credit: LLNL, Left Credit: LANL; both via Wikimedia Commons)

The University of California Lawrence Radiation Laboratory, California, as it looked when it opened in 1953. The AEC wanted two laboratories competing against each other to encourage weapons development. (Credit: LLNL, cropped)

Herbert York (left), LRL's first Director and Ernest Lawrence (right), for whom it was named. Lawrence developed the Calutron, which the Oak Ridge facility used to enrich uranium. (Left Credit: NNSA, Right Credit: US Government, both via Wikimedia Commons)

A 500-foot-high shot tower at the NTS with a shot cab at the top. The towers were completely vaporized if their tests were successful. Placing a nuclear device on top of such a tower reduced fallout and allowed instruments a clear view of the explosion. (Credit: NNSA)

A nuclear shot cab hangs from a tethered balloon. This test method was considerably cheaper than building towers, and it allowed nuclear devices to be detonated at higher altitudes. (Credit: NNSA)

Setting up for an underground shot at the Nevada Test Site. Note the numerous diagnostic vehicles, cabling, and the adjacent subsidence craters from previous tests. Underground tests eliminated fallout. (Credit: NNSA / DOE Photo NF1679; Color Original)

The Shrimp shot cab located on a sand spit at the Pacific Proving Grounds in preparation for the Bravo thermonuclear test. Diagnostic vacuum pipes (background) crossed a 1.4-mile gap to a data bunker. The towers to the right (foreground) reflected light from different locations around the bomb to streak cameras in an instrument bunker located some miles away. Testing large nuclear devices in the Pacific eliminated fallout in the US. (Credit: LANL Photo D-55-8; Color Original)

which Rutherford gave the name "nucleus." He further concluded that electrons formed distant orbits around the nucleus. Rutherford correctly placed neutrons into the nucleus, along with the proton.

Rutherford's model helped to make sense of the "periodic table of the elements," which Dmitri Mendeleyev created between 1869 and 1871. Mendeleyev ordered the known chemical elements according to their atomic weights and produced a table that exhibited a periodicity of chemical properties. Once the nature of the atom was established, Rutherford could demonstrate that atomic weight was basically the sum of the neutrons and protons contained in the nucleus of an element. However, in a departure from Mendeleyev's reasoning, it was atomic number "Z" or the number of protons residing in an element's nucleus that was the cause of chemical periodicity. The protons attracted an equal number of electrons that formed shells around the nucleus and established bonds with other atoms to create "molecules." The electrons were held in place by an "electromagnetic force," while protons and neutrons were bound together by the "strong nuclear force."

The periodic table also showed that heavier elements required increasingly more neutrons to stabilize their nuclei. Even with the extra neutrons, the nuclei of some elements remained out of balance. The radioactive emission of photons and particles was their way of adjusting to a more comfortable energy state. Upon later investigation of radioactive processes, Sheldon Glashow and Abdus Salam discovered that a "weak nuclear force" controls neutron stability and the phenomenon of beta particle emission. Together with "gravity", the electromagnetic, strong, and weak nuclear forces constitute the four basic forces found in nature.

Studies of radioactive elements soon brought new discoveries. Rutherford and Frederick Soddy found that thorium gave off an inert gas that turned out to be radon. They had observed the spontaneous activity of a radioactive element, which encouraged them to identify the products of such elements as they changed their nature by radiating away alpha and beta particles. Loss of an alpha particle meant transmutation into the element two protons down the periodic table whereas the loss of a beta particle meant a change to the element one proton up.

Rutherford and Soddy further discovered that each radioactive substance had a radioactive "half-life." This was the time required for an element to reduce its radioactive intensity by half. Radioactive half-lives measure the rate at which particle emissions transmute radioactive elements into new elements. These half-lives differ for each element and for "isotopes" – variations of elements that differ by the number of neutrons contained in their nuclei. For example, U238 and U235, which are the most abundant isotopes of uranium, are the same element because both isotopes contain 92 protons. They differ isotopically because U235 contains 143 neutrons and U238 contains 146. The half-lives of radioactive isotopes range from billionths of a second to billions of years.

In 1938 came a momentous discovery. Otto Hahn wrote a letter to Lise Meitner informing her that he had found barium in purified samples of uranium that he had subjected to neutron bombardment. He was mystified how this could have happened since it was inconceivable that neutron bombardment could have knocked off enough particles from uranium atoms to form the lighter element. In subsequent discussions with her cousin, Otto Frisch, came the realization that some of the uranium nuclei had absorbed bombarding neutrons, grown unstable, and disintegrated. The barium that Hahn discovered was a byproduct of nuclear disintegration. In March 1939, Leo Szilard and Walter Zinn demonstrated that the process of atomic disintegration also released a number of neutrons.

Hahn called the newly discovered phenomenon "fission" after the biological process of cell division. It was only possible because neutrons are electrically inert and can easily slip into the nucleus of an atom, with which they can then react. Atomic fission or disintegration usually produces two daughter products, with one fragment approximately two-thirds the weight of the other. The energy released by fission amounts to approximately 200 million electron volts (MeV). The electron volt or eV is an extremely small unit of energy often used as a proxy for velocity or temperature. The 200 MeV released by a fission event is sufficient energy to make the movement in a grain of sand just visible to the human eye.

For U235 (total mean fission energy of 202.5 MeV per disintegration event), the kinetic energy of daughter fission products amounts to 167 MeV; the fission fragments flying apart at 3 percent of the speed of light due to Coulomb repulsion. The kinetic energy of U235's average 2.5 neutrons is 5 MeV or 2 MeV each. Fission of a U235 atom also releases about 5 MeV in prompt gamma rays and 10 MeV in capture gamma rays. This sums to a prompt energy release of 187 MeV, divided into 7.5 percent gamma ray energy, 2.5 percent neutron kinetic energy, and 90 percent as the kinetic energy of fission fragments. The fission reaction releases neutrons because the lighter daughter products do not require as much mass to stabilize their nuclei. U235's daughter products further release delayed radiation as beta particles (7 MeV), neutrinos (10 MeV), and gamma rays (8 MeV). The total prompt and delayed energy thus adds up to 202.5 MeV – neglecting neutrinos. Neutrinos do not interact with matter in any substantial way.

Some examples of uranium fission reactions are:

U235 + n > U236 -> Sr90 + Xe144 + 2n + 202 MeV
U235 + n > U236 -> Kr94 + Ba139 + 3n + 202 MeV
U235 + n > U236 > Rb90 + Cs142 + 4n + 202 MeV

An example of a plutonium fission reaction is:

Pu239 + n > Pu240 > Xe134 + Zr103 + 3n + 211 MeV

Weighing the products of fission and quantifying the energy released by a fission event confirmed Albert Einstein's famous formulation that $E = MC^2$. The energy from each fission event is millions of times larger than the energy produced by a chemical reaction. In his lectures to new arrivals at Los Alamos, Roger Serber noted that the complete fission of one kilogram (1,000 grams) of U235 would produce energy equivalent to the explosion of 20,000 tons (20 kilotons, 20 KT) of TNT. The weight of the matter remaining after fission would total 999.15 grams. The missing 0.85 grams represented the matter to energy conversion of the fission process.

Although not widely known at the time, Leo Szilard had already predicted the fission process in 1932. While crossing a street one evening in London, he considered what might result from the discovery of an element that emitted two neutrons for every atom split by a neutron. If other atoms in a sufficiently large or "critical mass" could capture the emitted neutrons, an atomic reaction might run away in geometric progression. Considering the amount of energy released for each interaction in such a "chain reaction," it might be possible to produce power on an industrial scale or to build an atomic bomb. To protect his idea, Szilard filed a patent application for an atomic bomb on July 4, 1934. In his application, he described the basic concept of the neutron-induced chain reaction needed to create an explosion and the concept of critical mass. The patent application was successful, making Szilard the legal inventor of the atomic bomb – even though the physical discovery of fission was four years away!

Thorium and uranium, elements 90 and 92 respectively, were both possible candidates for Szilard's hypothesized fissile element. Based on predictions from the periodic table, so was an element with 94 protons. This element did not occur in nature, so after hearing about fission, Glen Seaborg and Emilio Segré decided to try to make it artificially. They converted a cyclotron to produce neutrons and bombarded a uranium sample for several hours. Their premise was that some of the atoms in the uranium sample might capture a neutron. Then, by a process of beta emission or "beta decay," the uranium atoms would transmute themselves up the table of elements. This is in fact what happened! Atoms of U238, the principal isotope in uranium, captured bombarding neutrons to become U239. The nuclei of the new isotope then emitted a beta particle, transmuting themselves into neptunium 239 – a new, manmade element containing 93 protons. Np239, which has a half-life of only 2.5 days, soon emits a second beta particle to become plutonium 239, a more stable element with 94 protons.

Although Seaborg and Segré had created plutonium, it took some time before enough of this manmade element was on hand for a detailed analysis. In the meantime, the scientific communities at Columbia University, Berkeley, and the University of Chicago Metallurgical Laboratory started work to determine the atomic properties of uranium and thorium. Undertaking such a study, Neils Bohr discovered that high-energy neutrons could split the nuclei of U238 and

Th232 atoms in a process called "fast fission." Intermediate energy neutrons, however, were absorbed in a "window of resonance" and through a process of beta decay, Th232 turned into U233, and U238 became Pu239. Because the majority of neutrons produced by fission have intermediate energies, Bohr determined that these isotopes could not sustain a chain reaction.

Bohr made a critical breakthrough when he found that the rare U235 isotope of uranium behaved differently from its more abundant U238 sibling. Its nucleus easily absorbed neutrons across a wide energy spectrum or "cross-section" to become U236. Then, an internal atomic adjustment in the newly created isotope released energy into its nucleus. This energy exceeded the binding forces within and caused the nucleus to disintegrate in a process called "slow fission." Since U235 released an average of 2.5 neutrons per fission when interacting with neutrons of intermediate energy, Bohr had found an ideal candidate for power production or a bomb. Later, the physicists at Los Alamos found that Pu239 had similar properties, releasing an average of 2.9 neutrons per fission. This made it an even more attractive candidate for a bomb material.

After the surprise attack by the Japanese on Pearl Harbor, and the ensuing declaration of war by the United States on the Axis powers in December 1941, President Roosevelt approved the Manhattan Engineering District for the production of nuclear explosives and the fabrication of atomic bombs. To support this entity, a central atomic laboratory was set up during the winter of 1943 at Los Alamos, New Mexico. It operated under a contract to the Board of Governors of the University of California and Professor J. General Leslie Groves chose Robert Oppenheimer, a distinguished theoretical physicist from the university as its first scientific director. Following a whirlwind recruiting drive, Oppenheimer began operations at his new lab on March 15.

Among the tasks assigned to the scientists stationed at Los Alamos or Site "Y" was the determination of the nuclear properties of fissile materials. Most important among these properties was their critical mass. As previously hypothesized by Leo Szilard, neutrons are lost from the surface of a chain-reacting material. If the mass of material in which a fission event occurs is too small, the loss of neutrons from its surface will prevent a self-sustaining reaction from taking hold. However, if the mass is large enough and exceeds the critical mass, its nuclei will absorb enough neutrons to support a self-sustaining chain reaction of fissioning atoms. If the "super-critical mass" of nuclear material holds together for long enough, the ongoing chain reaction will release energy equivalent to the explosion of tens or even hundreds of thousands of tons of TNT. Establishing how much material constitutes a critical mass for U235 and for Pu239 was a crucial step required to finalize the design for a bomb.

Early in 1944, Los Alamos received its first quantity shipments of U235 from an enrichment facility constructed at Oak Ridge, Tennessee. To hide the true nature of the material, it was code named "ORALLOY," for Oak Ridge ALLOY. Otto Frisch immediately started the experiments necessary to determine its critical mass. Using metal hydride at first and then pure metal, Frisch dropped enriched uranium slugs through a central opening in a pile of enriched uranium blocks to detect bursts of criticality. Frisch performed his experiments at a facility in a canyon some distance from the main installation. Had a slug accidentally become stuck in the apparatus, the resulting uncontrolled atomic reaction would have been disastrous.

Following the arrival of plutonium samples from production reactors constructed near Hanford, Washington, the physicists at Los Alamos carried out a second type of experiment. In this experiment, a physicist gradually moved a pair of beryllium coated plutonium hemispheres together until he could measure neutron production. Richard Feynman called this dangerous experiment "tickling the tail of the dragon." Tragically, Louis Slottin received a fatal dose of radiation in an accident while recreating this experiment after the war. Eventually, critical values for both Oralloy and plutonium were determined. A bare uranium sphere enriched to 93.5 percent U235 had a critical mass of 52 kilograms. A bare sphere of α (alpha) phase plutonium, containing 93.5 percent Pu239, had a critical mass of 10.5 kilograms. For the δ (delta) phase allotrope of plutonium, which is less dense than the α-phase, critical mass increases to 16 kilograms of fissile material. Δ-phase plutonium must be alloyed with 3 weight percent of gallium to stabilize its structure at standard temperature and pressure.

The creation of an atomic explosion thus requires the formation of a supercritical mass of fissile material into which spontaneous fission or some artificial process introduces a neutron to start a chain reaction. However, because fissile materials have high neutron backgrounds arising from their natural radioactive decay, it is necessary to assemble a critical mass in a very short time or face the possibility of "premature detonation"; also, known as "pre-ignition." The production of spontaneously produced neutrons can initiate atomic reactions even before the sub-components of a super critical mass can come into full contact. The energy from the resulting chain reaction then drives the fissile masses apart before much of the involved material can undergo fission. This circumstance results in a severely reduced explosive yield.

As implied in the preceding paragraph, the most obvious method of creating a super critical mass is "gun-assembly." This requires two nearly critical fissile masses in the form of a small diameter solid cylinder known as the "insert" and a larger diameter, hollow cylinder known as the "target." For their gun-assembled bomb, the Los Alamos scientists sized the insert so it fit inside the target. The target was fired at the insert in a long-barreled gun loaded with cordite propellant. The assembly of the two components created a solid, super critical mass. Because the hollow and solid cylinders were relatively squat, their final geometry approximated a sphere, which is the most efficient shape for a super critical mass. The complete amalgamation of the hollow and solid cylinders took about half a millisecond, after which time the bomb's designers anticipated that a spontaneously produced neutron would start a chain reaction. This methodology was well suited to the low spontaneous neutron environment associated with an enriched uranium bomb.

Plutonium, the main other fissile material under consideration at the time, had such a high natural neutron background that the necessary velocities to avoid pre-ignition were hopelessly out of reach. To make use of plutonium as bomb material, Los Alamos scientists developed another method that compressed (rather than assembled) a spherical, sub-critical, "core" to criticality by means of an inward directed or "convergent" explosive-driven shock wave. This technique is termed "implosion." An explosive shock wave generates pressures of 500,000 kilobars or 500,000 times the pressure at the earth's surface and temperatures that can soar to 5,000 °C. Since critical mass increases with the square of density, doubling the density of a near critical core produces a configuration of almost four critical masses. As demonstrated by early weapons, a chain reaction initiated with this arrangement can produce yields of 20 - 30 kilotons, enough power to destroy a city of 250,000 inhabitants.

A special property of high explosives is a "detonation wave" that forms in front of the chemical reaction that radiates outward from an "initiation point." Moving at very high velocity, expanding gases produced by the chemical reaction accompany the detonation wave. When the detonation wave reaches the fissile core in a nuclear device, it is no longer driven by the chemical reaction of explosives but continues to travel inwards as a "classical" or mechanical shock wave, with an abrupt transition from shocked to un-shocked material. A zone of compression accompanies the mechanical shock wave that results from the inward-directed flow of matter in the area immediately behind the shock wave. A material's "equations of state" or "Hugoniot" describe the difference in the conditions (temperature, pressure, density, particle velocity, etc.) on either side of a shock wave. An implosion-based nuclear device avoids pre-ignition because an explosive driven shock wave can compress a core to criticality in about four microseconds, 100 times faster than a gun-assembled mechanism can achieve assembly.

The initial approach to implosion employed a large number of detonation points spread over a uniform explosive envelope that surrounded a solid, spherical pit. This method produced undesirable shaped-charge like jets along the interfaces where the various waves merged. In the spring of 1944, a number of prominent British physicists arrived to assist in the work at Los Alamos and one of these men, James Tuck, suggested using three-dimensional "explosive lenses" to shape a convergent shock wave in the same way that the curved surface on a pair of glasses bends light rays to sharpen an optical image. The explosives group manufactured Tuck's 3D lenses by combining specially shaped pieces of high and low-speed explosives. An assembly of such lenses could produce the desired continuous, spherical, convergent shock wave. In perfecting this method, the explosives group started with work on cylindrical implosion. Ultimately, they fabricated some 20,000 explosive lenses into assemblies and detonated them in a trial and error program intended to perfect spherical implosion. The final design of the explosive assembly used cones made of low-speed explosives inserted into the base of high explosive blocks.

In a typical implosion weapon, the period of maximum compression is limited to less than a microsecond before the material in the core begins to rebound to its normal volume and density. This precludes reliance on spontaneous fission as a source of neutrons to start the chain reaction. Instead, it was necessary for the Los Alamos scientists to devise a mechanism that produced neutrons during the extremely short period when the core was approaching its maximum density. The bomb's designers accomplished this feat by placing a walnut-sized mechanical "neutron initiator" into a spherical cavity at the center of the core. Mechanical initiators contained radioactive polonium 210 (code-named Postum) and beryllium, which were separated by gold and nickel layers until violently mixed together by the implosion process. For every million alpha particles released by polonium, beryllium collisions produce 30 low-energy neutrons. The number of alpha particles released is dependent on the polonium content of the initiator.

In the early 1960s, Sandia Laboratories devised electronic neutron initiators to trigger chain reactions more precisely. Known as "pulsed neutron tubes," these advanced initiators made use of a powerful current surge to fuse heavy isotopes of hydrogen and produce a burst of neutrons. Although this type of initiator is located outside of the core, its neutron burst is so intense that it can penetrate right through implosion debris. For a brief period, tens of thousands of neutrons per nanosecond flood the core. These neutrons initiate the chain reaction and because of their number, shorten its duration by about 15 percent.

The chain reaction in a super critical core begins as soon as an initiating neutron ruptures a fissile nucleus. The distance that the resulting neutrons travel between successive fission events is their "mean free path" and this measurement is different for each isotope. MFPs are dependent on the distance between fissile nuclei and the number of times a neutron is scattered before it is absorbed. An average "path for scattering" is 2 or 3 centimeters long and in typical circumstances, a neutron reflects about five times before it is absorbed. At normal densities, the mean free path for U235 is 16.5 centimeters and for Pu239 it is 12.7 centimeters. Using an average neutron speed of 1,400 kilometers per second, the time between succeeding generations in a chain reaction calculates to about ten nanoseconds, a term that nuclear physicists refer to as a "shake."

In a nuclear reaction, a "multiplication coefficient" determines its growth rate. This coefficient is the number of neutrons released by each fission event, less the fraction of neutrons absorbed in non-fission events by U238, and the fraction of neutrons that escape from the core. Assuming a multiplication coefficient of two, in a bomb like the one dropped on Nagasaki at the end of WW II, a geometric progression (2, 4, 8, 16, 32, 64…) for 81.7 generations produces an energy output of 20 kilotons. A multiplication coefficient of at least two is necessary for an efficient explosion. The maximum multiplication coefficient for U235 is 2.5, which corresponds to the average number of neutrons released by this isotope in a fission event. For Pu239, the maximum multiplication coefficient is 2.9. As it turns out, some neutrons are lost, some neutrons are absorbed by U238, some neutrons emerge promptly, and some neutrons have a slight delay, so that successive fission events tend to overlap in a continuous process. The chain reaction for plutonium with its 2.9 multiplication factor takes about 60 shakes to run to completion in a Fat Man type nuclear pit. This is the time required for 56 generations in a geometric progression. Significantly, the last four generations of a chain reaction release more than 95 percent of the energy in an atomic explosion.

Because of the overlap in the neutron multiplication process, physicists prefer to quantify nuclear processes with expressions that describe their neutron density. This is convenient because reaction rates and energy release rates are proportional to the number of neutrons per unit volume. It follows, then, that the optimization of this process is central to the design of efficient weapons.

A particularly useful term in the mathematical analysis of a chain reaction is its "alpha." It turns out that this term is the reciprocal of the doubling or effective multiplication time. Physicist measure alpha in fission events per microsecond on a logarithmic scale to indicate the rate of neutron build up. When the effective multiplication time has a negative value, as for a sub-critical mass, it states that the rate of neutron production and energy release in the system is declining. The term is exactly zero for a critical mass that maintains a constant neutron density and it is positive for a super-critical mass in which the neutron density and rate of energy release are increasing. Counts of 25 to 200 events per microsecond are typical for implosion weapons. During full-scale nuclear tests, alpha is determined by the time taken to convert gamma rays to light in a mechanism that contains a fluorescing organic compound.

A system's alpha is dependent on a number of factors that include the type of nuclear material under consideration and its density. Because compression is a function of convergence, a core will always achieve its greatest density and alpha at its center. This means that the chain reaction proceeds most quickly here and that a diminishing energy gradient establishes itself from the center of the compressed core outwards to its edge. Since compression and disassembly are dynamic processes, the continually changing density of the core contributes to changes in alpha throughout the implosion process. Physicists at Los Alamos carried out a number of criticality experiments to determine effective multiplication rates for nuclear materials in various states of compression. The resulting alphas were equivalent to doubling times of 3 to 30 nanoseconds.

Implosion has several advantages with respect to gun-assembly – and one big disadvantage. The disadvantage is the complexity of the explosive sphere and the exacting tolerances required for its explosive lenses. The advantages of implosion result from compression, which shortens the mean free path of neutrons in the core and reduces the time between successive generations. This allows the chain reaction to proceed more quickly than in a critical uncompressed mass. Neutrons also find it difficult to escape from a compressed core because of the reduced spacing between its atomic nuclei. Both of these mechanisms contribute to enhanced "efficiency," which is the ratio of the energy produced from a nuclear fuel mass divided by the total energy possible. The original gun-assembled U235 bomb achieved an efficiency of only 1.5 percent, whereas the first plutonium implosion bomb had an efficiency of 16 percent (11 percent if one assumes that 30 percent of its energy came from the fissioning of uranium in its tamper.) Considering the enormous cost of fissile cores, it is no surprise that designers discarded gun-assembled bombs as mainstream weapons.

In practice, designers never use explosives to compress an atomic core directly. Instead, they place a tamper and / or a neutron reflector between the explosive sphere and the core. In the case of a tamper made of a dense substance, like natural uranium, the mechanical and thermal processes following compression and the initiation of the chain reaction come into play. The rebounding mechanical shock wave in association with energy liberated by the fission chain reaction create very large core pressures, which produce an outward acceleration at the core-tamper interface, which is a free surface. This produces slip lines along which highly compressed tamper material piles up just ahead of the expanding core, in what is referred to as a "snowplow region". Technically, the structure that results is known as a "duplex."

The inertia of the material in the duplex delays expansion of the core, so that a pressure gradient builds up from the center of the core to its outer surface. As a result of the delayed expansion, the volume of the compressed core remains essentially constant during the first 50 or so generations following initiation of the fission reaction. After this interval, almost the whole of the core's energy is released within 5 generations - during which time the core expands until it becomes subcritical. The liberation of energy may be regarded as over when the dimensions of the core reach criticality. The brake on expansion provided by the moment of inertia in the duplex zone adds several generations to the length of the atomic reaction. The net effect of this phenomena is to increase critical mass without having to add expensive nuclear material.

Surrounding a fissile core with a neutron reflecting material or "neutron tamper" is another method used to increase critical mass. This technique reduces the loss of neutrons from the surface of the core and effectively increases the reactivity of the contained fissile material. The lightweight metal, beryllium, was found to have the best qualities for this purpose. Despite the fact that expansion and disassembly are not retarded in this configuration, the increased reactivity provided by the neutrons reflected from a 2-inch thick beryllium shell provide the same benefit as a bulky U238 tamper. Beryllium tampers thus contributed to the development of lighter and more compact weapons. Los Alamos scientists also discovered that beryllium could augment neutron flux through a neutron multiplication reaction.

Depending on the benefit desired, there is room for considerable compromise in tamper design. U238 is a good all-around choice because it is both dense and a moderately effective neutron reflector. It also augments yield because the higher energy neutrons escaping from the core will cause some of the tamper material to undergo fast fission. For these reasons, designers chose U238 as the tamper material for use in the first implosion weapons. Because U238 has a moderate neutron background that can contribute to premature detonation, a dense, non-radioactive material like tungsten carbide is a better choice for gun-assembled weapons.

Once an atomic chain reaction is set in motion inside a compressed core, it releases energy. When the temperature at the center of the core reaches 10 million °C, the core begins to expand. At this time, the total energy yield is about 0.2 kilotons, an efficiency of 0.15 percent in a Nagasaki type bomb. The temperature then rises continuously until the inner core reaches a maximum temperature of 50 million °C and the outer core reaches 20 million °C. At this time, the energy yield of the atomic reaction is about 2 kilotons, an efficiency of 1.5 percent, and its power output is approaching 20 billion watts per square centimeter. From this time forward, the core's temperature decreases as some of its thermal energy turns into kinetic energy or transfers to the tamper or reflector.

Expansion of only a few inches over a period of 40 - 50 nanoseconds is enough to stop the ongoing chain reaction. This turns out not to be a problem because this time frame corresponds to the last 4-5 generations of the atomic chain reaction, which release almost all of the bomb's energy. This rapid release of energy creates an outgoing "radiative" shock wave, which differs from a classical shock wave in that it contains the vast store of energy released by the chain reaction. As it moves outward, the radiative shock wave releases photons that continually preheat the material in front of it – hence the name.

At the center of the core, where the release of energy has raised temperatures to 50 million °C, physicists term the radiative wave "super-critical." This is to say that the outward flow of heat associated with photon emission at the wave front dominates energy transport. The kinetic energy of the vaporized bomb material is only of secondary importance. By the time the shock wave has expanded to the edge of the core, it will have cooled and may only be "critical" or "sub-critical." The amount of photon preheating in the region in front of the shock front will no longer produce temperatures higher than that of the material behind the front.

When a radiative (or a classical) shock wave reaches a low-density interface, like the one that is found at the edge of a nuclear pit, the energy of compression turns into kinetic energy or energy of motion. This causes the material in the pit to "disassemble" from the outside in, typically at velocities well above the speed of sound. Because the uranium tamper of the pit is of similar density to the core, disassembly of the core is delayed until disassembly of the tamper reaches the core - tamper interface. The time between the transmission of the firing signal to the detonators in a fission device and the beginning of observable nuclear reactions is called the "transit time." The period between neutron injection and disassembly is the "incubation period." Disassembly typically takes between 0.5 and 2 microseconds.

After the war, Sandia undertook a good deal of research to improve the engineering of atomic bombs. The high thermal output from the natural decay of the solid plutonium cores used in the first implosion bombs, and the short shelf life of their electronic components, required the assembly of bombs just before use – a process that took several days! Much of Sandia's engineering effort was, therefore, concentrated on producing weapons that could be stored ready for use in a fully assembled format: the so-called "wooden bomb" project. Another focus of post war research was adapting enriched uranium for use in implosion weapons and in creating composite cores to take better advantage of available stocks of enriched uranium and plutonium. Composite cores had an inner plutonium sphere surrounded by a shell of enriched uranium. This arrangement produced a much lower thermal output than an all-plutonium core and did not damage delicate bomb components. A drawback to composite cores was the increase in mass required by the use of enriched uranium. This increased the composite core's energy requirement for compression.

A complimentary concept used with composite cores was the "levitated pit," in which an air gap separated the core of a nuclear device from its tamper. The term "pit" refers to the fissile core, the tamper, and any associated hardware. As soon as the inward propagating shock wave from the detonation of a bomb's explosive sphere reached the air space between the tamper and the core, the conversion of compressive energy into kinetic energy accelerated the tamper across the pit's air gap. The expanding gases from the chemical explosion also contributed to the acceleration of the tamper. Ted Taylor, a renowned Los Alamos weaponeer, likened this process to swinging a hammer at a nail, rather than just pushing on it with the hammer's head.

The collision of the flying tamper with the levitated core converted the tamper's kinetic energy back into a compressive shock wave. The extra energy imparted to the tamper as it accelerated over the air gap produced enough additional compression to double the 16 percent efficiency of a first-generation weapon. A typical composite-levitated strategic atomic bomb produced yields varying from 100 to 150 kilotons and could demolish the heart of a city with a population of 1,000,000 persons. Some advanced versions of the levitated pit placed a shell of fissile material on the inner surface of the tamper. The collision of this material with the motionless fissile sphere at the center of the assembly created both ingoing and outgoing shock waves. The pressure in the area between these waves was initially constant and led to even compression throughout the combined fissile mass. Pits of even greater complexity incorporated several shells of tamper or fissile material. Each succeeding shell had half the thickness of the preceding shell. The interval between shells also fell by one half. This arrangement seems to have created the highest shell velocities, producing the greatest degree of compression.

Los Alamos and Sandia Laboratories designed the new, composite-levitated pits to allow the removal of their nuclear cores, which were stored inside of special climate-controlled containers known as a "birdcages." These cages accompanied operational flights during which a crewmember would have transferred the core to the bomb if SAC authorized an attack. For safety reasons, the crew carried out this procedure only after the bomber was airborne. Removable cores, coupled with longer life components produced by Sandia Laboratory (the non-nuclear arm of Los Alamos), allowed atomic bombs to be stored in a fully assembled format (but without their nuclear cores) after 1949.

Beginning in 1957, fusion boosting began to improve the yield of levitated pits. This technique injects a high-pressure mixture of deuterium and tritium gas into a hollow core or the headspace between the core and a fissile shell before detonation. The rising temperatures caused by fission ionize the hydrogen isotopes, which undergo fusion with the release of high-energy neutrons at a temperature of 10 million °C. Pure deuterium, a possible alternate fusion fuel, ignites only at temperatures where the chain reaction is very close to its end. The deuterium-tritium reaction produces neutrons early enough in the chain reaction for them to have a significant effect on its ultimate yield. It also occurs before significant Rayleigh Taylor mixing of the plutonium and the boost gas can take place. Tritium and deuterium are thus important components in modern nuclear weapons.

An important property of neutrons produced by fusion is energy that is much higher than the energy of neutrons produced by fission (14.1 MeV vs. 2 MeV). This serves to reduce the average time between succeeding generations in the chain reaction. They also produce a larger than average number of neutrons for each fission event in which they are involved. Alphas for boosted weapons are typically about 1,000 per microsecond, 5 - 40 times higher than for un-boosted warheads. Thus, in a fissile core, the flood of high-energy neutrons from fusion can produce fission efficiencies as high as 80 - 90 percent.

The idea of using fusion as an energy source came from studies of the processes that produce power in the interior of our sun and the stars. In the late 1800's, lacking any knowledge of the sun's makeup, Lord Kelvin and Hermann von Helmoltz found that the gravitational collapse of a large body of gas could raise its internal temperatures to tens of millions of degrees. It would then take 100,000,000 years to radiate away this energy, providing a long-term source of energy for a body such as our sun. Unfortunately, it was not long before geologists discovered that life on earth had existed for billions of years and that a much longer-lived source of energy was required to explain solar and earthly processes.

Sir Arthur Eddington took the first significant step on the path to the discovery of the sun's true energy source in 1926 following the discovery of atoms. Sir Arthur noted that an atom of helium is 0.7 percent lighter than four hydrogen atoms. Based on Einstein's equation for mass and energy this mass deficit could provide a suitable source of solar energy given that astronomers could discover a large quantity of hydrogen in the sun. Cecilia Payne-Gaposchkin provided this evidence through her work in the science of spectrography. Based on her data, astrophysicists determined that a star like our sun has been around for 4.6 billion years and contains enough hydrogen fuel to keep it burning for another 5.5 billion years.

There was still the problem of how protons got close enough to bond, since the electromagnetic repulsion of their positive charges should have prevented this from happening. George Gamow, Fritz Houtermans, and Robert Atkinson provided the answer to this enigma in 1928. They based it on the newly emerging science of quantum mechanics. In accordance with Heisenberg's uncertainty principle, the scientists hypothesized that atomic constituents behaved simultaneously like a particle and a wave. If two protons approached each other in a high temperature gas, their wave fields could overlap to the extent that the particles could "tunnel" through the electromagnetic barrier. The strong nuclear force would then bond them together into an atomic nucleus. They determined that in the sun, this process takes place at temperatures greater than 15,000,000 °C.

Despite the discovery of quantum tunneling, it took many more years before Hans Bethe and Charles Critchfield were able to describe the fusion process in detail. The first step in the fusion process takes place when two protons bond by means of the tunneling process. One proton immediately ejects a neutrino and a bit of antimatter known as a positron, transmuting itself into a neutron in a process called reverse beta decay. This creates a nucleus of deuterium - an isotope of hydrogen comprised of one proton and one neutron. The newly formed deuterium nucleus then fuses with another proton to form He3 with the emission of a gamma ray. Finally, the collision of two nuclei of He3 form a nucleus of He4 with the emission of two protons and the process is complete. The amount of energy released by fusion is significant – about 1/10 of the energy produced by the fission of a heavy, radioactive atom. However, there are many more hydrogen atoms in an equal mass of material so that on a mass for mass basis, fusion releases about four times more energy than is released by fission. It also releases 24 times as many neutrons.

In the practicalities of bomb design, it is necessary to use the heavy hydrogen isotopes deuterium (H2, D) and tritium (H3, T) as fuel. Garden-variety protium (H) is not reactive enough to undergo fusion in the artificial conditions created inside a weapon. Theoretically, the fusion of 12 kilograms of deuterium can yield an energy equivalent of 1,000,000 tons (1 megaton, 1 MT) of TNT. Unlike atomic bombs, which are constrained by the amount of fissionable material that can be practically placed in close proximity, a hydrogen bomb's destructive capacity is limited only by the amount of fusion fuel that can be conveniently compressed by the force of an atomic explosion. The larger the atomic explosion, the more fusion fuel it can compress. In early weapons, design considerations necessitated producing weapons with multimegaton yields, vastly more energy than was needed to wipe out the largest metropolitan areas on earth. Nobel Laureate Enrico Fermi suggested the idea for the "Super" to Edward Teller at Columbia University in September 1941. The weapon so fascinated Teller that he spent the rest of his life pursuing its development in one form or another.

Calculations to demonstrate the feasibility of the hydrogen bomb proved very difficult to carry out and the urgency of the war years kept the staff at Los Alamos focused on the atomic bomb. It was not until 1950, after the Soviets had occupied Prague, precipitated the Berlin crisis, and exploded their first atomic bomb that President Harry Truman directed the scientific agency to produce a hydrogen bomb. In a discussion with David Lilienthal on January 31, 1951, he categorically demanded that the Director of the Atomic Energy Commission immediately commence work on this fearsome weapon.

The initial design for the "Classical Super" hydrogen bomb called for the heat from an atomic explosion to ignite a fusion reaction in an adjacent mass of deuterium fuel:

$$D + D + D > He4 + n + p + 21.62 \text{ MeV}$$

Since fusion reactions are primarily temperature dependent, scientists expected that the continuous heating of deuterium next to an ignition point would propagate a wave of burning through the entire fuel mass. For this reason, the hydrogen bomb was termed a "thermonuclear" weapon. The continuous release of energy in the fusion process is effectively a chain reaction but not in the same sense as a fission reaction. It more closely resembles the process of flame propagation. As a result, the term "burning" came into popular usage to describe thermonuclear reactions.

A deuterium-tritium mixture, which reacts 100 times faster and at much lower temperatures than pure deuterium, is an even better choice for a thermonuclear fuel:

$$D + T > He4 + n + 17.59 \text{ MeV}$$

Despite its advantages, the bomb designers at Los Alamo rejected this combination because of tritium's high cost and its short half-life. As well, it was necessary to produce tritium in specially designed reactors that would have constrained plutonium production. In consideration of this factor, Los Alamos' bomb designers chose pure, liquid (cryogenic) deuterium as the fuel for the Classical Super. As research progressed, the designers discovered that the Classical Super was a dead end because its deuterium fuel ionized quickly to transparency, allowing radiation to escape. This prevented the development of the high temperatures needed to maintain an ongoing wave of burning. Even adding quantities of tritium as an accelerant, a suggestion made by Emil Konopinski, could not solve the Classical Super's problem.

Fortunately, a meeting between Stanislaw Ulam and Edward Teller in January 1951 provided a critical intellectual breakthrough. Together these physicists conceived of the "Equilibrium Super," a bomb in which the x-rays emitted by an atomic explosion could be used to first compress and then heat a deuterium-filled fuel cylinder to the extreme conditions needed to produce an ongoing wave of fusion burning. The radiant energy emitted by an atomic explosion forms a continuous "blackbody" spectrum of photons whose energy distribution is determined by the temperature of the fireball. These x-rays reach the fuel cylinder in about 1 nanosecond whereas it takes neutrons 20 nanoseconds or more to reach the cylinder. The shock wave from the primary takes a full microsecond to arrive. Thus, there is ample time for the x-rays to affect the cylinder before neutrons preheat it or the shock wave disrupts it. Teller also realized that radiation compression of several hundred-fold would alter the equations of state in the mass of thermonuclear fuel contained within the cylinder and slow its rate of thermal emission to acceptable levels.

As pressures increase above several megabars (millions of atmospheres) during the compression of thermonuclear fuel, the fuel's electronic structure begins to break down. The "Coulomb Force" becomes so strong that it displaces electrons from their atoms and the fuel begins to resemble individual atomic nuclei floating in an electron gas. Because quantum mechanical laws govern the properties of electrons (fermions), physicists refer to an electron gas as a "Fermi Gas." Deuterium at 1,000 times its normal liquid density becomes a true Fermi gas when all of its atoms are fully ionized.

The ignition of deuterium plasma releases neutrons and photons. During the early stages of burning, when deuterium nuclei control the fusion reactions in the bomb, thermonuclear reactions deposit approximately two thirds of their yield as kinetic energy in neutrons. Deuterium nuclei are very light, which gives them a strong moderating effect that rapidly absorbs the kinetic energy of the neutrons as heat. Later in the fusion process, interactions between the nuclei of the helium byproduct and deuterium nuclei transfer back about 20 percent of the produced energy to the neutrons. In addition, "three-body" reactions between photons, electrons, and nuclei scatter photons to prevent them from escaping. In effect, the degenerate nature of matter in the highly compressed thermonuclear fuel renders it opaque to the escape of radiation or neutrons, promoting self-heating of the plasma's various components to an "equilibrium" temperature.

The basic components of the Equilibrium Super Bomb are a radiation source or "primary," a physically separated "secondary" container of thermonuclear fuel and a "radiation duct" or "inter-stage" that channels the radiation from the primary to allow radiation induced compression of the secondary. With this type of configuration, the most important diagnostics for thermonuclear efficiency are inter-stage (transit) time and radiation channel temperature.

The earliest thermonuclear weapons incorporated a moderately powerful atomic bomb placed at the end of a thick walled, hollow metal cylinder as a radiation source or "primary." Designers lined the cylinder with a dense or "high-Z" material such as lead or natural uranium to control the flow of radiation from the atomic primary after it was exploded. Inside, separated by a gap from the outer radiation lining, was a second cylinder that contained the thermonuclear fuel, which in the case of the first hydrogen bomb was a vacuum flask of cryogenically cooled liquid deuterium. Separated from the fuel cylinder by a standoff gap was a high-Z (uranium) tamper. Together, the fuel cylinder and the tamper become the "secondary." The tamper forms a shield between the fuel cylinder and the atomic primary, an arrangement that prevents preheating of the fuel by the flow of neutrons that emerge from the atomic primary ahead of the expanding debris produced by the detonation of its high explosives. Preheated fuel is impossible to compress to the density required to sustain fusion. The terms "isentropic" and "adiabatic" compression describe an increase in density without heating.

After radiation emerges from an exploding primary, it flows into the duct between the radiation liner and the secondary, a process that takes about a nanosecond. This "radiation channel" is filled with polyethylene or some similar, "low-Z," material. The intense radiation from the atomic primary quickly vaporizes this filling, turning it into plasma. This process scatters a lot of energy to bring the interior of the bomb into "thermodynamic equilibrium" at a temperature of 10 - 25 million °C, a process that takes only 1 or 2 shakes. It also renders the channel transparent to the flow of radiation. Without the use of this technique, partially ionized high-Z material, vaporized from the surfaces of the tamper and the radiation liner, would distort or block the radiation path.

As x-rays scatter throughout the radiation channel, they create "radiation pressure." They also create "kinetic pressure" when they heat the polyethylene filler into plasma. More importantly, the tamper surface absorbs x-rays, which can penetrate 0.1 millimeter into uranium at a temperature of 2 KeV or 20 million °C, and 2 millimeters into uranium at a temperature of 10 KeV or 100 million °C. This process heats and vaporizes material from the tamper surface in a process called "radiation driven ablative implosion." Although the thermalization of the polyethylene plasma in the radiation channel takes about 1 shake, ablation takes from 10 - 50 shakes, the latter process providing the primary compression of the fuel in the thermonuclear secondary. The force exerted on the tamper is a function of the mass ablated from its surface and the square of its ejection velocity. For those with a mathematical interest, Carey Sublette provides a mathematical treatment of thermonuclear processes in his *Nuclear Weapons Frequently Asked Questions*. Andre Gsponer and Jean-Pierre Hurni also provide a mathematical treatment of this process in their paper on *Fourth Generation Nuclear Weapons*.

Ablation is not continuous but proceeds in discrete steps, with each step inducing an inward propagating shock wave as radiation absorption blows off a layer of tamper material. The closely spaced shock waves traverse the tamper where they are converted to kinetic energy on its inner surface. This process progressively accelerates the tamper inwards to collide with the fuel cylinder in the same manner that the levitated tamper in a fission bomb collides with its fissile core. The collision of the tamper with the fuel container imparts sufficient energy to compress its contents to near stellar conditions. In early weapons, the degree of compression was several hundred-fold, whereas in modern weapons, compression of a thousand-fold is possible. (In the inertial confinement fusion experiments used to develop fusion power reactors, a compression of ten-thousand-fold is possible, with fusion fuels achieving a density one hundred times that of lead.) Even so, Teller was unsure if temperatures sufficiently high for ignition (about 1 KeV or 10 million °C) would obtain after compression. He therefore employed a clever subterfuge.

Teller placed a hollow rod of plutonium containing a small amount of tritium – essentially a boosted core – at the center of the secondary. Designers later replaced the fissile plutonium with enriched U235. When compressed by the implosion of the surrounding tamper and thermonuclear fuel, the explosion of this inner "sparkplug" raised the fuel's temperature. Since thermonuclear reaction rates are sensitive to pressure as well as temperature, this arrangement insured that the fuel reached the necessary conditions to initiate a fusion reaction. A thermal or "Marshak" wave then formed and moved away from the area of ignition, continuously heating the fuel in front of it to the critical temperature needed to sustain the fusion process. Since photons moving at the speed of light dominate the thermal wave, this process is very swift, taking place in 10 - 20 nanoseconds. Unlike the highly spherical geometry needed to support a fission chain reaction, a thermal wave is capable of sustaining fusion in a compressed, cylindrical mass with considerable distortion. This feature gave early weaponeers a degree of freedom in producing thermonuclear weapons.

The energy produced by fusion causes the compressed thermonuclear fuel mass to disassemble in the same way that the core of a fission bomb disassembles. The ablation-induced mechanical shock wave also supports disassembly after it reaches the center of the fuel mass and rebounds. Since this wave is travelling at several hundred kilometers per second (several millimeters per nanosecond), it traverses the fuel very quickly. The reduction in fuel density associated with the rebounding shock wave and thermal expansion allows radiation to escape the fuel mass. When the density of the fuel drops below a critical value, the fuel mass cools to a point where it can no longer support an ongoing wave of fusion. The Marshak wave then progressively transforms into a mechanical shock wave and finally into a sonic wave that dissipates in the remaining unburned fuel.

There are three ways to delay this disassembly and to improve the efficiency of a burning mass of thermonuclear fuel. The first is to insure a very high initial state of compression. The fuel will thus react faster and take longer to disassemble. The second is to ensure that the mass of the tamper remaining after ablation and compression is sufficient to retard disassembly. The third method is to increase the pressure and (or) temperature at the boundaries of the thermonuclear fuel. Increased boundary pressure will result if the thermonuclear tamper is composed of fissionable or fissile material. After compression to a very high-density, the secondary's tamper will be extremely effective in absorbing the sudden flood of neutrons produced by fusion. The absorption of these neutrons will induce fast fission in elements such as U238 or slow fission in fissile elements such as U235.

Because the secondary's tamper is in close proximity to the thermonuclear fuel, it takes less than a shake before neutrons arrive to initiate the fission process. The conditions at the tamper boundary immediately soar to very high temperatures, with an associated rise in pressure that retards disassembly in a process called "secondary atomic compression." In some early weapons with extremely massive tampers, the energy released through secondary fission produced most of the explosive power of the bomb – with fission energy exceeding fusion energy by a factor of five or so. In this regard, first generation thermonuclear weapons were nothing more than big, dirty fission bombs burning large quantities of cheap, natural uranium.

Liquid deuterium, as used in the first thermonuclear device, is not a particularly practical fuel. As a result, the AEC produced only a few, short-lived, emergency capability versions of this device (the EC16). Follow-on weapons used a solid thermonuclear fuel that significantly reduced their bulk and improved their handling characteristics. The Y-12 plant produced the new solid fuel, known as lithium deuteride or LiD, by reacting deuterium at high temperature with the light element lithium. Y-12 molded the resulting compound into a hard ceramic that machinists shaped to the exacting tolerances needed to satisfy the design requirements for advanced weapons.

Fortunately, diluting deuterium with lithium does not prevent fusion from taking place. In fact, neutron interactions convert lithium into the complimentary thermonuclear fuel, tritium. The ratio of lithium and deuterium in LiD is ideal, lithium breaking down to provide one tritium nucleus for every deuterium nucleus! Physicists discovered the lithium decomposition phenomenon at the Cavendish Laboratory in the 1930s through the medium of the cyclotron. After absorbing a slow neutron, they found that Li6 decomposed into He4 and tritium according to the reaction:

$$Li6 + n > He4 + T + 4.78 \text{ MeV}$$

The more abundant Li7 isotope reacted similarly, to produce He4, tritium and an excess neutron according to the reaction:

$$Li7 + n > He4 + T + n - 2.47 \text{ MeV}$$

The need for a flood of neutrons to initiate tritium production in the thermonuclear fuel mass may help to explain the use of boosting in the secondary's sparkplug, since fusion reactions produce an abundance of neutrons. Some authors indicate that this is not a requirement. Fuel compression is essential in propagating the lithium decomposition process by increasing the likelihood that lithium nuclei will interact with a neutron. Because reactions involving Li6 are more energetic than those involving Li7, lithium deuteride enriched in Li6 is a superior choice for a thermonuclear fuel. Fusion reactions involving the tritium produced from lithium contribute about half of the energy derived from solid thermonuclear fuel.

Contractors manufactured early thermonuclear bomb casings from cast iron and steel. As testing provided knowledge about the length of time and the amount of material needed to contain radiation in its channel, the AEC had casing wall thicknesses thinned. Eventually, steel casing walls gave way to aluminum alloy rolled about one half of an inch thick. A casing had to retain its integrity only long enough for a bomb to achieve efficient compression and ignition of its secondary. A significant advance in hohlraum design took place when thin sheets of polished uranium replaced lead-lined aluminum casings. Properly aligned and finished to optical tolerances, uranium sheets reduce the rate at which radiation is absorbed by the outer casing and minimize hot spot development. This in turn reduces the rate at which the casing suffers ablation and allows for thinner designs. A number of sources attribute the arrangement of the nested mirrors used in NASA's Chandra x-ray observatory to research carried out to improve thermonuclear weapons.

By the late 1950s, the accurate delivery of nuclear weapons had improved to the point that the multimegaton yields of early hydrogen bombs were no longer required for many military applications. The Air Force also required lighter warheads to increase the range of its ICBMs. The use of boosted atomic warheads as primaries made this requirement possible. Only a dozen or so inches in diameter and weighing a few hundred pounds, these primaries allowed the scaling of thermonuclear weapons to the single megaton range. The explosives used in boosted thermonuclear primaries contain only low atomic number components (H, C, O, N), making their chemical cloud transparent to the radiation flowing into the secondary. The same flow of radiation ionized their low-Z beryllium tampers.

A final important innovation in the evolution of thermonuclear weapons was the spherical thermonuclear secondary developed by the British and the Lawrence Radiation Laboratory. Its geometry provided a design superior in its ability to use the energy curve from a boosted primary to obtain a high order of compression and thus improve efficiency. Newer weapons have only been able to achieve incremental improvements over this arrangement.

Until 1973, blast and heat provided the primary military effects of nuclear and thermonuclear explosives. In that year, a new class of "enhanced radiation" weapon emerged that was popularly known as the "neutron" bomb. Conceived by Samuel Cohen, the neutron bomb was a tactical warhead deliberately designed for minimal blast characteristics and a high radiation output. It made "the people and their pets disappear," while minimizing collateral damage to infrastructure. The penetrative neutron radiation produced by this warhead could immediately incapacitate troops caught in the open or sheltering in armored fighting vehicles at ranges outside of its blast effects. The weapons laboratories also found neutron radiation to be effective against the electronics of hostile warheads in space, where heat and blast are quickly dissipated.

Enhanced radiation warheads are modified thermonuclear devices that use a high-pressure container of tritium and deuterium gas in place of solid thermonuclear fuel. Because this fuel is highly reactive, an enhanced radiation warhead does not require a central sparkplug. Compression of the fuel is the same as for a hydrogen bomb but with one exception. The fusion tamper surrounding the high-pressure gas container is a non-fissionable material such as tungsten carbide. This prevents the capture of high-energy fusion neutrons by uranium nuclei. Made of a neutron transparent material, the outer casing allows neutrons to escape into the environment as the primary effect of this weapon. Because neutrons are highly attenuated in air, the effects of a radiation weapon are limited to a few hundred meters. This limits the effective yield of enhanced radiation to the sub-kiloton range.

The weapons laboratories could not have developed the technology that lies at the heart of modern weapons without significant testing. In order to probe the concept of implosion, the Los Alamos laboratory began a series of "hydrodynamic" tests in 1944. These tests used inert materials in place of fissile cores to evaluate the effects of explosive envelopes. Scientists compressed ferrous cores in magnetic fields whose distortions they recorded to assess the efficiency of the implosion process. An advantage of this method was its suitability for use with full sized explosive assemblies. Another hydrodynamic method made use of lanthanum recovered from spent reactor fuel. Strategically arranged ionization detectors measured the patterns of radioactivity produced by radioactive lanthanum "RaLa" cores as they were compressed. Gamma rays from the lanthanum induced currents that were proportional to the density of the imploding core in strategically placed ionization chambers.

In addition to the aforementioned hydrodynamic procedures, technicians recorded high-speed x-ray movies and photographs to reveal the characteristics of imploding assemblies containing substitute cores made of natural uranium or thorium. Natural uranium is an excellent substitute for enriched uranium and thorium is useful as a plutonium substitute. Using these metals, hydrodynamic experimentalists avoided concerns over accidental atomic explosions and radio-toxicity. Because x-rays reveal density differences, it was possible to follow the development of shock fronts as they burned their way through the explosive blocks surrounding the cores. They also captured the evolving shapes and the degree of compression in the substitute cores.

The continuing importance of hydrodynamic experiments in evaluating nuclear weapons has resulted in the construction of a new complex at Los Alamos. DARHT, a state of the art, "dual axis, radiographic hydrotest facility," contains two x-ray machines that are used for three-dimensional imaging of components from the aging stockpile. In addition to x-rays, DAHRT has linear accelerators that can probe experiments with electron and proton beams. Assisting DAHRT in its hydrodynamic evaluations is the Flash X-Ray Facility at LLNL's Site 300, which has the capability of producing two time-exposures for test events.

In 1955, Los Alamos began "hydronuclear" or "low yield" testing at the Nevada Test Site, (NTS) now referred to as the Nevada National Security Site (NNSS). Hydronuclear tests supplemented the data produced by hydrotests using fissile cores. Early versions of such tests assessed "one-point safety" by triggering the explosives surrounding nuclear assemblies at a single point to ensure that the resulting nuclear yield did not exceed the equivalent of four pounds of TNT. Some of these tests were quite spectacular when they over shot their anticipated yields by hundreds of tons of TNT equivalent. In order to facilitate this program, the NTS dug a complex of 16 tunnels under Rainier Mesa. An important part of this complex was the 1,800-foot long Horizontal Line of Site (HLOS) pipe that tapered from 30 feet in diameter at the test chamber to a few inches at the entrance to the instrumentation chamber. Immediately after a test emitted its initial pulse of radiation, explosive driven doors closed the tunnel to prevent blast damage to instruments. This system also evaluated the radiation hardening of warheads, reentry vehicles, and guidance systems. It was even possible to evacuate air from the tunnel environment to simulate the effects of weapons in space!

The hydronuclear tests carried out at NTS revealed quite a bit about the performance of atomic weapons. Hydronuclear tests with yields less than one half pound TNT equivalent provided information about criticality and alpha. Yields of less than four pounds TNT equivalent provided information about pressures and temperatures after the onset of criticality. Tests that produced yields from a few pounds to several hundred pounds TNT equivalent provided an accurate estimate of the final performance for a weapon, as well as all of the previous kinds of information. Tests with yields from a few tens of tons to a few hundred tons TNT equivalent generated data about the performance of boosted fission and thermonuclear weapons.

In order to supplement data on atomic primaries generated by DAHRT, the DOE built the "National Ignition Facility" (NIF) for laser-based fusion testing at Lawrence Livermore National Laboratory. Laser-based fusion is a way of gathering information about the performance of thermonuclear weapons. In a process called "inertial confinement fusion" (ICF), technicians train a number of powerful lasers on a pellet containing deuterium / tritium gas, lithium dueteride or lithium tritide to create miniature thermonuclear explosions. The properties of nuclear materials can thus be determined at temperatures of tens or even hundreds of millions of degrees. These properties can be quite different from room temperature properties and are often not directly scalable. Although inertial confinement fusion does not operate at the same "energy levels" as nuclear tests, it operates at the same "energy densities." Scientists at the NIF carry out two different types of ICF experiments: "direct" and "indirect." In direct experiments, NIF scientists use lasers to ablate a polyamide or beryllium / copper-jacketed pellet of fusion fuel. In indirect experiments, the scientists shine ultraviolet lasers through windows onto the inner walls of a gold cylinder (hohlraum) that contains a coated pellet of fusion fuel. The gold walls absorb and re-radiate the ultraviolet laser light as high-energy x-rays that ablate the pellet's coatings and compress its fuel. In recent experiments, scientists have pressurized gold hohlraums with 1 - 5 atmospheres of a low-Z gas. Silicon nitride windows placed over the entrance holes contain the gas but allow entry of the laser beams. Calculations show that plasma blown off from the hohlraum's inner walls would have filled the hohlraum to a high enough density to absorb most of the incident laser light. Since laser light has a lower absorption coefficient in low-Z gas, it can more easily propagate to the hohlraum walls. Present codes predict a 70 percent conversion of laser light to x-ray energy with a temperature of about 300 eV. Data from NIF experiments confirms the purpose of the low-Z foams used to fill the radiation duct in thermonuclear weapons.

It is interesting to note that NIF experiments use modulated laser energy. The laser bursts typically have a 12 - 16-nanosecond low-energy foot followed by a 4-nanosecond high-power pulse. The duration of the "foot" (assembly time) is determined by the time it takes lower energy x-rays to drive an initial shock through the pellet shell, establishing a smooth pressure gradient. If a strong shock were to preheat the shell, a high degree of compression would not be possible. In the larger world of weapon physics, weapons designers achieve much the same effect by subjecting the fuel container to multiple ablative shock waves. The final, short, high power pulse generates a central "hot spot" that exceeds the conditions necessary to start a fusion reaction. Although very high compression obtains throughout the fuel mass, the collision of this second shock wave at the center of the fuel mass (shock ignition) creates pressures and temperatures that correspond to the use of a sparkplug in thermonuclear weapons.

In addition to their scaled experiments, the weapons laboratories carried out full-scale weapons nuclear testing during the years 1945 to 1992, right up to the implementation of the Comprehensive Test Ban Treaty. "Physics Tests," normally of fractional kiloton yield, supplied information about nuclear and thermonuclear processes. "Weapons Development Tests," which comprised the vast majority of nuclear tests, confirmed the engineering principles assumed in a weapon's design. The DOE also conducted weapons tests in their quest to produce so-called third generation nuclear weapons. Scientists hoped that these advanced weapons could exploit nuclear explosions to power optical and x-ray lasers, produce particle and directed microwave beams, and accelerate hypervelocity pellets. "Production Verification Tests" of warheads taken randomly from the production line insure their proper manufacture, and "Confidence Tests" evaluate the effects of aging on the materials in older weapons.

In order to evaluate the performance of full-scale tests, LASL's scientists had to calculate their yields and efficiency. To meet this need, six methods presented themselves as diagnostic tools: the intensity of the fireball's gamma ray emissions, the intensity of fast neutron emission, the intensity of heat and electromagnetic radiation, the peak pressure impulse as a function of distance from the hypocenter, the characteristics of the growth of the fireball, and the radiochemical analysis of fission products.

As it turned out, the analysis of fission products proved to be the most diagnostic tool for determining yield. This knowledge caused the Air Force to collect samples of radioactive residue by flying its aircraft through cooling fireballs during early atmospheric atomic tests! After collection of the airborne samples, LASL scientists measured the samples radionuclides in terms of type and quantity. From this information, they could determine the ratio of daughter products to the amount of Pu239 or U235 remaining in the sample. This ratio determined the efficiency of the atomic explosion. Since the scientists knew the amount of fissile material used for the atomic device, it was also possible to calculate yield. The scientists then calibrated the rate of fireball growth, the final diameter of the fireball, and the fireball's integrated gamma ray output so they could also calculate yield by optical means. As well, scientists found they could correlate yield to the strength of the rarefaction wave that followed the main shock wave.

Scientists most commonly use optical parameters relating to fireball growth and fireball diameter to calculate the yield of a thermonuclear explosion. They still have to calculate the fission component of the yield from collected samples. Given that thermonuclear weapons are three-stage devices, it is necessary to factor the fission yield of the primary, the uranium tamper and the sparkplug into the total yield equation. Fortunately, scientists found it possible to isolate the primary's fission yield as a component of total thermonuclear yield. This calculation required the reaction history curve for the device and the predicted hydrodynamic and neutronic states of the device at the time of its maximum yield. The results of this calculation are accurate to within plus or minus 15 percent standard error.

Los Alamos conducted the first atomic test at Alamogordo, in a relatively unpopulated part of New Mexico. Alamogordo was conveniently located within driving distance of the Los Alamos Laboratory. One of the most important instruments used to record this test was the Fastax camera that could expose 8,000 frames per second. It recorded the blast effects and the development of the fireball for later examination. Eventually, the EG&G (Edgerton, Germeshausen, and Grier) Company developed Rapidtronic cameras capable of taking millions of frames per second to record these events.

Following World War II, nuclear testing moved to the "Pacific Proving Grounds" in the Marshall Islands. Between 1946 and 1958, the United States conducted 23 tests at Bikini Atoll and 43 tests at Enewetak Atoll. The tests at Enewetak included the first thermonuclear device. Testing eventually moved offshore to barges because much of the archipelago's real estate was disappearing due to the power of the blasts. Following a two-year moratorium from 1959 - 1960, testing was transferred to Johnston and Christmas Islands in a more remote part of the Pacific. Most of the 36 tests conducted here were airdrops. In addition, the AEC conducted a number of high-altitude and space-based tests using missiles launched from Johnston Island and from an aircraft carrier in the South Atlantic.

In the continental United States, the Nevada Proving Grounds (later the Nevada Test Site) opened for business in 1951. Here, the AEC conducted 828 underground and 100 atmospheric tests. The early tests focused on tactical and strategic atomic weapons, as well as boosted warheads and the primaries for thermonuclear weapons.

It is important not to overlook the importance of testing. Because of the two-year test moratorium authorized by President Eisenhower in 1959, the AEC issued the Armed Forces a number of dud thermonuclear warheads. This resulted from a misunderstanding of the interactions between tritium and its He3 decay product, which rapidly accumulates in the gas reservoirs associated with boosted primaries. Other warheads were equipped with defective implosion systems when the composition of their explosive lenses was changed or deteriorated. The AEC only discovered these warhead problems after the resumption of testing in 1961.

The grave results of the post-moratorium tests caused the DOE and the DOD to formalize Stockpile Stewardship programs that the Department of Energy refers to as New Material and Stockpile Assurance and the Department of Defense refers to as Quality Assurance and Reliability Tests (QART). The information discussed here derives mainly from sealed pit weapons, which first appeared with the W25 air-to-air warhead in 1958. Since this time, the Stockpile Stewardship program has continuously evaluated 45 different sealed pit warheads, bombs, and artillery shells, from the W25 through the W88. The program also evaluated the W33, which was not a sealed pit weapon. Altogether, through 1995, the Stockpile Stewardship program randomly sampled and tested 13,800 weapons and devices in one form or another. It continues to do so, albeit at reduced levels. This is in part because there are only seven bomb or warhead types remaining in the active stockpile. These weapons are the B61, B83, W76, W78, W80, W87, and the W88.

Some of the data used in Stockpile Stewardship programs came from hydrotests and full-scale tests as previously described. Laboratory and "joint test assembly" evaluations produced the bulk of the data. During the laboratory evaluation of a nuclear device, technicians test all of its fuzing modes at the highest level of fuzing assembly possible. They expend one-time devices and apply electrical loads as required. During system-level tests of bombs, the technicians generate artificial signals to certify pre-arming functions and employ a centrifuge to evaluate warhead trajectory-sensing devices. After the technicians complete their system-level tests, they evaluate the warhead's firing-set and conventional explosive package. The conventional high explosive package undergoes a variety of tests, which in some cases involves its detonation. In others, explosive samples have their physical and chemical properties examined. The technicians then remove permissive action links and command disable equipment for bench testing. They also examine plastics for age effects and for exposure to radiation. Finally, they examine the nuclear components with a variety of physical, radiographic, acoustic, electrical, and metallographic techniques.

Flight surveillance testing evaluates non-nuclear components of nuclear weapons in operational flight environments. Since these tests employ missile flights or bomb drops, they require coordination between the DOD and the DOE. Joint flight-testing employs a nuclear device in which DOE personnel replace fissile materials with data recording and telemetry equipment (Joint Test Assembly). The DOE conducts some tests with parts of the nuclear explosive package installed and other parts simulated to obtain data directly on the nuclear package.

The results from the first 37 years of Stockpile Stewardship testing produced 1,800 "findings" with regard to nuclear weapons. (Six hundred findings came from other sources.) Of the 2,400 total findings, 848 were "distinct." Of the distinct findings, 416 were "actionable." That is, they required changes in the weapon or its stockpile to target sequence. Of the actionable findings, DOE and DOD personnel found problems with the primary in 34 weapon types, with the secondary in eight weapon types, and with non-nuclear components in 38 weapon types. In some cases, a weapon type had overlapping actionable findings. These can be further broken down into actionable findings in arming and safing systems, cables and connectors, use control (PAL) devices, neutron generators, gas transfer systems, radar equipment, parachutes, structure and assembly, the primary, the detonators, and the secondary.

In the 1960s the AEC compiled the data from nuclear and experimental tests into a series of computer codes to support the Stockpile Stewardship Program. Only the National Security Agency has more computing power than that available to the nuclear agencies. A state-of-the-art IBM Blue Gene \ L supercomputer is in operation at LLNL while LANL maintains IBM's comparable Cielo. Programmers anticipate these computers will be able to provide reliable designs without the need for physical tests. (A warhead designed with the aid of a supercomputer in the mid-1960s required 23 field tests for its development whereas a warhead developed in the 1970s required six tests.) Of course, the reliability of computer predictions is greater for warheads that approximate the design of derivative weapons. The second thermonuclear test conducted by the United States in 1954 ran away to 15 megatons against a predicted yield of 6 megatons when its designers overlooked a neutron reaction involving Li7!

With the implementation of the Limited Test Ban Treaty in 1963, nuclear testing moved underground. This ended the release of hazardous radioisotopes into the atmosphere from atmospheric tests by the United States and the Soviet Union. At the Nevada Test Site, drill rigs with bit diameters ranging from three to twelve feet bored holes up to 5,000 feet deep to emplace nuclear devices. Fortunately, the need for large, multimegaton tests was largely over. Deployed weapons were moving to smaller, multiple independently targetable reentry vehicles carrying warheads with yields of several hundred kilotons, making this program practical. Nevertheless, the AEC carried out one final, high yield experiment. In 1971, the AEC conducted the 5-megaton Cannikin proof test of the W71 ABM warhead in Alaska. There was considerable opposition to this underground test over concerns that the blast might trigger a devastating series of earthquakes in this seismically active region. The signing of the Threshold Test Ban Treaty in 1976 subsequently limited the maximum size for subsequent underground tests to 150 kilotons and the Comprehensive Test Ban Treaty, signed in 1991, banned tests with a nuclear yield altogether.

Since the end of nuclear testing, the DOE has focused on safety and on developing an improved set of simulation codes to model the nuclear processes in atomic weapons. In conjunction with personnel attached to the DOD Nuclear Weapons Safety Standards Program, members of the DOE Nuclear Explosive and Weapons Surety Program are working to prevent accidents and the inadvertent or unauthorized use of nuclear weapons. A revised set of safety standards issued in 2005 as DOE Order 452.1C emphasizes the DOE's responsibilities for nuclear explosive operations. The five Nuclear Explosive Surety Standards outlined in this order provide controls that:

- Minimize the possibility of accidents, inadvertent acts, or approved activities that could lead to fire, high explosive deflagration or unintentional high explosive detonation
- Minimize the possibility of fire, high explosive deflagration, or high explosive detonation through accidents or inadvertent acts
- Minimize the possibility of deliberate unauthorized acts that could lead to high explosive deflagration or high explosive detonation
- Insure adequate security of nuclear explosives
- Minimize the possibility of or delay unauthorized nuclear detonation
- Four complimentary DOD Nuclear Weapon System Safety Standards provide positive measures that:
- Prevent the production of a nuclear yield in nuclear weapons that are involved in incidents or accidents
- Prevent deliberate pre-arming, arming, launching, or release of nuclear weapons except upon execution of emergency war orders
- Prevent inadvertent pre-arming, arming, launching, or release of nuclear weapons in all normal and credible abnormal environments
- Insure adequate security of nuclear weapons

Established in 1995, the DOE's "Advanced Simulation and Computing (ASC) Program" is intended to refine the codes that implement the algorithms used to solve hydrodynamic and radiation transport equations; build the supercomputers needed to implement these codes; and develop improved data, materials and scientific models for "high fidelity" weapons simulation. The program is a consortium of LANL, LLNL, and Sandia in partnership with computer manufacturers and an alliance of five major universities. Designers use the codes from this program to certify the safety and reliability of the existing stockpile and to answer questions about aging components.

Los Alamos certifies the stockpile of nuclear weapons using a concept known as "quantification of margins and uncertainties." At each stage in the sequence of events that simulates the explosion of a stockpiled weapon, researchers in the Applied Physics Division at Los Alamos National Laboratory assess the reliability margins of the physical quantities that enable the weapon to perform as designed. The QMU process further addresses variations in the data and the physics models underlying the weapon to determine their impact on the accuracy of full simulation results. Statistical analysis of data from the past, and data from contemporary measurements at the Los Alamos Neutron Science Center (LANSCE), play a crucial role in reducing the uncertainties of nuclear cross-sections. As well, designers use the small-scale, integral experiments performed at the Los Alamos Critical Experiment Facility (LACEF) to validate and reduce of uncertainty in nuclear data.

At LLNL, the Computational Nuclear Physics Group provides similar support for the analysis of thermonuclear processes. The CNPG collects and evaluates internal and external experimental data that they compile into specially tailored formats used as inputs for interactive simulation. Outputs from the simulation process include improved models of thermonuclear reaction rates, radiation transport processes, and deterministic particle transport mechanisms.

After nuclear testing moved from the field to the laboratory, the DOE consolidated its various atomic proving grounds at the 864,000-acre Nevada Test Site. The NTS is still active, although not in its original role. Currently, Bechtel Nevada operates the "Device Assembly Facility" at the NTS. This facility manufactures and tests the explosive components incorporated in implosion assemblies. Los Alamos tests its products at "Complex Able" whereas Lawrence Livermore National Laboratory tests its products at "Complex Baker". The NTS still maintains a mandate to resume nuclear testing, should this be required.

The Nellis Air Force Range borders the NTS on three sides. Nellis is the site of the 525-square mile, Sandia operated Tonopah Test Range used for evaluating the ballistic properties of nuclear weapon casings. A wide variety of tracking, communications, and data acquisition equipment supports operations at the test range. Major systems at the TTR include air traffic control, multiple radar and optical trackers, telescopic high-speed cameras, telemetry stations, and computers that provide trajectory analysis. Tonopah provides separate areas for firing artillery shells, dropping bombs, and launching missiles.

CHAPTER 2
PRODUCING AND FINISHING "SPECIAL MATERIALS"

President Franklin D. Roosevelt approved the development of atomic bombs on October 9, 1941. Early the next year, the Office of Scientific Research and Development followed Roosevelt's directive with preliminary studies for the weapon, while the military watched from the sidelines. At this early stage in the development of the atomic bomb, its feasibility was still in great doubt. By June 1942, the OSRD had made such little progress that the Army Corps of Engineers assumed control of the project under the command of Colonel James C. Marshall. Although Colonel Marshall identified a 57-square mile tract of land along the Clinch River near Oak Ridge, Tennessee, as an ideal site for the production of nuclear materials, progress still did not meet expectations.

The Army made a critical decision on October 23, 1942, when it promoted Colonel Leslie J. Groves to Brigadier General and placed him in charge of the bomb project. Deputy Chief of Construction for the US Army, Colonel Groves had recently completed the Pentagon building in Washington and was recognized as a no-nonsense administrator with the ability to get results. Although the bomb project, now known as the "Manhattan Engineering District," was officially a military venture, the federal government supplied its finances and scientists seconded from public institutions staffed it. A group of public and private corporations provided the wherewithal to build, maintain, and operate the plant and equipment that became the heart of this project. By the spring of 1945, when sufficient "special material" was on hand to build the first atomic bomb, General Groves had invested two billion dollars in the project. At the time, the MED was one of the most expensive military undertakings in history!

The cornerstone of the MED and the successor programs of the AEC was its special materials – uranium, plutonium, lithium, deuterium, and tritium. Uranium, a moderately common element, occurs in ore bearing veins found in igneous and sedimentary rocks. Important uranium minerals include pitchblende, uraninite, coffinite, and carnotite. The Belgian Congo and Canada supplied the pitchblende ore from which the United States extracted the uranium for its first bombs. The AEC later supplemented these sources with domestic production from the Colorado Plateau
and by purchases from Portugal, Australia, and South Africa. The United States government stimulated domestic exploration when it artificially raised the price it would pay for uranium ore. After it is mined, uranium ore is milled and leached using various processes to produce concentrates suitable for refining. These concentrates are collectively called "yellowcake." The Manhattan Engineering District constructed America's first uranium refinery on the site of the St. Louis Sash and Door Works. Here, the Uranium Division of the Mallinckroft Chemical Works converted yellowcake into uranium oxide, uranium tetrafluoride, and then metal. The Manhattan Project used uranium metal and compounds processed here as fuel for the plutonium production reactors at Hanford and for the Oak Ridge U235 enrichment program. The St. Louis uranium refinery closed in 1957, after 15 years of operations. The AEC demolished the uranium-contaminated building in 1996.

The Oak Ridge K-25 power plant with the S-50 liquid diffusion enrichment site shown in the background. The S-50 plant doubled the U235 content of natural uranium. (Credit: AEC-Edward Westcott)

The Oak Ridge K-25 gaseous diffusion site. The plant enriched the product from the S-50 site to about 25 percent U235. At the time, it was the largest building in the world. (Credit: AEC-Edward Westcott)

A row of gaseous diffusion cells at the K-25 plant. Because of its attractive economics, gaseous diffusion ultimately became the enrichment process of choice in the United States. (Credit: AEC-Edward Westcott)

The Oak Ridge Y-12 plant. It reprocessed the partially enriched uranium from the K-25 plant using an electromagnetic process. During the Cold War, Y-12's role transitioned from uranium enrichment to producing secondaries for thermonuclear bombs. (Credit: AEC-Edward Westcott)

A Calutron racetrack at the Y-12 Plant. It electromagnetically enriched U235 from partially enriched uranium supplied by the Y-12 Plant. (Credit: AEC-Edward Westcott)

A blank made of enriched uranium metal. It is formed into a hollow tube to increase neutron loss, which enables safe handling and storage. (Credit: DOE)

The B reactor at the Hanford, Washington site (left). It irradiated uranium rods to transmute some of their U238 content into plutonium. It was built in a year. (Credit: DOE)

The face of the B reactor showing its fuel rod arrangement. In total, there were 2004 aluminum tubes that held uranium fuel rods. (Credit: DOE, Photo D 8320)

The S Plant or Canyon, a chemical processing site at Hanford that used the PUREX process to extract plutonium nitrate from irradiated uranium fuel slugs. The plant protected its workers with thick concrete walls and was fully automated because of the high radioactivity associated with processing. It went into production in 1951. (Credit: DOE; Color Original)

Inside the PUREX facility was one long gravity feed system that moved the product down a series of cells to facilitate production. (Credit: DOE; Color Original)

The Hanford Plutonium Finishing (Z) Plant (PFP) where plutonium nitrate recovered from fuel rods in the canyons was converted into plutonium metal. (Credit: DOE; Color Original)

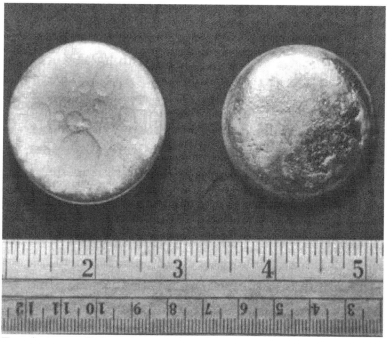
Two buttons of plutonium produced by Hanford. This plutonium was used in the Fat Man bomb dropped on Nagasaki. (Credit: DOE)

Plutonium containers stockpiled at the Rocky Flats Plant in 1988. The plutonium was for the production of weapons parts. (Credit: NPS Photo Rocky Flats 800; Color Original)

A worker prepares to cast a plutonium part in an electric induction furnace while standing outside a protective glovebox (left). A precision plutonium casting mold. (right). (Credit Left: LANL Photo C2178-05; Credit Right Photo: NPS; Color Originals)

The heavy water production plant at the Alabama Ordnance Works. This was one of three plants assigned to the Manhattan Project's P-9 program. The other two plants were at the Wabash River Works and the Morgantown Ordnance Works. Heavy water can be used as a moderator in reactors or it can undergo electrolysis to produce deuterium. (Credit: National Archives, Photo MED 38-4)

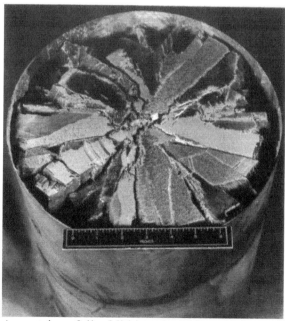

A container full of lithium hydride, which cracked during a machining operation. Its properties are very similar to lithium deuteride, which is used as thermonuclear fuel. (Credit: NASA)

Depending on the nature of the solutions used, ore-leaching processes precipitate yellowcake containing anywhere from 60 - 89 percent uranium by weight. Historically, yellowcake was refined using two processes. By means of a "dry process," a series of chemical operations converted sodium diuranite (Na_2U_2O) yellowcake into uranium hexafluoride gas (UF_6), which refiners purified by means of fractional distillation. In a second "wet process" refiners dissolved uranium octaoxide (U_3O_8) yellowcake in nitric acid. Solvent extraction with tributyl phosphate and a refined kerosene additive then precipitated high purity uranyl nitrate ($UO_2(NO_3)_2$). The refiners further decomposed the nitrate in a furnace to produce orange oxide (UO_3), which a hydrogen process further reduced to brown oxide (UO_2). Wet processing eventually replaced dry processing because it precipitated purified uranium compounds earlier in the refining sequence. This reduced the amount of equipment needed for processing.

The final steps undertaken in the production of uranium metal require the conversion of brown oxide to green salt (uranium tetrafluoride, UF_4) by treatment with anhydrous hydrofluoric acid. Uranium hexafluoride treated with hydrogen gas can also be converted to UF_4. Refiners then react the UF_4 with magnesium in a refractory crucible. The resulting exothermic process achieves temperatures in excess of 1,300 °C, producing an ingot or "button" of high purity uranium metal and a slag of fluorite. In order to manufacture weapon parts, technicians place the hockey puck sized buttons of pure uranium into crucibles and then insert the crucibles into special induction furnaces where the uranium is melted. They then bottom pour the molten uranium into graphite molds. After the uranium castings cool, machinists turn them into their final shapes.

Uranium metal has a melting point of 1,132 °C and a density of 19.1 grams / cubic centimeter in its stable α-phase. It is malleable, ductile, and slightly softer than steel. The metal oxidizes easily, reacts with hydrogen and water vapor, and quickly coats itself with a colorful film. Uranium is pyrophoric, igniting in air at temperatures of 150 °C to 200 °C. Apart from its density and pyrophoric nature, uranium does not require special metal working techniques. Important compounds of this element include uranium tetrachloride (UCl_4) and uranium hexafluoride (UF_6), which the Oak Ridge plant used in U235 enrichment processes for the production of reactor fuels and weapon parts. An important physical property of UF_6 or "hex" is its ability to form a vapor at temperatures above 56 °C.

Although modern weapons are vacuum-sealed, uranium may still react with oxygen, hydrogen, and water vapor generated by internal warhead components such as explosives, plastics, and organics. LANL scientists have, therefore, investigated a large number of binary and tertiary alloys to produce more workable and stable uranium components. In particular, they have researched uranium alloyed with tungsten, titanium, tantalum, and niobium. They found that a 6 percent niobium-uranium (U-6Nb) alloy has characteristics particularly relevant to the production of radiation cases in thermonuclear weapons. Since the NNSA is extending the working lives of modern weapons well beyond their intended retirement dates, it is now necessary to monitor the stability of the crystal phases of the uranium alloys that the NNSA's facilities used in earlier warhead construction.

The isotopic composition of natural uranium is approximately 99.3 percent U238 and 0.7 percent U235, with the actual ratio dependent on the specific deposit from which it comes. It also contains trace amounts of U234. Metallurgists have identified eleven other uranium isotopes. Of these, only the artificial isotope U233 is of interest in weapons applications. This is because it can form a critical mass and has a low spontaneous fission rate.

U238, uranium's most common isotope, is "fissionable" rather than fissile. This is to say that although a high-energy neutron can split U238's nucleus, the isotope cannot form a critical mass, nor can it sustain a chain reaction. U238 is, however, an important nuclear material because atomic reactors can transmute it into plutonium. Its high-density and its neutron reflecting characteristics also make it an excellent tamper. The primary mode of decay for U238 is by alpha particle emission. A secondary mode of decay is by spontaneous fission at a rate of 5.5 fissions / second / kilogram, which is about 100 times higher than the rate found in U235. This equates to 10.8 neutrons / second / kilogram and is a drawback to its use in gun-assembled weapons, which have low insertion rates. The half-life of U238 is about 4.5 billion years.

U235 is the only fissile material known to occur in nature and can form a critical mass from a bare, 48-kilogram sphere of pure metal. Critical mass increases to 52 kilograms for a sphere of "weapons grade" metal. The term weapons grade refers to uranium enriched to 93.5 percent U235, a level of purity that is not necessarily a physical requirement for a bomb. It derives from production economics and the fact that U235 criticality increases very rapidly at high purity making it difficult to handle.

A 5-centimeter (2-inch) thick beryllium shell can reduce the mass of weapons grade uranium necessary for criticality to only 21 kilograms, whereas a theoretically infinite beryllium tamper reduces the critical mass to 17.2 kilograms. U235 has the lowest spontaneous neutron emission rate of any fissile material and for this reason, designers used it in gun-assembled and other weapons sensitive to prevent pre-detonation. Like U238, its primary mode of decay is through alpha particle emission. Its half-life is 700,000,000 years.

Establishing quantity production of U235 for use in nuclear weapons was the first big hurdle for the Manhattan Engineering District. Like all isotopes of a common element, U235 and U238 have virtually identical physical properties. Because of this similarity, it is impossible to separate these two isotopes chemically. Instead, it was necessary to consider four physical processes – thermal diffusion, gaseous diffusion, electromagnetic separation, and gas centrifuge separation – to "enrich" the U235 isotope content in natural uranium. Since no one had ever used these processes on an industrial scale, it was far from certain if they could be adapted to support the needs of a weapons program. Despite this concern, General Groves decided in early 1943 to establish a giant military-industrial complex known as the Clinton Engineering Works (Site "X") at Oak Ridge, Tennessee, to pursue the two most promising enrichment processes.

The Kellex Corporation designed a gaseous diffusion plant with 2,000,000 square feet of floor space at the K-25 Site on the Oak Ridge Reservation. The J. A. Jones Company constructed it. Inside the world's largest building (at the time), a 3,024-stage system passed pressurized uranium hexafluoride gas through a cascade of semi-permeable barriers to separate fast moving molecules of U235 from slow moving molecules of U238. Harold Urey of Columbia University, the world's leading expert on isotope separation at the time, designed the K-25 diffusion system. Typically, only about half of the gas in the system passed through the barriers to their low-pressure side. The uranium hexafluoride gas was then re-circulated with successively reduced concentrations of U235 until it reached a cutoff level of 0.1 - 0.3 percent and was discarded.

Houdaille-Hershey manufactured the barriers used in the diffusion process from a combination of compressed nickel powder stabilized by an electro-deposited nickel mesh. The barriers were capable of withstanding attack by the incredibly corrosive uranium hexafluoride feedstock. The microscopic pores through which the molecules passed were only 10 nanometers, or 10 billionths of a meter in diameter. The Chrysler Corporation built the special tanks that housed the barriers and DuPont produced the hexafluoride feed stock for the diffusion process. Almost 600 miles of high-pressure stainless-steel piping was required to connect the barrier tanks and DuPont had to create a new wonder material called "Teflon" to seal their connections. Union Carbide contract-operated the K-25 Plant.

Stone & Webber designed the second enrichment facility and the Tennessee Eastman Division of Eastman Kodak constructed it at Site Y-12. Here, inside of special scaled-up mass spectrometers known as California cyclotrons or "Calutrons," an arrangement of electromagnetic fields separated streams of U235 bearing molecules from a feed of uranium tetrachloride ions. Ernest O. Lawrence at the University of California, Berkeley designed the instruments and Allis-Chalmers manufactured them. Due to wartime copper shortages, Allis-Chalmers used 300,000,000 dollars' worth of silver bullion borrowed from the Treasury to manufacture their electrical windings. General Electric supplied the high voltage electrical equipment used to power this system and Westinghouse provided the vacuum equipment.

Set up in oval cascades known as "racetracks," the calutrons provided weapons grade enrichment in a two-stage electromagnetic separation process. Nine alpha racetracks, each composed of 96 large (four-foot high) calutrons, produced a 15 percent partially enriched product. This output was fed into six beta racetracks, each of which contained 72 smaller (two-foot high) calutrons. The second racetrack produce a final product enriched to 90 percent U235. In contrast to the continuous gaseous diffusion process, electromagnetic enrichment was a batch process and due to inefficiencies in the design of its equipment, required prodigious quantities of feedstock, which was produced onsite by the hydrochlorination of brown oxide. Tennessee Eastman contract-operated the Y-12 facility after its completion.

The Manhattan District initially rejected liquid thermal diffusion as inefficient but the Navy did not. The Navy built a thermal enrichment plant at the Philadelphia Naval Yard where waste heat from a boiler laboratory was available to power the process cheaply. The Navy Yard piped steam up tubes placed inside of pipes containing a uranium hexafluoride solution. The lighter U235 molecules in the solution diffused toward the warmer inside of the pipe and, as the uranium bearing solution convected upwards, a mechanical arrangement skimmed enriched fluid off at the top. A sequence of columns produced a continuous cascade like the setup adopted for the K-25 diffusion process. The Navy's purpose in producing low enriched uranium was to evaluate its potential as a reactor fuel for shipboard propulsion.

When difficulties arose in getting the electromagnetic and gaseous diffusion plants into full production, the Manhattan District borrowed the Navy's technology and had H. K. Ferguson build a 2,100-column liquid thermal diffusion plant at Site S-50 in only 90 days. The speed with which H. K. Ferguson placed thermal diffusion into operation was only possible because of the availability of waste heat from a thermal electric plant used to power the Y-12 calutrons. The 2 percent marginally enriched uranium produced by this process fed the K-25 diffusion operation, which in turn passed its 23 percent enriched output into the Y-12 calutrons.

Electromagnetic separation provided most of the enrichment for the components of the first uranium bomb. This was due to delays in obtaining reliable barriers for the K-25 Plant. Gaseous diffusion proved to be the superior technology in the end. After the war, the high maintenance and power requirements needed for the Y-12 calutrons priced the electromagnetic separation process off the market. Oak Ridge shut its alpha racetracks down in September 1945 and followed suit with the beta racetracks at the end of 1946. By this time, Los Alamos had built a new K-27 diffusion facility that could enrich uranium to weapons grade on its own. The K-29,

K-31, and K-33 facilities followed K-27. Together, these operations added another 1,540 stages to the diffusion operation. Despite the immense size of the Oak Ridge Complex, the Manhattan District measured its daily production of fully enriched uranium in ounces up to the war's end.

After World War II, the deepening climate of the Cold War caused President Truman and the National Security Council to ask Paul Nitze, the Director of Planning for the State Department, to study the "United States Objectives and Programs for National Security." The State Department published this study as document NSC-68 in April 1950. Nitze concluded that the main objective of the Soviet Union was world domination and that the United States had a moral obligation to defeat the Soviet threat. He further concluded that the United States could afford to increase military spending to meet this obligation.

Although the bulk of the increased military spending recommended by NSC-68 went to arming conventional forces in the United States and in Europe via Military Assistance Programs, a direct result of NSC-68's recommendations was the decision to increase the rate of production for enriched uranium by 150 percent and the rate of plutonium production by 50 percent. For this purpose, the AEC contracted the Goodyear Atomic Corporation and Union Carbide to build and operate two additional gaseous diffusion plants. These came on line in 1956. Located at Portsmouth, Ohio, and Paducah, Kentucky, the new gaseous diffusion plants provided another 5,892 stages of enrichment. Of the new facilities, Paducah was the feed point. It produced and shipped low enriched uranium to Portsmouth and Oak Ridge for upgrading to weapons grade metal. Electric Energy Inc. and the Tennessee Valley Authority constructed the additional generating capacity needed to power the plants. At peak enrichment, the power consumed by the Nuclear Weapons Complex totaled just over 60 billion kilowatt-hours per year or 12 percent of national electricity production.

The AEC initially purchased uranium hexafluoride feedstock for its enrichment programs from a variety of commercial operations in the northeastern United States. It discontinued these purchases in 1947 when Oak Ridge began refining its own feedstock at the newly constructed F-27 Plant. It also constructed additional refining capacity at Fernald, Ohio and St. Louis, Missouri. The AEC further increased feedstock production by providing Paducah and Portsmouth with their own refineries after they reached operational status. In 1962, the AEC consolidated all of its feedstock production when it had Honeywell Specialty Chemicals construct a large uranium conversion plant at Metropolis, Illinois.

Because U235 is a recyclable product, weapons grade material from the expansion programs accumulated rapidly. In 1964, the AEC discontinued enrichment of uranium for military purposes when production reached 645 tons. The gaseous diffusion plants at Paducah and Portsmouth then switched their production to reactor fuels. The United States Enrichment Corporation now owns and operates both plants. Should additional stocks of weapon grade uranium be required, the DOE could arrange for USEC to modify these plants to produce weapons grade material once again. The DOE could also build new plants to take advantage of more sophisticated enrichment technologies.

Foremost among new enrichment technologies was gas centrifuge separation. In this process, centrifugal force separates lighter, U235 containing molecules from a stream of natural uranium hexafluoride gas. Like the diffusion process, gas centrifuges require a cascade of stages to produce a final product. However, the power requirements to run a centrifuge operation are lower than for diffusion and the setup is more amenable to industrial scaling. The DOE constructed a gas centrifuge plant at Portsmouth in the early 1970's, but never placed this installation on full production. Gas centrifuge facilities have produced slightly more than half of the world's enriched uranium supply.

Aerodynamic separators, cousins of the gas centrifuge, use the forces created by spinning high-pressure gas out of a specially designed nozzle to separate uranium isotopes of different atomic weights. This process is less attractive than a centrifuge operation because aerodynamic processing requires significant dilution of its UF_6 feed gas with hydrogen or helium to attain the velocities necessary for efficient isotopic separation. The processing of the diluted gas significantly increases this method's power utilization.

Uranium processors have also developed three laser processes for isotope enrichment. Atomic vapor laser isotope separation (AVLIS) uses tuned lasers to ionize U235 atoms in a stream of uranium vapor. After the atoms are ionized, a negatively tuned plate attracts and collects them. Molecular vapor laser separation (MVLS) takes advantage of atomic excitement induced in U235 hexafluoride molecules by shining a tuned infrared laser on a stream of natural uranium hexafluoride gas. Shining a second laser in the gas stream frees a fluorine atom from the excited U235 molecules to precipitate uranium pentafluoride. Separation of isotopes by laser excitation (SILEX) is a classified process rumored to be an order of magnitude more efficient than other processes in the enrichment of U235.

Pu239, which is also commonly used in nuclear weapons, does not occur naturally. Scientists created it artificially by bombarding uranium with neutrons in an atomic reactor, a device that Enrico Fermi first pursued at the University of Chicago in 1942. He was interested in testing the concept of criticality and in researching the reactor's potential for power generation and plutonium production. Fermi realized that despite natural uranium's low concentration of U235, a sufficiently large mass of such fuel might be used to produce a controlled, self-sustaining chain reaction.

The neutrons necessary to start up a reactor fueled with natural uranium come from the spontaneous fission of its U238 content. There are three possible outcomes for these neutrons. They may be lost through the surface of the fuel mass; absorbed by another nucleus of U238 to produce plutonium by way of beta particle emissions; or absorbed by a nucleus of U235 to undergo fission with the release of energy and more neutrons. Enriching the U235 concentration of the fuel or moderating the speed of produced neutrons to prevent them from being absorbed in the U238 resonance window increases reactivity.

The best moderators are elements with light nuclei such as hydrogen or carbon. Engineers chose graphite as a moderator in early reactors because it is an inexpensive naturally occurring carbon solid. Light water, which contains two nuclei of ordinary hydrogen, is also a good moderator. Unfortunately, it has the undesirable property of absorbing neutrons, which impairs criticality. A better choice is heavy water that contains deuterium in place of ordinary hydrogen. Deuterium has a much lower cross-section for neutron absorption than hydrogen has. Each time a neutron strikes a light nucleus, it loses a little bit of energy until it is "thermalized." U235 has a much larger cross-section for absorbing thermal neutrons than U238. For this reason, the AEC designed its reactors so that their nuclear fuel was interspersed in a mass of moderating material to take advantage of its inexpensive boost to reactivity.

When newly created neutrons exactly balance the number of neutrons causing fission, a reactor has a "reproduction factor" of one. If the reproduction factor is smaller than one, the reactor eventually shuts down. If it is larger than one, reactivity can run away and produce a meltdown from overheating, such as happened at Chernobyl in the Ukraine. For the purposes of controlling the reproduction factor, technicians insert special control rods into a reactor. In early reactors, control rods were made of cadmium. Hafnium and boron steel have recently become popular alternatives. All three of these materials have a voracious appetite for absorbing neutrons. Designers incorporate sufficient control rods into a reactor to produce an excess of neutron absorption with which to control reactivity. When a reactor starts up, technicians oversee the gradual withdrawal of these rods until it "goes critical" and a "steady state" supply of neutrons is established.

Enrico Fermi had the world's first nuclear reactor built on an unused squash court under Stagg Field at the University of Chicago. For its construction, it required 771,000 pounds of graphite moderator in combination with 12,400 pounds of uranium and 80,590 pounds of uranium oxide fuel. The graphite was machined into blocks and piled in layers; one layer of solid graphite alternating with two layers in which the graphite blocks were embedded with a pair of 5-pound uranium spheres. Leo Szilard provided the arrangement of the uranium-filled blocks. Key to the design of the reactor was acquiring pure graphite because boron, a notorious absorber of neutrons, is a common contaminant of this mineral.

The final pile, named "Chicago Pile Number One," had an elliptical shape with a major axis of 25 feet and a minor axis of 20 feet. It was 76 layers or 20 feet high. Technicians controlled CP-1 by means of neutron absorbing cadmium sheeting nailed to 13-foot long wooden strips interspersed throughout the core. After their insertion, technicians padlocked the strips in place as a safety measure. Fermi first ordered his assistants to withdraw the strips by hand on the night of December 2 - 3, 1942, at which time the reactor went critical with a power output of 1/2 watt.

After its first successful test, Fermi operated the CP-1 reactor at power levels up to 200 watts. He also built and operated the CP-2 and CP-3 reactors at less public locations. These early reactors had a number of engineering deficiencies. They could not adequately cool themselves at higher thermal outputs and technicians had to tear them apart, block by block, to gain access to their fuel. The 3.5-megawatt X-10 experimental reactor, subsequently built at the Oak Ridge Reservation in Tennessee to evaluate the requirements for full-scale plutonium production, was a much better design.

The scientists at Oakridge constructed the X-10 reactor with a graphite core that measured 24 feet on each side and surrounded it with a 7-foot-thick concrete radiation shield. Spaced on 8-inch-centers, 1,248 aluminum-lined fuel channels penetrated through the core. In these channels, technicians placed five tons of natural uranium metal in the form of aluminum-jacketed slugs, whose inner uranium cores measured 8 inches long by 1.36 inches in diameter. The aluminum jackets prevented the corrosion of the uranium fuel and prevented the release of fission products. The channels were large enough that fans were able to blow a cooling flow of air around the fuel elements. The technicians loaded the reactor through its front face in a process facilitated by an elevator. Four horizontal shim rods, two horizontal regulating rods, and six vertical safety rods provided control. In addition, a safety system could pour boron steel shot into the reactor in the event of an emergency.

The neutrons produced by the decay of U238 in the Oak Ridge X-10 reactor slowed as they repeatedly bounced off molecules of graphite. Some of these neutrons provided reactivity by fissioning U235 atoms, whereas others transformed U238 into U239 through a process of absorption. In less than a week, beta particle emissions transmuted the U239 into Np239 and then into Pu239. When sufficiently irradiated, technicians ejected the fuel rods through the rear face of the reactor into cooling ponds where their short-lived fission byproducts decayed. An underwater monorail, which provided radiation shielding, transferred the spent fuel rods to an adjacent demonstration plant, set up for the chemical separation of the slugs' plutonium content. Oak Ridge then sent its first Pu239 samples to Los Alamos for analysis.

The results were bitterly disappointing. Metallurgists described the physical and nuclear properties of plutonium as "fiendish." It occurs in six different crystal phases of which α-phase plutonium is stable at room temperature and up to 122 °C. This phase has a density of 19.84 grams per cubic centimeter but is almost impossible to work due to its brittle nature. Fortunately, metallurgists can alloy the δ-phase, normally found in the temperature range 319 °C to 476 °C, with 0.8 percent by weight of gallium to stabilize it at room temperature. Δ-phase plutonium has a density of 15.92 and is quite ductile, making it easy to work. Δ-phase plutonium has the rather dangerous characteristic of permanently reverting to the denser α-phase if subjected to a shock. Weaponeers must account for this property in their weapon designs.

Plutonium melts at 641 °C and has low thermal conductivity. It oxidizes in air and even more violently when in contact with moisture. A few shavings of plutonium ignited at the Rocky Flats Plant in 1969 starting a major fire in the weapons fabrication area. Over two tons of plutonium metal eventually went up in smoke and if the fire had burned through the containment, the city of Denver would have become a radiotoxic wasteland. The incident had the potential to become one of America's most serious industrial accidents.

The radiochemistry of plutonium is just as interesting as its physical properties. Metallurgists have identified 15 isotopes of plutonium and all are capable of forming a critical mass! Nevertheless, only a few isotopes are useful to weaponeers. The most important is Pu239, the predominant isotope produced in nuclear reactors. Depending on the period of time that fuel remains in a reactor, Pu239 can absorb low-energy neutrons to become Pu240. Pu240 can yet again absorb neutrons to become Pu241. modern weapons grade plutonium is therefore composed of approximately 93.5 percent Pu239, 6 percent Pu240, and 0.5 percent Pu241. It also contains traces of Pu238 and Pu242. The actual composition is dependent on the irradiation period of the fuel rods and the neutron flux specific to the irradiating reactor's design.

Plutonium isotopes decay primarily by alpha particle emission. They also have high spontaneous fission rates. In particular, Pu240 has a spontaneous rate of 478,000 fissions / second / kilogram, which produces just over 1,000,000 neutrons per second. Unavoidable contamination by even small amounts of Pu240 renders plutonium unsuitable for use in gun-assembled weapons because of their low insertion rates. The scientists at Los Alamos were in a quandary about how to weaponize plutonium until they developed their scheme for high-speed implosion to overcome its spontaneous neutron background.

A 10.5-kilogram sphere of weapons grade, α-phase plutonium forms a bare, critical mass. Surrounding a plutonium sphere with 5 centimeters of beryllium reduces its critical mass to 5.5 kilograms. American designers have directly bonded plutonium to beryllium in operational nuclear weapons. In its more usual gallium stabilized, δ-phase, a bare, critical mass of plutonium constitutes a 16.6-kilogram sphere. This reduces to 5.8 kilograms with a theoretical, infinitely thick, beryllium tamper. A core will feel warm to the touch because plutonium's nuclear decay rate produces energy at a rate of 1.9 watts / kilogram. Because of this energy production, an insulated core can reach its melting point in only a few hours, an effect that weaponeers must address in the design of weapons. Plutonium's high nuclear decay rate, the result of its 24,000-year half-life, is the reason for its extreme radio-toxicity.

Because of the hazards associated with plutonium, the Manhattan Engineering District located its plutonium production reactors on a remote, 560-square-mile reservation near Hanford, Washington. The MED designated this area as "Site W," and called it the Hanford Engineering Works (HEW). In order to test materials for use in Hanford's planned, full-sized, production reactors, MED engineers built the 305 Test Pile. The 305 Reactor was a graphite cube that measured 16 feet on each side. It was shielded by five feet of concrete, air-cooled, and fueled with 27 tons of natural uranium placed in horizontal tubes with a lattice spacing of 8.5 inches. Horizontal regulating and shim rods controlled its power. In addition, vertical and horizontal safety rods and a steel-shot-filled vertical safety tube could SCRAM the reactor in case of an emergency. In addition to fuel rods, 20 horizontal openings accommodated test stringers for irradiating samples.

Hanford engineers began operating the 305 Reactor at 50 watts of power in March 1944. They used it to test graphite, aluminum, uranium, and other materials that the Manhattan Engineering District planned for use in the site's full-scale production reactors. These tests allowed Hanford engineers to verify which materials met their specifications for use in the production reactors. The 305 Reactor also provided radiation that the engineers used for instrument development. After the war, the 305 Reactor operated through 1972, at which time the AEC converted the building that housed it to a fuel fabrication facility. The AEC disposed of the reactor's core between 1976 and 1977.

Situated on the Columbia River, the Hanford Engineering Works used water for cooling its production reactors. Unlike the X-10 experimental reactor built at Oak Ridge and Hanford's 305 Reactor, the 250 MW DuPont plutonium production reactors required more effective cooling than fan-driven airflow. DuPont began construction of the reactors in the latter half of 1943, with the idea of bringing three units on line, spaced at intervals of six miles. A 1,200-ton graphite block penetrated by 2,004 aluminum tubes formed the basic design for the site's B, D, and F reactors. Inside the reactor's aluminum tubes, technicians placed 200 tons of natural uranium in the form of aluminum-jacketed slugs. The slugs were about 8 inches long, 1.4 inches in diameter and weighed 8 pounds. Operators controlled each reactor by means of 9 vertical control rods and 29 horizontal safety rods. Every 100 days, new slugs inserted into the front of the reactor pushed irradiated fuel slugs out of the back.

A 10-inch thick cast iron thermal shield and a 4-foot thick biological shield made from laminated steel and Masonite surrounded the reactor cores. A welded steel box, equipped with expansion joints, encased the reactor's shields. Built as a "single pass" design, a throughput of 75,000 gallons of water per minute cooled the reactors before returning directly to the Columbia River. A closed-loop cooling circuit serviced the thermal shield. Unlike the X-10, which was open to the atmosphere, Hanford's reactors operated inside sealed, helium-charged systems. This simplified leak detection and allowed for the recovery of gases vented from the core.

A variety of companies in the Midwest manufactured fuel slugs for the Oak Ridge and Hanford reactors. In the spring of 1945, the Manhattan District centralized much of this work at Hanford's Building 314 in Area 300. Here, machinists cut uranium rods to length and sealed them in aluminum jackets. Between 1946 and 1948, Hanford also undertook the extrusion and rolling of uranium rods, but in the end, returned to purchases from commercial suppliers. Starting in 1952, FMPC of Fernald, Ohio, took over the machining of uranium slugs. In 1956, a second plant built at Weldon Springs, Missouri, supplemented FMPC's production. The late 1950s saw enriched uranium replace natural uranium metal, hollow tubes replace solid uranium stock and the introduction of zircalloy cladding.

HEW carried out the complex aluminum-silicon cladding process in the 313 Metal Fabrication Building located in Area 300. The fuel slugs differed from the design adopted for X-10 by having their jackets bonded to their uranium cores to improve heat conduction. At the start of the cladding process, technicians cleaned rolled uranium stock in a trichloroethylene vapor degreaser before dipping it in a nitric acid bath and two rinse tanks. Following cleaning and drying, the technicians placed lengths of the stock in steel cylinders, added aluminum end caps and dipped the cylinders in a molten bronze (53 percent tin and 47 percent copper) bath to raise their temperatures to 660 °C - 770 °C, uranium's β-phase. They then dipped the cylinders in a molten tin bath to remove any traces of bronze from the stock. The tin bath reduced the cylinder's temperatures below 660 °C, to uranium's α-phase. This process randomized the uranium stocks' grain structure to prevent expansion when it was irradiated. Centrifuging removed any traces of tin.

The transfer of the cores to an aluminum-silicon brazing bath took place while the cylinders' temperatures were still in uranium's α-phase. After a dip in the cladding solution, the cores were water quenched. The heating, dipping, and cooling process took place three times to ensure that an adequate thickness of aluminum-silicon had accumulated on the cores. After completing the triple dip process, technicians removed the fuel rods' steel jackets and cleaned the rods with sodium hydroxide. After cleaning, a fluoroscope determined the proper lengths of the fuel rods and a technician marked them with a punch to show the amount of machining required to bring them to their specified length. Finally, the fuel rods' end caps were tungsten welded to the porous cladding using an inert-gas process that insured a good seal.

After machining, the fuel rods arrived at Hanford's reactors for insertion and irradiation. Hanford kept the fuel rods' irradiation period short to prevent the newly formed Pu239 from transforming itself into Pu240. Because of this decision, Hanford's early plutonium cores contained less than 1 percent Pu240 – a composition that is now designated "super grade." Technological advances allow modern bombs to use less expensive "weapons grade" plutonium, with a 6 or 7 percent Pu240 content. As with uranium, the term weapons grade refers primarily to production economics. Following the removal of the irradiated fuel rods from the reactor, remote handling equipment transferred them to cooling ponds in which their short-lived fission byproducts decayed. Additional equipment sealed the rods in casks, which a rail system took to a pair of processing areas, discretely located on the other side of Gable Mountain.

The Manhattan District designed the HEW's processing capability around a pair of bismuth phosphate processing plants, located at each of its two processing areas: 200 East and 200 West. Hanford built Plants T and U at site 200 West and Plant B at site 200 East. It cancelled Plant C when it determined that only two plants were required to process its entire output of irradiated fuel rods. HEW then turned Plant U into a training facility for personnel assigned to Plants B and T. In addition to plutonium reprocessing, Hanford used Plant B for the separation and purification of cesium, and for strontium encapsulation. Plant T provided support as a decontamination and repair facility.

HEW personnel called the B and T processing plants "canyons" or "Queen Marys" because of their size. These were massive concrete containment structures, 800 feet long by 65 feet wide by 80 feet high. Each plant had 20 separate cells that accommodated the progressive stages of the bismuth phosphate chemical separation process developed by Stanley Thompson and Glenn Seaborg. Engineers used corrosion resistant stainless-steel to build the processing equipment located inside of the cells, which technicians operated by remote control. After startup, lethal levels of radioactive contamination prohibited human entry to the cells. Seven-foot-thick, reinforced concrete walls with 35-ton steel lids protected plant personnel from the deadly contents of the cells.

At the start of reprocessing, a concentrated sodium hydroxide solution removed the aluminum fuel jackets from Hanford's irradiated uranium slugs, which contained about 250 parts per million of plutonium. The processing sequence next called for the dissolution of the fuel slugs in a combination of nitric and sulfuric acids to keep uranium and other fission products in solution. The addition of bismuth nitrate and phosphoric acid precipitated a combination of plutonium nitrate and bismuth phosphate solids, the latter compound giving its name to this process. Centrifuging separated the precipitate from the solution.

The addition of nitric acid again dissolved the plutonium precipitate. The further addition of potassium permanganate and sodium bismuthate oxidized the plutonium content to keep it in solution and precipitated bismuth as bismuth phosphate. After a number of purification steps, the addition of hydrogen peroxide, sulfates, and ammonium nitrate precipitated the hexavalent plutonium as plutonium peroxide. The dissolution of the plutonium peroxide in pure nitric acid allowed the aqueous plutonium to precipitate in its nitrate form. The technicians then placed the wet plutonium nitrate in small shipping cans and boiled it with heated air. This process reduced the plutonium to a paste, which was stored in the 213-J and 213-K vaults at the southeast end of Gable Mountain. Each shipping-can held about 1 kilogram of plutonium.

In order to produce plutonium metal, Hanford experimented with a REDOX extraction process at its "S" Plant, beginning in 1951. It also implemented a plutonium uranium recovery by extraction (PUREX) process in 1956. The PUREX process was similar to the methodology used for the wet extraction of uranium. It has since become the standard for plutonium extraction because it also recovers reusable uranium. Variations of the PUREX process include uranium extraction (UREX), transuranic extraction (TRUEX), diamide extraction (DIAMEX), selective actinide extraction (SANEX), and universal extraction (UNEX).

The conversion of plutonium nitrate to plutonium metal took place in Hanford's "Z" or Plutonium Finishing Plant (PFP). This agglomeration of 60 buildings was the end of the line for plutonium processing at Hanford. Here, technicians carried out metal production behind appropriate shielding using glove boxes and pneumatically operated mechanical manipulators. Plutonium nitrate arrived at the Plutonium Finishing Plant in special "shipping bombs," sealed inside of protective cases. The shipping bombs normally contained 80 or 160-gram batches of plutonium nitrate slurry. After receipt at the PFP, a number of processes produced plutonium tetrafluoride in preparation for its final reduction to a metallic phase.

One such fluoridation process called for the dissolution of plutonium nitrate in acidified water. PFP personnel kept their processing batches smaller than 320 grams to prevent criticality accidents. The addition of hydrogen peroxide maintained the plutonium in solution as plutonium (IV). The solution was next heated to 60 °C, at which time oxalic acid was added to precipitate plutonium oxalate ($Pu(C_2O_4)_2$). After recovery and air-drying, the plutonium oxalate was heated to a temperature of 450 °C in a hydrogen fluoride and oxygen atmosphere to form plutonium tetrafluoride ($PuF4$). During this process, the plutonium first decomposed to its oxide (PuO_2) with the emission of carbon dioxide (CO_2). It then formed its tetrafluoride (PuF_4) through the absorption of fluorine and the emission of water vapor. During the course of fluorination, the neutron flux increased rapidly through alpha bombardment of the fluorine absorbed by the plutonium compound. The neutron flux of a completely fluorinated 45-gram batch of plutonium tetrafluoride reached 14,000 neutrons / second / centimeter^2 at a distance of 30 centimeters. The calculated total neutron flux of the batch was 97,000,000 neutrons / second.

After fluorination, technicians allowed the contents of the furnace to cool to 300 °C under a flow of hydrogen fluoride and oxygen gas. Below 300 °C, a flow of dry argon flushed excess reactant gases from the system and into a scrubber that contained potassium hydroxide. Here, the KOH neutralized excess hydrogen fluoride before technicians pumped the resulting products out of a radioactive gas vent. When the temperature of the system dropped to 100 °C, the technicians terminated the flow of argon gas and opened the furnace. They then used mechanical manipulators and a pneumatically operated powder transfer unit to move the plutonium tetrafluoride to a fused magnesium fluoride crucible. A Lucite barrier in front of the operators reduced the neutron flux below the maximum safe limits set by the AEC.

After the plutonium tetrafluoride powder reached room temperature, PFP technicians mixed it with high-purity powdered calcium in a crucible that they placed into a reduction bomb made from 304-grade stainless-steel. They then placed a perforated, fused, magnesia disc on top of the crucible to support a quantity of iodine. The iodine initiated a chemical reaction when the stainless-steel bomb and its contents reached a temperature of 720 °C inside an induction furnace. Considerable heat evolved during the course of the chemical reaction, which raised the temperature of the mixture by several hundred °C, sufficient to form a molten mass. In this molten state, pure plutonium separated and sank to the bottom of the crucible where it formed buttons that weighed up to 750 grams. In response, a slag of fluorite (CaF_4) floated to the surface of the crucible. A rapid decrease in neutron flux accompanied the reaction and indicated the exact time of reduction. An argon gas atmosphere prevented oxidation of the plutonium during the heating steps of the refining process.

Before plutonium buttons were manufactured into weapon parts, they were cast into feed ingots and then into war reserve ingots. Production control personnel calculated the precise mixture of feed ingots needed to produce a war reserve ingot of specified purity before the second casting. The plutonium casting process consisted of weighing the metal, placing it in tantalum crucibles, and melting it in electric induction furnaces under a vacuum. During the melting and casting of plutonium, the addition of agents such as gallium stabilize it in its δ-phase. Melting also vaporized and drove off low atomic weight impurities that were more volatile than plutonium. The purified molten metal was next poured into chilled aluminum or steel molds coated with oxidized tantalum. At Rocky Flats, technicians rolled, forged, heat-treated, and cut war reserve ingots in a blanking press. After cutting, the technicians annealed and homogenized the blanks. Following this treatment, a hydroform press formed the blanks into hemispherical shells.

After forming, plutonium parts were annealed a second time and then measured on a density balance. The hemi-shells underwent a final shaping with lathes, mills, drills, and hydraulic presses. Due to plutonium's highly oxidative nature, Rocky Flats plated finished plutonium components with nickel in a special nickel carbonyl atmosphere, after which the components were marked with a serial number, cleaned, weighed, and inspected. Technicians welded and leak-tested the plutonium hemispheres before forming them into boosted cores.

Rocky Flats routinely encased finished plutonium cores in beryllium tampers. Beryllium operations at Rocky Flats include foundry casting and the production of blanks. A "wrought" beryllium process developed in 1962 involved the casting of beryllium ingots, sawing of the ingots, "canning" the ingots in stainless-steel, rolling the ingots into sheets, and then cutting the cans away. Beryllium ingots are very brittle and in order to roll them, Rocky Flats had to encase them in stainless-steel and heat them to 900 °C to 1,000 °C. After the removal of the stainless-steel cans, technicians cut the beryllium sheets to their final shapes and formed them in presses.

Finished plutonium or beryllium-clad plutonium parts were collected into subassemblies, assemblies, and finally into complete triggers (primaries). Assembly activities included machining, cleaning, parts matching, brazing, welding, heating under vacuum for trace contaminant removal, marking, weighing, monitoring for surface contamination, and packaging for shipment. Assembled triggers underwent a final inspection prior to shipment.

Hanford ran its plutonium production reactors flat out until the end of the war. At this time, engineers detected structural irregularities. The intense neutron flux from continuous operation had caused the graphite in their cores to swell, distorting the fuel channels in a process that operators named "Wigner's disease", after the physicist who identified the problem. Hanford reduced one reactor to 80 percent of rated power and shut down a second reactor to serve as a reserve to breed Po210 for initiators in case of a war emergency. During the war, Hanford had bred Po210 for neutron initiators in its reactors by placing special targets made from bismuth 209 into areas of high neutron flux. Neutron capture transmuted the targets into Bi210 and thence into Po210 via beta decay. The short, 138.5-day half-life of Po210 required the frequent replacement of initiators as their potency declined. At the end of 1 year, only 18 percent of the active product remained, which was further reduced to 2.6 percent at the end of 2 years.

In order to increase plutonium production at the behest of the AEC, Hanford had DuPont built five more reactors between 1948 and 1955. It added a ninth reactor in 1963. DuPont then refurbished and upgraded the three original reactors. To supplement production from Hanford, the AEC approved the construction of a second plutonium production site in 1950. It was located at Savannah River, North Carolina. Here, DuPont built five reactors to a completely new design that allowed them to "breed" a variety of nuclear materials. The contractor accomplished this feat by incorporating enriched uranium fuel slugs in the reactors' design to increase their reactivity. The increase in reactivity required enhanced cooling. For processing, the AEC constructed facilities similar to the PUREX operation at Hanford.

DuPont built Savannah River's reactors inside large stainless-steel tanks that contained 600 fuel and target assemblies. Instead of graphite, the reactors were equipped with closed circuit, heavy water moderating systems that operated at lower temperatures than Hanford's graphite reactors. Technicians used 491 safety and control rods in each reactor to adjust its reactivity. Water drawn from the Savannah River provided cooling and a helium atmosphere maintained over each reactor simplified the recovery of gases vented from the core. Because of their design, the Savannah reactors could produce 0.86 neutrons for breeding as opposed to each neutron that was either lost or used to maintain the chain reaction. The reactors had special openings to allow for the insertion of breeding materials as "targets" or as "blankets."

A lesser known element produced in conjunction with plutonium by the irradiation of uranium fuel rods is neptunium. Neptunium 237 can be produced through the interaction of neutrons with U235 or U238. It can then be separated based on chemical techniques, although this has never been done commercially. Neptunium 237 is a hard, silvery, ductile, radioactive actinide metal, and is similar to uranium in terms of its physical workability. It has a density of 20.25 grams per cubic centimeter and a melting point of 639 °C. It has four known allotropes.

Recent measurements at LANL show it to have a critical mass of some 60 kilograms, similar to U235. However, it is radioactive, poisonous, and pyrophoric. Because of these traits, and its high critical mass, it has never been employed in US weapons. Approximately, 1,000 critical masses of neptunium are produced each year in commercial reactors.

In addition to plutonium, one of the special isotopes manufactured at Savannah was tritium for use in the eagerly anticipated hydrogen bomb. In the event that this weapon lacked feasibility, the AEC and the Department of Defense had a backup plan to produce long-lived and highly radioactive isotopes such as cobalt 60. The DOD intended to use these isotopes as radiological weapons to render an enemy's cities and military installations uninhabitable.

One of the special breeding materials used at Savannah River was thorium. It occurs naturally in the phosphate mineral monazite, which commercial operators mine from pegmatitic and detrital deposits. Natural thorium consists almost entirely of the isotope Th232. The metal is silvery white in color, ductile, melts at a temperature of 1,750 °C, and has a density of 11.72 grams per cubic centimeter. Its primary use in the nuclear industry is to produce U233 by means of neutron irradiation. The transformation begins when Th232 absorbs a slow neutron to become Th233. Then, by a process of beta decay, it transforms into palladium (Pa233) and finally U233. A slightly more complicated side chain produces small amounts of U232.

Uranium enriched to 98.25 percent U233 has a critical mass of 16 kilograms. Surrounded by 4 centimeters of beryllium, the critical mass drops to 7.6 kilograms. In this regard, U233 is much more like plutonium than is its U235 sibling. Before geologists established that sufficient supplies of uranium existed for the production of U235, the AEC thought that quantity production of U233 might be necessary to build up the nuclear stockpile. To supplement reactor production, Luis Alvarez built an experimental linear accelerator as a test bed for the transmutation of Th232 to U233.

To evaluate the potential of U233 as a weapons material, Los Alamos tested the Teapot MET device using a composite U233 / plutonium core. Los Alamos also tested a number of thermonuclear assemblies with sparkplugs made from U233. This latter concept assumed that U233 would be less likely to affect surrounding lithium deuteride than plutonium, which is a strong neutron emitter. Neutrons produced by the plutonium sparkplug could potentially break down adjacent lithium deuteride in the secondary to deuterium and tritium gas. This gas would then be free to form unwanted hydrides with the sparkplug or migrate into other parts of the bomb to react with its uranium case walls, the primary's core, or electronics. modern warheads incorporate hydrogen scrubbing organic materials to avoid this problem. The conversion of lithium deuteride to gas can also cause unwanted changes to the structure of the thermonuclear fuel and the spontaneous decay of tritium into He3 has unwanted effects on thermonuclear reaction rates. For these reasons, thermonuclear sparkplugs are now made of enriched U235 rather than plutonium.

It has never been declassified whether a production weapon incorporated U233. A serious drawback to its use was its unavoidable contamination with small amounts of U232, an energetic gamma emitter. Since gamma rays are much more penetrative than alpha particles, serious safeguards need to be in place for working with this material or in handling weapon pits constructed from it. As well, the production of U233 is much less efficient than the production of plutonium. In total, Savannah River produced about 625 kilograms of weapons grade U233, of which 500 kilograms remain in the inventory. Current plans call for its down blending with depleted uranium to render it unusable as a fissile material. Before Savannah River started its reactors, it had to supply them with large quantities of heavy water for their moderating circuits. Heavy water is water in which ordinary hydrogen or protium has been replaced by deuterium, a hydrogen isotope that contains a both a proton and a neutron. Deuterium was predicted in 1926 by Walter Russell based on his understanding of a "circular" periodic table. Five years later, Harold Urey, a chemist at Columbia University, detected it spectroscopically. A collaborator, Ferdinand Brickwedde, next distilled five liters of cryogenic hydrogen to produce a single milliliter of pure deuterium that he used to determine its properties. The discovery of deuterium, which came before the discovery of the neutron in 1932, was an experimental triumph that resulted in the award of a Nobel Prize for Chemistry to Urey.

In the early 1940s, Karl-Herman Geib and Jerome Spevack independently invented a process that concentrated deuterium from natural water, making use of deuterium's presence at a ratio of 1 part in 5,000. Commercialized by the Girdler Company, the "Girdler Sulfide Process" is still in use today and relies on the affinities of deuterium to displace normal hydrogen from hydrogen sulfide (H2S) or water (H2O) at specific temperatures. By passing a combination of water and hydrogen sulfide through a series of towers with varying temperatures, it is possible to end up with water enriched in deuterium at a concentration of 15 percent. This marginally enriched product is then fractionally distilled based on heavy water's boiling point, which is 1.4 °C higher than that of normal water. The result is "heavy water" with a deuterium content of 99.8 percent. Heavy water production plants constructed at Savannah River, Georgia, and at Newport, Indiana, began operating in 1952.

In addition to using heavy water as a moderator, the AEC used it as a feedstock for the electrolytic production of deuterium. An electric current passed between two electrodes disassociates heavy water into oxygen and deuterium gas. The deuterium gas can then be compressed or liquefied at a temperature of -249.5 °C for transport or storage. A stable isotope of hydrogen, deuterium is colorless and odorless. It can be reacted with lithium to form the solid, ceramic compound: lithium deuteride, which is the preferred fuel for use in thermonuclear weapons. The Y-12 facility machined LiD into its final shapes in special dry rooms to prevent it from reacting with even the tiniest trace of moisture. The fuel shapes were vacuum-sealed before installation in secondary assemblies.

Most of the lithium used in the preparation of lithium deuteride came from lithium carbonate or lithium chloride recovered from brine deposits by commercial operators. Mining operations also produced small amounts of ore from pegmatite deposits containing the minerals lepidolite, petalite, and spodumene. The lightest of known metals and an alkaline earth, lithium's main method of production is the electrolysis of molten lithium chloride mixed with potassium metal. Lithium has a density of 0.534 grams per cubic centimeter and a melting temperature of 180.5 °C. Its only two isotopes are Li6 and Li7, with the heavier isotope more abundant at 92.6 percent. One of lithium's important properties is its ability to bond with hydrogen isotopes to form metallic hydrides that can store 10 percent more deuterium per unit volume than is contained in solid deuterium. Lithium deuteride has a density of 0.8 grams per cubic centimeter.

In thermonuclear weapons, lithium is more than just a convenient storage medium for deuterium because fusion reactions rapidly convert it into tritium. When struck by a neutron produced by a fusion reaction, lithium decomposes according to the equations:

$$Li6 + n > He4 + T + 4.78 \text{ MeV}$$
$$Li7 + n > He4 + T + n - 2.47 \text{ MeV}$$

Since tritium is a highly reactive fusion fuel, both of these reactions ultimately make an enormous contribution to thermonuclear energy production. The superior energetics of Li6 has led to its enrichment in thermonuclear fuel as a standard procedure. The AEC has used levels of 40, 60, and 95.5 percent enrichment in secondary assemblies. It is likely that the 95.5 percent enrichment level is an economic rather than a physical limit.

The AEC chose the Y-12 Plant at Oak Ridge, Tennessee, as its lithium enrichment site. Y-12 had ceased the electromagnetic enrichment of uranium in 1946. Here the Carbide and Carbon Chemical Corporation, later the Union Carbide Corporation - Nuclear Division, developed three lithium enrichment processes through to the demonstration stage. OREX was an organic exchange process that Union Carbide did not pursue due to a variety of technical difficulties and ELEX was an electrochemical vat process that it operated on a trial basis between 1953 and 1956. Union Carbide discontinued the ELEX operation because of various inefficiencies after it adopted the COLEX process.

COLEX was a column exchange process that enriched the bulk of the lithium used in American weapons at two Y-12 operated facilities between 1955 and 1963. These made use of an affinity of Li6 to dissolve in mercury. Lithium - mercury amalgam was agitated with a lithium hydroxide solution into which Li7 preferentially migrated. By passing the amalgam through a continuous series of cascades or columns, Y-12 could enrich Li6 content to any desired level, after which it recovered and reused the mercury from the amalgam. The two COLEX Plants had access to 12,000 tons of mercury to produce their lithium amalgam. President Eisenhower personally authorized the transfer of this material from the National Strategic Stockpile. Union Carbide enriched approximately 450 tons of lithium in various grades for military purposes.

The Y-12 Plant passed deuterium gas over enriched lithium to produce lithium deuteride according to the equation:

$$2 \text{ Li} + D_2 > 2 \text{ LiD}$$

The reaction proceeds rapidly at temperatures above 600 °C. It could proceed, however, at temperatures as low as 29 °C. Chemical reactions yield LiD in the form of lumped powder, which can be formed via compression. More complex shapes can be cast from a melt.

After its production, Y-12 technicians finished LiD into weapons parts. The primary elements of this process included powder production, forming, finishing, and inspection. Lithium deuteride taken from storage, recycled weapons parts, and manufacturing scrap was first broken, crushed, and ground to produce a powder of specified granularity. The powder was then loaded into molds, isostatically cold pressed into solid blanks, and outgassed in a vacuum furnace. After the blanks had cooled, technicians packed them into form-fitting bags for reheating and warm pressing. After a second cooling period, technicians removed the pressed forms from their bags and radiographed them to detect any high-density inclusions. The technicians performed all of these operations inside of dry glove boxes to minimize the reaction of lithium deuteride with atmospheric moisture. Single-point machining and finishing operations reduced the pressed forms to their final dimensions. Before installation in a weapon, the technicians inspected all finished components with precision balances and contour measurement machines. Certified parts received a final vacuum outgas treatment before installation in weapons. Y-12 found it necessary to provide long-term storage for pre-produced lithium deuteride billets and interim storage for lithium deuteride components from disassembled or retired weapons and rejected components.

A third and final fusion fuel (already discussed in conjunction with lithium) is tritium, which weaponeers use in combination with deuterium to boost atomic primaries. Tritium is a radioactive isotope of hydrogen with a nucleus containing one proton and two neutrons. A colorless and odorless gas at standard conditions, it has a half-life of 12.3 years and for this reason it is rarely found in nature. Tritium's primary mode of decay is by beta particle emission to He3. The utility of this isotope derives from its reaction rates with deuterium, which take place at lower temperatures than the rates for pure deuterium and are 100 times faster. When large quantities of tritium were required for boosted primaries and thermonuclear sparkplugs in the mid-1950s, the AEC chose to produce it via the light water reactor irradiation of tritium-producing burnable absorber rods at Savannah River. Each TPBAR assembly contained columns of Li6 enriched lithium aluminate ceramic pellets. The neutron flux in Savannah's reactors converted the lithium into tritium with the same in situ reaction that takes place in thermonuclear weapons. Following irradiation, SRS chemically processed the pellets to remove their charge of tritium gas.

Although a declining need for plutonium saw two of Savannah River's five production reactors shut down in the 1960s, its last three reactors continued to produce tritium until 1988. Hanford produced subsidiary amounts of tritium, along with production targets, until its last reactor ceased operation in 1971. Because tritium decays rapidly into He3, it is periodically necessary to process the continuously diminishing national supply to remove its helium buildup. The DOE restarted tritium production in February 2007 to insure a continuous supply of gas for the boosted primaries in the present generation of thermonuclear warheads.

Rather than restart a number of aging reactors, the DOE put forward a plan to produce tritium by means of a large accelerator it planned to build at Savannah River. At this facility, high-energy protons generated by the accelerator would spall tritium fragments from tungsten or lead targets. The DOE subsequently cancelled this project in favor of the traditional irradiation method of TPBAR assemblies due to economic reasons. To this end, the DOE modified the Tennessee Valley Authority's Watts Bar Reactor to accept TPBAR assemblies. After irradiation and removal, the DOE transports the assemblies to the Savannah River Site for processing.

During the closing months of WW II Los Alamos was the first facility to produce weapons using material supplied by the national nuclear production infrastructure. In order to meet a demand for increased production, the Manhattan District implemented a division of labor in July 1945 to free Los Alamos to pursue its scientific activities. It set up a support facility known as Sandia Base at Kirtland AFB, near Albuquerque, New Mexico, to assemble atomic bombs. A year later in 1946, Monsanto's Mound Laboratory began production of polonium-beryllium initiators at Miamisburg, Ohio and the Manhattan District transferred the manufacture of explosive assemblies to the Salt Wells Pilot Plant (code-named Eye) at the China Lake Naval Ordnance Test Station in California. The Burlington, Iowa, Ordnance Plant (Sugar) supplemented this production in 1947. Over a 2-year period, Burlington next assumed the assembly responsibilities initially assigned to Sandia Base. Together with the plutonium pit manufacturing facility at Technical Area TA-21 in Los Alamos, these facilities produced the components for the 120 handmade MK III bombs that provided deterrence until 1950.

On September 1, 1949, the AEC reassigned Sandia Base as an independent branch of Los Alamos Scientific Laboratory and gave it a mandate to carry out research, design and limited production of non-nuclear weapon components. A top priority of Sandia Laboratory was developing storable atomic bombs with longer shelf lives. At this time, Hanford assumed sole responsibility for the manufacture of plutonium pits at its Plutonium Finishing Plant. The production of enriched uranium components for pits and tampers went to the Y-12 facility at Oak Ridge.

In 1952, the Dow Chemical opened the Rocky Flats Plant (Apple) near Golden, Colorado, for the production of plutonium, oralloy and depleted uranium components. Honeywell took up the manufacture of bomb casings and electronics at the Kansas City Plant (Royal) in Missouri and Mason and Hangar – Silas Mason Company converted the Pantex conventional munitions plant (Orange) near Amarillo, Texas, into a modern assembly line weapon facility. At the same time, Edward Teller and Ernest O. Lawrence created the Lawrence Radiation Laboratory at Livermore (LRL) as a branch of the University of the California Radiation Laboratory (UCRL) that Lawrence founded in 1931. Authors use the terms Lawrence Radiation Laboratory and University of California Radiation Laboratory interchangeable up until 1971. In that year, the laboratory's name changed to the Lawrence Livermore Laboratory, and then to the Lawrence Livermore National Laboratory. The laboratory quickly affiliated itself with the AEC to provide additional science for the development of hydrogen bombs. Sandia located a second facility near the Lawrence Radiation Laboratory to provide non-nuclear support. Ultimately, the weapon designs from Lawrence Livermore and Los Alamos competed to fill Department of Defense requests.

At Pantex, a WW II type munitions facility, the AEC produced nuclear physics packages, also known as nuclear explosive assemblies. The AEC designated this activity a Class II Hazardous Operation. In 1956, Pantex transferred the assembly and modification of nuclear weapons to purpose-built structures known as "Gravel Gerties." The design of the new structures prevented the propagation of a blast to adjacent facilities in the event of an accident. There was, however, no provision for the survival of personnel inside such a structure. A typical assembly cell or "round room" occupied a cylinder 10.4 meters in diameter and 6.6 meters high. It had 0.3-meter thick reinforced concrete walls surrounded by a gravel berm. An additional 6.7 meters of gravel covered the entire structure. The assembly room opened into a staging area, to which technicians gained entry through massive, blast resistant, revolving steel doors that created an air lock. The interiors of assembly facilities had filtered air, temperatures controlled between 21 °C and 24 °C, and humidity that did not exceed 15 percent.

Pantex also mated or "mechanically assembled" nuclear physics package to bombs and warheads. Technicians conducted this activity inside munitions facilities that had 0.3-meter thick reinforced concrete walls and cantilevered concrete slab roofs covered by 0.6 meters of earth. The buildings used the "separated bay" concept, in which 0.5-meter thick reinforced concrete walls and interlocking blast doors isolated individual assembly areas. Temperature and humidity controls were the same as for the Gravel Gerties. Typical employment levels at the main modification and assembly plants in the mid-1960s were 1,300 personnel at Burlington, 1,050 at Pantex, 200 at Clarksville and 1,100 at Medina.

The production of hydrogen bombs and fusion-boosted atomic warheads in the mid to late 1950s required many changes and upgrades to the Nuclear Weapons Complex. The Y-12 Plant began manufacturing lithium deuteride and uranium parts for thermonuclear secondaries and the Rocky Flats Plant took over the final assembly of components shipped from Hanford and Y-12. Savannah River began loading tritium into weapons in 1955, and the Westinghouse operated Pinellas Plant at Largo, Florida, began producing external neutron initiators in 1957. In 1958, Rocky Flats took on the additional task of manufacturing beryllium shells for boosted warheads and the South Albuquerque Works, New Mexico, assumed sole responsibility for the fabrication of stainless-steel parts. Following these changes, the Mound Plant switched over its production lines to the manufacture of detonators, cable assemblies, and firing-sets. Further reorganization in the early 1960s saw Hanford's pit manufacturing capability terminated and Y-12 given the sole responsibility for producing highly enriched uranium components.

Rocky Flats began a unique program in 1957. The small amount of Pu241 unavoidably incorporated in plutonium decays to Americium 241 by beta decay within a few years. Am241 emits hazardous gamma radiation and has the undesirable trait of absorbing neutrons. As a result, the AEC found it necessary to reprocess the entire inventory of plutonium to remove its americium ingrowth. Rocky Flats performed this function in Building 771, selling the extracted americium to commercial users. Technicians melted plutonium metal in crucibles along with a mixture of sodium chloride, potassium chloride, or calcium chloride salts, into which the americium migrated. After cooling, the crucibles were broken open to retrieve the purified plutonium, which segregated and settled to the bottom. Americium recovery ceased in 1980, by which time Rocky Flats had cycled the entire US stock of weapons grade plutonium.

Despite the vast sums of money invested in plant and equipment for the production of fissile and fusible materials, the level of warhead production anticipated in the 1980s, created a "material gap." No less than 9,071 bombs and warheads were in serial production during 1982. In addition, the AEC planned for the production of 10,070 bombs, warheads, and artillery shells during the period 1982 - 1987. Further, the AEC planned for the production of 10,440 bombs and warheads from 1987 through the early 1990s, with 7,500 SLBM warheads under consideration. Together, this totaled 37,091 warheads! The so-called material gap illustrates the ease with which planning for a never-ending arms race created both real and artificial shortages. Near the end of the 1970s, the AEC obtained much of the material used to build tactical and smaller strategic warheads from recycled multimegaton warheads that it had deployed in the 1960s. By the 1980s, this source of material had dried up. While the projection for 37,091 warheads was unrealistic, the MIRVing of the nation's ICBM and SLBM forces created a very real shortage of enriched uranium. This shortage resulted in a decision to produce the Navy's Trident W88 SLBM warhead at a yield of 475 kilotons but to reduce the Air Force's Peacekeeper W87 ICBM warhead from a yield of 475 kilotons to a yield of only 300 kilotons.

In addition to a requirement for large quantities of nuclear material, the production of increasingly sophisticated nuclear and thermonuclear warheads required the adoption of new and improved manufacturing techniques. The weapons complex had to machine weapon parts precisely and accurately for reliable function. This led to the creation of three-axis machine tools in 1952, followed by five-axis tools in 1954. The five-axis tools were able to produce extremely complex and accurate shapes because the repositioning of work was no longer necessary. Computer numeric control, which combined high speed with precision, came later. Computer control allows the production of preprogrammed shapes by continually varying the motions of multiple axes along predetermined paths.

Metrology, the dimensional inspection of parts, also had to be improved. Tolerances for the first atomic bombs were about 1 millimeter. The national weapons complex now measures components by means of lasers and other non-contact techniques during and after manufacture. Tolerances have increased to about ten millionths of a meter. At the extreme, diamond turning and ion beam polishing to an accuracy of a nanometer prepare special optical surfaces in modern weapons!

Some authors have suggested that America's world leadership in manufacturing came about because of the need to produce nuclear weapons. In addition to multi-axis machine tools with precise numerical control, a variety of other processes was developed. These included cryogenics and the high temperature electric furnaces and hot isostatic presses used to form uranium, plutonium, and tungsten carbide. Special equipment to fabricate thin-walled, hollow bodies and superplastic forming / diffusion bonding processes for the fabrication of sheet metal structures made of exotic alloys were also developed. In addition, weapon designers developed a wide variety of adhesives, foams, plastics and other advanced materials for use in modern weapons. The fields of power management, storage, and switching; fiber optics; and electronics have also significantly advanced.

As the basis for weapons production, the AEC and the DOD signed an "*Agreement between the AEC and* the *Department of Defense for the Development, Production, and Standardization of Atomic Weapons*," in 1953. More commonly known as "The 1953 Agreement" or the "Missiles and Rockets Agreement," this Memorandum of Understanding outlined the responsibilities of the AEC and the DOD in the production of weapons. The agreement made the AEC responsible for manufacturing the warhead, including its nuclear components, detonators and firing system. The agreement in turn made the DOD responsible for building rocket or missile parts, including fuzes. Grey areas of responsibility were reentry vehicles and the production of the "adaption kits" required to mate warheads to missiles. The Army delegated this responsibility to Picatinny Arsenal while the Air Force assigned it to contactors like AVCO or General Electric. The Navy left the work of developing its reentry vehicles to Sandia National Laboratory. In 1983, the DOD and the DOE signed a further Memorandum of Understanding, *Objectives and Responsibilities for Joint Nuclear Weapon Activities* that provided more detail for the interagency division of responsibilities.

Further to the original 1953 agreement, the DOE assigned nuclear weapons production to six phases as outlined in DOE / AL Supplemental Directive 56XB: *Nuclear Weapons Production and Development*. These six phases are:

- Phase 1: The DOE and / or DOD jointly or independently produce weapon concept studies. They then publish their study results as a *Concept Study Report* that determine whether further formal study is required. There is no automatic commitment to proceed to Phase 2.
- Phase 2: The DOE and DOD jointly produce program feasibility studies to determine the "Military Characteristics" and the "Stockpile to Target Sequence" for approved weapon concepts. Military Characteristics outline the nuclear design requirement for a particular weapon system, detailing what is desirable and what is achievable. They detail the purpose of a weapon, its competing characteristics, operational considerations, physical characteristics, required weapon systems or aircraft, vulnerability, reliability, safety, maintenance, monitoring, transportation, command and control. The stockpile to target sequence supplements military characteristics by outlining logistical and operational considerations, and the environments the weapon will encounter from fabrication to delivery (or decommissioning).
- Prior to the completion of Phase 2, the DOE issues a *Major Impact Report* that provides a preliminary evaluation of the resources required to continue the program and any impact that the program might have on other nuclear weapon

systems. If warranted, a design definition and detailed cost study known as a *Weapon Design and Cost Report* (WDCR) is undertaken (Phase 2A). This report identifies baseline design and resource requirements, establishes tentative development and production schedules, and estimates warhead costs.

- Phase 3: The DOE produces a *Developmental Engineering Study* for all weapons deemed desirable of deployment. This phase entails the final definition of a tested, manufacturable design along with any special or acceptance equipment. The DOE also evaluates warhead component prototypes to ensure that they meet the MC and STS requirements of a warhead able to enter initial production. The warhead's production schedule, technical risks and life cycle cost are then refined. Phase 3 typically includes at least one nuclear test to confirm that the warhead will meet requirements. If significant redesign is required, the DOE carries out additional developmental nuclear tests. Prior to the completion of Phase 3, the DOE issues a *Preliminary Weapon Development Report* (PWDR) and a preliminary Design Review and Acceptance Group (DRAAG) Evaluation is conducted to determine if the warhead characteristics meet DOD requirements.

- Phase 4: To prepare the weapon for production engineering, the DOE next adapts the weapon design for manufacturing. This includes product engineering, process engineering, production line tooling, and prototype procurement. Non-nuclear testing and evaluation of prototype components continues through this phase. A Program Officers Group (POG) meets as needed to solve problems and choose between design tradeoffs. At the end of Phase 4, DOE issue a *Complete Engineering Release* (CER) for each component, assembly, and sub-assembly.

- Phase 5: The DOE performs early production oversight to identify manufacturing issues, establish quality control measures, and determine inspection procedures. During this phase, the DOE conducts testing and evaluation of warhead components taken from the production line. The Program Officer Group continues to meet as required to solve problems concerning competing characteristics and trade-offs. The DOE uses most of the warheads produced in Phase 5 for Quality Assurance (QA) testing, although the DOD uses some as War Reserve warheads to meet its Initial Operating Capability. Also, during this phase, operational reviews the DOE and DOD conduct safety studies to determine the efficacy of safety features. Prior to the completion of Phase 5, the DOE issues a *Final Weapon Development Report* (FWDR). Based on this report, the DOD conducts a final DRAAG evaluation to determine if the warhead ha met its requirements. Phase 5 culminates with the issuance of a *Major Assembly Release* (MAR), and formal delivery of the First Production Unit (FPU) to the DOD.

- Phase 6: The DOE begins quantity production to supply the DOD with a specified number of weapons. During Phase 6, the rate of War Reserve (WR) warhead production is increased. For important weapon systems, this process lasts 10 years or more. Phase 6 continues beyond final production until all the warheads of a given type are retired. During deployment, the testing and evaluation of warhead components continues as part of the Quality Assurance and Reliability Testing (QART) Program that includes Stockpile Laboratory

Tests (SLT) and Stockpile Flight Tests (SFT). Normally, the DOE extends component production beyond the requirement for WR warheads to establish an inventory of parts intended for future surveillance item rebuilds under the QART program. DOE also perform stockpile maintenance, such as the replacement of Limited Life Components or LLCs.

Although the development and production of nuclear weapons was at first assigned to six phases, the DOE later added a seventh phase as weapons began to reach obsolescence:

- Phase 7 begins when the first warhead of any given type is retired. The DOE defines retirement as the reduction in quantity of a warhead type in the stockpile for any reason other than to support the QART program. The DOE may however find it necessary to initiate Phase 7 activities to perform dismantlement and disposal activities for surveillance warheads destructively tested as part of the QART program. The Phase 7 program continues, often over a period of several years, until all warheads of a given type are retired and dismantled. From the DOD perspective, a warhead type beginning retirement may remain in the Active or Inactive Stockpiles for a period of years. Currently, the DOE divides Phase 7 into Phase 7A, Weapon Retirement; Phase 7B, Weapon Dismantlement; and Phase 7C, Component and Material Disposal.

Assuming that the first three phases for a new warhead design are completed, the President makes the decision to place a weapon in production when he signs the annual *Nuclear Weapons Stockpile Memorandum*, submitted by the DOD and DOE. This memorandum covers the establishment of infrastructure within the Nuclear Weapons Complex, the production of nuclear materials, and the number and types of weapons the DOE will manufacture or retire. After the president approves it, DOE Headquarters converts this document into a *Production and Planning Directive*, which it sends to its Operations Offices. These offices follow up with Weapon Program Management Documents that they send to the Area Offices that directly oversee the manufacture of weapons.

President Reagan placed the entire nuclear program and its infrastructure under review in January 1985 when he established the "Blue Ribbon Task Group on Nuclear Weapons Program Management." Reagan chartered this task group to examine the procedures used by the DOD and the DOE to research, develop, test, produce, monitor and retire nuclear weapons. The Task Group issued its final report in July 1985. It found the relationship between the DOD and the DOE to be sound but asked for a number of administrative and procedural changes intended to produce closer integration between nuclear weapons programs and national security planning.

One of the suggested administrative changes was the formation of a joint DOD - DOE body tasked with coordinating nuclear weapons programs. The Military Liaison Committee, originally formed for this purpose in 1954, was by this time reduced to an intra-agency DOD group that no longer supported interagency communication. Consequently, the DOD - DOE created a joint, senior task group to coordinate nuclear weapons acquisition issues and matters related to joint activities. They named it the Nuclear Weapons Council. Responsibilities assigned to the NWC included preparing the annual *Nuclear Weapons Stockpile Memorandum* (NWSM), developing stockpile options and costs, coordinating programming and budgetary matters, identifying cost effective production schedules; safety, security, and control issues; and monitoring the activities of Project Officers Groups.

The NWC is currently composed of five members: the Under Secretary of Defense for Acquisition, Technology, and Logistics; the Under Secretary of Defense for Policy; the Vice Chairman of the Joint Chiefs of Staff; the Commander of the US Strategic Command and the Under Secretary of Energy for Nuclear Security / National Nuclear Security Administration Administrator. The Under Secretary of Defense for Acquisition, Technology, and Logistics serves as the Chairman of the Council. The Assistant to the Secretary of Defense for Nuclear and Chemical and Biological Defense Programs serves as the Staff Director.

Authorized by Section 179 of Title 10 of the United States Code, the NWC serves as both an oversight and a reporting body that is accountable to both the Legislative and Executive branches of government. It must fulfill four annual reporting requirements: production of the *Nuclear Weapons Stockpile Memorandum / Requirements and Planning Document* (NWSM / RPD); production of the *NWC Report on Stockpile Assessments* (ROSA); production of the *NWC Joint Surety Report* (JSR); and production of the *NWC Chairman's Annual Report to Congress*. Presidential direction, Congressional legislation, and agreements between the Secretaries of Defense and Energy create additional requirements. Much of this work is coordinated at the subordinate level and then finalized and approved by the NWC.

A *DOD / DOE Memorandum of Agreement* redefined the agenda of the Nuclear Weapons Council in 1997. Its new mandates included:

- Establishing committees to provide senior staff support within the limits of the Council's responsibilities
- Acting on committee recommendations relating to the nuclear stockpile
- Working with the National Nuclear Security Agency on matters of nuclear weapons safety, security, and control
- Authorizing studies affecting the nuclear stockpile; reviewing, approving and providing recommendations to DOD and NNSA authorities
- Providing guidance to the DOD and the NNSA on matters regarding the nuclear weapon life cycle
- Reviewing nuclear program matters as directed by the Secretaries of Defense and Energy
- Fulfilling the annual reporting requirements required by its charter

The NWC conducts its operations through a number of subordinate organizations. Two such organizations were the Nuclear Weapons Council Standing Committee (NWCSC) and the Nuclear Weapons Council Weapons Safety Committee (NWCWSC). Established in 1987 the NWCSC served as a joint DOD - DOE flag level committee to coordinate routine matters put before the Council. Established in 1989, the NWCWSC was also a senior level joint committee. It provided advice and assistance on matters of weapons safety to the Council, to its Staff Director and to the NWCSC. In 1994, the NWC merged the two Committees into the Nuclear Weapons Council Standing and Safety Committee (NWCSSC).

In 1995, the Assistant to the Director for Atomic Energy delegated responsibility for the supervision of NWC staff to the Deputy Assistant to the Secretary of Defense for Nuclear Matters. He also provided an Action Officers Group and additional NWC staff to provide support to the council and its subordinate bodies. The Chairman of the NWC then established the Nuclear Weapons Requirements Working Group (NWRWG) to prioritize weapons requirements. This group stopped meeting in November 2000 when its members voted to transfer their functions to the NWCSSC. Also, in November 2000, the DOE / DOD formed the Compartmented Advisory

Committee (CAC) to provide information and recommendations to the NWC concerning nuclear weapons surety upgrades. In 2005, the NWC created the Transformation Coordinating Committee (TCC) to coordinate the development and implementation of a strategy for the transformation of the National Nuclear Enterprise to meet the needs of the 21st century.

As previously indicated, The Nuclear Weapons Council has considerable contact with the DOE's National Nuclear Security Agency. The DOE created the NNSA in 1999 as a response to the Wen Ho Lee scandal, in which the government accused a Chinese American scientist of supplying information relating to advanced thermonuclear weapons to the Communist Chinese. From its inception, the NNSA had responsibility to carry out four specific missions with respect to National Security: providing the United States Navy with safe, reliable nuclear propulsion plants, promoting international nuclear safety and non-proliferation, reducing global danger from weapons of mass destruction and supporting American leadership in science and technology.

From the late 1980's onward the *Intermediate Range Nuclear Forces Treaty* (*INF*) and the *Strategic Arms Reduction Treaty* (*START I*) have reduced the need for nuclear warheads. This resulted in a halt to pit manufacturing in 1989 and warhead assembly in 1990. Severe inroads in the nation's nuclear infrastructure followed the cessation of weapons manufacturing. The DOE closed the Burlington, South Albuquerque, Mound, and Pinellas Plants and environmental concerns resulted in the cessation of operations at Rocky Flat and the Hanford Reservation. Supporting fuel fabrication operations have long since ceased operation.

The Los Alamos and Lawrence Livermore National Laboratories have both retained their mandates to design nuclear weapons at the behest of the Department of Defense. LANL specializes in the analysis of neutron scattering, enhanced surveillance techniques, pit production, and plutonium science and engineering. It also oversees the refurbishment of the nuclear and nonnuclear components of stockpiled weapons and handles diagnostics for plutonium pits. The DOE has assigned LANL with the oversight of the B61-3/4 and B61-7/11 bombs, and the W76, W78, W80-0, W80-1, and W88 warheads. Together with LANL, LLNL supports NNSA's integrated surveillance program, specializing in techniques to predict aging phenomena and the diagnostics for nuclear secondaries. LLNL also provides high explosives research and manages the National Ignition Facility for weapon physics experiments. LLNL is specifically responsible for oversight of the B83-0/1 bomb, and the W84 and W87 warheads. In the future, it was to assume responsibility for the W80-2/3, programs which were cancelled in 2006. It is now responsible for the W80-4 LEP, an update of the original W80-1.

DOE assigned Sandia National Laboratory the responsibility for supporting LANL and LLNL to insure the safety, security, and reliability of the nuclear stockpile. To this end, the lab maintains two separate facilities. Sandia, New Mexico, is active in the systems engineering of nuclear weapons and in the design, development, manufacture and testing of nonnuclear components for LANL. The New Mexico facility also handles the production of neutron generators and their targets, a task the DOE transferred to Sandia after the closure of Pinellas, Florida. Sandia, California provides mechanical, electrical, structural, and chemical engineering for the program at LLNL. This facility is also the principal producer of non-nuclear components for stockpile management. The Kansas City Plant, currently managed and operated by Honeywell Federal Manufacturing & Technologies, also produces bomb components. Its products include electrical, electronic, electromechanical, plastic, and non-nuclear metal components for weapons.

At Oak Ridge, the K-25 Site is currently in the process of environmental remediation, whereas Y-12 is the national storage site for special nuclear materials. Programs at Y-12 include manufacturing and refurbishing weapon components, dismantling retired weapon components, and providing production support for various other programs. Y-12 is responsible for uranium components, salt components, and secondary assembly. Operated by BWXT, the facility has maintained its ability to produce and assemble uranium and lithium components and to recover materials used in their fabrication.

The Savannah River Site is responsible for limited life component exchanges, reservoir surveillance, and tritium extraction. These missions involve the filling and shipping of new and reclaimed reservoirs containing tritium, deuterium and non-tritium gases and the surveillance of gas transfer components. Tritium operations include the purification and enrichment of tritium, the mixing and compression of tritium, deuterium, and non-tritium gases, pinch welding gas filled reservoirs, reclamation of returned reservoirs, function testing, reservoir surveillance; quality inspection, packaging, shipping and tritium extraction from irradiated targets.

Pantex is nearing the end of the surplus-weapon dismantlement process and has resumed the limited production of W88 warheads using cores produced by the Los Alamos TA-55 facility. Managed and operated by BWXT, DOE has charged Pantex with maintaining the safety, security, and reliability of the nation's nuclear weapons stockpile. The plant has five primary missions: evaluating, retrofitting and repairing weapons in support of life extension programs and certification of weapon safety and reliability; dismantling weapons that are surplus to the stockpile, sanitizing components from dismantled weapons; developing, testing and fabricating high explosive components; and providing interim storage and surveillance of plutonium cores or plutonium bearing pits.

CHAPTER 3
STOCKPILE LOGISTICS

The enormous casualties incurred in the Pacific Island-Hopping Campaign created tremendous domestic pressure to end World War II. For this reason, the Manhattan Engineering District did not at first stockpile any of the atomic bombs produced by the Los Alamos Scientific Laboratory. Rather, the MED immediately used Los Alamos's first bomb in a proof of concept test and the USAAF dropped the next two bombs on Japan.

Right after the completion of "Gadget," the implosion mechanism for the first atomic device, Los Alamos technical personnel bagged, crated, and lashed it to the bed of a five-ton army truck. Camouflaged under a tarpaulin, the technicians and a detachment of military police took it on an overnight drive to a desolate test site located near Alamogordo, in southern New Mexico. The sedan filled with military police that preceded the truck unwittingly drew attention to the secret mission when the police sounded a siren as the small convoy drove through the darkened streets of Santa Fe. No one wanted an accident with 2 1/2 tons of sensitive high explosives.

The Los Alamos technicians next secured Gadget's plutonium core inside a ventilated, bumper equipped, magnesium field case that they secured with bailing wire. Army personnel then placed the field case in the back seat of a sedan for its trip to the MacDonald Ranch House near the atomic test site. A car full of military police preceded the core-bearing sedan and another vehicle transporting the scientific team brought up the rear. Los Alamos technicians and scientists used the MacDonald ranch house as a temporary "clean room" for assembly of the core. During this procedure, a detachment of Army guards surrounded the house. After Los Alamos scientists and technicians completed the core assembly operation, Army personnel lugged it back out to the sedan and drove it to the test site for installation in the Gadget, which sat atop a 100-foot-high tower adapted from an oil derrick. Shortly before dawn on July 16, 1945, Los Alamos cleared the last of its personnel from the test site in preparation for "Trinity," the world's first atomic explosion.

While the Los Alamos Laboratory was preparing for Trinity, the Manhattan Engineering District was preparing to ship atomic bomb components to the Marianas Islands in the Pacific Theater. For this purpose, it activated Operations "Bowery" and "Bronx." The Navy began preparations to receive the atomic components when it sent Commander Walden Ainsworth to Tinian Island to select an operational site in February 1945. Colonel Elmer Kirkpatrick, the MED's Deputy Engineer, followed Commander Ainsworth to Tinian in March. In a meeting with Pacific Fleet Commander, Chester Nimitz, Kirkpatrick made the final arrangements for space to accommodate the USAAF's 509th Composite Group, which would drop the bombs. Kirkpatrick then flew to Guam in order to obtain the necessary materials to build the world's first atomic base. He also arranged for a Battalion of Seabees to carry out construction and to provide security. Kirkpatrick's first priorities were to build a restricted flight apron and construct the facilities needed to assemble and to test the bombs.

Tinian Island to select an operational site in February 1945. Colonel Elmer Kirkpatrick, the MED's Deputy Engineer, followed Commander Ainsworth to Tinian in March. In a meeting with Pacific Fleet Commander, Chester Nimitz, Kirkpatrick made the final arrangements for

The original five civilian Atomic Energy Commissioners (AEC), who took control of nuclear weapons in 1946. Left to right: Robert Bacher, David Lilienthal, Sumner Pike, William Waymack, and Lewis Straus. Bacher was formerly head of G Division, which developed the Fat Man bomb at LANL. (Credit: LANL)

The Manhattan Engineering District established what is now Sandia National Laboratory in 1945 at Kirtland AFB to support nuclear weapons development, testing, and assembly for LASL. In 1980, its mission changed to research and development for the non-nuclear components of nuclear weapons. Sandia opened a second lab to support LLNL in California in 1956. (Credit: SNL; Color Original)

The Pantex Plant (Zone 12) is the only nuclear weapons assembly and disassembly facility maintained in the United States today. It oversees the safety, security, and reliability of the nuclear stockpile. Pantex is located on a 16,000-acre site northeast of Amarillo, Texas. (Credit: NNSA; Color Original)

World War II era munitions bunkers still used for the temporary and long-term storage of nuclear weapons and components at the Pantex facility in Zone 4. Pantex has the ability to permanently store up to 20,000 nuclear cores. The storage area is located some distance from the main plant area. (Credit: NNSA / DOE; Color Original)

"Gravel Gerties" used for the assembly and dis-assembly of nuclear weapons at Pantex. Each is a circular reinforced concrete room, 33 feet in diameter, capped with 17 feet of sand and gravel. This is sufficient to absorb the power of an accidental conventional explosion, trapping gasified plutonium in the rubble. There are no provisions for the survival of the occupants. (Credit: NNSA / Pantex Press Kit; Color Original)

A cutaway view of the Kirtland Underground Munitions Storage Complex (KUMSC), the successor to National Site Able (Manzano Base). It began operation in 1992 and has storage for over 3,000 nuclear weapons. It is the main American nuclear storage site. (Credit: USA; Color Original)

Nuclear storage area aboard an aircraft carrier in the 1950s. In the center is a MK 8 bomb and to the rear is a MK 5 bomb. Both weapons are fastened to cradles that were bolted to the floor. (Credit: USN)

Rail sidings at the Pantex plant filled with ATMX nuclear transport cars. Rail transport was used exclusively from 1951 - 1976, augmented with road transport from 1976 - 1987, and curtailed in 1988. The cars were painted white for easy identification. (Credit: NNSA / OST; Color Original)

An innocuous looking, safe, secure, transport truck (SST). Transport by road is favored over rail or air and the routes are carefully planned. These armored trucks are referred to as a "bank vaults on wheels" and are accompanied by heavily armed security detachments. The drivers are trained combatants. Note the thickness of the rear door. (Credit: SNL)

space to accommodate the USAAF's 509th Composite Group, which would drop the bombs. Kirkpatrick then flew to Guam in order to obtain the necessary materials to build the world's first atomic base. He also arranged for a Battalion of Seabees to carry out construction and to provide security. Kirkpatrick's first priorities were to build a restricted flight apron and construct the facilities needed to assemble and to test the bombs. The Seabees built four air-conditioned 20 x 48-foot, steel arch-rib buildings of the type normally used for Navy bombsight repair: two for the fuzing team, one for the electrical detonator team, and one for joint use by the pit and the observation teams. They next provided three air-conditioned 20 x 70-foot buildings for the assembly of the atomic bombs. Other miscellaneous buildings constructed by the Seabees included steel arch-rib warehouses, a variety of explosives magazines, and a modification shop. Because the atomic weapons were too large to fit under the fuselage of a B-29, the Seabees emplaced loading pits equipped with hydraulic lifts into the special taxi areas they prepared. Last on Kirkpatrick's construction list were accommodations for the bomber crews, aircraft maintenance staff, and the team of science specialists from Los Alamos who would assemble and test the bombs.

The Seabees built four air-conditioned 20 x 48-foot, steel arch-rib buildings of the type normally used for Navy bombsight repair: two for the fuzing team, one for the electrical detonator team, and one for joint use by the pit and the observation teams. They next provided three air-conditioned 20 x 70-foot buildings for the assembly of the atomic bombs. Other miscellaneous buildings constructed by the Seabees included steel arch-rib warehouses, a variety of explosives magazines, and a modification shop. Because the atomic weapons were too large to fit under the fuselage of a B-29, the Seabees emplaced loading pits equipped with hydraulic lifts into the special taxi areas they prepared. Last on Kirkpatrick's construction list were accommodations for the bomber crews, aircraft maintenance staff, and the team of science specialists from Los Alamos who would assemble and test the bombs.

Captain W. S. Parsons, aided by Norman Ramsey as his Scientific and Technical Deputy, commanded the Scientific Team at Tinian. General Groves selected Commander F. L. Ashworth as the Team's Operations Officer and as the alternate to Captain Parsons. Roger Warner directed the Fat Man Assembly Team, Francis Birch headed the Little Boy Assembly Team, E. B. Doll led the Fuzing Team, Commander E. Stevenson commanded the Electrical Detonator Team and Phillip Morrison and C. P. Baker controlled the Pit Team. Luis Alvarez and Bernard Waldman oversaw Airborne Observation and Sheldon Pike led the Aircraft Ordnance Team. Robert Serber, W. G. Penney, and Captain J. F. Nolan acted as Special Consultants. The 1st Aviation Ordnance Squadron under the command of Captain Charles Begg assisted the Scientific Team by preparing the bomb loading pits and related equipment. Although Begg's military contingent hoped to play a role in bomb preparation and assembly, the Scientific Team considered their training inadequate for this purpose.

The first atomic parts shipped to Tinian were part of a gun-assembled, enriched uranium bomb named "Little Boy." The Army trucked three boxes containing the bomb, its uranium projectile, and some scientific test instruments to Kirtland AFB, New Mexico, accompanied by six carloads of military police. Here, the boxes were transferred to a pair of C-47 aircraft and flown to San Francisco, California. At Hunter's Point Naval Yard, the Navy took the boxes and loaded them aboard the cruiser USS Indianapolis (CA-35). They also loaded some spare bomb units for use in drop tests. Aboard the cruiser, sailors padlocked the lead box containing the uranium projectile to eyebolts welded to the deck of the Flag Lieutenant's cabin. During the sea voyage, a pair of armed Army Officers ceaselessly watched over the box in an early version of the two-man rule. The bomb's enriched uranium insert followed separately by air, separated into three shipments placed on board C-54 aircraft belonging to the 509th's "Green Hornet" Logistics Squadron. On July 28, these aircraft landed at North Tinian Airfield to complete the inventory of Little Boy's parts.

Little Boy was an all up bomb that required only a charge to its lead-acid batteries and the installation of its uranium components, neutron initiators, and propellant to make it operational. Apart from the installation of its propellant charges, the Los Alamos Scientific Team completed the assembly and checkout of Unit L-11 on July 30. The delay in loading the propellant was a necessary safety precaution. The impact of a crash on takeoff would most likely have set the weapon off in a full yield explosion, vaporizing the airfield. To prevent this from happening Captain Deak Parsons, the weaponeer for Little Boy, installed the bomb's propellant charges only after the B-29 was safely in the air. He spent the night of August 5 - 6 practicing the procedure until he was certain he could carry it out in the confines of the B-29's unpressurized bomb bay. Parsons successfully loaded the cordite charges early on the August 6 mission. Morris Jepson, the weapon test officer, then installed the live arming plugs that connected the bomb's firing circuits to its batteries.

The Air Force flew the two hemispheres for the Fat Man implosion bomb to Tinian on board a C-54, also from Kirtland AFB. Three B-29 bombers, each carrying a Fat Man bomb casing, followed thereafter. After the bomb casings reached Tinian on August 2, the Scientific Team moved the casings to air-conditioned buildings where they started assembling their internal mechanisms. The implosion bombs came as a collection of parts that had to be tested and then laboriously hand-assembled in a procedure that took several days. In order to meet the planned attack schedule, the Scientific Team omitted a number of tests during the assembly process. They completed Unit F-31 on August 8. That same day, Unit F-33 was practice dropped and exploded with a dummy core to ensure that it was in proper working order. Because Fat Man would have its plutonium core and its internal neutron initiator installed when it took off, the bomb was just as dangerous as Little Boy. A crash would undoubtedly have resulted in an atomic explosion that might have come close to full yield. For this reason, Frederick Ashworth, the weaponeer, installed the bomb's arming plugs only after the mission was safely in the air on August 9. He then used aircraft power to charge the x-unit and verified that the bomb was operational.

Los Alamos completed the core for a second implosion bomb on August 13 and released a schedule that established future rates of core production. Starting with three cores per month in September, the lab expected to increase productions to a rate of seven cores per month by December. Robert Oppenheimer, the lab's Director, planned to make some of the new cores from enriched uranium instead of plutonium. Implosion was so much more efficient than gun-assembly that the uranium used in Little Boy could have produced four implosion cores. Following the success of Trinity, Oppenheimer made this suggestion to General Groves who, nevertheless, vetoed the idea because it would have delayed the bomb's use in combat. Groves did delay the shipment of the second plutonium pit to Tinian. He anticipated the Japanese surrender, which came on August 14, local time.

The end of the war resulted in significant changes to core production when the MED cancelled the advanced designs upon which Los Alamos had based its production schedule. Following this action, many of the scientists and engineers working at Los Alamos returned to their civilian jobs. Hanford shut down one of its reactors due to technical problems and allowed plutonium nitrate from fuel reprocessing to accumulate without converting it to metal. Oak Ridge shut down its S-50 Plant and wound down the Y-12 Plant, ceasing electromagnetic enrichment of uranium at the end of 1946. Because of these actions, the national atomic stockpile contained only nine plutonium cores and seven operable neutron initiators in July 1946. Los Alamos increased the number of plutonium cores to thirteen in July 1947, although there was enough nuclear material on hand to fabricate about 100 more.

Part of the reason for the delay in stockpiling cores came from a discussion on how to proceed at the beginning of the nuclear age. Suggestions ranged from providing the world with unilateral nuclear assistance to giving the newly formed United Nations control over atomic programs through an International Atomic Energy Agency. At the end of the day, President Truman decided to place both the commercial and military development of atomic power under the auspices of a national Atomic Energy Commission. This decision left the awesome power of nuclear energy in civilian hands – directly responsible to the President. Senior government personnel thought that the scientific community might balk at military control, which was in any case poorly equipped to manage the new and technically complex atomic weapons.

On January 1, 1947, the Truman Administration disbanded the Manhattan Engineering District and created the civilian Atomic Energy Commission to assume control of its weapons and weapons infrastructure. The *Atomic Energy Act* (Public Law 585) of August 1, 1946, authorized this change in custody. Headquartered in Germantown, Maryland, five civilian commissioners managed the AEC, with one acting as Chairman. As constituted, the AEC comprised four Divisions: Research, Production, Engineering, and Military Application. The AEC subsequently assigned the research, testing, production, and storage of atomic weapons to its Santa Fe Operations Office, which was located at Los Alamos, New Mexico. Except for some restrictions on research, the Truman Administration allowed the Commission to define the functions of each of its Divisions. It also allowed the Commission to designate each Division's Director, with the proviso that the Director of Military Application be a member of the Armed Services. To provide legislative oversight, the House of Representatives and the Senate created a Joint Committee on Atomic Energy (JCAE), with nine members chosen from their ranks. To assist the JCAE in planning, the AEC created a General Advisory Committee of senior scientists, chaired by Robert Oppenheimer.

Due to the inherent conflict in a single agency promoting and regulating nuclear power, the Ford Administration dissolved the AEC in 1974. In its place, Public Law 93-438 created a pair of new agencies: the Nuclear Regulatory Commission and the Energy Research and Development Administration. ERDA assimilated the federal government's energy research and development activities into a single unified agency. These included the nuclear defense activities carried out by the Santa Fe Operations Office and the AEC's former R&D functions, which ERDA's Chicago Operations Office assumed. The Nuclear Regulatory Commission (NRC) acquired the remaining parts of the AEC's former functions. These included reactor safety and security, reactor licensing and renewal, radioactive material safety, and spent fuel management. The Carter Administration subsequently incorporated both of these agencies into the Department of Energy (DOE) on October 1, 1977. Currently, five Operations Offices, acting as contracting agencies for onsite service providers, control the nuclear operations of the DOE. In addition, there are a number of Area Offices responsible for day-to-day operations at individual sites.

After its creation, the AEC assumed a number of responsibilities, the most important of which was the sole control of fissionable material and the facilities that produce it. With respect to mining, a license from the Commission was required to "transfer or deliver any source material from its place of deposit." A second important area of control was the utilization of atomic energy. A license from the Commission was required to "manufacture, produce, or export any equipment utilizing fissionable material or atomic energy." A third grant of power revoked (subject to compensation) all patent rights previously granted for inventions in the production of fissionable material and the utilization of fissionable material for atomic energy or for a military weapon. A fourth area of control covered the development and production of weapons, where the Commission was to concentrate its efforts.

The Truman Administration directed the Commission to produce atomic weapons subject to "the express consent and direction of the President obtained at least once a year." In the interest of national defense, the President could direct the Commission to deliver quantities of fissionable material to the Armed Forces as deemed necessary and to authorize the Armed Forces to manufacture atomic weapons. Finally, the President gave the Commission control of all restricted data pertaining to the manufacture and utilization of atomic weapons, the production of fissionable material or the use of fissionable material in the production of power. As the sole repository for restricted data, the AEC was required to obtain an FBI clearance check of all its employees and to obtain agreements from licensees and contractors not to divulge any restricted data. In fact, the predecessor Manhattan Engineering District required FBI clearance inspections for all of it employees from its inception.

Having spent a significant effort on security, it came as a shock in the late 1940s when the VENONA investigations of the FBI revealed that Communist spies had infiltrated the nuclear project and the government! Decryptions from Soviet one-time code pads showed that major Soviet espionage campaigns had targeted the United States, Britain, and Canada as early as 1942. Identities soon emerged for American, Canadian, Australian, and British spies. These included Klaus Fuchs, a senior Los Alamos scientist; Alan Nunn May, a British physicist; and David Greenglass, a Los Alamos employee who passed atomic secrets to couriers Harry Gold and Julius and Ethel Rosenberg. The FBI found other spies within the workings of the US Government. These included Ager Hiss, a State Department lawyer involved with establishing the United Nations; lawyer Harry Dexter White, the second-highest official in the Treasury Department; Lauchlin Curry, a personal aide to Franklin Roosevelt; and Maurice Halperino, a Section Head in the Office of Strategic Services.

A "witch hunt," instigated by Senator Joseph McCarthy, followed the FBI revelations. In February 1950, McCarthy claimed to have a list of "members of the Communist Party and members of a spy ring" that were employed in the State Department. McCarthy was never able to prove his sensational charges, but for a while, he enjoyed widespread publicity that gained him a powerful following that included J. Edgar Hoover, the Director of the FBI. While acknowledging that there were inexcusable excesses during the period of "McCarthyism," the depth of Soviet espionage in America needed the national exposure it received, resulting in the tightening of security around state and military secrets. Later espionage cases increased security to the point that President Lyndon Baines Johnson had to authorize the *Freedom of Information Act* (Title 5 U.S.C. Section 552) to curb government excesses. It is through this act that the Author derived much of the information used in this book.

As the sole repository for restricted data, and because of national security concerns, the AEC asked the DOD to conduct background checks on its personnel before it released Restricted (Atomic) Data to them, a suggestion at which the DOD balked. The DOD, nevertheless, recognized the need for a security program. To this end, it issued a Statement of Principles regarding security clearances on January 6, 1948. These principles were:

- Clearances were to be based on "compelling, official necessity"
- Individuals would receive clearance only for lengthy assignments
- Clearances were to be explicit and limited to "need to know"
- Central screening services had to be sufficiently familiar with the scientific or engineering field in question to understand the implications of specific requests
- After some discussion with the AEC, the DOD proposed three types of formal clearance
- A "Q" clearance for Restricted Data was to be granted by the AEC based on an investigation by the FBI as prescribed by Section 10 (b) (5) (B) (1) of the *Atomic Energy Act*
- A "Z" clearance for Restricted Data was to be granted by a Military Department or Joint Agency after an investigation by the FBI and subject to disapproval by the AEC.
- An "M" clearance for Restricted Data granted by a Military Department or Joint Agency after an investigation by one of the Departments, subject to disapproval by the AEC.

The DOD also made provision for granting temporary clearances after a favorable check by a concerned Department or Agency pending the completion of a background investigation and for granting of emergency clearances at the discretion of the AEC pending a background investigation.

The AEC agreed on both Q and M clearances but eliminated the Z clearance by expanding the M clearance to bridge the gap created by its cancellation. This status was to remain in effect until May 1951, by which time the DOD was to have secured agreement with the AEC that a standard military clearance could stand in the stead of an M clearance. The new procedures came into effect on September 30, 1948, with Air Force issuance of AFR 205-3. Their implementation cleared up a data transfer problem that had resulted in the use of improper ballistic coefficients for the airdropped atomic bomb tested in Operation Crossroads.

Although the AEC - DOD agreement sorted out restricted data management issues, the production of clearances for thousands of civilian employees and hundreds of military personnel threatened to overwhelm the system. In March 1950, about 75,000 Air Force personnel were in the clearance pipeline. Strategic Air Command had a requirement for 14,000 clearances but found that 7,000 were in continual backlog. The situation was serious enough that SAC estimated there were not enough trained and cleared personnel to staff its atomic fleet and called into question the national security of the United States! The Air Force did not satisfactorily solve the problem of security clearances until it, and the other Armed Services, received the right to store and maintain nuclear weapons. The DOD then developed its own security clearance system.

The fight to wrest atomic weapons from the custody of the civilian AEC began with the Army. It objected that for purposes of military readiness, atomic bombs needed to be in the hands of end users. Although the Army was unsuccessful in its quest to gain control of atomic weapons, Senator Arthur Vandenberg sponsored an amendment to the *Atomic Energy Act* to create a Military Liaison Committee (MLC) as a means of communication between the AEC's Board of Governors and the DOD. This amendment passed.

The MLC consisted of seven members and three official observers. Following the signing of the *National Security Act* on July 27, 1947, its Chairman became the Assistant to the Secretary of Defense for Atomic Energy. In addition to its Chairman, two flag level representatives from each Military Service served on the MLC's Committee. A representative from the AEC, the Joint Chiefs of Staff, and the Armed Forces Special Weapons Project participated as observers. An Action Officers Group composed of AOs representing each of the seven members and the three official observers supported the MLC. Other organizations with a direct interest in nuclear weapons, such as the National Weapons Laboratories, frequently participated in Action Officer level activities.

A memorandum issued from the Secretary of War and the Secretary of the Navy to the Chiefs of Staff for the Army and for Naval Operations created the Armed Forces Special Weapons Project (along with the MLC). General Groves had solicited this memorandum, which anticipated the future Air Force, although it would not come into being until after the approval of the *National Security Act*. The Secretaries placed General Groves, the former head of the Manhattan Engineering District, in charge of the AFSWP, and made it the operational link between the branches of the Armed Services and the AEC.

Previously, in September 1945, General Groves had approved the establishment of a logistical area known as Sandia Base at Kirtland AFB near Albuquerque, New Mexico. Los Alamos transferred most of Z Division, its new engineering and production arm, to this base.

Groves assigned Z Division the responsibility for testing, designing, developing, stockpiling, and assembling atomic bombs. Groves also made Z Division responsible for the engineering, procurement, and storage of all bomb components except the nuclear pits. He then established Technical Area 1 for Z Division's classified operations. Since Sandia was now designated the main national stockpile and assembly area for nuclear weapons, it was no surprise when the AFSWP was also located here.

The Secretary of War, Robert P. Patterson, assigned the AFSWP responsibility for "all military service functions of the Manhattan Project as are retained under the control of the Armed Force. Of these, he specified three: the training of Bomb Commanders, weaponeers, and specially trained personnel required by the project; military participation in the development of atomic weapons; and the development of radiological safety measures with other agencies. General Groves further outlined the responsibilities of the AFSWP on April 4, 1947, when he issued a memo proposing that Secretary Patterson grant the agency ten special functions:

- Training of special personnel
- Military participation in the development of atomic weapons
- Coordination for the development of joint radiological safety measures
- Storage and surveillance of atomic weapons in military custody
- Recommending security procedures within the military
- Assistance to the services in providing training material and courses
- Assistance in the preparation of staff studies and war plans as requested
- Production of material for public education
- Command of military units assigned the storage
- Surveillance of atomic weapons
- Furnishing staff assistance to the MLC

Even before Sandia received the AFSWP, Kirtland AFB had developed an interest in atomic weapons. A small group of Kirtland's Officers formed the Armstrong Committee in 1946, with the goal of developing training programs to teach Army Air Force personnel how to handle nuclear weapons. The Armstrong Committee produced a series of reports on the strategic applications of atomic bombs and the organization and types of military units that should handle them. This thinking agreed with discussions held between Norris Bradbury, the post-war Scientific Director at Los Alamos, and General Groves, who had determined that the most pressing need in the field of atomic weapons was for teams that could assemble the bombs on short notice. In July 1946, General Groves ordered Colonel Gilbert M. Dorland to create these teams in conjunction with the scientific staff at Sandia and to take command of the base. Lieutenant Colonel John Ord joined Colonel Dorland in August 1946 as his second in command. Together these men activated the 2761st Engineer Battalion (Special) on the 19th of that month. The 2761st had an authorized strength of 70 commissioned officers, 1 warrant officer and 397 enlisted personnel.

The Army officially assigned 77 personnel to each Assembly Team, although this number varied considerably. Of these personnel, 35 performed specialty functions with the remainder provided support. The separation of specialty functions into three categories served to reduce the need for training and to streamline the work of atomic assembly. Los Alamos trained the nuclear specialists to maintain and assemble the bomb's cores, which consisted of neutron initiators, plutonium hemispheres, uranium tampers, and aluminum pushers. The nuclear trainees then learned how to install the cores in assembled bombs by practicing with stainless-steel facsimiles. The electrical specialists took a "crash" course in electronics and learned how to test and assemble the bomb's batteries, fuzing systems, firing-set, and connections. The mechanical specialists learned how to assemble the 96 explosive blocks of the implosion sphere, insert its detonators, and to attach the ballistic envelope and support cones.

It was not long before the specialty teams learned to create storable sub-assemblies with which to speed their work. At the end of 1946, a team assembled and loaded a Fat Man bomb onto a B-29 in less than a day. In 1947, Dorland's engineering teams were absorbed into the AFSWP. At this time, the Army Air Force decided to test the competence of the new organization with a maneuver (Operation AJAX) simulating overseas deployment to a forward base. This was necessary because the B-29's limited range did not permit strikes originating from the continental United States. In order to facilitate this exercise, Dorland had portable structures designed and built to provide the environmental and safety standards needed for forward basing. He also recruited noncommissioned officers and had them trained to complete the teams. Finally, in November 1947, the AFSWP activated Operation AJAX in coordination with the newly created Strategic Air Command.

As part of Operation AJAX, Dorland's teams pulled the parts for six Fat Man bombs from their storage areas and loaded them aboard B-29's for a circuitous trip to Wendover AFB, Utah, which represented the forward base for the exercise. Portable assembly buildings accompanied the B-29's in cargo aircraft belonging to the 1st Air Transport Unit. Shortly after their arrival at Wendover, Dorland's teams erected their portable buildings, assembled and tested their MK III bombs, and delivered them to the 509th Bombardment Group. The 509th then flew the bombs on simulated missions before it returned five of the bombs to Sandia for disassembly and storage. The sixth bomb was drop tested to ensure that it was fully functional. Although it did not contain a nuclear core, it produced a gratifying high order explosion at its intended detonation altitude.

Eight months after AJAX, the Air Force conducted a second atomic exercise. It concluded from this exercise that it could increase atomic bomb assembly to a rate that exceeded one bomb per team per 24-hour period. It also concluded that the use of the complete assembly method at forward bases was more economical of staff than the rear-forward method used in AJAX, and that use of the recently developed C-97 flying assembly room (code named Chickenpox) was effective if auxiliary power was available.

Based on AJAX, the AFSWP determined that the principal requirements to prepare an atomic bomb for use at a forward base consisted of 39 Assembly Team personnel, one assembly kit, nine C-54 cargo aircraft, one airfield, one assembly building, and a strategic aircraft and its crew. Together, the necessary material for this task occupied a volume of 3,800 cubic feet and weighed 40,000 pounds. Organizing a strike with 25 atomic bombs had much larger requirements: 975 Assembly Team personnel, 25 equipment kits, 9 airfields, 225 cargo aircraft, 25 assembly buildings, and 25 strategic aircraft and their crews. Early war plans that envisioned dropping 75 - 125 atomic bombs required thousands of assembly personnel, about one thousand support aircraft, and millions of pounds of equipment – a non-trivial undertaking!

In parallel with the development and training of Assembly Teams, the AFSWP developed training courses for Bomb Commanders, Weaponeers, and Staff Officers through its Operations and Training Division. In 1947, the unit charged with this program was the 2761st Engineer Battalion (Special), which had previously trained the Assembly Teams. The AFWSP later delegated the training function to the 38th Engineer Battalion (Special) before it re-delegated these activities to 8470th Technical Training Detachment on June 25, 1948. The 8460th Special Weapons Group; the 1140th Special Reporting Wing; and Army, Navy and Air Force assembly units under the control of the 8460th SWG assisted the 8470th SWG.

The AFSWP began the training of Senior Weaponeers or Bomb Commanders (atomic pilots) in the spring of 1946. The early trainees formed the nucleus of the team responsible for Operation Crossroads. After two weeks of preparatory work, successful candidates participated in a three-week course with six students assigned to a class. During 1948, the AFSWP increased the rate of training until 20 Bomb Commanders a month were graduating. By the end of the year, there were 222 Air Force, 37 Navy, and 2 Army alumni. The training of Junior Weaponeers (bomb technicians) occupied 24 weeks, with entry to the course set at twelve-week intervals. Enrollment was ten officers per course. The principle components of the course were nine weeks of mathematics, electricity, and electronics; four weeks of atomic bomb components and their assembly; four weeks of maintenance and the use of electronic test boxes; three weeks of disaster procedures and Silverplate installations; and four weeks of flight training in B-29s. By the end of 1948, 24 students were graduating per month and 162 had completed training – 144 Air Force and 18 Navy personnel. The Staff Officer training course lasted three days and could accommodate 25 students per session. By the end of 1948, 300 officers had graduated from this course. In terms of assembly teams, the AFWSP could turn out one team per month by the end of 1948 and did not envision any difficulty in meeting a goal of ten fully trained teams.

In order to support a strategic nuclear capability, the Navy desired the same training that the AFSWP was providing for the Army Air Force and in 1947 requested assistance from Sandia to achieve this end. The Chief of Naval Operations intended to upgrade three aircraft carriers for the storage, maintenance, and transportation of atomic bombs. Sandia was immediately able to help because it had recently completed its design exercise in creating temporary assembly facilities for use at forward air bases and was active in training Army recruits for the AFSWP.

Navy trainees underwent a program very similar to the Army Air Force program. After completion, the graduates joined Special Weapons Units composed of 70 - 80 officers and enlisted personnel. The first such unit was NSWU 471. NSWU 802 and NSWU 1233 followed shortly thereafter. These units first served at National Storage Sites before assisting engineers and technicians from Sandia's Z Division with the conversion of aircraft carriers and sea trials at Norfolk NB. Navy nuclear technicians from NSWU 471 assembled a MK III Fat Man bomb, a Little Boy bomb, and a MK IV mockup on board the USS Midway (CVB-41) in November 1948. The Navy achieved its desired nuclear strike capability on August 31, 1950, when it declared the AJ-1 medium bombers of flight VC-5 operational on board the USS Coral Sea (CVA-43).

After 1953, when the national nuclear support units were decommissioned, the Navy transferred its atomic personnel to Special Weapons Unit Atlantic (Norfolk) and Special Weapons Unit Pacific (San Diego) under the respective commands of Naval Air Forces Atlantic and Naval Air Forces Pacific. As more advanced strategic weapons and tactical weapons became available, the size of the Special Weapons Units shrank to five or six officers and 18 - 20 enlisted men. The modernized weapons required less maintenance. The Navy assigned smaller CVA carriers a single SWU whereas it assigned the larger CVBs two such units.

Since atomic bombs were in short supply when first released to Navy custody, they were off loaded from a carrier as soon as it came into its homeport, transferred to a special weapons storage site for any required modifications or maintenance, and immediately loaded onto another carrier leaving on patrol. Entry into the nuclear storage or "Special Aircraft Service Stores" spaces for CVA carriers was from the third deck amidships. Larger CVB carriers had SASS areas located fore and aft. A small entry area guarded by an armed marine contingent contained a ladder that led downwards to a cross deck passageway. Off this passage, the Special Weapons Unit Office, the instrument shop, the nuclear shop, the battery locker, a pair of storerooms, and the ladder that led further down to the special weapons magazine were accessible.

The magazine on a CVA surrounded a trunk that ran from the third deck down to Damage Control Central. Detonators were stored in a special locker where they were tested and inspected. The atomic weapons were stored aft of the trunk, next to assembly stations where the electrical section's test equipment was stowed. SWU crewmembers stowed tactical weapons such as the MK 5 and MK 7 in special holders that aligned them for attachment to the ammunition hoist. They bolted MK 8 weapon containers directly to deck plates through pre-positioned threaded holes. An overhead hoist transported electronic cartridges removed from atomic weapons for disassembly and testing in the electrical section. Batteries were transferred one deck up to the Battery Locker for charging or testing. Separable nuclear components (cores) were stored in a special compartment further aft and below in their birdcages. From this location, a system of manual dumbwaiters transferred them to the SASS space.

Navy Special Weapons Unit staff hoisted bombs aboard carriers in "roadable" containers that they towed with the same vehicles used to position aircraft. The tow-vehicles first moved the containers to a transfer passage where seamen removed the container's outer coverings. The vehicles then towed the wheeled portions of the containers to the magazine elevator where the lower part of the container and its bomb descended to the SASS magazine. Once in the magazine, a rail mounted ammunition hoist picked up the bombs and positioned them in their holders. NSW staff then pushed the bomb containers back to the transfer passage, reassembled them, and returned them to the special weapons storage site onshore.

The Navy implemented similar safety and handling procedures for cruisers when it began arming them with nuclear tipped surface-to-air missiles and rocket-assisted nuclear depth bombs in 1955. A Talos armed cruiser received both conventional and nuclear warheads in standard warhead containers, which dockside personnel brought alongside in trucks or barges at Naval Weapons Depots. Only a serial number stenciled on its exterior identified a container's contents. After acceptance, the Talos crew moved the warhead containers, or "cans," to warhead strikedown hatches on the cruiser's O2 deck over the port and starboard test areas. The crew then used a davit to lift the warheads out of their containers and to lower them to the Special Weapons Office on the second deck. Here, sailors attached the warheads to receiving stands. From the Special Weapons Office, the guided missile crew transferred the warheads to a magazine on the second platform at the base of the ship. Both the missile house and the warhead storage magazine were designated nuclear weapons spaces. Access to nuclear weapons spaces was only by authorized personnel who abided by the two-man rule at all times.

The ship carried a Marine detachment that guarded its nuclear weapons. Entry to any nuclear designated area required the presentation of an authorization notice to the Marine Sergeant of the Watch. If the authorization was valid, the Marine Sergeant dispatched armed guards to the appropriate door or hatch. Only after the guards informed the Sergeant of the Watch that they were in place did the Sergeant open the door or hatch to allow entry of the authorized personnel. All hatches and doors into the missile house and warhead magazine were alarmed. When an alarm sounded in the Marine Detachment Office, armed Marines immediately proceeded to the alarm site. The guards had authorization to shoot anyone attempting to enter a nuclear weapons space without proper authorization. In the event of a fire or similar such emergency in a nuclear designated space, the missile crew acted as first responders because of their authorization level. If necessary, damage control crews without authorization could enter under the supervision of marine guards. This exception was necessary because of the large amounts of explosives and propellant stored in nuclear designated areas.

While Sandia Base was assisting the Navy in achieving its storage and transport capability for atomic bombs, the AEC established Technical Area II at Kirtland AFB to accommodate Z Division's expanding operations. Designed as a weapons assembly facility, Technical Area II became the primary assembly site for America's nuclear weapons between 1948 and 1952. Technical Area II was renamed Sandia Laboratory as a separate branch of the University of California in 1949.

In conjunction with establishing Technical Area II, the AEC undertook the building of three onshore National Storage Sites at which to keep the atomic stockpile dedicated to the Army Air Force. The Truman Administration authorized these sites in 1946 after General Groves stated his concerns about unsafe storage practices and the deteriorating condition of the stockpile. David Lilienthal, the first director of the civilian Atomic Energy Commission, was shocked to find the nation's supply of nuclear pits sitting in chicken wire cages during his first fact-finding tour of Los Alamos and Sandia. In the event of war, Air Force bombers could arrive at the three widely dispersed Storage Sites to obtain atomic bombs for transport to forward overseas bases. The overseas bases would serve as the launching platforms for their final attacks. The Army Air Force had significant concerns about the locations of the three sites, which the AEC chose for physical security rather than operational suitability.

The three AEC / AFSWP operated National Storage Sites (NSS) were designated Able, Baker, and Charlie. They were located at Manzano Base (Able) near Kirtland AFB, New Mexico; Killeen Base (Baker) located between Gray AFB and Fort Hood, Texas; and Clarksville Base (Charlie) adjacent to Campbell AFB and Fort Campbell, Tennessee and Kentucky. The Army Corps of Engineers completed Site Baker in 1948, with Sites Able and Charlie following in 1949. Site Able had 41 underground storage structures and two underground assembly units, Site Baker had 49 underground storage structures and an airstrip for B-36 operations, and Site Charlie had 24 underground storage locations and an above ground assembly plant. For security purposes, the Army Corps of Engineers carried out the construction of these sites in three phases. In the first phase, it proceeded with surface work and tunneling that did not reveal the nature of the projects. In the second phase, it undertook the construction of the assembly and storage areas. In the third and final phase, it completed the installation of all necessary equipment. the assembly rooms had a minimum rock cover of 100 feet.

Each base had approximately 150 administrative and 200 security personnel, as well as two maintenance and assembly teams; each composed of 77 technically trained personnel. In addition to their regular maintenance duties, the assembly teams convoyed bombs to the flight line for training exercises or in the event of war. They also provided instruction for Air Force and Navy Bomb Commanders and Weaponeers. By rotating personnel through AFSWP training, the DOD had approximately 1,500-trained technicians by 1950.

The concept of the civilian custody of nuclear weapons stored at a restricted number of national sites came under scrutiny in late 1949. By this time, it was common knowledge that the Soviet Union had established a strategic bomber force and had exploded its first atomic bomb.

American cities and military bases could become vulnerable to attack if the Soviet Union decided to send their bomber fleet on a one-way mission. Planners considered the possibility that the Soviets might capture Canadian airports as refueling points. The Korean War, which started in June 1950, brought further worries. A number of influential figures were concerned that the war might be a feint to draw American Forces away from the real target, which was Europe!

In order to further disperse the nuclear stockpile and increase the nation's offensive readiness, the AEC expanded the number of National Stockpile Sites to six locations, adding Bossier Base (Dog) near Barksdale AFB, Louisiana; Medina Base (King) near San Antonio, Texas; and Lake Mead Base (Love) near Nellis AFB, Las Vegas, Nevada. It also gained approval for the construction of five smaller Operational Storage Sites. It is not coincidental that the OSS or "Quick Alert" facilities were located at Loring (Easy), Ellsworth (Fox), Fairchild (George), Travis (How) and Westover (Item) Air Force Bases, where the Air Force stationed its B-36 intercontinental bombers. The storage of nuclear weapons on base greatly reduced the time needed to mount a strike. This had previously required 36 hours, during which time the AEC and the Air Force had to distribute atomic bombs to arming points. The AEC also built Stockpile Sites at North Depot Base (Yoke) near the Seneca / Romulus Army Depot, New York; and Skiffes Creek Annex (Jig), Yorktown Naval Weapon Station, Virginia.

Sandia Corporation, the successor to Sandia Base, undertook construction of the new sites. Ultimately, Air Materiel Command assumed the sites' operation on behalf of SAC and the AEC. An Aviation Depot Squadron equipped with Boeing C-97 Stratofreighters supported each of the various operational sites. Because these aircraft had B-29 airframes, they had the necessary room and lifting capacity to transport nuclear weapons. The bombs had to be loaded in pieces, though. Ground crews loaded the nose cap and forward bomb components in the front of the cargo compartment, placed the explosive sphere in chocks amidships, and stored the box tail and aft bomb case in the rear of the cargo compartment. It was not long before the larger C-124 Globemaster II replaced the Stratofreighter. These aircraft had a 74,000-pound carrying capacity and a range of 2,175 miles.

Black and Veatch of Kansas City designed the new storage sites, known as "Q" areas for their high level of security clearance. The firm had a long history with both the Army and President Truman. Black and Veatch also designed other installations such as the Burlington, Iowa, Assembly Plant. They laid out each Q Site along similar lines with a series of specialized buildings and bunkers or "igloos" for storage of the non-nuclear and explosive bomb components. The Q Sites adjoined major military installations for security. Three rows of fencing, topped by barbed wire surrounded them. The middle row of wire carried a fatal charge of electricity, but death for an intruder was just as likely to come from a bullet since the area between the fences was a designated "kill zone." Concrete pillboxes and lighted patrol roads on the Q Sites' perimeters added further security.

Security at Q Sites was very strict, with the "two-man rule" enforced to the letter. This rule required at least two qualified individuals to be present around any nuclear or non-nuclear component of a bomb, in weapons storage or in the servicing areas. Buildings were equipped with electronic sensors and motion detectors monitored by an Air Police Operations Center. Teams wishing to enter a building or storage igloo had to use an exterior phone jack to call the Operations Center and give an access code based on a continually changing matrix. The Operations Center then sent an Air Police team to authenticate that at least two individuals with proper identification were present. The Air Police had authority to deactivate the alarms and unlock the structure to provide access. For maintenance buildings, a second inner door was remotely unlocked from the Operational Control Center only after the Air Police relocked the outer door. On weekends and after hours, guard dog patrols supplemented security.

In addition to a group of administrative buildings, the Q Sites contained "A" Structures that housed the nuclear cores or capsules. These were stored individually in M102 "birdcages," special cylinders charged with dry nitrogen gas to prevent corrosion of the atomic pits. The A structures also provided storage for neutron initiators. The buildings were two-storied, with 10-foot thick reinforced concrete walls. The upper story was a 17-foot thick dummy composed of reinforced concrete. The lower story was either below ground or partly recessed and protected by bermed earth. The whole effect was to simulate an innocuous office building. The Black and Veatch designed "A1" Structures had a pair of single-entry rooms on each side of a central corridor. Stored on specially constructed steel racks, 30 per room, each structure accommodated 120 birdcages. The AEC had to undertake special studies to determine the minimum spacing between the birdcages to prevent unwanted nuclear interactions. Black and Veatch later designed "A2" Structures to house the DT booster reservoirs used in fusion weapons. In order to safeguard their contents, A Structures were fitted with heavy, bank-vault type doors.

The AEC used another type of building, known as a "C" Structure for the maintenance and inspection of nuclear pits and initiators. It did not build these structures to the same blast proof specifications as A structures but gave them the same unassuming appearance. A common activity carried out in this facility was the removal and replacement of threaded plugs from plutonium and uranium cores as part of the inspection process. This procedure often produced radioactive shavings, which were disposed of onsite.

The Q Sites also contained "Plants," which were used for the maintenance and inspection of an atomic bomb's non-nuclear components. Type I Plants had arched reinforced concrete roofs with earthen bermed walls that tapered from 2 feet thick at the base to 1-foot-thick at the spring. They were 1.5 feet thick at the crown. During construction, workers mixed metal filings with the concrete used to construct their floors. This product insured the dissipation of static electrical charges. Each plant had six bays and tunnels that led to adjacent plants. The bays were designated mechanical (M-bay) or electrical (E-Bay) depending on the type of activity conducted inside. These activities included the inspection, testing, and maintenance of various sub-assemblies, mechanical systems, and electronics. To facilitate in the final assembly and movement of weapons, the plants were equipped with heavy-duty forklift trucks and overhead cranes.

Early hydrogen bombs were so large that Q Sites required specialized Type II Plants to service them. These plants had two bays that contained heavy lifting equipment. They also had a special alarmed room in which to service the tritium components of boosted primaries and warheads. Because tritium gas is highly radioactive and therefore carcinogenic, the rooms had special vacuum intake devices installed in their ceilings to vent any release of tritium gas. The move to boosted warheads equipped with electronic neutron initiators resulted in the cessation of operations in C structures when the AEC withdrew the last weapon featuring a mechanical neutron initiator from service 1962.

A late type of building constructed at Q Sites was the Surveillance or "S" Structure. The Quality Assurance and Inspection Agency, staffed by AEC personnel from Sandia, carried out developmental surveillance and quality assurance inspections in these buildings. Quality assurance inspections were distinct from the routine maintenance operations undertaken in the electrical and mechanical bays of the Plants. Surveillance Structures contained their own electrical and mechanical facilities, a calibration room, and a photographic lab.

The formation of the North Atlantic Treaty Organization (NATO) on April 4, 1949, brought further requirements for the dispersal of nuclear weapons. The fundamental role of NATO was to safeguard the freedom and security of its twelve member-nations (increased to fourteen in 1952) by both political and military means. *Article 5* of the *NATO Treaty* regarded an attack on any member in Europe or North America as an attack upon all. Using secret agreements, the United States obtained permission to station bombers, missiles, and atomic weapons at NATO bases for the mutual protection of the Alliance. The success of NATO resulted in the 1954 formation of a similarly structured Manila Pact or Southeast Asia Treaty Organization (SEATO) after the fall of French ambitions in Viet Nam. In 1958, the United States joined the Baghdad Pact or Middle East Treaty Organization, which was renamed the Central Treaty Organization (CENTO). By these means, the United States surrounded the Communist Bloc with a series of containment treaties consistent with the philosophy of document *NSC-68*.

President Truman authorized the first foreign shipment of nuclear weapons on June 11, 1950. This was for 89 sets of MK IV non-nuclear components to Great Britain. On July 1, 1950, after the Communist invasion of South Korea, Truman authorized the further deployment of MK IV non-nuclear components to Guam and to the carrier USS Coral Sea in the Mediterranean. The bomb's nuclear capsules followed later to make them operational. By the end of August, the AEC had authorized the construction of seven foreign nuclear operational sites. The Truman Administration anticipated that the forward deployment of bulky, non-nuclear components would speed up reaction time in the event of war; the AEC and the Air Force need only disperse small atomic cores to place the bombs in service.

In October 1952, the Secretary of Defense asked the Joint Chiefs' opinion on the deployment of nuclear cores to Armed Forces controlled sites where non-nuclear weapons components were already in storage. The Joint Chiefs recognized that this question was the wedge they needed to pry loose nuclear weapons from the AEC. They once again stated their opinion that this option was critical to maintaining operational readiness and to insure flexibility of response. They also stated that disruptions to communications and the difficulty in bringing nuclear weapons forward to units under attack might prove disastrous. They therefore requested that the President approve the disposition of nuclear materials, and on June 30, 1953, Eisenhower authorized the deployment of nuclear components to the Air Force in numbers equal to the non-nuclear components already in its custody.

The distribution of nuclear and non-nuclear components to Air Force storage sites and the relentless training pursued by SAC in search of operational excellence kept nuclear technicians busy. By 1953, the MK IV had superseded the MK III and the MK 6 strategic and MK 5 lightweight strategic bombs were waiting in the wings. The MK 7 tactical bomb and the MK 90 depth charge were also entering service. Although the stockpile contained only 369 warheads in 1950, four years later it had expanded by a factor of more than 5 to 2,063 weapons! All the Branches of the Armed Services were by this time equipped with nuclear weapons. Ominously, the enormously destructive hydrogen bomb was poised for mass production.

In order to maintain the new weapons arriving at storage sites, the DOD had to develop formal training schools and curriculums. The Air Force officially recognized three specialty classes of Air Depot technicians at its operational bases: 463X0 Mechanical Specialist, 331X0 Electrical Specialist and 332X0 Nuclear Specialist, Training for mechanical specialists was conducted at Lowry Air Force Base, Colorado and then at Sandia Base. The electrical specialists received their training at Keesler AFB, Mississippi and then at Sandia. Upon graduation, Sandia selected top 331X0 graduates for further training as 332X0 nuclear specialists. Course durations for these specialties ranged upwards to about 1,200 hours of instruction.

The nuclear specialists were responsible for servicing neutron initiators, assembling nuclear pits, and maintaining the environmentally controlled M102 storage cylinders in which the pits were stored. They had to check neutron initiators periodically to ensure that their radioactivity levels were strong enough to start a chain reaction. They also had to clean each pit or core also with trichloroethylene and examine them for signs of spalling and corrosion. They then carefully weighed the cores and measured their radioactivity levels measured. Finally, the nuclear specialists checked their readings against a table for each type of core to certify that no one had tampered with removed material from them. After specialists checked the cores, they added desiccant to and recharged with nitrogen the birdcages used for the cores storage and transport.

The electrical and mechanical specialists performed yearly inspections for each weapon, as well as an inspection every time SAC flew a bomb. They removed the bomb's NiCad batteries, topped them up with water, and charged them on a regular schedule, about every 3 months. During a full inspection, the mechanical techs removed the tail and service ports from bombs, allowing their electrical cartridges to be removed for transportation to an electrical bay. Here, the electrical technicians partially disassembled and inspected the bomb's cartridges to make sure their vacuum tubes and circuits were operating properly. The electrical techs also maintained the test equipment necessary to perform their various functions.

The specialists also inspected detonators, tested firing-sets, and calibrated barometric fuzes in a vacuum chamber to ensure that they would function at the proper altitude. If the bomb had an automatic in-flight insertion mechanism, they serviced and cycled it to make sure that it operated properly. The bomb was then reassembled and the results of the inspection recorded on a "Weapon Summary Sheet" that was stored inside the nose section. This sheet also provided information on radar and barometric settings for the bombardier and Aircraft Commander. Finally, the specialists sealed a container of desiccant inside the bomb to maintain a desired level of humidity. Following the inspection, they recorded the results were on a *Weapon Inspection Report*, of which "Special Copies" were forwarded to the Air Force and the AEC.

In addition to inspections, specialists performed modifications and upgrades to weapons as part of improvement programs. For example, when Los Alamos discovered that the MK 6 MOD 6 strategic atomic bomb had an electrical fault that might trigger its detonating system if unintentionally grounded, the weapons had to have specifically numbered wires severed and properly capped. Another problem identified with W31 Nike warheads required specialists to drill holes though the fiberglass covers of their high explosive implosion systems in order to reroute wiring. They then had to clean cut wire ends and fiberglass and metal shavings from the interior of the weapons. The continual modification or retirement of thousands of weapons eventually took so much time that in 1958, the AEC built the Clarksville Modification Center at Fort Campbell, Kentucky and the Medina Modification Center at Medina, Texas. Clarksville was primarily responsible for the dismantlement of retired weapons whereas Medina concentrated on modification and stockpile evaluation testing. The centers operated until 1965 and 1966 respectively, at which time the Pantex and Burlington Assembly Plants assumed their functions.

During an alert, ground crews had to retrieve weapons from Munitions Storage Areas, which required the opening of 5-ton blast doors on storage igloos. After retrieval from their igloos, the ground crews prepped their nuclear bombs for strike at maintenance and inspection plants. In the case of the MK 6, specialists gave a final charge to the bomb's batteries, which they installed in insulated and heated boxes for their high-altitude ride to the target. They also removed desiccant from the bomb casing and stripped tape from barometric fuze ports. Early parachute-retarded hydrogen bombs, such as the MK 15, had to have a nuclear core installed in their in-flight insertion mechanisms. This necessitated the removal and replacement of the parachute package through the rear of the bomb's fuselage.

After the specialists had their bombs prepped for strike, they towed them to the flight line; providing security with .45 caliber semi-automatic pistols. The drive to the flight line could take some time because maintenance and inspection areas were often located several miles away from munitions storage areas and munitions towing was limited to a speed of 5 MPH. In the early years, the ground crews simply towed Fat Man type strategic atomic bombs on their maintenance dollies. These vehicles gave way to balloon tire trailers that could traverse increasingly longer distances to flight lines. With the advent of the 22-ton MK 17 hydrogen bomb, ground crews used a beefed-up version of the Ross straddle carrier, rated at 50,000 pounds, to transport the bombs. Later, Air Materiel Command developed specialized handling equipment such as the hydraulically operated MHU-7M loading / transport trailer. Hauled by a 2 1/2-ton truck, the ground crew could operate the trailer either electrically or manually. Transported along with the strategic weapons were 20mm cannon shells and .50 caliber machine gun cartridges with which to equip the bomber's defensive armament.

At the flight line, the specialists turned their atomic weapons over to specially certified SAC crews (462X0), who loaded the bombs into each aircraft under the direction of the aircraft commander. The aircraft commander had the final responsibility to ensure that the nuclear bombs loaded on his aircraft were properly stowed and that their monitoring, control, and release systems were connected.

Every 30 days, the loading crews had to satisfactorily complete an upload / download for each type of bomb and each type of aircraft for which they were responsible. They had to jack up B-29 and B-50 aircraft at the nose and remove a bomb bay door to center a bomb for loading. The B-36 simplified loading procedures because it had enough clearance for even the largest weapons. Nuclear bombs were at first winched into bomb bays using a modified version of the pneumatically operated U-2 system developed to load 22,000-pound "Grand Slam" bombs into British Lancaster bombers during WW II. Later, special "clip-in" assemblies facilitated the loading process. Ground crews winched bomb pallets into bomb bays and mated them with suspension systems. Different types of pallets could accommodate a single large bomb or four smaller bombs.

A team located at Kirtland AFB, New Mexico, designed and implemented aircraft modifications as well as the equipment used to load bombs. The Air Force transferred this facility to Air Materiel Command as a Logistics Center in 1946. At this time, Air Materiel Command tasked the Logistics Center with providing flight test support for the operations at nearby Sandia Laboratory. As Los Alamos produced new weapons, the Logistics Center carried out the necessary modifications to allow the SAC bomber fleet to carry them. In December 1949, the Air Force designated Kirtland AFB as the home of Special Weapons Command, a division of Air

Research and Development Command. Special Weapons Command researched the hardening of aircraft, missiles, and missile sites against the effects of nuclear attacks.

All of the weapons activity up until 1955 was merely the tip of the iceberg; the main period of weapons production took place between 1956 and 1967. In order to manage the delivery and storage of new weapons, the AEC drew up and activated a long-range deployment schedule. By early 1956, the AEC had shipped about 20 percent of its available nuclear components overseas and had stored another 20 percent stored at CONUS bases as part of the "Bombs on Base" program. The remaining nuclear components remained under AEC control at its national storage sites. In order to recognize the changes in these storage arrangements, the DOD and the AEC entered into a new agreement regarding the *"Responsibilities for Stockpile Operations,"* which both parties amended in 1959.

In 1957, the number of nuclear weapon storage facilities stood at seven national sites and eight domestic operational sites, with another seven operational sites located worldwide. Nine Strategic Air Command bases had received nuclear weapons and follow-on plans were in place for nuclear weapon storage at 10 Naval Bases, 21 Nike Sites, and 35 Air Defense Interceptor Bases. As well, the Navy deployed nuclear weapons on a number of ammunition ships and aircraft carriers. Eventually, nuclear weapons would find their way onto ballistic missile submarines, into missile silos, and onto many more air, naval, and army bases. In addition to the United States, nuclear weapons were stored at sites in the United Kingdom, Canada, Greenland (Denmark), Iceland, Spain, Belgium, the Netherlands, West Germany, Italy, Turkey, Greece, Morocco, Guam, Okinawa, the Philippines, South Korea, and Taiwan. At these international sites, SAC maintained an alert force of its B-47 bombers, which by this time numbered some 1,300 aircraft. SAC required each of its 29 Air Bombardment Wings to rotate 15 fully armed B-47s through these locations on a continuous basis.

There was no use in pretending to any sort of civilian control with this type of dispersal. Therefore, in 1962, the AEC yielded operational control of its weapons – apart from those that it was required to modify or repair. With full delegation of control, SAC disbanded its Air Depot Squadrons and assigned the maintenance of its nuclear weapons to Munitions Maintenance Squadrons (MUNMS) attached to Air Logistics Command, the new name assigned to Air Materiel Command in 1961. Over the course of the next decade, the AEC shut down its National Storage Sites as the Army, Navy, and Air Force moved the stockpile of nuclear weapons onto their bases and into upgraded or new storage areas. The Armed Services also assumed full responsibility for the storage, security, maintenance, transportation, and developmental surveillance of these weapons.

Unlike the Army, the Navy, and the Air Force, the Marine Corps never received direct custody of its nuclear weapons. The Navy held the Marine's atomic weapons in custody until such time it deemed that the Corps required them. The Marine's nuclear weapons were for the most part the same as those used by the Army. However, because of Marine force mobility requirements, the Navy did not supply heavy missiles for use by the Corps. Instead, Marine aircraft armed with tactical weapons filled the heavy weapon / long-range nuclear niche. Marine Wing Special Weapons Units supported Marine Air Wings, and Marine Nuclear Ordnance Platoons supported Marine ground forces. Within a Marine Amphibious Unit, a special weapons shop was responsible for supply and maintenance of nuclear weapons. The Navy certified its nuclear weapons for transport on board amphibious assault ships, amphibious transport docks, dock landing ships, and tank landing ships. The Navy also authorized Marine CH-46 and CH-53 helicopters to transport nuclear weapons from ships to the beach.

In Europe, the United States built Special Weapons Depots in support of NATO. At the request of the Army, a plan was prepared in 1958 to provide advanced weapons logistical support from 116 Basic Load Sites and 17 Special Weapons Support Depots. The Corps of Engineers began construction of these sites the following year. The Engineers provided both Type II and Straddle reinforced magazines for the storage of nuclear warheads and missile bodies. Mutual Assistance Program (MAP) ownership accounts were then set up for the purpose of managing component repair, parts, and ancillary equipment for NATO missile systems. The Commander in Chief of the US Army in Europe (CINCUSAREUR) assumed operational control of all units involved in the monitoring, storage, maintenance, modification, operational readiness, custody, and security of atomic weapons designated for non-US NATO forces.

Shortly before the final transfer of nuclear weapons to the Department of Defense, the Air Force modified the mechanical 463X0 AFSC designation by adding an A to designate tactical bombs or a B for strategic bombs. Specialists working on warheads received a missile badge. The introduction of missiles and tactical bombs caused major changes to the duties and working arrangements for many of these technicians. Instead of residing on base or in the relative comfort of an aircraft carrier, they could now find themselves servicing W27 warheads for Regulus cruise missiles on board submarines or mating a W7 warhead to an Honest John missile a few miles away from a Soviet Division.

The exclusive use of sealed pit nuclear weapons after 1962 rendered the duties of 332X0 nuclear specialists obsolete. The Air Force rolled up the remaining responsibilities for this position with the electrical designation. Then, in 1992, the Air Force introduced a single 2W2X0 AFSC general classification for nuclear weapons specialists. This decision acknowledged the introduction of solid-state electronics and self-testing circuits, as well as the general reliability of new weapons. It also reflected the reduced number of weapon types that remained in the stockpile after the implementation of the *INF Treaty* and *START*. In addition to warheads, 2W2X0 specialists inspect and service reentry vehicles, launching pylons, penetration aids, and Permissive Action Links (PALs). The Navy has its own designation system for nuclear specialists. The Army retired its 55G Nuclear Weapons Maintenance Specialist designation when it ceased operating these weapons in 1994.

By mid-1962, there were more than 25,000 bombs and warheads in the Armed Forces inventory. Transporting these weapons from AEC production plants to national, operational, and base storage sites was a considerable undertaking. The AEC moved the weapons using a combination of air, rail, road, and ship. In the beginning, atomic bombs were so large that they were simply loaded into B-29 or B-50 bombers and flown to their final destinations. This was also true for the MK 17 hydrogen bomb, which SAC could most easily transport by means of B-36 bombers. Later on, as shipping volumes increased and bombs became smaller, railways became an important means of transportation within and between bases. For this purpose, "Safe Secure Trains" that towed specially constructed ATMX rail cars with removable roofs were developed. AEC personnel could lower weapons, secured to their handling dollies, into the cars and then bolt them to the floor for security. They would next fasten down the roofs of the cars from the inside and exit through a hatch that was too small for the removal of the weapons. Rail transportation was never a preferred mode of transportation because of security concerns.

However, to maintain volume, it became a "necessary evil." The AEC made its last rail shipment of nuclear weapons in 1987.

Air Materiel Command and its successor, Air Logistics Command, also engaged in the mass transport of nuclear weapons. Aircraft such as the C-141 Star lifter were fitted with special racks that could secure several dozen warheads or freefall bomb. The air transport of nuclear weapons from one base to another was favored as a mean of maintaining the strictest security, especially for overseas deployments. The movement of warheads to individually dispersed missile silos required the use of road transport. Heavily armed SAC security teams accompanied the vehicles moving these weapons. In Europe, the Army used armored personnel carriers for the transport of nuclear artillery rounds. For its part, the Navy transported nuclear weapons onboard ships. It never used commercial vessels and it preferred ammunition ships over combat vessels.

The AEC initially transported nuclear bombs as is, but security and safety concerns soon dictated the development of specialized shipping and storage containers. The Sandia Corporation manufactured a roadable storage container for the MK 6 bomb that also served as a dolly and a maintenance bed for assembly operations at forward air bases. Because of their size, the provision of such containers was an exception for freefall bombs. Sandia produced shipping and storage containers mostly for warheads and nuclear artillery shells. In the case of later model nuclear artillery shells, Sandia built important accessories such as programming units into shipping containers. It also fitted the containers with limited try electronic locks (PALs) that rendered a warhead inoperable in case of tampering. The DOE currently transports warheads in sealed, aluminum, two-part drums. These have foam padding and spacers that secure the warheads to insure their safety.

Although the Department of Defense often uses its agencies to transport finished nuclear weapons, the AEC and its predecessors have moved special nuclear materials and nuclear weapons ever since the AEC transported the plutonium core for Gadget from Los Alamos to the Trinity test site. It was not until after the nuclear core had arrived safely at the McDonald ranch house that General Thomas Farrell signed a receipt taking control of the material used for Trinity.

The threat of terrorism in the 1960s caused the AEC to design special vehicles for the transport nuclear weapons, materials and radioactive waste – but never together. The Transportation Safeguards Division (now the Office of Secure Transportation (OST) adopted these vehicles after its formation in 1975. The eighteen-wheeled tractor-trailer used by the OST is designated the "Safe Secure Trailer." It is of extremely robust construction and capable of withstanding enormous impact. The storage area within the vehicle has additional crumple zones to improve crash resistance. Heavily armored, the SST is a "bank vault" on wheels. Its defenses include sophisticated electronic locks as well as gas and explosive booby traps with which to delay unlawful entry. Special armed couriers accompany SST shipments along transportation routes carefully planned with attention to weather conditions and the availability of safe havens at major military installations. A Secure Communications Network (SECOM) stays in contact with the SST and its accompanying security, ready to dispatch assistance if necessary. Currently the TSD handles all shipments of weapons to and from military bases and has an impeccable safety record.

Department of Defense regulations require the shipment of nuclear weapons in a safe manner. Because of this regulation, the Air Force and the Navy transported their early weapons without their nuclear capsules installed. They also shipped them with inert firing plugs installed and without their detonators. Qualified personnel installed the bomb's detonators only after they had arrived at an approved storage site. The Air Force and Navy authorized the installation of capsules and live arming plugs on board aircraft only in the event of an actual strike.

After the weapons laboratories developed boosted, sealed pit weapons with permanently installed cores, Sandia developed safety switches to arm or safe these weapons. Ground crew wired the safety switches to the safe position before shipment. The current crop of nuclear weapons has smaller implosion systems, insensitive explosives, and boosted warheads that make them inherently safe. Sandia also designed them with fusing and arming systems that will not activate unless preceded by a special set of circumstances surrounding their deliberate use.

As an example, the safety rules from *Air Force Instruction 91-111* (dated October 1, 1997) for a B-52H carrying B61 or B83 hydrogen bombs are as follows:

12.2.2. Common Strategic Rotary Launcher Mated B61-7, B83-0, and B83-1 Gravity Bombs or AGM-86B / W80-1 Missiles:

12.2.2.1. Pilot's Missile / Munitions Consent Panel:
- Off / Pre-arm switch in the OFF position with the cover down, safety wired, and sealed
- Lock / Unlock switch in the LOCK position with the cover down, safety wired, and sealed

12.2.2.2. Weapon Control Panel:
- Nuclear Lock / Unlock switch in the LOCK position with the cover down, safety wired, and sealed
- Nuclear Pre-Arm Enable (PA ENBL / OFF) switch in the OFF position with cover down, safety wired, and sealed
- Weapon Jettison Select (SEL / NORM) switch in the NORM position with cover down, safety wired, and sealed

In addition to safety concerns and secure storage, the Department of Defense places great importance on inventory control. For instance, the Air Force performs a number of redundant checks before it flies a nuclear weapon on a strike aircraft. In the case of nuclear-armed cruise missiles, munitions maintenance personnel prepare sheets that record the pylon number and the pylon's missiles and warheads' serial numbers before a breakout team enters a storage facility to verify the presence and status of the weapons. (Training, test, and live devices are currently stored together requiring identification by readily visible means.) After verification is complete, a convoy team again verifies the status and payload of all weapons before they tow them to a strike aircraft. After arrival at the flight line, the Crew Chief accepts the load of missiles only after again verifying their identities and payloads. The loading crew then installs the missile pylon and its missiles and yet again verifies the status of each missile and its warhead after loading. Finally, the strike aircraft's Bomb Commander is required to verify the payload in each missile and the safety status of each missile before accepting the load.

Despite this emphasis on safety and inventory control, a remarkable incident occurred in 2007. Minot AFB shipped a consignment of advanced cruise missiles to Barksdale AFB on a launch pylon attached to the wing of a B-52H bomber. The Air Force had slated the weapons for retirement. Imagine the surprise of technicians at Barksdale when they discovered that the missiles still had their W80 thermonuclear warheads installed! The key safety risk in this incident was that the aircraft commander might have ordered weapons jettisoned during an airborne emergency. Ground crew had not connected the missiles or their warheads to the B-52's onboard arming and launching systems. This incident generated significant interest in reinvigorating the nuclear management process.

In addition to stringent safety and storage regulations, the Armed Forces certify both military personnel and units connected with the use of nuclear weapons to carry out their duties safely and effectively. The Armed Forces began these certification procedures in the 1960s. The NNSA must also annually certify that the nuclear stockpile will function safely and as predicted. This process is based on technical evaluations that are summarized in statements made by the directors of Lawrence Livermore, Los Alamos, and Sandia National Laboratories and findings reported by the joint DOE National Nuclear Security Administration / Department of Defense Project Officers Group, the Commander in Chief of Strategic Command and the Nuclear Weapons Council and its Standing and Safety Committee. The Secretaries of Energy and Defense, who receive these reports, then make a recommendation on the safety and reliability of the stockpile in a memorandum to the President.

The remarkable ends to which safety is adhered means that although there have been a number of accidents, an accidental nuclear explosion has never occurred. Nevertheless, the DOE and DOD developed a coded system for reporting potential nuclear incidents on a graduated scale of severity. These are as follows:

> NUCFLASH – An incident involving the possible detonation of a US nuclear weapon that could provoke a risk of nuclear war with the USSR
>
> BENT SPEAR - A significant incident involving a nuclear weapon / warhead, nuclear component, or vehicle when nuclear loaded
>
> BROKEN ARROW - An accident involving a nuclear weapon or nuclear component
>
> DULL SWORD - A nuclear weapon safety deficiency
>
> EMPTY QUIVER - The seizure, theft, or loss of a US nuclear weapon
>
> FADED GIANT - An event involving a nuclear reactor or a radiological accident

Although an accidental nuclear explosion has never happened, accidents involving nuclear weapons have produced some rather spectacular conventional explosions and unwanted attention. The AEC documented approximately 32 major accidents between 1945 and 1980. For example, a B-36 dropped a MK 17 hydrogen bomb through its bomb day doors while on approach to Kirtland AFB on May 22, 1957. Although the bomb did not have its nuclear capsule inserted, the high explosives in its primary exploded on impact, leaving a 25-foot diameter crater. The explosion spread debris over a one-mile radius. On January 25, 1961, a B-52 suffered catastrophic structural failure near Goldsboro, North Carolina, releasing two MK 39 hydrogen bombs. One bomb parachuted safely into a tree whereas the other bomb fell into a bog where it broke up. The Air Force had to undertake a substantial excavation to recover the parts from the broken bomb.

On January 17, 1966, a high profile international nuclear accident occurred over Palomares, Spain, when a B-52 collided with a KC-135 refueling tanker and the crippled bomber released four MK 28 hydrogen bombs. The conventional explosives in two of the bombs exploded on impact with the ground and a third bomb landed safely. Months later, the Navy recovered the fourth bomb from the floor of the Mediterranean Sea with the help of a deep diving research submarine. A second incident with a B-52 carrying four MK 28 hydrogen bombs occurred at Thule AFB in Greenland on January 21, 1968. The aircraft in question crashed and burned, setting off a conventional explosion that caused extensive radioactive contamination requiring substantial cleanup. The accident was embarrassing because neither the Danish nor the US governments had been entirely forthcoming about their nuclear arrangements. On a final note, a technician accidentally dropped a socket down a Titan II missile silo on September 18, 1980. The wrench bounced and punctured a fuel tank, ultimately resulting in an explosion of the fully fueled missile. The W53 thermonuclear warhead in its MK 6 reentry vehicle was blown clear of the silo, which was sealed with a 740-ton blast door – to land 200 yards away!

The *Intermediate Nuclear Forces Treaty* that went into effect in 1987, and the *Strategic Arms Reduction Treaty* of 2001, greatly reduced both the number and type of nuclear weapons in the stockpile and this has simplified storage, maintenance and safety. As of January 2006, the United States had an active stockpile of 5,235 strategic and 500 tactical warheads and bombs. It held another 4,225 warheads in reserve or in the inactive stockpile.

A great many of the reserve warheads are stored at the Kirtland Underground Munitions Storage Complex (KUMSC), which was activated in 1992. This facility provides storage, shipping, and maintenance for the United States Air Force and Navy. It is operated by the 898th Munitions Squadron (898 MUNS) and the 377th Security Forces Squadron (377th SFS). The facility occupies more than 300,000 square feet and is located entirely underground. Total storage capacity for warheads exceeds 3,000, and includes the B83 and B61 gravity bombs, and W80, W88, and W87 warheads. Kirtland AFB serves as one of two main Air Force nuclear weapons general depots in the United States (the other is at Nellis AFB in Nevada). Because of its 300-mile proximity to PANTEX, Kirtland serves as a transshipment base and storage point supporting the disassembly facility.

The DOE divides the active and inactive stockpiles into a number of subcategories. This arrangement provides the flexibility to accommodate a variety of contingencies and to maintain operational warhead quantities. Active stockpile (AS) warheads consist of strategic and tactical weapons kept ready for use with their boost gas bottles and other limited life components installed. The NNSA regularly assesses these warheads to insure their reliability and safety. The active stockpile includes operationally deployed warheads, augmentation warheads and logistics warheads. Operationally deployed warheads are stored at operational bases and are ready for immediate employment. Augmentation warheads are stored at operational bases or depots, maintained in an operational status and are ready to serve as operationally deployed weapons in less than six months. Augmentation warheads are never loaded onto delivery vehicles or launch platforms. Logistics warheads are stored at either operational bases or depots, maintained in an operational status, and used to replace operationally deployed or augmentation warheads undergoing maintenance or testing for quality assurance.

Inactive stockpile (IS) warheads are strategic or tactical weapons maintained in a non-operational status with their boost gas bottles and other LLCs removed. The inactive stockpile includes augmentation warheads, logistics warheads, quality assurance and reliability testing (QART) warheads, and reliability replacement warheads. Augmentation warheads are stored at a depot, maintained in a nonoperational status and require at least six months for conversion to AS operationally deployed weapons. The Army and Navy maintain logistics warheads in a nonoperational status until authorized for reactivation to serve as AS logistics warheads associated with reactivated augmentation weapons. The DOE maintains QART replacement warheads in a non-operational status until authorized for reactivation to replace AS warheads used as QART samples. The DOE may replace AS or IS weapons that develop reliability or yield problems with QART replacement warheads stored at depots. Reliability replacement warheads are maintained in a non-operational status until used as replacements for AS or IS warheads that develop safety, reliability or yield problems.

Currently, only the Air Force and the Navy maintain operationally deployed and active stockpile weapons. The Air Force deploys a force of 450 Minuteman missiles armed with W78 and W87 strategic warheads around F. E. Warren AFB, Wyoming, Malmstrom AFB, Montana, and Minot AFB, North Dakota. The Air Force deactivated the last Peacekeeper heavy strategic missile in 2005 and moved its boosters to storage. The Air Force keeps B61 bombs in storage at Kirtland AFB, New Mexico and at Nellis AFB, Nevada. B61 bombs and B83 bombs are also stored at Whitman AFB, Missouri, for use by B-2A stealth bombers. The Air Force stores ALCMs, armed with W80 warheads, at Barksdale and Minot AFBs for their B-52 wings.

NATO also deploys the tactical version of the B61 bomb. Sandia designed a special storage system unique to NATO's quick alert requirements for use with this weapon. Starting in 1990, the Bechtel Corporation built a subterranean weapon storage vault (WSV) into each of 249 protective aircraft shelters (PAS) located at air bases throughout Europe. The mechanical part of the WSV elevates a platform assembly, designed to hold four B61 bombs, out of a reinforced concrete and steel foundation inside of each shelter. Air Force technicians carry out maintenance on bombs removed from WSVs inside their PAS shelters, following the two-man rule. By 2000, there were 204 of these vaults in operation at eight bases in six European countries with a stockpile of 520 B61 bombs. By means of reductions or elimination from a number of European bases that included Lakenheath, UK, and Spangdahlem, Germany, the United States has reduced the NATO stockpile to 100 - 200 tactical bombs.

The Navy maintains most of its nuclear warheads on missiles kept aboard its Fleet Ballistic Missile submarines. It bases these submarines at Strategic Weapons Facility Atlantic, (SWFLANT) King's Bay, Georgia, and at Strategic Weapons Facility Pacific, (SWFLPAC) Bangor, Washington. The Navy has decommissioned its last Trident I missile to leave the SSBN fleet universally equipped with Trident II missiles. The Navy arms its Trident II missiles with a combination of W76 and W88 warheads that it services at the fleet strategic facilities. The Navy also maintains a stock of W80-0 tactical warheads for its Tomahawk Land Attack (nuclear) cruise missiles, which are currently in storage. Formerly located at a number of Naval Weapons Stations (NAVWPNSTA), the W80-0 warheads are now stored at King's Bay and Bangor.

CHAPTER 4
INTELLIGENCE, TARGETING, AND EARLY WARNING

In 1947, former British Prime Minister Sir Winston Churchill gave a speech in which he declared that Soviet Dictator Joseph Stalin had brought down an "Iron Curtain" across Europe. Behind this wall of secrecy, Stalin was busy consolidating control over Eastern Europe, which his armies had overrun on their way to defeat Hitler and the Nazis. The Soviet leader was wary of Western intentions and wished to create a buffer zone against possible military intervention into Soviet affairs by Europe's democracies. For a while, the United States gained intelligence about Stalin's activities by interrogating repatriated Soviet internees and prisoners of war. Almost 2,000 specially trained military and civilian personnel attached to Project Wringer carried out this activity. Far East Command activated project Wringer in December 1946, which European Command soon joined. The intelligence gained from this project indicated that Stalin was in the act of assimilating Eastern Europe and planned to absorb Western Europe in the near future.

This disturbing intelligence caused the United States to formulate a comprehensive Soviet "threat estimate" known as Pincher. The Joint War Plans Committee (JWPC) carried out this estimate in 1946 on behalf of the Joint Staff Planners (JSP). Pincher not only evaluated the Soviet threat, but also considered how the Soviet Union might be defeated if it attacked Europe, the Middle East, or the Far East. For the purposes of the Pincher estimate, planners estimated the composition of the Soviet Armed Forces at 113 Army Divisions equipped with modern armor and supported by 20,000 ground attack aircraft. In contrast, they considered the Soviet Navy and its Naval Air Force to be "ineffective." Because American, British, and French forces had largely demobilized after WW II, American analysts considered their military forces incapable of stopping a Soviet invasion of Europe. Thus, to counter a Soviet attack, the JCS had no choice but to rely on strategic airpower.

"Crushing the enemy's will to resist" was to be accomplished with precision, conventional attacks that were intended to destroy Soviet war making capacity in "vital areas." In order of importance, these vital areas were Moscow, the Caucasus, Ploesti, the Urals, Stalingrad, Kharkov, Lake Baikal, and Leningrad. Any future war plan had to remain provisional until reconnaissance missions could gather additional information. To conduct precision air campaigns, such as the Army Air Force prosecuted against Germany and Japan in WW II, planners needed detailed information about the Soviet economy and its Armed Forces. For a start, SAC needed information about Soviet transportation networks, electric power infrastructure, factories, raw materials, and military bases. In order to hit these kinds of targets, bomber crews needed the detailed maps, aerial photos, and weather information that went into WW II vintage "target folders." Because this type of information was lacking for its estimate, Pincher simply designated urban areas as targets. Thirty cities thus became the centers for a projected air campaign. Pincher's planners concluded that in the future, a lack of strategic information might prove disastrous and recommended a peacetime reconnaissance program to gather much needed data.

The Air Force never converted Pincher to a war plan – it simply did not have enough resources to conduct a campaign on the scale envisioned by this study. Instead, it put together Emergency War Plan Makefast. Air Force planners directed EWP Makefast at Soviet access to

A U-2A reconnaissance aircraft on display at the USAF Museum. The U-2 was at first used for photographic missions. Later versions added antennae for electronic eavesdropping. (Credit: USAF; Color Original)

An ERB-47H reconnaissance aircraft. The weapons bay was refitted to hold three monitors who controlled its electronic monitoring equipment, in a very cramped environment. The aircraft could also be fitted for photo reconnaissance missions. These aircraft would have flown damage assessment missions during a nuclear war. (Credit: USAF)

A U.S. Air Force C-130 modified with poles, lines, and winches from its rear cargo door captures a film capsule and parachute ejected from a Discoverer satellite. This space-based espionage program was very successful. (Credit: USAF Photo 57-0257- via the 6594th Test Group; Color Original)

A KH-9 satellite equipped with a 72-inch diameter telescope. It had an infrared capability, and carried between four and six film capsules. The NRO equipped it with SIGINT antennas and a double-lens secondary camera for high resolution and wide area surveillance. The KH-9 transmitted SIGINT data and secondary images electronically. (Credit: NRO; Color Original)

The "Big Board" at SAC's underground strategic headquarters, Offutt AFB, Nebraska in 1961. The screens showed information on weather conditions, force deployment, aircraft, and missiles to aid SAC's staff in making vital decisions. (Credit: NSA)

Radar Station LIZ-2, one of 30 stations under USAF control along the Distant Early Warning (DEW) Line which ran 3,600 miles, from Alaska, across Northern Canada, to Greenland. The DEW Line provide warning of an air attack. The original Dew Line has been replaced by the North Warning System, which is under Canadian control. (Credit: USAF - Tech. Sgt. Donald L. Wetterman, Photo DFST8803446; Color Original)

Three BMEWS AN/FPS-50 detection radars at Clear AFB, Alaska. In the late 1950s through the 1990s, BMEWS provided warning of a ballistic missile mass attack over the North Pole. (Credit: USAF - Photo by C. Henry, HAER AK-30-A-96F)

An AN/FPS-120 Solid-State Phased Array Radar System (SSPARS) in operation at the Thule (BMEWS) AFB, Greenland. Similar radars were installed at Clear AFB, Alaska, and Flyingdales Moor, England. The radars have now been upgraded to the AN/FPS 132 standard. (Credit: USAF; Color Original)

The AN/FPS-115 Pave PAWS Site at Cape Cod, Maine. Pave PAWS Sites protected the US by detecting SLBM and fractional orbital bombardment attacks. The radars at Cape Cod, Maine and Beale AFB, California have now been upgraded to the AN/FPS 132 standard (Credit: USAF - Cmsgt. Don Sutherland, Photo DF-ST-93-03907; Color Original)

The NORAD complex at Cheyenne Mountain, Colorado. Below 2,000 feet of rock, buildings made of battleship steel sit on springs isolating them from the impact of a nuclear blast outside. NORAD analyzed data provided by radar sites to provide advance information of a nuclear attack to SAC, SAGE and the NCA. The Cheyenne Complex has closed and NORAD's headquarters have moved to Petersen AFB, Colorado. (Credit: USAF - Staff Sgt. Andrew Lee; Color Original)

petroleum in order to shut down its war machine. This was one of the strategies used to defeat Germany in WW II. It would probably have been effective against the Soviets because the US still had target folders for the Ploesti oil fields that now supplied petroleum to the Soviet Union. It also had target folders for some other parts of the USSR, which it had captured from Germany. Makefast did not envisage the use of nuclear weapons.

As the number of returnees to Western Europe began to dry up in the late 1940s, it became necessary to devise new methods to discern Soviet political and military intentions. The only practical way to gather the intelligence needed to expand the available set of target folders was through espionage. Recognizing the potential for a clash between Communist and Western ideologies, the US Army Air Force had begun overhead reconnaissance of Europe the month after Hitler was defeated. In an operation dubbed Casey Jones, the 306th Bomb Group photographed much of western and south-central Europe from their B-17s to determine which routes an invading Soviet Army might take. As east - west tensions transformed into the Cold War, photoreconnaissance bombers and patrol planes belonging to the Air Force and Navy flew missions along the margins of the Soviet Bloc in an attempt to peek as far inland as possible. This program was termed the Peacetime Airborne Reconnaissance Program. The objectives of PARBRO were three-fold. In addition to ascertaining and evaluating the military capabilities of the Soviet Union, targets of strategic importance were marked for possible destruction, and early warning and air defense sites were marked for avoidance.

Following the 1949 explosion of a Russian atomic bomb and the fall of China to Mao Tse Tung's revolutionaries, it became even more important to learn the intentions of the Communists. Because of this need, the Army Air Force and the Navy stepped up over-flights of Soviet and Chinese territory. These over-flights had begun in 1948 when SAC Reconnaissance Wings began to fly "Big Safari" missions using a variety of military aircraft. Airborne platforms included RF-100 Super Sabre jet fighters, the RB-45 Tornado, and RB-47 Stratojet bombers, which had been adapted to carry cameras. The immense RB-36 only made a few over-flights out of concerns MiG fighters might shoot it down. The most dangerous missions involved head on passes at Soviet radar stations in an effort to force them to reveal their operating modes and frequencies. The Air Force used the data from these missions to devise jamming procedures and to plot the safest paths for attack routes in the event of war. Unfortunately, for reconnaissance aircrews, once a radar site began to track their aircraft, its operators could vector in air defense interceptors.

During the period 1945 to 1950, the Soviet Air Force shot down approximately 40 Air Force and Navy reconnaissance aircraft around the margins of the Soviet Union and along the air corridors to Berlin. Despite the risks, a wholesale invasion of Soviet air space was in progress by the mid-1950s and only the overwhelming threat of the American nuclear arsenal prevented war. During a 2-month period in the spring of 1956, sixteen RB-47E photo reconnaissance and five RB-47H electronic reconnaissance aircraft flew 156 missions over Siberia. Project "Home Run," the finale to this operation, was a massed flight by no less than six RB-47s! There was nothing to distinguish a snooping RB-47 from a B-47 on a nuclear bomb run to a Russian radar operator.

Intelligence gathering gained new heights when the Central Intelligence Agency joined the fray in 1956. The CIA had the Lockheed Skunk Works build the revolutionary U-2 spy plane, capable of level flight at altitudes in excess of 70,000 feet. It was equipped with interchangeable noses, one of which contained a model 73 B camera developed by Edwin Land of Polaroid fame. A special lens equipped with an extremely sophisticated light meter compensated for the speed and vibration of the aircraft and set a performance benchmark for subsequent reconnaissance cameras. Another type of U-2 nose carried side looking radar. The aircraft eventually sprouted so many aerials for signals intelligence that they were nicknamed "antenna farms." The Soviets did not have any aircraft, artillery, or missiles with which to reach the U-2. This shortcoming allowed CIA sponsored aircraft, flown by SAC pilots with false civilian identities, to crisscross the Soviet Union photographically recording 100-mile wide swaths in unprecedented detail. Ultimately, SAC assumed the responsibility for U-2 operations from the CIA.

The Target Planning Staff categorized all of the material collected by intelligence operations and placed it into special databases. Categorization consisted of assigning a five-digit numeric "functional classification code" that indicated the products, capability, or activity associated with each target. Examples of these codes are 43900 for dams, 80052 for ground-attack fighter airfields, 8224X for command bunkers and 604XX for nuclear weapon storage depots. Cities had the functional classification code category 70210 and towns were designated 70220. Apparently, the Air Force planned to spare villages and hamlets from direct nuclear attack since it did not classify them. The *Standard Coding System Functional Classification Handbook* helped guide classification activities.

The *Target Intelligence Handbook* supplemented the *Standard Coding System Functional Classification Handbook*. The TIHB contained specific guidance and procedures for use in the analysis, production, and processing of target intelligence by the Defense Intelligence Agency (DIA), Unified and Specified Commands, Services, and other organizations. It included information on numbering and naming procedures, security classification guidance, coordinates, abbreviations, target intelligence programs, products, and geographic areas of coverage. The Department of the Air Force published the handbook to facilitate the production, maintenance, and use of the *Basic Encyclopedia, Geographic Installation Intelligence Production Specifications*, and related target intelligence documentation programs. The *Basic Encyclopedia* (BE) described every worldwide installation of interest to operational staffs or intelligence agencies. This information guided the selection of targets or, conversely, excluded installations such as hospitals.

The Air Force based the relative importance of each installation using six criteria. The most import of these was the ability to carry out an attack on the continental United States, while the least important was war support. For instance, planners rated Soviet air defenses lower than Soviet offensive bombing capability, except where Soviet air defenses might interfere with SAC's destruction of strategic airfields. Planners now store categorized data in the *Modernized Integrated Database* before they transfer it to the BE. The MIDB is a standardized system designed for data exchange between intelligence and operational assets from the national to tactical levels. It is the basic source for data describing installations, military forces, population concentrations, C^2 structures, significant events, and equipment.

The Air Force uses information from the *Basic Encyclopedia* to prepare the target folders that SAC (now STRATCOM) bomber crews need to identify their targets. The BE gives each installation a unique, ten-digit number. The first four digits of the ten-digit code identify the World Aeronautical Chart on which the target is located and the final six digits are the target's installation number. BE numbers are not reissued if a target is deleted. The *Basic Encyclopedia* covers the entire world in five areas: Latin America, Western Europe, Eurasia, the Pacific, and the Middle East with Africa.

The Air Force placed a subset of material from the Basic Encyclopedia in a *Target Data Inventory* that it maintained at the Air Force Intelligence Center in Washington. The Air Force updated the TDI three times a year with a standardized set of data it used for target planning, coordination, and weapons application. *The Target Data Inventory Handbook*, a user's manual, supported the TDI. From its inception through 1960, the TDI grew to more than 20,000 items listed on over 1,000 pages in two volumes covering three areas: Eurasia, Southeast Asia, and the Middle East. *Volume I* listed targets alphabetically whereas *Volume II* arranged them by category. Planners listed target data sequentially across each page of computer printout. After the target's location and general area, came its name, BE number, latitude, longitude, and target significance.

Established by the DIA, *Geographic Installation Intelligence Production Specifications* (GIIPS) have now replaced the TDI *Contingency Planning Facilities List* (CPFL). GIIPS provides comprehensive documentation about the product specifications for various MIDB programs and contains specific guidance and procedures for use in analysis, production, and processing of target intelligence by the DIA, Armed Services, and other organizations responsible for maintaining the DOD installation intelligence database. The *Point Reference Guide Book* (PRGB) provides guidance for selecting the reference points needed to derive geographic coordinates. Such reference points locate critical functional elements of installations in the various target categories. Each photograph or sketch depicts a sample installation, annotates the reference point at the recommended location, and explains briefly how to locate the reference point.

Another type of information included in the TDI and GIIPS is a vulnerability number (VN), or in the case of hardened targets, a physical vulnerability (PV) number. The Air Force calculated this number for the susceptibility of a target to moderate damage and to severe damage. The Vladivostok radar station had a PV of 09Q5 for severe damage. The first two digits expressed relative hardness to a 20-kiloton blast, the letter identified the sensitivity of the structure to peak overpressure (P, L), dynamic pressure (Q), or cratering (Z), and the final digit was the target's K factor, or ductility. An L designation describes a target that becomes resistant to peak overpressure at smaller distances to ground zero than does a P designation. The K factor indicates a target's response to a 20-kiloton blast wave and the capability for vulnerability adjustment to weapons with higher or lower yields. As might be expected, the Q designation for the Vladivostok radar station indicates the installation was primarily sensitive to dynamic pressure.

Soviet Missile Silos have (or had) physical vulnerabilities as follows:

Missile system	Silo Type	Physical Vulnerability to Severe Damage
SS-4	N/A	31P1
SS-5	N/A	31P1
SS-7	III-A	37P6
SS-8	III-B	37P6
SS-9	III-C	37P6
SS-11	III-D	46L8
SS-13	III-E	44L7
SS-17	III-H	51L7
SS-18	III-F	52L7
SS-11 / 19	III-G	52L8
SS-11 / 19	III-G MOD	55L8

The physical vulnerability number increased from an 800-PSI hardening for an SS-7 silo deployed in 1958 to 15,000-PSI for an SS-18 silo deployed in 1974. Modified SS-19 silos may have had hardening as high as 25,000 PSI but some knowledgeable analysts disagree with this comment.

Russian TEL mobile launcher shelters have physical vulnerabilities as follows:

Missiles System	Shelter Type	Physical vulnerability to Severe Damage
SS-25 / SS-27	N/A	11Q9
SS-25 / SS-27	Steel Framed	13Q7
SS-25 / SS-27	Earth Mounded	26P3

A Russian railway mobile launch system would have physical vulnerabilities as follows:

Missile Systems	Vehicle Type	Physical vulnerability to Severe Damage
SS-27	Locomotive	21Q5
SS-27	Launcher car	13Q5

The Air Force now keeps physical vulnerability data in its own *Physical Vulnerability Handbook*. Supplementing the PV Handbook is a *Nuclear Damage Handbook* that contains nuclear damage definitions, damage criteria, and recuperation time estimates for TDI categories and subcategories. The Air Force uses a *Target Value System Manual* to determine the relative importance of enemy installations and a *Bomb Damage Assessment Handbook* to assist in the assessment of battle damage, either by nuclear or conventional weapons.

General Vandenberg summed up American nuclear policy in 1953 when he stated that the purpose of Strategic Air Command was to destroy enemy industrial capacity, prevent a nuclear attack on the United States, and support ground forces engaged with the enemy. To prevent duplicate attacks on identical targets after the Navy and Theater Commands acquired nuclear weapons, the Air Force established a Joint Pentagon War Room and Joint Coordination Centers in two theaters of operation. SAC Zebra in Europe and SAC X-ray in the Far East, and the Pentagon War Room all received identical target information to eliminate the potential of duplicate attacks.

In order to reduce the possibility of overlapping attacks by different commands, Secretary of Defense Thomas Gates Jr. set up the Joint Strategic Target Planning Staff in 1960. The JSTPS was composed of 227 Air Force representatives (219 from SAC), 29 Navy representatives, 10 Army representatives, and 3 Marine representatives. A SAC General commanded the JSTPS and a Navy Admiral directed day-to-day operations. The formation of the new agency ended a decade of strategic bickering that had previously characterized relations between the Air Force and Navy. One of the first products issued by this organization was *SIOP-62*, the *Single Integrated Operational Plan* for an all-out, coordinated attack on the Sino-Soviet bloc.

At the lower end of its multi-option range, *SIOP-62* envisioned an alert force strike composed of 1,004 delivery vehicles carrying a total of 1,685 nuclear bombs and warheads. The size of the strike force increased with the remaining options to a preemptive strike in which 2,344 delivery vehicles delivered 3,267 bombs and warheads. When President Kennedy asked for more flexibility, the JSTPS conceived *SIOP-63* to accommodate his request. It consisted of attacks built around five primary attack options. These options were soviet strategic nuclear forces, military forces located away from urban areas, conventional forces located near urban areas, command and Control Centers, and a category that combined all of the above. Despite Kennedy's wish for flexibility, the outcomes offered by *SIOP-63* still used the majority of the nuclear weapons available, because the JSTPS based the *SIOP* on their capabilities rather than objectives.

In November 1962, Robert McNamara, the Secretary of Defense, presented a study on the number of weapons required for "Assured Defense" – the destruction of the Soviet Union's population and urban industrial infrastructure. Essentially, this was the secure strategic force the United States needed for a second strike. McNamara had the study based in megaton yields and warheads because this was the most common yield deployed on American strategic delivery systems. He found that a force of 400 megatons / warheads would destroy 55 million people (40 percent of the Soviet urban population) and 67 percent of Soviet industrial capacity. He then found that a force of 800 megatons / warheads could destroy 77 million people (or 71 percent of the urban population) and 77 percent of the industrial capacity. Thus, there was little use in a strike that exceeded 400 warheads because a strike of this size could destroy the Soviet Union's war making capacity. More warheads could only be justified to limit damage to the United States. The Navy would eventually take on the job of assured destruction with a fleet of 41 ballistic missile submarines. The fleet's size allowed the Navy to keep 25 submarines continuously on station. The submarines on station carried an aggregate of 400 missiles armed with megaton range warheads.

In the meantime, changes to the National Strategic Command and Attack Policy turned SIOP into a combination of attacks to provide outcomes around three core categories. These were:

- (Alpha) attacks prosecuted against Soviet and Chinese political and military command centers along with nuclear delivery capabilities located outside of urban areas
- (Bravo) attacks against conventional military centers located outside of urban areas
- (Charlie) attacks against nuclear capable installations located inside of urban areas, together with 70 percent of the urban areas themselves

This targeting strategy was an update of the Bravo, Delta, and Romeo priorities first established in 1950. These former priorities were for blunting a Soviet nuclear offensive (Bravo), the destruction of Soviet industry and urban areas (Delta) and retarding a Soviet attack across Europe (Romeo).

The attack options envisioned by *SIOP-71* were:

- A preemptive strike against 1,700 Alpha targets with 3,200 bombs and warheads (including redundant targeting by multiple independently targetable vehicles (MIRVs)
- A preemptive strike against 2,200 Alpha and Bravo targets with 3,500 bombs and warheads
- A preemptive strike against 6,600 Alpha, Bravo, and Charlie targets with 4,200 bombs and warhead (some targets were co-located)
- A retaliatory strike against 2,100 Alpha and Bravo targets with 3,200 weapons
- A retaliatory strike against 6,400 Alpha, Bravo, and Charlie targets with 4,000 weapons

The Nuclear Planning Process that produced the *SIOP* has gradually evolved since its inception. It currently consists of:

- Guidance from the President, the Secretary of Defense, and Force Commanders on the strategic objectives of the United States

- Target development, validation, nomination, and prioritization
- Capabilities analysis to describe the expected results of the plan's implementation
- Allocation of specific weapon systems to targets
- Mission planning and force execution analysis to initiate unit preparation, produce a tasking order, and to receive authorization for the plan's execution
- Assessment to determine whether the attack achieved its military objectives

A *Presidential Directive* that incorporates advice from the National Security Council supplies primary guidance for the Nuclear Planning Process. The Department of Defense focuses the *PD* into military employment objectives to meet outlined policy. In support of the *Presidential Directive*, the Secretary of Defense provides *Policy Guidance for the Employment of Nuclear Weapon* in a document called *NUWEP* that supplies information about targeting philosophy, objectives, and constraints to the Chairman of the Joint Chiefs, who in turn produces a *Joint Strategic Capabilities Plan - Nuclear Supplement*. The *JSCP-N* provides direction to war planners for the development of a nuclear *OPLAN* that codifies specific *SIOP* objectives.

The nuclear *OPLAN* directs a specific Attack Structure that has gradually evolved from a single, all out "spasm" into a number of scenarios. In the second step of the planning process, the JSTPS establishes targets and aim points known as desired ground zeroes (DGZs) to meet the scenarios. Planners look to the *Target Data Inventory* to select the targets used to fulfill *OPLAN* requirements. The JSTPS selects DGZs to achieve a desired level of damage expectancy (DE) with a designated weapon. DE is determined by multiplying together the probability of damage (PD) and the probability of arrival (PA) for a given weapon system. PD is determined by calculating the weapon yield, accuracy, height of burst, target characteristics, and desired level of damage. PA is determined by multiplying together pre-launch survivability, weapon system reliability, and probability of penetration. Depending on desired objectives, the strengths and weaknesses of a weapon system determine its mission usage. The results of this process are target files allocated to meet OPLAN objectives.

Allocation is the process by which targeting personnel select the best weapon, or group of weapons, for use against a target. Each weapon system has advantages and disadvantages such as range, yield, preparation time, accuracy, recallability, and vulnerability to defensive systems. Planners require inputs from service components on the number of assets available for nuclear tasking and their capabilities at any given time. The result of this determination is a *Force Commit* document that directs compliance with an *OPLAN* to ensure that the assigned nuclear force is capable of meeting the objectives of the defined attack structure.

The viability of the attack structure begins with establishing sortie availability. Planners must balance their targeting requirements with sortie maintenance schedules to determine how they will cover an allocated group of targets. After the planners determine where they will direct each sortie, they assign individual weapons to specific DGZs. In this process, the planners apply and reapply weapons until they can account for all required targets and accommodate asset availability. Planners then time targeted sorties against all the other missions in the overall *OPLAN* to integrate attack options and to prevent fratricide of friendly weapons. After successfully timing missions, planners perform an internal quality check to ensure that they have optimized sortie capabilities.

The last step in nuclear planning insures that target effects are aligned with strategic or theater objectives. This process is composed of three interrelated procedures: battle damage assessment, munitions effectiveness assessment, and re-attack recommendation. In peacetime, planners carry out the assessment process by means of simulations and war games. They subject the results to a rigorous analytical process to determine attack effectiveness. Computer modeling determines if the target damage from simulated attacks was consistent with campaign objectives. In the event of nuclear combat, intelligence data would have to be physically collected to determine if an attack has met its desired objectives.

The *SIOP* envisioned both first strike and retaliatory options. In order to preserve American nuclear forces in the event of an enemy strike, it was necessary to develop a radar early warning network. The United States began the development of this capability shortly after the Soviets deployed strategic bombers in the late 1940s. The British used radar networks with great success during World War II. Just prior to the war, they had built two sets of radar stations, Chain Home and Chain Home Low along the country's east and south coasts. These networks were very successful in detecting incoming waves of Luftwaffe bombers and directing RAF fighter squadrons to intercept them. Had it not been for radar, superior German numbers might have overwhelmed the thinly held ranks of the RAF. As it was, the RAF rose to the occasion in numbers sufficient to achieve eventual victory, even after the Germans switched to night attacks.

Radar operates by emitting a pulse of electromagnetic energy and then detecting an echo when the energy bounces back off a distant object. The azimuth of the returning pulse reveals the direction of the object, and its distance can be determined by establishing the time between transmission and detection. Dividing the time lapse in half and multiplying by the speed of light yields distance. The first successful radars used the multi-chamber, resonant-cavity magnetron developed by John Randall, a British physicist. After the war, radars based on electronic tubes known as klystrons supplanted the magnetron.

The electromagnetic energy generated by early radars had meter length waves. Engineers gradually reduced radar wavelengths, allowing the resolution of much smaller objects. Radar waves are now categorized into "bands" ranging from the 30-centimeter L-Band through S, C and X to the centimeter K-Band. An undesirable effect of shorter wavelengths is "clutter" or spurious reflections returned from small, extraneous objects. Shorter waves are also susceptible to absorption by water vapor in the atmosphere. Longer wavelengths produce less clutter and have greater range so that aircraft warning systems operate primarily in the L-Band.

After receiving a magnetron from the British Government, the United States made great strides in developing radar. Advances included the very successful SCR-584 aircraft tracking radar and bombing radars such as the AN/APQ-13 used in the atomic attack on Nagasaki. The US Navy, in particular, used a wide assortment of radars. These included surface search, airborne early warning, and fighter direction systems. Based on radar's success in WW II, General Gordon Saville, the Author of *Air Defense Doctrine*, proposed building a "radar fence" of 233 permanent stations around the United States in 1947. Congress rejected Seville's proposal, designated SUPREMACY, as too expensive. The Air Force then built a temporary system called "LASHUP" using AN/CPS-5 search radars left over from WW II. The system had a number of problems. Like other contemporary radars, the AN/CPS-5 did not function well against low-altitude targets and the radio communications gear the Air Force used for reporting was susceptible to ionospheric interference.

The Air Force emplaced LASHUP stations around major metropolitan centers and important military installations including Oak Ridge, Hanford, and Los Alamos. In June 1950, SAC tested the capability of the system by flying a series of simulated attacks against the city of

Seattle. At altitudes of 20,000 to 25,000 feet, LASHUP easily detected the bombers, but below 5,000 feet, the bombers always got through. The day after the exercise was completed; the North Korean army crossed the 38th parallel in an all-out invasion of the South. The Air Force placed LASHUP on 24-hour alert.

Encouraged by the success of LASHUP, the US Air Force proceeded with a "Permanent System" of eighty-five radar-warning stations and eleven Control Centers distributed throughout the continental US and Alaska. It also entered into an agreement with the Royal Canadian Air Force to jointly build and operate a series of radar sites along the International Border known as the Pinetree Line. A *Joint Statement on Defense Collaboration* issued in 1947, linked the United States and Canada in mutual air defense.

The Pinetree Line consisted of thirty-three manned and six automatic gap-fill sites, all of which were in operation by 1953. The radar and communications systems employed in its construction were designed to provide both warning and interception control. At the main sites, AN/FPS-20 L-Band pulse radars searched the horizon out to a range of 200 miles and AN/FPS-6 S-Band radars estimated target altitudes to a height of 75,000 feet. AN/FPS-14 gap-fillers were S-Band search radars optimized for low-altitude coverage.

Farther north, the Canadians built a second warning network known as the Mid-Canada Line. This unique system was a chain of 90 automatic Doppler radars controlled by eight section-stations. Instead of measuring the direction and length of time for a radar pulse to return from a target, Doppler radars emit a continuous radar wave that changes its frequency when reflected by a moving object. The same principal causes the pitch of a train whistle to change as it approaches a listener. Because this technique is useful in determining the velocity of a target and not its range or direction, the Mid-Canada line only triggered a warning if it detected movement from an aircraft as it flew overhead.

Offshore, the US Navy operated a series of radar picket ships to provide advanced early warning during the period 1955 to 1965. These were converted WW II Liberty ships equipped with SPS-17A detection systems. ADC also built four Texas Towers based on oil platform technology on the continental shelf off the northeast coast of the United States. The Texas towers were originally equipped with an AN/FPS-3 search and two AN/FPS-6 height finding radars. ADC shut the towers down after TT-4 was lost along with a 28-man crew during a winter storm on January 15, 1961.

In addition to the Navy's offshore activity, the Air Force operated RC-121 early warning aircraft that flew racetrack patrol patterns off the east and west coasts of the United States. The RC-121 Warning Star was a Lockheed Constellation modified to carry AN/APS-20 search and AN/APS-45 height-finding radars. RC-121s flew out of Otis AFB, Maine; McClellan AFB, California; and McCoy AFB, Florida. The replacement of the AN/APS-20 with the longer-range AN/APS-95 in 1962 greatly increased the effectiveness of airborne early warning. Upgraded versions of this aircraft re-designated EC-121 served with distinction in Vietnam to become the progenitors of modern Airborne Warning and Control Systems (AWACS).

For the purpose of co-coordinating early warning systems, President Eisenhower announced the formation of Continental Air Defense (CONAD) Command in 1954. Only a few months before this announcement, he had issued Directive NSC-139 approving the construction of the Distant Early Warning Line, which the Air Force strung across Alaska and Canada's North-West Territories, a few degrees north of the Arctic Circle. After its inception, CONAD took responsibility for the new system. DEW Line stations were located in one of the least hospitable environments known and CONAD did not achieve an initial operating capability until 1957. At that time, the Canadian and American governments announced the integrated operational control of their joint air space by means of North American Air Defense Command (NORAD).

When completed in 1962, the DEW Line consisted of 6 main, 23 auxiliary, and 28 intermediate sites located about 100 miles apart. The primary detection system at main and auxiliary sites consisted of an identical pair of AN/FPS-19 L-Band search radars that fed a pair of fixed, back-to-back antennas. Two plan position indicator monitors provided 360-degree coverage out to a range of 160 miles. An automatic alarm sounded if a radar installation detected a target, at which time a scope camera photographed a full sweep of its PPI to provide a permanent record of the event. Intermediate sites were equipped with unmanned AN/FPS-23 Fluttar (Doppler) radars to fill in low-altitude gaps between the main and auxiliary stations. Remote or bi-static receivers were located at both main and auxiliary sites to monitor AN/FPS-23 returns. In order to complete the DEW Line, NORAD equipped four sites in Greenland with powerful AN/FPS-30 systems.

The Air Force equipped DEW Line sites with a number of communication systems with which to exchange information with NORAD, SAC Headquarters, and SAC aircraft. For east - west communications along the line, NORAD employed the White Alice tropospheric lateral scatter system. The original equipment was subject to fading, so NORAD eventually upgraded it with the Surestop I and Surestop II programs. The Vietnam War siphoned off equipment intended for Surestop III, which resulted in a long delay before Alaskan sites received their final upgrades.

The DEW Line had two primary purposes: to alert NORAD that an attack on the United States and Canada was in progress and to launch a retaliatory strike. Before March 1958, the CINCSAC could not launch his alert forces until senior political and military authorities issued him with direct orders. With concerns about the vulnerability of the bomber force to a Soviet missile attack, the Air Force developed a new operational concept, the Positive Control Launch (PCL), or "fail-safe" procedure. Using this procedure, CINCSAC could order US bombers to take off and fly toward the Soviet Union on prearranged routes, as he deemed necessary. After the alert aircraft reached a position known as the "Positive Control Turn Around Point" (PCTAP), a position outside Soviet territory, the bombers were to orbit as long as possible waiting receipt of a "go-code" ordering an attack. A special SAC SSB radio network known as Short Order transmitted the go-codes. The system "fail-safed" in the event of a communications loss – the SAC bombers could only enter Soviet Air after receiving authenticated go-codes.

In order for SAC to communicate with bombers orbiting overhead at their fail-safe points, the main DEW Line sites were equipped with the 487-L Survivable Low Frequency Communications System. SLFCS could operate in an environment disturbed by nuclear explosions allowing DEW Line sites to receive one-way teletype messages from SAC. DEW Line operators could then rebroadcast SAC's messages on one of two UHF frequencies used by the SLFCS system. The 488-L Greenpine Communications System connected the main DEW Line sites to the commercial telephone grid by means of a landline. Using Greenpine, SAC operators in the continental United States could broadcast directly to their aircraft by remotely operating UHF equipment located at the main sites.

Although it gave the DEW Line state of the art radar and communications equipment, the Air Force did not equip it to Air Defense assets for use against an attack in progress. Instead, the radar stations of the Pinetree Line and the Permanent System, located further south, supplied ground-control intercept direction. This capability took a giant leap forward when the first of 43 planned Direction and Combat Centers for the Semi-Automatic Ground and Environment System came on line at McGuire AFB, New Jersey in 1958. The hardened SAGE Centers collected information over telephone lines from assigned radar installations and after evaluation, transmitted the results to Ent AFB at Colorado Springs, where NORAD's Headquarters was located. SAGE did this by means of a specially developed electronic data transmission device known as a "modem." SAGE could also provide air intercept data directly to ADC fighter squadrons and launch coordinates to BOMARC and Nike surface to air missile batteries. SAGE Centers were able to communicate directly with associated Air Defense District Headquarters.

Beginning in 1964, a semi-automatic 416M Backup Interceptor Control System projected radar data onto rectangular map like plots, instead of the polar projections used by cathode ray scopes. BUIC was equipped with AN/GSA-51 computers that allowed operators to call up weapons information by means of an interactive light pen, request solutions for possible intercepts, and transmit launch codes and intercept data over appropriate connections. Although NORAD deactivated SAGE in the early 1980's, its lineal descendant lives on. A chance meeting between the CEO of American Airlines and one of IBM's SAGE program managers resulted in the SABRE Airline Reservation System.

NORAD only built 26 of SAGE's 43 planned centers. After the launch of Sputnik in 1957, NORAD began building the Ballistic Missile Early Warning System (BMEWS) and gave air defense a back seat. It completed three special missile warning stations at Thule AFB, Greenland (1960), Clear AFS, Alaska (1961) and RAF Flyingdales Moor, England (1963). NORAD supplied Thule and Clear with General Electric L-Band detection radars, each equipped with three enormous fence antennas, 165 feet tall and 400 feet long. Together, the antennas provided 120-degree coverage of the horizon, out to a distance of 3,000 miles. In the case of Thule, Greenland, Navy ships anchored offshore initially supplied the base with power.

Four separate radars received power from pairs of gigantic klystron tubes dedicated to each antenna. The system could probe different parts of the sky depending on how an arrangement of feedhorns connected the system's antennas. Radar beams were projected as two parallel lobes, one several degrees above the other. As a Soviet missile moved through the radar lobes, measurements of range, angle and Doppler shift provided an estimate of velocity and trajectory. After detection, the AN/AFPS-50 passed targets over to AN/FPS-49 tracking systems. RAF Flyingdales Moor relied solely on three AN/FPS-49 radars for both detection and tracking.

Located inside of huge radomes, Flyingdales' AN/FPS-49 RCA radars made use of conical scanning technology and a fully steerable 84-foot diameter dish to paint targets with tight bursts of radar pulses. At the center of the radar cone thus painted, the target returned a constant signal that decreased in intensity if it moved away from the axis. Converted to direct current, the voltage drop from the signal controlled the electrical servomotors that kept the antenna trained on its target. In this manner, extremely accurate trajectories for space objects could be determined. Instead of conical scanning, Flyingdales Moor used a monopulse feed whereby computers analyzed single pulses. which were detected by a pair of dishes, to determine target trajectories.

The three BMEWS sites were equipped with redundant communications systems. Commercial lines carried communications from Alaska and Great Britain to the continental United States but Thule was isolated and required the installation of a pair of special communications networks. When the radar began operating in 1960, communications took place through a single, 1,950-mile-long, underwater cable from Greenland to Newfoundland. From

Newfoundland, a commercial microwave radio system relayed data to NORAD Headquarters in Colorado Springs. NORAD passed this information to SAC Headquarters in Omaha, Nebraska. During the Cuban Missile Crisis, Soviet trawlers cut the underwater cable on three occasions.

In 1961, a high-powered tropospheric scatter facility named Dew Drop joined the BMEWS communications systems. Dew Drop bounced radio signals off the troposphere to a receiving station on Baffin Island, located 700 miles away. This transmission capability was an impressive feat that required a quadruple diversity, SSB radio utilizing two 50-KW klystron amplifiers that fed a pair of 120-foot-high parabolic reflectors at either end of the facility. Another tropospheric scatter system on Baffin Island retransmitted data to a site in Labrador, which sent the data through commercial lines to NORAD Headquarters at Colorado Springs. Thus organized, BMEWS could provide a 15-minute warning of a missile attack on the United States. Unfortunately, for the United Kingdom, that country's close proximity to the Soviet Union limited an attack warning to 5 minutes. Although BMEWS was so powerful that during early trials, radar signals returned from the rising moon generated a false alarm, the system could only identify a mass ICBM attack in its early configurations. One or two missiles did not reflect enough radar energy for detection.

Because it was located extremely far forward and directly under the path of a missile attack, planners considered Thule's BMEWS site a tempting target for the Soviets. Thus, from SACs perspective, it could not rely solely on NORAD systems for the survivability of its bomber force. SAC instead wanted communications systems that were not under the control of another organization and that did not have common failure modes. It, therefore, assigned one of its Chrome Dome airborne alert B-52's on the northern bomber route to become a BMEWS monitor. The aircraft orbited in a 300-mile long oval above Thule. The monitoring B-52 had to maintain hourly HF radio, radar, or visual contacts with the BMEWS site. In the event that a monitor bomber aborted a sortie, SAC orders required that the KC-135 tanker used to support the Chrome Dome missions had to monitor the BMEWS site.

With this operational innovation in place, Strategic Air Command created its own round-the-clock warning system. In the event that communications with Thule were lost, SAC Headquarters could contact its airborne B-52 through the Short Order radio network. The B-52 monitor aircraft could then report directly to SAC on whether it had witnessed a nuclear detonation at the radar site, signaling that a major Soviet attack had begun, or whether BMEWS had merely suffered a temporary communications failure.

The existence of this special B-52 mission was highly classified. Those with proper security clearance considered the B-52 monitor mission an important backup for clarification of what otherwise might be an ambiguous attack warning. Combined with other warning and communications systems connecting Thule to southern military commands, the B-52 monitor aircraft and NORAD warning systems enabled SAC to differentiate between false alarms and real attacks.

During normal peacetime circumstances, the B-52 monitor aircraft provided redundant confirmation that the BMEWS system was up and operating. In times of international stress, it provided the capability to distinguish between a communications failure, a Soviet conventional attack, or sabotage at the base. A report from the B-52 of a nuclear detonation provided independent confirmation of a Bomb Alarm report of a nuclear attack on the base. The Bomb Alarm System consisted of nuclear detonation detecting sensors located at some distance from the base. SAC considered the BAS a particularly important asset, because the early BMEWS system could not detect just one or two Soviet ICBMs launched in a precursor surprise attack.

Other possible attack scenarios envisioned Soviet conventional attacks utilizing long-range bombers, saboteurs, or submarines against the Thule Base and the Short Order radio transmitters, and fighters against the orbiting B-52 monitor in an effort to blind US warning systems just before launching a nuclear attack. In this case, the inability to contact the B-52 would provide confirmation of an attack. Finally, if the Soviet Union used a barrage attack of SLBMs and / or depressed trajectory ICBMs to blanket the area, destroying both Thule and the B-52 monitor, the bomb alarm system would report the detonation and SAC'S inability to contact the B-52 would provide an important confirmation that a major nuclear attack had indeed occurred at Thule.

In parallel with the deployment of BMEWS, the Air Force constructed the 440-L Over-the-Horizon Forward Scatter Network and the Navy developed the Naval Space Surveillance System. OTH-N was a series of four AN/FRT-80 radio frequency transmitters located in Japan, Okinawa and the Philippines coupled with five AN/FSQ-76 receivers located in Italy, Germany, England, and Cyprus. The system could detect perturbations in an electrical field projected around the world by bouncing it off the ionosphere. OTH-N could identity high-altitude bombers and ICBM missiles within minutes of launch. It could also detect Soviet missiles launched over the South Pole in a fractional orbital bombardment attack that would have been invisible to BMEWS.

For its part, the Navy activated NAVSPASUR, a space surveillance system, to detect and track space objects. It was a modification of the Minitrack System it had formerly used to support the NRL Vanguard Space Program. The Navy's space surveillance system made use of three continuous wave transmitters that projected an electronic fence across the United States and out into deep space. Six antenna arrays, acting as interferometers, determined the angles and angular rate of arrival for space objects. Data interpolated from multiple target passes and from different sites allowed the accurate determination of orbital parameters for space objects. When integrated with the Space Detection and Tracking System, in which BMEWS played a role, NORAD could also monitor the activities of Russian spy satellites. Although the Navy operated NAVSPASUR, the system became a significant NORAD asset.

The debut of phased array radar systems in the late 1960s substantially improved early warning. This type of radar adjusts the timing of signals emitted by a large number of miniaturized transmitters to steer an electronic beam without physically having to move a dish. Because a phased array has no moving parts, the United States could build larger and more powerful radars. The first such system was the Bendix AN/FPS-85, which reached operational capability near Eglin AFB, Florida, in 1968. It employed 5,928 transmitters and 19,500 receivers arranged in two separate arrays across the face of a fourteen-story high building. The radar operated in the UHF band and provided detection in a 120-degree arc to a range of 22,000 nautical miles. Phased array systems, depending on their computing power, can simultaneously detect and track a large number of independent targets. Although its original purpose was to keep track of space objects, NORAD pressed the AN/FPS-85 radar into service for monitoring Cuban air space and detecting the possible launch of Soviet SLBMs from the Caribbean.

The success of the Eglin phased array radar resulted in the gradual replacement of older systems. Four PAVE PAWS (Phased Array Warning System) AN/FPS-115 radars replaced six AN/FSS-7 radars that NORAD had previously deployed along the Atlantic and Pacific coasts.

The AN/FSS-7 was a stopgap measure based on modifications to the AN/FPS-26 height finder. The PAVE PAWS radars took over the responsibility for detecting the launch of SLBMs. Of the original PAVE PAWS sites located in the continental United States, only Cape Cod AFS, Maine, and Beale AFB, California, are still active. A relocated PAVE PAWS system, transferred from Eldorado AFS, Texas, now monitors the Arctic Ocean from the Clear AFS BMEWS site.

Like BMEWS, PAVE PAWS could detect and track objects out to a range of 3,000 miles and had a predictive capacity with which to ascertain simultaneously, the time of arrival and point of impact for large numbers of missile warheads. NORAD has now upgraded all of the original radars to the AN/FPS-123 configuration. The AN/FPQ-16 Perimeter Acquisition Radar System (PARCS), originally built for the defunct Safeguard ABM base in North Dakota, supplements PAVE PAWS by searching for Soviet SLBM launches out of Hudson Bay, Canada.

In the 1990s, the BMEWs sites at Thule, Greenland, and RAF Flyingdales Moor, England, received new, Raytheon / Cossor Aerospace and Control Data Corporation solid-state phased array radars (SSPARs) at a cost of 100,000,000 dollars each. Flyingdales Moor acquired a three-face, AN/FPS-126 antenna to provide 360-degree coverage whereas Thule had a 240 degree, AN/FPS-120 two-face antenna. Operating in the UHF band, these systems currently provide long-range detection and tracking. Clear AFS, Alaska, did not receive one of the new SSPAR systems. Instead, NORAD gave Clear AFS an AN/FPS-92 upgrade to replace its original tracking radar. This has since been upgraded to a single face AN/FPS-120 system.

Despite the post-Sputnik emphasis on missile detection, NORAD also improved its air warning capability. It re-designated the DEW Line as the North Warning System and in 1990 turned its operation over to the Royal Canadian Air Force. In its new incarnation, it consists of 15 long-range AN/FPS-117 and 39 AN/FPS-124 short-range automated radars controlled from five logistics sites. The reason for this turn of events was the deployment in the United States of OTH-B, an over-the-horizon radar that employed an ionospheric backscatter technology similar to that used for communication with BMEWS and along the DEW Line. Developed by GE, AN/FPS-118 OTH-B radars were some of the most powerful ever built. They employed megawatt transmitters to bounce their signals off a layer, 200 miles up in the ionosphere, and could track aircraft flying over the Soviet Union. The development and construction of three Atlantic and three Pacific stations cost more than two billion dollars. In 1991, NORAD cancelled the construction of three Alaskan OTH-B sites and mothballed the rest of the system because of the Soviet Union's demise.

Joint Surveillance System (JSS) Regional Operations Control Centers (ROCCs) located at Griffiss AFB, New York; March AFB, Arizona; McChord AFB, Washington; and Tyndall AFB, Texas now provide air warning. Together, these Operations Control Centers receive data from 46 radar sites. Fourteen more radar sites feed data to an ROCC at Elmendorf AFB, Alaska, and two radar sites supply data to an ROCC in Hawaii. Twenty-Four Canadian air defense network radars supply data to a pair of ROCCs in North Bay, Ontario. The first JSS region entered operation with Air Defense Tactical Air Command (ADTAC) in March 1983. The remaining regions then joined one at a time until all continental US ROCCs were operational in March 1984. The first Hawaiian site followed later that year and Iceland received an interim site in 1988. The JSS has recently deployed 44 ARSR-4/FPS-130 unattended radars under the FAA / Air Force Radar Replacement Program with which to replace older radars. The ARSR-4/FPS-130 is a 3D, solid-state radar with an effective detection range of 200 - 250 nautical miles.

Regional Operations Control Centers automatically process information supplied from their radar networks, which they then display for identification, tracking, and the assignment and direction of interceptor aircraft. ROCCs can transfer their command and control functions to E-3A AWACS aircraft and provide backup for AWACS. In peacetime, the Air Force assigns six AWACS aircraft to co-operate with the JSS. ROCCs can also pass information to NORAD's Combat Operations Centre (COC).

In order to supplement its early warning radars, the United States developed the previously mentioned Bomb Alarm System (BAS), and the space-based Missile Detection and Alarm System (MIDAS). The BAS became operational in 1961. It evaluated the intensity, timing, and pattern of a nuclear attack. To do so, three infrared sensors were located at distances of 11 miles around each of 97 strategic targets. The Air Force connected BAS sensors to signal generators linked by telegraph lines to six remote Control Centers. These centers in turn fed their data to NORAD, which leased the whole system from Western Union.

The space-based MIDAS infrared sensors detected the hot exhaust plumes of Soviet ICBMs during their boost phase. The Air Force placed the first sensor bearing MIDAS satellite into orbit in 1960, which it judged a qualified success. Although it worked as intended, the satellite's low earth orbit caused it to reenter the atmosphere and burn up after only a year of operation. A continuous series of launches were required to keep the system operational.

In 1966, Defense Support Program (DSP) satellites placed into geosynchronous orbits replaced MIDAS. The Air Force linked the DSP satellites directly to NORAD at Cheyenne Mountain to provide a 30-minute warning in the event of a land-based missile attack. DSP-SED and improved DSP-I satellites later replaced the original DSP satellites. DSP-I makes use of improved sensors that are resistant to laser attack. Three operational and two backup satellites are currently in orbit. Together, with NORAD's radar facilities, DSP satellites provide the dual detection required to confirm that a nuclear attack is under way. They can also detect the surreptitious launch of missiles from an orbital platform. Currently, the Air Force is implementing the Space Based Infrared System. SBIRS is a consolidated system that will use ground-based data processing to evaluate information from the current crop of DSP satellites with IR sensors on new satellites placed in geosynchronous and highly elliptical orbits.

MIDAS and DSP were only a small part of the Cold War space program. To identify unauthorized nuclear detonations, the Air Force launched a pair of satellites code-named Vela Hotel into geosynchronous in 1963. Stationed at an altitude of 70,000 miles, the Vela satellites carried gamma ray and neutron detectors that could pinpoint a nuclear explosion at a distance of 100,000,000 miles. Subsidiary sensors could detect the unique double flash of light associated with a nuclear explosion as well as its electromagnetic pulse. The original Vela satellites were only capable of detecting explosions in space but the Advanced Vela satellites that followed could detect atmospheric explosions. The Integrated Operational Nuclear Detection System (IONDS) superseded Vela. Currently, a variety of sensors placed on Global Positioning Satellites can detect, locate, and measure the intensity of nuclear explosions. By quickly determining the locations and extent of a nuclear attack, the National Command Authority can authorize an appropriate response.

The Air Force activated Space Command in 1982 to manage a constellation of 24 ground-based and space-based surveillance and missile warning sensors. It established a telemetry, tracking, and command facility to coordinate this operation at the Consolidated Space Operations Center (CSOC) located at Falcon AFS, Colorado. One of Space Command's first programs was the development of a Mobile Ground System that could provide survivable data acquisition and assessment from DSP satellites. AFSPC's 4th Space Warning Squadron operated this equipment.

Space Command also assumed responsibility for AN/FPS-17 and AN/FPS-79 radars located at Pirinçlik AFB, Turkey, and the AN/FPS-108 COBRA DANE station at Shemya, Alaska. These three radars monitored Soviet missile tests. Space Command also managed the Space Detection and Tracking System (SPADATS) and the Pacific Barrier System (PACBAR). The latter system maintained stations at Kwajalein Island, the Philippines, and Saipan. Its C-band radars could detect Soviet satellites during their initial orbit. Presumably, the system would have provided an alert to NORAD if it detected activity consistent with a fraction orbital attack. The construction of the Ground-based Electro-optical Deep Space Surveillance System, which could image objects the size of a basketball at a range of 10,000 miles, followed PACBAR. Four GEDS locations at Maui, Socorro, Choe Dong-San, and Diego Garcia were equipped with meter diameter telescopes for this purpose.

In addition to their early warning satellites, the United States developed satellites to provide visual surveillance and targeting information. The Research and Development (RAND) Corporation, a non-profit think tank spun off from the Douglas Aircraft Company to evaluate scientific and technological issues, first suggested the use of spacecraft for this purpose in 1946. RAND continued to develop the space based photographic intelligence (PHOTINT) theme and, in 1954, produced a two-volume report for the CIA and Air Force entitled: *An Analysis of the Potential for an Unconventional Reconnaissance System*. The study resulted in a 1956 contract awarded to Lockheed Missile Systems to develop Weapon System 117L. Codenamed "Pied Piper," WS-117L consisted of two satellite reconnaissance systems and the MIDAS surveillance program. One type of reconnaissance satellite was designed to transmit line-scanned images of negatives developed in orbit, while the other was equipped to send back undeveloped film by means of a capsule ejection system.

The Air Force chose electronic transmission over film retrieval because it was interested in real time reconnaissance. It launched the first Satellite and Missile Observation System (SAMOS) into space in late 1960, as part of a second stage Agena rocket boosted by an Atlas ICBM. Although the system functioned well enough, its low resolution prevented it from capturing very effective information.

The CIA pursued the film return system since it was only interested in high quality imagery. This program had an inordinate number of failures at its outset. Orbited under cover as Discoverer satellites, success did not come until the thirteenth launch on January 10, 1960. The next day, Navy frogmen retrieved an empty "film bucket" after it parachuted into the Pacific. Shortly thereafter, a C-119 aircraft snagged the parachute of a bucket containing exposed film dropped from Discoverer 14. This event established aircraft retrieval as a standard operating procedure. The success of the CIA program caused the Air Force to abandon electronic data transmission and adopt film return for later SAMOS missions. It terminated this program after eleven launches.

In order to acquire the highest resolution pictures possible, the Air Force launched the CIA's Discoverer satellites into low earth orbit. In this orientation, air friction limited the amount of time the satellites could stay aloft, so their use was target specific. The program was so successful that that it was quickly hushed up. The NRO gave reconnaissance satellites the program designation Keyhole, with Discoverer designated KH-1. It further referred to Discoverer by the code name Corona. Advanced KH-2 through KH-4b satellites followed KH-1, accounting for 121 launches of which 93 qualified as successful. The KH-4a used twin film canisters to extend its time in orbit and reduce launch costs. Two special missions followed Corona. Argon or Super SAMOS was a sequence of twelve KH-5 Air Force sponsored wide-area mapping satellites, whereas Lanyard comprised three KH-6 satellite launches (one successful) that the Air Force used to image a suspected Soviet ICBM site at Tallinn. NASA later used the KH-5 camera system for Mars and Lunar missions and in commercial LANDSAT earth-imaging satellites.

The reconnaissance satellites took their pictures on 70mm film exposed in 24-inch focal length Itek cameras. Early reconnaissance cameras had a resolution of 7.5 meters but improved lenses increased this feature to 1.5 meters in the KH-4b. Vehicles parked in Red Square were discernable, as well as the shadows of visitors waiting in line at Lenin's tomb. Although there were some early issues with electrostatic fogging, the intelligence catch quickly exceeded that produced by the U-2. It also became apparent that a "missile gap" did not exist, and this realization may well have contributed to a reduction in Minuteman missile deployments.

Prior to the launch of Argon, the Air Force knew military and industrial targets lying to the east of the Ural Mountains only to a precision of 10 or 20 miles. This accuracy was insufficient to aim ICBMs, even if armed with multimegaton warheads. Following the Argon missions, the list of targets uncovered by interpreters grew so fast that it became necessary to deploy MIRVed warheads to cover them all. Photoreconnaissance satellites imaged Soviet launch complexes, test ranges, and air defense batteries in detail. Also imaged were military airfields, where analysts provided fighter and bomber inventories. This helped to dispel the notion of a "bomber gap." Argon imaged the first deployments of anti-ballistic missiles, detected their characteristic radars, and helped verify *SALT Treaty* provisions. KH-5 mapping satellites also helped to define attack routes for SAC bomber that avoided air defense sites. The enormity of the Air Force intelligence take eventually caused the Army and Navy to accuse it of "bootstrapping" or creating targets in order to appropriate the lion's share of defense budgets.

A significant threat identified by space borne reconnaissance during the 1960's was an increase in deployed Soviet missile submarines, which were detected as they lurked in the waters off the continental United States. The most dangerous of these were "Yankee" SSBNs. The Russian answer to Polaris, the Yankee carried 16 R27 (NATO SS-N-6, Serb) missiles with a range of 1,440 miles. Serb missiles had an average flight time of only 8 minutes to their targets and carried a megaton yield warhead or three independently targetable warheads with 200-kiloton yields. In order to counter this development, the Navy deployed the Sound Surveillances System (SOSUS) on continental shelves and in ocean basins around the world.

The practice of passively detecting submarines using low-frequency sound waves transmitted through oceanic sound ducts originated with Project Hartwell in 1950. In support of this project, the Navy contracted Western Electric Company to build and place 1,000-foot long arrays of 40 hydrophone elements in 240 fathoms of water for tests at Sandy Hook, New Jersey and Eleuthera Island, Bahamas. They also adapted a speech analysis device or "sound spectrograph" into LOFAR, a low frequency analysis and recording device for real time analysis of the data collected by SOSUS. Post war diesel submarines were relatively noisy, especially when they snorkeled, and LOFAR could discern identifiable submarine noises from the ocean background at a distance of 1,000 miles! The Navy used multiple stations to triangulate and locate the source of noise.

Following the success of Project Hartwell, the Navy constructed SOSUS arrays along the East and West Coasts of the United States, as well as on the sea floor around Hawaii. It transferred SOSUS data over multi-conductor armored cables to shore stations known as Naval Facilities or NAVFACS for analysis. Later, the Navy expanded the system worldwide to cover the approaches to Soviet Naval bases. It could thus track Soviet submarines from the time they left port. Eventually, SOSUS analysis was concentrated at central locations using data retransmitted via Fleet Satellite Communications (FLSATCOM) from shore receiver stations. This allowed the Navy to bring greater computing power to bear on the problem of locating and tracking the entire Soviet submarine fleet. Supplementing SOSUS in the mid-1980s was a small fleet of civilian operated surveillance craft that towed surface arrays known as SURTASS. The Navy had the data from this program transmitted via satellite to its central processing locations. Together, the surface ships, fixed bottom arrays, and evaluation centers were designated the Integrated Undersea Surveillance System. Although the Navy has reduced the scope of the IUSS since the end of the Cold War, it gained access to an "Advanced Deployable System" for in theater use in 1996.

The IUSS passed its intelligence to American nuclear attack submarines (SSNs) that shadowed Soviet missile submarines using spherical bow mounted sonar and passive towed arrays, which were the descendants of SOSUS. At the first signs of a missile launch, American SSNs could have torpedoed their adversaries with nuclear tipped MK 45 Astor or SUBROC torpedoes. Although the Soviets did not possess this high level of sophistication, a United States Navy mole named John Walker provided them with codes and information about submarine patrol activity until he was unmasked in 1985. The 18-year long spy operation cost the Soviets the piddling sum of 1,000,000 dollars and seriously compromised the entire American FBM program!

Despite the advantages inherent SLBMs, the technological challenges involved in producing reliable undersea weapon systems caused the Russians to place their main strategic reliance on land-based missiles. These showed great improvement after they were equipped with solid fuel boosters and MIRVed warheads. The improved acceleration of solid fuel missiles condensed the time available for American command authorities to respond to an attack. In the 1950s, the DEW Line provided a 12-hour warning for a bomber strike, but the ICBMs deployed in the late 1960s took no more than 30 minutes to arrive at their targets after detection. By 1980, there was a perception that increasingly accurate Soviet missiles had compromised the security of the American Minuteman force in its protective silos. In order for this force to remain effective, it became necessary to "launch on warning," or face the possibility of annihilation. For this reason, President Jimmy Carter issued Presidential Directive PD-59 that called for modernization of a variety of intelligence systems to supply real time information about possible attacks.

As part of Carter's intelligence modernization program, the Air Force launched a third generation of reconnaissance satellites code-named Gambit. These satellites provided a combination of detailed and wide area intelligence. Although the Air Force launched the satellites, the National Reconnaissance Office operated them, supplying their data to various users like the CIA, the Defense Intelligence Agency (DIA), and the Air Force. A joint CIA and Air Force staff operates the NRO under the direction of the Undersecretary of the Air Force, who reports directly to the Secretary of Defense. NRO staff use the BYEMAN control system to determine the physical security requirements of the product, assign codes, and grant access to information. Separate programs for the Air Force (A), the CIA (B), and the Navy (C) prioritize the NRO's operations. Restructuring at the end of the Cold War has resulted in the agency's partition into Directorates for Advanced Systems, Communications, Image Intelligence (IMINT), and Signals Intelligence (SIGINT).

Gambit satellites were notable in possessing both infrared and multi-spectral scanners (MSS). Itek, the company that had manufactured spy cameras during the startup of space-based reconnaissance, developed this new equipment. The NRO eventually replaced scanners with the more versatile thematic mapper (TM). As used in LANDSAT satellites, both multi-spectral scanners and thematic mappers photograph scenes that are approximately 185 kilometers square. Multi-spectral scanners produce images in four spectral bands while thematic mappers record in seven bands. Analysts can digitize these images and use them to identify strategic materials with the aid of mathematical algorithms. For instance, an analyst could determine whether the hull of a submarine was made of steel or titanium, allowing inferences about its depth of operation.

KH-7, the first Gambit satellite, was an aerial surveillance platform with wide-angle cameras, whereas the KH-8 had narrow angle cameras with the remarkable resolution of 6 inches in later models. Provided with an Aerojet re-startable engine for maneuvering, KH-8 was so large that it required a Titan booster to take it into orbit. The KH-8 program spanned 22 years and 60 launches. Many of its operations targeted events of specific political interest such as the Arab - Israeli conflicts.

Advances in technology quickly replaced the KH-7 with KH-9 Hexagon satellites. Launched in 1971, the KH-9 Big Bird platform was equipped with a Perkin-Elmer Cassegrain focus telescope having a 72-inch diameter mirror. The Hubble Space Telescope is really just one of these satellites turned the other way around to look at the stars. NASA shipped Hubble to its launch pad in a container identical to the ones used for the transport of KH-9s. The KH-9 telescope could direct its sharply focused images through a number of different sensors for recording on film. These included greatly improved multi-spectral and thematic mappers. It also had an improved infrared capability and a photomultiplier to help it see in the dark. The satellites had between four and six film capsules and could eject them as required. The NRO also equipped the KH-9 was with a number of SIGINT antennas and a double-lens Kodak secondary camera for high resolution and wide area surveillance. The KH-9 electronically transmitted its secondary camera's images to a series of worldwide receiving stations along with any signals intelligence collected.

The KH-11 Crystal satellite superseded Hexagon and Gambit in the late 1970s. The Air Force had planned the KH-10 as a Skylab-based manned reconnaissance platform but cancelled this concept because of KH-11's remarkable capabilities. The NRO supplied the KH-11 Crystal with a charge coupled device (CCD) camera that did not use film and could supply its electronic images in real time by means of dedicated relay satellites. This system was very cost effective because the Air Force had to launch satellites supplied with film canisters on a regular schedule. The KH-11 and Improved KH-11 likely used adaptive or active optics in which actuators bend a primary mirror to account for atmospheric distortion. Adaptive optics can produce a resolution of a few inches under optimum conditions. Like its predecessor, the KH-11 has a host of sensors that record in both visual and infrared wavelengths.

Notable accomplishments of late model reconnaissance satellites were observations on the development and deployment of Soviet UR-100MR (NATO SS-17, Spanker), R-36 (NATO SS-18, Satan), and UR-100N (NATO SS-19, Stiletto) silo-based missiles. The satellites also followed the development of the RSD-100 mobile MRBM. In 1976, a KH-9 discovered the

Soviets MIRVing 600 IRBMs facing China and Europe. The KH-9 also identified the giant phased array radar located at the Sary-Shagan ABM test range. Space reconnaissance has continually monitored the construction of naval vessels in Soviet shipyards, including ballistic missile submarines and aircraft carriers. Improved Crystal satellites are able to mark images so that mappers can use them directly.

In 1982, radar-imaging satellites joined photoreconnaissance satellites with the launch of Indigo. Radar satellites have the desirable quality of being able to see though cloud cover or in total darkness. However, the long wavelengths (compared with light) at which they operate impair their ability to provide high-resolution images. To help overcome this problem, satellites make use of synthetic aperture (SAR) imaging, which combines signals received over a portion of the craft's flight track to approximate a very long antenna. It is also likely that radar satellites can broadcast and receive over a range of frequencies in order to provide detailed or wide area coverage. VEGA, the current standard in radar imaging satellites has a resolution of about 1 - 2 meters from an altitude of 400 kilometers. This is sufficient to distinguish between various types of armored fighting vehicles.

Although the United States developed the analytical capability to locate and assess Soviet targets, it then had to strike them accurately. As Soviet air defenses grew in sophistication, the Armed Forces could no longer accomplish this mission reliably with bombers. Instead, it was necessary to strike targets with warheads delivered from space by ballistic missiles. These weapons rely for their guidance on inertial navigation systems that make use of sensors that are responsive to acceleration. The guidance systems correct for deviations to the anticipated acceleration supplied by propulsion systems and for such extraneous factors as the effect of high-altitude winds. Because gravity acts as an acceleration and varies globally, it is necessary to have an accurate model of the earth's geoid or gravity field. Only after accounting for the varied effects of gravity, can specialists determine and correct for propulsion deviations and external effects.

In order to generate a more accurate model of the earth's geoid, the NRO began optically tracking the orbital deviation of bright space objects such as the Echo and Pageos satellites in 1960. The data derived from this program improved geodetic accuracy from 200 meters to 10 - 15 meters. Radio Doppler tracking of Transit communications satellites further increased geodetic accuracy, which the Navy refined by laser tracking corner reflectors on BE-B and DIADEME satellites. Skylab, Geos, and Seasat satellites, launched in the 1970s, also carried altimeters that detected such subtleties as undulations in the marine geoid due to the topographic expression of the seafloor.

Data from SIGINT or signals intelligence gathering satellites supplements data from IMINT satellites. The first of these satellites, code named GRAB, went into space on June 22, 1960, during the closing years of the Eisenhower Administration. The success of this small satellite in characterizing Soviet Air Defense radars was encouraging and resulted in its follow up by a number of "Ferret" satellites. KH-4, KH-7, and KH-8 launches could carry a single one of these small, secondary satellites into orbit whereas a KH-9 launch could carry two. The secondary ferrets gathered intelligence about Soviet antiaircraft and ABM radars. Heavier ferrets, independently launched into higher orbits, eavesdropped on communications within the Soviet command and control system.

The field of satellite-based signals intelligence is very sophisticated and even more secretive than the photo intelligence program. In 1952, the Eisenhower Administration created the

National Security Agency to manage SIGINT operations. It currently has a ten-billion-dollar budget for this purpose. The NSA is so secretive that a number of wags have declared its acronym to stand for "No Such Agency." The actual operator of signals intelligence satellites is the National Reconnaissance Organization, although the NSA bears the satellite's design and costs. Data reception and processing facilities for SIGINT satellites are located at Pine Gap, Australia; Menwith Hill, UK; and Fort Meade, Maryland: the last of which is the NSA's headquarters.

The NSA currently has at least three types of SIGINT satellites in orbit. It has tasked these with different missions. It launches Trumpet satellites into highly elliptical inclined orbits approximating the orbits of Soviet Molniya communications satellites. The parallel orbits suggest that Trumpet's primary mission is to intercept government, military, and diplomatic communications. The NSA places the two other kinds of SIGINT satellite into geostationary orbits with at least three satellites in each class required for global coverage. Vortex type birds intercept global UHF traffic whereas Orion type birds intercept microwave communications.

The Vortex and Orion programs date back to 1970 when the TRW Corporation developed Rhyolite. This NRO designed the satellite primarily for TELINT or the capture of telemetry information from Soviet missile programs and made use of a 20-meter diameter mesh dish that unfolded after it reached orbit. Interpreting the information captured by Rhyolite and its successors has allowed American intelligence to estimate the performance of succeeding generations of Soviet missiles. Although the secret of this system was betrayed to the Russians by a disgruntled TRW employee and an accomplice who were known as the "Falcon and the Snowman," TELINT satellites have been continuously updated.

Following a change in cryptographic techniques that occurred in 1958, the NSA found itself unable to crack Soviet encrypted messages. For the entire period of the 1960s, the NSA was essentially deaf to Soviet communications and this loss in ability undoubtedly contributed to the Cuban Missile Crisis. In the 1970s, when the Soviets began to use scrambled telephones, the NSA scored a major victory by capturing their communications using COMINT receivers on satellites. A series of transmission busts and retransmissions allowed Anne Caracristi's A Group to break the encryption scheme on these scrambled messages. For this success, the NSA promoted Caracristi to Deputy Director. Chalet / Vortex satellites with a 38-meter antenna and Magnum / Orion satellites with antennae up to 50 meters in diameter are currently used to intercept UHF and microwave communications, a secondary task of the original Rhyolite satellites.

The United States Navy scored one final intelligence coup in conjunction with the CIA. They took older nuclear attack submarines and modified them so that divers could exit and re-enter at depth. The Sealab Program of the 1960s was a cover used to develop this capacity. Divers exiting from the spy subs were able to tap submarine communication cables in Soviet waters, some of which were direct links between military bases. Analysts have described the resulting data haul as the "Crown Jewels" of intelligence for the depth and breadth of insight provided into Soviet military capabilities, operations and doctrine. Over the years, the combination of space based and undersea intelligence has been of great importance in evaluating and anticipating a potential attack by Soviet strategic forces.

CHAPTER 5
THE EFFECTS OF NUCLEAR WEAPONS AND NUCLEAR WAR

Due to the large amount of energy released by a nuclear explosion, scientists measure atomic yields in thousands of tons (kilotons) or millions of tons (megatons) of TNT equivalent. They divide this energy into four categories: the kinetic energy (velocity) associated with the motion of vaporized bomb materials, the thermal energy (heat) associated with the fireball, prompt ionizing energy (radiation), and the delayed energy (radiation) released by fallout. At an altitude below 50,000 feet, a nuclear explosion expends approximately 50 percent of its yield on developing a shock wave in the surrounding atmosphere. It releases another 35 to 40 percent of its yield as thermal energy and 5 percent more as prompt radiation. At higher altitudes, where the air is thin, a nuclear explosion releases less energy in its shock wave and more energy in the form of electromagnetic radiation. A nuclear explosion releases its remaining energy (5 - 10 percent) over a longer period as the delayed radiation emitted by fallout.

All of the energy in an atomic explosion is at first contained in a super-critical radiative wave that forms at the center of the bomb's core and then propagates outwards in a radial fashion. When it reaches the low-density interface at the edge of the pit, the wave divides into two components. One is a slow, mechanical shock wave driven by the vaporized material containing the bomb's kinetic energy and the other is a faster, radiation driven component that becomes the nuclear fireball. The fireball's rate of expansion slows as it cools and this ultimately allows the mechanical shock wave to catch up, at which time the two components merge to form a single wave once again.

The fireball created by a nuclear airburst forms in the microsecond after the radiative wave separates from the core. It has a temperature of about 100,000,000 °F and pressures of many millions of pounds per square inch. Two scientific principles govern the history of this fireball. The Stefan-Bolzman Rule dictates that the fireball will emit a spectrum of electromagnetic radiation with intensity proportional to the fourth power of its surface temperature. Because the temperature of the early fireball is thousands of times hotter than the sun's surface, this rule further implies that the amount of radiation emitted must be truly prodigious. Depending on the nuclear yield, the thermal pulse associated with the fireball lasts for a period of 3 - 15 seconds. The Wien Law then requires that the dominant wavelength of the electromagnetic radiation emitted by the fireball be inversely proportional to its temperature. The Stefan-Bolzman Rule has already established this temperature as much hotter than the sun so the photons emitted by the fireball must have very short wavelengths. Because short wavelength photons are very energetic, the radiation dominating early emission packs quite a wallop. The photons are primarily x-rays produced by orbital changes of electrons, rather than the type of nuclear interactions that produce gamma rays. They are further termed "soft" x-rays because they have energies at the lower end of the x-ray spectrum.

The fireball's internal temperature is at first nearly homogenous or "isothermal" because the capture lengths of its photons are about the same magnitude as its diameter. A glowing envelope, brighter than the sun, forms around the fireball because cool air absorbs x-rays and then re-emits them at longer, visible wavelengths. The continuous absorption and re-emission of

The first atomic test (Trinity) near Alamogordo, New Mexico. The "Gadget" nuclear device produced a 22-kiloton fission yield in July, 1945. (Credit: National Archives; Color Original)

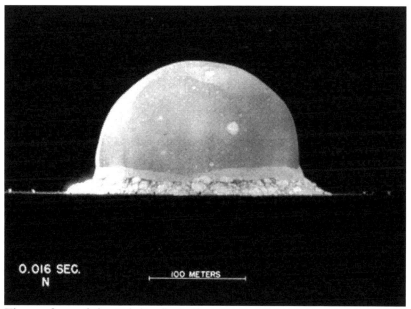
The surface of the Trinity fireball at T plus 0.016 seconds. This amazing photo was shot with a Rapidtronic camera able to capture photos with a shutter speed of 1/1,000,000 of a second. Note the interaction of the shock wave with the ground surface. (Credit: National Archives; Color Original)

The Castle Bravo thermonuclear test at the PPG in March 1954. Fueled with solid lithium deuteride, the test ran away to a yield of 15 megatons. This yield was typical of early aircraft delivered thermonuclear bombs. The photo is from a height of 12,500 feet and the mushroom cloud is some 75 nautical miles distant. (Credit: DOE; Color Original)

Hardtack Orange, the test of a 3.8-megaton W39 warhead at an altitude of 141,000 feet. Orange caused communications blackouts over a wide area of the Pacific Ocean due to the injection of a large quantity of fission debris into the ionosphere. (Credit: DOE; Color Original)

Detonation of an ASROC delivered 10-kiloton nuclear depth bomb. The USS Agerholm (DD 826) is in the foreground. (Credit: USN)

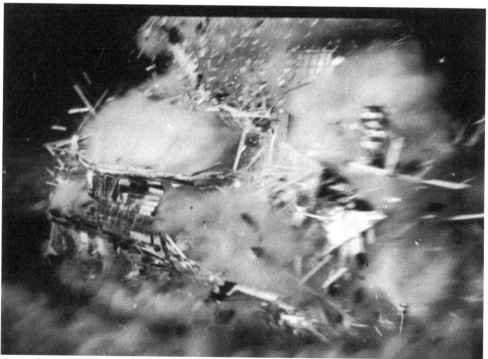
A nuclear shock wave strikes a house. This blast was part of a series of weapons effects tests. (Credit: DOE)

Left, the mushroom cloud that rose over Hiroshima. Right, the mushroom cloud that rose over Nagasaki. Both mushroom clouds quickly dissipated. (Hiroshima Credit: National Archives Photo 58189 AC; Nagasaki Credit: Charles Levy / USAAF)

An enormous pyrocumulus cloud rises above Hiroshima's firestorm 3 hours after the atomic attack. It formed when water nucleated on particles that were carried into the atmosphere by thermal updrafts. A black radioactive rain then fell from the cloud. The firestorm released energy equivalent to 1,000 times the energy of the bomb! It was not related to the nuclear mushroom cloud. (Credit USAAF)

The remains of Hiroshima after suffering a 15-kiloton atomic explosion and a massive firestorm. Only a few brick or reinforced concrete buildings remain standing. The firestorm incinerated non-ambulatory casualties, greatly increasing the death toll from the attack. (Credit: DOE)

The remains of the Mitsubishi Torpedo Works in Nagasaki after the 22-kiloton atomic attack. Nagasaki suffered multiple conflagrations due to high winds at the time of the attack, a situation that burned out much of the city and its industrial base. Only earthquake resistant smoke stacks and crushed steel frame buildings survived the attack. (Credit: DOE)

electromagnetic radiation at progressively longer wavelengths allows the fireball to expand and to cool in a process called radiative transport. Because of its high initial temperature, the fireball is relatively buoyant and rises in the cool atmosphere After the rapidly expanding fireball has cooled to a temperature of 300,000 °F, a shock wave separates and moves away from its surface much faster than radiative transport can keep up. Scientists term this phenomenon "hydrodynamic separation."

The shock front is at first opaque and hides the isothermal sphere from sight. However, after a short period of cooling, the shock front becomes transparent in a process termed "breakaway," and the glowing fireball once again reveals itself. This millisecond long process characterizes the distinctive double flash of a nuclear explosion. Although too fast for the human eye to distinguish, nuclear detection (NUDET) satellites in geosynchronous orbit can recognize it and sound an alert. Finally, at a temperature of 3,000 °F, radiative transport becomes ineffective and can no longer contribute to cooling. The diameter of the fireball at this time ranges from 100 yards for a 20-kiloton tactical warhead to more than 3 miles for a 1950s type 20-megaton strategic bomb.

The shape of the fireball changes as it ascends, evolving from a sphere into a violently rotating toroid because of friction with the surrounding air. Large amounts of air, sucked into the fireball, provide further cooling and contribute to its final dissipation. As the fireball cools, small drops of water derived from the air form a mushroom shaped cloud. The color of the cloud is initially reddish brown, due to the presence of nitrous acid and oxides of nitrogen synthesized from the atmosphere. As the fireball cools and condensation occurs, the color changes to white, mainly due to the formation of water droplets.

Depending on the height of the nuclear burst and its energy, an updraft phenomenon known as "afterwinds" can carry varying amounts of soil into the fireball at speeds up to 200 miles per hour. Afterwinds form the stem of the developing mushroom cloud. During the updraft process, fission products that condense from the fireball mix with soil debris to form highly radioactive particles. Gravity eventually separates these particles from the cloud, at which time they may combine with water droplets to fall as rain.

The final height of the mushroom cloud is dependent on the yield of the explosion. The clouds produced by large thermonuclear tests conducted in the 1950s reached heights of 25 miles but heights of less than 75,000 feet are more typical for modern strategic warheads. Once the mushroom cloud has topped out, it becomes "stabilized." At this time, it begins to spread laterally until it "dissipates."

Ultimately, the shock wave that separated from the isothermal sphere reaches the ground and produces a secondary blast wave through reflection. This reflected wave eventually catches up with and reinforces the primary wave in what is termed a "Mach Front." The point at which this phenomenon occurs is a function of yield and burst altitude. For the typical range of yields associated with nuclear weapons, this process takes from 0.5 to 5 seconds. Thereafter, the Mach Front continually merges with the primary shock wave at progressively higher altitudes. The pressure in the Mach Front is generally double that of the primary shock wave. Both fronts behave like a wall of compressed air and initially move at supersonic speed. Following behind the Mach Front are transient winds that may have speeds in excess of several hundred miles per hour.

The phenomena associated with a surface burst differ with respect to an airburst primarily through interactions of the fireball with the ground. The proximity of the fireball to the ground induces radioactivity in soil materials through a process of neutron capture. Enhanced afterwinds then carry these materials aloft in large quantities, leading to intense radioactive fallout. The blast wave forms a crater, the size of which depends on the energy of the explosion and the physical nature of the substrate. At the Pacific Proving Grounds, large multimegaton surface tests created craters more than a mile across and 250 feet deep in porous coral rock. If a nuclear explosion occurs near the surface of a water body, large amounts of vaporized water contribute to cloud formation through condensation.

In the subsurface, the effects of nuclear explosions are dependent on whether the fireball breaks through to the surface. If this happens, the shock wave throws large quantities of hot gas and debris upward in the form of a hollow plume. At shallow depths, the movement of material away from the center of the explosion produces a crater surrounded by a thick blanket of ejecta. A highly radioactive cloud is usually associated with this type of explosion, even though debris falling back into the crater buries large amounts of radioactive material. At intermediate depths, correspondingly more of the ejecta will fall directly back into the crater. Finally, at depths where the confining pressure prevents the explosion from reaching the surface, a nuclear blast forms a cavity with fused rock walls. A substantial amount of material falls from the ceiling of the cavity to bury a portion of the radioactive products from the blast. In all cases, subsurface nuclear explosions produce ground waves similar in nature to those generated by shallow focus earthquakes.

Underwater, the heat from a nuclear explosion forms a gas bubble by instantly vaporizing millions of tons of seawater. If the blast is shallow, the bubble bursts through to the surface and forms a truncated hollow column of water, similar to the effect produced by a shallow underground burst. It also induces an outward propagating shock wave in the water with a velocity higher than that produced in the air. A condensation cloud of water droplets may briefly form after the passage of the air blast. Some of the radioactive gases will vent through the top of the water column to form a granular crown before the column collapses back into the sea. The force of millions of tons of radioactive water pouring down creates a base surge that follows the underwater shock at a speed of several hundred feet per second and is associated with waves that initially may have a trough to crest height of more than 100 feet. Finally, large amounts of radioactive water begin to condense and rain out of the mushroom cloud.

A nuclear explosion in space produces effects that are quite different from those produced by an explosion within the confines of the atmosphere. In the first microsecond after a nuclear weapon explodes at orbital height, high-energy gamma rays spread out and strike neutral molecules in the tenuous upper atmosphere. These collisions produce energetic electrons that in turn generate a strong electromagnetic pulse. During the next second, the remaining energy of the blast spreads out in the form of lower energy x-rays. The ionized bomb debris then expands upward over a distance of several hundred kilometers where it begins to interact with the earth's magnetic field to produce low frequency field emissions and an artificial aurora. The field emissions spread outwards, reflecting off the surface of the earth and the underside of the ionosphere to propagate their way around the planet, inducing telluric currents as they pass.

Blast, heat, electromagnetic pulse, prompt radiation, and the delayed effects of radioactive fallout produce the military effects of nuclear explosions. Of these, the most significant is blast, which can damage structures at excess pressures (overpressures) as low as 1/2 pound per square inch (PSI) above atmospheric. The main characteristic of blast is a shock wave that creates a sharp pressure front as it expands outward from its point of origin. After the passage of the shock front, the strength of the associated overpressure quickly decays, eventually becoming weakly negative before a brief, final positive phase. Although the destructive effects of a nuclear blast are generally associated with peak overpressure, the "following winds" can also cause substantial damage. The dynamic pressure of these winds is proportional to the square of their velocity and to the density of the air behind the shock front. At an overpressure of 70 PSI, the accompanying wind speed is 1,150 MPH and dynamic pressure is equal to overpressure. Above 70 PSI, dynamic pressure is in excess of overpressure, and at lower pressures, the reverse is true.

Shock fronts produce damage by means of "diffraction loading" whereas the destructive effects of afterwinds result from "drag." When a shock front encounters a structure, overpressures rapidly build to levels at least twice that of the incident front due to the transfer of kinetic energy from the following mass of air. As the shock front bends or diffracts around the structure, it exerts an inward force from all sides. The implosion of windows and doors rapidly equalizes this pressure, but if these are lacking, diffraction loading at an appropriate level of overpressure can cause a building to collapse. Following the passage of the shock front, the accompanying winds strike the face of any structure left standing. Rather than the near instantaneous effect of a shock wave, the following winds exert an extended dynamic pressure over a period of several seconds. This effect causes drag sensitive structures to collapse.

Buildings and box shaped structures feel the effects of diffraction loading most strongly, whereas the effects of drag are most noticeable on transmission towers, telephone poles, trees, and truss type bridges. Building size, design and construction quality are important factors in determining resistance to blast. Earthquake proof structures show the best performance, while large buildings with a substantial moment of inertia hold up well under dynamic loading. The ability of a structure to yield elastically is very important. Load bearing structures made from non-reinforced stone or masonry fail immediately. Wood frame structures that can yield by flexure show better resistance. Finally, buildings constructed with ductile steel frames or reinforced concrete show the highest degree of resistance. For this reason, the Air Force constructs its missile silos from reinforced concrete and large amounts of closely spaced steel rebar to withstand overpressures of several thousand PSI.

In order to optimize the effects of blast, the Air Force planned to explode its nuclear weapons at an altitude that extended a desired level of damage over the greatest possible area. For instance, a 5-PSI overpressure is sufficient to cause widespread damage to most residential and commercial buildings but a 10-PSI overpressure is necessary to demolish reinforced concrete structures. Strategic Air Command targeted industrial areas for a 30-PSI overpressure to insure the utter destruction of both factories and the equipment contained therein. The requirements for high overpressures reduce the area over which a bomb can project its effects.

In addition to structural damage, shock waves produce biological effects, although these are relatively unimportant because most organisms are quite resistant to blast. Direct exposure to an overpressure of at least 40 PSI is required to kill a human being. Much more dangerous than shock effects are collapsing buildings and flying debris. In 1962, a "nuclear effects calculator" in the form of a circular slide rule was available from the Superintendent of Documents for the princely sum of one dollar. It calculated many types of destructive effects, including the optimum height of burst for a given yield to ensure that a blast would drive glass fragments at least one-centimeter-deep into human flesh!

The thermal radiation from a nuclear explosion can add significantly to blast damage. An explosion releases approximately 1 percent of its heat during the millisecond period of the initial flash. At this time, the fireball is small but extremely hot. The explosion releases the remaining 99 percent of its heat from the fireball after breakaway, when the fireball is cooler but much larger. This period lasts for several seconds but is generally over before the arrival of the shock front. Since the spectrum of the light and heat released in a nuclear explosion is similar to very intense sunlight, atmospheric conditions strongly affect thermal emissions. Reflection off snow or low cloud cover can amplify these effects, and fog and haze can attenuate them.

Although the intensity of blast falls off with the square of distance, the thermal emission from a nuclear explosion is so great that it can cause harm at a range where blast is ineffective. The destructive effects of thermal radiation are primarily a function of proximity and exposure. Since thermal radiation can be scattered by atmospheric particles, it can also affect objects not directly exposed to the fireball. The absorption of large amounts of thermal radiation in a very short time prevents its dissipation by conduction. This serves to produce very high surface temperatures. Dark colored or low reflectivity objects and thin kindling-like materials such as leaves, paper, and wood splinters are prone to catch fire.

Intense light and heat can be particularly injurious to organisms. Temporary flash blindness is common in those directly exposed to the fireball. Severe burns can also occur, even at extreme distances. A third degree burn causes tissue death to a depth where the stem cells necessary for re-growth are affected. As a result, this type of burn is extremely difficult to heal and poses a serious threat of infection. At close range, immediate heating can cause the moisture in flesh to flash into steam. The subsequent shock wave can easily tear this injured flesh from the bone.

Along with blast and heat, a nuclear explosion produces large amounts of prompt and delayed ionizing radiation. The initial radiation pulse is composed of neutrons and gamma rays emitted during the fission process. Thermonuclear weapons supplement these products with neutrons and gamma rays produced by fusion. Most of the gamma rays from the initial burst are absorbed in the surrounding cloud of bomb debris. However, a short period of intense radiation produced by the decay of short-lived fission products follows the initial pulse. This subsequent radiation is a spectrum of neutrons and gamma rays mixed with alpha and beta particles. The atmosphere strongly absorbs the particles; what reaches the ground in the first minute after an airburst is an intense flux of gamma radiation accompanied by delayed neutrons.

Radiation disrupts chemical bonds in body tissue. This can impair cell function and in extreme cases, cause cell death. As well, chromosomal damage can produce various types of cancer. The basic unit of radiation is the roentgen, which measures the exposure of organisms to ionizing energy. Medical units are the "radiation absorbed dose" (rad) and the "radiation equivalent man" (rem). The rad is a measure of the amount of radiation deposited per unit mass whereas the rem modifies this measure by considering the relative abilities for different types of radiation to cause harm. Because the body provides a certain amount of shielding for its internal organs, whole body exposure to radiation does not necessarily apply to an individual organ.

An "acute" (24-hour) whole body radiation dose of 200 rems produces moderate fatigue and nausea but no significant degradation of physical performance. These symptoms can last from a few days to several weeks. At doses of 400 rems, nausea becomes universal and vomiting is prevalent within an hour of exposure. Hair loss happens frequently, along with hemorrhages of the mouth and kidneys. Mortality for untreated patients approaches 35 percent. This rate jumps to almost 90 percent for exposures of 600 rems, with the rapid onset of symptoms. A dose of 1,000 rems is almost universally fatal. Death results from internal bleeding and infection within a month. A dose of 5,000 rems interferes with the nervous system, causes immediate incapacitation, and death within 48 hours.

Injuries from prompt radiation are only associated with low yield weapons. The blast and heat effects of larger weapons are fatal at ranges far exceeding the reach of their radiation pulse. However, a second consequence of nuclear explosions is delayed radiation produced by fallout. A nuclear explosion creates approximately 350 isotopes of 36 elements, many of which are viciously radioactive. The explosion generates about two ounces of fission for every kiloton of yield. The unit of radioactive measurement for these products is the curie, which corresponds to the activity of a gram or radium – one of the first radioactive substances discovered. For example, the two ounces of fission product from a kiloton yield explosion have an activity of 30 billion curies. If spread evenly over an area of a square mile, this material would produce an exposure rate of 2,900 rads per hour in the early period after an explosion. A fatal dose of radiation would be absorbed within a very few minutes.

Fortunately, the effects of fallout decrease with time. To calculate this decay, the "Rule of Sevens" acts as a guide. For every seven-fold increase in time after a nuclear explosion, the radioactive intensity of fallout decreases by a factor of 10. Thus, after 7 hours, the radioactivity is only 1/10 as strong as it was immediately after the blast. In just over 2 days or 49 hours, it drops to 1/100 of its initial levels. After 14 days or 2 weeks, it has dropped to just 1/10 of 1 percent of its initial intensity. Finding adequate shelter during this period after an attack is essential for survival.

Radiation shelters provide protection from fallout. They also protect against the initial pulse of radiation. Engineers measure the efficacy of radiation shielding in "tenth-thicknesses." This represents the amount of shielding needed to reduce the intensity of a given flux of radiation by 90 percent. It follows that a radiation shield, equivalent to three "tenth-thicknesses", will reduce levels inside a shelter to about 1/1000 that of the rate outside. Depending on the density of the shielding, more or less material will be required to provide an effective barrier. Some representative tenth-thickness values are 3.3 inches for steel, 11 inches for concrete and 16 inches for rammed earth. Adequate filtering is also necessary to prevent airborne fallout particles from penetrating the shelter.

The effects of fallout and prompt exposure are roughly equivalent. However, the body receives prompt radiation in a concentrated burst whereas the radiation from fallout is absorbed over time. The symptoms for the same radiation dose will therefore be slightly more severe from prompt exposure. The body can engage in some healing during an exposure period to fallout. Since fallout occurs as fine particles, humans may breathe fallout into their lungs, or ingest it into the body with food and drink where internal organs can absorb it. Alpha and beta particles, which are harmless if shielded by even a simple layer of clothing, become dangerous when they gain direct access to delicate tissues.

Among the most dangerous fission products are the isotopes strontium 89, strontium 90, cesium 137, and iodine 131. Strontium is chemically similar to calcium and shows a strong preference for absorption into bone tissue. The short half-life of Sr89 is prone to cause immediate damage whereas Sr90, with its 28-year half-life, produces longer-term effects. In the human body, there is a steady turnover in body tissue material that eliminates ingested radioisotopes with time. Physicians have established a "biological half-life" that characterizes residency time in the body for many radioisotopes. The 50-year biological half-life of strontium indicates that the human body cannot readily eliminate it. The primary effect of strontium is to produce bone cancer.

Cesium is chemically very similar to potassium. Thus, both plants and animals readily absorbed it. After a human has ingested Cs137, it is concentrated in blood electrolytes and distributed throughout the entire body. For this reason, it exposes both the blood and a wide range of tissues to the effects of highly penetrative gamma rays. Radioactive cesium is responsible for a wide range of cancers including leukemia. Fortunately, it has a biological half-life of 100 days, allowing the body to cleanse itself over a period of a year.

Radioactive iodine concentrates in the thyroid gland and because of its high activity poses a substantial threat for cancer to this organ. Its biological half-life does not play a significant role because I131 decays in a matter of weeks. Typically, humans do not ingest radioactive iodine directly but consume it in milk from cows that have eaten contaminated fodder. Iodine is a particular threat to children because they have small thyroid glands and consume large amounts of milk.

Radiation also poses a threat to electronic equipment, which it can affect temporarily or permanently, depending on the level of exposure. A common, temporary effect in microchips and integrated circuits is the generation of spurious currents. This happens when host atoms displace loosely bound electrons through x-ray absorption. In transistorized circuits, energy absorption produces excess amplification in direct proportion to the intensity of the radiation flux. As well, tubes used for high-speed current switching in nuclear weapons can exhibit spurious firing behavior if their gas content becomes ionized.

Permanent alteration of semiconductor materials can result from atomic displacement. This phenomenon occurs when high-energy neutrons kick loose atoms from molecular structures. In most cases, the most serious side effects of this phenomenon are the degradation of gate voltages and amplifying capability. Tube failure brought about by breakage of the glass envelope or deterioration of glass - metal seals is a more serious consequence. A high neutron flux can also result in the temporary or permanent alteration of materials used in the manufacture of resistors, wiring, and insulation.

Radiation induced changes to electronic components can cause performance degradation or complete failure in communications equipment, GPS units, radar, gyroscopes, computers and the like. These circumstances can in turn produce malfunctions in the guidance systems for nuclear delivery vehicles, as well as the arming and fusing mechanisms of their warheads. Induced currents can also cause detonators to spontaneously fire. The presence of neutrons produced by a nuclear explosion may cause premature initiation of the chain reaction in nuclear weapons intended to explode near a preceding burst.

Many scenarios for WW III begin with a high-altitude nuclear explosion over a target country. The purpose of this stratagem is to disrupt communications by means of a powerful pulse of electromagnetic energy known as a high-altitude electromagnetic pulse (HEMP). At a sufficiently high-altitude, the x-rays from a nuclear explosion descend into the atmosphere to produce a large, pancake-shaped region of ionization, hundreds of miles across and 50 miles thick. The electrons released by this action spiral along the lines of the earth's natural magnetic field. The resulting radial acceleration produces a strong downward directed pulse of electromagnetic energy. As this energy travels away from the region of ionization, metallic conductors collect it in the same way that antennas collect radio waves. Depending on the strength of the electromagnetic pulse, the current it induces can short circuit electrical generating and transmission equipment, disrupting regional power supplies. It can also destroy computers, communications gear, and radar stations, rendering a nation blind to follow-on attacks by bombers and missiles.

Equipment containing solid-state components is particularly sensitive to the effects of EMP. Electrical equipment and vacuum tubes are much more resistant to its effects. However, any type of electrical equipment connected to long runs of cable is vulnerable to powerful EMP induced currents. An example of this type of vulnerability was the failure of breakers on 30 streetlight strings on the island of Oahu, 800 miles distant from a thermonuclear test conducted above Johnson Island. Disconnecting equipment from a regional power supply can reduce this type of damage. Good grounding, power surge arrestors, amplitude limiters, band pass filters, breakers, and fuzes can also provide some degree of protection, especially if used in concert. Surrounding particularly vulnerable areas with a continuous shield of metallic mesh can protect them from EMP.

EMP is not just restricted to high-altitude explosions. In a surface burst, interactions with the ground polarize the expanding cloud of ionized bomb debris. This effect is termed source region electromagnetic pulse (SREMP). Lighter, negatively charged particles move upwards faster than positive particles and produce a charge separation. The field potential attains its maximum value in about one shake and induces a strong vertical current to equalize the disparity in charge. The current flow results in the release of a radial electromagnetic pulse. The effects of EMP for a surface burst, while significant, are restricted to a small area of the battlefield.

Los Alamos scientists examined in detail the effects of the two atomic attacks carried out against Japan, in 1945. The Air Force dropped the first atomic bomb on the city of Hiroshima, which is located on the broad delta of the Ota River near the southern tip of Honshu Island. At the time of the attack, the city had a population of 255,000 civilian inhabitants, down from a peak level of 380,000 due to wartime evacuations. It also billeted 40,000 soldiers assigned to 2nd Army Headquarters, which was responsible for the defense of Southern Japan. Hiroshima covered an area of 26 square miles, of which seven square miles constituted the densely-occupied city core. There was no differentiation into residential, commercial, or industrial areas. Construction consisted of tile roofed wood frame homes interspersed with a dense collection of wooden workshops and a number of reinforced concrete structures designed to withstand earthquakes. The harbor was a major troop embarkation area for overseas deployments.

The weapon used in the attack was a gun-assembled uranium bomb with a nominal yield of 15 kilotons. It exploded at an altitude of 1,900 feet at 8:16 AM on August 6, 1945, directly over the city core. Initially, it produced a bright flash that lasted 4 or 5 seconds. Immediately after the blast, a controller at the Japanese Broadcasting Corporation headquarters in Tokyo noticed that the Hiroshima radio station had gone off the air. Communications could not be re-established because the main telephone trunk line into the city had ceased to operate. Along with the flash came a wave of intense heat. At the hypocenter, the point directly below the fireball, temperatures reached 3,000 °C and out to a distance of 1,300 yards, tile roofs melted at an estimated temperature of 1,700 °C. At distances up to 4,000 yards, the heat from the fireball charred wooden telephone poles and ignited flammable objects. Exposed skin was flash heated to temperatures well above the boiling point.

The shock wave arrived after the initial period of intense heating. Starting out at supersonic speed, the shock wave threw up a dense cloud of smoke and dust as it moved. By the time the crew of the Enola Gay had a chance to look back from their diving turn at a distance of

11 miles, the afterwinds from the explosion had sucked up much of this debris to form the stem of a mushroom cloud, 20,000 feet high. The shock wave collapsed the city's few masonry buildings and blew frame buildings apart to a radius of a mile. Earthquake-proof, reinforced-concrete buildings, nevertheless, showed remarkable resistance. Structural damage extended to a radius of approximately two miles from the hypocenter. The bomb destroyed 48,000 of Hiroshima's 76,000 buildings and damaged another 22,000. Much of the debris from the collapsed buildings fell into the streets to form a major obstacle to movement. This debris quickly became the focus of fires ignited by overturned stoves, sparking electrical equipment and debris left smoldering after the thermal pulse.

The consequence of the numerous fires that sprang up throughout the heart of the city was the creation of a firestorm. Firestorms develop under conditions of low humidity and light winds. They require large quantities of fuel, which was available in abundance from the debris of collapsed frame homes and wooden workshops. Firestorms also require multiple sources of ignition that coalesce to produce a rapidly rising column of superheated air. This serves to draw in strong winds, which provide the necessary oxygen to feed the firestorm. Once ignited, a firestorm will continue to burn until all available combustible material is exhausted. The Hiroshima firestorm release 1,000 times the energy of the bomb that ignited it.

The destruction of almost 80 percent of the city's fire fighting forces and equipment contributed to the formation of the firestorm. Surviving fire units from outside the city center were unable to render significant assistance because of the debris that blocked their entry into the devastated area. Temperatures in the firestorm area climbed to hundreds of degrees. Survivors who plunged into the shallow rivers that cut through the city perished when the heat from the storm caused the water within their banks to boil. In all, the firestorm razed 4.4 square miles of the city core. The firestorm started only 20 minutes after the bombing.

The Los Alamos scientists were able to correlate Hiroshima's casualties to population density and the distance from the hypocenter of the blast using three concentric rings:

A) Within 0.6 miles of the hypocenter
B) Between 0.6 and 1.6 miles of the hypocenter
C) Between 1.6 and 3.1 miles of the hypocenter

These distances approximate the ten, three, and one-PSI overpressure contours. Population densities within built up areas A and B were about 25,000 persons per square mile, while the density within residential area C was only 3,500 persons per square mile. Area A suffered 26,700 killed and 3,000 injured, area B had 39,600 killed and 53,000 injured and Area C had 1,700 fatalities and 20,000 injured. The casualty rate was 95 percent in area A, 64 percent in area B, and 27 percent in area C. The high number of deaths in area A and the inner part of area B were in part a result of the rapidly ensuing firestorm that incinerated non-ambulatory casualties within the 4.4 square miles affected.

A second contributing factor to the overall number of deaths at Hiroshima was a lack of medical aid. Over 90 percent of the city's medical staff were themselves casualties of the bomb and were unable to help. Seventy percent of the survivors had mechanical injuries, 60 percent had burns, and 30 percent had radiation effects. These figures add up to more than 100 percent because many victims suffered multiple injuries. Although survivors began to recover from their injuries, a mysterious series of deaths began after about a week. These were the result of lethal radiation doses, now beginning to take effect. American scientists had grossly underestimated the number of radiation casualties because they had concluded that those individuals close enough to receive a serious radiation dose would perish from of the blast.

Although the Air Force exploded the atomic bomb at an altitude thought sufficiently high to reduce fallout, the updraft produced by the firestorm sucked large amounts of ash and moisture up into the atmosphere. The moisture then nucleated on the ash particles to create a rainstorm, with drops colored black from the inclusion of the particles. Desperate survivors drank the water from the rainstorm, not realizing that it contained fission products that had adhered to the ash particles.

The Air Force prosecuted a second attack against the city of Nagasaki on August 9. This city straddled the length of the Urakami River valley where it entered a long natural harbor on the southern island of Kyushu. At the time of the bombing, the city had a population of 195,000 people and, except for the city core near the harbor, did not have as dense a population as Hiroshima. Due to its low population density, Nagasaki sprawled over an area of 35 square miles. The city's construction was mostly flimsy wooden homes and shops with a small number of reinforced concrete structures. Near the northern end of the city was a large, modern, steel and ordnance manufacturing facility that belonged to the Mitsubishi Corporation.

Due to heavy cloud cover, the B-29 carrying the 22-kiloton plutonium implosion bomb dropped it away from the city core, over the Mitsubishi manufacturing complex. It exploded at an altitude of 1,650 feet. The attack came at 11:02 AM, following a failed attempt to bomb the city of Kokura. The bomb's shock wave demolished the steel-frame buildings of the Mitsubishi Plant, along with the northern end of Nagasaki. Casualties at the end of the year were 39,000 killed and 25,000 injured.

One significant difference in the two atomic attacks was the absence of a firestorm at Nagasaki. This was due to natural firebreaks and a strong 35 MPH wind blowing from the sea. Instead of a firestorm, a number of separate conflagrations swept along until they ran out of combustible materials. The combined casualty numbers for both atomic attacks constitute 1/4 of the total bombing casualties suffered by Japan during WW II and are unprecedented in the history of war. Previously, the most severe firebombing attack on Tokyo produced a 10 percent casualty rate, which was in itself a record.

In the period 1953 - 1955, the AEC conducted a series of military effects tests at the Nevada Proving Ground to establish the criteria needed to estimate the damage and casualties inflicted by a nuclear attack. Megaton range effects tests at the Pacific Proving Grounds supplemented the earlier low yield effects tests in 1956 by. The AEC then used the results of both test series, as well as the data from the Japanese attacks to establish an expected range of damage for the *Single Integrated Operational Plan*, which the JSTPS established for fiscal year 1962. Military planning dictated that since the other side now possessed a considerable stockpile of nuclear and thermonuclear weapons, it would be necessary to destroy their nuclear delivery capability in order to prevent or at least minimize retaliation.

In the language of *SIOP*, strikes against military installations are "counterforce" attacks, whereas the strikes against population centers are "countervalue" attacks. The Air Force favored a counterforce campaign because it reduced an opponent's ability to carry out retaliatory operations against the United States and its weapon systems had the accuracy to get the job done. Counterforce had the negative connotation of implying a first strike and became less attractive as the size of the Soviet arsenal increased. The Navy, on the other hand, favored countervalue targeting as a means of deterrence, in part because its nuclear delivery systems lacked the accuracy necessary for counterforce. Politicians preferred a countervalue policy because deterrence required a smaller stockpile of nuclear arms and did not imply a first strike policy.

The Nuclear Effects Project Staff attached to the Office of Technical Assessment in 1979 estimated the impact of a countervalue attack on a major population center. They produced three scenarios for an attack on St. Petersburg (Leningrad), Russia's second largest city. These included the explosion of a 1 megaton MK 28 bomb such as would be delivered by a B-52 bomber, the explosion of ten W68 warheads, each of 40 kilotons yield, as would be delivered by a MIRVed Poseidon SLBM and a 9-megaton explosion from a B-52 delivered MK 53 bomb or a Titan II's W53 warhead.

St. Petersburg is a major industrial and transportation center lying along the Neva River where it enters the Gulf of Finland. The city has an industrialized harbor area surrounded by 10 - 12 story apartment buildings constructed from pre-cast concrete, overlaid on steel frames. The city has little in the way of single residence homes. At the time of the study, St. Petersburg occupied an area of almost 200 square miles and had a uniform population density of 25,000 persons per square mile, for a total population of 4,300,000.

The Air Force expected that a 1-megaton MK 28 bomb, exploded at an optimum height over the city center, would kill 890,000 people and injure another 1,260,000. This compares with a MIRVed Poseidon attack, which planners expected would kill 1,020,000 people and injure 1,000,000 more. The differences are due to a larger area covered by 5-PSI overpressures for the Poseidon attack and a greater area between the 2 PSI and 5 PSI overpressure contours for the megaton yield explosion. These figures are only half of the 2,460,000 killed and 1,100,000 injured expected for a 9-megaton blast. Total casualties for this final scenario are 3,560,000 or 83 percent of the total population – for a single warhead!

As horrific as these scenarios are, they do not tell the real story. The JSTPS applied redundant targeting to important cities or military bases, to insure their destruction. JSTPS based the degree of redundancy on several factors. These included the ability of SAC pilots to hit a target based on yearly bombing competitions. They also included the results of regular SAC and Navy test firings of strategic missiles to determine their reliability and accuracy. The JSTPS then combined reliability with an estimate of Soviet defensive capabilities before assigning a unique combination of weapons for the destruction of each target. This process, when coupled with conservative estimates of nuclear effects led to significant over-targeting. For example, planners assumed the need for three bombs of 80-kiloton yield to destroy a Nagasaki size target, when the Air Force had demonstrated during WW II that a single 20-kiloton bomb was more than adequate!

A more realistic scenario for a countervalue attack on St. Petersburg would have been an initial strike against surrounding radar installations, airfields, and defensive missile sites with a number of 170-kiloton W62 warheads delivered by Minuteman missiles. The main attack would have followed the annihilation of St. Petersburg's air and missile defenses. The attack would have been composed of Poseidon missiles, each armed with ten W68 warheads of 40 kilotons yield. Multiple submarines, stationed at widely separated locations would have launched the missiles. One or more Titan II missiles, armed with 9-megaton W53 warheads and aimed at hardened targets, would have arrived simultaneously or shortly thereafter. The "layering" provided by a large number of warheads and decoys would have overwhelmed any surviving defenses.

After a delay of several hours to improve visibility, SAC would have prosecuted a "cross-targeting" attack with a flight of B-52s supported by FB-111A fighter-bombers. At least one B-52 would have carried a pair of 9-megaton MK 53 bombs. The remaining B-52s would have each carried four short-range attack missiles armed with 200-kiloton warheads and four megaton range MK 28 bombs. SAC would have armed the FB-111A's with a combination of SRAMs and B43 or B61 bombs. The SRAMs would have helped these aircraft blast their way through any surviving defenses in a low-level attack with bombs fuzed for laydown delivery. Upon detonation, the resulting surface bursts would have encompassed the city in a shroud of highly radioactive fallout. The total nuclear yield of the weapons directed at St. Petersburg would likely have been in the range of 50 - 70 megatons and it is unlikely that anyone would have survived the attack!

Moscow would have fared even worse than Leningrad because of its Galosh A-35 ABM system. A projected US ABM suppression strike in 1968 involved the use of 130 Minuteman and Polaris warheads with a combined yield of 115 megatons to take out the Galosh System. The Air Force and Navy would have used many more warheads to attack the city's air defense system and its political, military, and industrial targets. In the late 1980s, planners estimated that an ABM suppression attack against Moscow's upgraded A-135 ABM system required 210 highly accurate warheads with a combined yield of 68 megatons. A retired American STRATCOM Commander reportedly commented that in total, the United States targeted Moscow with 400 nuclear warheads!

Although counterforce attacks supposedly minimized civilian casualties, their results would still have been catastrophic. A 2001 study by the National Defense Research Council estimated the results for an attack on the Kozelsk SS-19 missile field that lies about 120 miles west of Moscow. To insure the destruction of the 60 super-hardened missile silos in this field, the NRDC considered surface bursts by 120 W88 warheads (combined yield of 50 megatons) necessary. The expected fallout pattern from the attack would have produced an integrated 48-hour radiation exposure greater than 10,000 rems for the citizens of Moscow. Anticipated casualties in the area of fallout were in excess of 16,000,000 persons, of which about 13,000,000 would have been fatalities!

A full *SIOP-62* strike would have expended 7,847 megatons while a retaliatory strike by alert forces would have delivered 2,164 megatons. Planners expected the retaliatory strike to kill 175,000,000 Sino-Soviets whereas the all-out strike would have killed 285,000,000 people with 40,000,000 more as serious casualties. By 1967, *SIOP* envisioned a pre-emptive strike with 8,489 megatons. Despite the tonnage, the JSTPS still estimated that Soviet retaliation would produce 102,000,000 American and 112,000,000 European and Asian fatalities. In 1982, the World Health Organization estimated that a full 10,000-megaton exchange by the superpowers would result in a billion deaths and another billion injured. The results of a study entitled "Global Atmospheric Consequences of Nuclear War" suggested even greater casualties. The high yield bombs in use at this time would have destroyed so much ozone and injected so much dust into the stratosphere that the resulting drop in temperatures would have produced a nuclear winter on a scale to rival the climatic upheaval created by the asteroid impact that wiped out the dinosaurs. There was a distinct possibility that the human race would not survive the holocaust of thermonuclear war.

After their first review of the *SIOP*, President John F. Kennedy and his Secretary of Defense, Robert McNamara, became strong proponents of "flexible response." Both men had personal difficulties with the levels of casualties inflicted by the plan. At their request, the JSTPS provided a selection of attack options for *SIOP-63*. The planners also included "withholds" of specific targets or countries to produce options for negotiation, but the level of force employed still remained very high. President Richard M. Nixon and his National Security Advisor, Henry Kissinger had similar difficulties with the plan. Nixon, therefore, issued National Security Decision Memorandum NSDM-242: *"Policy for Planning the Employment of Nuclear Weapons"* that required the Joint Chiefs to provide a series of limited responses. Although these options were forthcoming, *SIOP* still relied heavily on countervalue strikes. This caused President Jimmy Carter to rescind NSDM-242 and issue Presidential Directive PD-59 *"Nuclear Weapons Employment Policy."* PD-59 required a shift away from cities to military assets. These assets included leadership bunkers, command, control, and communications facilities, and strategic targets.

The Joint Chiefs of Staff have never been very comfortable with options that could lead to massive American losses. In their consideration, the risks associated with limited strikes are not acceptable. Fortunately, the retirement of high yield weapons from world stockpiles has reduced the likelihood of mass extinction in the event of an all-out nuclear war. The smaller weapons currently in use are not capable of injecting the large amount of dust into the upper atmosphere that would produce a nuclear winter. On the negative side, much of the radioactivity that would have accompanied that dust into the stratosphere will now remain behind to settle as fallout. While megatonnages have fallen and the ratio of fusion to fission in deployed weapons has increased, there will be a substantial problem with fallout in an all-out nuclear exchange. We can only hope that if a nuclear war does come about that it will not produce the doomsday scenario envisioned by Nevil Shute's novel *On the Beach*.

CHAPTER 6
INITIAL DEPLOYMENTS: NUCLEAR FORCES AND STRATEGIC DOCTRINE FROM 1944 - 1968

The United States never used atomic weapons for their intended purpose – to defeat Germany. Hostilities in Europe ended before the United States was able to test its first atomic device. Rather than dropping an atomic bomb on Berlin, the United States instead dropped atomic bombs on two Japanese cities, thus ending WW II in the Pacific. These acts came at the end of a long and bitterly fought campaign that began after the December 7, 1941, sneak attack on Pearl Harbor. Although analysts have attempted to second-guess the motives of the men who made the decision to use atomic weapons, these analyses are of little value. Just as the simple act of observation can change the nature of subatomic particles, the events of the war with Japan altered the emotional balance and thought processes of America's political, military, and scientific leaders in ways unimaginable to the modern mind. A Gallup poll conducted from Aug 10 - 15 in 1945, found that 85 percent of Americans approved the use of nuclear weapons.

As the war in the Pacific ground toward its inevitable conclusion, the Japanese fought with increasing desperation. Recognizing that they were losing, Japanese military leaders hoped to negotiate a settlement by demoralizing the United States with the human price of an invasion. This was not to be. The attack on Pearl Harbor had crystallized American anger. Public sentiment convinced the Roosevelt Administration that the only acceptable outcome to the Pacific war was unconditional surrender.

In order to prepare for the invasion that would finish off Japan, American leaders decided to weaken that country's war economy. The means to accomplish this goal was the B-29 bomber, an aircraft capable of return trips to Japan from bases located 1,500 miles away in the Pacific Archipelago. The man placed in charge of the bombardment operation was General Curtis. LeMay, a central figure in the reduction of Germany's cities and industrial infrastructure. LeMay commenced operations against Japan from the Marianas Islands on January 23, 1945.

Initially, the results of high-altitude bombing raids were disappointing. Recognizing that Japan's fighter assets were limited, LeMay had the defensive armaments removed from his aircraft to increase their bomb loads and directed their pilots to attack at low-level. On March 9, 1945, a massive raid against Tokyo at an altitude of 8,000 feet produced the results LeMay was looking for. Cluster incendiary bombs started a firestorm that incinerated almost 15 square miles of the city. The raid left one million people homeless and destroyed more than a quarter of a million buildings. It also produced 83,793 deaths and 40,918 serious injuries.

In the months that followed the historic raid on Tokyo, the 20th Army Air Force burned down another 105 square miles in the heart of industrialized Japan; at which time, it ran low on bombs. Because the Japanese showed no willingness to surrender after this campaign, American authorities held a targeting meeting from May 10 - 11, to plan for the use of atomic bombs. As an outcome of this meeting, the USAAF left five cities: Kyoto, Hiroshima, Kokura Arsenal, Yokohama, and Niigata, intact from firebombing. Their destruction would clearly demonstrate the effects of atomic weapons to the recalcitrant Japanese.

Air Force General Curtis Lemay with President John F. Kennedy and General Power during the Cuban Crisis. General LeMay built the Strategic Air Command into a credible strike force. General Power, who succeeded him, produced the first *SIOP*, a plan for an all-out nuclear strike against the Communist Bloc. (Credit: Harry S. Truman Presidential Library and Museum Photo 2006-135)

Army Generals Bernard Medaris and Holger Toftoy flank German scientist Werner von Braun, who developed the V2. The Army was the first service to get tactical nuclear missile development rolling with home grown and German help at the Redstone Arsenal. They also produced the Saturn booster to establish a military outpost on the moon. (Credit: Cropped USA Photo)

A photo of Assistant Secretary of the US Air Force for Research and Development Trevor Gardner and General Bernard Schriever. These men were relentless in building the modern missile-based strategic deterrent and reconnaissance programs. (Credit: USAF)

Admirals Arleigh Burke (left) considered submarines to be the key to modern naval warfare – powerful attack submarines to pursue the enemy and ballistic missile submarines for a secure second-strike capability. (Credit: USN) Admiral Rickover (right) gave him nuclear propulsion so his submarines could remain submerged for months. (Credits: USN Photo70134-10-65)

Robert McNamara was Sec Def from 1961 to 1968 under Presidents Robert Kennedy and Lyndon Johnson. McNamara represented a watershed in the acquisition of nuclear weapons. Formerly Generals and Admirals had been in control, which then shifted to Congress. The Secretary of Defense became a critical intermediary between these two groups. (Credit: US Government)

The National Military Command Center (NMCC), located in the Pentagon, is responsible for sending Emergency Action Messages (EAMs) to launch control centers, nuclear submarines, recon aircraft and battlefield commanders. Shown is the Emergency Conference Room. A hardened Alternate National Military Command Center (ANMCC) lies under Ravenrock Mountain, not far from Camp David. (Credit: National Archives - R. D. Ward, Photo 6342760; Color Original)

A Looking Glass KC-135 Flying Command Post, which made its debut in 1960. On board is a senior SAC officer – known as the Airborne Emergency Action Officer (AEAO) - who took command of the SIOP if communications were lost with the National Command Authority, the Pentagon, and SAC headquarters. (Credit: USAF)

An early version of the "Nuclear Football." An aide accompanying the President carries this briefcase at all times. The president will use the contents to launch a nuclear strike if necessary. (Credit: SI – Jamie Chung, via Smithsonian Institute Magazine)

Recognizing the horror that atomic bombing would inflict on the populations of Japanese cities, American authorities at first considered demonstrating the bomb's power on an uninhabited island. In the end, the Atomic Bombing Committee decided that only a full demonstration over an inhabited area (Operation Centerboard) would bring about the desired result of unconditional surrender. This was in preference to the 250,000 - 1,000,000 American casualties expected from Operation Olympic, the planned invasion of Kyushu Island set for November 1, 1945. In anticipation of this invasion, the Army had 500,000 Purple Heart medals manufactured, a stock that would last through the end of the Vietnam War.

Immediately after he had two atomic bombs available, General Groves drafted a directive to use the bombs and sent it to the Potsdam Conference in Yalta, USSR, where Secretary of State George Marshal and Secretary of War Henry Stimson signed it. President Truman also authorized the attack, setting a precedent requiring presidential approval for the release of nuclear weapons. This level of approval remains in effect to this day. The directive provided a final list of four targets: Hiroshima, Kokura, Niigata, and Nagasaki. After the directive's approval, the Secretaries cabled bombing orders to General Carl Spaatz, Commander of the Strategic Air Force in the Pacific. He immediately directed Colonel Paul Tibbets, the Commander of the 509th Composite Group, to carry out the orders.

The 509th was the world's first atomic strike force. The Army Air Force created it by combining the 320th Troop Carrier Squadron, the 390th Air Service Group, the 603rd Air Engineering Squadron, the 1027th Air Materiel Squadron, the 393rd Bombardment Squadron and the 1395th Military Police Company, in September 1944. The 1st Ordnance Squadron, Special Aviation and the 1st Technical Services Detachment, which administered the Los Alamos personnel attached to the project, later joined these units. Located at Wendover AFB, Utah, for training, the 509th was equipped with specially modified or "Silverplate" B-29 bombers capable of hauling a heavy atomic bomb to Japan. Wendover AFB was known as Site K or "Kingman."

By May 1945, the 509th had mastered the techniques needed to deliver the bombs and moved to North Tinian Airfield. Shortly thereafter, on August 6, a B-29 piloted by Colonel Tibbets dropped the first atomic bomb on the city of Hiroshima. Three days later, the 509th dropped another atomic bomb on Nagasaki. Earlier that same day, 1,600,000 Soviet troops attacked Japanese forces in Manchuria, opening up a two-front war. General LeMay followed the Soviet offensive on August 10 with another incendiary attack on Tokyo. Recognizing that further resistance was futile, Japan announced its unconditional surrender on August 14, local time.

The demobilization of America's Armed Services came swiftly after the war's end. During this period, the Truman Administration reduced personnel and materiel to levels well below that which the Joint Chiefs of Staff considered prudent. The Army requested a peacetime force of 25 Divisions but because of budget cuts, had to reconstitute itself in eleven under-strength and under-funded Divisions. The Navy suffered a similar plight. It had to scrap or mothball half of its vessels, leaving a two-ocean force of 277 ships that retained 11 aircraft carriers. The Army Air Force in turn deactivated 68,000 aircraft and reorganized its combat assets into Air Defense Command, Tactical Air Command, and Strategic Air Command. ADC was tasked with protecting the continental United States from air attack and TAC was made responsible for air-to-air operations, ground support, and theater reconnaissance. SAC, which ultimately received the bulk of Air Force funding, had the responsibility for strategic operations and reconnaissance. AAF Regulation 20-20 called for SAC to establish a "global striking force in constant readiness." Despite is global mission; the Air Force had to make do with 48 Air Groups instead of the 70 Air Groups that it requested.

The Truman Administration based its rationale for American military reductions on lessons learned in fighting World War II. Of particular importance were the concepts of the "Unified Command" and "Integrated Theater Operations." Before World War II, an Army and Navy Board had managed inter-service relationships based on "mutual cooperation." The catastrophe at Pearl Harbor revealed this concept was useless. The effectiveness of Unified Commands and integrated theater operations first showed their effectiveness when these concepts were used to organize Allied Forces in the European Theater. General Dwight Eisenhower, who served as Supreme Commander of the Allied Expeditionary Forces in Europe, later wrote the guidelines for operations in the Pacific Theater based on this experience. In undertaking post war reorganization, the Truman Administration anticipated that a plan for mutual support between the various armed services would reduce the need for personnel and materiel thereby generating significant cost savings. The reorganization assumed a durable, lasting peace.

At Truman's request, General Eisenhower produced an *Outline Command Plan* for post war reorganization, which the President approved in December 1946. Successive versions of this plan have been designated *Unified Command Plans* and continue to govern the organization of the Armed Services. Unified Commands consist of elements from two or more Military Departments. As an example, European Command was formed from the 6th Mediterranean Fleet, the 7th US Army in Europe and the 8th US Air Force in Europe. Unified Commands operate with joint staffs and as originally constituted, were responsible to a specified member of the Joint Chiefs. Their Commanders communicated with the various Service Headquarters regarding training, logistics, and administration.

In 1947, the *National Security Act* abolished the two Armed Services and, in their place, created independent Departments of the Army, the Navy, and the Air Force. These made up a National Military Establishment. The *Act* also placed a Secretary of Defense in "general" control of the NME and in charge of three civilian Departmental Secretaries. The three Military Departments reported to the Secretary of Defense through the three Departmental Secretaries. The Secretary of Defense also formed one-half of the National Command Authority, together with the President of the United States. The NCA has the constitutional authority to direct the execution of military action and the inter-theater movement of troops. (*National Security Council Document NSC 30*, issued in 1948, confirmed the use of atomic weapons as a means of national defense subject to the approval of the NCA.)

The *National Security Act* also formalized the positions of the Joint Chiefs of Staff and provided for a separate Chairman. The Act called for the Chiefs to continue in their roles as advisors to the President and their Departmental Secretaries, with whom they were accountable for planning and budgeting. They also retained their authority to create Unified Commands in strategic areas important to national security, thus making them responsible for a "national strategic direction." The President could also create Unified Commands for specific military operations, as well as "Specified Commands." Specified Commands are composed of assets drawn from a single Military Department but act as unified commands with regard to all or part of their mission. Good examples of Specified Commands include SAC, TAC, and ADC. The Specified Command concept insured that assigned missions in critical areas such as strategic bombing or air defense were free from interference by other command elements.

At a conference held during March 1948, the three Departments of the Armed Services attempted to establish their areas of responsibility. This was in response to a policy paper entitled: *The Function of the Armed Forces and Joint Chiefs of Staff*, which James Forrestal, the first Secretary of Defense, drafted. As outcomes from the conference, the Air Force assumed control of long-range missiles and strategic aircraft, the Army continued work on medium range missiles, and the Navy prepared to conduct such air operations it needed to prosecute a successful naval campaign. This agreement unraveled in 1949 with the so-called "Revolt of the Admirals", when the Navy's planned 65,000-ton super carrier USS America (CVA-66) was deleted from the budget in favor of B-36 bombers for the Air Force.

An "anonymous document" circulated in Washington stated that the B-36 was a "billion-dollar blunder." It also charged Secretary of Defense Louis Johnson and Secretary of the Air Force Secretary Stuart Symington with influence peddling on behalf of Consolidated Vultee, the aircraft's manufacturer. A special committee convened to investigate these allegations quickly traced the letter back to Cedric Worth, the civilian assistant to Under Secretary of the Navy. There followed an investigation of the House Armed Services Committee in which the two Secretaries were cleared, Worth was fired, and the B-36 vindicated. The Air Force got its 70 Air Groups and with them a preeminent role in the strategic atomic mission. This was not surprising. The capabilities of Soviet submarines and the limited range of Navy patrol aircraft rendered its carriers vulnerable.

In 1949, Truman amended the *National Security Act* to resolve a number of outstanding issues. The National Military Establishment was renamed the Department of Defense and the Secretary of Defense was given "full" authority for military planning and budgeting. In order to prevent interference with the functions of the Military Departments he was, however, denied power to rearrange, consolidate, or abolish any of their established functions. This action demoted the Department Secretaries from their former cabinet level positions but, in view of their reduced powers, Truman allowed them to take Departmental matters directly to Congress after first informing the Secretary of Defense.

That same year, Bernard Brodie established the basis for nuclear deterrence when he published his influential book: *The Absolute Weapon: Atomic Power and World Order*. Although Brodie characterized atomic weapons as very dangerous, he realized that threatening their use would have a pacifying effect on belligerent nations that were not so armed. The threat had teeth because the United States had recently employed them against Japan to end World War II. After the Berlin Crisis in 1947, the emergence of the Cold War caused President Truman and the National Security Council to ask Paul Nitze, the Director of Planning for the State Department, to study the security of the United States. The State Department published this study as document NSC-68 in April 1950. Nitze concluded that the main objective of the Soviet Union was world domination and that the United States had a moral obligation to defeat the Soviet threat. This would require substantial military spending.

In order to defend against a possible Soviet attack on Europe, a threat made credible by continued Soviet militarization, the United States, Canada, and a number of European Nations created a self-defense pact known as the North Atlantic Treaty Organization in 1949. In terms of their ability to contribute forces to a war in Europe, the British and the French had reserves of personnel that if armed, could match Soviet forces. The only problem was that they were still rebuilding from WW II and were not yet finically able to do so. Germany at this time also had resources of personnel, but post war treaties restrained it from rearming. Instead, American,

Canadian, British, and French Divisions garrisoned it. At this time, the Soviet's had not yet built up the necessary three to one advantage needed to overrun Europe, so an attack on their part would have led to a manufacturing contest to see which side could rearm faster. Stalin believed the USSR could win such a race and planned for a European war in 1952.

Despite the limited effectiveness of NATO, President Truman was still reluctant to increase military spending. It took the outbreak of war in Korea in 1950 for him to change his mind. It was now apparent that nuclear deterrence might protect the United States from attack but this deterrence did not carry over to nations in Eurasia. The United States needed a strong conventional force as well as a nuclear deterrent to establish global stability. The war in Korea also caused concern that the Soviet's might ignite a major war in Europe while the United States was distracted elsewhere. Thus, Truman approved expansions of the Army, Navy and Air Force that would stymy Stalin's plans for a European War. Even during the most critical periods of the Korean War, the bulk of American armament production went to Europe.

Upon taking office in 1953, President Eisenhower established the Rockefeller Committee to make further recommendations for DOD reorganization. This resulted in an order designating Military Departments as executive agents for Unified and Specific Commands. It also had the effect of de-emphasizing the Joint Chiefs' role as service representatives. Finally, in 1958, Eisenhower authorized the Defense Reorganization Act, which established a stable, ongoing command structure. Commanders of Unified and Specified Commands had full operational control of their forces and were responsible to the President of the United States through the Secretary of Defense for their missions. The Joint Chiefs became responsible to the Secretary of Defense, and the three Military Departments assumed the administration, training, and support of component forces.

Although Eisenhower's military reorganization established a stable command structure, a critical shortcoming was its failure to specify roles and responsibilities. This perpetuated the military power struggle that emerged after the development of the atomic bomb. Although National Security Council document NSC 30 confirmed the use of atomic weapons as a means of national defense, it did not apportion responsibility for their use. Instead of cooperating in the development of a national atomic strategy, the Military Departments quickly fell to squabbling over how to divide a greatly diminished budget and a limited number of atomic bombs. In particular, the Air Force and the Navy fought for funds with which to establish strategic atomic forces. The Air Force wanted the expensive, intercontinental B-36 bomber for SAC, and the Navy wanted flush deck super carriers to project power in the form of long-range patrol bombers.

President Eisenhower enshrined "massive (atomic) retaliation" as a national doctrine in 1953, with document NSC-162/2: *A Review of Basic National Security Policy*, in which he discarded the concept of a durable and lasting peace. Secretary of State John Foster Dulles publicly announced the new policy in a speech on January 12, 1954. NSC-162/2 was the product of three blue ribbon panels comprising Project Solarium – so named for the room in the White House where the project originated. The intent of Project Solarium was to provide a long-term strategy for conducting the Cold War. Panel A, chaired by Soviet expert George Kennan, recommended a steady military buildup that was intended to outpace the Soviet Union's military machine and thus nullify its influence on the world stage. Panel B, conducted by General James McCormack, recommended drawing a "line of aggression" around the Soviet Union. The United States would meet Communist expansion beyond this line with an atomic attack. This plan had the advantage of limiting military spending, but also had the difficult task of defining the line that had to be crossed. Panel C, led by Vice Admiral Richard Connolly, advocated a "roll back" of Communism in which military, economic, political and covert methods would be used to "swiftly eradicate" the menace. NSC-162/2 was a combination of recommendations made by the first two panels.

Prior to any war planning, the United States carried out a number of geographic and strategic studies dubbed Pincher. Pincher found that targeting Soviet industry would be the most effective way of defeating the USSR. The Pincher estimates were followed by "Joint Outline War Plan" Makefast, which envisioned a massive conventional bombing campaign. Makefast was followed by JOWP Broiler, which was formulated in 1947. Broiler presaged the type of atomic deterrence espoused by NSC-162/2. In contrast to the conventional bombing campaign associated with JOWP Makefast, JOWP Broiler viewed the atomic bombing of Soviet war-making centers as a key to victory. The United States hoped that the destruction of the Soviet Union's war-making capacity would compel it to withdraw to its 1939 boundaries, force it to abandon political and military aggression, and establish international stability.

Broiler's atomic air offensive reflected an important shift in nuclear strategy. Whereas JOWP Makefast targeted Soviet industry, JOWP Broiler targeted the Soviet cities in which the industry was located. Broiler saw the destruction of urban areas as inseparable from the destruction of the industries themselves. This shift in strategy resulted from American military weakness and a lack of Soviet target information. The targets espoused by Broiler included government facilities, urban industrial areas, the petroleum industry, submarine bases, construction and repair facilities, transportation systems, the aircraft industry, the coke, iron and steel industry and electric power systems. Due to a lack of resources, planners transferred never transferred Makefast or Broiler to the Armed Services for implementation.

Emergency War Plan Halfmoon expanded on Broiler in 1948 with the inclusion of more nuclear weapons. Halfmoon and a successor plan designated Fleetwood called for "massive" conventional strikes against the Soviet Union after an invasion of Europe, and the use of about 20 atomic bombs. General Omar Bradley regarded Halfmoon as the inspiration for the strategy of massive retaliation. In the face of the 1949 Berlin crisis, the Air Force formulated EWP Trojan. Trojan envisioned dropping 133 atomic bombs on 70 of Russia's main population centers. It is doubtful if SAC could have carried out Plan Trojan. The command anticipated a 30-day delay in assembling its nuclear weapons and moving them to forward bases before it could prosecute a coordinated attack. Strike folders with photographic and radar image detail to find targets were not available. Further compounding SAC's lack of strike folders was the need to attack in bad weather or at night to avoid Soviet air defenses.

The concept of withdrawing from Europe in the face of a Soviet attack remained in effect until Emergency War Plan Offtackle replaced Trojan in 1949. Offtackle called for the defense of Western Europe and a strategic withdrawal only as far east as the Rhine, or the Pyrenees if necessary. The United States formulated Offtackle after the signing of the North Atlantic Treaty Organization Pact on April 4, 1949. Article V of the pact formally committed the United States to the defense of Canada and Western Europe. Because NATO could not match Soviet strength in conventional forces, plans to defend Western Europe led to increased reliance on nuclear weapons. With each passing year, the operational plans drawn up for this purpose added more targets as the availability of atomic bombs and nuclear capable aircraft increased.

The first practical atomic campaign to emerge after the formation of NATO was EWP 1-51. This war plan had sufficient modern weapons and aircraft to carry it out, including the B-36 intercontinental bomber. EWP 1-51 called for a force of B-36 bombers, launched from Maine to drop 20 bombs in the Moscow Gorky area. After the attack, they would land in the UK. B-29's and B-50s launched from Labrador would drop 12 bombs in the Leningrad area, also recovering in the UK. Medium bombers based in Britain would drop 52 bombs along the Volga and recover in Egypt and Libya. Bombers from the Azores would attack the Caucasus with fifteen bombs, landing in Dhahran. Finally, bombers from Guam would attack Vladivostok and Irkutsk with 15 bombs. A first strike on the part of the Soviet Union would trigger the retaliatory strike, although there was no way to detect a strike until the first Soviet bombs exploded. SAC expected the Soviet attack to last 30, days so it was essential for SAC to disrupt this attack through superior performance.

Despite NATO plans that called for an increased reliance on nuclear targeting, General Curtis LeMay found Strategic Air Command's ability to implement a war plan to be woefully inadequate after the Air Force promoted him to Chief of SAC in 1948. He ran a simulated all out atomic attack from his headquarters in Omaha, Nebraska, in which not a single aircraft was able to carry out its assigned mission. Instead of getting mad, LeMay got down to business. He reorganized the Command and worked it tirelessly, until it was capable of carrying out what he believed to be its true mission. This was a pre-emptive strike with SAC's full complement of aircraft and atomic weapons to prevent a foreign aggressor from doing the same to the United States. He firmly believed that Japan had lost World War II because it did not possess a strategic bombing capability. During WW II, the United States had the luxury of "time and space" to gear up its massive industrial base for making war. Now, however, atomic bombs had compressed time and space to the point that America's industrial base could be wiped out in a single surprise attack.

At the time of LeMay's simulated all out atomic mission, SAC was in possession of only 32 atomic-capable B-29s. These included 13 of the 46 Silverplate B-29s modified during WW II and 19 that SAC converted to the Silverplate standard in the post war era. These aircraft belonged to the 509th Bomb Group assigned to the 58th Air Wing. In order to upgrade SAC's capabilities, the Air Force found it necessary to improve the 509th's equipment. It accomplished this through Projects Gem and Saddletree. Project Saddletree focused on modifying a large number of aircraft as atomic bomb carriers, with a target of 227 operational aircraft by December 15, 1948, 290 operational aircraft by June 30, 1949, and 492 operational aircraft by mid-1950. Project Gem upgraded SAC's operating abilities through the provision of in-flight refueling equipment, winterization, and electronics. SAC designated the aircraft that received these modifications "Ruralist."

Project "On Top" succeeded Saddletree and Gem in 1951. It involved the modification of all MK III capable aircraft to carry the MK IV atomic bomb by means of the universal bomb suspension system. In parallel with On Top, Operation Back Breaker focused on the conversion of F-84 and B-45 aircraft for tactical atomic support. The President's Air Policy Commission fully supported the expansion of SAC and the introduction of deterrence as a strategic concept, declaring that any foreign power threatening the United States could expect "a counterattack of utmost violence."

The improved capabilities of Strategic Air Command allowed the formulation of Emergency War Plan 1-1953. Consistent with the philosophy of massive retaliation, EWP 1-1953 envisioned atomic strikes against 500 industrial Soviet targets by six heavy bomber wings originating in the continental US, and 23 medium bomber wings situated throughout the UK,

North Africa, the Far East, and Alaska. Both SAC and TAC would make additional atomic strikes against Soviet nuclear delivery capabilities and the advancing Soviet Army in Europe. The Commands would launch these strikes in four waves and would vary their tactics according to the season of the year, the amount of resistance encountered, and the depth of Soviet penetration. SAC aircraft were to penetrate Soviet airspace at low-level and then rise to 30,000 feet on approach to their assigned targets. Recovery of aircraft was to take place at Allied bases. The Air Force expected minimum losses of 25 percent in such an engagement.

Although the Air Force located Strategic Air Command at Offutt AFB, Nebraska, it distributed SAC's growing fleet of atomic capable aircraft throughout the United States, Europe, North Africa, and the Far East. In spite of its expansion, SAC's atomic bombs remained in civilian depots and the National Command Authority that controlled SAC's use of atomic weapons resided in Washington. These circumstances forced the Air Force to devise improved methods of communication. As early as 1947, the Air Force attempted to include the Western Union Defense Organization in its command arrangements. The Air Force followed this overture in 1949 with a teletype system known as AIRCOMNET, which was dedicated for administrative and operational traffic. The AIRCOMNET system fell far short of expectations and in 1950; a combination telephone and teletype Strategic and Operational Control System replaced it. SOCS paralleled the SAC chain of command with the RAMROAD telephone network, which fanned out to subordinate headquarters in the United States and overseas.

In 1950, the Air Force opened a Command Post in the Pentagon building to manage its battle functions for the Korean War. The War Department built the Pentagon in 1941 to house its senior military personnel. The Air Force Command Post expanded on a "War Room" that the Air Force had previously established for intelligence displays and briefings. Since the Air Force was in charge of Air Defense, it added dedicated phone lines to connect the AFCP with the National Command Authorities. The AFCP was the direct predecessor of the National Military Command Center.

The Air Force also had an Alternate Joint Command Post constructed at Fort Ritchie, Maryland. The various Armed Forces Battle Staffs were to relocate to the AJCP if their Washington facilities became untenable. In 1951, the AFCP moved to a permanent area in the Pentagon, which had a communications center and an operations area. An Emergency Air Staff Actions Office at this location prepared a series of messages for transmission to meet anticipated hostile situations as required. At the Joint Chiefs request, the Air Force also set up a Joint War Room Annex for the coordination of all atomic capable forces.

In 1952, Secretary of the Air Force, Thomas K. Finletter, and Air Force Vice Chief of Staff, Nathan F. Twining, undertook a review of Air Force doctrine that resulted in a paper entitled: *The New Phase*" Among the recommendations of this paper was a call for a standing nuclear deterrent force. It also requested that the Joint Chiefs designate nuclear operations as a key element of NATO, a strategy subsequently adopted as NATO Military Committee decision MC-14/1. The ultimate result of the New Phase was a plan endorsed by President Eisenhower to redevelop the Armed Forces around a "New Look" based on nuclear deterrence. The United States expected that atomic weapons and the threat of "massive retaliation" would restrain the Communists from attempting to tie down a limited number of American forces in local wars in order to allow an overwhelming attack elsewhere.

As part of the New Look, the Air Force requested an expansion to 143 Air Wings. The heart of this updated force was to consist of ten B-36 heavy bomber wings, to give it a true intercontinental capability. The Air Force had received its first B-36 in 1948 and, following its award of the strategic atomic mission in 1949, the Truman Administration and substantially increased appropriations for this aircraft. Although some thought it cheaper and more prudent to wait for an all jet bomber, General LeMay was determined to defend America at any cost. He was quite willing to equip SAC with the B-36 and then re-equip it with all-jet bombers as they became available. He believed that SAC had to close a "window of vulnerability" and jet bombers were not yet available to meet this threat.

As it turned out, the issues surrounding the B-36 proved anti-climactic when the Korean War erupted in June 1950. The annual Defense Budget then jumped from 14 billion to almost 50 billion dollars. The Truman Administration sustained this increase for the duration of the War, allowing the Air Force to expand toward its target of 143 Air Wings. By August 1954, it had six Heavy Strategic Wings (B-36), four Heavy Reconnaissance Wings (B36), 17 Medium Strategic Wings (2 B-50 and 15 B-47), and six strategic Fighter Wings (F-84F & G). The Army established a force of 20 Divisions and the Navy expanded to 409 combat ships, including 12 aircraft carriers. It purchased it first super-carrier, the USS Forrestal (CV 59), in 1955. An omen of things to come, the Navy launched its first nuclear powered attack submarine, USS Nautilus (SSN 571), in 1954.

Although the Air Force and the Navy held onto their gains, the Army lost ground as soon as the Korean conflict ended. In order to make itself more relevant, the Army began to re-arm with tactical atomic weapons, which the AEC had placed on a crash development basis at the start of the Korean War. In a secret study known as Project Vista, the RAND Corporation sent atomic physicist Samuel Cohen to Korea, so that he could witness the recapture of Seoul. RAND intended Cohen's study of urban warfare to provide insight into the best uses for tactical weapons.

As a basis for the most effective use of tactical weapons, Chief of Staff General Maxwell Taylor organized the Army into "Pentomic" Divisions in 1956. The Army, however, had a rather naive understanding of such weapons, considering them only as a more powerful form of artillery. A contemporary estimate of atomic ordnance requirements made by General James M. Gavin, the Army's Deputy Chief of Staff for Research and Development, required 151,000 nuclear weapons to fight a protracted war with the Soviet Union. He based his estimate on 106,000 warheads for battlefield use, 25,000 warheads for air defense, and 20,000 warheads for Allied support. General Gavin estimated daily consumption during a hard day's nuclear combat at 423 atomic warheads, exclusive of air defense requirements!

The core of each Pentomic Division was composed of five "Battle Groups," which replaced the predecessor Regimental Combat Teams. Battle Groups had an aggregate strength of just under 1,500 personnel, large enough to fight independently, but small enough to be expendable if targeted by an enemy nuclear warhead. Commanded by a Colonel, Battle Groups were composed of four Combat Maneuver Companies, each of which had five Platoons. For support, the Battle Group's Headquarters Company had increased reconnaissance, intelligence, maintenance, and signals assets. Nevertheless, the Battle Group had to rely on Divisional artillery and armored support, and to the Corps level for support with theater weapons. The Honest John rocket was the only tactical weapon deployed at the Battle Group level. An organizational drawback equipped each Pentomic Division with only enough armored personnel carriers to move a single Battle Group at a time. This lack of mobility significantly hampered the concept of small, highly mobile forces maneuvering on the nuclear battlefield.

In order to support the Pentomic Army in its new role, Tactical Air Command also received theater weapons. These were primarily tactical bombs for TAC's force of F-84G fighter-bombers and lightweight strategic bombs for its B-45 Tornado bombers. TAC also received all-weather cruise missiles armed with nuclear warheads. Originally known as pilotless bombers, TAC based its Matador Missile Squadrons in both Europe and Southeast Asia.

The reorganization of the Army and TAC coincided with the deployment of aircraft and atomic weapons to overseas bases by SAC. Secret agreements with NATO and other American allies made these international deployments possible. The agreements moved SAC's medium range B-47s closer to their targets, thereby reducing the time needed to mount an atomic attack. Significant benefits of the new Air Force deployments were the dilution of a Soviet attack on the United States by diverting some of its resources to nullify American forward air bases and the denial of the countries containing American air bases to Communist encroachment.

The United States decided to supply nuclear weapons to its NATO allies as a force multiplier. These weapons were (and still are) governed by Atomic Stockpile Agreements that provide for the introduction, storage, custody, security, safety and release of the weapons by a user nation. They also contain a cost-sharing component to cover the US contribution to foreign defense. A further Service Level Agreement implements the stockpile agreements. Third Party Stockpile Agreements provided for the storage of atomic weapons in third nations for use by the United States or NATO Forces. In 1967, a NATO Nuclear Planning Group (NPG) met for the first time to become the forum for defining the conditions and procedures for nuclear use. Article MC-14/2 established massive retaliation as the basis for NATO military doctrine at this time.

Placing nuclear weapons in the hands of NATO necessitated a heightened measure of American security to prevent their unauthorized use. At first, the United States stationed guards next to quick reaction fighter-bombers, armed with atomic bombs. When an inspecting officer asked a young guard what he would do if a pilot began an unauthorized take off, he replied that he would shoot the pilot. The officer told him to shoot the bomb instead. This is not as crazy as it sounds. The pilot sat in an armored cockpit with a bulletproof Plexiglas canopy against which the guard's carbine was unlikely to be effective. However, a bullet fired through the bomb's ballistic cover was likely to wreck its delicate electronics, rendering it inoperable. The level of safety incorporated in the design of atomic weapons made an accidental nuclear explosion unlikely.

This situation was hardly satisfactory, so Sandia developed a new means of security known as the Permissive Action Link or PAL. This piece of equipment prevented the unauthorized use of nuclear weapons. President Kennedy issued National Security Action Memoranda NSAM 160 *Nuclear Permissive Action Links for NATO* and NSAM 230 *Assignment of Highest Possible National Authority to Project PAL* to deal with this issue. The earliest type of Permissive Action Link was a three or five-digit combination lock fitted to warhead containers. PAL A was an electromechanical switch requiring the input of a four or five-digit code to arm a weapon. Later PAL designs, such as Category D or Category F PAL, are equipped with six or twelve-digit electronic keys. If someone makes more than a limited number of incorrect attempts to arm a weapon, the PAL device permanently disables it. Some nuclear capable aircraft cockpits were fitted with an Aircraft Monitoring and Control (AMAC) system that allowed the pilot or bombardier to control a nuclear weapon's safety systems as well as its arming and fusing. Bomb Commanders currently transmit PAL codes to nuclear weapons through AMAC.

The Air Force focus on offensive operations meant that Air Defense got off to a slow start. SAC, in particular, was worried that expenditures allocated to ADC might detract from its strategic build up. As well, the Army felt that the air defense mission rightly belonged to its antiaircraft assets, for which it was developing surface-to-air missiles. The outcome of this combined opposition resulted in the inactivation of Air Defense Command on July 1, 1950. This state of affairs did not last for long. The recognition of improved Soviet strategic capabilities resulted in the Command's reactivation on January 1, 1951. Definitely, the junior Command in the Air Force organization, ADC at first had to arm itself with WW II vintage propeller driven aircraft. It only received jet aircraft when castoff F-86D Sabres became available from TAC. Nevertheless, by 1955, ADC had more than 1,000 Sabres in service. Because the F-86D was more of a fighter than it was an interceptor, ADC held a design competition in 1954 to select its successor. This was the supersonic F-102 Delta Dagger.

Starting with the F-102, ADC began arming its interceptors with nuclear tipped Genie air-to-air missiles. Using this weapon, individual interceptors could shoot down flights of bombers "en masse" and a miss by a quarter of a mile could still prove lethal. Since it was neither practical nor cost effective to have packs of ADC fighters scouring the sky looking for airborne enemies, the Eisenhower Administration activated Continental Air Defense, the forerunner to North American Air Defense, on September 1, 1954. CONAD's purpose was to provide both an alert and air intercept direction for a planned force of 25 Air Defense Squadrons. Backing up the ADC interceptors was a force of long-range BOMARC surface-to-air missiles located at eight American and two Canadian Air Force sites. For point defense, Army Air Defense activated 365 Nike Hercules short-range missile batteries. Like ADC's Genie missile, BOMARC and Nike had nuclear warheads. With Nike, the United States acquired a superb high-speed and high-altitude defensive weapons system that it exported to Europe.

The transmission of real time data from distant radar-warning networks provided the basis for the integrated command of America's nuclear forces. In 1955, the Air Force Command Center in the Pentagon became the National Air Defense Warning Center. Upon the arrival of a significant event at this center, CINCNORAD (who also commanded ADC) contacted CINCSAC, alerted the Joint Chiefs of Staff, and declared an Emergency Condition (EMERGCON). This action had the result of sending all US forces to Defense Condition (DEFCON) 1 or maximum readiness.

Defense conditions vary from DEFCON 5 for normal peacetime conditions to DEFCON 1, in which the United States has mobilized all of its forces and placed them on full alert. Upon discussion and confirmation that the United States was under nuclear attack, the Joint Chiefs would have notified the National Command Authorities – the President of the United States and the Secretary of Defense. After deciding on a course of action, the nation's leaders would have communicated their decision back to the Joint Chiefs. The Joint Chiefs in turn would have let the various Army, Navy, Air Force and Theater Commands know the orders they were to follow by means of Emergency Action Messages. The National Military Command Center (NMCC), which was set up in August 1959, transmitted EAMs. The NMCC was located at the Pentagon in the Joint Chiefs War Room. The system for handling nuclear alerts had relatively few links and apparently operated successfully since no accidental nuclear actions ever happened. NORAD did almost start World War III when someone put a training tape in a computer and the big warning boards lit up with hundreds of imaginary inbound missiles.

Declassified documents reveal that President Eisenhower pre-delegated authority for the release of nuclear weapons to subordinate commanders in the event of specified emergency conditions, such as the death of the National Command Authorities in a sneak attack. The authority to release nuclear weapons takes a different path than the devolution of the Presidency, which the Presidential Succession Act of 1947 and the 25th Amendment govern. Those legislative documents anticipate a normal peacetime transition. Pentagon Directive No. 13 states that the authority to release nuclear weapons passes to the Vice President or the Secretary of Defense, the Deputy Secretary of Defense, and finally to the Chairman of the Joint Chiefs in an emergency situation. Eisenhower's concern was that the Soviets might simultaneously wipe out this small cadre in a surprise nuclear attack.

In March 1956, President Eisenhower asked the Departments of State and Defense to prepare a policy statement on the pre-delegation of nuclear authority. Later that year, he approved instructions authorizing the unrestricted use of nuclear air defense weapons against Soviet bombers in the event of an attack. Following additional study, planners presented Eisenhower with an enhanced set of options. These options resulted in further pre-delegation instructions from the Joint Chiefs to the Commanders in Chief of Atlantic Command (CINCLANT), European Command (CINCEUR), and Strategic Air Command (CINCSAC) in December 1958. Presidents Kennedy and Johnson, while expressing concerns over pre-delegation upon taking office, chose to retain and even update *Instructions for Expenditure of Nuclear Weapons in Emergency Conditions*.

The Soviet launch of Sputnik on October 4, 1957, had serious repercussions for the survival of American nuclear forces and the national command and control structure. This event clearly demonstrated the ability of the USSR to target thermonuclear tipped ICBMs onto the United States, with a flight time of about an hour. At a stroke, the Soviets had rendered NORAD's alert abilities and ADC's defensive capabilities obsolete. A Weapons Systems Evaluation Group (WSEG) Report published in 1960 found that 35 Soviet ICBMs could easily knock out 90 percent of the political and military leadership of the United States in a single, decapitating strike.

The precarious situation outlined by the WSEG seemed confirmed when Nikita Khrushchev made outrageous claims about nonexistent Soviet strategic capabilities. Although the Soviet Premier had begun to downsize the Soviet military, come clean about Stalin's murderous excesses, and repudiate the inevitability of military conflict between communism and capitalism, he tried to cover up Soviet weaknesses with bluster. In 1957, he announced that thermonuclear tipped missiles were rolling off assembly lines in Russian factories like "sausages." The result was panic in America, which did not have a counterbalancing missile force because of early concerns over the accuracy and throw weight for this type of weapon.

Because of Khrushchev's rhetoric, the Air Force moved NORAD's command center into a tunnel complex bored beneath Cheyenne Mountain in Colorado. This complex was, and most likely still is, the most secure place in the United States. It is located beneath 2,000 feet of solid granite and the portals to the 4 1/2-acre complex are equipped with multiple, 25-ton blast doors. The buildings within the facility are made of continuous-weld, 3/8-inch-thick steel, and sit on spring-loaded suspension systems. The metal walls serve to attenuate the effects of EMP. In the event that the complex should become unusable, the Air force installed a complete backup at SAC headquarters. The Army Corps of Engineers constructed a third, hardened, underground command center at the Olney Special Facility located near Gaithersburg, Maryland. With the end of the Cold War, NORAD's mission has been de-emphasized and Olney now serves as a satellite tele-registration facility.

SAC also anticipated the need to protect its command personnel from a nuclear attack when it moved its headquarters underground in 1958. SAC's original headquarters were located in the disused administrative offices built for the Glenn L. Martin bomber plant at Offutt AFB, Nebraska. SAC built its underground command center, often referred to as the Pentagon West, at this location. Designed by the architectural firm of Leo A. Daly, SAC's new headquarters consisted of a contemporary above ground office building and a hardened subterranean command complex. The underground command center, colloquially referred to as the "molehole," was completely self-sufficient. It had 24-inch thick reinforced concrete walls and 10-inch-thick intermediate floors. The roof varied in thickness from 24 to 42 inches, also constructed of reinforced concrete. The Pentagon West featured the latest in computer technology as well as the "big board" that plotted worldwide developments along its 264-foot length.

In 1960, SAC activated Project Looking Glass. Looking Glass comprised five KC-135 tankers that the Air Force converted into airborne command posts. It placed these aircraft on continuous airborne alert in July 1961. The Looking Glass aircraft stayed in constant communication with the Joint Chiefs, SAC Headquarters, and SAC bases, as well as SAC aircraft on the ground and in the air. Even if the Soviets managed to knock out all of America's underground command centers, SAC's airborne command posts would have been able to control retaliatory operations. At this time, General Powell, CINCSAC, ordered a force of B-52 bombers placed on continuous airborne alert to provide a retaliatory capability for use by Looking Glass. The Navy followed the Air Force lead in 1963 with TACAMO airborne command posts. These were propeller driven Hercules EC-130 aircraft with specialized low frequency communication equipment to contact the submerged ballistic missile fleet.

In order to provide further support for its command structure, the DOD constructed a hardened underground alternate to the National Military Command and Joint Communications Center at Ravenrock Mountain near Waynesboro, Pennsylvania. Backed up by internal power, Site R was equipped with a large number of communications systems that supported various Army, Navy, and Air Force Emergency Operations Centers. Such was the importance of the National Emergency Command Post (NECAP) that the DOD implemented airborne (NEACP) and floating (NECPA) backups.

The political leadership in Washington mirrored the military by also building underground facilities. The White House had a shelter constructed for the President and his staff at Mount Weather, a special underground "Continuity of Government" facility operated by the Federal Emergency Management Agency (FEMA). The underground facility also housed FEMA's 200,000 square foot National Emergency Co-ordination Center. The site contained a complete radio and television studio connected to the Emergency Broadcasting System. Another COG facility codenamed Casper was located at the Greenbrier in Sulfur Springs, West Virginia. The Greenbrier was the relocation bunker for the government of the United States and included separate facilities for the House of Representatives and the Senate. A special room was included for joint sessions. The Army Corps of Engineers constructed additional underground shelters for government officials near Hagerstown, Boonsboro, Mercersburg, and at more than 90 other locations. It also built special hardened radio transmission facilities at Davidsonville and the Brandywine Receiver Site to insure communication between the government and Armed Forces as part of the National Communications System (NCS), authorized by NSAM 252.

In order to facilitate the execution of a nuclear response under emergency conditions, a military aide with a special briefcase known as the "nuclear football" accompanies the President.

The nuclear football is a Zero Halliburton metal briefcase carried in a black leather jacket. It weighs about 45 pounds with its communications gear and has a small antenna protruding near the bag's handle. The nuclear football contains a "Black Book" of retaliatory options, a list of classified site locations where the President can be taken in an emergency, a manila folder that describes procedures to activate the Emergency Broadcast System, and a 3 X 5 plastic card with authentication codes. The DOD apparently arranges the options in the loose-leaf Black Book in sections labeled "Rare, Medium, or Well Done."

In the event that the President opens the nuclear football, it sends an alert signal to the Joint Chiefs of Staff. The President then contacts the Secretary of Defense, reviews their available response options, and selects a preset war plan. It is not clear if the football is equipped with AFSATCOM, VLFT, or VHFT communications. After the President makes a final decision, the military aide contacts the National Military Command Center. The President then provides positive identification by entering codes from a special plastic card" The Secretary of Defense, who is the second member of the National Command Authority, then confirms the President's orders. After the National Military Command Center confirms his codes, it issues Emergency Action Messages. The NMCC re-verifies the orders for authenticity. There are three footballs: one that travels with the President, a spare that is stored at the White House, and a third that accompanies the Vice President.

Fortunately, the perceived missile gap with the Soviets was not as serious as the stampede to move underground suggested. The Air Force had received information from Edward Teller and John von Neumann in 1951 that powerful thermonuclear warheads would soon be available to compensate for the ICBM's lack of accuracy. Coupled with advances in missile guidance and intelligence on military developments in the Soviet Union, the Air Force had revised its standing position on long-range missiles and issued contracts for preliminary research into an American ICBM named Atlas before the launch of Sputnik.

Missile development had begun in earnest following the delivery of a report to the White House by a special committee chaired by Herbert Hoover Jr. in 1954. In his report, Hoover declared that the avowed objective of the USSR was nothing less than "world domination." Responding to this report and to a special lobby by the influential senators Clinton Anderson and Henry "Scoop" Jackson, President Eisenhower assigned ICBM development a DX status of one (highest national priority). A year later in 1955, Secretary of Defense Charles Wilson approved additional programs for the development of Intermediate Range Ballistic Missiles. The Defense Department's Science Advisory Committee had made it clear through the release of the Killian Report that the Air Force needed IRBMs to provide an interim offensive capability before it could deploy ICBMs. American technology and production methods, pioneered by General Bernard Schriever at the Western Development Division of Air Research and Development Command, soon closed the missile gap. By 1963, the deployment of 123 Atlas and 64 Titan ICBMs in 21 Squadrons was complete. SAC equipped these missiles with city-busting thermonuclear warheads, with yields ranging from 1 - 4 megatons.

Because of his extensive military background, Eisenhower was uniquely suited to make decisions about strategic forces. Soon after he approved the development of Atlas and Titan, Eisenhower approved work on a solid fuel missile. As a hedge, however, he continued to support the development of heavy intercontinental jet bombers. This resulted in an order for 400 Boeing B-52 bombers, which Eisenhower ultimately expanded to 744 aircraft. The Air Force deployed the bombers in Bomb Wings of 45 aircraft supported by 20 KC-135 air-refueling tankers. From

1956 onwards, air bases were equipped with special alert aprons and adjacent crew facilities to get their bombers airborne within the 15-minute timeframe provided by early warning. The Air Force also arranged for the dispersal of wings in 15 aircraft allotments to subsidiary facilities during times of crisis.

The Russians had only themselves to blame for SAC's increased deployments. In 1955, they had flight after flight of Bison heavy bombers fly over their Air Force Day parade in a brazen attempt to intimidate the West. In fact, it was the same group of bombers flying in a great circle. This action precipitated the notion of a "bomber gap" and the ratcheting of strategic production by the United States. To arm American bombers, President Eisenhower approved the mass production of hydrogen bombs.

Eisenhower also supported the Navy in its desire to acquire strategic nuclear weapons. In particular, the Navy was interested in submarine launched ballistic missiles, a concept it acquired after reading German documents that it captured after WW II. Ordered to participate with the Army in the Jupiter IRBM program, the Navy began independent research on a Jupiter "S" solid fuel variant. Solid fuels were much safer for shipboard applications. Since the large size of the Jupiter missile did not permit its use in submarines, Admiral William Raborn secured permission from President Eisenhower and Secretary of Defense Charles Wilson to begin development of a smaller missile known as Polaris. Because the Navy had access to a substantial body of research carried out by the Army and the private sector, the missile's design proceeded quickly, and delivery was set for 1965. Following the launch of Sputnik, the Navy decided to accelerate the development of its launch platform by slicing in half the keel of a nuclear-powered attack submarine under construction and adding a center section housing 16 vertical missile tubes. Named USS George Washington (SSBN 598), the new ballistic missile submarine began her first combat patrol on November 15, 1960.

President Eisenhower is thus the father of the "strategic triad," which consists of long-range bombers, land-based intercontinental ballistic missiles (ICBMs), and submarine launched ballistic missiles (SLBMs). The first offensive plans based on this concept were *Basic War Plan 60-1* and the *Strategic Integrated Operational Plan* (*SIOP*) introduced in 1962. The United States still bases nuclear deterrence on the strategic triad concept. Following Eisenhower's example, Presidents Kennedy and Johnson eventually rode herd over the strategic deployments of the 1960s and ultimately defined the triad's final composition.

President Kennedy entered office with a very different perspective from that of his predecessor. The writings of General Maxwell Taylor greatly influenced him in developing his view. Appointed Army Chief of Staff in 1955, General Taylor disagreed strongly with the Eisenhower backed doctrine of massive retaliation and instead called for "flexible response." He wished to see less emphasis on nuclear intimidation and more emphasis on conventional war fighting. Maxwell was aware that the American Administration had failed to authorize the use of nuclear weapons during the Korean War and to support the French in Indochina in 1954. Thus, the doctrine of massive retaliation had a hollow ring. Taylor attempted to sway the National Security Council and the Department of Defense toward his viewpoint, but the establishment rebuffed him. Disillusioned, he retired and documented his views in a book entitled: *The Uncertain Trumpet*. President Kennedy both read and appreciated this book.

After his inauguration, President Kennedy brought General Taylor out of retirement and appointed him Chairman of the Joint Chiefs of Staff. Kennedy then proposed increasing conventional forces to provide the ability to fight 2 1/2 wars. By this, he meant sufficient men and materiel for a 90-day conventional holding action in Europe, a simultaneous action with Chinese Forces in the Far East and one other smaller engagement. Kennedy projected an increase from eleven to 16 Army Divisions by 1963 and an increase in supporting Tactical Air Wings from 16 to 24 by 1968. Kennedy realized that while nuclear deterrence would prevent any sort of attack on the United States or NATO, it could not prevent the Soviet Union or China from undertaking military adventures in the third world. Kennedy's decision to increase America's conventional forces was timely because of developments in Southeast Asia.

Kennedy also formulated new roles and responsibilities for all three Services. In 1961, the Army created the "ROAD Division" under the Reorganization Objectives, Army Divisions. The Army intended these new units to meet the requirements of flexible response and combat on the nuclear battlefield. Presumably, the re-organization incorporated lessons from the Desert Rock exercises carried out in conjunction with testing at the Nevada Proving Grounds during the period 1951 - 1957. Operations Desert Rock I - VIII integrated maneuvers by elements of the Army and the Marines with live nuclear tests. Troops in the open and in trenches observed the effects of explosions with yields in the tactical range. They then maneuvered to distances measuring only hundreds of meters from the explosion's hypocenters, to witness the effects of the devastation.

The ROAD Division emphasized interchangeable, Battalion sized combat maneuver units within and between Divisions as a means of task organization. Each Division consisted of three brigades and associated support units. The Brigades acted as Tactical Headquarters, controlling ten or eleven maneuver Battalions, depending on their assigned task. Different types of ROAD Divisions had their own specific make up. The Infantry Division had eight infantry and two tank Battalions; the Mechanized Division had six Tank and five Mechanized Infantry Battalions; and the Airborne Division had nine Airborne Infantry and an Airborne Gun Battalions. Later, in 1965, the Airmobile Division joined this structure. Organizationally, the Airmobile Division was equipped with 450 aircraft and 1,100 vehicles, whereas a typical ROAD Division had 3,400 vehicles and only 100 aircraft.

The Army staffed ROAD Divisions with approximately 15,000 officers and men and equipped them with four Artillery Battalions armed with 105mm, 155mm, and 8-inch guns. By this time, nuclear artillery shells had begun to supplement Missile Battalions armed with Honest John or Little John rockets. Nevertheless, the Army still attached the bulk of theater nuclear forces at the Corps level. These included Sergeant SRBMs that could carry a 200-kiloton warhead to a range of 85 miles. A six-man crew could erect and launch this solid fuel missile in the unheard-of time of an hour and a half. The Army also introduced Pershing, a two-stage MRBM that could carry a 400-kiloton warhead to a range of 460 miles. Although these weapons had powerful thermonuclear warheads, the Army considered them "tactical" because they did not have sufficient range to reach the central Soviet Union from their deployments in Europe and South Korea.

Kennedy also reorganized the Navy. After being relieved of its strategic bombing responsibility, he ordered it to focus on the Fleet Ballistic Missile Program, which he greatly expanded. At first 24 and then 31 improved Lafayette and Ben Franklin SSBNs were approved, the last of which reached operational capability in 1968. This number was in general agreement with Chief of Naval Operations Arleigh Burke's estimate of a credible, survivable nuclear deterrent force of 25 missile submarines on patrol in two oceans at all times.

Like General Taylor, Admiral Burke had his own view of deterrence. Burke believed in a concept he termed "limited" or "finite" deterrence. This doctrine maintained that the only requirement to prevent a Soviet first strike was a survivable strategic force that could inflict an unacceptable level of damage in return. Twenty-five missile submarines, launching 400 missiles at Soviet cities, could kill 1/3 of the Soviet population and destroy 1/2 of its industrial capacity. After such an attack, the USSR would cease to function as a social and political entity. The Kennedy Administration could transfer the savings in defense spending brought by limited deterrence back to the American people.

RAND Corporation studies undoubtedly influenced Burke's view of deterrence, which by this time focused on the complexities of nuclear war. RAND's investigations applied game and systems theory to economic and military strategy. Herman Kahn subsequently published the findings in a massive work entitled: *On Nuclear War*. A significant conclusion offered by Kahn was the possibility of an unwanted pre-emptive strike as the result of nuclear blackmail. Considered far more important than a massive nuclear arsenal was the Navy's secure, second strike capability that could completely devastate any aggressor that launched an attack. As a result, the Navy targeted its missile force against cities. Their low accuracy in any case prevented them from competing with land-based ICBMs against military targets.

In 1961, the Soviet Union shot down a trespassing American U2 spy plane piloted by Colonel Francis Powers. This demonstrated to President Kennedy and his Secretary of Defense, Robert McNamara, that the day of the high-flying bomber had ended. Believing that the future of deterrence now lay in ballistic missiles, Kennedy and McNamara made significant changes to the strategic weapon systems planned and operated by the Air Force. They cancelled the experimental XB-70 Valkyrie bomber and terminated production of the B-58 and B-52H. They then had the Air Force reassign its existing B-52s to low-level attacks after improving their penetration capabilities with Hound Dog standoff missiles. Aircrew could use the megaton range warheads on these missiles to destroy radar and missile sites along the B-52's penetration routes, as well as to attack terminal air defenses. The B-47 entered phased retirement and the Air Force injected the cost savings generated by this measure into the accelerated deployment of ballistic missiles.

At the height of the Cold War in 1962, the US military stockpile consisted of more than 27,000 bombs and warheads. Strategic delivery systems included 10 solid fuel ICBMs, 129 liquid fuel ICBMS, 144 submarine-launched IRBMs, and 1,300 heavy and medium bombers. During the Cuban Missile Crisis in October of that year, CINCSAC General Thomas Power placed 2,952 nuclear weapons on alert. Although the Soviets had reached operational readiness with four regiments of R-12 (NATO SS-4, Sandal) MRBMs in Cuba, America's overwhelming nuclear advantage helped resolve this threat, along with a side deal to remove Jupiter IRBMs from Turkey. Polaris armed SSBNs prowling the depths of the Baltic and Mediterranean Seas had already made the vulnerable Jupiter missiles obsolete. A positive outcome of the Cuban Crisis was the installation of a telex "hotline" between Washington and Moscow to allow for direct communication between American and Soviet leaders.

In the final days of the Cuban Missile Crisis, SAC missile crews struggled to ready ten Minuteman solid fuel ICBMs for launch. The Air Force saw these thee-stage solid fuel missiles as less complicated, more compact, and safer than liquid fueled missiles, making them ideal for silo based or mobile deployment. As an added bonus, the Air Force could launch these missiles instantly. By 1959, the Eisenhower Administration had given the Minuteman Program a DX status of one. While the parallel Polaris program in the Navy focused on IRBMs because of size restrictions arising from submarine hull dimensions, the Air Force had the luxury of developing a full-sized ICBM. The Air Force planned for both mobile and silo-based deployments until Kennedy cancelled the mobile requirement in favor of cheaper silos.

The sites chosen for Minuteman were in the northernmost states of the Continental United States in part because of the missile's range and in part to shorten their trip to the USSR. Prior to the final selection of the sites based in the northern states, the DOD considered basing modified Minuteman missiles under the Greenland Ice Cap to further shorten the range to their targets. Conceived in the late 1950s by the Army Engineering Studies Center, Project Iceworm envisioned boring 2,500 miles of tunnels through the icecap to set up thousands of firing positions to house 600 two-stage Iceman MRBMs that would be moved on a regular basis. Iceman was to have a 3,300-nautical mile range, a CEP of 0.8 nautical miles, and a 2.4 megaton (W39?) warhead, potentially upgradeable to 4 megatons. The system would have a 20-minute response time for a first strike and 40 to 60 minutes if it had to launch under attack.

Launch Control Centers would be located below the ice surface and hardened to 100 PSI, with the Launch Centers hardened to 30 PSI. Stealth was the primary element for protection of the LCCs and LCs with clusters of LCs spaced 4 miles apart. It was expected the Soviets would have to launch 3,500 warheads to disable the system. To test the viability of this concept, Camp Century was constructed 150 miles east of the Thule AFB on the northwest flank of the North Greenland ice sheet. After ice cores demonstrated that the flow of the ice sheet was too severe too support the project, it was cancelled. Nevertheless, the Camp Century ice cores have proven invaluable in providing environmental data going back over thousands of years.

The accuracy of Minuteman was so great that Kennedy and McNamara decided against continuing with an all-out countervalue program against Soviet cities. Instead they called for attacks against Soviet military infrastructure, which they termed a "no cities first" strategy. The National Command authority would only authorize the Air Force to attack Soviet populated areas if the Russians attacked American population centers first. The Air Force embraced Kennedy and McNamara's new philosophy because by 1970 they projected the need to attack 10,400 counterforce targets. To be effective, a counterforce strategy required targeting an opponent's military installations with at least two warheads each to insure their efficient destruction. The Air Force, therefore, wanted a force of 3,000 Minuteman missiles, 150 Atlas missiles, 110 Titan missiles, and 900 heavy bombers! Many influential leaders in Congress and the Senate became alarmed at what appeared to be a blatant first strike policy.

The first Minuteman missiles acquired by the Air Force had a range of 5,000 miles and carried megaton range thermonuclear warheads. Because the missiles' solid fuel boosters had no need of heavy pumps, piping, or fuel tanks, their thrust to weight ratio was double that of contemporary liquid fuel missiles. This gave them tremendous acceleration and a 30-minute flight time to their targets. Minuteman reached official operational capability on July 3, 1963, although the Air Force had previously rendered a number of missiles operational during the Cuban Crisis. The Air Force deployed it in Squadrons of 50 missiles divided into five Flights of ten missiles each. Two officers located in a hardened underground Launch Control Center controlled the Flights. The Air Force used underground cables to connect the Control Centers to their missile silos, which were spaced three to seven miles apart. The Air Force connected each Launch Control Center in a Squadron with its other LCCs so that in an emergency, each LCC could launch the Squadron's entire complement of missiles.

An improved Minuteman IB missile armed with a 1.2-megaton warhead reached service in late 1963, at which time the Air Force began the phased retirement of Atlas and Titan. By 1965, the Air Force had emplaced 150 Minuteman 1A and 650 Minuteman 1B missiles in silos at four bases in the American west. Fifty-four Titan II second-generation liquid fuel missiles supplemented Minuteman. Titan II's range with a 9-megaton W53 warhead allowed it to attack the most hardened of Soviet facilities. Its propellants were internally storable, allowing for the missile's safe emplacement in hardened silos.

For the Soviets, the global humiliation suffered during and after the Cuban debacle made them vow to never again fall into such a tenuous position. Leonid Brezhnev deposed Nikita Khrushchev in 1964 and adopted an armaments program aimed at seizing strategic superiority from the United States. The centerpiece of the program was increased ICBM production that eventually peaked at an emplacement rate of 200 super-hardened silos per year. Important new missiles were the R-36 (NATO SS-9, Scarp) and UR-100 (NATO SS-11, Sego). The SS-9 was a liquid fuel ICBM with internally storable propellants. Generally similar to Titan II, the Soviets armed this missile with a single 10 - 20 megaton yield warhead. They emplaced about 270 launchers, of which 18 made up a fractional orbital bombardment (FOB) system. The SS-11 was a solid fuel equivalent to Minuteman, of which the Soviets emplaced 990 launchers. Armed with a megaton range warhead, it was the most numerous of Soviet missile types.

After Kennedy's assassination in 1963, Vice President Lyndon Johnson succeeded him, going on to serve a full term as President in 1964. Johnson was a Social Democrat with a vision of a "Great Society" based on a broad range of social legislation. Instead of being able to concentrate on this cherished dream, he found himself more and more absorbed by a war in Vietnam and a defense budget that were spiraling out of control. With Soviet strategic capabilities on the rise, Johnson and McNamara (who had stayed on as Secretary of Defense) had to alter Kennedy's counterforce strategy. It was becoming apparent that the United States could no longer economically maintain its massive imbalance in strategic weapons. The two leaders, therefore, adopted a strategy of "Assured Destruction" that had been developed for the RAND Corporation by Colonel Glen Kent of the Pentagon's Deputy Directorate of Defense Research and Engineering (DDR&E).

"Assured Destruction" and "Damage Limiting" attacks were accounting procedures that Kent used in his studies of nuclear deterrence. His damage limiting strategy implied a first strike against counterforce targets and implicitly assumed a full-blown nuclear war, which even after a first strike could lead to unacceptable damage to the US. Assured destruction was a war avoidance strategy that assumed unacceptable damage to the Soviet Union should they choose to attack the United States with a first strike. Assured destruction included attacks on both strategic infrastructure and population centers with a smaller force than was previously envisioned for massive retaliation.

Although the Johnson Administration attempted to reduce defense spending, it was able to carry out many of the strategic deployments called for by Kennedy. Between 1965 and 1968, it authorized the emplacement of 500 Minuteman II missiles in new silos. In accordance with Kennedy's earlier insistence, the new missiles were equipped with guidance and firing circuits resistant to the effects of EMP, and solid-state memory that could store the coordinates for eight individual targets. The improved guidance systems allowed Circular Error Probable (CEP), or the chance that one out of two warheads will fall within a circle of stated radius, to drop from 3,000 to 1,200 feet. In addition, the Air Force lined the casings of its warheads with neutron and x-ray absorbent materials to protect them from the radiation flux of nearby nuclear explosions. The first 200 of these missiles brought the Minuteman force to a final authorization of 1,000 missiles, while the remainder replaced aging IA and IB models. The Johnson Administration used the savings brought about by reducing strategic deployments to finance social spending and the Vietnam War.

A program of major importance instituted late in the Johnson Administration was a comprehensive group of studies that examined future US missile basing concepts and missile performance characteristics with a view towards anticipating Soviet strategic developments that might arise during the period 1975 - 1990. Known as the Strat-X, the studies were sponsored by Lloyd Wilson, the Deputy Director of Defense Research and Engineering and carried out by the Research and Development Support Division of the Institute for Defense Analysis. A variety of industrial, governmental, and research agencies furnished the information used to prepare the studies. The studies explored about 125 options that included larger missiles, mobile basing, improved accuracy, multiple independently targetable warheads, and stealthier submarines.

Exhausted by the struggles of his first term and disheartened by a massive decline in popularity over the war in Southeast Asia, President Johnson elected not to run for a second. As he left office in 1968, Soviet ICBM programs were achieving parity with American land-based missile forces. In the 6 years following the Cuban Missile Crisis, the United States had deliberately allowed its commanding strategic lead to lapse.

In achieving their strategic goals, the Soviets had to increase their defense spending to 25 - 30 percent of their gross domestic product. In America, defense spending averaged only about 5 percent of GDP. Russian military doctrine centered on large numbers of basic, dependable weapons whereas the United States based its defense on smaller numbers of more sophisticated weapons. Soviet armament factories were churning out strategic weapons at a time when the United States looked to produce more consumer goods. Americans had opted for butter while the Russians had decided on more guns. In a few years, the butter would smell a bit rancid as Americans stared at a forest of Russian missiles.

CHAPTER 7
DEFEATING THE EVIL EMPIRE: NUCLEAR FORCES AND STRATEGIC DOCTRINE FROM 1969 - 1991

On July 1, 1968, fifty-nine nations signed the "Treaty on the Non-Proliferation of Nuclear Weapons." Three of these countries: the United States, the United Kingdom, and the Soviet Union, were nuclear powers. The treaty signing followed 10 years of negotiations and was inspired at an early stage by the almost fatal confrontation over the Soviet transfer of nuclear tipped Medium Range and Intermediate Range Ballistic Missiles to Cuba. After ratification by Congress, the United States deposited its copy of the *Nuclear Non-Proliferation Treaty* with the United Nations on March 5, 1970. Similar actions by the UK, the USSR, and the non-nuclear signatories brought the treaty into effect.

The *Nuclear Non-Proliferation Treaty*, which is still in effect, calls for acknowledged nuclear weapon signatories to refrain from transferring either nuclear weapons or nuclear weapons technology to non-nuclear weapon states. The non-nuclear weapon signatories are in turn required to refrain from acquiring or developing nuclear weapons. The Treaty grants all signatories the rights to the peaceful use of nuclear power but requires non-nuclear weapon states to allow the periodic inspection of their nuclear facilities by agents of the International Atomic Energy Agency (IAEA) to prevent clandestine weapon development. Individually negotiated "*Safeguards Agreements*" govern these inspections.

A banner year for peace initiatives, 1968 also saw American President Lyndon B. Johnson announce that Soviet Premier Alexei Kosygin had agreed to discuss limits on strategic launchers. This action was consonant with wording in the *Nuclear Non-Proliferation Treaty*, which called for nuclear capable signatories to reduce and liquidate their arsenals through continued dialog. Disarmament talks were not a new idea, in the interwar period leading up to World War II, the major powers attempted to restrict the size and number of warships by means of the Washington Treaty. After Johnson's retirement, the success of disarmament negotiations created a temporary era of détente in which President Richard M. Nixon and General Secretary of the Soviet Union Leonid Brezhnev signed the *Strategic Arms Limitation Treaty* (*SALT I*) and the *Anti-Ballistic Missile (ABM) Treaty*.

Before the signing of the ABM Treaty, an early 1960s perception of a missile gap had driven the need for an anti-ballistic missile network in the United States. This need took on urgency when the Soviets preempted the United States by deploying the "Galosh" ABM system around Moscow. As a response, Richard Nixon proposed a nuclear-tipped, twelve-site ABM network to protect Washington and the Minuteman ICBM force. Known as "Safeguard," the United States expected that the proposed ABM system would also provide a hedge against a limited attack by China, which had exploded its first atomic bomb in 1964. Congress and the scientific community, which were skeptical over the efficacy of such a system, opposed Nixon's proposal. They were concerned that Safeguard would touch off a new round in the strategic arms race.

Melvin "Bom" Laird was Sec Def from 1969 to 1973 under President Richard Nixon. Laird prune conventional budgetary requests in order to proceed with strategic systems such as the B-1 bomber, the Los Angeles SSN, the Trident SSBN, and cruise missiles. To do this, he cut military personnel from 3.5 million in 1969 to 2.3 million January 1973. (Credit: US Government)

James Schlesinger was Sec Def from 1973 to 1975 under Presidents Richard Nixon and Gerald Ford. Schlesinger believed that credible deterrence depended on maintaining equivalent force effectiveness with the Soviet Union, a survivable force to deter coercive or desperation attacks against US population or economic targets, and a fast-response force to deter additional enemy attacks. (Credit: US Government)

Harold Brown was Sec Def from 1979 to 1981 under President Jimmy Carter. Brown considered it essential to maintain the strategic nuclear triad of ICBMs, SLBMs, and strategic bombers as a deterrent. He backed the development of the Peacekeeper missile, the accelerated development of the Trident submarine and carried out the conversion of Poseidon submarines to a fully MIRVed system. (Credit: US Government)

Caspar "Cap" Weinberger was Sec Def from 1981 to 1987 under President Ronald Reagan. Weinberger oversaw a massive rebuilding of US military strength including the B-2 bomber and a 600-ship Navy. This stressed the Soviet centralized economy to the point that it collapsed – ending the Cold War. He, nevertheless, believed that the USSR had not given up its long-term aggressive designs. (Credit: US Government)

After its ratification in 1972, the ABM Treaty became the underpinning for "Mutually Assured Destruction." MAD was the natural extension of the "Assured Destruction" doctrine adopted by Lyndon Johnson and his Secretary of Defense, Robert McNamara, in the mid-1960s. McNamara's detractors added the adjective "Mutual" after the Soviets built up their strategic forces. It was the hope of many that an ABM Treaty would prevent the superpowers from launching nuclear attacks against each other because they would have no defense against a return strike. Treaty provisions limited the opponents to a pair of ABM sites each, later reduced to a single site when the US and USSR ratified a Treaty Protocol in 1976. *SALT I* then froze the number of strategic launchers to prevent a compensating spiral of offensive arms acquisition.

Prior to the signing of the *Strategic Arms Limitation Treaty*, the Soviet Union pulled ahead of the United Stated in strategic missile deployments and, by 1970, enjoyed a 40 percent advantage in land-based ICBMs. In order to sell the idea of a strategic freeze that seemed to perpetuate the Soviet advantage, President Nixon issued National Security Decision Memorandum NSDM-16: *Criteria for Strategic Sufficiency*. Strategic Sufficiency called for high confidence in the survival of a second strike nuclear force that could inflict unacceptable damage on the Soviets after a first strike. NSDM-16 was a combination of finite deterrence and assured destruction but did not call for a return to nuclear ascendance. This return was in any case moot because of the general outlook in Congress and the cost of the Vietnam War, which provided a brake on the Nixon Administration's strategic spending. Nevertheless, Nixon and his first term Secretary of Defense, Melvin Laird, put forward a modernization program for each leg of the nuclear triad. If quantity was out of the question, quality was not.

When the United States signed the *SALT I Treaty* in 1972, it had a force of 1,054 land-based ICBMs and 41 ballistic missile submarines carrying 656 SLBMs. Due to disparities in the makeup of Soviet and American Forces, the United States was allowed to expand its ballistic missile submarine fleet by three additional vessels, increasing its complement of SLBMs to 710. Due to the overwhelming number of heavy bombers possessed by the United States, the Treaty allowed the Soviet Union 1,618 ICBMs and 950 SLBMs. The Treaty barred the relocation of land-based missile launchers (silos) and the upgrading of land-based launchers to support heavier missiles.

President Nixon and his second term Secretary of Defense, Melvin Laird, nevertheless sidestepped the fixed number of missile launchers mandated by *SALT* by deploying multiple warheads in independently targetable reentry vehicles (MIRVs). The Air Force added the first of these delivery systems to the Minuteman force, even as negotiations proceeded. The new technology had the ability to overwhelm ABM defenses, attack an increased number of targets from existing launchers, provide a high assurance of target destruction through redundant targeting, and reduced the number of manned bombers required to carry out the *Single Integrated Operational Plan*.

In implementing MIRV, the Air Force had Minuteman III missiles equipped with an enlarged third stage, a maneuverable trans-stage, and three independently targetable warheads. It later replaced the 170-kiloton warheads supplied for the initial deployment with 335-kiloton warheads in reentry vehicles endowed with superb accuracy for counterforce targeting. The Minuteman III MIRV initiative replaced 500 Minuteman IB missiles in existing silos.

The Navy, in its turn, replaced the Polaris Fleet Ballistic Missile with Poseidon, a MIRVed SLBM that carried ten reentry vehicles, each armed with a 40-kiloton warhead. The relatively low yield of Poseidon's warheads made them unsuitable as counterforce weapons and reduced the provocation offered to the Soviets by the MIRVing of (most of) America's strategic missile force. The United States nevertheless made it clear to the Soviet Union that if they launched a first strike, the Navy would use its 3,040 warheads to target population centers and infrastructure. Unfortunately, Poseidon never achieved its planned operational range and this shortcoming required the Navy's Fleet Ballistic Missile Submarines to patrol within reach of Soviet anti-submarine defenses.

Nixon's defense initiatives also included modernization of the outdated bomber leg of the Strategic Triad, which had seen the cancellation of the B-58 and the XB-70 bombers by the Kennedy Administration in 1962. President Johnson's Secretary of Defense, Robert McNamara, had then proposed an "Advanced Manned Strategic Aircraft" to replace the aging B-52 – because of concerns over structural fatigue and engine life. When the technical difficulties associated with the AMSA program militated against any sort of resolution in a reasonable period, McNamara, proposed an "Interim" Strategic Bomber. The designs for this bomber included a reintroduction of the B-58, a major upgrade of the F-111 serving with TAC and a slightly modified F-111. In the interests of funding and time saving, McNamara selected the latter alternative as the FB-111, based on the concept of "commonality." Then, in a sudden turnabout, he ordered 263 FB-111As and testified to Congress that the aircraft was a "permanent" long-term replacement for the B-52.

The FB-111A flew in 1967, and General Dynamics delivered the first production aircraft to the Air Force in 1968. The FB-111A was equipped to carry both freefall nuclear bombs and short-range attack missiles – the forerunners of air-launched cruise missiles. Despite its nuclear capability, the FB-111A was not a suitable replacement for the B-52. It was a medium range, medium bomber that was severely restricted in the number of conventional weapons it could carry. Thus, in 1972, President Nixon and his first term Secretary of Defense, Melvin Laird, cancelled FB-111A production at 76 aircraft and reinstituted the AMSA bomber as the B-1A.

The "interim bomber" incident was symptomatic of changes that affected strategic procurement in the 1960s. During the 1950s, virtually unlimited capital was available for the acquisition of strategic weapons. This outlay was in large part the result of nuclear weapons' novelty and the awe in which the public held them. By the early 1960s, the House of Representatives and Congress had begun to balk at the staggering sums requested by DOD and, in the period 1962 - 1966, it cut the amount of capital for strategic spending from 23 billion to 9 billion in adjusted 1981 dollars. McNamara was simply trying to make ends meet with his FB-111A.

Another fundamental change to the nuclear procurement process came with an improved understanding of these weapons and their strategic milieu. In the 1950s, the government officials associated with the control of these weapons were relatively ignorant about nuclear technology and willing accepted requests made by the DOD and DOE. By the mid-1960s, the Secretary of Defense, Congressmen, and Senators had become much more sophisticated in their understanding of strategic technology. They were now willing to propose their own programs and to subject military proposals to intense scrutiny. Unfortunately, the proposals put forward by the military, the Secretary of Defense, Congressman, and Senators were often at odds, generating a great deal of acrimony during the selection and funding processes.

In planning for future strategic requirements, the Nixon Administration espoused a policy of "Essential Equivalence" as outlined by James Schlesinger, the second term Secretary of Defense. Through this doctrine, Schlesinger argued that the United States and Russia did not need identical strategic forces, as long as the makeup of their forces provided for equivalent capabilities and that both sides saw them as such. Weapon systems proposed by Melvin Laird and later promoted by Schlesinger included a second-generation cruise missile. They also included the MX (Peacekeeper) ICBM and the Undersea Long-range Missile (Trident) System.

As Nixon reviewed employment plans for the strategic forces under his control, he became concerned that his only nuclear option was an all-out attack on the Soviet Union and its allies. He, therefore, issued NSDM-242: *Policy for Planning the Employment of Nuclear Weapons* as a guide to the creation of a more flexible *SIOP* in 1974. As a means of escalation control, NSDM-242 called for a selection of nuclear options to terminate conflicts with the lowest possible level of force. This resulted in *SIOP 5A*, which included a number of Regional Nuclear Options (RNOs) for areas such as the Korean Peninsula. It also provided Major Attack Options (MAOs), Selected Attack Options (SAOs), Limited Attack Options (LAOs), and withholds to deal with China, the Soviet Union, and their allies on a joint or individual basis.

In conjunction with this new strategic policy, Defense Secretary Schlesinger issued *NUWEP-1*, a document that provided collateral damage guidelines relating to *SIOP* inflicted fatalities and damage. *NUWEP* added a layer of civilian oversight to what was essentially a military undertaking. The combined result of Nixon and Schlesinger's policies was a return to Kennedy's doctrine of flexible response, with a reduced emphasis on conventional forces. Nixon downsized conventional forces to fight a 1 1/2-war scenario because he believed that a clash in Europe or Asia would rapidly escalate into a full nuclear exchange.

In an effort to obtain DOD funding in support of Nixon's defense policies, Schlesinger frequently clashed with Congress. The Secretary of Defense was particularly concerned that *SALT II* negotiations might result in the relegation of the United States to an inferior strategic position with respect to the Soviet Union. This opinion brought him into conflict with Secretary of State Henry Kissinger during the Ford Administration. Both Kissinger and President Gerald Ford were strong supporters of the *SALT* process and were concerned that strategic modernization might upset negotiations. They were also willing to compromise with Congress on defense issues. Schlesinger's intransigence in the face of Ford's opposition to defense spending ultimately resulted in his dismissal.

Following the Carter Administration, General Secretary Brezhnev and President Jimmy Carter signed the *SALT II Treaty* in 1979 but because of the 1980 Soviet incursion into Afghanistan, the United States Congress never ratified it. Nevertheless, both sides complied (more or less) with its terms. These included a ceiling of 2,400 strategic nuclear delivery vehicles, including ICBMs, SLBMs, and heavy bombers. Neither side was to MIRV more than 1,320 of these systems. There was also a ban on the construction of new ICBM launchers and constraints on the deployment of various types of offensive arms. Negotiating victories by the US required the USSR to dismantle its fractional orbital bombardment system and delayed a ban on cruise missiles until at least 1981.

As a basic strategic philosophy, President Carter inherited a "launch on warning" doctrine formulated during the Nixon Era. The Nixon Administration thought this doctrine was necessary because land-based missiles were becoming increasingly vulnerable to counterforce targeting. Launch on warning was a "use it or lose it" approach to nuclear war that assumed that the Soviets could wipe out the US land-based ICBM force in a first strike. The launch on warning approach to retaliation left many politicians queasy about a false alarm starting WW III. The Carter Administration, therefore, modified this doctrine was to "launch under attack," in which a retaliatory strike would not be authorized until a nuclear detonation over the United States had been confirmed by IONDS and eyewitness reports. Because the launch under attack policy required the opening of missile silos in a trans-attack environment that compromised their integrity, many considered this concept flawed.

Rather than launch under attack, the National Command Authorities considered riding out and evaluating a first strike before formulating a "measured response." This approach was possible because of the security afforded by America's fleet of ballistic missile submarines. Hidden in the ocean depths, analysts expected most of these launch platforms to survive an initial attack by the Soviet Union or China. The United States would then have the opportunity to evaluate and respond with SLBMs and its residual ICBM force in the post-attack environment. The doctrine of measured response appealed to both President Carter and his Secretary of Defense, Harold Brown, who were in any case planning to overhaul the Department of Defense and update the strategic doctrines pertaining to the use of nuclear weapons after they assumed office.

As the new Secretary of Defense, Brown brought an impressive set of technical credentials to the table. He had previously held positions as Director of the Lawrence Livermore National Laboratory and as Director of Defense Research and Engineering for the DOD. By 1979, he had developed an approach to nuclear defense that he codified as a "countervailing strategy." By this, he meant the use of nuclear forces to deny success to an enemy's plan for an attack on the United States or American interests – no matter the scale. Although Brown did not rule out assured destruction, he indicated that this would not be an automatic response. As a strategic doctrine, he sought a broadened base of proportional and flexible responses with which to seek "war termination under the best possible conditions." The countervailing strategy was essentially a policy of scaled, selective, counterforce.

Carter made Brown's strategic position official in 1980, with the release of Presidential Directive PD-59: *"Nuclear Weapons Employment Policy,"* at which time he rescinded NSDM-242. PD-59 especially called for upgrades to command, control, and communications. Brown intended the upgrades to provide the flexibility, enduring survivability, and performance necessary to survive an enemy attack. In conjunction with PD-59, Brown released *NUWEP-2*, which called for a switch from urban targets to command installations, conventional and nuclear military bases, and supporting industrial infrastructure. The idea was to spare the Soviet population from direct attack but hold the Soviet political and military leadership at personal risk for an attack on the United States.

Under previous strategic doctrines, the command, control, and communications infrastructure had to remain operable only long enough to support a retaliatory strike, after which time the Joint Chiefs considered it expendable. The new countervailing strategy required C^3 to stay fully operational through an extended trans-attack period that might consist of a series of nuclear exchanges and well into post-attack recovery. Key vulnerabilities to the system in the early 1980s were a general susceptibility to the effects of EMP and a lack of redundancy. There were only two national level command centers backed up by four SAC operated Boeing 747 E-4A aircraft that comprised the National Airborne Command Center. Other key installations included 15 hardened bunkers for nuclear force commanders, eight VLF ground-based transmitters for submarine communication, and 27 KC-135 SAC "Looking Glass" flying command posts. These represented a rather small but select target for Soviet strategic planners.

Carter and Brown not only had American communication assets hardened against the effects of EMP but had the E-4A fleet upgraded to the E-4B Nightwatch configuration. The latter aircraft modifications provided accommodation for the National Command Authorities and a staff of 114 personnel, including a general officer, with which to direct US nuclear forces. Originally flown from Andrews AFB, Maryland, the Air Force relocated Nightwatch aircraft to Offutt AFB, Nebraska, to keep them out of reach of Soviet SLBMs. One aircraft, however, remained on alert for the President and the Secretary of Defense at Andrews AFB. The Navy followed the Air Force lead by replacing its aging VLF transmitters with new ELF facilities in Wisconsin and Michigan. ELF transmissions send a simple one-way message telling submarines to come to shallow depth, at which time higher frequencies from TACAMO aircraft transmit orders. In conjunction with improvements to submarine communications, the Navy upgraded TACAMO aircraft to the E-6B configuration based on a Boeing 707 airframe with 15 hours of un-refueled endurance.

While the Carter Administration's stance on command, control, and communications was timely, its stewardship of nuclear forces was less impressive. One troubled area was Air Defense Command, which the Air Force demobilized in 1980. Over the years, the composition of the Soviet Air Force had remained heavily weighted toward medium bombers for use against China and Western Europe. As a result, the air defense role fell to Air National Guard Units in the face of a downgraded threat. The introduction of Sidewinder and Sparrow air-to-air missiles then resulted in a gradual withdrawal of nuclear tipped missiles, leaving ADC armed with conventional weapons. The Air Force, therefore, delegated ADC's role to Tactical Air Command and divided its assets between SAC and TAC. The demobilization of BOMARC and a decline in Nike missile sites paralleled the demise of ADC. By 1975, only four semi-mobile Nike training units remained active with European Command.

The delayed ban on cruise missiles was significant because new and improved versions of this weapon were under development in the United States. Although ballistic missiles were extremely capable weapons, they were also expensive. In order to manage budgets, the various branches of the Armed Forces had begun to reevaluate more cost-effective cruise missiles in the late 1960s. The sinking of an Israeli destroyer by a Soviet built Styx anti-ship cruise missile in 1967 motivated the Navy to look at a number of possibilities that included long-range cruise missiles launched from Polaris missile tubes and short-range versions launched from torpedo tubes. The Navy chose the latter alternative and proceeded to develop the Tomahawk sea-launched cruise missile. The Air Force in its turn considered the alteration of airborne decoys into air-launched cruise missiles.

The success of the cruise missile projects caused President Carter to approve both Naval and Air Force versions of this weapon. The deployment of Tomahawk on attack submarines, destroyers, frigates, cruisers, and battleships greatly increased the number of available naval nuclear capable platforms and caused great consternation among the Soviets. The American cruise missile strategy effectively countered their "First Salvo" policy of taking out a limited number of fleet ballistic missile submarines and nuclear capable carriers. Tomahawk was also adapted as the Gryphon ground launched cruise missile and deployed to Europe with TAC in 1983. Targeted at fixed sites, Gryphon freed tactical aircraft to hunt Soviet mobile missile launchers. TAC armed the GLCMs with variable yield thermonuclear warheads that allowed tailored attack options.

At the cost of its cherished B1-A penetration bomber, Carter ordered the Air Force to equip its aging B-52 fleet with air-launched cruise missiles. This was both a strategic and an economy measure that extended the life of these aircraft. On airborne alert and safe from a first strike, air-launched cruise missiles (ALCMs) gave B-52s an ability to attack targets deep inside the Soviet Union without having to confront its formidable air defenses. Equipped with under wing pylons and a rotary launcher in its bomb bay, the B-52 could carry an aggregate of 20 ALCMs. A key weakness of second-generation cruise missiles was their subsonic air speed, which made them vulnerable to attack by faster and more maneuverable fighters. The Air Force, therefore, procured an advanced cruise missile (ACM) with stealth characteristics to improve its penetration abilities. Like the navy with Tomahawk and the Army with Gryphon, the Air Force armed air-launched cruise missile with variable yield thermonuclear warheads.

Although Carter cancelled the B-1A, the possibilities of stealth so intrigued him that he had the Air Force issue contracts for an Advanced Technology Bomber in 1981. The delay associated with the B-2 left SAC with an aging fleet of B-52s and a handful of FB-111s. Whether Carter's intent was genuine or a ploy to reduce defense spending has been a matter of some debate. The same reasoning that had previously led to the demand to drop the B-36 in favor of the B-47 may have influenced Carter and Brown's thoughts. There are always new technologies around the corner and one needs a good deal of insight about when to decide on implementation. In the face of the decision to cancel the B-1A, the Soviets took the opportunity to accelerate development of their Tu-160 Backfire bomber, which was in many respects its equivalent. In so doing, they took advantage of America's weakened air defenses and stole a march on the B-52.

Another hi-tech weapon championed by Carter and Brown was the MX (Peacekeeper) ICBM. The MX had been in development since the Nixon Administration, which had called for a mobile successor to Minuteman. The intent of MX was to offset future Soviet capabilities, including an opposing mobile ICBM force. By this time, a pervasive view equated mobile basing with a survivable land-based ICBM force. Under *US Air Force Directive 22*, MX was to employ either an air or a ground-based mobile system. To satisfy these requirements, the Air Force evaluated a variety of concepts during Nixon's presidency. In 1979, President Carter decided to deploy 200 MX missiles on mobile rail launchers, which would shuttle between 4,600 protective shelters. This arrangement would have forced the Russians to use a large number of their strategic warheads to insure the destruction of the MX force. The price tag for this plan was so large and the amount of land required so vast that Congress rejected it. Although the Carter Administration authorized full-scale engineering development of MX, it never achieved a satisfactory basing plan.

Carter had much better success with the Trident I SLBM. Deployed in 1979 this missile proved a tremendous success by greatly exceeding Poseidon's range limitations. It was actually capable of reaching targets inside the Soviet Union when launched from homeports. The Navy quickly approved follow-on backfits to all twelve Ben Franklin Class subs, enabling them to carry the new missile. The Navy then moved the patrol areas for Trident I missile submarines further back from the margins of the Soviet Union to reduce the risk of their detection. This allowed the Navy to discontinue the foreign basing of its SSBNs and to begin servicing them entirely from American ports. Each Trident missile carried six reentry vehicles armed with 100-kiloton warheads. The accuracy of these reentry vehicles was sufficient to provide a limited counterforce capability.

The Carter Administration also accelerated the development of Ohio Class Fleet Ballistic Missile Submarines, allowing the first such vessel to reach operational deployment in 1981. The new submarines had 24 vertical launch tubes and displaced twice the tonnage of earlier SSBNs. This increased firepower allowed two Ohio submarines to replace three submarines from earlier classes and reduced operating costs. With improved speed and low noise levels, the Navy designed its Ohio submarines for exceptional survivability. In addition, their spacious facilities allowed them to undertake extended patrols, the first of which commenced on October 1, 1982.

A significant political stand late in the Carter administration was a show of force in the face of European developments. Throughout the period of détente, the Warsaw Pact (Soviet equivalent to NATO) continued to build its conventional forces. Events took an ominous turn in 1977, when the USSR supplemented these forces with large numbers of road mobile SS-20 MRBMs, each armed with three independently targetable 250-kiloton warheads. These missiles had the capacity to wipe out NATO's capital cities in a matter of 5 minutes. To deal with this situation, a carrot and a stick approach attempted to reduce nuclear deployments and tensions in the area by means of diplomacy. In the event that the diplomatic carrot failed, Carter intended the introduction of advanced Pershing II MRBMs and Gryphon Cruise Missiles to entice the Soviets to rethink their position.

Although Carter built an effective blueprint for a new and invigorated strategic force, his termination of Air Defense Command, the launch of only one ballistic missile submarine, and the cancellation of highly visible programs like the B-1A bomber had a negative impact on his image. Nightly news about the Soviet intervention in Afghanistan and the Iran Hostage Crisis placed the public in a militant mood and they voted him out of office. In his stead, the American public elected President Ronald Reagan, in no small part because of his promise to upgrade America's military, which he declared to be a "hollow force." Upon taking office, Reagan officially repudiated détente, which he viewed as a one-way street for Soviet militarism and expansionism.

American Administrations were well acquainted with the makeup of Soviet forces through "*National Intelligence Estimates*," produced under the supervision of the Director of National Intelligence. Through these, Reagan was aware that the Soviets had deployed new and improved ICBMs, including the notorious SS-18 Satan, which could carry either a single 25-megaton warhead or ten MIRVed warheads with megaton range yields. The MIRVing of the Soviet missile force had increased that nation's warhead count from 1,600 to nearly 6,000. The Soviets had also improved accuracy to the point that their warheads threatened the Minuteman ICBM force. Some of the new missile designs used a cold launch technique that allowed for rapid reloads, further threatening American military facilities. Soviet naval and tactical forces had seen equivalent improvements. The Russians doubled the size of their submarine fleet and MIRVed its SLBM complement. In Europe and Asia, the Russians deployed conventional forces and theater weapons out of all proportion to available targets.

The Soviet military presence in Eastern Europe served a two-fold purpose. It provided a basis for defensive and offensive operations against NATO while serving to keep the Eastern Bloc nations under control. The Russians made their intentions to maintain control over Eastern Europe abundantly clear when they crushed the Czech government of Alexander Dubcek after it began to experiment with social and economic reforms in 1968. In order to justify this repressive action, the Soviet Union announced the *Brezhnev Doctrine*, which gave it the right to interfere in the governance of socialist states to insure their adherence to orthodox communist principles. Unlike American Forces that were invited guests in NATO nations, Soviet forces were a de facto army of occupation. The USSR further extended its influence to the third world by providing client nations with military advisors and massive shipments of conventional arms. The Soviets armed terrorists and guerilla movements to foment unrest where they could not obtain an open invitation for entry.

The policies through which Reagan intended to combat expanding Soviet influence were set out in National Security Decision Directive NSDD-32 *US National Security Strategy* and amplified in NSDD-75 *US Relations with the USSR*. Their strident tone was a hearkening to the earlier trumpet call of NSC-68. Reagan vowed to contain and reverse Communist expansion and to compete with the Soviet Union in military balance and regional influence. In order to carry out his plans, he called for a two-trillion-dollar injection into military spending. This program included a Strategic Defense Initiative popularized by its opponents as "Star Wars." SDI envisioned the creation of a defense umbrella using "third generation" nuclear weapons to shield America from a missile attack. At the end of the day, President Reagan planned to "leave Marxism-Leninism on the ash-heap of history" by surpassing it in strategic capability or by spending it to death if it tried to match American capital outlays.

Caspar Weinberger, Reagan's Defense Secretary, ably assisted him in his endeavor. Although Weinberger did not have the expertise possessed by his predecessor Brown, he was an able administrator who had previously worked with Reagan in California. At that time, he had acquired the nickname, "Cap the Knife", for his ability to cut costs. He now applied his talents to extracting funds from Congress with which to fuel Regan's defense programs. During his tenure, he increased the defense budget from 175.5 billion dollars in FY 1981 to 287.8 billion dollars in FY 1988. The new Secretary quickly developed a close working relationship with the military staff at the Pentagon.

One of Reagan's campaign promises was to place the B-1 bomber back in development. It was to be an interim, low-level, penetration bomber prior to the deployment of the B-2A stealth bomber. This necessitated a number of design changes to the original B-1A design concept. Although the redesigned bomber could still fly at supersonic speed, the Air Force deleted a Mach 2 dash capability and ordered its airframe strengthened for low-level flight. The Air Force also ordered its bomb bay modified to accept rotary cruise missile launchers and its skin coated with radar absorbing materials to reduce its radar signature to 1/100 that of a B-52. The first of 100 B-1B Lancer bombers reached operational readiness in 1985.

The subsequent development of the B-2A Spirit bomber ended in an order for 100 aircraft at a billion dollars apiece. A novel design, the B-2A is a flying wing that uses advanced composites and titanium to reduce its weight and give it stealth characteristics; although the primary reason for its low radar signature is a computer designed shape that deflects and minimized its radar returns. Northrop also gave the B-2A special engine exhaust ducts that reduce its infrared signature. Because it can penetrate heavily defended air space, the Air Force did not have Northrop equip the B-2A for cruise missile delivery. Its primary nuclear armament consists of B83 freefall and B61-11 earth penetrating bombs.

Although President Reagan quickly sorted out the Air Force's bomber problem, his biggest difficulty lay in establishing a basing plan for the Peacekeeper ICBM. His first proposal, Dense Pack, called for the Air Force to place 100 missiles in super hardened silos spaced 1,800 feet apart. Any missile attack would lead to "fratricide" as incoming warheads blew themselves up in attempting to destroy the concentration of silos. This guaranteed a surviving force sufficient for retaliatory purposes. The audacity of this plan led to its rejection by Congress. Determined to solve the Peacekeeper basing dilemma and to develop a national missile policy, Reagan turned to retired Air Force General Brent Scowcroft to chair a bi-partisan committee to sort out the problem.

The Scowcroft Commission delivered its *Report on Strategic Forces* in 1983. As a preamble, the report recognized that the accuracy and quantity of Soviet ICBMs constituted a very real threat to the United States. The Soviet Union had the capability of striking the entire American land-based missile force twice over, with plenty of missiles held in reserve for use against other targets. The 2,000 American Minuteman warheads had no such capability and faced Soviet silos hardened to very high standards. The same was generally true for other Soviet strategic targets. The report, therefore, recommended the immediate deployment of 50 Peacekeeper missiles in refurbished Minuteman silos in order to increase the number of American ICBM warheads. Since each Peacekeeper was equipped with ten warheads, the 50-missile deployment provided a 25 percent increase in operational land-based warheads. In agreement with the Scowcroft Commission, Congress approved the Commission's recommendations, the first Peacekeeper missile achieving operational capability at F. E. Warren AFB on December 22, 1986.

The Air Force planned to follow Peacekeeper's silo deployments with "Rail Garrison," a basing plan in which it would base 50 more missiles on railroad cars stored in protective shelters. In times of international tension, the missiles could disperse onto the American rail network, after which they would be impossible to target. To backup Peacekeeper, the Air Force placed a small road-mobile ICBM referred to as "Midgetman" in development. The Air Force intended to arm it with a single warhead and hide it in the mountains of the Western United States, where it too would be impossible to target.

In a continuation of the Carter policy that called for the hardening of C^3, Reagan approved the construction of a new underground headquarters for Strategic Air Command. The headquarters was located at Offutt AFB, Oklahoma, and connected to SAC's original Command Post. The addition featured 16,000 square feet of floor space in a two-story underground structure that was hardened to the effects of EMP. Some authors indicate that SAC built the new headquarters as a response to the portrayal of its facilities in thriller movies – the war rooms of Hollywood's fictional headquarters were superior to the reality!

In order to test the readiness of his reconstituted command structure, Reagan approved Operation Ivy League in 1982. It was the largest war game held in 30 years and focused not so much on readiness as survivability. At the end of 5 days, although much of the United States lay theoretically in ruins, an impaired American command structure was still operating with sufficient ability to continue a fight. This validated Reagan's investments in Strategic Defense and confirmed the strategy of measured response.

Since measured response was to a large degree dependent on America's ballistic missile submarines, Reagan continued building the SSBN fleet of multi-billion-dollar Ohio class SSBNs at the rate of one per year. These were equipped with super-sized launch tubes in anticipation of the arrival of the ULMS. Even before the deployment of the Ohio submarine, the Navy realized it could achieve a hard target kill capability by building an SLBM that carried an ICBM sized warhead. The result of this insight was the 65-ton Trident II D5 missile. Twice the weight of Trident I, the new missile could carry between eight and fourteen W88 warheads, each with a 475-kiloton yield. The warhead's RVs have an astounding CEP of 300 feet! First deployed in

1990, the accuracy of second-generation Trident missiles is on par with the most accurate and most powerful land-based ICBM.

Based on his call to invested trillions of dollars in defense, it would be easy to characterize Ronald Reagan as a hawk. This was not strictly true. His primary aim was to reduce world tensions through the elimination of nuclear stockpiles. In order to manage this task, he took the same carrot and stick approach with the Russians that Jimmy Carter had adopted for his negotiations on theater forces in Europe. Reagan not only continued theater negotiations where Carter left off, but also-called for more comprehensive talks aimed at reducing strategic arms.

Reagan's contradictory actions and rhetoric confused the Russians. On the one hand, he was calling for disarmament, whereas on the other hand, he was rebuilding America's strategic forces. The timing of Reagan's announcements could not have come at a worse time for the Russians, who were deeply embroiled in an ill-conceived military adventure to Afghanistan. This left them with little in the way of resources to prop up the failing economies of the Warsaw Pact countries and to meet a new round of expenditures in the arms race. Essentially, they were in the same position as Johnson and Nixon during the Viet Nam War era. Although the Russians did not want to give up a position of military superiority that had taken 20 years to achieve, there simply was not enough money to go around. It was a bitter pill to swallow.

The stress of this situation led to severe paranoia in Russian military circles. They interpreted a NATO exercise (Able Archer) scheduled for late 1983 as the prelude to an all-out attack by the west and subsequently geared up for this eventuality. Fortunately, a double agent was able to warn of the Russian fears and NATO scaled back the exercise, averting a crisis. While Russian military leaders took Reagan's 1983 missile defense initiative seriously, they were sufficiently astute to recognize that no technology is perfect. In the words of the pre-WW II Italian strategist, Giulio Douhet, "the bomber will always get through." However, they were acutely aware that if Star Wars proved even partially successful, the preponderance of missiles to "get through" would most likely be American.

Taking advantage of international opinion that condemned the Russian invasion of Afghanistan, the Reagan Administration pressed hard to attain its goal of arms reduction. On November 18, 1981, it placed a *Zero-Option* proposal on the table with regard to intermediate-range nuclear weapons. The United States would cancel its planned INF deployments if the Soviet Union agreed to eliminate its R-12, R-14 (NATO SS-5, Skean) and RSD-10 mobile missiles. The Soviets immediately rejected this proposal and so, on November 30, at the commencement of formal negotiations in Geneva, Switzerland, the United States tabled a new proposal that called for the complete elimination of long-range intermediate nuclear forces and mutual constraints on short-range intermediate nuclear forces. The Soviet Union made a counter proposal that provided for a ceiling of 300 medium range missiles and tactical aircraft (including British and French forces) for each side. The United States in turn rejected this proposal with the insistence that the agreement exclude third parties.

In 1982, while INF negotiations were still under way, President Reagan engaged the Soviet Union in Strategic Arms Reduction Talks, also held in Geneva. The original American proposal limited each side to 5,000 warheads, with no more than half of the warheads mounted on ICBMs. The proposal also called for a limit of 850 deployed strategic ballistic missiles, including a sub-limit of 110 heavy missiles (Titan II or SS-18). A second phase of talks tackled constraints on heavy bombers and other strategic systems. At this time, the United States removed China from the *SIOP*, reflecting that country's potential as an ally in any future clash with the Soviet

Union. The JSTPS reduced Chinese nuclear targeting to a few contingency options that involved strategic reserve forces and conventional weapons.

In the interest of expediency, Reagan announced in March 1983 that the United States and NATO were willing to sign an interim INF agreement to establish equal numbers of warheads on Soviet and American missiles. Although a zero option was still desirable, a warhead count in the range of 50 to 450 weapons would be acceptable. The Soviets rejected this overture in April, in part because Reagan announced an American Strategic Defense Initiative. In response to the Soviet rejection, the US and NATO began to deploy Pershing II IRBMs and Gryphon GLCMs in Europe. Because of this action, the Soviet delegation pulled out of formal nuclear negotiations in Geneva.

After taking a breather, the United States and the Soviet Union announced in November 1984 that they would enter into new negotiations, known as the Nuclear and Space Talks. The Russians planned to use these talks to kill Star Wars, whereas Reagan planned to use the talks as a lever to reach an agreement on arms reduction. In the interim, the domestic situation in Russia had deteriorated due to the Afghan war. The United States had shipped "Stinger" hand held surface-to-air missiles and Milan anti-tank missiles to Afghani guerillas via Pakistan. These took an appalling toll of Soviet armor, ground support aircraft, and helicopters. In response to the carnage in Afghanistan, a reform movement began to take shape.

In early 1985, US Secretary of State, George Schulz, and Soviet Foreign Minister, Andrei Gromyko, agreed to include discussions on INF as part of the NST talks, getting negotiations back on track. Behind the scenes, the Soviet people elected Mikhail Gorbachev as the General Secretary of the Communist Party on a reform ticket. He soon had his hands full with dissent over the Afghan war, unrest in Eastern Europe, and a stalled economy. In order to fix his economic problems Gorbachev instituted limited economic reforms known as Perestroika or "restructuring." He also instituted Glasnost or "openness" that removed the suffocating restrictions of the Brezhnev era. Gorbachev intended Glasnost to mend fences with the Chinese and with Soviet client states.

Despite Glasnost, Gorbachev still tried to maintain a strong Soviet military position through continued defense spending. In 1986 at the Reykjavik Summit, he insisted on linking negotiations with an end to the American experiment with strategic defense. Although the United States reiterated its position on Star Wars, Gorbachev and Regan were able to agree on limits to 1,600 Strategic Nuclear Delivery Vehicles (SNDVs) and 6,000 ICBM, SLBM and ALCM warheads. They also had discussions regarding the reduction of intermediate nuclear missile systems.

The collapse of international oil prices in 1986 cut off the main source of hard currency supporting the Soviet economy. Officially, this was a result of market forces and the inability of Saudi Arabia to maintain OPEC discipline. Conspiracy theorists believe that this event was actually the result of a pact between the Saudi Royal Family and the Reagan Administration. The United States apparently guaranteed the security of Saudi interests in the face of Soviet expansionism, while cheap oil was used to kick-start a stalled American economy.

A side effect of this conspiracy ended funding to the Soviet military machine when the supply of hard currency from foreign oil sales dried up. With his economy in tatters, Gorbachev finally de-linked INF negotiations from Star Wars and announced that the USSR was prepared to eliminate all of its INF missile forces. Because of this pronouncement, Secretary of State Shultz declared that the United States and its allies would stop deployment of GLCMs as soon as the

Soviets signed a mutually agreeable treaty. On December 8, 1987, Ronald Reagan and Mikhail Gorbachev signed the *Treaty on the Elimination of Intermediate Range and Sorter-Range Missiles* and the US Senate ratified the Treaty on May 27, 1988. In addition to the elimination of intermediate missile forces within a specified period, it also called for onsite inspections to verify the removal and disposal of intermediate nuclear weapons.

The signing of the INF Treaty let Gorbachev withdraw both conventional and nuclear forces from the nations of the Warsaw Pact. He also began a phased withdrawal of Soviet forces from Afghanistan. These actions allowed him to increase domestic spending with which to stimulate the Russian economy. However, Glasnost had in the meantime led to unforeseen repercussions with client states. These included the strengthening of the Solidarity Movement in Poland and the fall of the Berlin wall in late 1989. A year later on October 3, 1990, East and West Germany reunified after 45 years of separation. All of this activity provided a backdrop to the continuing negotiations for the reduction of strategic weapons.

In the United States, improved relations with the Russians resulted in the removal of Titan II from alert and the suspension of continuous Looking Glass Operations for the first time in 29 years. The ratification of the *Conventional Forces in Europe Treaty* in 1990 further reduced military buildups and political tension. The CFE Treaty was a complex document that provided for equal deployments of major weapon systems for both East and West in the area from the Atlantic Ocean to the Ural Mountains. The new military balance imposed limits for 20,000 tanks, 20,000 artillery pieces, 30,000 AFVs 6,800 combat aircraft, and 2,000 attack helicopters to each side. Within a year of its implementation, on July 1, 1991, the Warsaw Pact officially disbanded and the Soviet Union withdrew its forces from client states. The countries of Eastern Europe were free to choose their own political futures.

George H. W. Bush and Mikhail Gorbachev signed the *Treaty on the Reduction and Limitation of Strategic Arms"* on July 31, 1991. It called for the two opponents to reduce the number of their strategic launchers to 1,600 with no more than 6,000 "accountable" warheads over a period of 7 years. Of the allowed warheads, each side could not place more than 4,900 on ballistic missiles. The Soviet Union also agreed to eliminate half of its heavy SS-18 missiles in a gradual process and to reduce its overall ballistic missile throw weigh to 3,600 metric tons. The cuts ranged from 25 to 35 percent of existing warheads. Like the INF Treaty, stringent provisions provided for onsite inspection and verification.

As a side benefit of improved relations and the reduction of strategic and conventional forces, *Presidential Nuclear Initiatives* (*PNI*s) by both sides in 1991 provided for the complete elimination of tactical nuclear deployments, apart from modest stockpiles of freefall bombs. President Bush required that:

- The United States Armed Forces eliminate its inventory of ground-launched, intermediate range, theater nuclear weapons
- The United States Navy remove all of its tactical nuclear weapons from surface ships, attack submarines, and land based naval aircraft bases. That it then eliminates these weapons except for SLBMs, which it will place in storage
- The United States Air Force stand down its strategic bombers from their alert postures, remover their nuclear weapons and store them in secure areas
- The United States Air Force de-alert its intercontinental ballistic missiles scheduled for deactivation under the terms of the *Strategic Arms Reduction Treaty*

- The United States Air Force terminate the development of the mobile Peacekeeper ICBM rail garrison system and terminate the mobile portions of the small ICBM Program
- The United States Air Force terminate the nuclear Short-Range Attack Missile Program (SRAM II and SRAM T)
- The United States Strategic Command submit a new Unified Command Plan

On December 26, 1991, the Soviet Union dissolved itself, allowing former Soviet Republics to choose their own destinies. After almost 45 years, the Cold War was over.

CHAPTER 8
PEACE AND WAR 1992 - 2016: AMERICAN ARMS REDUCTION AND REARMAMENT

Following the collapse of the Soviet Union, the DOD consolidated America's strategic nuclear forces under the aegis of United States Strategic Command (USSTRATCOM) on June 1, 1992. The transition to the post-Cold War era created a period in which American political and military leaders developed an opinion that the nation's strategic nuclear forces could (and should) be reduced. In support of this opinion, General George Butler, the last CINCSAC, and the first CINCSTRATCOM, proposed a reorganization of the Armed Forces Unified Command Structure in November 1990. General Butler envisioned a reduction in the number of Unified Commands from ten to six, one of which was to be a Joint Strategic Command assigned direct responsibility for the nuclear Triad and the anti-satellite mission. In support of his view, General Butler initiated a Strategic Force Structure Review, most often referred to as the *Phoenix Study*.

Strategic Air Command completed the *Phoenix Study* in September 1991, a few weeks after the signing of the *Strategic Arms Reduction Treaty* (START), and a few weeks before President Bush announced sweeping changes to US nuclear forces. The *Phoenix Study* then became the basis for several follow-on studies as STRATCOM struggled to define itself and its mission during the 1990s. In determining the size and composition of US strategic forces, the study examined who should be targeted, what targets should be held at risk, how many aim points or "desired ground zeroes" were required, what quality and how many weapons and weapon reserves were needed, and how to "hedge" against an uncertain future. At the time of the *Phoenix Study*, the number of DGZs in *SIOP* was in decline – from a level of 16,000 targets maintained in the mid-1980s down to 7,000.

Since the Soviet Union was the only nation with nuclear forces capable of threatening the United States, it was designated the "adversary" for the *Phoenix Study*. SAC back calculated the number of Soviet installations it could hold at risk from a formula that defined the number of weapons available for a "basic attack force" less the number of weapons required for a "reserve force." The reserve force was for future contingencies and limited re-strikes in case part of the basic attack force failed in its mission. The reserve force included weapons assigned to a "hedge force" for possible immediate use.

As part of the *Phoenix Study*, SAC assumed that not all of the warheads used in a nuclear engagement would reach their targets; some would fail because of technical malfunctions, some would be subject to prelaunch losses, and some would be lost to attrition by local defenses. In the case of aircraft, some would be lost due to adverse weather conditions. Considering these losses, the *Phoenix Study* established "rules of thumb" for calculating the number of weapons required to destroy a given number of installations. To this end, SAC calculated separate probabilities of arrival (PA) for warheads delivered by bombers, ICBMs, and SLBMs. PA was a function of pre-launch survival (PLS), weapons system reliability (WSR), and the probability to penetrate (PTP) enemy defenses.

Air Force General George Lee Butler, the last Commander of Sac and the first Commander of USSTRATCOM. General Butler oversaw the consolidation of the three legs of the nuclear triad into a single command. (Credit: USAF)

Admiral James Ellis, who assumed command of a merged USSPACECOM and USSTRATCOM on October 1, 2002. The merged command retained the US Strategic Command title and remained headquartered at Offutt AFB, Nebraska. (Credit USN)

On January 10, 2003, President George Bush signed Change Two of the Unified Command Plan that tasked USSTRATCOM with four previously unassigned responsibilities: global strike, missile defense integration, DOD Information Operations, and C^{4}ISR. (Credit: US Government)

USSTRATCOM's new, 1.2-billion-dollar headquarters under construction at Offutt AFB, Nebraska. The old headquarters can be seen in the background. (Credit: USSTRATCOM; Color Original)

Since the US defense posture at this time was launch on warning, and submarines have a very low detection threshold, the PLS for ICBMs and SLBMs was defined as very high, whereas for bombers it was set lower. The WSR for ICBMs and SLBMs was set quite high, based on their operational histories, whereas that of bombers was slightly lower. PTP for ICBMs and SLBMs was very high, but the PTP for bombers was much lower because of the Soviet Union's extensive antiaircraft defenses. Although SAC has never declassified the actual figures for PLS, WSR and PTP, it assigned ICBMs and SLBMs a PA of 0.75 as a class and assigned bombers a PA of 0.5. For the purposes of the study, SAC assumed that any warhead that reached its DGZ was capable of destroying the associated installation(s), although some hardened or geographically extensive targets actually required multiple DGZs.

In assigning weapons to the Phoenix force structure, the DOD dedicated 25 percent of the SLBM force as a reserve. One hundred percent of the ICBM force, 75 percent of the SLBM force, and 50 percent of the bomber force comprised the basic attack force. The remaining 50 percent of the bomber force formed the basic attack's hedge force. This disposition of delivery systems resulted from dramatic improvements in the capability of submarine launched ballistic missiles and their launch platforms and a less prominent posture attributed to bombers. The study described the extensive use of missiles for the basic attack as a "Twin Triad" posture, with bombers assigned mostly to backup.

Having established the preceding parameters, the study enumerated how many warheads it would take to destroy eight individual installations, along with the number of associated warheads required for the hedge and reserve forces. This worked out to four ICBM warheads, three SLBM warheads, and six bombs for the basic attack, one SLBM warhead for the reserve, and six bombs for the hedge – a total of 20 warheads, or 2.5 warheads per installation. From this point, it was possible to go forward and calculate the number of warheads needed for an attack on a fixed number of targets taken from the Target Index or the number of DGZs / installations that SAC could attack given a specific number of warheads. Given the 6,000 warheads allowed by *START I*, it would have been possible to attack 2,400 DGZs, which translates into 3,000 installations assuming the destruction of 1.25 installations per DGZ targeted.

In addition to its numerical calculations, the *Phoenix Study* established a number of force-planning principles that later influenced *START II* negotiations, the *1994 Nuclear Posture Review*, and the organization and composition of STRATCOM. Foremost among these principles was the concept that the Soviet Union (later Russia) was the only nation capable of destroying the US. It should therefore remain the focus of strategic force planning. A corollary to this concept was the need to maintain a credible nuclear force to reduce the need for proliferation among American allies. Deterrence would remain focused on a Triad of ICBMs, SLBMs, and bombers. This concept derived from the recognition that the future Strategic Triad would contain fewer warheads, fewer types of warheads, and a reduced number of strategic nuclear delivery vehicles. In addition, American ICBMs and SLBMs, the so-called Twin Triad, would carry the weight of deterrence. Separate from the forces previously discussed, STRATCOM made recommendations for the maintenance of a stock of inactive warheads that the DOE could return to operational status if required.

Prior to the completion of the *Phoenix Study*, General Butler's views and calls for a "peace dividend" at the end of the Cold War combined to erode the longstanding inter-service rivalries that had previously prevented the formation of a single nuclear command. During a conference in 1990, General John T. Chain, CINCSAC and Director of Strategic Target Planning, offered an inclusive view of a Unified Strategic Command that combined the strategic bomber force, ICBMs, the Navy's ballistic missile submarines, the Space Defense Initiative, and elements of United States Space Command.

After evaluating alternative Navy and Space Command models, the Joint Chiefs' Chairman – General Colin Powell – put forward the final structure for the new Unified Strategic Command. This was a consolidation of Air Force strategic missiles and aircraft with the Navy's ballistic missile submarines and the Joint Strategic Target Planning Staff. President George Bush approved and publicly revealed the new organization as United States Strategic Command during a speech on the evening of September 27, 1991. At this time, the President also announced the removal from alert of all bombers and missiles not scheduled for destruction under the *Strategic Arms Reduction Treaty*. He then held up Strategic Command as a way to manage the command and control of the nation's strategic forces more effectively. Shortly thereafter, Secretary of Defense Richard Cheney reinforced the president's message by stating that USSTRATCOM emphasized DOD commitment to Unified Military Commands and Joint Service arrangements.

Following USSTRATCOM's official announcement, organizational implementation came quickly. General Robert Linhard, the Strategic Air Command Deputy Chief of Staff for Plans and Resources, and General Albert Jensen, the JSTPS Deputy Director for Analysis, Concepts, and Systems, established a transition team that they co-chaired. They released their organizational plan for service coordination in early November 1991, and by mid-January 1992, had activated a Provisional Command Headquarters at Offutt AFB, Nebraska. In February, President Bush nominated General George Butler as the first commander of USSTRATCOM and by May, all necessary requirements for establishing the new command were in place. The DOD published details of the organization in March 1992, and on April 7, President George Bush approved a revised Unified Command Plan. Senate confirmation of General Butler as CINCSTRATCOM took place on May 22, and set the stage for the Command's inaugural ceremonies, held on June 1, 1992.

In parallel with the process that led to the formation of US Strategic Command, the Air Force reorganized itself along new lines proposed by General Butler. The reorganization of the Air Force involved the replacement of SAC, TAC, and MAC with two new commands – Air Combat Command and Air Mobility Command. Secretary of the Air Force Donald Rice announced this reorganization on September 17, 1991, at the Air Force Association's national convention. Beginning in early October, SAC's forces began to transfer to the newly established commands. Headquartered at Langley AFB, Virginia, Air Combat Command consisted of SAC and TAC combat assets that placed air power at the disposal of the theater / JTF war fighter. Headquartered at Scott AFB, Illinois, Air Mobility Command consisted of airlift and air refueling assets that supported Air Combat Command. An important goal of Air Force reorganization was the retention of qualified personnel.

While SAC's drawdown was taking place, General Linhard established STRATCOM's Headquarters using a traditional joint staff structure. STRATCOM's Commander in Chief was a four-star billet and its second in command was a three-star billet that served as both Deputy Commander in Chief and Chief of Staff. STRATCOM later separated the second in command's staff responsibilities and placed them in an Air Force Colonel's position. The two senior staff positions rotated between Air Force and Navy personnel. Initially, STRATCOM's Chief of Staff supervised five Joint Staff directorates:

- Manpower and Personnel (J1)

- Intelligence (J2)
- Operations and Logistics in a combined office (J3 / J4)
- Plans and Policy (J5)
- Command, Control, Communications, Computers, and Intelligence (J6)

STRATCOM later added:
- Joint Exercises and Training (J7)
- Capability and Resource Integration (J8)
- Mission Assessment and Analysis (J9)
- Joint Reserve Directorate (J10)

STRATCOM also assumed the Joint Strategic Target Planning function – development of the *SIOP* – to form the core of its Plans and Policy Directorate. This was necessary because one of STRATCOM's initial assignments was to develop a new strategic war plan. Concurrent with the formation of STRATCOM, the Air Force formed the Strategic Joint Intelligence Center at Offutt AFB from Strategic Air Command's 544th Intelligence Wing. STRATJIC fell under the operational control of STRATCOM's Director of Intelligence.

In June 1992, representatives of STRATCOM briefed Secretary of Defense Dick Cheney and Joint Chief of Staff Chairman Colin Powell on the combined implications of the *INF Treaty*, the *START Treaty*, and the *Presidential Nuclear Initiatives*. The *INF Treaty* and the *PNI*s dealt primarily with tactical nuclear weapons whereas the *START Treaty* focused on the strategic nuclear arsenal. *START I* reduced the number of strategic delivery vehicles to 1,600 ICBMs, SLBMs, and heavy bombers. The Treaty allowed Russia to retain 154 heavy ICBMs. Inventories were not to exceed 6,000 accountable warheads, of which the signatories could mount no more than 4,900 on ballistic missiles. Ultimately, the agreement reduced the size of the Soviet and American strategic arsenals by 30 - 40 percent through the end of 2001. In order to comply with the *START Treaty*, the United States inactivated thee Minutemen Strategic Missile Wings between 1994 and 1998, bringing its total deployment of land-based missiles down to a level of 500 vehicles.

The *START II Treaty*, signed on January 3, 1993, called for the elimination of heavy intercontinental ballistic missiles and required the downloading of all remaining ICBMs to a single warhead. It was thus a "de-MIRVing" agreement. It also reduced the number of accountable strategic nuclear warheads to one third of pre-*START* levels. The signatories were to accomplish this reduction in two phases. In the first phase, they were to reduce total deployed warheads to a level of 3,800 - 4,250 units and in the second phase, reduce total warheads to no more than 3,000 - 3,500 units. They could MIRV SLBMs to a total level of 1,700 - 1,750 warheads. The Treaty attributed all heavy bomber types with potential payloads. This capped the number of heavy nuclear bombers at a force that did not exceed the total allowable warheads minus deployed ICBM and SLBM warheads. In addition, the Treaty capped total heavy conventional bomber forces at one hundred aircraft. The Treaty required the completion of its specified arms reductions by January 1, 2003.

Although the US and Russia signed the *START II Treaty*, it never entered force due to concerns on the part of the Russian Duma. Nevertheless, both sides tacitly abided by its terms. On April 14, 2000, the Duma finally ratified the Treaty, contingent on the preservation of the *ABM Treaty* and agreement on an addendum that identified the differences between strategic versus tactical missile defense. On June 13, 2002, the US withdrew from the *ABM Treaty* and on the following day, the Russia Federation announced that it would no longer consider itself bound by *START II* provisions.

Responding to the *START Treaties*, STRATCOM held a number of conferences with the Joint, Air and OPNAV Staffs, Air Combat Command, and the Commanders of the surface and submarine fleets in the Atlantic and Pacific. These meetings produced a "preferred USSTRATCOM force structure" for America's nuclear posture. The preferred strategic force structure, presented to Secretary of Defense Cheney and Joint Chief of Staff Powell, included:

- Nuclear certification for the new B-2A bomber
- Retention of B-52H bombers
- Transition of the B-1B bomber force to a conventional role
- Assignment of Air Reserve units to nuclear bomber functions
- Retention of the heavy Peacekeeper ICBM through 2001
- The modernization and life extension of Minuteman III ICBMs
- Transfer of W87 warheads from retiring Peacekeepers to Minuteman III
- Maintenance of a two-ocean SSBN force of 18 vessels
- The retention of SLBM MIRV capabilities

The force structure study was STRATCOM's first chance to define itself after replacing the independent Air Force and Navy strategic forces that had dominated the Cold War. Yet STRATCOM was not satisfied with just determining a structure compatible with *START*. It began a review intended to meet the needs of the warfighter by creating an affordable, survivable, and flexible strategic nuclear force. These principles became the basis for the *Sun City* and *Sun City Extended* "Alternate Force Structure" Studies carried out in 1993 and 1994.

Like the *Phoenix Study*, the 1993 *Sun City Study* focused primarily on Russia as a nuclear adversary and compared the options it generated with *Force Structure Option 1*, the preferred USSTRATCOM force structure that had been presented to Dick Cheney and Colin Powell the previous year. In all, STRATCOM examined nine options, six of which were at the *START* limit of 3,500 accountable warheads and three that fell well below. STRATCOM also examined four target bases. These included the Russian threat as well as several smaller threats. STRATCOM analyzed each option for its ability to hold a target base at risk while maintaining flexibility and affordability. The study considered STRATCOM's newly enforced reliance on a non-MIRVed ICBM force and a reduction in the number of Russian strategic installations. One benefit of a non-MIRVed ICBM force was extended missile range.

The *Sun City Study* also concerned itself with a calculation of "strategic stability" in US - Russian relations. Strategic stability examined "advantage ratios" that compared American and Russian delivery vehicles, weapons, megatonnages, and hard target kill capabilities. It also examined "stability measures" such as a stability index, sensitivity to generate missions, sensitivity to prompt retaliatory launch, second-strike dialect, incentive index, and drawdown curves. Stability index expressed the cost of initiating a nuclear strike versus the cost of waiting to strike. Cost was considered as either damage to the US or loss of damage to Russia, with a higher index considered to be more stable. That is, the maintenance of a highly capable nuclear posture that included an American capability to strike first reduced Russia's temptation to launch a first strike.

Because STRATCOM closely linked stability analysis to the question of alert level, it saw the stability index as increasingly important as it reduced its forces. De-alerting nuclear forces jeopardized stability by creating a premium for a preemptive first strike. It followed that any act that restored, or that the signatories perceived to restore, de-alerted forces was destabilizing. This concern later became moot with the development and deployment of new mission-generating and re-targeting software. At the individual delivery platform level, the time needed to generate a B-2A stealth bomber sortie shrank to 1 day, whereas the time needed to generate adaptively planned missions fell to 8 hours. Implemented in 1996, the Minuteman Rapid Execution and Combat Targeting system allows retargeting of missiles in the space of a few minutes. In similar fashion, the SLBM re-targeting system installed on all ballistic missile submarines permits fast, accurate, and reliable re-targeting in support of *SIOP* (*OPLAN*).

Russia's selection as the major target, and a residual cold war mentality at STRATCOM, resulted in a *Sun City* conclusion that the most preferable force structure was the most capable force structure. STRATCOM analyzed the smaller force structures and target set options mainly for "parametric purposes." The study further concluded that the United States should retain the traditional Nuclear Triad and that additional force reductions beyond *Option 1* were undesirable, at least in the short term. It also maintained that Triad flexibility, capability, and affordability should be of paramount importance to planners, especially in light of a thinning target base.

STRATCOM used the outcome of the 1993 *Sun City Study* to promote Force Structure Option 1 as its choice during a Nuclear Posture Review conducted by the DOD from October 1993 through September 1994. During this period, STRATCOM carried out the *Sun City Extended Study* in further support of the Nuclear Posture Review. The *Sun City Extended Study* generated about 40 more nuclear attack options that focused on small target bases. In particular, it generated a number of Chinese options, two of which were declassified. One of these options looked at a confrontation over North Korea whereas the other was a Major Attack Option for a full-scale US / Chinese confrontation. The increased focus on China resulted from China's modernization of its long-range strategic nuclear forces.

The *Nuclear Posture Review* was the first DOD study of any kind to incorporate a comprehensive review of policy, doctrine, force structure, operations, infrastructure, safety, security, and arms control. As opposed to the *Sun City* Studies, the *NPR* focused on the deterrent, rather than the warfighting capabilities of the strategic arsenal. The review was necessary because of the significantly different visions for future force structures that key figures in the Administration and the DOD championed. While a number of civilians favored complete denuclearization, many in the military wished simply to maintain the status quo.

A five-man steering committee, co-chaired by Major General John Admire, the Vice Director for Strategic Plans and Policy at Joint Staff, and Ashton Carter, the Assistant Secretary of Defense for Nuclear Security and Counterproliferation, led the Nuclear Posture Review. The other three members of the committee came from the intelligence, nuclear and space agencies. DOD organized the review around six working groups comprised of military and civilian experts from OSD, the Joint Staff, the Services, and the Unified Commands. It assigned each group a specific topic for review:

- The role of nuclear weapons in US security strategy
- The structure of US nuclear forces
- US nuclear force operations
- Nuclear safety and security measures
- The relationship between US Nuclear posture and counter proliferation policy

- The relationship between US nuclear posture and threat reduction policy with the former Soviet Union

The Deputy Secretary of Defense and the Vice Chairman of the Joint Chiefs of Staff presented the final report to the Secretary of Defense and the Chairman of the Joint Chiefs of Staff after 10 months of deliberation. The DOD then forwarded the *NPR* to the President, who approved its recommendations on September 18, 1994.

The Nuclear Posture Review affirmed that the United States had both national and international nuclear deterrent postures that required retention of the maximum number of warheads allowed under *START II*. STRATCOM, however, had to reallocate these warheads to a nuclear force with a reduced number of strategic delivery vehicles. Strategic policy adjusted US strategic nuclear forces to:

- 14 Trident submarines each carrying 24 D-5 Trident II missiles MIRVed with five warheads
- 20 B-2A bombers carrying gravity bombs
- 66 (later 76) B-52H bombers carrying air-launched or advanced cruise missiles
- 3 Missile Wings containing 450 / 500 Minuteman III missiles; each Minuteman missile armed with a single warhead

In addition to its force reductions, STRATCOM removed strategic bombers from alert, de-targeted ICBMs and SLBMs, and reduced its command post structure and its systems endurance.

The DOD then reconfigured its non-strategic nuclear forces. It eliminated nuclear-capable, carrier-based aircraft and eliminated the option to deploy nuclear-capable Tomahawk cruise missiles on surface ships. Because of requests by Japan and South Korea, it retained the option to deploy nuclear-capable Tomahawk cruise missiles on attack submarines (with all missiles stored in port). It also retained a NATO commitment to maintain European and CONUS based nuclear-capable aircraft.

Other outcomes of the *NPR* included improvements to the nuclear infrastructure. The DOD ordered the Air Force to re-motor and upgrade the guidance systems of its remaining Minuteman III missiles and the Navy to continue Trident D-5 missile acquisition to maintain the strategic ballistic missile industrial base. It also ordered the Navy to install a Trident coded control device on all D-5 missiles and to install permissive action links on all its nuclear weapons prior to 1997. The DOD further committed funding to sustain guidance and reentry vehicle development. The DOE ordered its laboratories to:

- Maintain their ability to design, fabricate, and certify new warheads without underground nuclear testing
- Develop a stockpile surveillance engineering base
- Demonstrate the capability to refurbish and certify the weapons in the stockpile
- Maintain the science and technology base needed to support nuclear weapons
- Decide on a production program to maintain the national supply of tritium

A significant result of the *Sun City* Studies and the preparation for the Nuclear Posture Review was a STRATCOM generated *SIOP*. The *SIOP* was at this time a deterrent against Russia's extensive nuclear arsenal and the use of weapons of mass destruction (WMDs) by third world states. In order to cover a multitude of global scenarios with a declining weapon base in a rapidly shifting political landscape, STRATCOM adopted "adaptive planning" and automation to maintain flexibility in its planning process. This change resulted in the "*Living SIOP*," first produced as *SIOP-95*. The development of a highly-automated War Planning System allowed changes to the manner in which STRATCOM's Planning Directorate formulated *SIOP*. With the *SIOP* adoption of highly automated planning technology, the time required for a complete overhaul of *SIOP* shrank from 18 months to 6 months, the generation of a Major Attack Option shrank to 4 months, and the production of Limited Attack options dropped to as little as 24 hours!

The DOD had begun automation during the 1960s with the "Strategic Warfare Planning System" that supported Strategic Air Command and the Joint Strategic Target Planning Staff. Although STRATCOM refers to the SWPS as a planning system, it also supports the data analysis used in the planning process. The current SWPS is distributed in a client-server architecture that draws on global data (data used by more than one application) that is processed by powerful graphics workstations. The primary database accessed by SWPS is the Strategic War Planning Systems Enterprise Database (SWPS-EDB).

The SWPS planning process is a sequential effort that still follows guidelines developed prior to automation. It begins with target selection, which planners follow with DGZ construction, resource allocation, aircraft and missile applications, penetration analysis, timing and resolution allotment, weapon review, and production. An Allocated Windows Planning System (AWPS) supports this process by insuring that weapon allocations achieve the intended damage expectancy thresholds for all installations targeted. It also performs analysis and a quality review of the *SIOP* during the plan's build. Currently, the AWPS acts as a pre-processor to an Automated *SIOP* Allocation (ASA) System that allows for the continuous production of war plans. In addition to supporting the strategic nuclear mission, SWPS now supports theater forces nuclear planning and non-strategic nuclear forces (NSNF) planning.

During the active *SIOP* planning process, operators manipulate computer results in a dynamic database. At the end of the process, the final output is stored in a number of static databases that are accessible by other applications. This arrangement eliminates redundant databases and allows for the standardization of analysis applications. It also reduces personnel, operating, and maintenance costs. A global server manages external links to the SWPS through a number of guard processors. Four network rings provide access to Top Secret *SIOP* / Extremely Sensitive Information, Top Secret Information, Secret Information, and Unclassified information. In 2003, STATCOM replaced the *SIOP* designation for its strategic war plan with the new designation: *OPLAN 8044 Revision 3*. The final number of the *OPLAN* changes each year to reflect the revisions made over the previous 12 months.

Following the signing of *START II*, STRATCOM initiated a Whitepaper to prepare for further arms reduction in the post *START II* era. While it was preparing this study, STRATCOM took on a number of new missions, some of which it actively solicited. These missions included theater nuclear planning, counterproliferation, stockpile stewardship, an expanded role with the Nuclear Weapons Council, and the creation of a new readiness posture.

STRATCOM actively solicited the theater planning mission in 1993 but met significant resistance from theater commands that preferred to retain this function. This situation forced STRATCOM to provide briefings to justify its request to the Joint Staff and Theater CINCs. These briefings took place from November 1993 through the spring of 1994. Fortunately, the downsizing of the DOD strengthened STRATCOM's position because its request centralized and affordably retained theater nuclear expertise. With the issuance of *Change 4 to Annex C of the Joint Strategic Capabilities Plan* (JSCP) in May 1994, STRATCOM assumed the nation's theater nuclear planning function. The new mission required a change to STRATCOM's planning style, because theater nuclear planning inherently requires more flexibility than does the strategic mission. To demonstrate its competence in nuclear planning, STRATCOM produced the theater nuclear support plans required for the exercise: Global Archer 94-3.

A second mission solicited by STRATCOM was countering weapons of mass destruction. Planning against WMDs resembles theater planning because it involves mounting limited strikes using both conventional and nuclear means. STRATCOM began work to attain this mission at a Counterproliferation Conference it hosted in September 1993 but encountered opposition from the Joint Staff and the theater CINCs. In the fall of 1994, further discussions with the Joint Staff and CINC staffs resulted in acceptance of STRATCOM for the Counterproliferation Mission.

Acceptance of the Counterproliferation mission resulted in the production of "SILVER books," an acronym that stood for *Strategic Installation List of Vulnerability Effects and Results*. The SILVER books were plans for military strikes against WMD facilities in a number of countries that included Iran, Iraq, Libya, and North Korea. The project involved the planning of "SILVER bullet" missions aimed at nuclear, chemical, and biological; and command, control, and communications installations. STRATCOM analyzed attacks against six WMD facilities with nuclear, conventional, and unconventional weapons and by late 1994, a proposed SILVER book was ready for the European Command and a prototype Book was in production for Pacific Command. Then, in early 1995, the JCS ordered STRATCOM to drop the SILVER project. This reflected a decision by the Defense Acquisition Board to use only non-nuclear weapons for attacking hardened and deeply buried WMD targets.

Stockpile Stewardship was another area in which STRATCOM took a role. Presidential Decision Directive PDD-15 *Stockpile Stewardship* and *Fiscal Year 1994 National Defense Authorization Act* (Public Law 103-160) established the Stockpile Stewardship Program as a national requirement. Because of a "zero yield" nuclear testing policy imposed by President Clinton, the program was designed to better understand the nuclear stockpile; predict, detect, and evaluate (potential) problems with aging weapons and weapon components in the stockpile; allow the refurbishment and re-manufacture of weapons and their components; and maintain the science and engineering institutions that support the nation's nuclear deterrent. To meet these goals, agencies of the DOD collaborated with agencies from the DOE to address civilian and military surety and safety concerns. STRATCOM further strengthened its role in stockpile stewardship by soliciting a voting membership on the NWSSCS's board in 1993, a role it received in 1994.

In response to a June 1994 Operations and Logistics Directorate initiative, STRATCOM began developing an improved readiness model for its strategic forces. Originally produced by SAC, this tool evaluates the availability of strategic forces. The SAC model was, however, labor intensive to produce and difficult to update on a continuous basis. The computerized graphics that emerged from re-development of the SAC readiness system were colorized displays that not only showed the disposition and status of all STRATCOM's assets, but how long it would take to recall them to generate a mission. Operation Bulwark Bronze tested the new readiness system in 1995 with great success. As a result, STRATCOM has since used it to provide daily force status assessments during command center exercises. The system is also used by the Joint Staff and other unified commands as a tool to evaluate force readiness, and to generate multiple attack scenarios.

After the US and Russia signed *START II* and the DOD completed the first Nuclear Posture Review in September 1994, STRATCOM prepared a Whitepaper on force structure for the post *START II* era. It carried out this study in compliance with Presidential Decision Directive PDD-37: *Post START II Arms Control*, which provided guidance to American arms control agencies. PDD-37 called for further arms reductions in compliance with Article VI of the *START II Treaty* and included a list of four "first principles" – deterrence, stability, hedge, and equivalence – with which to guide the US approach to arms control.

The *Sun City Studies* had already evaluated three of PDD-37's principles – deterrence, stability, and hedge. STRATCOM linked the fourth principle, equivalence, with stability. Using these principles, STRATCOM established a number of guidelines for the implementation of post *START II* arms control. STRATCOM devised these guidelines to:

- Protect the US strategic nuclear delivery vehicle force by retaining as many delivery types as possible
- Retain US warheads at a level consistent with warfighting needs
- Minimize the impact of Russian systems posing the greatest threat to American interests
- Reduce and eliminate US and Russian non-deployed warheads and fissionable material
- Address non-strategic nuclear forces as part of the overall effort to stem nuclear proliferation

In establishing a post *START II* strategic force structure, STRATCOM's Whitepaper recommended against reducing strategic offensive weapons below 2,000 - 2,500 deployed warheads; STRATCOM noted that whatever limit that was finally chosen, it would drive the future composition of US strategic forces. The Whitepaper indicated that STRATCOM should leave its B-52s in the strategic mix and that a reduction of the ICBM force below 350 SNDVs would significantly affecting the *SIOP*. In fact, any reduction to the ICBM force would erode the number of strategic targets in the US, a potentially de-stabilizing strategic measure. Within these constraints, the Whitepaper concluded that the large number of air-launched cruise missiles allocated to B-52H bombers would have a significant impact on how many SSBNs could be retained. Since SSBNs had been designated an increasingly important role in the Twin Triad, the less valuable bomber was held up as a candidate for future arms reduction.

STRATCOM's Whitepaper also analyzed potential reductions to Russian strategic forces, in order to understand how that country might implement its strategic forces. Since the Russians had not provided definitive information regarding their post *START II* structure, this analysis was mostly broad brush. Three trends that emerged from the Whitepaper were a reduced Russian emphasis on heavy bombers, an increased emphasis on mobile ICBMs, and an assumption that SLBMs would eventually comprise the single largest component of Russian strategic forces. This last assumption proved incorrect.

A final topic of the study was the disposition of fissionable material. STRATCOM deemed this issue the least important aspect of the study. The Whitepaper did conclude, however, that it was not desirable to proceed with warhead elimination until STRATCOM could establish detailed, verifiable information on strategic and nonstrategic stockpiles of fissionable material. Related to this issue were the needs to establish a method of data exchange, mutual reciprocal inspections, and a chain of custody covering the total inventory of warheads and fissionable material. For the process to work, it would also be necessary to establish safeguards, transparency, and an irreversible methodology for disposing of unwanted fissionable material.

The Whitepaper concluded that the failure to reach an agreement on the disposition of fissionable material from eliminated warheads should not prevent deeper cuts.

In December 1996, a few months after the completion of its post *START II* arms control Whitepaper, STRATCOM completed a second force structure study. This study addressed STRATCOM's nuclear warfighting philosophy, as evidenced by its title: *The Warfighter's Assessment*. The DOD has not declassified much of this study but it very likely considered the *START III* framework of 2,000 - 2,500 accountable warheads. The study acknowledged that its guidance for employment of nuclear weapons remained unchanged. Thus, it not only relied on President Clinton's PDD-37, but also relied on Presidential Guidance dating back as far as Ronald Reagan's National Security Decision Directive NSDD-13, issued in October 1981.

As in several previous force structure studies, the Warfighter's Assessment concluded that the characteristics of the strategic force would become increasingly important for both deterrence and warfighting as the DOD reduced the number of weapons in the stockpile. That is, STRATCOM had to maintain flexibility under future arms control treaties to create and maintain a credible, effective deterrent that incorporated the warfighting principles developed during the Cold War. These characteristics included modernization, stockpile stewardship, survivability, a robust planning capability, C^2 connectivity, and timely threat warning.

As identified in T*he Warfighter's Assessment*, the issues of command, control, and communication become increasingly important with reductions to the size of the strategic stockpile. This exacerbates the urge to "use or lose" available forces. STRATCOM thus removed its strategic weapons from alert and persuaded the Russians to do the same in order to address American concerns about an unintentional launch. The de-alerting of American weapon systems, ICBMs and bombers in particular, had the effect of increasing the vulnerability of these nuclear triad components to a first strike. Thus, it became necessary to improve C^3I so that STRATCOM could quickly generate and transmit emergency action messages to permit a launch prior to the arrival of incoming warheads. The need to upload target coordinates to delivery vehicles prior to their launch now compressed this period.

To improve its C^3 capabilities, STRATCOM conducted annual Global Archer exercises that it used to evaluate command connectivity with its strategic forces. Beginning in 1996 with the Global Guardian command post and field training exercise, STRATCOM Headquarters interacted closely with all of its task forces, field units and outside organizations to confirm their ability to generate and execute warfighting missions as directed. Global Guardian saw the first use of the Mobile Consolidated Command Center (MCCC) and validated STRATCOM's plans, policies, strategies, and decision-making processes in a stressed environment.

Following the Global Guardian exercise of April 1998, STRATCOM conducted the first operational flight of the Navy's E-6B, the "Take Charge and Move Out" or TACAMO aircraft that inherited the "Looking Glass" nuclear command post mission from the Air Force's EC-135. Not only could the E-6B command, control, and communicate with all three legs of the nuclear Triad, it could remotely launch Minuteman III ICBMs, a capability it demonstrated in June 1998. The E-6B officially assumed STRATCOM's command post mission on September 25, 1998, followed by the transfer of the National Airborne Operations Center (NAOC) organization from the Joint Staff to Strategic Command on October 1, 1999.

A new Unified Command Plan, issued by President Clinton in 1997 required CINCSTRATCOM to upgrade the security of the nation's nuclear forces in light of global terrorist attacks. These included STRATCOM's headquarters, SSBNs, ICBMs, bombers, and C^3

aircraft. To meet this requirement, STRATCOM initiated new force protection measures. It heightened security around the LeMay Building Headquarters with a card Reader to control access and issued distinctive color-coded badges to all personnel. Nuclear command and control facilities and the bomber leg of the triad also received security upgrades.

In the same period that the Air Force prepared its post *START II* studies, the DOD conducted its first *Quadrennial Defense Review*. The origins of the *QDR* date to 1995 when the Congressional Commission on Roles and Missions recommended "a comprehensive strategy and force review at the start of each new administration – a *Quadrennial Strategy Review*." In 1996, the DOD accepted the Congressional recommendation and initiated a *QDR* modeled after a "bottom up" review in fulfillment of the Lieberman Amendment to the *FY97 Defense Authorization Bill*. The Lieberman Amendment also directed the establishment of a National Defense Panel.

The *1996 QDR* was concerned primarily with tactical air power, although it also examined a number of readiness postures that had the potential to reduce operational costs. In terms of the overall military force, the DOD made only minor adjustments. It reduced the Air Force from thirteen to twelve active Fighter Wings and cut F-22 procurement from 438 to 339 aircraft. The Marine Corps sustained minor cuts and the Navy retained its twelve Carrier Battle Groups, although the DOD reduced the number of its surface combatants from 128 to 116 vessels and capped F/A-18E / F procurement at 548 aircraft. The Army retained 19 active Divisions and 2 Armored Cavalry Regiments, with the loss of 15,000 active duty personnel and 45,000 guardsmen and reservists. Along with these changes, the *QDR* introduced the slogan – "shape, respond, and prepare."

In terms of nuclear forces, the *QDR* reaffirmed the need for a robust and flexible nuclear deterrent, an appraisal that resulted in President Clinton issuing Presidential Decision Directive PDD-60, *Nuclear Weapons Employment Policy Guidance* in 1997. Although STRATCOM regularly updated its nuclear plans to accommodate circumstances such as the *Presidential Nuclear Initiatives* and the *Nuclear Posture Review*, PDD-60 was the first revision of nuclear guidance in over 15 years. PDD-60 directed that the DOD maintain an assured response capability able to inflict "unacceptable damage" against a potential enemy's valued assets. It also directed that STRATCOM plan a range of options allowing it to respond to aggression in a manner appropriate to the provocation, rather than an "all or nothing" Major Attack Option. The new guidance indicated that the US would not necessarily "launch on warning" but would respond promptly to any attack. Thus, it eliminated the Cold War concept of winning a protracted nuclear war. The Directive reaffirmed the need for a survivable triad of strategic deterrent forces maintained at the level of 2,000 to 2,500 accountable warheads envisioned for *START III*.

While the 1990s saw the continuous removal and destruction of warheads to meet the requirements of START agreements, the makeup of strategic nuclear delivery vehicles remained little changed. The DOD made a minor change in targeting when STRATCOM returned China to the *SIOP* in 2000. Both *Russian Integrated Strategic Operational* and *Chinese Integrated Strategic Operational Plans* (RISOP and CHISOP) were now included in the *Single Integrated Operational Plan*. Overall, however, the makeup of America's strategic nuclear forces remained little changed until affected by four important events in 2001. These were the election of George W. Bush, the *2001 Quadrennial Defense Review*, the 2001 Nuclear Posture Review, and the September 11 terrorist attack on the United States.

The DOD completed both the *2001 Quadrennial Defense Review* and the *2001 Nuclear Posture Review* after the September 11 terrorist attack. Although the attack influenced these reviews, the planning groups for these documents did not have sufficient opportunity to address the new reality in full. The DOD intended the *2001 QDR* to reflect the president's desire to transform the US defense posture from an outmoded Cold War instrument to a flexible force that could meet emerging threats, maintain stability in critical regions, and preserve American leadership and freedom of action for the future. It was based on four strategic priorities: to "assure" allies and friends of American resolve, to "dissuade" adversaries from developing threatening military forces or ambitions, to "deter" threats and coercion against the US, its allies and interests and to decisively "defeat" any threatening adversary in a manner, place and time of American choice. The DOD was to organize, train, and staff its forces to meet a so-called "1-4-2-planning construct" to:

- Defend the homeland
- Operate in the four areas of Europe, Northeast Asia, the Asian littoral, and the Middle East / Southwest Asia
- Swiftly defeat two adversaries in near simultaneous campaigns
- Win decisively in one of those campaigns while conducting limited small-scale contingency operations

In order to meet the president's goal of a transformed military, the *QDR* formulated what it called a "capabilities based" defensive Triad. The old "threat based" defensive Triad consisted of SLBMs, ICBMs and bombers and been sized to meet a particular threat (the nuclear arsenal of the former Soviet Union). It had little flexibility and considered a missile defense destabilizing. The new Triad comprised a mixed deterrent force (the old nuclear Triad and conventional assets armed with precision guided munitions); an active ballistic missile defense backed up with passive civil defense measures; and a responsive defense infrastructure better able to field new generations of weapon systems. The DOD predicated the success of the new Triad on improved command and control, more effective intelligence gathering, and adaptive planning. Although it still had to deal with Russia's nuclear forces, it also had to contend with multiple, non-country specific threats, reduce US dependency on nuclear deterrence, and deal with threats for which the political stakes might be varied and not equal in size. The *QDR* also required that the DOD conduct the *2001 NPR* in concert with its recommendations.

In terms of force structure, the 2001 *NPR* stayed true to the recommendations of the 1994 review. It reiterated the importance of nuclear weapons in warfighting and reaffirmed the importance of Russia as the most serious threat to the United States. In addition to Russia, the United States named six other countries that could present a serious threat and therefore might warrant a nuclear strike: China, Iran, Iraq, Libya, North Korea, and Syria. In terms of accountable warheads, the *NPR* acknowledged that 1,700 - 2,200 strategic warheads would comprise the active stockpile in 2012. Where it differed significantly from the 1994 report, was in its adherence to the recommendations of the *2001 QDR*. In this aspect, it reasserted the *QDR*'s commitment to flexibility through the capabilities-based framework of the New Strategic Triad.

In addition to its commitment to the New Strategic Triad, the *2001 NPR* advocated the development and deployment of low yield nuclear weapons, including a nuclear bunker buster that would provide the primary means of destroying underground facilities supporting weapons of mass destruction. In line with the *QDR*, the *NPR* rejected the need for binding arms control agreements as an essential component of US foreign policy. In fact, the *NPR* rejected the *ABM*

Treaty and the Comprehensive Test Ban Treaty in the new security environment. The *NPR* made the radical assumption that the United States should base future Russian relations on increased cooperation and friendship, rendering Cold War treaties redundant. Thus, the United States formally withdrew from the *ABM Treaty* in 2002, calling for a National Ballistic Missile Defense System able to defend against limited attacks. The Russian's were not impressed.

Following the terrorist attacks against New York and Washington on September 11, 2001, President George W. Bush declared that he would establish a unified military organization dedicated to defending the domestic United States from attack. To this end, Secretary of Defense Rumsfeld announced on April 12, 2002, that a new unified command, NORTHCOM, would be responsible for the defense of North America and up to 500 nautical miles offshore. He also gave hints about a merged US Strategic / US Space Command. Shortly thereafter, President Bush met with Russian President Vladimir Putin at a May 2002 Moscow Summit during which the leaders signed a *Strategic Offensive Reduction Treaty* (*SORT*). This was just as well since the Russians had withdrawn from *START II* after the US repudiated the *ABM Treaty*. *SORT* promised bilateral reductions to reduce deployed strategic warheads to 1,700 to 2,200 units for each country by the year 2012. It left the mix of strategic nuclear delivery vehicles to the discretion of the individual powers.

The DOD officially announced Rumsfeld's rumored merger on June 26, 2002. It reassigned USSPACECOM's mission to USSTRATCOM and nominated Admiral James Ellis as its new commander. The merged command retained the US Strategic Command name and remained headquartered at Offutt AFB, Nebraska. The activation of the new USSTRATCOM under Admiral Ellis took place October 1, 2002, and shortly thereafter on January 10, 2003, President Bush signed Change Two of the Unified Command Plan that tasked USSTRATCOM with four previously unassigned responsibilities: global strike, missile defense integration, DOD information operations, and C^4ISR. This unique combination of roles, capabilities, and authorities, brought new opportunities to the strategic arena and improved support for theater combatant commanders.

The 2003 change from *SIOP* to *OPLAN* highlighted STRATCOM's new role in US military strategy and coincided with an Air Force publication on Joint Operations. In order to be effective, the Air Force found that it had to carry out joint operations planning simultaneously at several levels. At the uppermost strategic level, joint operations planning involved the formulation of military objectives to support the national security strategy. At the middle operational level, joint operations planning linked the tactical employment of forces to strategic objectives, and at the lower tactical level, joint operations planning concerned itself with units in combat.

Although joint operation planning begins with the National Command Authorities, it is primarily the responsibility of the Chairman of the Joint Chiefs of Staff and senior Combatant Commanders. It includes the preparation of operation plans (*OPLAN*s), operation plans in concept format (*CONPLAN*s), functional plans, campaign plans, and operational orders (*OPORD*s) by joint force commanders. Joint operation planning includes the full range of activities required for conducting joint operations activities that include mobilization planning, deployment planning, employment planning, sustainment planning, and redeployment planning.

Mobilization planning assembles and organizes national resources to support national objectives during times of war (and for military operations other than war). Primarily the responsibility of the individual Armed Services, mobilization planning is intended to bring selected units to the states of readiness required to deal with specific contingencies. Deployment planning is concerned with the movement of forces and force sustainment resources from their bases into and around theaters to conduct joint operations. This type of planning is the responsibility of CMDRSTRATCOM and / or theater combatant commanders. The use of tactical nuclear weapons or the theater use of strategic weapon systems (conventional or nuclear) requires further coordination between STRATCOM and Theater Commands.

Employment planning by CMDRSTRATCOM and combatant commanders prescribes the application of only enough force to attain specified military objectives. This planning determines the scope of mobilization, deployment, sustainment, and redeployment planning. Sustainment planning provides and maintains the levels of personnel, materiel, and consumables required for combat activity of an estimated duration and for a desired level of intensity. This function is also the responsibility of CMDRSTRATCOM and combatant commanders. Redeployment planning transfers units, individuals, or supplies for further employment after achieving an objective.

Joint Operation Planning and Joint Operation Planning Execution Systems are the principal means through which the DOD turns policy into *OPLAN*s. These planning and execution systems provide a means to respond to emerging crisis situations or the transition to war through rapid and coordinated planning and implementation. To achieve effective responses requires command and control techniques and processes supported by computerized information systems as previously described.

The main war plans used by STRATCOM are the *OPLAN*; the *CONPLAN* with or without time phased force and deployment data (TPFDD); and the *Functional Plan*. An *OPLAN* is a complete and detailed plan that contains a full description of the concept of operations. It identifies the specific forces, support, deployment sequence, and resources required for execution. A *CONPLAN* without TPFDD is an abbreviated operations plan that requires considerable expansion to become an *OPLAN*, *Campaign Plan*, or *OPORD*. A *CONPLAN* with TPFDD contains more detailed planning for the phased deployment of forces. Detailed planning may be required to support contingencies that are critical to national security but not likely to occur in the near term. A Functional Plan involves the conduct of military operations in a peacetime environment. STRATCOM traditionally develops these plans for specific functions or discrete tasks, such as nuclear weapons recovery, logistics, or communications.

The categories that define the availability of nuclear forces and resources for planning and conducting joint operations are termed assigned, apportioned, and allocated. During normal peacetime conditions, the Secretary of Defense places assigned forces and resources under the authority of a Unified Commander in his *Forces for Unified Commands* memorandum. The Secretary of defense makes apportioned forces and resources available for deliberate planning as of a certain date. The JSCP apportions these forces for use in developing deliberate plans and may vary from the forces actually allocated for execution planning. The NCA provides allocated forces and resources for execution planning or implementation through procedures established for Command Authorities planning (CAP). During actual implementation, the NCA attaches augmenting forces to a receiving combatant commander.

In order to focus STRATCOM on strategic integration and Unified Command Plan missions, General James Cartwright, the second CMDRSTRATCOM, delegated authority for operational and tactical planning, force execution, and day to day management to four Joint Functional Component Commands and a number of Functional Components in January 2005.

Cartwright delegated the nuclear mission to a Joint Functional Component Command for Space and Global Strike that he headquartered at Offut AFB, Nebraska. He then split this Command into Space (JFCC-SPACE), and Global Strike and Integration (JFCC-GSI) Component Commands on July 19, 2006. The divided commands allowed their individual commanders to focus better focus on their primary missions, thereby optimizing planning, execution, and force management.

Currently, STRATCOM's Joint Functional Component Commands and Functional Components are:

- Joint Functional Component Command - Global Strike, Offut AFB, Nebraska. JFCC-GS develops strategic plans and assists in the development of theater plans to combat adversary weapons of mass destruction worldwide and provides integrated global strike capabilities to deter or defeat adversaries through decisive kinetic (nuclear and conventional) and non-kinetic combat effects. It is capable of rapid force execution against threats to the United States, its territories, possessions, and bases. The Commander of the 8th Air Force (AFSTRAT-GS) serves as the Joint Functional Component Commander for Global Strike.
- Joint Functional Component Command - Space, Vandenberg AFB, California. JFCC-SPACE coordinates, plans, integrates, and controls space operations to provide tailored, responsive, local and global support to other STRATCOM components, theater combatant commanders, the DOD, and non-DOD partners in support of military, national security, and civil operations, while denying the same support to an enemy. JFFC-SPACE missions include communications, intelligence, navigation, missile warning, and weather forecasting. After the amalgamation of SPACECOM with STRATCOM, The Air Force moved its headquarters from Cheyenne Mountain, Colorado, to Vandenberg AFB, California. This action resulted in NORAD moving its operations to nearby Petersen AFB, Colorado, in 2006 to reduce duplication between the sites. The Air Force has since mothballed the Cheyenne facility and renamed it the Cheyenne Mountain Directorate. The Air Force currently maintains it on "warm stand-by" for use on short notice. The Commander of the 14th Air Force (AFSTRAT-SP) serves as the commander for JFCC-SPACE.
- Joint Task Force - Global Network Operations, Arlington, Virginia. JTF-GNO directs the operation and defense of the global information grid to assure timely and secure net-centric capabilities across strategic, operational, and tactical boundaries in support of warfighting, intelligence and business missions. The Director, Defense Information Systems Agency, serves as Commander for JTF-GNO.
- Joint Functional Component Command - Network Warfare, Fort Meade, Maryland. JFCC-NW plans and executes operations in cyberspace to assure the US and its allies of freedom of action while denying this same freedom to adversaries. It works with other national organizations in Computer Network Defensive and Offensive Information Warfare as part of the Global Information Operations mission. Through its Special Technical Operations (STO) branch, it can penetrate and disrupt enemy computers to steal or manipulate data and to

crash vital systems. The Director, National Security Agency, serves as Commander for JTF-NW.
- Joint Functional Component Command - Integrated Missile Defense, Schriever AFB, Colorado. JFCC-IMD carries out planning, execution, and force management to deter attacks against the United States, its territories, possessions and bases. It is responsible for operational support, tactical level plan development, force execution, and day-to-day management of assigned and attached missile defense forces. The Commander, US Army Space and Missile Defense Command / Army Forces Strategic Command, serves as the commander for the JFCC-IMD.
- Joint Functional Component Command - Intelligence, Surveillance and Reconnaissance, Bolling AFB, Washington. JFCC-ISR develops strategies and plans, integrates national and allied capabilities, and executes DOD ISR operations to satisfy combatant command and national operational and intelligence requirements. It plans, executes, and integrates ISR activities in support of strategic and global missions. Its area of interest includes transnational threats, weapons of mass destruction, and the Global War on Terror. The Component's four Divisions are: Operations, Plans and Strategy, Assessments, and Special Activities. The Commander, Defense Intelligence Agency serves as the Commander for JFCC-ISR.
- Joint Information Operations Warfare Center, Lackland AFB, Texas. JIOWC plans, integrates, and synchronizes information operations in direct support of Joint Force Commanders. It executes both offensive and defensive intelligence operations that involve the integrated use of operations security (OPSEC), psychological operations (PSYOP), military deception (MILDEC), electronic warfare (EW), and computer network attack (CNA). A member of the Senior Executive Service commands the JIOWC.
- Center for Combating Weapons of Mass Destruction, Fort Belvoir, Virginia. SCC-WMD provides the Defense Department with expertise in contingency and crisis planning to interdict and eliminate the proliferation or use of Weapons of Mass Destruction. The Director, Defense Threat Reduction Agency commands the SCC-WMD.

Joint Functional Component Commands execute their missions by means of Task Forces that a number of Service components in turn support. Task Forces used for the execution of the Global Strike missions include:
- Task Force 124 – Airborne Communications. The Navy's E-6B TACAMO aircraft assigned to Strategic Communications Wing One, Tinker AFB, Oklahoma, provide survivable communications between national decision makers and the nation's strategic forces. The E-6B Mercury enables the President and the Secretary of Defense to make direct contact with crews on board ballistic missile submarines, land-based ICBMs, and long-range bombers.
- Task Forces 134 (Pacific) and 144 (Atlantic) – Ballistic Missile Submarines. The most survivable leg of the strategic Triad, Navy ballistic missile submarines provide a global launch capability using Trident missiles. The Navy bases its Atlantic SSBNs at Kings Bay Submarine Base, Georgia, with headquarters at

Norfolk NB, Virginia. It bases its Pacific SSBNs at Bangor NB, Washington, with headquarters at Pearl Harbor NB, Hawaii.

- Task Force 204 – Strategic Bomber and Reconnaissance Aircraft. The most flexible Leg of the Strategic Triad, the bombers assigned to 8th Air Force located at Barksdale AFB, Louisiana, are capable of deployment around the globe. B-1B Lancer conventional heavy bombers are stationed at Dyess AFB, Texas, and Ellsworth AFB, South Dakota; nuclear cruise missile armed B-52H Stratofortress heavy bombers are based at Barksdale AFB, Louisiana, and Minot AFB, North Dakota; and nuclear gravity bomb armed B-2A Spirit stealth bombers are stationed at Whiteman AFB, Missouri. Worldwide reconnaissance aircraft assigned to the 8th Air Force include the RC-135 Rivet Joint at Offut AFB, Nebraska, and the U-2S Dragon Lady at Beale AFB, California.
- Task Force 214 – Land-based Intercontinental Ballistic Missiles. The most responsive leg of the nuclear Triad, Air Force ICBMs dispersed in hardened silos provide a fast reacting and reliable component of STRATCOM's forces. Minuteman III missile launch Control Centers are located at F.E. Warren AFB, Wyoming; Malmstrom AFB, Montana; and Minot AFB, North Dakota. ICBM crews report to the 20th Air Force located at F.E. Warren AFB. The 625th Strategic Operations Squadron provides targeting and strategic communications for this task force.
- Task Force 294 – Aerial Refueling / Tankers. STRATCOM's ability to conduct global combat and reconnaissance operations with its aircraft are dependent on Air Mobility Command tankers assigned to the 18th Air Force, Scott AFB, Illinois.

The previously described re-organization demonstrates just how fast the DOD converted STRATCOM from a strategic command implementing the nuclear deterrent mission into a complex organization tasked with global strike; space operations; intelligence, surveillance, and reconnaissance; integrated missile defense; information operations; global network operations; and combating weapons of mass destruction. This dramatic change resulted in six broad, accelerating trends that have affected US nuclear enterprise and the Air Force in particular. These were:

- Reduced funding
- Reduced advocacy for the resources supporting nuclear capabilities
- Embedding nuclear missions in organizations whose primary focus was not nuclear
- An overwhelming emphasis on conventional operations
- Lowered grade levels for personnel whose daily business involved nuclear weapons
- The devaluation of the nuclear mission and those who perform it

In addition to the deficiencies identified with STRATOM's nuclear mission, a review entitled *DOD Nuclear Weapons Management*, undertaken by a Task Force led by James Schlesinger, made it clear that personnel involved in the Air Force nuclear mission did not properly understand the new Triad concept as introduced in the *2001 QDR*. The Task Force found that the Navy, had a higher level of commitment to its nuclear mission – consolidating it in two functional elements: an operational force and an acquisition / lifecycle support organization. To be fair, much of the Navy's equipment (fourteen Trident submarines and the Trident II missile) was relatively new and it managed its strategic operations from only two naval bases. The Air Force nuclear enterprise was instead comprised of ageing equipment it had distributed across the United States and Europe.

Schlesinger's Task Force also noted that Air Force problems extended beyond the issues previously identified. The Air Force desperately needed to address underinvestment in the nuclear deterrent mission, fragmentation of nuclear related authority and responsibility, ineffective processes for addressing nuclear compliance and capability issues, erosion of nuclear expertise, and a critical lack of self-assessment. Lacking a proper understanding of the nuclear mission's importance, some Air Force personnel developed a view that the nuclear mission was no longer critical and a lack of discipline ensued.

A number of events in 2006 and 2007 highlighted this trend. In one incident, STRATCOM personnel unwittingly transferred a number of W80 warheads installed in ACMs from Minot AFB, North Dakota, to Barksdale AFB, Louisiana. This incident resulted from a breakdown in the accounting, issuing, loading, and verification procedures established for the handling of nuclear weapons. In another incident, Air Force personnel shipped four AF&F assemblies (for use with the W78 warhead on Minuteman III MK-12A reentry vehicles) to Taiwan in place of helicopter batteries. This incident was again the result of procedural inattention. The Air Force only recovered the warhead components in March 2008 when notified by the Taiwanese government. Finally, the Air Force has reported incidents involving a lack of discipline and procedural awareness on the part of Minutemen crew personnel.

The problems encountered by the Air Force relate back to the increased complexity of STRATCOM, a situation created by the lack of focus of its leadership on the nuclear mission. Senior STRATCOM personnel incorrectly assumed that nuclear enterprise could sustain itself with minimal oversight while they established the Command's new missions. Nuclear enterprise was also a casualty of the Base Realignment and Closure Program that shut down the San Antonio Air Logistics Center, the sole location for Air Force nuclear sustainment. The Air Force followed this change by distributing the Special Weapons Directorate's (SWD) responsibilities to six conventional weapons centers and consolidated ICBM reentry system components with other missile system components at the Air Logistics Center in Ogden, Utah. These actions ended the specialized management of 12,000 nuclear weapons related components, which were then treated as ordinary commodities. This led to the handling of nuclear weapons components by personnel who lacked an appreciation of their new and heightened responsibilities.

Along with base closures, the DOD cut funding for Strategic Forces by roughly 65 percent during the period 1990 to 2007 – far more than any other segment of the Armed Services. It instead directed funding toward conventional forces to deal with conflicts in the Middle East. This redirection led to severe funding cuts in nuclear forces personnel and for nuclear infrastructure. Consequently, bomber and ICBM forces are currently understaffed, possessed of inadequate training, inadequate resources, and unable to meet sustainability goals. In light of the complex demands of the nuclear mission, this situation could not be more serious.

In addition to staff reductions, the Air Force reduced the seniority level of personnel within the nuclear enterprise. For instance, it replaced general officers and members of the senior executive service with colonels and mid-level civilians, leaving the nuclear enterprise without a four star general for oversight and advocacy since the early 1990s. Schlesinger's Task Force also observed that nuclear experienced officers were disadvantaged in comparison to their non-nuclear peers in selection for promotion, a clear indication that the maintenance of nuclear-trained officers has not been an Air Force priority.

In order to remedy these problems, United States Air Force formed a new major command based on the recommendations of former Secretary of Defense James R. Schlesinger's investigation into the status of Air Force nuclear surety. Secretary Schlesinger's recommendation was the creation of a single command under which the Air Force was to place all of its nuclear assets for improved accountability. Secretary of the Air Force, Michael Donley, announced the creation of Air Force Global Strike Command on October 24, 2008. AFGSC began operations on August 7, 2009. It assumed responsibility for the nuclear assets of Air Force Space Command on December 1, 2009, and the nuclear assets of Air Combat Command on February 1, 2010. Its headquarters are located at Barksdale AFB, Louisiana.

Following the release of Schlesinger's recommendations, President Obama issued new guidance to the DOD regarding the development of United States nuclear policy on June 19, 2013. The issuance of this guidance followed the detailed analysis of United States nuclear policy and deterrence requirements by the Department of State, the Department of Defense, the Department of Energy, and the intelligence community. The new guidelines implied that the United States could ensure its security and that of its allies while safely pursuing a one-third reduction in deployed strategic nuclear weapons from the level established in the *New START Treaty*. Obama declared that his intent was to seek negotiated cuts with Russia, so that both countries could move beyond Cold War nuclear postures.

The principal components of President Obama's new strategic guidelines were:
- The maintenance of a credible deterrent that will convince adversaries that the consequences of attacking the United States or its allies outweigh any benefit they might obtain through an attack
- The alignment of strategic defense with the policies of the *NPR*, which states that the United States will only use nuclear weapons in extreme circumstances to defend the vital its interests and those of its allies
- Strengthening conventional capabilities to reduce the role of nuclear weapons in deterring non-nuclear attacks
- Reducing the role of launch under attack in contingency planning since the President considered the potential for a surprise, disarming, nuclear attack exceedingly remote (The United States will nevertheless retain a launch under attack capability)
- Developing an alternative to hedging against technical or geopolitical risk with nuclear weapons leading to more effective management of the nuclear stockpile
- Retaining nuclear weapons for as long as they exist in the arsenals of other nations to guarantee the safety, and security of the United States and its allies

In maintaining a credible deterrent, President Obama believed it was necessary to update the currently deployed nuclear delivery systems, some of which dated back to the 1950s. The Obama administration then identified a trillion-dollar program that the DOD was to carry out over the next three decades. This program included:
- A replacement for Minuteman III - dubbed the "Ground-Based Strategic Deterrent" - with an updated Minuteman or a totally new missile, which may

adopt a new launch mechanism and share hardware in common with a future submarine launched ballistic missile for the Navy
- The replacement of the B-52H and the B-2A with a stealth, long range strike – bomber that will be engineered to fly unmanned conventional and manned missions with existing weapons as well as emerging and future weapons while defeating processors and sensors designed to track stealth aircraft at long range
- Development of a Follow-on Long-Range Stand-off Vehicle, or LRSO to replace the ALCM and equip the long-range strike bomber
- An SSBN-(X) replacement for the Ohio FBM submarine featuring sixteen missile launch tubes
- Maintenance of the current deterrent force during the update process

In concert with President Obama's nuclear guidelines, the DOD summed up the strategic defense plans of the United States in the *2014 Quadrennial Defense Review*, released on March 14. The *Defense Review* calls for:
- The deterrence and defeat of attacks on the United States and the support of civil authorities in mitigating the effects of potential attacks and natural disasters
- Building global security to preserve regional stability, deter adversaries, support allies, and cooperating with others to address common security challenges
- Projecting power to win military contests decisively, to defeat aggression, disrupt and destroy terrorist networks, and provide humanitarian assistance and disaster relief

Across the three pillars of the 2014 Defense Strategy, the DOD committed itself to finding creative, effective, and efficient ways to make strategic choices to achieve its goals. The DOD is identifying new defense paradigms, including the forward positioning of naval forces in critical areas, deploying new combinations of ships, aviation assets, regionally aligning or rotating ground forces, and providing crisis response forces, all with the intention of maximizing effects and minimizing costs. With its Allies, the DOD says the United States will coordinate planning to optimize their security. Reflecting the requirements of this defense strategy, the Armed Forces of the United States plan to defend the homeland, conduct counterterrorist operations, deter aggression, and assure allies through the forward deployment of its military forces simultaneously.

Although President Obama's strategic guidelines called for the maintenance of a credible deterrent and the DOD planned to project power to win military contests in a decisive manner, the *Budget Control Act (BCA)* of 2011. The *BCA* imposed caps on annual defense appropriations from 2013 through 2021, and automatic spending reductions that began in March 2013. This cap means that in future, the DOD will have to operate with a budget that is lower than that of 2010. Under the *BCA*, the DOD's budget will fall to the amount that the Department received in 2007 and will remain flat over the nine-year period during which the BCA is in effect. The timing of the *BCA* is unfortunate, because of military actions on the part of the Russian Federation and unwarranted territorial claims on made by the PRC made in 2014.

Thus, the United States decided to reduce defense spending at a time when the Russian Federation and the PRC were increasing their defense spending. This is eerily similar to the situation that existed just prior to the Korean War, but in a world filled with nuclear weapons is much more serious. Looking at the situations in Southeast Asia and Eastern Europe, the world is turning into a nuclear powder keg with Vladimir Putin and his dupes repeatedly threatening nuclear war. A similar campaign resulted in the withdrawal of nuclear forces and the non-deployment of enhanced radiation weapons by NATO in the 1980s. In order to maintain an effective deterrent and the ability to project conventional power, Congress and the Obama Administration need to rescind the BCA and increase nuclear defense spending so as not to fall behind the Russian Federation and the PRC in quality and quantity of arms and equipment. This approach was successful in forestalling an attack on Europe by Joseph Stalin in the late 1940s and 1950s and the invasion of Taiwan by Mao Tse Tung in the 1950s. Only after the United States restores a credible nuclear deterrent and upgrades its capabilities to win decisively in any military contest, will the citizens of the United States be able to enjoy their Constitutional entitlement of "life, liberty, and the pursuit of happiness".

CHAPTER 9
THE COSTS, BENEFITS, AND CONSEQUENCES OF NUCLEAR DETERRENCE

Stephen Schwarz itemized the costs of building, maintaining, and operating America's nuclear deterrent from 1940 through 1996 in *Atomic Audit: The Costs and Consequences of US Nuclear Weapons since 1940*. He put this figure at a minimum of 5.5 trillion inflation-adjusted 1996 US dollars. A subsequent estimate based on the Atomic Audit accounting methodology found that costs through 2005 totaled 7.5 trillion inflation-adjusted 2005 US dollars. Dividing by 65 years, the period over which the DOD expended this capital, yields an average yearly investment of 120 billion inflation-adjusted 2005 dollars for nuclear deterrence.

Over the period 1945 - 2009, the United States averaged 600 billion inflation-adjusted 2009 dollars per year in total defense spending. It spent approximately 470 billion dollars per year on conventional defense spending and 130 billion dollars in inflation-adjusted 2009 dollars on nuclear deterrence. Assuming that without nuclear weapons the US would have had to double the size of its conventional forces to afford the same level of deterrence provided by its combined conventional and nuclear forces, conventional defense spending through 2009 would have increased by 69 years x 470 billion dollars per year to a total of 32.4 trillion dollars. Less the outlay for nuclear deterrence in inflation adjusted, 2009 dollars, the savings to the United States for having pursued a nuclear deterrent since 1945 exceeds 25 trillion dollars. This is a vast sum of money and does not count the number of American (and Allied) lives potentially saved, since without a nuclear deterrent, regional conflicts in Europe and Southeast Asia would have been inevitable. These conflicts might well have turned global. Thus, the benefits of America's nuclear deterrent are a 25 trillion dollar saving in defense spending, a 65-year interval in which WW III did not erupt, and a 65-year period in which no sovereign power attacked the United States or its NATO allies.

To the American people, the most tangible benefits of the nation's nuclear weapons program were education, employment, and an improved standard of living. This is because a significant fraction of the trillions of dollars invested in nuclear weapons went towards salaries and benefits. The DOE and the DOD paid these salaries out to highly trained scientists, engineers, and technicians in the Nuclear Weapons Complex, to the skilled engineers and tradesmen who developed and produced nuclear delivery systems, and to the highly-trained personnel in the Armed Forces who maintained and operated the weapons and their delivery systems. Producing so many trained scientists, engineers, tradesmen, officer specialists, and technicians required a major expansion of America's educational system that continues to pay dividends today. In addition to creating a highly motivated and educated work force, industry adapted the manufacturing techniques associated with nuclear weapons to produce consumer goods that increased the nation's standard of living.

The consequences of developing and deploying nuclear weapons were not all beneficial, however. Local military conflicts replaced the global wars that nuclear weapons deterred. As well, the mining and refining of uranium, and the development, testing, production, maintenance,

The tailings pile at the Moab, Utah, uranium mill, situated alongside the Colorado River. In August 2005, the Department of Energy announced that 11 million tons of radioactive tailings would be moved by rail and buried in a lined hole some distance from the river. There are hundreds more sites such as this, albeit smaller. (Credit: DOE Photo 7631911338; Color Original)

Twelve of Hanford's 177 nuclear waste holding tanks awaiting drainage and clean up. (Credit: DOE Photo N1D0001315)

A few of the 60,000 cylinders filled with depleted UF6 lying in storage yards in the US. The cylinders weigh from 2 - 14 tons. They are stacked two high with the bottom cylinder resting on concrete or wooden chocks. (Credit: DOE / NRC; Color Original)

Low level radioactive waste being buried at the Hanford Site. Low level waste from military programs is typically buried at the sites where it was produced. (Credit: DOE via Hanford Nuclear News; Color Original)

Waste storage canisters arrive at the Waste Isolation Pilot Plant (WIPP), New Mexico, for deep underground burial in a thick layer of salt. The WIPP is reserved for the disposal of military waste. Commercial nuclear waste has yet to find a permanent home. (Credit: WIPP / DOE; Color Original)

Drums of high-level radioactive waste being moved into a salt cavern at the WIPP. Note the rock bolts and steel mesh in the ceiling, which are used to stabilize the salt layers until the tunnel is backfilled. (Credit: WIPP / DOE; Color Original)

and storage of nuclear weapons and their delivery systems caused environmental damage, created waste disposal problems, and exposed nuclear workers to radiation and toxic hazards.

After the United States stymied the USSR's intentions to annex Europe, Stalin tried to outflank the West by exporting revolutionary ideology and conventional arms to developing nations around the globe. This "Cold War" strategy replaced a head-on confrontation that could have led to the mutual annihilation of the "Superpowers." The Communist Chinese assisted their Soviet allies by exporting arms and revolutionary ideology. The Sino-Soviet strategy of outflanking the West ultimately ignited major wars in Indochina and Korea, as well as minor wars and revolutions in Africa, the Middle East, Central and South America, and the Caribbean.

To counter the Sino-Soviet strategy, the West also began exporting arms to the third world. This allowed indigenous peoples to fight proxy wars, even though they did not necessarily understand the ideologies of Communism and Democracy. They were, however, eagerly willing to accept arms and financial inducements to promote their own aims. Unfortunately, indiscriminate arms exports destabilized large regions of the third world, resulting in deaths by the millions. The international instability and misery created by the Cold War still plague us today and will do so for many years to come. Hundreds of millions of small arms and billions of rounds of ammunition litter the developing world.

The effects of the Cold War most noticeable to Americans were the casualty counts from the Korean and Vietnam wars, in which US Forces engaged the Communists alongside domestic nationalist forces. The Korean War cost 53,500 US dead and 92,000 US wounded, and the Vietnam War cost 58,000 US dead and 153,500 US wounded. Financially, the Korean War cost the United States 341 billion inflation-adjusted 2011 dollars and Vietnam cost 738 billion inflation-adjusted 2011 dollars – exclusive of military aid. The conventional war for control of the Middle East has so far cost the US between 4 trillion and 6 trillion dollars.

During the Cold War, the United States distributed some 1.6 trillion, inflation-adjusted, 2004 dollars in foreign aid over the period 1945 - 2004. This spending was in five categories:

- Bilateral assistance programs designed to foster sustainable long term, broad based economic progress and social stability in developing countries
- Humanitarian aid intended for the immediate relief of humanitarian emergencies and for refugee relief
- Multilateral assistance that was combined with contributions from other donor nations to finance multilateral development projects
- Economic aid supporting US political security interests
- Military aid composed mostly of conventional arms to developed and developing nations

The development / economic / humanitarian component of foreign aid was designed to win "hearts and minds," whereas the political / military aid component was designed to directly curb the spread of Communism.

As a percentage of total foreign aid, development / humanitarian aid totaled 40 percent, economic / political aid totaled 30 percent, and military aid totaled 30 percent. Thus, the US skewed foreign aid towards economic development as opposed to military assistance. Whereas some countries, particularly those in Africa and South America, squandered this aid, other countries like Germany, Japan, Taiwan, and South Korea used it to establish thriving economies. Thus, economic aid, which the United States distributed as an outcome of the Cold War, has provided substantial benefits to countries that invested it wisely. It has also helped build good will toward the United States in select instances.

Apart from the costs associated with foreign aid, the United States is facing an expensive cleanup associated with the production of the fissile materials used to feed the nuclear weapons complex. Concerns about the toxic and radioactive waste emanating from the nuclear weapons program emerged just after the civilian Atomic Energy Commission took charge of this operation in 1947. At this time, the AEC tasked a committee with assessing disposal methods for both radioactive waste and toxic substances. Although there was some recognition that the environmental management of nuclear waste was a serious issue, it was not until the 1970s that it became clear that the nation had to tackle the nuclear waste disposal problem. This realization resulted in the Comprehensive Environmental Response, Compensation, and Liability Act of 1980 (CERCLA), a federal law designed to clean up sites contaminated with hazardous substances. Also known as the "Superfund," CERCLA has broad national authority to deal with releases, or threatened releases, of hazardous substances that could endanger public health or the environment. The law authorizes the Environmental Protection Agency (EPA) to identify parties responsible for contamination and gives them authority to compel the responsible parties to clean up their sites and to clean up legacy sites using a special trust fund. This includes government owned and government owned - contract operated agencies.

In the spirit of environmental protection, the Department of Energy has set up a special DOE - Environmental Management (EM) Office. The mission of the Environmental Management Office is to clean up the legacy left by 50 years of nuclear weapons development and government sponsored nuclear research. The DOE-EM is currently:

- Constructing and operating facilities to convert radioactive liquid tank waste into a safe, stable form to enable ultimate disposition
- Securing and storing nuclear material in a stable, safe configuration in secure locations to protect national security
- Transporting and disposing of transuranic and low-level wastes in a safe and cost-effective manner
- Decontaminating and decommissioning facilities that provide no further value to reduce long-term liabilities and maximize resources for cleanup
- Remediating soil and ground water contaminated with the radioactive and hazardous constituents
- Fulfilling its commitments to reduce risk and complete the cleanup of all contaminated sites for generations yet to come

Not surprisingly, every major site of the Nuclear Weapons Complex has made it onto the Superfund cleanup list. Thus, the DOE-EM has a long task ahead of it. Currently, this office works with 5-year budgets and spends about 5 - 6 billion dollars a year. The DOE does not expect a full cleanup of the Nuclear Weapons Complex until the end of the 21st Century.

Probably the most widely distributed and neglected of the sites requiring cleanup are the mines and refineries that produced uranium metal for the weapons complex. There are two kinds of mines: underground and open pit. As a rule, open pit mines have greater exposure to modification by wind and rain, and underground mines have greater exposure to modification by groundwater. Due to a lack of records, it is difficult to differentiate those mines that produced uranium for the weapons industry and those mines that produced uranium for the power industry. The purpose of each mine notwithstanding, the mine sites are in deplorable condition. Again, due to a lack of reliable records, environmentalists are still stumbling across previously unknown sites.

Uranium tailings are of particular environmental concern because:
- They contain long-lived radioactive materials derived from uranium ore
- They contain toxic heavy metals and other dangerous compounds derived from the uranium ore and from processing operations
- Their porous nature makes them susceptible to leaching, erosion, and collapse
- They may contain sulfide minerals that acidify groundwater, a process that accelerates the leaching and transfer of radioactive materials and heavy elements to groundwater
- They are often thinly spread over large areas, which increases the risk of their interaction with the wind and surface water systems
- The large area occupied by tailings deposits renders the underlying land unfit for other uses

The need for uranium early in the weapons development program saw many new mining districts developed with both underground and open pit mines. Mills disposed of their tailings as close as practically possible for economic reasons, using practices common to the mining industry at that time. There were no special efforts made to improve containment security, reduce radon flux, isolate the tailings from the wind, or minimize interaction with surface and groundwater systems. The small size of many mines saw central mills service groups of mines, with tailings disposed of close to urban centers where the mills were located.

Remediation of uranium production facilities consists of activities intended to restore relict mines, mills, waste management facilities, and tailings piles to an acceptable state. At the outset of a remediation program, it is necessary for all stakeholders to agree on the use of the site following remediation. At the same time, remediation must be consistent with financial constraints. Environmental officials must consider the principles of environmental protection, sustainable development, and intergenerational equity at all stages of the remediation process.

A remediation program typically includes the following
- Plans and specifications that comply with relevant laws, regulations, license provisions, and established criteria
- Minimization of the sites residual impact taking into economic and social factors
- The control and containment of residual contaminants for as long as necessary
- Control of radon and radioactive dust emissions
- Protection of all water resources, above and below ground, from contamination
- Assessment of radiation exposure and exposure pathways to individuals who might visit, live at, or work on the remediated site
- Remediation in such a way that future maintenance requirements are minimized

The work to remediate the legacy of the uranium mining and refining industry is daunting. Between the late 1940s and early1990s, thousands of mines operated in the United States, mostly in the western half of the country. In order to keep track of this massive inventory, the Radiation Protection Division of the Office of Radiation & Indoor Air has compiled mine location information from federal, state, and tribal agency partners into a GIS accessible database. The current number of locations associated with uranium, as identified in the EPA database, is around 15,000. Of these mines, over 4,000 have documented uranium production. Three uranium mines are presently on the Superfund list, whereas others are in the *EPA Comprehensive Environmental Response, Compensation, and Liability Information System* (CERCLIS) hazardous waste database.

Turning to major industrial sites, the Superfund lists two that still require substantial cleanup. These are the Paducah and Portsmouth gaseous diffusion plants. The Paducah Site, occupying 3,400 acres, is located in rural western Kentucky near the confluence of the Ohio and Mississippi rivers. The Portsmouth Site, comprising 3,714 acres, is located approximately 75 miles south of Columbus, Ohio, in the foothills of the Appalachian Mountains. Historic operations at Portsmouth and Paducah have contaminated areas both onsite and offsite. Principal contaminants include uranium, technetium, trichloroethylene, and polychlorinated biphenyls. Through spills and disposal operations, these contaminants have entered groundwater aquifers, formed plumes, and in some cases, have migrated offsite to contaminate private drinking water wells. Since its inception, the Paducah Site has generated, stored, and disposed of hazardous and nonhazardous waste that include radioactive transuranic waste, mixed waste, and large quantities of scrap metal.

Since Portsmouth and Paducah (and the East Tennessee Technology Park, also known as Oak Ridge) came on line in the 1950s, depleted uranium hexafluoride (hex) from their enrichment operations has been stored in large steel cylinders onsite. The DOE is consequently responsible for the management of approximately 700,000 metric tons of depleted uranium hexafluoride stored in about 60,000 cylinders. These thin-walled, steel cylinders stand in open-air yards, exposed to the elements. Each cylinder contains approximately 14 metric tons of hex. If exposed to the environment, uranium decomposes to hydrofluoric acid (HF) and water-soluble uranyl fluoride (UFO_2), both of which are highly toxic. In 1986, a leaky cylinder ruptured releasing 29,500 pounds of hex into the environment. This accident resulted in one fatality and dozens of injuries. Because of this accident, plant personnel now regularly inspect hex cylinders for corrosion and seal integrity.

In July 2004, the DOE began construction of depleted uranium conversion facilities at both Paducah and Portsmouth. The intent of these facilities is to convert the sites' depleted uranium hexafluoride inventories into a more stable form of uranium for reuse or disposal. The DOE expects these conversion facilities to operate over the next 20 years. The United States Government will ultimately hold the DOE-EM responsible for the sites' deactivation and decommissioning upon completion of the uranium hexafluoride reprocessing program.

In addition to hex, metallic uranium is stored onsite at the Paducah and Portsmouth facilities. Part of this inventory belongs to the DOE and part to the United States Enrichment Corporation, which operates the facilities. Technetium 99, a beta particle emitter, contaminates a significant amount of the surplus uranium, thus eliminating its value for commercial sale. The United States Enrichment Corporation operates the only US facility for removing Tc99 at Portsmouth, sending the resultant product for further processing at Paducah. As of September 2006, 52 metric tons of the United States Enrichment Corporation's uranium inventory and 4,727 metric tons of the DOE's uranium inventory awaited technetium reprocessing to meet American Society for Testing and Materials (ASTM) standards for purity.

The Fernald Feed Materials Production Center, which produced uranium metal products in support of the national weapons program during the Cold War, is a formerly contaminated Superfund site. The Fernald Environmental Restoration Site occupies 1,050 acres in southwestern Ohio. During 37 years of operation, Fernald produced 462 million pounds of uranium metal products for use in Hanford's and Savannah River's production reactors. When operations ceased in 1989, Fernald had 15,500 tons of uranium product, 1,250,000 tons of waste, and 2.75 million cubic yards of contaminated soil and debris onsite. In addition, 223 acres of the underlying Great Miami Aquifer had uranium levels above safe drinking water standards. The most hazardous waste was radium-bearing residue from the uranium extraction of pitchblende ore. Fernald stored this waste in Silos 1 and 2 and radium bearing cold metal oxides in Silo 3. There were miscellaneous pits containing other low-level radioactive waste. The DOE declared cleanup operations at Fernald complete on October 29, 2006.

The most contaminated DOE Superfund Sites are Savannah River, Georgia, and Hanford, Washington, where nuclear reactors burned uranium and chemical plants recovered plutonium from spent fuel rods. Forty percent of the one billion curies of manmade radioactivity that exists across the American nuclear weapons complex resides at Hanford. The Richland Operations Office is managing the cleanup at Hanford with the exception of the site's liquid waste tank farms, which the US Office of River Protection manages. Over 40 years of operations, the Hanford site produced 74 tons of plutonium – about 2/3 of all the plutonium acquired by the DOE. Between 1943 and 1963, the MED / AEC built nine plutonium production reactors alongside the Columbia River. Associated plants recovered plutonium and reusable uranium from irradiated fuel using various chemical processes.

The 1,533-square-kilometer (586-square-mile) Hanford Site comprises the Central Plateau and the River Corridor. The Central Plateau, also known as the 200 Area, is an elevated region that lies above the Columbia River's water table between Hanford's 100 and 300 Areas. The 200 Area is where Hanford chemically processed spent fuel from its production reactors to separate plutonium. The plants in the 200 Area also recovered several other valuable isotopes, such as polonium. The Manhattan Engineering District initially built three separation plants in the 200 Area: T Plant, B Plant, and U Plant. The S Plant (Reduction-Oxide) and the Plutonium Uranium Extraction Plant followed later. During the 1950s, the AEC tasked the U Plant with recovering uranium that was stored in waste tanks during the rush of World War II. This area also housed the Plutonium Finishing Plant used for the production of plutonium nitrates, oxides, and metal from 1950 through early 1989. From these operations, Hanford produced a vast inventory of radioactive waste and toxic chemical materials. Historic releases of these materials have contaminated Hanford's facilities, groundwater, and soil. Hanford buried over 625,000 cubic meters of solid waste onsite and discharged more than 1.7 billion cubic meters of liquid waste containing radioactive and toxic chemicals to the ground. The current cleanup operation involves about 800 waste sites and 1,000 facilities.

The Central Plateau has ongoing waste management activities that include the storage of 2,300 tons of spent nuclear fuel at the Canister Storage Building, storage of cesium and strontium capsules in the Waste Encapsulation and Storage Facility, and storage of transuranic waste, mixed low-level waste, and low-level waste. The DOE ships transuranic waste processed at the Waste Receiving and Processing Facility to the Waste Isolation Pilot Plant in New Mexico. Non-transuranic waste is permanently disposed at the Environmental Restoration and Disposal Facility. Other activities include the burial of mixed low-level waste in trenches, the treatment of mixed low-level waste to meet regulatory requirements, the disposition of over 200 defueled naval reactor compartments in a dedicated trench, and treatment of generated liquid wastes at the Effluent Treatment Facility, Liquid Effluent Retention Facility, and Treated Effluent Disposal Facility.

In addition to the aforementioned sites, 177 underground storage tanks (149 single-shell tanks and 28 double shell tanks) in 18 tank farms store approximately 190 million curies of radiation in approximately 200,000 cubic meters of chemically hazardous radioactive waste on the Central Plateau. The DOE suspects that 67 of the 177 tanks have leaked waste to the environment. Fortunately, Hanford has completed the transfer of 12,500,000 curies and 1,100,000 gallons of waste from its 149 single shell tanks to safer double shell tanks. The DOE is constructing a Waste Treatment and Immobilization Plant to vitrify this waste. It will be the largest radioactive / chemical processing facility in the world. Retrieval of sludge / salt cake waste from the remainder of single shell tanks continues. Construction of the Hanford Integrated Disposal Facility, which will dispose of mixed low-activity and low-level wastes, is complete.

The River Corridor was home to the Hanford Site's nuclear production reactors. The MED / AEC constructed these in 100 areas. Nine nuclear reactors designated B, C, D and DR, K-west and K-east, H, F, and N were constructed and operated between 1944 and 1985 when the DOE shut down the N reactor. The DOE has now placed all of these reactors in safe storage to cool off. They await demolition and recovery, a task the DOE will not complete before the end of this century. In addition to the remediation of Hanford's reactors, the DOE must reclaim cooling ponds used to store coolant water irradiated by spent fuel. The 300 Site, also located in the River Corridor, was home to a fuel fabrication facility and laboratories that focused on improving the efficiency of plutonium production.

Following the contamination of forty-two workers with plutonium during demolition of the plutonium finishing plant from 2017, The DOE found that plutonium contamination was far more widespread than anticipated. As a result, the DOE halted demolition of the plant. Fixatives sprayed on rubble piles to reduce the movement of dust were apparently ineffective. Apparently, Hanford officials placed too much reliance on air-monitoring systems that failed to detect the spread of radioactive particles. The DOE is now reviewing its plans before further work will be carried out at the site.

Savannah River, the DOE's other plutonium production site, encompasses 310 square miles with 1,000 facilities divided into 18 site areas according to their previous missions. The DOE has currently tasked Savannah River with stabilizing, treating, and disposing of legacy nuclear materials, spent nuclear fuel, and waste. This includes the deactivation, decommissioning, and remediation of all Savannah River DOE-EM operated facilities and inactive waste units. As cleanup activities are completed, the DOE will move its operations to the site's core. The land surrounding the central core provides a protective buffer.

As part of the site's reclamation, the DOE plans to construct a facility that will treat the large quantities of waste created during plutonium production and refining. Approximately 37,000,000 gallons of this waste is being stored in 49 underground waste storage tanks. Of this volume, approximately 3,000,000 gallons are sludge waste and 34,000,000 gallons are salt waste, consisting of 16,500,000 gallons of solid salt cake and 17,500,000 gallons of salt supernate. Volume estimates are subject to change because of new waste the DOE is transferring to the tanks, the evaporation of supernate, and the removal of sludge for processing and vitrification. The DOE expects that the remediation of this waste will ultimately require the processing of 84 million gallons of salt and supernate solution. Continued long-term storage of this waste in underground storage tanks poses an unacceptable environmental risk.

Savannah River is managing low-level solid radioactive waste and solid mixed low-level waste in accordance with hazardous waste regulations and DOE Order 435.1 "*Radioactive Waste*

Management." As part of the solid waste program, hazardous mixed wastes such as lead, solvents, paint, and pesticides are stored until the DOE can arrange their transport to offsite treatment facilities. There are presently no federal disposal facilities for such substances as low-level radioactive liquid waste or spent nuclear fuel, although the Nevada Test Site and the Hanford Site may be made available for this purpose in the future, Typical low-level wastes at the Savannah River Site are radioactively contaminated job control waste, equipment, plastic sheeting, gloves, and soil. The DOE has transported some low-level wastes to NTS, Hanford, and commercial sites. The continued operation and cleanup of Savannah River depends on the site's ability to continue the shipment of waste products to offsite facilities.

As of June 2006, the Savannah River Site has received and stored the contents from 372 spent nuclear fuel casks, both foreign and domestic. It has produced more than 2,132 vitrified waste canisters in conjunction with the removal of radioactive liquid waste from the site's storage tanks. Technical improvements have gradually increased the waste loading per canister. The Site also makes shipments of transuranic waste to the DOE Waste Isolation Pilot Plant (WIPP) in Carlsbad, New Mexico. Through its soils and groundwater project, Savannah River has reclaimed 334 of 515 waste sites and decommissioned over 200 facilities. Nevertheless, recent reductions in funding have placed the Savannah River reclamation program well behind schedule and the State of Georgia has threatened to sue the DOE.

Another nuclear waste repository remains on Runit Island (Enewetak Atoll) a part of the Marshall Islands. In the late 1970s, the DOE deposited an estimated 85,000 cubic meters of contaminated topsoil and debris from the Bikini and Rongelap Atolls in the Cactus nuclear test crater beneath an 0.46-meter thick concrete dome. The dome covers the 30-foot deep, 350-foot diameter crater. This waste disposal site was supposed to be a temporary measure. However, the site remains and the DOE has yet to address leakage around its margin. Although the DOE conducts comprehensive radiological monitoring of people living on Enewetak Atoll, the exposure risks posed by the Cactus Dome have been limited to catastrophic release scenarios. Evidence, however, indicates open hydraulic communication between waste and intruding ocean water, with migration pathways leading to local groundwater and circulating lagoon waters. Colloids and dissolution / complexing reactions under low–pH anoxic likely facilitate radionuclide migration from the contaminated topsoil contained in the dome.

Apart from nuclear waste, the United States has declared 174.3 metric tons of highly enriched uranium as surplus to future weapons needs. In order to make this material unsuitable for use in nuclear weapons, the DOE will dilute approximately 85 percent of the surplus inventory for use in commercial reactors. The remaining highly enriched uranium will be disposed of as waste. A Tennessee Valley Authority vendor is handling the dilution by the extended operation of Savannah River's H Canyon through Fiscal Year 2019. This program is in accord with the mandate of the Enriched Uranium Disposition Project.

The previously mentioned Waste Isolation Pilot Plant is under the management of DOEs Carlsbad, New Mexico, Field Office and will be the final resting place for much of the transuranic waste produced by the nuclear weapons complex. Located 26 miles south of Carlsbad in Eddy County, the plant is the national site for the permanent disposal of defense-generated transuranic waste. The WIPP covers an area of 10,240 acres, with the fenced surface portion of the active site occupying 35 acres. The area surrounding the facility has a low population density that is involved in grazing and the development of potash, salt, and hydrocarbon resources.

The waste disposal area is a mined geologic repository located 2,150 feet underground in the 250-million-year-old Salado Formation, a 3,000-foot-thick strataform salt bed. Because salt is plastic and thus self-sealing, it was chosen as a host medium for the WIPP. The baseline repository layout comprises eight panels, each hosting seven disposal rooms. Each room is 4 meters high, 10 meters wide and 91 meters long. The completed repository's footprint is approximately 0.5 square kilometers. The WIPP is digging the panels sequentially because of salt's rheological (creep) characteristics. It will excavate additional panels as disposal demands arise to minimize cost and ground control requirements.

The current disposal scheme involves the stacking of contact-handled transuranic waste (CH-TRUWa) in 208-liter drums and larger standard waste boxes (SWBs) in disposal rooms and the placement of remote handled (RH-TRUWa) containers in horizontal holes drilled into the walls between disposal rooms. The WIPP places bags containing granulated magnesium oxide (MgO) on top of the stacked containers and in the void space between the containers and the walls. The MgO will minimize actinide solubility during the project's 10,000-year long regulatory period. Actinide stability is an important consideration for safe long-term storage at the WIPP, as its waste is stored in short-life containers. The MgO has other benefits, such as brine absorption.

Disposal activities at the Waste Isolation Pilot Plant have currently ceased due to a number of radioactive spills. This may disrupt plans to close the site by 2030. Decommissioning of surface facilities and permanent closure of the underground system was to take place between 2030 and 2035, after which time the placement of markers would deter inadvertent human intrusion for at least the next 10,000 years. The facility processed 9,000 shipments of radioactive waste through 2010.

In addition to sites associated with the production of nuclear materials, the Environmental Management Office is conducting cleanups at weapons design and production facilities. These include Oak Ridge; Los Alamos, Lawrence Livermore, and Sandia National Laboratories; and the Idaho, Mound, Rocky Flats, and Pantex facilities. The Oak Ridge Reservation is comprised of three facilities: The East Tennessee Technology Park, the Oak Ridge National Laboratory, and the Y-12 Plant. The East Tennessee Technology Park occupies approximately 2,000 acres adjacent to the Clinch River, 13 miles west of Oak Ridge, Tennessee. The majority of the 125 major buildings on this site have remained unused since uranium enrichment ceased in 1985. The 3,300-acre Oak Ridge National Laboratory historically supported defense production operations and civilian energy research efforts. Cleanup at this site includes environmental remediation; decontamination and decommissioning of hazardous and radioactively contaminated facilities; and the disposition of legacy low-level, mixed low-level, and transuranic waste.

Y-12 covers 811 acres and is located 2 miles southwest of Oak Ridge, Tennessee. The site was originally a uranium processing facility but now dismantles nuclear weapons components and serves as one of the nation's secure storehouses for special nuclear materials. Y-12 has numerous operating units within three areas: Chestnut Ridge, Upper East Fork of Poplar Creek, and Bear Creek Valley. Types of contamination at Y-12 include radioactive, hazardous, and mixed wastes. The sanitary landfills for all of the Oak Ridge Reservation are located here as well. Since the end of its cleanup in 2015, the Oak Ridge National Laboratory continues to operate as a world-class research facility. In addition, Y-12 continues to fulfill its national security mission. The DOE will make the East Tennessee Technology Park available for use as a private sector industrial park.

The Los Alamos National Laboratory is located in north-central New Mexico, in Los Alamos County, approximately 60 miles north-northeast of Albuquerque and 25 miles northwest of Santa Fe. The site occupies approximately 40 square miles on the Pajarito Plateau. The surrounding land is largely undeveloped. Four Native American Pueblos border Los Alamos National Laboratory, and other Federal agencies hold large tracts of land to the north, west, and south. Since its founding in 1943, the Los Alamos National Laboratory has carried out nuclear weapons research and development. In performing this mission, the Laboratory released hazardous and radioactive materials to the environment through outfalls and stack releases. Beginning in 1989, cleanup at this site characterized the disposition of legacy waste and assessed the decontamination and decommissioning requirements of its nuclear facilities. At the same time, workers staged mixed low-level and transuranic waste in preparation for off-site disposition.

There are four distinct geographical areas associated with the cleanup of Los Alamos National Laboratory: the town site, Technical Area 21, Technical Area 54, and the area's watersheds. The town site contains potential release areas associated with the Manhattan Project and early Cold War operations. These areas occupy property currently owned by private citizens and local governments. Technical Area 21 contains material disposal areas A, B, T, U, and V; former process waste lines; and a broad category of environmental sites that served process facilities in Delta Prime West and Delta Prime East including the Tritium Systems Test Assembly Decontamination and Decommissioning Facility. Technical Area 54 is both a former and the current waste disposal area for the Laboratory. The scope of work for decontamination and decommissioning requires the cleanup of major material disposal areas G, H, and L. Sites scheduled for investigations and cleanup other than that G, H, and L are included within the watersheds. There are eight watersheds across the laboratory, which collectively drain into the Rio Grande. Within these watersheds are more than 650 sites that still require investigation and remediation.

Los Alamos National Laboratory has developed a plan to cleanup of its EM Legacy Waste Sites. This plan integrates the retrieval and disposition of legacy transuranic waste, decommissioning, and decontamination of excess facilities at Technical Areas 21 and 54, and the remediation and site restoration at 760 remaining Solid Waste Management Units. The current transuranic waste disposition strategy ships legacy transuranic waste and mixed low-level waste to the Waste Isolation Pilot Plant in New Mexico. The laboratory's environmental restoration strategy complies with all regulatory requirements thus providing for future land use scenarios, protecting and monitoring the regional aquifer, and installing long-term surveillance and monitoring systems. The estimated end date for cleanup operations for Los Alamos is currently 2015.

Like Los Alamos, Lawrence Livermore National Laboratory is a multi-disciplinary research and development center focusing on weapons development, nuclear stewardship, and homeland security. The Environmental Management program for this site accomplished the complete disposition of its legacy waste in 2005; the transfer of the Newly Generated Waste Program to the National Nuclear Security Administration in 2006; completion of the Lawrence Livermore National Laboratory Main Site remedial activity build-out in 2006; and completion of Site 300 remedial activity build-out in 2008. Starting in 2007, the National Nuclear Security Administration became responsible for the long-term stewardship of the Lawrence Livermore National Laboratory Main and 300 Sites.

The only Environmental Management Program awaiting completion at Lawrence Livermore National Laboratory is the Site 300 Environmental Restoration Project. Site 300 is an 8,000-acre preserve located about 15 miles east of Livermore, California. LLNL used it primarily for explosive hydrodynamic testing and analysis of weapons components. The surrounding area has limited agricultural development and a sparse population. Site 300 has onsite soil and groundwater contamination and limited groundwater contamination offsite. The major contaminants are volatile organic compounds, tritium, depleted uranium, and perchlorate, nitrate, and high explosive compounds. To date, the project has completed construction, installation, and operation of 20 treatment systems.

Sandia National Laboratories, New Mexico, carries out a broad range of scientific and technical research and development programs. It is located in Bernalillo County, New Mexico, 6.5 miles east of downtown Albuquerque. The Laboratory consists of five technical areas and several remote sites covering 2,820 acres in the east half of Kirtland AFB. The base sits on two broad mesas bisected by the Tijeras Arroyo. The Manzano Mountains border it to the east and the Rio Grande River to the west. The Sandia National Laboratories Environmental Restoration Project is focused on the remediation of inactive waste disposal and release sites at Albuquerque and other offsite locations. These sites all have known or suspected releases of hazardous, radioactive, or mixed waste.

Except for three sites, environmental remediation at Sandia National Laboratories is complete. Two hundred and sixty-eight sites have been subject to investigation and potential corrective action. Three of these sites will remain as "deferred active mission sites" that will require future remediation. All remaining remediation activities, includes installing a cover and a rock bio-barrier at the Mixed Waste Landfill, and groundwater sampling as prescribed by the Corrective Measures Evaluation process. Long-term stewardship includes all activities necessary to ensure continued protection of human health and the environment after remediation, disposal, or stabilization at this site.

Prior to generating the aforementioned wastes, approximately 30,000 people received exposure to radiation in the course of mining and processing uranium ore in the United States. Worldwide, possibly twice this number of people received exposure because the United States received substantial shipments of uranium ore or yellowcake from the Belgian Congo, Australia, and Canada. Because uranium deposits are often associated with much more dangerously radioactive elements such as radium, even short-term exposure to its ores can cause health problems. In Canada, some deposits had such high radioactivity that they could only be mined using robotic equipment. In enclosed areas without proper ventilation, radioactive radon 222 gas can build up as a decay product of uranium. Since very little was known about the human effects of radioactivity until the 1950s, intense exposure was received by many early miners who inhaled this gas. It has subsequently been determined that a statistically significant number of these men developed small cell carcinoma after Ra222 exposure in unventilated mines or buildings. Some of the survivors from this era or their descendants have received compensation under the "1990 *Radioactive Exposure Compensation Act.*"

In producing the aforementioned wastes, approximately another half million scientists, engineers, and workers received exposure from radioactive substances and toxic chemicals through their employment in the Nuclear Weapons Complex. In addition to the exposure of these workers, technical personnel, military personnel, and the world's population received exposure to fallout from nuclear testing at the Pacific and the Nevada test sites and by the tests conducted by other nuclear powers. The MED carried out the first test explosion of a nuclear weapon, Trinity, in New Mexico on July 16, 1945. The Soviet Union tested its first nuclear weapon at near Semipalatinsk, Kazakhstan, in September 1949. The US, the USSR, and the UK continued atmospheric nuclear testing nuclear until 1963, when they signed the Limited Test Ban Treaty. France and China, neither of which signed the 1963 treaty, pursued atmospheric testing from 1960 through 1974 and 1964 through 1980, respectively. Altogether, these nuclear powers exploded 504 nuclear devices at 13 test sites, yielding the equivalent explosive power of 440 megatons of TNT.

As evidenced by the focus of science fiction during the 1950s, the earliest concerns about health effects from exposure to fallout focused on genetic alterations among the offspring of the exposed. However, medical scientists have not observed inheritable effects from radiation exposure during follow up studies of Hiroshima and Nagasaki survivors. Instead, radiation exposure studies have demonstrated a risk of leukemia and thyroid cancer within 10 years of exposure and increased risk of solid tumors later on. Thus, the studies of populations exposed to fallout point to an increased cancer risk as the primary health effect of radiation exposure. These studies also identify the specific radionuclides implicated in fallout related cancers and other late exposure effects.

Scientists classify the radiation risk from nuclear fallout in terms of distance from a detonation or test site. "Local fallout" is within 50 - 500 kilometers from ground zero, "regional fallout" 500 - 3,000 kilometers from ground zero and "global fallout" describes worldwide patterns. Because fallout disperses with time and distance from the explosion, and radioactivity decays over time, the highest radiation exposures are generally in areas of local fallout. The radiation risk is both external and internal if fallout contaminated food is consumed. Individuals who spend more time outdoors receive the highest doses of radiation exposure. Finally, the isotopes to which an individual is exposed vary with time and distance, based on the radionuclides' specific half-lives.

The activity of fallout is currently measured in Becquerels (Bq), defined as the number of radioactive disintegrations per second. As described in Chapter 5 the radiation-absorbed dose (RAD) is the energy per unit mass imparted to a medium (such as tissue). The basic international unit used to characterize radiation dose is the gray (Gy), defined as the absorption of 1 joule of energy per kilogram of tissue. The Sievert (S) measures the equivalent dose of radiation having the same effect as an equal dose of gamma rays. (*The International System of Units* (ISU) is gradually supplanting the previous system based on dose units of RAD and radiation equivalent man (REM).) Conversion is, however, straightforward: one Gy equals 100 RAD. In order to compare the effects of fallout with natural sources of radiation, it is useful to know that the annual exposure from natural radiation sources is 1 milligray (mGy, one-thousandth of a Gray). In addition, the dose from a whole-body computer-assisted tomographic (CAT) scan is about 15 mGy and the dose received from cosmic rays during a transatlantic flight is 0.02 mGy.

The primary source of fallout exposure to Americans was radioactive products released by atmospheric nuclear testing at the Nevada Test Site. Secondary sources of fallout exposure were products from atmospheric tests conducted at the Pacific Proving Grounds and from atmospheric tests conducted by foreign powers. Recently, radioactive doses received by Americans from atmospheric testing were estimated using mathematical models and historical fallout concentration data.

The NTS conducted atmospheric nuclear tests from early 1951 through 1962. It exploded 86 of these tests at or above ground level. In addition, it conducted fourteen underground tests that released significant amounts of radioactive material to the atmosphere. The DOE ascribes doses from internal exposure within local and regional fallout areas to the inhalation of radionuclides and the consumption of food contaminated with radionuclides. Doses from internal irradiation were smaller than doses received from external irradiation, the only significant exception being the thyroid. Estimated thyroid doses were mainly due to consumption of milk and milk products contaminated with iodine 131. The DOE estimates the average thyroid dose from fallout for adults living within 300 kilometers of the NTS at 5 mGy, with 2 mGy coming from exposure to iodine 131. For children, who consumed large amounts of milk products, the DOE estimates total exposure at 35 mGy, with 30 mGy coming from exposure to iodine 131. This exposure level is consistent with 49,000 related thyroid cancer cases in the United States, almost all of them among persons who were children during the period 1951 - 1957.

For global fallout, the ratio of radionuclides that contributed to exposure in the United States differed from NTS fallout, because radioactive debris injected into the stratosphere takes several years to deposit, during which time short-lived radionuclides like I131 decayed. Of greatest concern from global fallout are two longer-lived radionuclides: strontium 90 and cesium 137, which are associated with the development of leukemia. The average internal dose to red bone marrow from global fallout received by adults living within the continental United States is 0.6 mGy and the average external dose is 0.7 mGy. This compares with 0.1 mGy internally and 0.7 mGy externally from NTS derived fallout. About 22,000 radiation related cancers, half of them fatal, might eventually result from red bone marrow external exposure to NTS and global fallout.

CHAPTER 10
NON-PROLIFERATION AND THE MANAGEMENT OF THE US NUCLEAR STOCKPILE

President Barack Obama highlighted the problem of nuclear proliferation in a 2009 Prague speech in which he declared that to meet 21st Century nuclear dangers, the United States must "seek the peace and security of a world without nuclear weapons." He emphasized that implementing such an ambitious undertaking would be a slow process – likely unattainable in his lifetime. The President, however, expressed his determination to follow a path of disarmament by reducing the absolute number of nuclear weapons and their role in US national security. As a caveat, the President pledged that as long as nuclear weapons exist, the United States would continue to maintain a safe, secure, and effective nuclear arsenal to deter adversaries and to assure allies and other security partners that they can count on America's security commitments.

The President demonstrated his commitment to a nuclear free world when he signed the *New START Treaty* with Russia's President Medvedev on April 8, 2010. With the ratification of the *New START Treaty*, the US and Russia must significantly reduce their strategic nuclear arsenals within seven years of the treaty entering force. As with *SORT*, each signatory has the flexibility to determine the structure of its strategic forces within limits set by the Treaty. For its part, the United States based these limits on a study conducted by DOD planners in support of the *2010 Nuclear Posture Review*. The limits are 74 percent lower than those established for the *1991 START Treaty*, and 30 percent lower than the deployed strategic warhead limit established for the *2002 SORT Treaty*.

The *New START Treaty* limits aggregate nuclear deployments to a total of 1,550 strategic warheads. All of the warheads on deployed ICBMs and SLBMs count toward this limit, but each deployed nuclear-capable heavy bomber counts as only one warhead. Each side is also limited to a combined total of 800 deployed and non-deployed ICBM launchers, SLBM launchers, and nuclear-capable heavy bombers. The signatories are further limited to a total of 700 deployed ICBMs, SLBMs, and nuclear-capable heavy bombers. The difference between deployed and non-deployed launchers takes into account the servicing requirements for ballistic submarines and silos. Verification measures under the treaty include onsite inspections, data exchanges, notifications related to the strategic offensive arms and facilities covered by the Treaty, and provisions to facilitate the use of national technical means for monitoring. To increase confidence and transparency, New START also provides for the exchange of telemetry.

The Treaty's duration is ten years, unless superseded by a subsequent agreement. The Parties can agree to extend the Treaty for a period of no more than five years. The Treaty includes a withdrawal clause that is standard in arms control agreements. The *2002 SORT Treaty* terminated after the US Senate and the Russian Legislature approved the *New START Treaty*. It entered force on February 5, 2011. The *New START Treaty* contains no constraints on the testing, development, or deployment of current or planned US missile defense programs or long-range (GLOBAL) conventional strike capabilities.

Under New START, the United States will deploy a force of 400 Minuteman ICBMs, some of which will be armed with two warheads. It will further deploy a force of 12 Ohio SSBNs

One of two linear-induction electron accelerators arranged at right angles in the Dual-Axis Radiographic Hydrodynamic Test Facility, DARHT. Focused on a metal target, the electron beams are deflected, converting their kinetic energy to powerful x-rays used to probe implosion hardware. (Credit: LANL via Los Alamos Press release; Color Original)

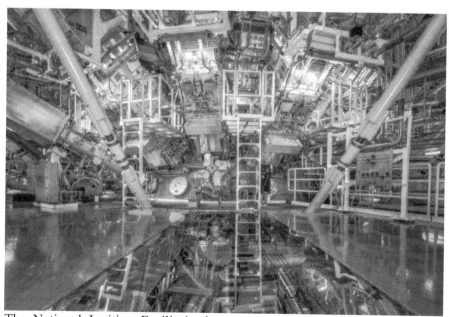
The National Ignition Facility's, laser target chamber where temperatures reach 100 million degrees and pressures up to 100 times the density of lead. Tests at the NIF pursue data concerning fusion power and fusion weapons. (Credit: LLNL - Damien Jemison, Photo NIF-875; Color Original)

The Los Alamos Neutron Science Center, LANSCE, is one of the world's most powerful linear accelerators. The Proton Radiography Facility, seen here, uses 800 MeV protons to investigate dynamic experiments in support of national weapons science and stockpile stewardship programs. (Credit: LANL - Rob Kramer; Color Original)

LANL's Atlas generator consists of twelve capacitor banks that surround a central target chamber. It provides the largest current pulse, 30 million amps, of any capacitor bank in the world. LLNL has a similar device called Jupiter. (Credit: LANL)

armed with 240 SLBMs, each of which will carry four warheads. Finally, it will field about 40 B-52G and 20 B-2A heavy nuclear bombers.

President Obama revealed his future course of action regarding nuclear forces, nuclear weapons, and nuclear infrastructure in the 2010 Nuclear Posture Review. As will be seen, his course of action represents a significant change from his predecessors. In the 2010 Nuclear Posture Review, President Obama reaffirmed his commitment to nuclear disarmament and focused on five key objectives relating to nuclear weapons, policies, and posture:

- Sustaining a safe, secure, and effective nuclear arsenal
- Reducing the role of US nuclear weapons in the national security strategy
- Maintaining strategic deterrence and stability at reduced force levels
- Strengthening regional deterrence and reassuring US allies and partners
- Preventing nuclear proliferation and nuclear terrorism

With regard to nuclear proliferation, President Obama declared that the United States would refrain from developing any new types of warheads in his 2010 Nuclear Posture Review, noting that the United States "can ensure a safe, secure, and effective deterrent without the development of new nuclear warheads or further nuclear testing". This is very likely true but may require both the retirement and the continued manufacture of some existing types of warheads in order to maintain an effective deterrent.

With regard to maintaining the current nuclear arsenal, he agreed with the objectives of the Science Based Stockpile Surveillance Program, which the NNSA implemented in 1995. The purpose of this program is to preserve "the core intellectual and technical competencies of the United States in nuclear weapons." In order to support this program, the NNSA has further developed a National Security Strategic Plan for nuclear weapons stewardship in the absence of nuclear testing under the Comprehensive Test Ban Treaty. In addition to monitoring the state of warheads in deployment or storage, this plan recognizes the need for improving the understanding and modeling associated with the reduced numbers of warheads and warhead types that remain in the stockpile. Individual program elements of the Science Based Stockpile Surveillance Program include:

- Hydrotesting using the Dual-axis Radiographic Hydrotest Facility (DARHT) and the Advanced Hydrotest Facility (AHTF) by LANL
- Inertial confinement fusion (ICF) studies carried out at the National Ignition Facility (NIF) by LLNL
- Neutron scattering studies for stockpile surveillance at the LANL Center for Neutron Scattering Studies (LANSCE)
- Pulsed power for weapons diagnostics and effects programs at the ATLAS (LANL) and JUPITER (SNL) facilities
- Special nuclear materials processing evaluation
- Advanced computing studies for improved stewardship

Despite the Science Based Stockpile Surveillance Program's ability to detect problems with nuclear weapons, the implementation of Life Extension Programs to correct these problems has become problematic because of facility closures within the national nuclear weapons complex. Depending on the urgency for new components, increased demand for the production of refurbished components may require new or expanded NNSA and commercial facilities. The lack of certain essential materials, coupled with changes in environmental and occupational safety standards has created sunset technologies for which the nuclear laboratories must find substitutes – without the benefit of testing. All of these factors affect the ability of the nuclear weapons complex to perform its functions.

Since nuclear testing is not permissible under the Comprehensive Test Ban Treaty, the DOE's nuclear laboratories must make design calculations for nuclear warhead Life Extension Programs using assumptions about how materials behave under the extreme conditions of a nuclear explosion. Scientists have to base many of these assumptions on classical data or on new, theoretical, Science Based Stockpile Surveillance Program predictions that they cannot verify under the CTB Treaty. The phrase "thoroughly tested" thus does not apply to warheads developed in the post CTB era or to those warheads that have been LEP modified.

For reasons that become apparent below, any attempt to produce a replacement warhead identical with the original is impossible. The problems with replica manufacturing arise from three considerations:

- Material batches are never the same, materials become unavailable, equivalent materials are not fully equivalent, vendors have gone out of business or stopped producing products, and new health and safety regulations prohibit the use of some critical materials and processes
- The documentation for many manufacturing processes is often no longer available – no one knows the manufacturing processes used for some materials or parts
- Only a nuclear test can provide the type of reliable data needed for certification

Recall the problems associated with the B61, the W68, and the W84 warheads. In these weapons, measured yields fell short of the predictions made by experienced weapon designers. This happened because nuclear weapons are extremely complicated; they operate under conditions that include material velocities of millions of miles per hour, temperatures and pressures that are hotter and denser than the center of the sun, and time scales as short as a few nanoseconds. Also, recall the W76 warhead, for which designers spent several years solving the problem of remanufacturing the Fogbank material used in its interstage region.

The assessment of weapon performance, whether for stockpile inspection, design, or remanufacture, thus depends to a large degree on scientific and engineering judgment. It takes years for designers to gain the experience on which they base their judgment. In the past, junior designers gained experience when senior designers passed on their knowledge as part of the weapons development process. Since the United States stopped testing, this continuity of experience has ended. The scientists and engineers whose judgment was honed by continued weapons development and testing have retired or moved on to other fields. The NNSA now faces the prospect of designers who lack practical experience when making judgments about the remanufacture and the maintenance of the nuclear stockpile.

Since the current nuclear stockpile will be the primary instrument of nuclear deterrence for the near future, the NWC approved *Phase 6.X Procedural Guidelines* to manage its life extension process in April 2000. These guidelines mirror the original DOE Joint Nuclear Weapons Life Cycle Process. The NNSA intends the *Phase 6.X guidelines* to govern legacy warhead refurbishment programs that may be concerned with a single component or many components in any given warhead type. Depending on the extent of refurbishment required, the NNSA labels the resulting warhead as an alteration (ALT) or a modification (MOD). It defines a refurbished warhead as an ALT, if modification has not changed its operational characteristics, control, or safety features, and these changes are "transparent" to the DOD user. If end use is affected, it labels the converted weapon a MOD. The DOD / DOE *Phase 6.X Guidelines* follow.

- Phase 6.1. The need for conducting life extension program arises from maintenance procedures, component testing, and ongoing studies conducted by the DOD, NNSA, and various POGs. These activities may identify a warhead or warhead component that warrants a *Program Study*. The DOD, NNSA, and POGs are each free to develop their own refurbishment ideas subject to impact analysis by their fellow agencies. The DOD conducts phase 6.1 activities jointly with the NNSA, after these agencies inform the NWCSSC about the planned activity in writing. After the NNSA / DOD complete the concept assessment phase, they may submit a written recommendation to the NWCSSC to proceed to Phase 6.2. The NWCSSC then makes a final determination whether it should authorize a Phase 6.2 Study.

- Phase 6.2. Typically, the DOE's various laboratories develop a number of life extension design options at the request of the NNSA. The POG then evaluates the options for feasibility, considering design, technical risk, life expectancy, R & D requirements, safety, qualification and certification requirements, production capabilities and capacities, maintenance and logistics issues, and delivery system and platform issues before deciding on whether to conduct refurbishment. The Program Officers Group assigned to a weapon type manages all feasibility and down-select studies carried out during the second stage of the life extension process. Phase 6.2 also includes a review of fielded and planned support equipment, and the technical publications associated with the refurbished weapon system. Military considerations, which the POG evaluates in conjunction with the design process, include operational impacts and the benefits derived from identified design options, operational measures, and requirements for joint non-nuclear testing. Various agencies update the refurbished weapon system's military characteristics, its stockpile to target sequence, and interface control documents during this phase. The POG then uses the weapon's refurbishment options to prepare a "down-select" package. This package includes impacts on the nuclear weapons complex, which the NNSA documents in a *Major Impact Report*. After releasing the MIR, the NNSA and the Lead Service coordinate regarding the down-selection of preferred option(s), after which the NWCSSC authorizes Phase 6.2A.

- The POG coordinates the Phase 6.2A design definition and cost study by incorporating NNSA and Service inputs into a *Joint Integrated Project Plan* (JIPP). The NNSA, the DOE's labs, and the production facilities then develop cost estimates for the design, testing, production, and maintenance activities for the planned LEP refurbishment. (The *Weapon Design and Cost Report* (WDCR) enumerates these estimates). The POG presents its information together with the estimated DOD costs to the NWCSSC along with a recommendation about whether to proceed to Phase 6.3. The NWCSSC then determines whether it should authorize Phase 6.3.

- Phase 6.3. Development engineering begins when the NWCSSC prepares a letter requesting joint DOD and NNSA participation. If the DOD and the NNSA agree to participate, the NNSA, in coordination with the DOD, conducts experiments, tests, and analyses to validate the design option(s) and the proposed MCs and STS. At this time, the DOE's production facilities assess the producibility of the design, initiate process development, and produce test hardware. They then formally update the WDCR into a *Baseline Cost Report*. They also prepare a draft addendum to the *Final Weapon Development Report* (FWDR). The report outlines the status of the refurbished design along with proposed qualification procedures, ancillary equipment requirements, and project schedules. The DOD Design Review and Acceptance Group then reviews the *Draft Addendum to the FWDR* and publishes its recommendations regarding the project. The appropriate Service forwards this material to the NWCSSC for approval. At the end of Phase 6.3, the POG updates the JIPP and prepares a final *Product Change Proposal* (PCP). At this time, the design is ready for release to the production facilities for stockpile production preparation activities.

- Phase 6.4. When development engineering is sufficiently mature, the NNSA authorizes production engineering. The NNSA intends this phase to adapt the developed design into a producible weapon, as well as prepare the production facilities for component production. Tooling, gauges, and test equipment must be defined and qualified; process development, process prove-in and process qualification are completed; materials are acquired and trainer components fabricated. Phase 6.4 also defines the procedures needed for refurbishment and the production of components. The NNSA updated production cost estimates based on preliminary experience from the PPI and product qualification. At this point, the DOE makes provision for spare components in conjunction with the DOD, and the Laboratory Task and Joint Task Groups update technical publications. The NNSA updates its *Stockpile Evaluation Program* (SEP) plan and the POG updates the JIPP. Generally, Phase 6.4 ends when the NNSA indicates that all production processes, components, subassemblies, and assemblies are qualified.
- Phase 6.5. DOE facilities produce the first refurbished weapons known as "First Production" weapons. The DOD and the NNSA then evaluate these rebuilt weapons as to their suitability and acceptability. Except in an emergency, the preliminary evaluation by the NNSA does not find a weapon acceptable for operational use. The NNSA and the DOE Laboratories only make a final determination after have completing a detailed engineering evaluation program. The POG then informs the NWCSSC that the LEP refurbishment program is ready to proceed to IOC and full deployment of the refurbished weapon.

The Lead Service for the refurbished weapon conducts a *Pre-Operational Safety Study*, 60 days before IOC or first weapon delivery. As part of this study, the NWSSG examines system design features, hardware, procedures, and operational aspects to determine whether the refurbished nuclear weapon will meet DOD nuclear weapon system safety standards. The responsible labs prepare a *Final Draft of the Addendum to the FWDR* and submit the document for final DRAAG review. If the addendum meets DRAAG approval, it informs the NNSA that the weapon meets the requirements of its MCs through the NWCSSC and the Lead Service. The responsible labs then complete a final addendum to the FWDR and a certification letter and the

POG updates the JIPP. After the evaluation of a limited production run and other reviews are completed, the NNSA issues a MAR releasing the first refurbished weapons to the Service. With the MAR, the NNSA advises the DOD that the refurbished weapon is suitable for use and the POG requests approval to proceed to Phase 6.6.

- Phase 6.6. After NWC approval, the POG requests that the NNSA undertake the full-scale production of new components and refurbish weapons as DOD removes them from the stockpile. After refurbishment, the POG prepares an end of project report for the NWCSSC to document the activities carried out during the Phase 6.X Process. Phase 6.6 ends after the refurbished weapons have re-entered the stockpile and all necessary certification and documentation is complete.

In order to assess President Obama's pronouncement on nuclear enterprise, Congress had William J. Perry and James R. Schlesinger chair a Select Committee that produced *America's Strategic Posture: The Final Report of the Congressional Commission on the Strategic Posture of the United States.* The Committee found that the President's program was feasible, providing that the DOD made wise choices in its selection of future weapon systems, that the US Government provided adequate funding to upgrade the existing nuclear manufacturing infrastructure, and that the government provide adequate funding to purchase new nuclear weapon systems. This the Obama Administration has not done.

To begin rebuilding the nuclear stockpile, the NNSA and the DOD require funding. This means that Sequestration (the limits placed on military spending by Congress) should be cancelled, military spending increased, and part of the funding for conventional weapon systems be diverted to nuclear weapon systems.

President Eisenhower invoked the New Look to save defense dollars by replacing a conventional Army with a nuclear Army and General Curtis LeMay observed that had nuclear weapons been used in a single strike at the outset of war in North Korea, that country might have ceased hostilities. He further observed that the USAF burned down all the cities targeted for nuclear destruction with incendiaries – killing their inhabitants anyway. Thus, his nuclear strike might have saved a million Chinese, North Korean, South Korean, and UN soldiers and civilians in the miserable years of war that followed North Korea's ill-advised invasion. Had the United States, Britain, and France launched a limited nuclear strike on Iraq in 2001, they likely could have forced a quick closure to unrest in the Mideast. The current war in Syria and the rise of ISIS are the end result of not taking a firm hand before Middle Eastern social and religious unrest ran out of control. The last, 25 years of fighting in the Middle East has cost the US 6 trillion dollars, one third of the national Debt! Reinvigorating the nuclear deterrent is estimated to cost only 1 trillion dollars.

Neither the Russian Federation nor the People's Republic of China are interested in non-proliferation, whether they have signed the *Nuclear Non-Proliferation Treaty* or not. Prior to the collapse of oil prices, the Russian Federation planned to spend 20 trillion rubles over the next 5 years updating its military forces, both conventional and nuclear. In fact, the Russian Military is introducing new technologies into its nuclear arsenals such as maneuvering reentry and hypersonic reentry vehicles, a fractional orbital bombardment system, an IRBM, and ground launched cruise missiles. It is also deploying improved strategic missiles such as the Sarmat heavy ICBM and the road-mobile Yars. Finally, it has Moved its Euro-bomber force into Crimea and East Prussia where satellite photos show TU-22M nuclear bombers parked on runways. It has moved its SS-26 nuclear capable SRBMs into positions much closer to Europe and has deployed dual capable shore defense missiles to Crimea and the Kuril Islands which it seized from Japan at the end of WW II.

The People's Republic of China is moving in the same direction as the Russian Federation. It is completing its fifth second-generation SSBN and has begun building its first third-generation SSBN. It likely intends to operate a fleet of ten SSBNs by 2020. To arm these vessels, the PLAN has been steadily updating the capabilities of its SLBMs, which can now reach Hawaii and Alaska when launched from the sheltered waters around coastal China. The PRC has shown a predilection for developing cruise missile of all types, many of which are nuclear capable. It continues to deploy new and improved ICBM, including road-mobile versions. The PRC has also begun building aircraft carriers with which to project its influence, possibly through the employment of nuclear-capable naval aircraft.

Looking at the military preparations by America's adversaries, the United States needs to act on its recent commitments to overhaul and update the nuclear triad. Although President Trump has increased military budgets and is supporting projects to produce modern nuclear bombers, ICBMs, SSBNs, and SLBMs, he has yet to address the inadequate yields of America's nuclear bombs and warheads and the need for new weapons production facilities. NATO nuclear members, Britain and France, are apparently following in the United States' footsteps with Britain intent on developing larger yields for it SLBM warheads. These actions are long past due.

CHAPTER 11
BIRTH OF THE ARSENAL: THE CLASSIC FAT MAN ATOMIC BOMB

America's nuclear weapons program began with a meeting between President Franklin Roosevelt and lobbyist Alexander Sachs at the White House on October 11, 1939. During this encounter, Sachs presented Roosevelt with a letter from Albert Einstein and a joint memorandum from noted Hungarian physicists Leo Szilard, Eugene Wigner, and Edward Teller. The subject of these documents was the use of atomic fission to build a bomb of unparalleled destructive power. Sachs tactfully outlined the peaceful aspects of fission before indicating that Germany, which had just invaded Poland, was pursuing the atom's military potential. Roosevelt was motivated to take immediate action.

Within a month of his meeting with Sachs, Roosevelt set up the Advisory Committee on Uranium. It consisted of Lyman Briggs, the Director of the Bureau of Standards, and noted ordnance experts Lieutenant Colonel Keith Adamson and Commander Gilbert Hoover. Roosevelt specifically directed the Committee to pursue the matter of atomic bombs. By the summer of 1940, the threat of armed conflict with Germany had become so acute that Roosevelt established the National Defense Research Committee with a mandate to appraise a broad range of new military technologies. Because the NDRC and the Uranium Committee had overlapping objectives, the NDRC absorbed the Uranium Committee to form the nucleus of its (S1) Atomic Sub-Committee.

The incorporation of the NDRC and various military laboratories into an umbrella organization known as the Office of Scientific Research and Development brought an expanded mission. In addition to conducting scientific investigations, the Roosevelt Administration empowered the OSRD to pursue the engineering development of new military technologies. Dr. Vannevar Bush, the former president of the Carnegie Institution and head of the NDRC, became the director of the OSRD on June 28, 1941. He immediately ordered the National Academy of Sciences to determine the state of knowledge on atomic fission. Supplemented with information supplied by the British Military Application of Uranium Detonation, or MAUD Committee, the NAS found that the radioactive elements uranium 235 and plutonium 239 had potential as nuclear explosives.

Following the entry of the United States into the Second World War, the Roosevelt Administration created the Manhattan Engineering District (MED) to produce nuclear explosives and to fabricate the explosives into bombs. This program informally became known as the "Manhattan Project," even in official circles. The Army promoted Colonel Leslie J. Groves, Deputy Chief of Construction for the US Army, to Brigadier General and ordered him to oversee the MED on October 23, 1942. As his scientific deputy, he chose Dr. J. Robert Oppenheimer, a brilliant theoretical physicist and self-made experimentalist from the University of California at Berkeley. Not everybody's choice as scientific director, Groves insisted on "Oppie" because of his general knowledge and his affability. Oppenheimer immediately demanded the establishment of a central atomic laboratory for the design of atomic weapons. At such an institution, scientists and engineers would be able to share their findings and come to grips with the many difficult technical problems that lay ahead.

A Silverplated Boeing B-29 bomber – Bock's Car – which dropped the Fat Man atomic bomb on Nagasaki. (Credit: USAF: Color Original)

The Model 1561 "Fat Man" atomic bomb being readied for loading on Bock's Car for the attack on Nagasaki. The four "headlights" are impact detonator connected to the explosive sphere with det cord. The arming plugs can be seen top right. (Credit: National Archives - Cropped)

The plutonium core for a second wartime Fat Man bomb is shown resting in an arrangement of tungsten carbide blocks for a criticality experiment. This core was detonated in the bomb used for the Crossroads Able shot, in 1946. Note the ruler for scale. (Credit: LANL Photo CIC-9:2498)

Shown in the center of the photo is a cylinder of uranium. It ft into a uranium sphere (the tamper), which had a cylindrical hole cast in it. The cylinder was split in halves and had an internal receptacle for the the bomb's plutonium core and internal neutron initiatorr. The cylinder completed the bomb's tamper when inserted into the sphere. Surrounded by a spherical aluminum pusher, these components formed the bomb's "pit". (Credit: Cropped from LANL Photo TR 229)

A partially assembed exposive sphere for a Fat Man bomb. It measured 55 inches in diameter and weighed 2 3/4 tons. The exterior of the bomb's nuclear pit (cork gasget surrounding aluminum pusher).can be seen nestled amonst the explosive blocks. (Credit: LANL)

A detonator location on the assembled Gadget device. The detonator lies between two metal cylinders, which contain diagnostic equipment. The two white wires provide redundant power to fire the detonator, which fits into a copper well located in its explosive block. In the lower left and right are redundant power distrbution boxes that fire the detonater network. (Credit: Cropped from LANL Photo TR-311)

238

A Model 1561 Fat Man bomb assembly on Tinian Island. On the right side of the explosive sphere are a drum and cone, which faced forward inside the bomb. Batteries were attached to the A Plate at the front of the drum and the x-unit was attched to the B Plate at the rear of the drum. The bomb's radar, barometric, and arming and fuzing elements were attached to the C Plate, which was held by a cone on the rear side of the explosive sphere (not shown). (Credit: LANL)

Two Convair B-36F Bombers flying in formation. Note the outboard jet engines. A B-36 could carry four Fat Man bombs. (Credit: Cropped USAF Photo)

A Boeing B-47E medium bomber. It had the performance characteristics of a jet fighter. Note the white anti-flash paint on the bomber's belly. (Credit: USAF)

A MK III "Fat Man" atomic bomb. Each weapon was handmade and hand assembled. It required a 39-man crew almost 2 days to assemble it. (Credit: DOD)

A MK IVN "Fat Man" atomic bomb. The N stood for Northrop, the company that manufactured the riveted casing. Note the circular core insertion panel and the four radar widows in the nose of the bomb. As well, note the changes to the tail configuration. This was the first assembly line manufactured nuclear weapon. (Credit: SNL)

A MK 6N "Fat Man" atomic bomb being loaded onto a B-50 bomber. Note the core insertion panel in the nose of the bomb and the spoiler bands. This type of casing was also used for the 500-kiloton MK 18 fission bomb and the atomic demolition munition. (Credit: SNL Photo D18043)

A type A atomic demolition munition created by giving a MK 6 atomic bomb a waterproof casing. It was fired by remote control or by a radio signal. The left end (not seen) was rounded and had attachments for firing cables. (Credit: LANL)

In selecting a location for the new atomic laboratory, General Groves specified that it have a climate suitable for year-round operations, adequate transportation, a local supply of labor, and isolation for safety and security purposes. In order to meet Grove's requirements, Oppenheimer acquired the Los Alamos Ranch School on a remote plateau in New Mexico. Despite the idyllic setting, the isolation of the laboratory did not impress the majority of the scientists invited to work on the bomb project. Even so, Oppenheimer was able to staff the laboratory following a brief, cross-country recruiting drive. He began operations at Los Alamos on March 15, 1943. Governed by a contract signed by the Board of Regents at the University of California, Los Alamos has since grown into a city and continues to provide innovative research on nuclear energy.

Following the entry of the United States into the Second World War, the Roosevelt Administration created the Manhattan Engineering District (MED) to produce nuclear explosives and to fabricate the explosives into bombs. This program informally became known as the "Manhattan Project," even in official circles. The Army promoted Colonel Leslie J. Groves, Deputy Chief of Construction for the US Army, to Brigadier General and ordered him to oversee the MED on October 23, 1942. As his scientific deputy, he chose Dr. J. Robert Oppenheimer, a brilliant theoretical physicist and self-made experimentalist from the University of California at Berkeley. Not everybody's choice as scientific director, Groves insisted on "Oppie" because of his general knowledge and his affability. Oppenheimer immediately demanded the establishment of a central atomic laboratory for the design of atomic weapons. At such an institution, scientists and engineers would be able to share their findings and come to grips with the many difficult technical problems that lay ahead.

In selecting a location for the new atomic laboratory, General Groves specified that it have a climate suitable for year-round operations, adequate transportation, a local supply of labor, and isolation for safety and security purposes. In order to meet Grove's requirements, Oppenheimer acquired the Los Alamos Ranch School on a remote plateau in New Mexico. Despite the idyllic setting, the isolation of the laboratory did not impress the majority of the scientists invited to work on the bomb project. Even so, Oppenheimer was able to staff the laboratory following a brief, cross-country recruiting drive. He began operations at Los Alamos on March 15, 1943. Governed by a contract signed by the Board of Regents at the University of California, Los Alamos has since grown into a city and continues to provide innovative research on nuclear energy.

Oppenheimer organized the new laboratory into a number of divisions to carry out the development of the bomb. He made an Administrative (A) Division responsible for procurement, personnel, and the shops that supported the ongoing research and development. The Theoretical (T) Division, under the direction of Hans Bethe, explored nuclear and hydrodynamic criteria relating to the design of the bomb. A separate Experimental Physics (P) Division, led by R. F. Bacher, developed the instrumentation needed by T Division. Chemistry (C) and Metallurgy (M) were at first a loose association of groups before Oppenheimer amalgamated them into the CM Division under the direction of J. W. Kennedy. This division was concerned with the purification and fabrication of materials for the bomb's tamper, initiator, and its core. Finally, since preliminary thoughts for an atomic bomb anticipated a gun-assembled mechanism, Oppenheimer created an Ordnance (O) Division under the command of Navy Lieutenant Commander William "Deak" Parsons.

The Los Alamos scientists at first pursued a gun-assembled device because it was the simplest means of creating an atomic explosion. This concept envisioned a sub-critical fissile projectile fired down a gun barrel to combine with a sub-critical target to form a super-critical mass. The final shape of the gun-assembled fissile components was a squat cylinder, with its length equal to its diameter. This geometry approximated a sphere – the optimum configuration for a super-critical mass. With the adoption of ordnance-oriented research, Los Alamos built a firing range at the Anchor Ranch, which was a part of the initial land purchase. To serve the needs of the test program, Los Alamos equipped the firing range with a 20mm cannon and a 3-inch antiaircraft gun that could be fitted with smoothbore barrels of varying lengths.

From the outset, Oppenheimer planned to use both plutonium and uranium in gun-assembled bombs. The laboratory's initial research focused mainly on plutonium because of the uncertainty associated with uranium's enrichment. Los Alamos scientists generally accepted the possibility of creating $Pu239$ in an atomic reactor and chemically separating the element from spent fuel. Using a small quantity of $Pu239$ produced in a cyclotron, Emilio Segré determined that this isotope had a spontaneous rate of 5 fissions / kilogram / second. Assembly at a speed of 3,000 feet per second would prevent pre-detonation with this material. Using contemporary propellants, the desired velocity was attainable in a gun with a 17-foot long barrel. Because of the gun's dimensions, the lab named the bomb that incorporated this device, the "Thin Man." Some sources consider the name an oblique reference to President Roosevelt, who initiated the atomic program or to Robert Oppenheimer, its Director. Other sources suggest it was a reference to the lead character in a popular movie series with the same name.

In parallel with the development of a gun-assembled bomb mechanism, Seth Nedermeyer initiated a second line of weapons research. Based on an idea brought forward by Richard Tolman, Nedermeyer planned to manufacture a hollow sphere from a super-critical mass of fissile material. In this configuration, the sphere's large surface area would emit so many neutrons that a chain reaction would not be possible. By using a surrounding envelope of explosives, Nedermeyer intended to collapse or "implode" the hollow sphere into a solid ball to create the geometry (and critical mass) necessary for an atomic explosion. This method had the advantage of being more compact than the unwieldy design adopted for the Thin Man. Early problems encountered by Nedermeyer in his quest for an implosion bomb included spalling on the inner surface of collapsing metal spheres and undesirable shaped-charge like jets that impeded formation of the desired spherical geometry. Nedermeyer also encountered considerable difficulty in obtaining the simultaneous ignition of the detonators that were then available.

In September 1943, a visiting physicist named John von Neumann pointed out the advantages in not only collapsing a hollow, fissile sphere but in compressing it as well. Compression could lead to much higher efficiency and the use of smaller amounts of fissile material. Since this would speed up the bomb program, Oppenheimer invited von Neumann to join the lab. He also delegated theoretical physicists Hans Bethe and Edward Teller to begin investigating the further ramifications of this concept, and recruited George Kistiakowsky, the leading explosives expert at OSRD, to assist Nedermeyer. A small cloud appeared on the horizon at this time when Segré measured the rate of fission for U235. He found it to be lower than expected, reducing the anticipated yield of enriched uranium bombs.

Implosion research moved to center stage on April 5, following the evaluation of plutonium produced in Los Alamos' experimental X-10 reactor. Segré measured a spontaneous rate of 50 fissions / kilogram / second and announced that unavoidable contamination of Pu239 by Pu240 had rendered it unsuitable for use in gun-assembly. Even worse, he anticipated that the plutonium produced at Hanford would have even higher levels of Pu240 contamination. Shortly after Segré's pronouncement, the bomb program diverged along two separate paths: a uranium gun-assembled device and a plutonium implosion device. Plutonium was only reprieved as a bomb material because the microsecond speed of implosion could compress a core to criticality before pre-detonation became an issue.

In order to accommodate the necessary changes to the weapon program, Oppenheimer instituted a general administrative reorganization of Los Alamos in August 1944 that included the formation of three new divisions: Weapon Physics (G), Explosives (X), and Fermi (F). He placed the Weapon Physics Division under the supervision of Robert Bacher and made him responsible for the critical assembly of the implosion device (G-1), electric detonators (G-7), and a neutron initiator (G-10). A Research Division (R) directed by R. R. Wilson replaced Bacher's old Experimental Physics Division. The Explosives Division, created for George Kistiakowsky, took on implosion theory (X-1), development (X-2), and detonating circuits (X-5). Oppenheimer placed Enrico Fermi in charge of his own (F) Division, which he directed to pursue advanced work on alternatives to the atomic bomb.

Oppenheimer now placed Francis Birch in charge of the O-1 Uranium Gun Group of the Ordnance Division and ordered him to produce a working bomb by the summer of 1945. Since there would not be enough uranium to test this design, Birch had the gun-assembled uranium bomb overbuilt to insure its reliability. In addition, because it did not require the high insertion speed necessary for the Thin Man, Birch built it with a shortened barrel and dubbed it "Little

Boy" – the Thin Man's younger brother. Little Boy destroyed Hiroshima in the very first atomic attack on Japan. In spite of this success, Los Alamos eventually discontinued this design because of its poor efficiency. The story of strategic atomic weapons then became one of implosion.

The Los Alamos physicists predicated the development of a plutonium bomb on the successful implementation of implosion. Up to this point, the efforts by Seth Nedermeyer had proved unsatisfactory but Oppenheimer hoped that Kistiakowsky's expertise would provide the necessary breakthrough. Calculations and suggestions made by Hans Bethe, Rudolph Peierls, and John von Neumann aided Kistiakowsky in his work. Managing the implosion operation, which at its peak expended 50 tons of high explosives per month, required the establishment of a dedicated facility. Located at the Sawmill Site "S" (TA-16), the explosive manufacturing facility included an office building, steam plant, casting factory, and equipment for trimming and shaping explosive lenses.

Also included at Site S was an x-ray laboratory for quality control, and a number of magazines for storing explosives. Lacking sophisticated equipment, the Los Alamos scientists at first used commercial candy cookers to melt explosive ingredients before pouring them into homemade molds. The molds were finished with a low melting point castable alloy known as Cerrotru.

The final design for the implosion assembly was a 5,300-pound hollow sphere, 55 inches in diameter and 18 inches thick. It consisted of 32 explosive lenses that fit together in the same pattern as a soccer ball. The explosives group manufactured each lens to a tolerance of 1/32 inch. The Kellermatic, one of the world's first numeric milling machines, produced the molds used to cast the explosive blocks. The Chrysler Corporation had to suspend the production of B-29 bomber engines while it machined these molds at its Detroit, Michigan, engine factory. Because of its large diameter, the lab informally named the implosion bomb "Fat Man," a possible reference to British Prime Minister Sir Winston Churchill or General Groves.

The upper part of each explosive lens consisted of a cone cast from low-speed Baratol explosive. Baratol was a mixture of 24 percent TNT and 76 percent barium nitrate oxidizer. It had a detonation velocity of 15,978 feet per second. To manufacture the lens, technicians from the explosives group poured a slurry of high-velocity Composition B explosive around the cone. Composition B was a mixture of 60 percent RDX and 40 percent TNT. It had a detonation velocity of 25,723 feet per second.

Based on experiments with the apex angle of the cones and their depth of penetration into the lenses, Kistiakowsky's group was able to shape a detonation wave that formed an inward-converging sphere. A second piece of Composition B, bonded to the base of each lens, accelerated the shock wave to its maximum detonation rate. This enabled the converging detonation wave to compress a solid plutonium core to twice its normal density in four microseconds. In examining the completed lenses, Kistiakowsky found that some contained large air bubbles. To solve this problem, he used a dentist's drill to bore into the bubbles so he could hand-fill them with additional slurry.

In addition to precisely configured explosive lenses, the implosion assembly required a highly-synchronized detonating system. To shape the desired shock wave, it was necessary to have all the detonators explode within 10 microseconds of each other. This ruled out conventional detonators, which had initiation tolerances of several milliseconds. Conventional detonators work by electrically heating a metal wire to set off a small amount of sensitive primary explosive that in turn explodes a much larger amount of secondary explosive.

The detonator team pursued two lines of research in developing detonators suitable for atomic applications. The first line of research evaluated spark gap detonators. These transmitted a powerful electrical discharge between two electrodes to set off a secondary explosive charge directly. A voltage applied to an intermediate wire in the gap triggered the discharge. Spark gap detonators could provide the necessary power and timing for an implosion system but the team rejected them out of safety concerns. Spark gaps were too prone to accidental activation by stray currents.

The detonator chosen for use in Fat Man was the Model 1773 exploding bridgewire. Larry Johnston and Robert Alldredge, working under the direction of Luis Alvarez, developed and patented this device. In an exploding bridgewire detonator, a pulse of high-energy current vaporizes a thin gold, platinum, or tungsten wire to initiate a shock wave that sets off a charge of secondary explosive (PETN). Current rise times for this type of detonator are a few nanoseconds and timing accuracies are a fraction of a microsecond. In order to insure a high order of detonation, each Model 1773 EBW detonator was equipped with a pair of bridgewires, either of which produced enough energy to set off its explosive charge. The incorporation of secondary, rather than primary explosives in the detonators increased safety.

It was not the direct explosion of the detonators that activated the implosion sphere. Rater, each detonator was placed in a copper well in an explosive block and it was the shock wave from the detonator driven copper wells that triggered the implosion sphere,

A young physicist named Don Hornig designed the electrical unit that fired the detonators. This device, known as an "x-unit", consisted of two arrays, each containing 32 high voltage capacitors. Hornig designed each capacitor bank to discharge a 5,600-volt electrical pulse simultaneously into separate firing harnesses. Two banks of 28-volt lead-acid aircraft batteries charged the x-unit's capacitors, using a 400-cycle inverter, a transformer, and a rectifier to step up the voltage. The entire system weighed about 500 pounds and either set of batteries was capable of fully charging both capacitor banks. During flight, a special pump kept the case of the x-unit pressurized. This prevented accidental flashover of high voltage at altitude when the x-unit was charged.

The firing pulse from the capacitor arrays funneled through a pair of triggered spark gap switches produced by EG&G. Triggered, or three-electrode, spark gaps are sealed, gas filled devices that have a non-conducting gap that is located between a pair of electrodes. An electric potential, applied to a trigger wire in the gap, completes the circuit. At a potential difference of 100 kilovolts, a triggered spark gap can reliably switch 100 kiloamps, with a current rise time of 10 - 100 nanoseconds. Despite the x-unit's potential for reliable switching, static currents accidentally discharged one of these units during testing and revealed a previously undetected safety problem that Los Alamos never corrected. Raytheon built Mark II copies of Hornig's experimental x-unit for the bomb program.

Prior to Hornig's development of the triggered spark gap, Los Alamos staff considered explosive switches and another type of electronic device for switching power in the x-unit. However, explosive devices were one-time use items, and Los Alamos personnel did not have sufficient confidence to use an explosive switch if it could not be tested before it was installed in the bomb. This left the electronic device for consideration.

The electronic device, called a thyratron, had been around since 1928. Thyratrons come in several varieties. They are gas filled tubes of which the hydrogen filled variety exhibit the fastest switching times. They consist of a discharge chamber in which are arranged a hot cathode, one or several grids, an anode, and a gas filler. The thyratron's gate must be biased highly negative in the off state and then biased positive to achieve switching. Hydrogen thyratrons operate like a latching switch, once activated, they can only be turned off by cutting their power supply. Unfortunately, the types of thyratron available at the time were bulky and lacked the ability to switch the high current loads required. Nevertheless, they would make their appearance in later Marks of the Fat Man bomb.

In order to reduce inductance between the x-unit's spark gap trigger and the bomb's detonators, the bomb's 32 detonators were connected in parallel. The use of independently charged, redundant firing circuits minimized the possibility of an electrical failure. Specialists carefully balanced the impedance of each wire with a Bowden camera, a special piece of equipment originally developed to measure the speed of light. This procedure insured the simultaneous delivery of current to each of the detonators. As a safety precaution, the wiring system had interchangeable green (inert) and red (live) five-prong plugs that disconnected the firing circuits until the bomb was ready to drop. These plugs fit into a socket located on the upper forward hemisphere of the bomb's casing.

The fuzing system that exploded the bomb was a complex affair based on RCA's AN/APS-13 rear warning aircraft radar. Los Alamos incorporated the radar in a device known as an "Archie." The device closed a switch in the firing circuits when it detected that the bomb had reached a preset distance from the ground. Each bomb had four Archies and agreement by any two radar units was required to operate the firing mechanism. The Archie units transmitted their radar signals through Yagi-type directional antennas, four of which were located near the nose of the bomb at 90-degree intervals around its circumference. Ironically, this antenna was a Japanese invention. The antennas' positions allowed the radar signals to bounce off the ground as the bomb fell nose first. Each antenna was composed of three simple aerials, with one bent in the shape of an inverted U. The preset detonation altitude for the bomb was not determined until after the first atomic test established its yield.

When dropped (from an altitude of 31,000 - 32,000 feet), pullout wires switched the bomb to internal power and started a pair of spring driven clocks. After 15 seconds and a fall of 3,600 feet, the clocks closed two switches to allow the flow of current from the 28-volt batteries to the x-unit choke. Here, the current built up a magnetic field. The clocks also activated a bank of six barometric switches. Eight ports, located about 30 inches aft of the maximum diameter of the bomb, pressurized an annular manifold that the baro-switches used for a reference. As soon as any two baro-switches agreed that the bomb had fallen past an altitude of 7,000 feet, they sent a signal that closed two more relays. One of these connected the x-unit to the firing harnesses and the other activated the Archie radar fuzes. The timed fall prevented detonation of the bomb by a radar return from the aircraft if one of the barometric switches accidentally shorted on release. Finally, at a preset altitude of 2,050 feet, the Archies closed a firing switch that applied power to the x-unit choke. The rapid closure of the firing switch broke down the choke's magnetic field, transferring energy to a condenser. When the voltage in the condenser built up to 4,000 volts, the spark gap switch operated, allowing current flow to the detonators.

Before installation in the bomb, the Air Force physically tested the barometric switches and the radar altimeters that measured height of burst by flying them in small aircraft to determine the accuracy of their settings. The University of Michigan also developed a backup proximity measuring or "Amos" radar fuze. The U of M made the fuze with commercially available equipment that operated at a lower range of altitudes than the Archie radar fuze, which was only capable of exploding Fat Man in the interval between 700 and 2,000 feet. Until confirmation of Fat Man's yield by the Trinity Test, it was possible that a lower altitude setting might be required. The altitude setting used for Fat Man incorporated an allowance for a 400-foot time of fall, during which the condenser broke down the spark gap to allow current flow to the detonators. The actual height of burst for Fat Man was 1,650 feet.

In case of a fuzing and firing system failure, Fat Man was equipped with four mercury fulminate AN219 contact destruct fuzes. Also known as, "salvage" fuzes, Los Alamos intended these components to detonate the implosion assembly and destroy the bomb to keep its nuclear material and technical secrets from the enemy. There was also a chance that the last resort fuzes would trigger a low-yield atomic explosion at ground level to partially complete the mission. Each fuze was connected to the explosive sphere by a length of primacord.

As part of Project "A" (for Alberta), the ballistic shapes for Fat Man and Little Boy were determined by means of drop tests conducted at the Tonopah Test Range, Nevada, and the Salton Sea Range, California. The final shape determined for the Fat Man bomb was a steel ovoid, 128 inches long and 60 1/4 inches in diameter. The two halves of the bomb's 3/8-inch-thick steel casing bolted together to protect internal components from antiaircraft shell fragments during a bombing run. A "box kite" tail provided aerodynamic stabilization and steel plates located within the tail acted as air brakes to prevent the bomb's terminal velocity from exceeding the speed of sound.

An inch-thick duralumin sphere held the explosive assembly together. The prototype Model 1222 sphere consisted of twelve pentagonal sections fastened together with 1,500 bolts and was a nightmare to assemble. For the operational bomb, Los Alamos used the considerably simplified Model 1561. It consisted of two half-shells and five equatorial band segments. A 1/2-inch-thick layer of cork formed a cushion between the explosives and the aluminum sphere. Mounting units permitted attachment of the x-unit and fusing mechanisms to a cone fitted to the rear hemisphere of the aluminum alloy sphere. The batteries and transformers attached to a cone fitted to the forward hemisphere.

The atomic hardware for the bomb sat inside of the hollow explosive assembly. At its center was a 3.6-inch diameter, 13 1/2-pound, (almost) solid plutonium sphere that was made in two hemispheres by hot pressing plutonium blanks cast in a special mold. Robert Christy devised the core's conservative shape while serving on the neutron initiation committee with Hans Bethe and Enrico Fermi. He was concerned about the problem of internal spalling when compressing a hollow core.

Alloying the plutonium in the 13 1/2-pound core with 0.8 percent by weight of gallium stabilized it in its δ-phase. After casting, the Chemistry and Metallurgy Division annealed and degassed the plutonium hemispheres by means of a heat treatment. Together, the hemispheres represented just over one-half of a bare, critical mass. To prevent surface oxidation, CM Division silver-plated the Trinity core. (It nickel-plated subsequent cores.) It also coated the flat face of each hemisphere with a 0.1-millimeter thick gold coating to prevent explosive jets from penetrating this interface during implosion. The weapons establishment later plated plutonium cores entirely with gold in lieu of nickel. An inch-wide space, accessible by means of a plutonium plug, was located at the center of the core. A special walnut sized "initiator" placed inside produced the neutrons necessary to begin the atomic chain reaction.

James Tuck proposed the design for the bomb's internal neutron initiator or "Urchin" and Charles Thomas, an executive of the Monsanto Company, had it produced. Before the war,

Thomas had worked as a chemist for Delco / GM, where he acquired the necessary expertise needed to carry out this phase of the Manhattan Project. General Groves gave Thomas permission to establish the initiator project (Weapon Physics Division, G-10) in the Runnymede Playhouse on the grounds of the Talbot family estate in a wealthy residential section of Dayton, Ohio. The playhouse was sufficiently large to accommodate a laboratory and a manufacturing operation. The facility, known as Dayton Unit IV, remained in use until 1949, when the AEC replaced it with the Mound Laboratory in Miamisburg, Ohio.

The Urchin initiator produced the 100 or so neutrons required to start a chain reaction when implosion mixed its components together near the height of core compression. The initiator consisted of a 1-inch diameter, hollow beryllium sphere with a 1/4-inch-thick shell. Technicians at the Dayton Unit inscribed fifteen parallel, wedge-shaped grooves on the sphere's inner surface before plating it with layers of nickel and gold. They then covered the gold surface with 50 Curies, or about 11 milligrams, of Po210.

In order to generate neutrons, the nuclear isotope used in the initiator had to be a strong alpha emitter. As well, it had to be a weak gamma emitter because gamma rays can knock loose neutrons from beryllium nuclei. At full activity, the initiator's polonium charge produced about 2,000,000 alpha particles per microsecond. These in turn generated 1 - 2 neutrons every 10 nanoseconds, a rate that was sufficient to initiate the chain reaction reliably. The urchin's gold and nickel layers were sufficiently thick to prevent the alpha particles from interacting with the beryllium shell until mixed. A small bracket levitated and centered the Urchin in a space at the center of the plutonium core.

A small sphere of beryllium, also plated with nickel and gold, was levitated inside the initiator on small pins. When compressed, the interaction of the grooved shell and the sphere created Munroe-effect, or shaped-charge like jets that thoroughly mixed the polonium and beryllium together. Designers initially considered radium for use as the active ingredient in neutron initiators but eventually rejected it because its high gamma ray emission rate rendered it a biological hazard. Edward Condon proposed the use of polonium as a replacement for radium in 1944. The final initiator configuration was the result of an intensive program that tested a wide variety of designs.

Following the Trinity atomic test, Los Alamos physicists carried out evaluations of a number of internal neutron initiator types around the periphery of the atomic crater. For this purpose, they buried concrete chambers 1 foot below ground and covered them with 16-foot-thick earthen mounds. The octagonal chambers were 14 feet long, 12 feet wide, and had 2-foot-thick walls. This arrangement prevented the venting of radioactive material to the atmosphere when explosives compressed the initiators. Remotely operated equipment made observations on the tests from a distance of 200 yards.

A 265 pound, 2 3/4-inch-thick natural uranium tamper surrounded the plutonium core and initiator. The tamper reflected stray neutrons back into the core during the chain reaction and contributed to yield through the fast fission of some of its atoms. Its mass also delayed disassembly. Geoffrey Taylor, a visiting British hydrodynamicist, changed this design when he found a problem with the direct placement of explosive lenses around the tamper. When a light material accelerates against a dense material, the intervening boundary becomes unstable and exhibits turbulent behavior. The effects of low-density explosive products (gases) pressing against the high-density uranium tamper were not predictable and might have seriously affected the compression of the core.

To overcome the effects of what is termed Rayleigh-Taylor instability, Taylor added a 4 1/2-inch-thick, 120-pound, "pusher" shell of aluminum between the explosives and the tamper. This arrangement created two interfaces of intermediate density contrast that behaved in a predictable manner. A 1/8-inch-thick layer of boron-impregnated, acrylic thermoplastic separated the pusher and the tamper. This feature prevented the thermalization and back scattering of spontaneous neutrons by the aluminum pusher, thus reducing the possibility of pre-initiation.

In addition to mitigating the problem of Rayleigh-Taylor instability, the density interfaces located at the explosive-aluminum and the aluminum-uranium contacts had the beneficial effect of reflecting and intensifying the explosive driven shockwave. Convergence magnified the explosive shock pressure of Composition B by a factor of five and the two density interfaces each doubled it. This provided a 20-fold increase in its intensity at the core. The actual pressure was equivalent to 6,000,000 atmospheres, or 6 megabars, slightly more than the pressure at the center of the earth! The shock wave also set up a permanent transition of the plutonium to its denser α-phase leading to an overall density increase of 2 1/2 times in the core. Coupled with the neutron reflectivity of the uranium tamper, compression and phase transition increased the 0.55 critical mass of the core by a factor of ten.

The density interfaces also modified the inward propagating zone of compression associated with the shock front, which terminated with a gradual decline in pressure to zero. Physicists term the low-pressure zone behind a shock front a "Taylor Wave." Although convergence amplifies the intensity of the shock front to aid in compression of the core, it also amplifies the effect of the following Taylor Wave in such a way that by the time the zone of compression reaches the center of the plutonium core, its outer regions have expanded back to their normal density. The density interfaces solved this problem by widening the zone of compression and by suppressing the Taylor Wave.

For the Gadget (the Trinity prototype) device, the plutonium core and the aluminum pusher were cast in two hemispheric pieces. The uranium tamper, however, was cast as a sphere with a cylindrical opening that penetrated through the entire component. The diameter of the hollow cylinder was slightly larger than the plutonium core. A uranium cylinder known as the "tamper plug", fit in the uranium sphere to complete its geometry. The cylinder was sliced lengthwise and had a cavity into which the core was inserted. The cylinder was then sealed with gold foil and fastened together with uranium hinges and screws at its ends. This arrangement allowed assembly of most of the bomb before a technician inserted the cylinder containing the core through the top of the uranium tamper. This arrangement was likely an attempt to prevent explosive jetting from distorting the core during compression because no single seam in the pit aligned with the core. It also prevented an explosive jet from prematurely activating the neutron initiator at the core's center.

Recently, some comments by Robert Bacher, a war-time Los Alamos physicist, were found to describe an interesting change to the design of the core in the first Fat Man bomb dropped on Nagasaki. For this bomb, the uranium tamper was cast as two hemispheres and the plutonium core was made in three pieces rather than two. The core consisted of two hemispheres that were beveled on their rim. No one knows the depth of the bevel, or if the bevel was up or down, although the Author prefers a bevel starting somewhere on each hemisphere's flat that is angled down toward the hemisphere's rim. The bevel left an empty pie-slice shaped space around the circumference of the two hemispheres when they were placed together. A pie-shaped ring of plutonium filled in this space to create a solid, spherical core. When assembled, the ring locked the two hemispheres together, preventing movement. The blunt face of the core's middle segment also prevented explosive jets penetrating the aluminum's and uranium's joins from reaching the initiator.

A Russian document released in the 1970's, illustrates the Nagasaki bomb's configuration using stolen Los Alamos data. It indicates that the Nagasaki core had a three-part arrangement as discussed by Bacher. What's more, the document doesn't show a uranium cylinder assembly such as used in the Trinity pit. Likely the new core / pit design was calculated to be reliable "as is" and the uranium tamper was now cast in two simple hemispheres and did not include the tamper cylinder used for the Trinity device. This design was used as the basis for the first run of Fat Man MK III bombs.

Altogether, the various metal layers that made up the bomb's pit weighed about 400 pounds. Collectively, this arrangement was known as a Type "A" Pit. This mass of material is the reason why the implosion assembly contained so much high explosive. The energy required to double the density of a pound of tamper or fissile material is equivalent to the energy released by 2 pounds of TNT. In modern weapons, the implosion process is about 35 percent efficient, so that 6 pounds of explosive is required to compress every pound of pit material. In Fat Man, the process was even less efficient, requiring 2 1/2 tons of explosives.

Fully assembled, Fat Man weighed 10,300 pounds. Its first test (Trinity) at Alamogordo, New Mexico, ushered in the atomic age with a 22-kiloton yield. The test was necessary because of the many unanswered questions regarding the performance of the bomb's components. The scientists and soldiers who witnessed Trinity did so from a distance of about three miles. Exploded at 5:30 am on July 16, 1945, it produced a blinding flash that illuminated the landscape for miles around. After the flash, an immense fireball appeared, ascending majestically into the sky. Before long, those present heard the blast, echoing as it bounced between the mountains enclosing the Jornada Del Muerto. The watching men experienced the entire gamut of emotions from awe to amazement to fear and loathing. Oppenheimer remarked "it worked!" as General Farrell congratulated General Groves on winning the war. Groves demurred, indicating that at least two atomic attacks would be required to produce a Japanese surrender.

The Army Air Force had prepared well for its upcoming atomic role. In 1943, it had selected the B-29 Superfortress to carry the atomic bombs, which were then in development. The veteran British Lancaster could also have carried them but for logistical reasons, the Army Air Force preferred an indigenous aircraft. The B-29 began life as the Boeing Model 316 in 1934. This design called for a fuselage with a high-mounted wing that Boeing eventually adopted for the production bomber. In 1938, an improved Model 322 investigated the use of a pressurized fuselage for use at high-altitudes. Boeing updated this design later in the year to the Model 333A, in which a tunnel through the bomb bay connected the forward pressurized crew cabin to the rear pressurized gunners' compartment. The next model, the 334A had the high tail plane, 135-foot-long high-aspect-ratio wingspan, and four Wright R-3350 radial engines included in the final design.

The outbreak of WW II caused the Army Air Force to take its first serious look at "Very Long-Range" bombers. On November 10, 1939, Air Force General Hap Arnold asked for and received permission to contact a number of aircraft companies regarding a VLR bomber and soon after, the Army called for proposals. Boeing submitted reworked specifications for a Model 345 bomber in May 1940 and in June, the Army issued contracts to four manufacturers for engineering data. The Model 345 made the final cut as the XB-29, and in August, the Army ordered two prototypes and a static test model. It soon increased its initial order to fourteen YB-29 service test models. Following the Japanese attack on Pearl Harbor, the Air Force awarded production contracts to Boeing and by March 1943, it had placed orders for 1,000 aircraft! The first XB-29 flew on September 1, 1942, and a year later, the first production models rolled off the Boeing assembly line in Wichita, Kansas.

The B-29 was a remarkable aircraft for its time. Instead of a thin aluminum skin bent and riveted onto an airframe as was the custom with contemporary bombers, it had a thick aluminum shell that was resistant to battle damage. This feature allowed much of its fuel to be carried in self-sealing wing tanks. Its eighteen-cylinder, turbo-supercharged, R-3550 Wright engines developed almost 9,000 horsepower. This gave the B-29 a maximum cruising speed of 342 MPH at an altitude of 30,000 feet. The practical operational radius for the aircraft was almost 2,000 miles. Its defensive armament consisted of eleven .50 caliber machine guns mounted in five electrically driven turrets controlled from remote sighting stations. Because of the many defensive weapon stations, the B-29 required an eleven-man crew. Its offensive bomb load was a maximum of 16,000 pounds carried in a pair of bomb bays. However, a 5,000-pound ordnance load was more realistic for long-range missions. For navigation, the Superfortress was equipped with an AN/APN-9 RCA LORAN guidance system and for bombing, it had an AN/APQ-13 search radar to supplement a Norden M-9 bombsight.

The B-29's Norden M-9 bombsight consisted of a sight head and a stabilizer. The sight head comprised an optical telescope and a "rate end" analog computer that controlled optical movement. The stabilizer, which attached to the sight head, used a vertical gyroscope to steady the bombsight's optics in pitch and roll, and a directional gyroscope to steady the sight body in azimuth. During a mission, the bombardier adjusted the sight with ballistic data from bombing tables that allowed for true airspeed, altitude, and the type of bomb carried. The only remaining information required was groundspeed, which the bombardier obtained by centering the target in the optical crosshairs and adjusting range and drift knobs until all apparent motion of the crosshairs was "killed." After this adjustment, the sight could track the target without further need for attention. From this moment on, the M-9's computer could set up a "drop angle" identified by a pointer in an index window on the sight head. Another pointer in the window moved toward the drop angle indicator to show the decreasing angle of the line of sight. When the two pointers overlapped, the bombardier could release the bombs manually or allow their automatic release. M-9 bombsights reached their peak of development shortly after the war when Norden added an automatic erection feature and replaced the telescope with a collimated optical system. Norden also rebuilt the bombsight's computer to allow the independent setting of offset aim points.

The AN/APQ-23 search radar followed the APQ-13. Combined with a CP-16 computer, it was the first synchronous radar bombing system installed in an operational aircraft. Western Electric began its development in March 1944, for installation in the B-17, B-24, B-29, and other heavy bombardment aircraft. The APQ-23 system supplied azimuth, distance, and drift information to both pilot and bombardier, and could perform direct and offset aiming. The main drawback of the system was the radar's resolution, which limited high-altitude bombing to large, clearly distinguishable targets. The APQ-23, nevertheless, operated satisfactorily at altitudes up to 30,000 feet and speeds up to 440 knots. The sight had a forward-looking range of 15 miles, could use offset aiming points 30,000 feet distant from the target, and at any azimuth bearing. Improved models saw service in the Korean War, during which Western Electric upgraded its resolution with a 60-inch diameter rotating dish antenna.

Despite the sophistication of the B-29, Boeing had to modify the aircraft significantly for its atomic missions. It called this procedure "Silverplating," possibly due to the cost of the alterations. Except for the B-29's tail guns, The Air Force removed the aircraft's defensive armament and armor to increase its payload and range. In addition, it had the engines tuned and upgraded with a fuel injection system to insure their performance. Wright had rushed the R-3550 engine into production and it was prone to overheating. This could start a fire, which, if it spread to the magnesium crankcase housing, would burn through the main wing spar. After the war, a modified aircraft known as the B-50 was equipped with Pratt & Whitney R-4360-35 radial engines to provide a more reliable means of propulsion. Essentially an improved B-29, the B-50 designation disguised the true nature of the aircraft for budget appropriations.

To permit the loading of large diameter atomic bombs, Boeing connected the B-29's bomb bays and modified the access tunnel that ran through them. It also rebuilt the hydraulic bomb bay doors with a pneumatic system that allowed faster opening and closing. Reversible Curtiss electric propellers enabled the aircraft to taxi over the recessed pits from which atomic bombs were loaded. The bombs were winched into the bomb bay with a heavy-duty hoist borrowed from the British Lancaster bomber. A special flight test box allowed an atomic weaponeer to monitor battery voltages, warm up the bomb's radars, charge the bomb's x-unit, and confirm the status of critical electronic components while in flight. An electronic countermeasures station was also available to scan for interference on the wavelengths used by the bomb's radar altimeter fuzes. Five pairs of pullout arming wires and monitoring cables ran from the weapon station to the bomb bay where they entered the bomb through ports mounted fore and aft of its central hoisting lug.

Boeing carried out the B-29's modifications as part of Project Alberta, the same program that evaluated Fat Man and Little Boy's drop shapes. A secondary goal of this project was the development of support systems needed to carry out atomic missions. Los Alamos formally established Project Alberta under Captain W. S. "Deak" Parsons in March 1945, who then chose Norman Ramsey as his scientific and technical deputy. Captain Parsons selected Commander F. L. Ainsworth as the Project's Operations Officer and as his military alternate. He then placed Commander Norris Bradbury and Roger Warner in charge of Fat Man assembly, Commander Birch in charge of Little Boy assembly, R. B. Brode in charge of fuzing, L. Fussell in charge of the electrical detonator system, and made Phillip Morrison and Marshall Holloway responsible for the pit. He also gave Luis Alvarez responsibility for aerial observations during the combat drops, placed George Galloway in charge of engineering, Lieutenant Colonel R. W. Lockbridge in charge of supply, Maurice Shapiro in charge of ballistics, and assigned Sheldon Dike to aircraft concerns. Commander T. J. Walker supervised 155 drop tests conducted at Wendover Army Airbase, Utah, and the Salton Sea, California.

After Sandia finalized the bombs' shapes and sizes, Boeing began Silverplating the B-29 bombers required for the atomic mission. As these aircraft started to roll off the assembly line in late 1944, the Air Force assigned them to the 509th Composite Group, commanded by Colonel Paul Tibbets. For training, it based the 509th at Wendover, Utah, where Project Alberta was conducting its drop tests and making aircraft modifications. The remote location helped to ensure that news of the Group's activities did not leak out. The official penalty for loose lips was a firing squad, although this did not become necessary. The Air Force transferred a few men to the Aleutian Islands, Alaska, for lesser transgressions.

A critical maneuver used to drop the bomb was a 60-degree diving turn that put as much room between the aircraft and the aim point as possible. The bomb's shock wave had the power to cripple or bring down any aircraft that remained too close. By May 1945, the 509th had mastered the techniques needed to safely deliver the bombs and was transferred to an American held airbase on Tinian Island, within striking distance of Japan. Shortly after receipt of President Truman's authorization to use the bombs, the B-29 Enola Gay dropped Little Boy on Hiroshima and the B-29 Bockscar dropped Fat Man on Nagasaki.

The end of hostilities brought a reduction to the pace of operations at Los Alamos and the laboratory put planned tests to develop an improved bomb on hold. There followed a yearlong hiatus before the decision was made to place the laboratory under the control of the civilian Atomic Energy Commission. During this period, many scientists and engineers returned to their prewar jobs or to academia. Others left because they had no desire to continue developing weapons of mass destruction.

Early in 1947, the first director of the AEC made a fact-finding tour of the Los Alamos and the Sandia facilities. David Lilienthal found a number of nuclear cores but parts for only one or two serviceable bombs. He then learned that it took 39 men two days to prepare an atomic bomb for service and that after only 72 hours, its batteries had to be recharged. Further use required the complete replacement of the batteries. The technicians had to dismantle the bomb within ten days of assembly or the heat flow from its plutonium core would damage critical components.

Upon hearing this news from Lilienthal, President Truman became deeply concerned. He had planned to use American atomic might to restrain an increasingly belligerent Soviet Union; the size and condition of the stockpile came as a shock. Like many Americans, the President thought that atomic bombs were off the shelf items, ready for instant use. Truman therefore encouraged the development of improved weapons and approved an atomic demonstration in a remote area of the South Pacific known as the Marshall Islands. A joint Navy / Air Force Task Force carried out operation Crossroads at Bikini Atoll in July 1946. The United States invited foreign observers, including some from the USSR, to witness the event. The operation's primary objectives were to gather data about atomic effects on ship concentrations and to impress the Russians.

Crossroads Able was an airburst test of a MK III MOD 0 implosion bomb, which was Fat Man's first post-war production model. The Air Force conducted the test over a group of 71 captured or obsolete warships moored in the atoll's lagoon. These included an aircraft carrier and a number of battleships. Unfortunately, the B-29, Dave's Dream, missed its designated ground zero by almost 2,000 feet and damage was not as severe as intended. The second test, Crossroads Baker, was a shallow underwater explosion. For this demonstration, the Navy placed a MK III MOD 0 bomb in a waterproof steel caisson suspended by a cable from a landing barge at a depth of 90 feet. The test was nothing less than spectacular and its film footage has since become a standard for demonstrating the effects of an atomic explosion. The Navy cancelled a third deep water test, Charlie, because of logistical problems and security concerns voiced by Los Alamos.

Strategic Air Command, the long-range arm of the US Air Force, conducted Crossroad's airdrop. Activated under the command of General George Kenney on March 21, 1946, at Bolling Field, Washington, DC, the Air Force assigned SAC the 15th Air Force, which it made responsible for operations west of the Mississippi River, and the 8th Air Force, which it made responsible for operations in the eastern United States. It armed these units with B-29 aircraft returned to the United States from the Far East 20th Air Force as part of Operation Sunset. SAC Headquarters moved to Andrews AFB, Maryland, on October 20, 1946.

Although Operation Crossroads did not intimidate the Russians, the atomic tests did impress Air Force General Curtis LeMay, who wrote his superior, General Carl Spaatz, that a concerted atomic attack with an obtainable stock of nuclear weapons would be able to destroy any nation's social fabric and economic infrastructure. It was now imperative that America develop the most effective atomic striking force possible. In 1949, the Air Force promoted LeMay to Commander SAC where he had 120 handmade MK III bombs at his disposal. The MK III bombs were carried primarily by B-29 and B-50 bombers.

Los Alamos based the MK III bomb on a formalized set of Fat Man specifications using a Type A pit. From its inception, Los Alamos produced solid plutonium cores in four types for the MK III (Types 030, 050, 080, and 090), providing it with four yields from 18 to 49 kilotons. It also supplied the MK III with improved detonators. The Department of Military Application designated the first version of the MK III as MOD 0. The term MOD is an abbreviation derived from the Mark-Modification-Alteration nomenclature used by the Atomic Energy Commission. This system indicates the overall design of a major assembly and any changes made to it. MOD "0" is the first version of a weapon design, with subsequent modifications and alterations numbered consecutively.

The AEC later stockpiled two improved versions of the MK III, the MODs 1 and 2, which had increased safety and reliability. By the end of October 1949, Los Alamos had converted all MK III MOD 0 weapons to the higher MODs. Nevertheless, the MK III was poorly suited for use by LeMay's planned strike force. In addition to many technical and logistical shortcomings, the bomb used an inordinate amount of fissile material. Considering the slow rate of growth in the fissile stockpile and the multi-million-dollar cost of atomic cores, there was a pressing need for better-engineered and more efficient weapons.

Fortunately, Los Alamos had resumed atomic development in the interim, with the intention of producing the weapons on SAC's wish list. After a short period, the AEC rescheduled the atomic tests it had cancelled at the end of WW II for the spring of 1948. Operation Sandstone, conducted at Enewetak Atoll in an area of the Marshall Islands that the AEC designated the Pacific Proving Grounds, evaluated a number of important concepts. The objectives of Sandstone were to:

- Prove the nuclear components and initiators in the stockpile
- Improve knowledge about the design of implosion type atomic weapons
- Test the principle of levitation
- Test a composite core
- Determine the most efficient design for the use of the available fissionable material in building the stockpile

Among these objectives, the most important were the development of the levitated pit and the development of the composite core.

A levitated pit enhances compression by accelerating an atomic bomb's heavy U238 tamper across an intervening air gap. The most common ways of achieving levitation were to suspend the core within an oversized tamper by means six truncated, hollow aluminum cones. (Los Alamos also used wires held under tension for this purpose.) Because the tamper is unsupported on its inner surface, it is necessary to overcome the effects of spalling that had plagued Seth Nedermeyer's original implosion experiments. This requires suppression of the

Taylor Wave to avoid excess tensile stresses during compression. Excess tensile stresses were avoided by means of a special core forging process.

The composite core enhanced efficiency. It consisted of an inner sphere of plutonium, surrounded by an outer shell of U235. This configuration, although larger than a solid plutonium core, had a lower spontaneous neutron flux and provided more consistent yields than plutonium cores. Composite cores generated much less heat than those made from pure plutonium, making them easier to store and handle. Finally, varying the ratio of plutonium to uranium in cores of roughly equivalent diameter provided a variety of yields without having to alter the dimensions of the pit and explosive assembly.

The AEC tested a Type B composite-levitated pit as shot X-Ray on April 15, 1948. An inner plutonium core greatly enhanced the efficiency of an outer core made from enriched uranium. X-Ray achieved a yield of 37 kilotons and efficiencies of 35 percent for the inner plutonium core and 25 percent for the enriched uranium outer core, a significant improvement over the original Fat Man bomb. Estimates of the fissile material used in the X-Ray core suggest it contained 14 pounds of enriched uranium and 7 pounds of plutonium – a 2:1 U235 / Pu239 ratio.

On April 30, Los Alamos conducted a second test with a core thought to contain 100 pounds of enriched uranium. This test achieved 49 kilotons, the largest yield to date, demonstrating that an all U235 implosion bomb was not only practical but could be made very powerful. On May 15, the Zebra test produced an 18-kiloton yield from a minimum-sized oralloy core. This bomb reputedly used less than 1/10 of the fissile material (about 16 pounds) incorporated in the Little Boy bomb. Los Alamos configured Yoke and Zebra to evaluate the upper and lower limits for enriched uranium cores. The Sandstone pits were all roughly identical in weight and volume.

The use of uranium in fissile cores had both drawbacks and advantages. As previously indicated, compression in a given amount of fissile material is proportional to the mass of the core, tamper, and pusher, and the amount of energy released by the implosion assembly. Uranium cores used five times more fissile material than a plutonium core in order to derive an equivalent yield. This extra mass subjected a uranium core to a compression penalty in an implosion assembly of fixed size or required a larger implosion assembly to achieve the same level of compression possible with a plutonium core. The mass of the tamper and pusher somewhat mitigated this effect. Despite this drawback, Los Alamos theoretician J. Carson Mark observed that Sandstone created the options to build more powerful weapons and to increase the size of the stockpile by re-fabricating the existing supply of solid plutonium cores. Since it is not possible to use more than a critical mass in a solid core, Los Alamos scientists calculated that the maximum yield for a solid plutonium core using weapons grade material was 100 kilotons. A solid uranium core contains much more fissile material than a plutonium core and, if suitably compressed, could produce a maximum yield of 500 kilotons. Solid composite cores would produce intermediate yields. The AEC began retiring MK III bombs in the spring of 1949, a task that was 35 percent complete by the end of that year. It dismantled its last MK III freefall bomb in late 1950.

Introduced in late 1949, the second-generation MK IV implosion bomb was a product of Division Z, which Los Alamos specially reorganized for this task. Under the overall direction of Jerrold Zacharias, Z1 performed experimental engineering, aerodynamics, airborne testing, ballistics, radar, and telemetry. Z2 was in charge of procurement, weapon assembly, and weapon transport. Z3 was responsible for the design of firing circuits and detonators, and Z5 handled fuze development. Z4, was responsible for general engineering and for overall project coordination of the MK IV "streamlined gadget." With the exception of Z1, the AEC ultimately transferred these groups to Sandia.

The MK IV MOD 0 incorporated the results of Operation Sandstone and 4 years of research at Sandia to become the first practical, deployable atomic bomb. It corrected many MK III deficiencies relating to size, weight, ballistics, the complexity of the fuzing and firing systems, assembly procedures, and aeronautical and structural weaknesses related to the bomb's tail assembly. Los Alamos intended the design concepts behind the MK IV to produce a significant improvement in the time required for field assembly, a more reliable and rugged bomb with lengthy storage characteristics, improved ballistics, and the ability for pre-assembled shipment without the unit's detonators and nuclear materials. The use of high-density explosives in its implosion assembly increased its weight to 10,800 pounds. The AEC produced the MK IV on assembly lines, delivering 550 bombs in four MODs during the period March 1949 to May 1951.

The shape used to improve the bomb's ballistics featured lift rather than drag stabilization. As well, streamlining recessed the loading lug and flush-mounted the radar antennas that had previously protruded from the bomb's envelope. Instead of the duct tape applied externally on the MK III, Sandia sealed the MK IV's casing with permanently fitted gaskets. Sandia also replaced the MK III's box kite tail with individual fins that attached to the bomb's 3/8-inch-thick mild steel casing. Since the strike aircraft's structure protected the front and the rear of the bomb, Sandia made the tail cone from 1/8 and 1/4-inch-thick mild steel and cast the nose and rear cover plates from 1/2-inch-thick aluminum alloy. These changes minimized vibration, improved ballistic coefficients for range and time, and reduced maximum ballistic yaw to 3 degrees from the 12 degrees associated with the MK III.

The MK IV initially used a Type C pit with three core types: the 49-LT-C (Sandstone Zebra levitated U-235), the 49-LC-C (Sandstone X-ray levitated composite) and the 50-LC-C (Ranger Fox levitated-composite). Prefaced by a two-digit numerical designation, LC-C stood for levitated, composite, Type C pit. T stood for test or experimental, since the 49-LT-C variant used an all U235 core. These pits gave it a range of yields from 1 - 31 kilotons with Type 110, 130, and 140 cores. Since the cores were interchangeable, the MK IV was essentially a variable yield weapon, as were the succeeding types of Fat Man bombs. A typical Type C pit / core arrangement consisted of an 11-centimeter diameter core, a 4-centimeter air gap, a 5-centimeter thick natural uranium tamper, and an 11-centimeter thick aluminum pusher. The explosive shell that surrounded the pit consisted of 21.5-centimeter-thick inner blocks and 21.5-centimeter-thick outer lenses. Installed in each outer explosive lens were two Model 1E20 detonators. These detonators improved simultaneous performance by a factor of two, featured a simplified explosives train, enhanced energy transmission from booster to explosive, and increased resistance to corrosion over the Model 1773 detonator used in the original Fat Man bomb.

In order to facilitate the insertion of its cores, Sandia equipped the MK IV with a front hatch. This feature allowed the bomb to be stored in a fully assembled format (minus its core) without fear of heat damage to electronic components. Inserting a nuclear core required removing the forward polar cap to expose the physics package. The removal of two outer pentagonal explosive lenses with their detonators and two inner explosive blocks, which together weighed 156 kilograms, exposed the pit. The pit's aluminum pusher had a removable 12-centimeter diameter, 1-kilogram weight trap door, and its uranium tamper had a removable 12-centimeter diameter, 3-kilogram weight trap door. After a weaponeer had removed these components, he could insert or remove the core with the use of a special vacuum tool. The whole process took about 1/2 hour during flight in a bomb bay. The nuclear core accompanied the bomb in a special, climate-controlled storage cylinder charged with dry, inert gas. This arrangement increased aircraft safety during takeoffs and landings – the aircrew only inserted the core during a War Emergency when a bomber was in level flight.

The MK IV MOD 0 implosion assembly remained relatively unchanged from the that of the MK III. It retained the Model 1E20 detonators used in the MK III MOD 1. In the MK IV, specialists could conveniently connect the 1E20 detonators to a wiring harness by means of bayonet mounts. The bayonet mounts replaced an installation procedure that had previously required crimping. The wiring harness was itself connected to a flange on the rear hemisphere of the implosion sphere casing. A special electronic cartridge that mounted almost all of the bomb's arming, fuzing, and firing components plugged into the flange and automatically connected to the firing harness by means of spring fingers. For the first time, it was possible to complete the wiring of the detonators before the installation of the firing-set.

The decision to consolidate the bomb's electronic components on a removable cartridge was an important innovation in the MK IV. In addition to the electronics, Sandia also relocated the bomb's mechanical clocks and batteries, previously incorporated in the x-unit assembly, to the cartridge. Because of these changes, it was only necessary to remove the antenna nose plate, rear cover plate, the cartridge, the split band, and a trap door to make a complete performance check before final assembly.

The MOD V x-unit used in the MK IV MOD 0 bomb was a repackaged version of the x-unit used in the MK III MOD 1. It was a great improvement over the MK II x-unit used in the MK III MOD 0. The energy used to fire the detonators was now stored in the magnetic field of a slim-loop ferroelectric generator. This feature made it unnecessary to pressurize the x-unit to prevent the flash over of high voltage at high-altitude. The impact of a rotor activated the slim-loop ferroelectric generator, which provided the sharp rise in voltage needed to charge the condensers prior to activation of a pair of M-26 thyratrons. The M-26 thyratrons had a radioactively stabilized breakdown voltage and were much less sensitive to the effects of static electricity than a standard spark gap. As opposed to the MOD II x-unit, in which each spark gap fired only half the detonators, the MOD V x-unit's M-26 thyratrons could each fire all 32 detonators using a 17-kilovolt electrical pulse. Because of the fast rise time of voltage in the ferroelectric firing unit, Sandia no longer incorporated a time delay relay. (Firing Units are discussed in detail in Chapter 16.)

Sandia upgraded the NT-6 the batteries that charged the x-unit in the MK III to the Willard ER-12-10, which had a gelatinized electrolyte to prevent sloshing. These batteries were still of lead acid design but had a conservatively rated shelf life of three weeks after a 20-hour charging cycle at a rate of 1.3 amperes. At the end of this time, a boost of 1.3 amperes for 5 hours extended the batteries' life for another two weeks. Battery box heaters maintained the batteries' temperatures above freezing at bomb bay temperatures of -40 °F. The Model ER-12-10 batteries showed reliable operation down to a temperature of -4 °F. The x-unit was itself operable over a range of temperatures from -22 °F to $+149$ °F and at pressures down to 7 PSI, which corresponded to an altitude of 19,000 feet.

The six barometric switches used in the MK IV MOD 0's fuzing system were the same BS-4 and BS-5 types used in the MK III MOD 1. The flow of air into the bomb's interior through six small ports placed near its nose activated the switches. The pressure across the ports produced nearly ambient conditions inside the bomb. Since the bomb case served as a manifold, no hose connections of the type used in the MK III were required. The barometric switches had a calibration sensitivity of 100 feet, a temperature and vibration sensitivity of 200 feet, error in prediction of barometric pressure of 300 feet, and a pressure system error of 600 feet for a total combined error of 750 feet. Based on this data, Los Alamos estimated that the barometric switches should be set to operate at 2,000 feet above the intended detonation altitude.

The MK IV MOD 0's Archie radars were very similar to those used in the MK III, but for the first time, Sandia equipped them with special anti-vibration mounts. Their flush mounted slot antennas were an entirely new kind that replaced the Yagi-type exterior antennas found on the MK III. Sandia mounted one slot antenna for each of the bomb's four Archie radar fuzes around the nose of the bomb in symmetrical fashion. The slot antennas were a relatively broadband type. In the MK III, specialists had to make height of burst modifications with separate delay lines that they soldered into the Archie to set a desired altitude range. This made it difficult to change the height of burst for new targets in a timely fashion. The modified AR-10A Archie used in the MK IV MOD 0 had convenient plug-in sockets.

Sandia provided improved radar fuzing for later MODs of the MK IV. Although nuclear research experienced a hiatus at the end of the war, developments in fuzing had moved ahead. In August 1945, Los Alamos had contracted the University of Michigan to produce an improved Amos unit known as the "Andy." The Air Force tested both Amos and Andy units to determine if they were susceptible to radar jamming, which might cause them to dud or explode prematurely. Jamming of radar fuzes was an enormous concern at this time so that Sandia also pursued a barometric fuzing project. In mid-1945, barometric fuzes had a potential error of 1,000 feet, which was excessive for the heights of burst then contemplated. Thus, the fuzes used in advanced MODs of the MK IV were "Abee" radar units that incorporated improved circuitry and packaging over the Archie. The Air Force did not consider the Abee fuzes sensitive to jamming and found they had acceptable range (altitude) settings.

Apart from the radar antenna cables, the bomb's firing and fuzing interconnections were located in a special junction box. This concept made it possible to disconnect any of the major subassemblies for replacement or repair without having to disassemble the electronics cartridge. The junction box contained the Archie integrating capacitors, relay networks, power fuzes, and pullout switches. The ten pullout wires that had previously been connected through five separate openings were reduced to two wires routed through a single aperture located just aft of the bomb lug. The pullout switches in the junction box were located so that specialists could insert pullout wires without having to remove the box cover.

Operation of the bomb, with the exception of a delay in the production of high voltages for the x-unit, was identical to that of the MK III. The biggest difference between the weapons was the time needed to prepare them for operation. As long as a specialist at a rear base had installed the bomb's detonators, the MK IV required only 2 hours for its preflight check. This compared with 8 hours for a MK III MOD 1, which also required a larger checkout team. The MK IV MOD 0 needed only 58 pieces of special handling and test equipment (excluding nuclear kits) as opposed to the 80 pieces of equipment required for the MK III MOD 0.

The MK IV MOD 1-C (Block 7) bomb followed the MOD 0. It was the first Fat Man bomb in which a crewmember could insert a Type C core while in flight. This improvement was followed by the MOD 2-C, which incorporated the first baro-manifold, a circular copper tubing construct installed in the rear of the ballistic case. This change greatly improved the accuracy of fuzing. The MK IV MOD 2-D used a Type 52LP-D pit that provided it with higher yields. Special handling equipment was required for the insertion and alignment of the core in this weapon's pit. By the end of 1951, Los Alamos had converted all Fat Man weapons to the Type D pit. The MK IV MOD 3-D was equipped with a Type 53LP-D pit and lightweight magnesium nose plate for ease of handling during in-flight insertion of its nuclear core.

In order to carry the MK III and the MK IV, SAC decided to convert the Consolidated Vultee B-36 Peacemaker from a conventional to a nuclear bomber. The B-36 originated in 1941 as a "hemispheric defense weapon." Its purpose was to attack Axis targets directly from the continental United States if Britain and Hawaii had fallen. A design competition called for an aircraft with a maximum speed of 450 MPH, a service ceiling of 45,000 feet, and the ability to carry a 10,000-pound bomb load a distance of 5,000 miles. Boeing and Consolidated Vultee each submitted plans for the aircraft whereas Douglas supplied specifications for its power plant. Of the two submissions, the Air Force chose the Consolidated Vultee Model 35 as the most promising and designated it the XB-36.

Various delays in finalizing the specifications for the bomber, and the continued availability of foreign bases from which to attack Germany and Japan, resulted in the USAAF focusing on the production of existing bomber types to win World War II. As a result, the first prototype B-36 did not roll off an assembly line until September 8, 1945. Its initial test flight did not take place until August 8, 1946. The Air Force was still interested in the aircraft because of its newly acquired, long-range atomic mission, but it had misgivings about the bomber's performance. After flight-testing by Air Materiel Command, and a return to Consolidated Vultee for modifications, the XB-36 returned for a successful evaluation as the YB-36 in June 1948.

The B-36A, which the Air Force used exclusively for crew training, did not receive a defensive armament. In 1950, these aircraft began conversion to the RB-36E reconnaissance mission. The first bomber variant of the Peacemaker was the B-36B, which the Air Force accepted in late 1948. With a length of 160 feet and a wingspan of 230 feet, it was almost twice the size of a B-29. Called the "magnesium overcast" because of its immense shadow, the B-36 was the largest operational aircraft ever flown by the US Air Force. Its fuselage incorporated many of the features found in the B-29. It had a forward pressurized crew compartment connected to a rear gunner's compartment by an 80-foot long, 25-inch diameter tunnel that ran along the side of four huge bomb bays. The aircraft had enough ground clearance to accommodate the largest available bombs.

The modestly swept wings of the bomber each mounted three Pratt & Whitney 3500 HP R-4360-41 Wasp air-cooled radial engines in pusher configuration. These gave it a maximum cruising speed of 381 MPH at an altitude of 34,500 feet. Its maximum, short-range conventional bomb load was 82,000 pounds. For the atomic mission, it could carry a 10,000-pound bomb to a range of 5,000 miles. Because of a shortage of atomic bombs, SAC originally configured the B-36 to carry a single MK III weapon in its number one (forward) bomb bay (Operation Saddletree). Then in 1950, it acquired the universal bomb suspension (UBS) system from North American to accommodate the forthcoming MK IV, MK 5, MK 6, and MK 18 bombs. As part of Project "On Top," Convair converted 30 bomber aircraft to carry atomic bombs in all four bomb bays and converted additional aircraft to carry atomic bombs in their number 1 and number 4 bomb bays. SAC would later convert all of its B-36 aircraft to carry four atomic bombs. SAC also converted all of its reconnaissance bombers to carry an atomic bomb in bomb bay 4. Later modifications to reconnaissance aircraft gave them the ability to carry atomic bombs in bomb bay 2 and in other bays.

Boeing equipped the B-36A with the APG-23A bombsight. The B-36B received the newer APQ-24 bombing-navigation system, composed of the APS-44 radar and the APA-44 bombing computer. The APS-44 was a high resolution, high-altitude radar that incorporated a 360-degree scan, a provision for "sector scan" (in which the antenna oscillated through any desired portion of the horizon), and a "displaced center" scan in which the vertex of the sweep started at or below the bottom of the cathode ray tube. Western Electric began development of the APS-44 in mid-1945 for use in a combined optical and electronic bombing system that a single individual could operate. However, when combined with an APA-44 computer in the APQ-24 bombing-navigation system, it formed a fully synchronous radar system without a requirement for an optical tie-in. At 30,000 feet, the APQ-24 bombing-navigation radar had a search range of 150 - 200 miles against large cities and could efficiently map an area with a radius of 75 miles. The system was able to distinguish inland targets without land-water contrast to aid in their identification. Boeing and Consolidated Vultee began system installations in June 1947. As of April 1, 1949, Western Electric had shipped the Air Force some 300 AN/APQ-24 systems.

The Air Force, nevertheless, considered the reliability and quality of the AN/APQ-24 substandard for an intercontinental bomber. In fact, the reliability of the AN/APQ-24 was so poor that SAC made a bombsight technician a permanent member of the B-36 crew. The technician's responsibilities included in-flight tube replacement, mechanical repairs, and calibration. Along with the bombsight technician, the B-36 carried several hundred pounds of tubes and spare parts on a mission.

Air Force and industry officials met during February 1951 to discuss this problem. SAC officials insisted on better scope quality to improve the problem of target identification and improved reliability; the many hundreds of tubes and the score or more of inter-connecting black boxes were a maintenance nightmare. A further caution from SAC concerned the new B-47 jet bomber, which had space limitations that make it impossible to carry a bomb system mechanic. Strategic Air Command also indicated an urgent demand for a radar resistant to jamming.

SAC found the solution to these concerns in the K-Series bombing-navigation system. The K-Series was a combination optical / electronic bombing system suitable for jet bombers on which the Sperry Gyroscope Company began development during the closing year of WW II. For this system, Sperry's standard radar bombing and navigation computer (SRC-1) seemed to offer the best possibilities. SAC intended it to form the basis for a one-man bombing system that provided both radar and optical aiming. The radar selected for this system was Western Electric's APS-23, which had a 5-inch diameter scope. Eastman Kodak initially built the optical periscopes for this system. SAC later split the optical part of the project between the Farrand Optical Company, which designed and fabricated a vertical (retractable) periscope (designated Y-1), and the Perkin-Elmer Corporation, which produced the Y-2 horizontal (non-retractable) periscope. The Massachusetts Institute of Technology also received a contract to build a "coordinate converter." This device transformed the Y-2 sight's two-axis data into three-axis data for use with the Sperry computer.

The Air Force installed the vertical periscope in aircraft like the B-36, which had considerable headroom at the bombardier's station. It installed the horizontal periscope in jet aircraft where space was at a premium. Periscopic sights overcame the need for a flat glass bombsight "window" in the aircraft's nose, which had a degrading effect on speed. Both the Y-1 and the Y-2 had sight lines stabilized in roll and pitch by means of remote signals from a stable platform. The first models of each sight were flight tested in B-29s. The entire bombing-navigation system was capable of continuously computing and displaying ground speed, ground track bearing, wind velocity, and wind direction, together with the latitude and longitude of the aircraft's position. This freed the bomber to engage in evasive maneuvers prior to and during the bomb run that included changes in altitude and azimuth.

The SRC-1 bombing computer received the military designation AN/APA-59 before Air Materiel Command renamed it the A1. AMC, had it designed with the assumption that navigation to and from the target was an integral part of any bombing mission and that bombing and navigation problems had many features in common. In both functions, the system had to determine the relative position of a recognizable landmark and track this landmark either optically or by means of radar. In mid-September 1946, the system's APS-23 radar and its interconnecting equipment were designated AN/APQ-31. The tracking range of the APS-23 radar reached out some 150,000 feet from the aircraft and projections indicated that the computer was capable of bombing at ground speeds up to 695 knots. The tests also proved that a "memory point" tracking feature could shorten bombing runs to 15 seconds. By March 1948, all the components of the 1,500-pound system were completed and installed in a B-29. One month later, the bombing-navigation system with the Y-1 vertical periscope received the designation K-1. On May 28, the first flight test of the K-1 system took place. Bombardiers reported that the resolution of objects on land and water at various degrees of magnification was very good. The same equipment with the Y-2 horizontal periscope for jet aircraft later emerged as the K-2 system. The K-1A bombing system allowed the radar detection of cities at a range of 200 miles and shipping at a distance of 50 to 100 miles.

In February 1950, the Air Force Armament Laboratory asked Sperry to study the possibility of using the AN/APN-81 Doppler radar from the AN/APN-66 Navigation System with the K-1 and K-2 systems, limiting extra weight to 200 pounds. The Armament Laboratory also asked Sperry to examine the feasibility of incorporating fully automatic navigation, "automatic crosshair laying", automatic target tracking until after bomb burst (for bomb damage assessment), and automatic navigation in Polar Regions. To get around the problem of radar jamming, Sperry designed an "inertial extrapolator" that could control an aircraft for one hour before bomb release. It was to supply the system with the rate and position information normally obtained by visual or radar tracking so that the bombardier had only to perform final positioning of the crosshairs for completion of the bombing run. The inertial extrapolator never became part of K-Series systems.

During 1950, changes in the K-system periscopes and computers led to a number of new designations. Farrand designed a non-retractable vertical periscope similar to the Y-1, which it called the Y-3. Sperry redesigned the A-1 computer, adding an improved amplifier, tracking computer, and navigation control. It designated this device the A-1A. The bombing-navigation system incorporating the Y-3 periscope and the A-1A computer (called the K-3A) became the standard system for B-36 aircraft starting with the B-36D. At the same time, Perkin-Elmer developed a Y-4 horizontal periscope that had binocular optics and was slimmer than the Y-2. It incorporated mounts for a camera that could photograph the radar and the optical displays. Together with the A-1A computer and the APS-23 radar, it became the K-4A bombing-navigation system installed in B-47 aircraft. The average error for these systems was less than 1,500 feet from a 40,000-foot bombing altitude, the majority of bombs falling within 1,000 feet of the target.

The B-36's defensive armament consisted of sixteen M24A-1 20mm cannon located in the nose, the tail, and six remotely controlled retractable turrets. An AN/APG-3 radar gun sight controlled the rear turret, whereas gunners using computing optical sights operated the retractable turrets. The large number of defensive positions necessitated a fifteen-man crew. This consisted of a pilot, copilot, radar operator/bombardier, navigator, flight engineer, two radiomen, three forward gunners, and five rear gunners. Despite its defensive capabilities, the Peacemaker became vulnerable after the Soviets introduced jet fighters.

To provide better performance, with which to counter Soviet jet fighter, SAC had Convair equip the B-36D with a pair of 5,200-pound thrust, General Electric J47-GE-19 turbojets mounted in pods underneath its outer wings. The jet engines decreased the bomber's takeoff run by 2,000 feet and increased its maximum speed to 439 MPH. Since these engines raised fuel consumption, they decreased the range of the B-36. To rectify this situation, Convair added bladder-type outer panel fuel cells that raised takeoff and landing weights to 370,000 and 357,000 pounds respectively. Convair also raised the aircraft's maximum bomb load to 86,000 pounds and fitted snap-action, split bomb-bay doors that could open and close in two seconds. The Air Force then had Convair upgrade the B-36B to B-36D specifications. This included the AN/APG-32 gun laying radar adopted for the B-36D's rear turret. Convair equipped the B-36F with slightly more powerful 3,800 HP Pratt & Whitney R-4360-53 engines. Produced in concert with the B-36H, The B36-F had a rearranged crew compartment and an AN/APG-41A radar system that aimed the two 20mm guns in the rear turret. The AN/APG-41A featured twin tail radomes.

In order to further increase the range of the B-36, SAC awarded Convair a contract to reduce its weight in 1954. The Air Force had originally intended the B-36 to fly in formation, each plane covering other aircraft with its heavy defensive armament. When the Air Force changed the B-36 mission profile to lone attacks with thermonuclear bombs, the heavy defensive armament became a liability. The Air Force had Convair remove much of the aircraft's armament, ammunition, and armor. Convair moved ahead with this program in three stages. In Project Featherweight I, Convair left only the front and rear firing guns on the B-36, removing some of the aircraft's armor. Removal of the aircraft's defensive armament reduced the crew from 15 to nine men. In Project Featherweight II, Convair removed the rear crew compartment's galley, which became unnecessary with the reduced crew. Project Featherweight III applied specifically to the B-36J, which Convair converted to the Featherweight standard on the assembly line. The last 14 B-36J-III aircraft had radar-aimed tail turrets as their sole defensive armament, extra fuel storage in their outer wings, and a landing gear that accommodated a maximum gross takeoff weight of 410,000 pounds. This increased the bomber's range and raised its operating ceiling to 47,000 feet.

Strategic Air Command accepted delivery of 383 B-36 aircraft, although no more than 250 aircraft were operational at any given time, even at the peak of its service. SAC organized these aircraft in six heavy bomb wings and four heavy reconnaissance wings. The B-36 was in a class by itself in comparison with contemporary American bombers, and way out in front of the vintage B-29 knock offs employed by the Russians. It served at bases in both the continental United States and overseas. Nevertheless, high-speed jet bombers placed in development by the Air Force soon rendered it obsolete.

Of long lasting benefit to the Air Force were the physical upgrades required to support the B-36. Its 300,000-pound - 400,000-pound takeoff weight necessitated the lengthening and strengthening of runways at many SAC bases. The costs associated with this program set off a major debate. The Air Force wanted the new runways made from reinforced concrete but government officials wanted to use less expensive asphalt. The interference of lobbyists only exacerbated this debate. The Air Force finally had the National Academy of Sciences conduct a fact-based investigation that recommended concrete. The NAS expected this material to prove more resistant to jet blasts from B-47s, fuel spillage and the high temperatures in southern states where SAC based these aircraft. To prove its point, the Air Force paved two runways, one with asphalt and one with concrete at Kelly AFB, San Antonio, Texas. The rapid disintegration of the asphalt runway vindicated the Air Force's choice.

In addition to the MK IV atomic bomb, the B36 could also carry the more advanced MK 6. The AEC successfully proof tested the MK 6 as part of Operation Greenhouse on April 8, 1951. With the success of this test, the AEC and the Air Force decided to place it in mass production as SAC's main strategic bomb. The new bomb had a welded aluminum casing that reduced its weight to 8,500 pounds. It also sported distinctive spoiler bands that had previously improved the ballistics of late model MK 4Ns. "N" stood for Northrop, the manufacturer of the lightweight aluminum casing. The spoiler bands were 1/2-inch-high and 3/4 inches wide. Sandia and Air Materiel Command repositioned the bomb's pullout wires and diagnostic connections to a single position behind the suspension lug at the top of this case to simplify loading. Air Materiel Command then initiated the "On Top" modification program to upgrade the fleet of B-29, B-50, and B-36 aircraft to handle the new specifications. By the end of 1951, almost 450 aircraft could carry MK 6 bombs.

The short-lived MK 6 MOD 0 and MOD 1 bombs were equipped with MK IV explosive assemblies. A more advanced, 60-lens "light sphere" implosion system manufactured at the Salt Wells Plant, California, reduced the weight of the MK 6 MOD 2 and all subsequent versions of the MK 6 to 7,600 pounds. The lightweight explosive sphere was 17.3 inches thick. The MK 6 had yields of 8, 22, 26, 31, 80, 154, and 160 kilotons using Type 110, 130, 150, 170, 210, 240, and 260 cores in Type D pits. The upper yield for the MK 6 was almost as powerful as the 170-kiloton W62 strategic warheads later deployed on Minuteman III missiles. These could flatten the heart of a city with a population of 1,000,000 inhabitants. For convenience, crewmembers could store a removable detonator and a cored sleeve of high explosives in a rotatable holder during in-flight insertion of the bomb's atomic core. The Bomb's Type D pit was equipped with a 120-pound natural uranium tamper. Rechargeable MC193 NiCad batteries that kept a 240-day charge provided the bomb with internal power.

Sandia also upgraded the internal neutron initiators for the MK 6, MK 5, and MK 7, to a smaller and more efficient "Tom" model. It based the Tom initiator on a design patented by Klaus Fuchs and Rubby Sherr of LASL in June 1946. Their initiator had a "timed neutron source" that was activated by the outgoing release wave (as opposed to the incoming implosion wave). This feature released neutrons at a more desirable degree of compression. Advanced initiator models had tetrahedral or conical pits on their internal surface to take advantage of the Munroe shaped-charge effect and did not require an internal ball like the Urchin. The Tom measured only about 1/2 inch in diameter and had a replacement life of 1 year whereas the Urchin had a replacement life of only 4 months. Los Alamos intended to use the Tom initiator in boosted warheads because of its longer life and diminutive size. It tested the new initiator in January 1951 during the Baker-1 shot of Operation Ranger. Monsanto then began producing the initiator in 1950. By May 1951, Los Alamos was adapting all new implosion cores to accommodate the Tom initiator and converting stockpiled cores at a rate slated for completion by the end of the year.

Los Alamos proposed the development of even longer-lived actinium 227 and polonium 208 based initiators in the summer of 1948. The lab first considered Ac227, which has a half-life of 21.7 years, as an alternate alpha particle source in the fall of 1947. Due to actinium's high production costs, Los Alamos discontinued its investigation for use in initiators during 1953. The Tom initiator likely used Po208 as its alpha emitter. Po208 has a half-life of 2.9 years but an alpha activity only 13 percent that of polonium 210.

Another new initiator design, designated Jonah, became available in October 1953. This initiator was cheaper than the Tom and did not require replacement on a regular basis. Jonah initiators may have incorporated uranium deuteride (UD3) as their active ingredient. China, and Pakistan, may have used this material in mechanical initiators for their first generation of implosion bombs. Although there was a small loss in yield associated with the Jonah initiator, the Air Force deemed the elimination of the complex logistics associated with replacing Po210 and Po208 charged initiators to be more important.

Although the MK 6 MOD 1 bomb was not equipped with the Jonah initiator, it was equipped with an improved Albert radar fuze that was more powerful and more reliable than the preceding Abee fuze. The Albert fuze incorporated solid-state electronics, resistance to jamming, and allowed the in-flight adjustment of the bomb's height of burst by means of a T-19 flight control box. The T-19 provided convenience of choice for the selection of secondary targets. A T-18 monitoring box supported the T-19 flight control box in the air and a T-23 circuit-testing box supported the T-19 on the ground. A crewmember initially selected height of burst on the ground from six preset altitudes. The aircrew could later adjust the altitude-setting while in flight.

In the MK 6 MOD 2 bomb, the lab moved the firing-set from the rear of the implosion sphere to the front of the bomb in order to maintain the bomb's center of gravity. The lab also moved the bomb's radar antennas to the sides of its casing. When this arrangement proved unreliable, the lab moved them back to the front of the bomb in the MK 6 MOD 3. The MK 6 MOD 4 featured heaters for the bomb's radars and batteries, and the MK 6 MOD 5 eliminated the bomb's radar fuzes in favor of all-barometric fuzes. The barometric fuzes accommodated the increased yields of this weapon, which required detonation at altitudes above which the radar fuzes did not function. The MK 6 MOD 5 was also equipped with banks of nickel-cadmium batteries known as a MK 2 power supply.

In 1956, Sandia converted all models of the MK 6 to the MOD 6 configuration with the installation of improved barometric and contact backup fuzes. These changes reduced concerns over radar jamming, although by this time studies had shown that the extensive jamming needed to prevent an atomic attack was impractical. The barometric fuze for the MOD 6 was designated MK 13 MOD 0.

As well as providing new fuzes, Sandia replaced the standard red and green arming plugs with a rotary Arm / Safe switch installed on the forward polar cap of the sphere's case. The aircrew activated the switch through an access door in the nose plate or remotely by means of an electrical circuit. Sandia also gave the bomb a faster firing x-unit, eliminated the choke used for resonant charging of the condensers, and added a thyratron trigger circuit to permit flow of power to the detonators when the proper signal was received from the fuze. The bomb featured barium titanite in its contact fuzes.

The mission sequence for a MK 6 MOD 6 bomb started before takeoff when air and ground crews set its arming baro-switch and applied power to its timer, battery, and trigger circuit heaters. Aircrew could remotely reset the arming baro-switch in-flight if required. Upon release from a B-29 or B-36, the extraction of the bomb's pullout wires switched the bomb to its internal batteries, which continued to supply power to the bomb's circuits. At the end of the bomb's preset separation time, a switch in its timer closed to connect the arming relay, which remained without power until the arming baro-switch closed. The switch also connected the inverter to the firing-set charging circuits, which completed the charging procedure in 0.5 seconds. Three seconds after safe separation time, a second switch in the timer closed the fuzing baro-switch ground circuit to ready the weapon for airburst detonation. When the fuzing baro-switch closed at its preselected altitude, it applied power to a pair of thyratrons that operated the bomb's dual firing circuits, discharging the condenser bank into the detonator's bridgewires. If airburst failed, detonation occurred upon impact. Setting the bomb for contact detonation isolated the baro-switches. Upon impact, the crystal contact fuzes applied a voltage to both thyratrons to activate them.

In addition to its use as a freefall bomb, the MK 6 was adapted for use as a nuclear land mine or atomic demolition munition. Designated a Project A type ADM, an existing stockpile weapon could be used in conjunction with a "Firing Device, Demolition, M22." The Special Weapons Project also studied the use of the firing device with MK 5, and Mk 7 bombs. The AEC conducted a full-scale test of this device during the Ess (Effects Sub-Surface) shot of Operation TEAPOT at the Nevada Proving Grounds on March 23, 1955. It buried the Ess device, containing a 1-kiloton Ranger Able core at a depth of 68 feet. The explosion created a crater 96 feet deep and 292 feet in diameter. Ten minutes after the blast, the crater was emitting a lethal dose of 8,000 roentgens per hour. The AEC forbade aircraft to fly within 2 miles of the crater for 3 days after the blast. The Ess test resolved uncertainties about the cratering effects of buried atomic demolition munitions and their effects on underground installations. The emplacement of the device served as a stockpile-to-target test sequence.

In order to produce a Project A type ADM, a MK6 physics package and electronics were installed inside a heavy cylinder which was rounded at one end and shaped into a truncated cone at the other. Additional electronics and batteries assisted the firing device in detonating the bomb. The rounded end of the ADM had connections that allowed for a number of different firing methods. A firing line of not more than 600 feet long was connected between the ADM's J8 junction and up to a floating buoy with a radio receiver antenna. This may give an indication of the bomb's maximum depth of use in water. The buoy's radio receiver could accept radio detonation commands sent over a range of 10 miles.

A second method of command detonation incorporated a cable not more than 8,500 feet long connected to the ADM's J6 junction to a shore box. The shore box was connected with a dedicated land line, that could be no more than 5 miles long, to the firing device. Presumably either method could be used on lad by connecting a radio receiver to the J6 terminal or a land line directly to the JE8 terminal.

There was also a J7 terminal for what purpose the Author is not clear. It may have had something to do with dedicate land use or for use with an alternate T46E2 remote control firing device. In 1974, the Army's Picatinny Arsenal released a Universal Firing Device (UFD) that was more rugged and compact than the M22 or T46.

Delivery of the new bombs coincided with the arrival of SAC's first all-jet strategic bomber, the Boeing B-47 Stratojet. The B-47 concept originated in 1943 when the USAAF made inquiries to manufacturers about a multi-jet airframe that it could use for reconnaissance or for medium bombing applications. A formal call for a medium jet bomber with a 3,500-mile range, a 45,000-foot service ceiling, and a 550 MPH speed followed on November 17, 1944. The Boeing

Corporation was already at work on a scaled down, jet powered version of the B-29 and along with three other manufacturers the Air Force awarded it a contract for advanced study. Although Boeing's plan originally called for straight wings, the evaluation of captured German aerodynamic studies by Chief Aerodynamicist, George Schairer, resulted in a design with wings swept back at an aggressive angle of 35 degrees. Almost all modern aircraft have this feature, which delays the formation of shock waves at high speeds. A reduction in wing thickness necessitated the relocation of fuel storage to the fuselage.

At the request of the Air Force, Boeing placed the B47's jet engines in nacelles and hung them under the wing on pylons to facilitate servicing and to reduce the risk of fire. Boeing also increased the number of jet engines from four to six. The final configuration placed two engines on an inboard pylon and one engine on an outboard pylon on each wing. The three-man crew sat in a bubble canopy, very similar to the ones found on fighter aircraft. In case of an emergency, late-model B-47s were equipped with fighter type ejection seats. The crew consisted of a pilot, a copilot, and a navigator / bombardier. The navigator could face rearward to operate the remote defensive armament in the tail when not occupied with other duties. Satisfied with the basic design of this aircraft, the Air Force ordered a pair of prototype bombers in April 1946.

Although Boeing delivered the first XB-47 on September 12, 1947, the aircraft did not make its maiden flight until December 17. Because the acceleration of early turbojets was slow, the fuselage had mounts for 18 solid fuel JATO units behind its wings. When the performance of the original J-35 GE engines was found disappointing, Boeing replaced them with the more powerful J-47 model in its second prototype. The increase in performance provided by the J-47 engines gave the XB-47 the necessary edge to win the bomber competition and on November 22, 1948, the Air Force awarded Boeing a production contract.

The B-47E was the most numerous version of the 2,042 Stratojets built. Its six General Electric J47-GE-25 turbojets each produced 5,970 pounds of thrust, dry and 7,200 pounds of thrust with water injection. They delivered a cruising speed of 500 MPH and a service ceiling of 33,100 feet. The aircraft's defensive armament consisted of a pair of 20mm cannon in the tail. An A-5 fire control system automatically detected and tracked threats and fired the guns. Boeing provided late model aircraft with AN/ALT-6 radar jammers to increase their defensive capability. The performance and maneuverability of the B-47 equaled that of contemporary fighters, giving it tremendous survivability. Air Materiel Command configured it to carry two atomic bombs although it was not until the mid to late 1950's that sufficient weapons were available for full load-out. The B-47B and E Model bombers had maximum payloads of 25,000 pounds, but the combat radius with a single 10,845-pound bomb was only 2,013 miles.

Following the installation of the K-4A bombing navigation system in early models of the B-47, SAC replaced the system's AS-23 radars with tunable AFS-64 radars equipped with pressurized 10-inch Motorola scopes. In this configuration, the K-4A was redesignated the MA-7A bombing-navigation system. The MA-7A entered production in November 1955, ending development of the B-47 bombing and navigation system. Bomber production continued through 1956 and in 1957, during which period the Air force retrofitted B-47B and B-47E early production aircraft with the MA-7A.

To improve the B-47's ability to carry out its atomic mission, the bomber was equipped with an in-flight refueling system. SAC also moved many aircraft to reconstituted WW II air bases located in Britain, Spain, French Morocco, and the Far East. SAC activated the 15th Air Force with headquarters at Madrid, Spain, in 1957. It operated the base through 1966 to oversee

B-47 activities in Spain and Morocco. The need for in-fight or aerial refueling in support of the B-47's mission forced SAC to make a substantial investment in tanker aircraft. Beginning in 1955, Boeing produced 816 KC-97 Stratotankers based on the C-97 cargo version of the B-29. Most of these aircraft could carry heavy cargo in addition to a 9,000-gallon load of jet fuel. The Stratotankers had refueling booms instead of the flexible hose arrangement found in the preceding KB-50J. In order to refuel much faster jet aircraft SAC had to resort to a technique known as "tobogganing." The tanker and B-47 met at altitude and then flew at a slight angle of descent to allow the tanker to pick up speed. The late model KC-97L was equipped with a J47-GE-23 jet engine in an under-wing nacelle on each wing to provide it with extra power. The use of aerial tankers meant that B-47s could now attack their targets directly from continental US airbases. SAC retained overseas bases to recover the aircraft after a strike.

Although the Air Force acquired the B-47 for high-altitude bombing, SAC split its mission in 1955 when it requested that Boeing retrofit 125 B-47s for low-level strikes. By this action, SAC intended to complicate the Soviet defensive posture through a combination of high and low-altitude attacks. Along with the B-47's split role, SAC implemented a 24-hour alert program for its strategic bomber force. Beginning in 1955 selected Bomb Wings had their B-47 bombers and KC-97 tankers armed, fueled, and parked on the runway, ready to go. Whereas the bomber crews only had to remain within a reach of a telephone call, the tanker crews had to remain on base. SAC launched the KC-97s ahead of the bombers to place them in the proper position for refueling. While the tankers were taking off, the aircrews of the bombers received their mission briefings.

By 1954, SAC could deliver a "Sunday Punch" of 750 strategic bombs from a combination of B-50, B-36, and B-47 bombers. Navy Captain William Moore, who attended a targeting session, reported that he saw a series of maps spaced at half hour intervals on which heavy lines showed the tracks of SAC's bomber wings converging on the heart of the Soviet Union. Stars representing Desired Ground Zeroes or "DGZs" liberally sprinkled the maps, demonstrating the speed with which SAC could utterly annihilate that nation. The Eisenhower Administration expressed its confidence in SAC when Secretary of State Dulles announced a policy of "massive retaliation" in 1952.

Not content with the MK-6 atomic bomb, Ted Taylor, one of Los Alamos' most prolific weaponeers, developed a follow-on device known as the TX13. The new bomb had a 56-inch diameter, 92-lens implosion system and a Hamlet "hollow core" in a Type D pit. This was the first use of a hollow core; all previous devices, whether experimental or operational, had solid cores, with just enough space for a modulated internal neutron initiator at their center. Aircrew could select contact burst inflight as the primary means of detonation or as backup. A safe separation timer for arming could be set from 5 - 45 seconds prior to takeoff. The TX 13 had an XMC356 rear inflight-insertion mechanism for its core. In the event of power failure, a crewmember could manually operate the mechanism by means of a flexible drive.

A pair of electronic betatron initiators, mounted external to the hollow core, started the chain reaction in the TX13. These initiators supplied a burst of electrons that induced photo-fission in the core as it imploded. The induced fission supplied the neutrons needed to start the chain reaction. External initiation was required because of the difficulty in centering an internal neutron initiator in a hollow core. Tested as Upshot-Knothole Harry in May 1953, the TX13 set an efficiency record for sub 100-kiloton weapons with a 32-kiloton yield. Even so, with the vastly destructive hydrogen bomb just around the corner, the concept of a heavy (relatively low-yield) strategic atomic bomb was on the verge of obsolescence. The Air Force cancelled the TX13 on August 5, 1954. It also cancelled a higher yield TX20 variant of this device.

Despite its cancellation, the TX13 explosive assembly and its inflight insertion mechanism managed a brief service life in the high-yield MK 18 atomic bomb from 1953 - 1956. The AEC deployed the MK 18, also developed under the direction of Ted Taylor, in a MK 6D casing as an interim measure while it stockpiled hydrogen bombs. D stood for Douglas, the bomb casing's manufacturer. Taylor used approximately 130 pounds of highly enriched uranium to replace a traditional composite core in the MK 18 and, for this reason, Los Alamos facetiously named it the Super Oralloy Bomb or "SOB." The lab chose uranium for the bomb's fissile fuel because of the time needed to implode the large diameter core. Plutonium, with its high spontaneous neutron production would have been prone to pre-detonate in a bomb with this yield. The largest pure plutonium warheads ever developed were the MR31 120-kiloton warheads produced by the French for their SSBS S2 strategic missiles. This was before the French weapons establishment had access to uranium enrichment. As previously discussed, the practical upper limit for a plutonium bomb using weapon grade material is about 100 kilotons. For almost pure or supergrade Pu239, the yield limit is about 200 kilotons.

The enriched uranium in the core of the MK 18 bomb exceeded the amount of fissile material required for a bare, solid, critical mass. Configured within its U238 tamper, the core material actually represented 3 critical masses and created a significant safety hazard that contributed to its short deployment. An 18-inch long "Boral" chain filled a hollow core in the central portion of the bomb to absorb neutrons and to prevent the core's collapse in the event of accidental detonation. For reasons of safety, the chain was only to be withdrawn when a crewmember activated the inflight core insertion mechanism.

As well as using enriched uranium in the MK 18's hollow core, the weaponeers at Los Alamos lined the inside of the innermost tamper shell with more enriched uranium, thus distributing the fissile material in such a way as to prevent criticality. The bomb likely had multiple levels of levitation in its tamper to enhance compression. Multiple levels of levitation would have led to an extremely efficient design. Los Alamos physicists were aware that a velocity increase results from the elastic collisions between bodies of decreasing mass. Insuring that the ingoing and outgoing shock waves arising from the successive collapse of multiple layers provided a uniform density throughout the core would not have been a trivial engineering exercise. The incorporation of "shock buffers" between the various layers may have minimized the conversion of kinetic energy to heat through successive impacts. Shock buffers are thin layers of low impedance materials like graphite. They can split a powerful shock wave into a series of smaller events to reduce the loss of energy through impact heating. The AEC asserted that the pit and the core of the MK 18 differed from those of the MK 6 bomb in such a way that and they were not interchangeable.

The method of initiation for the MK 18 is unknown. Los Alamos and Sandia may have supplied it with the same electronic initiators as the TX13. This would have improved efficiency and would have simplified the design problems associated with an internal initiator in such a complex weapon. The MK 18 MOD 0 had a MK 6 MOD 5 barometric fuze with a safe separation timer. The MK 18 MOD 1 bomb had the same insensitive MK 12 MOD 0 fuze used to arm the MK 6 MOD 6 bomb. The MK 12 fuze incorporated a contact option.

Los Alamos proof tested a MK 18 device as Ivy King on November 16, 1952, with a yield of 500 - 550 kilotons, the largest non-boosted atomic explosion on record. Since U235 delivers a yield of 17 kilotons per kilogram, 29.4 kilograms or 67.4 pounds of U235 had to fission in order to produce the MK 18's yield. This implies an efficiency of about 50 percent. Los Alamos manufactured 90 MK 18 bombs between March 1953 and February 1955. The Air Force retired its MK 18 bombs by converting them to the lower yield MK 6 MOD 6 prior to April 1956. This procedure was relatively simple because the MK 18 used the same case, implosion system, and fuzing as the MK 6.

      The kiloton yields of atomic bombs were no match for the megaton yields produced by hydrogen bombs and the nuclear hysteria of the late 1950s. By 1960, there were more than 5,000 of these super weapons in service and the need for strategic atomic bombs had faded away. The last of nearly 1,100 MK 6 bombs was retired from service in 1962, ending almost 2 decades of strategic atomic deterrence. The era of Fat Man was over.

CHAPTER 12
LIGHTWEIGHT STRATEGIC WARHEADS: ATOMIC WEAPONS IN TRANSITION

In June 1950, General Douglas MacArthur called for atomic artillery shells and freefall atomic bombs to support his operations against North Korea. This request was consistent with United States doctrine that called for the use of atomic weapons to support military elements in combat with the enemy. MacArthur also requested that the Air Force laydown a lethal radioactive belt of cobalt 60 between China and North Korea to prevent the transfer of troops and supplies! President Truman refused MacArthur's requests and ultimately replaced him because of his aggressive views on combat with North Korea. It did not take long before General Mathew B. Ridgeway, MacArthur's successor, repeated the request for atomic weapons. Truman refused Ridgeway, just as he refused MacArthur.

In the case of Korea, the United States did not abstain from using atomic weapons because this act might lead to a widespread nuclear war. Compared with the paltry 25 weapons in the Soviet arsenal, the United States had 450 operational atomic bombs. The United States abstained from using atomic bombs to avoid negative world opinion. Despite the fact that American military doctrine called for the use of atomic weapons in support of troops in combat, and despite the release of nuclear weapons to military custody, the employment of nuclear weapons was and remains a civilian (governmental) decision to this day.

The call for tactical atomic support came at an opportune time. Los Alamos had already begun work on a "light sphere" implosion system that it used in later MODs of the MK 6 strategic bomb, and a smaller implosion assembly for a "lightweight strategic bomb." The laboratory initiated these projects on October 31, 1947, when the Military Liaison Committee informed the Atomic Energy Commission "that current weapons did not lend themselves to wide or flexible employment." At a conference held at Los Alamos over September 2 - 3, 1948, LASL, the Air Force Office of Atomic Testing (AFOAT), the AFSWP, RAND, the Navy, and the aircraft industry decided that a new bomb with a 40 to 48-inch diameter and a weight of 5,000 - 6,000 pounds would significantly improve aircraft performance and increase the likelihood of successful delivery. They also decided that the new weapon could remain 128 inches long. The lightweight strategic bomb project ultimately produced a bomb that bridged the gap between the MK 6 and a true tactical bomb.

In designating the new bomb, the Military Liaison Committee asked the AEC to identify it with the prefix "X" for "Experimental." It also asked the AEC to add the letter "T" to indicate that the design was "Tentative." The AEC and its predecessors have since used the prefix "TX" for all experimental or developmental bomb projects. Thus, the lightweight strategic bomb became the TXV. The AEC later changed this designation to TX5 in order to eliminate use of the cumbersome Latin numeric system. The Air Force accepted the bomb into service as the MK 5 Bradbury, so named for Dr. Norris Bradbury who championed it.

The weaponeers at Los Alamos reduced the size and weight of the TX 5's implosion sphere by increasing the number of initiation points on its surface. This increased the number of lenses needed to form its explosive sphere and shortened the distance between initiation points.

A North American B-45 Tornado, the first Americn jet (tactical) bomber. It could carry a MK 5 atomic bomb internally. (Credit: USAF)

A MK 5 atomic bomb casing with its core insertion doors open. The white panels allow operation of the bomb's radar fuze throgh the bomb's casing. (Credit: NAM)

The USS Cusk (SSG 348) fires a JB-2 Loon missile. Data from the Loon program benefitted both the Matador and Regulus programs. The Loon was a modified V-1 cruise missile. Credit: USN Photo 08034807)

A Martin B-61A Matador missile on its mobile launcher. Despite their mobility, most of these missiles would have been launched from parking areas at their operational bases. Matador was armed with a 50-kiloton W5 warhead. (Credit: USAF)

A Chance Vought SSM-N-8 Regulus missile on the dek of USS Tunnny (SSG-282). The missile was stored with its boosters attached and armed with a 50-kiloton W5 warhead (Credit: USN, via All Hands Magazine)

USS Los Angeles (SSG 574) launches a Chance Vought SSM-N-8 Regulus missile. The cruiser carried three missiles, which were launched from her fantail. The Navy armed submarines, cruisers, and carriers with Regulus missiles. (Credit: USN Photo NH 97391)

A Martin MGM-13A (TM-61B) Mace missile on a "Teracruzer" TEL. Originally designated Matador B, the many changes to this missile resulted in its new Mace designation. Mace A and Mace B were equipped with megaton-range W28 warheads. (Credit: USAF; Color Original)

The launch of a Martin MGM-13B Mace missile at Cape Canaveral. This launch site simulated the hardened launch sites that the Air Force deployed to Germany and to Okinawa, Japan. In Europe, the Air Force also used simple environmental shelters. (Credit: USAF; Color Original)

A hardened Mace B missile site under construction by the Army Corps of Engineers on Okinawa. Each group of four shelters had its own underground control center. The heavy blast doors that protected the entranceways are seen in the lowered position. (Credit: USACE)

The smaller explosive lenses decreased the time it took for each shock wave to merge with neighboring waves. Since the shock waves travelled inwards as well as sideways, it was necessary to thin the lenses to achieve a symmetrical implosion wave. The implosion assembly thus had a smaller diameter due to the reduced thickness of its lenses. It also weighed less than preceding designs. The MK 5 implosion sphere comprised 92 lenses and measured 10.24 inches thick.

The Salt Wells Pilot Plant cast the lenses for the TX5 implosion assembly with the same Composition B explosive used in the MK III, MK IV, and MK 6 bombs. Composition B is a castable, high velocity explosive made by melting TNT and adding powdered hexahydro-1,3,5-trinitro-1,3,5-triazine or RDX $(CH_2-N-NO_2)_3$, a powerful nitramine explosive. Technicians stabilized the explosive mixture by adding a de-sensitizing wax. As formulated for early atomic weapons, Composition B consisted of 65 percent RDX, 34 percent TNT, and 1 percent wax.

Because the MK 5 atomic bomb was the leading candidate for use as a primary in early thermonuclear weapons, its designers could not use Baratol as a low explosive component. The barium contained in Baratol is a moderate-Z element that does not fully ionize in an atomic explosion. Partially ionized barium nuclei would have impeded the flow of radiation that imploded the thermonuclear secondary. This consideration resulted in the replacement of Baratol with Boracitol, an explosive formulated with 40 percent TNT. It also contained 60 percent boric acid as inert filler. The highest-Z element contained in boric acid is oxygen, which ionizes easily to become transparent to radiation. The production MK 5 implosion assembly measured 40 inches in diameter and weighed 2,700 pounds.

Intended for use with tactical bombers that did not have room for manual in-flight insertion of their nuclear capsules, the MK 5 freefall bomb required the implementation of a mechanical procedure for in-flight insertion. The new flight safety device was necessary because the reliability of even highly maintained nuclear bombers was suspect at this time and the chemical explosives used in atomic bombs were sensitive to heat and shock.

During the 1951 deployment of ten MK IV atomic bombs (minus their nuclear cores) to Guam from Fairfield Suisun AFB, California, two out of ten B-29s used for delivery had to turn back because of mechanical problems. A third B-29 crashed on takeoff when one of its engines malfunctioned and its landing gear failed to retract. Although rescue personnel saved some of the crew, 19 personnel died when the explosives in the MK IV bomb cooked off in the ensuing fire. With a nuclear core installed, the results could have been far worse. An atomic yield ranging from tens to hundreds of tons of TNT equivalent might have occurred, with a spray of lethal gamma and neutron radiation to a distance of 400 - 500 meters. Takeoffs with a nuclear core installed would have been a foolhardy practice indeed.

Concerns over the MK 5's safety thus resulted in the development of a remotely controlled, electro-mechanically operated screw type device to automatically insert or extract the bomb's nuclear capsule – along with a segment of its tamper and its explosive shell. The aircraft's Bomb Commander controlled in-flight insertion (IFI) or in-flight extraction (IFE) with a switch located in the cockpit. This operation took several minutes to complete, and green, amber, or red lights indicated the bomb's status (open, operating, or closed). Because the fissile core was stored safely outside of the bomb's explosive assembly, aircraft could takeoff or land with relative nuclear safety. Although a nuclear explosion was not possible, the detonation of a bomb's HE assembly was a big deal and could scatter radioactive debris over a wide area.

The MK 5 had many of the advanced features found in the MK 6. The bomb's electronics were cartridge mounted, attached to the rear of its implosion assembly, and provided with power from redundant banks of NiCad batteries. Although these batteries were rechargeable, they still had relatively short lives that required charging every 10 days. An arm / safe switch attached to the electronics cartridge was accessible through a panel in the rear housing. The MK 5 was also equipped with eight MC300 impact fuzes. Each impact fuze contained crystals of barium titanate, a material that has piezoelectric properties. When crushed on impact, the crystals could generate an electrical impulse able to discharge the x-unit and explode the bomb. The bomb's main fuzing system was a radar unit with two forward facing horn antennas mounted inside a streamlined, riveted, aluminum casing. The bomb had a radar transparent window located in front of each radar antenna.

The AEC released the MOD 0 version of the MK 5 in March 1952. The Air Force soon discovered that this weapon would explode prematurely if dropped below its intended release altitude with its low burst cables installed. Sandia carried out the modifications to rectify this problem in April 1952. Specialists also noted that the resistance of the barometric contacts increased with age, a problem that they corrected with regular cleaning. The upgraded MOD 1 version of the MK 5 was equipped with cables that connected the internal battery and radar heaters to an external power source and the MOD 2 version had an improved electronics cartridge that allowed its conversion from a bomb to a warhead.

Sandia equipped the MOD 3 bomb with a new type of fuze. This upgrade resulted from general dissatisfaction with the bomb's radar fuzing system. In consequence, Sandia evaluated several new designs as possible replacements. The new fuzes were already in use with the MK 7 and MK 12 but required as many as 30 different pieces of support equipment. After much debate and experimentation, Sandia installed a relatively simple baro-contact fuze with a safe separation timer known as "Fuze A."

A MK 5 MOD 0 bomb was successfully proof tested as Greenhouse Easy on April 20, 1951, with a yield of 47 kilotons. A number of additional tests followed in a range from 11 - 16 kilotons. Based on this data, many sources incorrectly indicate that this bomb had a variety of yields from 11 - 47 kilotons, identical with those established during its test program. Some sources indicate a single, 81-kiloton, yield for the MK 5. These latter sources incorrectly attribute the Greenhouse Dog proof test of the MK 6 bomb to the MK 5. It is possible to ascertain the use of a 60-inch device for the Dog test from photographs taken prior to the test. In fact, the MK 5's Type D pit was compatible with Type 110, 130, 150, 170, 190, 210, 240, and 260 cores, which the AEC had specified for substitution in all strategic bombs as a design characteristic. Some of the later high-numbered cores were replacements for earlier low-numbered cores. The AEC stockpiled the MK 5 in yields of 6, 16, 55, 60, 100, and 120 kilotons. The same core that produced a 160-kiloton yield in the MK 6 produced a 120-kiloton yield in the MK 5. It is thus possible to calculate that the MK 5 implosion assembly was 25 percent less efficient than the assembly used in the MK 6.

Los Alamos tested the following pits (A - F) for use in weapons produced from 1945 through the end of 1952

- Type A Christy (used in MK III)
- Type 48LP-B (used in MK IV)
- Type 49LC-C and 49LT-C (used in MK IV)
- Type 50LC-C (used in MK IV)
- Type 52LP-D (used in MK IV and MK 6)
- Type 53LP-D (used in MK 5 and MK 7)
- Type 59P-E (used in MK 7)
- Type 60P-E (used in MK 7)
- Type 61P-F (used in MK 12)
- a few special designs

The fissile cores used in MK 5 and MK 6 pits were compatible with the MK 7 pits discussed in the following chapter. The various core / yield options for the MK 5 resulted in a weight that varied between 3,025 and 3,175 pounds. The rapid development of the smaller MK 7, a true tactical bomb, ended production of the MK 5 after only 140 units.

The Air Force and the Navy flew the MK 5 primarily on first generation medium bombers. For this purpose, SAC had North American Aviation develop the B-45 Tornado. Unfortunately, secrecy and compartmentalization resulted in an aircraft that was unable to accommodate a full-sized strategic bomb. The Air Force thus relegated the B-45 to Tactical Air Command in 1952. At that time, TAC was building its aircraft inventory after regaining status as a major command. (The Air Force downgraded TAC to an Operational Headquarters, which it assigned to Continental Air Command in late 1948.)

The only notable characteristic of the B-45 was a 575 MPH speed provided by four GE J47A turbojets rated at 5,200 pounds thrust each. The bomber had a tactical range of 1,600 miles and a ceiling of 40,000 feet. As part of Project Backbreaker, TAC modified 41 B-45s in three squadrons to carry MK 5 or MK 7 atomic bombs. The AEC also used the Tornado to conduct airdrops of nuclear weapons at the Nevada Test Range. Like most emerging technologies, the aircraft had a short operational life and was retired from service in 1958.

The Navy's first post war bombers were the hybrid-powered AJ-1 and AJ-2 Savage, produced by North American Aviation. Each bomber had a pair of Pratt & Whitney R-2800-44W radial engines mounted under wing, and a 4,000-pound-thrust Allison J-33-A-10 turbojet located at the rear of the fuselage under the tail. The jet engine provided power assist for carrier-based takeoffs, evasive maneuvers, and bombing runs. The Savage's wide body provided room for a full-sized MK IV bomb, a lightweight MK 5 bomb, or a MK 8 penetrator. Delays in building a planned supercarrier relegated the Savage to smaller vessels that limited its numbers and effectiveness. By the time that the first supercarrier, USS America (CVA-66), entered service, the all jet A-3 Skywarrior had eclipsed the AJ-2.

The MK 5 bomb is notable for being the first nuclear weapon supplied for use by an allied nation. "Project E" was a 1958 arrangement by which the United States provided nuclear weapons for carriage by a force of 72 British "Valiant Force" bombers. These were located at RAF bases Marham, Waddington, and Honington. The United States later provided the MK 7 for British Canberra aircraft operating within Bomber Command and for RAF units posted to Germany. This arrangement continued through 1963 for the weapons in Bomber Command and through 1969 for the RAF units in Germany. Although available for use by the RAF, the United States retained custody of its nuclear weapons, which limited the ability of the RAF to disperse its assets in times of world tension. This custodial situation became critical during the Cuban missile crisis when the three Valiant Squadrons assigned to SACEUR at RAF Base Marham were to be loaded with their nuclear bombs. The Americans quickly discovered that there were insufficient custodial officers to maintain control of the weapons. This resulted in orders from the Commanding General, USAFE, to transfer the custody of the nuclear weapons to the British Base Commander.

Beginning in 1954, the Air Force and Navy acquired W5 warheads to arm cruise missiles. The design of nuclear warheads had lagged behind the development of freefall bombs in the post war era because of missile technology's slow progress. Influential figures such as Vannevar Bush, the head of OSRD, had made known their disbelief that anyone could fly a 5-ton atomic warhead on a ballistic missile. This situation resulted in curtailment of missile development in 1947 as part of a program to reduce defense spending. Nevertheless, the decision to implement the MOD 2 conversion of the MK 5 bomb to a W5 warhead allowed for the relatively speedy manufacture of 100 units.

Subsonic cruise missiles such as the Navy Regulus and Air Force Matador were only able to survive post war budget cuts and prioritization because analysts assumed that they were cheaper and faster to develop than ballistic missiles. The Air Force based its decision to develop cruise missiles on a ruling made by General Joseph T. McNarney, the Army Deputy Chief of Staff. In 1944, he assigned the development of aircraft-launched and winged missiles to the Army Air Force and the responsibility for ballistic missiles to the Army Service Forces. To reinforce their position that cruise missiles were exclusively in the realm of the Air Force, it called them "pilotless bombers." For its part, the Navy pursued missile technology because it did not want to give up control of a potential strategic weapon to the Air Force.

From an operational standpoint, cruise missiles offered several advantages over manned aircraft. The Air Force or Navy could fly them against heavily defended sites without concern for aircrew and at night or in bad weather using remote guidance. Contemporary light and medium bombers were not equipped with adequate navigational or bomb aiming equipment to be useful under these circumstances. Finally, TAC could disperse mobile missile launchers in times of rising tension, as opposed to its aircraft, which remained concentrated at fixed airbases.

American cruise missile development received a boost when the Army Air Force acquired a number of crashed Fi 103 / FZG-76 (V-1) cruise missiles from Great Britain during WW II. Designed by Robert Lusser for the Fiesler Company, the German vengeance weapon could carry a 1,870-pound warhead a distance of 250 miles. An Argus AS 014 pulse jet engine pushed it to a speed of 390 MPH. Air Materiel Command reverse engineered the captured missiles at Wright Field to produce the JB-2 Jet Bomb for the Army Air Force and the LTV-N-2 Loon for the Navy. In the event that the first atomic bombs proved to be duds, the Army Air Force planned a barrage of 75,000 jet bombs as a prelude to the invasion of Japan. Willys-Overland produced 1,200 test missiles under contract to Republic Aircraft.

Based on a positive recommendation by Theodore von Karman, a senior civilian advisor, the Army Air Force awarded a contract to Martin Marietta to develop the Matador cruise missile. A former professor of aerodynamics at the California Institute of Technology, von Karman had released a comprehensive technical report on military technology after the end of the war. "Toward New Horizons" recommended that the Air Force adopt an orderly program of missile development starting with a jet-powered cruise missile that might reach deployment in a period of five years. In much the same manner, the Navy was sufficiently encouraged with the performance of its submarine launched Loon that it engaged the Vought Corporation to produce Regulus, a cruise missile with sufficient power to carry a nuclear warhead.

Both the Navy and the Air Force selected jet engines to power their cruise missiles. In fact, they both chose the GE J-33 turbojet engine that powered the P-80 Shooting Star, America's first mass produced jet fighter. This engine differed significantly from the pulsejet engine used to power the Loon. The pulsejet engine is essentially a hollow cylinder, with shutters at the front end. When in flight, the flow of air opens the shutters and then moves to a combustion chamber where it mixes with fuel. When the air-fuel mixture ignites, the shutters slam shut, forcing the hot gases to leave through the engine's tailpipe, thus creating forward thrust. The cycle of ignition takes place about 60 times a second. Although this design is very simple, it has a low specific impulse and requires a secondary power source to start airflow through the engine.

The turbojet differs from a pulsejet in having a rotating compressor behind its air inlet to compress incoming air before passing it to a combustion chamber where it acts as an oxidizer. The heated exhaust gases from the combustion chamber spin a gas turbine before flowing out through an exhaust nozzle to provide thrust. The gas turbine operates the compressor. An Englishman, John Barber, patented a stationary turbine as early as 1791. Maxime Guillaume, a Frenchman, filed the first patent for using a gas turbine in a jet aircraft in 1921, and Alan Griffith published *An Aerodynamic Theory of Turbine Design* in 1926. This led to experimental work at the Royal Aircraft Establishment (RAE). Early problems with jet engine fabrication included, metallurgy, reliability, and sustained operation.

Although the United States did not pursue jet engine development prior to WW II, British aid in the early part of the war provided a 1,250-pound-thrust Whittle W.2B engine based on a centrifugal flow air compressor. A 1,650-pound-thrust Griffith H-1 engine with an axial flow compressor followed. Although axial flow was by far the better principle with the designs and materials available, mechanical problems with the H-1 engine rendered it unreliable. This caused GE to redesign the more robust W.2B into an all-new axial flow engine that culminated in the

4,600-pound thrust J-33. Axial flow engines compress airflow along the axis of a jet engine whereas centrifugal flow engines compress airflow around the outer internal edge of an engine.

The specific engine ordered for 15 Martin Matador XB-61 prototype missiles in late 1947 was a 4,600-pound-thrust J33-A-37 turbojet rated for only 10 hours of operation. There was no point in acquiring a more expensive version of this engine since both it and the missile it powered were one-time use items. Wherever possible, the Air Force encouraged Martin Aircraft to use "off the shelf" items to speed development and minimize production costs. The first XB-61 was launched at the Army's White Sands Missile Range in New Mexico in 1949, and by 1951, testing had progressed to the YB-61.

In the latter part of 1951, the Air Force moved the Matador flight test program to Patrick AFB near Cape Canaveral, Florida. Here, the 6555th Guided Missile Test Wing continued with evaluation of the YB-61, which for the first time was equipped with an operational guidance system. Eighty-four production launches followed 46 prototype firings. These resolved a number of problems that included structural defects found in the tail and wings. They also resulted in a change from a mid-wing design to a high wing that gave better launch clearance.

In addition to launching missiles, the 6555th trained personnel to staff the Pilotless Bomber Squadrons. It conducted this training in three-stages. In the first stage, guidance technicians received 43 days of instruction, and propulsion and missile assembly technicians received 13 days of training. In the second stage, 6555th integrated the technicians into assembly, checkout, and launch teams. This phase required approximately six weeks of training. In the final stage of instruction, the guidance, propulsion, and assembly teams learned to coordinate their activities under the direction of their future staff officers, a process that took about a month.

The final design for the B-61A (later TM-61A) Matador resembled a contemporary jet fighter. The 4 foot 6-inch diameter missile had a length of 39 feet 6 inches and a swept wingspan of 28 feet 7 inches. In order to simplify its guidance, Martin reduced the control surfaces to a horizontal stabilizer and snap action wing spoilers. The 6-ton missile could carry a 3,000-pound warhead over a maximum range of 700 miles with a 250-gallon load of jet fuel. This carrying capacity comfortably exceeded the 2,650-pound maximum weight of a W5 warhead. Matador had a cruising speed of 600 MPH and a maximum ceiling of 35,000 feet. A 57,000-pound-thrust Aerojet solid fuel booster that imparted a speed of 200 MPH in a span of 2.4 seconds propelled the missile into the air.

Because Chance Vought designed Regulus to fit inside a submarine, it was smaller than the Air Force's Matador. It was 32 feet long and had folding wings with a 21-foot span. The 56 1/2-inch diameter, 7-ton missile could also carry a 3,000-pound payload. However, because of its greater weight and the reduced lift provided by its smaller wings, its maximum range was only 575 miles with a 310-gallon fuel load. Its J33-A-14 turbojet had a much longer operational life than the engine that powered Matador. Regulus B followed the A Model with an improved J33-A-18A engine. Chance Vought equipped test vehicles of this model with a retractable landing gear for reuse. The Navy carried out its initial trials at Edwards Air Force Base with its first successful flight in March 1951. A novel feature of the test program allowed the control of missiles directly from chase planes. In late 1952, testing moved to surface ships, and in 1953, the Navy announced the start of serial production.

The two Armed Services initially operated their cruise missile with command guidance. In the case of Regulus, the Navy guided the missile's preplanned flight using a Sperry A-12 autopilot originally developed for manned aircraft. Sperry based its autopilot on a discovery made by Leon Foucault, a French experimental physicist who wanted to explore the Arctic by submarine. Unfortunately, for Foucault, a magnetic compass cannot work when surrounded by an iron hull and he remained without a means for navigation. He finally found a solution to his dilemma while experimenting with the gyroscope, a device he patented in 1852. A gyroscope consists of a spinning wheel that can maintain its orientation through the conservation of angular momentum. A child's top works on the same principle. Foucault discovered that a gimbaled gyroscope, confined to rotate in a horizontal plane, would gradually rotate until it pointed towards true (not magnetic) north. He had invented the gyrocompass, even if he never did get to explore the Arctic.

Elmer Sperry first applied the gyroscopic principle to the problems of aircraft navigation in 1914. Using a gyroscope, he built an artificial horizon indicator that demonstrated controlled flight based on an "autostabilizer." This device used the artificial horizon indicator and a magnetic compass to control the elevator, rudder, and ailerons of a test aircraft. The Navy supported Sperry in its research and by 1933, it had produced a hydraulic-pneumatic "autopilot" in which three orthogonally mounted gyroscopes detected and corrected deviations from straight and level flight. Such a unit was requested and obtained by Wiley Post, who flew his aircraft "Winnie May" around the globe in a record of 8 days.

The Sperry A-12 autopilot that flew Regulus appeared immediately after WW II. The Air Force demonstrated it to the world in 1947 when it equipped a C-54 transport with an A-12 unit, an automatic approach control, and a Bendix throttle. Programmed by means of punch cards, the A-12 piloted the "Robert E. Lee" from Stephenville, Newfoundland to a landing at Brize-Norton, England, unaided by human hands. Along the way, a pair of ships transmitted guidance updates via radio signals. The flight demonstrated the modern concepts of control and navigation about the axes of pitch, roll, and yaw, as well as location through the calculation of altitude, latitude, and longitude.

As implemented in Regulus, Sperry housed the A-12 autopilot in a box that measured 12 inches square by 30 inches long. The vacuum tube filled unit was so heavy that it required two technicians to carry it. An onboard telemetry system supported the A-12 by providing ground stations with tracking and flight data; including such information as airspeed, altitude, engine RPM, control surface movement, and hydraulic pressure. Using a "paired pulse" radar system known as Trounce, a ground unit evaluated the transmitted data and returned guidance commands to the missile's autopilot. The command system had two modes of operation. In the first, a selector switch with 20 positions could increase or decrease a selected function. The second mode provided continuous control of pitch, rate of turn, and throttle. Arming commands that activated the in-flight insertion of the nuclear capsule were command transmitted as well. Ground controllers were restricted to guiding a single missile at a time.

The Navy initially requested simple barometric fuzing for Regulus. After Sandia reported that this task would take 10 months to complete, the Navy asked Sandia to improve the MK 5's radar fuzing system as an interim measure before developing the barometric fuze. To fulfill the Navy's request, Sandia modified the MK 5 fuzing system with a pair of timers that provided 6 minutes of safe separation before they connected power to the warhead's arming and fuzing system. If the ground crew closed a command arming bypass switch prior to launch, automatic equipment inserted the nuclear capsule in the W5 warhead. If not, the receipt of a command nuclear arming signal initiated this process. When Regulus received the command signal to dive onto its target, a switch connected the radar fuze's arming circuits to the warhead's fuzing baro-switches. After the missile descended beneath 20,000 feet, an arming baro-switch closed to start x-unit charging. At an altitude of 700 feet above the desired height of burst, the fuzing baro-switch activated the missile's two radars. If these operated normally, detonation occurred when both radars agreed on reaching a preset altitude. If only one of the radars was operational, detonation occurred 300 feet beneath the preset altitude when a final element in the baro-switch closed. If both radars failed to operate, an impact fuze triggered detonation.

Matador was fitted with an Automated Guidance System (MARC) adapted from the AN/MSQ-1 system used to guide fighter-bombers during the Korean War. Three modified SCR-584 mobile ground radars tracked an AN/APW-11 control beacon mounted in the missile to determine its distance, direction, ground speed, and altitude. Using this data, technicians tracked the missile's position on an AN/MPS-9 board and transmitted corrections via the radar control. A heading error analyzer caused the missile to enter a fatal terminal dive if its flight path deviated more than a specified amount from the preset path. Once it was determined that Matador was over the target, a radio command armed the warhead and precessed the autopilot's vertical gyro to send it into a steep, terminal dive. The missile could not exceed the speed the sound in its terminal dive or it would disintegrate.

Sandia initially based Matador's fuzing on the radar-activated system it had designed for the MK 5 bomb. After a successful launch, an acceleration sensor closed a safety switch to allow receipt of an externally transmitted arming signal when the missile reached its target. The arming signal initiated the fuzing sequence and the missile's terminal dive to its target. When the missile reached a point in its dive where the external air pressure was equal to 1/2 sea level pressure, an x-arm baro-switch closed to start two inverters charging the firing-set. At a preset lower altitude, a fuzing baro-switch closed to place two radars in operation. The fuze initially connected the two radars in series. Both radars had to "range" in order to detonate the warhead. If this did not happen, slightly below the desired height of burst, the fuze reconnected the radars in parallel so that either could detonate the warhead. Sandia later incorporated an impact fuze as backup.

Because command guidance operated in line of sight mode only, Matador's guidance system was only effective to a range of 150 miles, thus limiting the performance of a missile. To maximize Matador's depth of penetration into Soviet airspace, its guidance sites were located on the tops of mountains near the border with East Germany, whereas its launch sites were located further to the rear for safety. Regulus had the same limitations as Matador, but its operators could transfer its guidance from one ship or submarine to another, thus increasing the distance of the launching vessel from shore-based defenses.

In order to improve the performance of Matador, TAC had Martin place a "B" Model in development. When Martin informed TAC that this model would require an extended period of development, the Command authorized an interim "C" Model. Martin equipped Matador C with a "Short Range Navigation Vehicle" guidance based on the SHORAN Navigation system. Two pairs of master / slave ground stations generated electronic azimuth and range signals that intersected over a target's location. The master stations synchronized the signals, which they automatically broadcast to Matador at intervals along its flight path. SHANICLE guidance was relatively insensitive to jamming when compared with more primitive command guidance units. SHORAN had a maximum range of 300 miles with a clear radio wave path.

The accuracy and reliability of early cruise missiles was nothing to celebrate. The experience of their controllers was the primary source of their accuracy. Regulus enjoyed a circular error probability (CEP), or the chance that one out of two missiles would land within a specified distance of its target, of 5 percent of the range to its target. Matador had a 71 percent reliability and a CEP of 1/2 mile. The Air Force considered this adequate for an atomic warhead capable of demolishing several square miles. Matador controllers situated at operational sites practiced their art by sending commands to specially equipped T-33 aircraft that simulated the missiles. These aircraft flew on a daily schedule often simulating 2 - 3 launches per flight.

The literature is far from clear on the yields of the W5 warheads used to arm the missiles. Most authors quote a yield from 40 to 50 kilotons, with specific mention of the 47-kiloton yield associated with the MK 5's proof test. This seems reasonable for Regulus because the Navy stockpiled only one core, the Type 210, for use with its warhead. TAC stockpiled two cores, Types 150 and 210, for use with Matador. The literature attributes 35 warhead conversions to Regulus and 65 to Matador. These warheads were equipped with a special "linear" in flight insertion mechanism and a fast firing x-unit. Initial fuzing was for airburst only. The Navy transferred 20 W27 warheads with a 2-megaton yield to Regulus IB missiles after it cancelled the supersonic Regulus II program. This required an enlarged weapon compartment that produced a characteristic "bulged chin."

The fuzing system for Regulus missiles equipped with W27 warheads differed from missiles that carried W5 warheads. For safety reasons, a breakaway relay prevented the 2-megaton warhead's x-unit from charging while the missile sat on its launcher. The warhead also featured an acceleration sensitive safety switch that prevented inadvertent charging of the x-unit until after launch. This system replaced the lanyard operated switch used with the W5 warhead and eliminated the need for pullout wires. The acceleration sensitive switch contained two sets of air pistons that closed after accumulating a 4G second impulse during launch acceleration. A technician set the safe separation time by means of a selector on the launch panel. The W5 had previously incorporated this function in its arming plugs. The timer contained two sets of motor driven gear and cam assemblies that closed electrical switches at the end of their timing intervals. The timers started when the missile took off.

A squib operated switch isolated the high voltage thermal batteries that charged the W27's x-unit. When the missile was in storage, the batteries were stored reversed in their holders. Prior to use, specialist had to open a hatch and reverse the missile's batteries so that their contacts connected to the electrical system. During safe separation time, an acceleration switch and a timer isolated the firing-set. A mission selector switch (command, airburst, contact) and an acceleration switch controlled low-voltage power.

Initially, Sandia considered a radar fuze for Regulus IB, but the lab discovered that a much simpler baro-fuze could provide the 600 feet of altitude accuracy required as a military characteristic for the W27. The firing-set could receive command arming and fuzing signals, which a contact fuze backed up. To minimize the possible effects of enemy antiaircraft fire on the fuze's impact crystals, the altitudes of the arming and firing fuzes were set very close. The baro-switch detonated the warhead during a dive from high-altitude and command detonation set off the warhead during a low-altitude level attack. The warhead's x-unit was identical to the one used for the W5.

In addition to nuclear warheads, TAC could equip Matador with blast / fragmentation or chemical / biological (CB) cluster warheads. The high explosive warheads would not have been very effective in engaging large targets but the CB warheads filled with Sarin bomblets would have influenced a much wider area. Sarin (Agent GB) is a virulent organophosphate compound related to a family of chemicals used as insecticides. Two German chemists discovered it in 1938.

The agent is highly volatile and can be inhaled or absorbed through the skin. Biological effects result from the inhibition of cholinesterase, which causes muscles to go into permanent spasm. Symptoms include nausea, vomiting, muscle spasm, and death by asphyxiation due to paralysis of the diaphragm. The chemical is rapidly fatal in minute doses. It has low persistence but a saturated area remains dangerous for several days. The CB warhead would have dispersed hundreds of ten-pound, cylindrical, M125 bomblets from an altitude of 5,000 feet.

Tactical Air Command operated Matador and the Navy-operated Regulus from zero length (ZEL) launchers. Regulus could also be launched from an aircraft carrier's catapults by means of an 11-ton, three-wheeled, "Steam Assisted Take-off for Regulus" or STAR dolly. Because the Navy secured Regulus on board combat vessels, it could leave the missile's warhead and its 33,000-pound thrust Aerojet General JATO units permanently installed. As previously mentioned, the missile was equipped with folding wings and tail surfaces that reduced its launch time to 15 minutes.

In the case of Matador, a 20-ton, 40-foot-long, truck-towed, ZEL launcher provided the missile with limited mobility. Matador's crew transported the missile to a remote launch site in four components carried on separate vehicles. These components consisted of the missile body, the warhead, the detachable wings, and the solid fuel rocket booster that propelled it from its launcher. The assembly and launch procedure for Matador took about 90 minutes. At L-90, the ground crew removed protective tarpaulins from the missile body as it lay on its transport vehicle and at L-80, they attached a cradle to the fuselage and transferred the missile body to the ZEL launcher using a 20-ton mobile crane. The ground crew lowered the missile's wings into position at L-70 and then connected ground supply power feeds. At L-60, nuclear specialists brought the warhead forward from the armory on a special trolley and winched it up the launching ramp into position. Electronic technicians then checked out the guidance system. At L-40, the ground crew hooked up the booster system and carried out a final checkout of the missile. Launch personnel elevated the missile to a maximum 70-degree angle and cleared the pad by L-05. Engine run up automatically started at L-03 and at L-01, the motor reached full power. At L-0, the missile's booster ignited, its umbilicals dropped free, and it roared off its launcher.

Tactical Air Command declared the 1st Matador Pilotless Bomber Squadron (later Tactical Missile Squadron) operational in late 1953 and deployed it to Bitburg AFB, Germany, on March 9, 1954. TAC followed the deployment of the 1st PBS with the deployment of the 69th PBS at Hahn AFB and the 11th PBS at Sembach AFB. TAC eventually rolled these Squadrons up into the 38th Tactical Missile Wing, which operated six Tactical Missile Squadrons and three Missile Maintenance Squadrons. TAC upgraded the 38th TMW with TM-61C (later redesignated MGM-1) Matador missiles, beginning with the re-deployment of the 11th Tactical Missile Squadron in 1957. It also armed Bundeswehr Flugkorpergruppe 11 with Matador TM-61C missiles. Unlike American Squadrons, the German Missile Squadron did not control the nuclear warheads for its ready missiles. A Special Weapon Depot under American control supplied these in time of war. A typical Squadron operated three "pads," each with a compliment of two launchers. In 1958, the TM-61C Squadrons were up gunned to eight launchers, presumably by adding a fourth pad. TAC conducted test firings and training twice a year at Patrick AFB, Florida, or at Wheelus AFB in the Libyan Desert.

The missile pads were located some distance from the main airbase, which improved their survivability. In effect, the pads were miniature airbases equipped with concrete aprons from which a Flight could launch its ready missiles. For security, each pad had a chain link fence topped with barbed wire and an Air Police Unit equipped with K9 dogs. In addition to an Administration Building, the miniature bases had accommodations for officers and enlisted men, as well as kitchens and mess facilities. A typical TM-61C Flight had 10 – 12-man crews assigned to each missile and total compliment numbering 80 men. Most of these were drivers. The ready missiles stood armed with their nuclear warheads on their launchers, with conventionally armed spares parked nearby. The pads had sufficient parking for the support equipment needed to take the Flight mobile as circumstances required.

Before the delivery of Matador missiles to their launch pads, technicians assembled and checked them out at missile support or depot maintenance areas located in the munitions area at each main AFB. These had a number of hangars that supported specialized functions such as guidance system or engine and airframe maintenance. Each base also had armament vans outfitted with all the equipment needed to mate warheads with missile bodies and to transport them from operational storage areas to launch sites. Martin Marietta supplied technical representatives to assist Air Force missile specialists in the performance of their duties at the maintenance facilities.

Matador's operational philosophy concentrated the missiles and launchers on their pads during peacetime for security. From these pads, each Flight could fire its missiles on command or disperse during periods of rising tension. The Flights did not randomly disperse their missiles. Instead, they transported the missiles to a number of carefully prepared sites they had previously surveyed. Each TM-61C system included a launcher, a transporter, special purpose checkout and targeting equipment, ground handling equipment, and servicing equipment. Support also included a commissary van. The Flight used standard, 5-ton Army tractors for transport. In order to reduce vehicle requirements, a Flight dispersed its launchers to pairs of sites served by a common guidance network. Total Squadron requirements worked out to 50 powered vehicles and 50 trailers. A Tactical Missile Squadron was theoretically capable of launching 20 missiles in 9 hours.

It is interesting to note that the 48 TM-61C launchers deployed to Europe after 1958 equated directly to the available number of atomic warheads. This most likely reflected a "use them or lose them" philosophy, since each launcher was kept on nuclear alert. Thus, the first flight of Matador missiles in any conflict would have sent a 5-megaton wave of destruction on its way to blunt a Soviet attack. The targets were most likely time-sensitive, fixed aim point locations as exemplified by military bases, airfields, and transportation hubs. After dispersal, surviving Squadrons could have supported further operations with conventional or CB warheads.

In 1957, the Communist Chinese began shelling the Nationalist Chinese held islands of Quemoy and Matsu. Fearing that an invasion of Taiwan was imminent, the Air Force dispatched the 17th Tactical Missile Squadron (later 868th) to Tainan AFB. It also sent the 310th Tactical Missile Squadron, assigned to the 58th Tactical Missile Group, to Kadena AFB on Okinawa. From these locations, the Tactical Missile Squadrons could have stopped cold any Communist Chinese advance across the Formosa Straits. The 310th TMS could also have attacked targets of opportunity in North Korea if hostilities had resumed in this region. The Air Force re-deployed the 310th Tactical Missile Squadron to Osan AFB, Seoul, and to Chinchon-Ni, Korea, in 1968. TAC armed the 310th with updated TM-61C missiles and replaced the 310th on Okinawa with the 498th TMS. Like their European cousins, these Missile Squadrons were equipped with only one nuclear warhead per launcher.

The Navy began deployment of the Regulus (SSM-N-8/RGM-6) missile on September 16, 1955, when it established Guided Missile Group One (GMGRU-1) at the San Diego NB, California. GMGRU-1's mission was to support detachments operating Regulus missiles aboard the aircraft carriers, cruisers, and submarines of the Pacific Fleet. Ten days later, the Navy established GMGRU-2 at Chincoteague NB, Virginia, to support the Atlantic Fleet. The Navy next converted two diesel-powered submarines, USS Tunney (SSG 282) and USS Barbero (SSG 317), to each carry a pair of missiles in externally fitted hangars. USS Tunney fired the first Regulus test missile in 1953 and USS Barbero conducted the first Regulus deterrent patrol in 1955. The Navy followed these conversions with the launch of USS Grayback (SSG 574) and USS Growler (SSG 577) in 1958. These second-generation, diesel powered submarines were purpose built to carry a pair of Regulus missiles in each of two internal hangars.

In 1960, the Navy commissioned the nuclear submarine USS Halibut (SSGN 587) with a hangar large enough to store five missiles. Originally built to carry the supersonic Regulus II missile, the Navy Refitted Halibut to carry Regulus I when it cancelled Regulus II in favor of Polaris. Operational planning required four submarine-based Regulus missiles to be on patrol at any given time. Thus, Tunney and Barbero had to patrol together, while the other submarines could patrol independently. In order to conduct a launch, the boats had to surface and remain at full stop in calm seas.

Along with the deployment of cruise missile submarines, the Navy converted four Baltimore class heavy cruisers, USS Los Angeles (CA 135), USS Helena (CA 75), USS Macon (CA 132), and USS Toledo (CA 133) to carry three cruise missiles apiece. The missiles were located in one of two stern hangars that the cruisers had previously used for the storage of scouting aircraft. The cruisers launched the missiles from an adjacent catapult. With a main armament of 8-inch guns mounted in three triple turrets, each cruiser displaced 17,500 tons and could make 33 knots. Built during the early period of WW II, this was the last class of heavy cruiser to see service with the United States Navy. Their original mission was to escort carrier groups. The Navy re-commissioned the USS Los Angeles as the first guided missile cruiser in 1955 and her sister ships the following year. Their time in service was short. The Navy phased out cruiser-based missile operations over the period 1959 - 1961, retiring one ship per year.

In addition to submarines and cruisers, the Navy converted ten Essex Class carriers to carry Regulus missiles. The Navy authorized the Essex Class in the summer of 1940. Twenty-for vessels were ultimately launched to make this class the most numerous fleet carrier constructed by any nation. Almost 900 feet long and displacing 27,100 tons, the carriers could steam at a maximum speed of 33 knots. As their main offensive weapon, they carried an air arm composed of 90 aircraft. With the advent of the jet age, the Navy rebuilt the carriers with new control islands and updated their launch and recovery gear. Each carrier could simultaneously operate four Regulus missiles. If required, the carriers could pass missile guidance to five specially outfitted submarines that allowed the carriers to operate at a distance from shore.

In order to extend the operating range of Regulus in the Western Pacific, the Navy outfitted six carriers with "Regulus Assault Missiles." The RAM program was an outgrowth of the original Regulus test program in which observers in accompanying aircraft controlled the missiles as drones. The AEC expedited this program in February 1953 when it had Sandia configure a small number of MK 5 MOD 2 bombs as W5 warheads to provide an interim offensive measure. These were retired from service when purpose built W5 warheads became available in mid-1954.

Single seat fighters, such as the F2H, F9F, and FJ-3, were equipped with special control equipment that the pilot operated, while still flying his aircraft! Having previously launched from a carrier, the pilot of the control aircraft timed an approach so that he could tuck in behind the missile as it left a catapult. Maneuvering in the right rear quadrant of the missile, the pilot flew at altitude until the missile / aircraft pair approached hostile radar coverage. The pilot then took the missile down low and transmitted a signal to charge the warhead's x-unit, which he radio-activated when Regulus was over its target. Navy Units conducted operational patrols up until 1964, after which the Navy expended its remaining Regulus missiles as target drones.

Approximately 1,200 Matador and 500 Regulus I missiles were produced. Together, these were the longest-range nuclear missiles in America's arsenal until the advent of Mace in 1959. Although the Air Force armed Mace with a W28 thermonuclear warhead available in a number of sub-megaton yields, the topic of Mace is included in this chapter because its technology and its mission were an extension of the Matador Program.

The AEC produced the W28 warhead in five MODs: the MOD 0 (1958) had an internal neutron initiator, the MOD 1 (1960) had an external neutron initiator, the MOD 2 (early 1962) was strengthened for laydown delivery, the MOD 3 (late 1962) was provided with CAT A PAL and the MOD 4 (1963) was provided with CAT B PAL. TAC stockpiled only the MOD 0 and the MOD 1 warheads for use with Mace.

When the Air Force compared Matador A with Regulus, it realized that its missile came up short. The weapon was both difficult to transport and slow to assemble and launch. For a time, the Air Force considered replacing Matador with the more compact Navy Regulus! They were, however, dissatisfied with the command guidance systems used for both missiles and, in the end, decided to develop an improved Matador B. This missile incorporated so many changes that the Air Force renamed it Mace and redesignated it TM-76A prior to its deployment. Martin began work on Matador B / Mace in 1954 and first tested the missile in 1956.

A 5,200-pound-thrust Allison J33-A-41 jet engine powered the prototype missiles launched at White Sands Missile Range. The new engine increased the size of Mace to 44 feet 9 inches long by 4 feet 6 inches in diameter. It weighed 13,800 pounds. In order to clear its launcher, Mace was equipped with a high, folding wing. It also retained Matador's 650 MPH air speed but had a larger fuel load that increased it range to 800 miles. Its W28 warhead reduced the missile's payload weight by about 1,000 pounds over the original W5.

The Mace launcher was also much improved. It hooked up to a dedicated MM1 Teracruzer that towed it in a fifth-wheel configuration. An eight-cylinder Continental engine powered the Teracruzer, which had a four-speed torque converter that could apply power to all of its wheels. The front suspension was manually adjustable to allow the vehicle to climb obstacles and the driver could vary pressure to tackle different road or ground conditions. In addition to a launcher trailer used to transport the missile body, three other trailers provided support. One of these was equipped with a crane and cradles to transport the booster rocket and the nose section that contained the missile's guidance unit and warhead. Another was a fuel hauler / power unit equipped with hoses and pumps; and the last was a preflight test equipment trailer that helped execute a launch.

The Mace ground crew had to emplace the missile launcher pointing at the target. This required a geographic survey. After the missile launcher was satisfactorily emplaced, the missile's guidance system had to be installed, the warhead attached, and the 97,000-pound-thrust Thiokol booster mounted underneath. After the missile crew completed fueling operations, they unfolded the wings and erected the launcher's ramp to the 19-degree angle required for launch. The ground crew then connected the missile to a preflight trailer that provided power, hydraulic pressure, and compressed air. After completing their servicing operations, the ground crew had to disconnect the umbilicals and hoses used for servicing and lay out a 150-foot cable to transmit the firing signal. Preparing the missile for launch required an officer, a crew chief, three avionics technicians, two airframe technicians, and conventional and nuclear warhead technicians. Despite sophisticated equipment and intensive training, the setup procedure took so much time that in real life, the missiles were pre-positioned at fixed alert sites in the same manner as Matador. Firing limits for Mace were in a temperature range from −50 °F to +103 °F, a 50 MPH head wind, 25 MPH tail or side winds, and an altitude below 5,000 feet.

Although Mace bore a strong family resemblance to Matador, its guidance system was a new concept that matured over an extended period. The delay in developing Matador B's (Mace A's) guidance system thus forced the interim deployment of Matador C. The Mace A guidance unit was an Automatic Terrain Matching and Navigation (ATRAN) system developed by Goodyear. After launch, a DPS-1 radar scanned the missile's path at 2 nautical mile intervals looking for preprogrammed topographic features. The guidance unit then compared the scans with radar images contained on a 35mm filmstrip and re-oriented the missile to stay on course. The use of radar allowed the missile to see through clouds and gave it an all-weather capability. The missile could not overfly featureless terrain such as water and its radar was susceptible to jamming. An initial difficulty lay in obtaining the necessary radar images to program fight paths. This shortcoming may have been one of the reasons that drove the over flight programs of the Eastern Bloc. Eventually the Air Force developed a system for converting topographic maps into suitable images.

The only fuzing option provided for Mace A was airburst. Sandia originally considered contact fuzing but dropped this option when TAC scrapped the missile's high-altitude attack profile. After launch, an acceleration of 3.5 Gs applied for at least two seconds closed a switch that activated a target-area timer circuit. Upon expiration of the timer interval, a second switch closed to connect the warhead's thermal batteries and its x-unit. High voltage activation, nuclear arming, and neutron initiator arming took place approximately two miles from the target upon receipt of a signal from the ATRAN guidance system. The ATRAN guidance unit provided an airburst detonation signal as Mace passed over its target.

TAC deployed its first Mace Missile Batteries to Germany in 1959, and by 1961, Mace had completely replaced its Matador units. TAC augmented the replacement Mace units in 1962 with the addition of the 89th TMS at Hahn and the 823rd TMS and 887th TMS at Sembach. At first, TAC provided each Squadron with eight launchers. It arranged these in groups of four at fixed, Rapid Fire Multiple Launch sites. The RFML sites featured concrete blockhouses measuring 8 feet by 10 feet. An officer and three enlisted personnel staffed the sites. They could launch the four missiles under their command in approximately 12 minutes. Later, TAC increased each Squadron to 12 launchers, deploying 72 launchers in six Squadrons. The missile launchers stood alert in the open until TAC provided weatherproof covers. The Air Force sent 300 of the 1,000 missiles produced to Germany. For these missiles, the AEC provided 100 W28 thermonuclear warheads.

Tactical Air Command was not entirely satisfied with the performance of Mace A. The missile had to operate at altitudes as low as 750 feet to guide accurately, and this reduced its range. Because of this problem, the Air Force began development of an upgraded Mace B missile in 1960. With the same outward appearance as Mace A, the follow-on missile incorporated a new inertial guidance system. Four main groups pioneered the development of inertial navigation and guidance systems in the United States: the MIT Instrumentation Laboratory, the Autonetics Division of North American Aviation, the Redstone Arsenal, and Bosch Arma. A fifth company, Northrop, worked on a modified version of the inertial concept.

Inertial systems had the same three-gyro arrangement as an autopilot but used them to create a "stabilized platform" as a fixed reference point. As the missile moved around the platform, sensors determined its orientation. Conversely, a technician could program inputs to orient the missile with respect to the platform and thus control its flight. In addition to its gyros, the stabilized platform was equipped with three orthogonally mounted accelerometers. As their name suggests, these instruments sensed accelerations applied to the missile – from its propulsion system or from outward forces such as wind gusts. This arrangement allowed the velocity of the missile to be determined in three-dimensional space. In conjunction with the gyros that determined orientation, electronic impulses sent to the flight control surfaces corrected any changes to the missile's preprogrammed path detected by the accelerometers. MIT developed the "Achiever" guidance system used in Mace B and the AC Sparkplug Division of General Motors manufactured it.

A significant complication in developing inertial guidance was separating the effects of gravity, which manifests itself as acceleration. The earth's gravity field is not uniform because of topographic effects and the variation of mass in the crust. In addition, since the reference platform remains in a fixed position, the orientation of the gyros to the gravity field changes as it moves around the curve of the earth. To minimize the effects of gravity on manufacturing imperfections, early guidance units were quite massive. The AC Achiever was nearly the same diameter as the Mace missile body and the whine from its gyros was so piercing that technicians had to wear special hearing protection during servicing and calibration. Nevertheless, the change in guidance increased Mace B's range to 1,500 miles, allowing its use against key Chinese industrial areas such as Chunking and Hankow. The Air Force intended Mace B to provide nuclear fire support in the Far East until Tactical Air Command could deploy its Composite Strike Force.

In order to attain a high degree of accuracy, TAC had to launch its inertially guided missiles from carefully surveyed fixed sites. For this reason, TAC decided to deploy Mace B in hardened shelters with a "Quick Reaction Alert" capability dictated by Secretary of Defense McNamara. Since TAC had not yet deployed Mace A to Southeast Asia, it constructed its first Mace B sites for the 489th Tactical Missile Group based on Okinawa. Each of the four semi-hardened sites constructed had eight openings for launching missiles. Considering that the missiles were pre-positioned and armed, The TMS could have carried out a launch in short order. The reinforced concrete used in the construction of the launch structures could withstand the overpressure of a nearby nuclear detonation. They were equipped with 100-ton blast doors that folded out and down to enable a launch. The two-story launch structures had conduits that directed exhaust from the missile's booster rockets down and out the back of the building. A launch center buried 60 feet beneath a paved loading and transport ramp located in front of the launch structure controlled each flight of four missiles.

The arming and fuzing system for Mace B was more complicated than that of Mace A. Mace B's inertial system could program different missions from three basic flight profiles. It could execute a terminal-dive high-level mission to a maximum range of 1,200 miles when flown at a Mach 0.9 speed and a 40,500-foot altitude. Severing the missile's wings with explosive charges initiated its terminal dive. It could also fly a terminal dive low-level mission to a maximum range of 500 miles at Mach 0.8 speed and a minimum altitude of 750 feet. Finally, Mace could fly a terminal dive, high - low mission that started with a high-level leg and ended with a low-level approach to the target. Mace could execute this mission to a maximum range of 1,100 miles.

The first of fifty Mace B launchers planned for Germany reached operational capability in 1965. These missiles replaced the Mace A units stationed at Bitburg. That same year, Secretary of Defense McNamara reassigned the Quick Reaction Role in Europe to Pershing MRBMs controlled by the Army. The most likely reason for this change was the vulnerability of the fixed Mace launch structures and the increasing vulnerability of subsonic cruise missiles to antiaircraft defenses. Pershing was a highly mobile ballistic missile that eventually attained a fire anywhere capability. TAC retired the remaining European Mace A missiles through 1966. With the deployment of Pershing IA in 1969, TAC stood down its European Mace B force. The Mace B missiles deployed to Okinawa continued in service until 1971 because the Army did not deploy Pershing missiles to the Pacific Theater. TAC only deactivated its Okinawan missile sites when the United States officially returned the island to Japan.

CHAPTER 13
THE MK / W7 TACTICAL ATOMIC WARHEAD

The MK 7 Thor tactical atomic bomb entered the nuclear stockpile only 2 months after Los Alamos delivered its first MK 5 lightweight strategic bomb. The MK 7 used many of the features incorporated in the MK 5, differing in only one significant aspect. Sandia designed the MK 7's 92-lens implosion system for use with much more powerful explosives than the Composition B used in the MK III, MK IV, MK 5, and MK 6 bombs. Some TX7 test devices contained Cyclotol, which increased the RDX in its castable high explosive mixture to 75 percent. Others used Octol, which replaced RDX with an even more powerful nitramine explosive, HMX ($C_4H_8N_8O_8$) or octahydro-1,3,5,7-tetranitro-1,3,5,7-tetrazocine. These changes reduced the diameter of the TX7 explosive assembly to 30 inches and its thickness to 5 inches. Contained within a fiberglass shell, MK 7 explosive assemblies weighed between 800 and 900 pounds. Because Los Alamos used the MK 7 physics package as a primary in a number of early thermonuclear weapons, it retained Boracitol as the slow explosive component for its implosion lenses.

The MK 7 bomb casing measured 15 feet 2.5 inches long by 2 feet 6.5 inches in diameter. The extreme length of the MK 7 has confused some authors who have declared it in error to be a gun-assembled weapon similar to Little Boy. Without its nuclear pit installed, the bomb weighed 1,485 pounds. With its pit and core, the MK 7 weighed between 1,645 and 1,700 pounds. The AEC produced the MK 7 in ten different MODs, not all of which saw concurrent service. The DOD retired the last (MOD 9} bomb in 1967. The Air Force stockpiled Type 110, 130, 150, 160, 170, 190, 210, 240, and 260 cores for the MK 7, which had at least six yields: 8, 19, 22, 30, 31, and 61 kilotons. Altogether, the AEC produced 1,700 - 1,800 MK 7 bombs during the period 1953 - 1962.

Douglas Aircraft built the MK 7's aluminum alloy case at a plant in El Segundo, California. Although streamlined, the MK 7 was not amenable to supersonic carry. This shortcoming handicapped the higher performance fighter-bombers to which TAC assigned it. Specialists could remove the forward end of the case to install batteries and detach the rear end of the case to install a nuclear capsule in its automatic IFI system. A three-fin system stabilized the bomb, the lowermost fin retracting to provide ground clearance on fighter-bombers. In the MK 7 MOD 0, 1, and 2, Sandia added small retractable airbrakes in the spaces between the fins. These maintained the subsonic velocity required by the bomb's MK 5 MOD 0 barometric fuze, which backed up the radar primary. Beginning with the MK 7 MOD 3, Sandia deleted barometric fuzing along with the airbrakes. The MOD 3 and later versions were spin stabilized by means of small roll tabs located on the tips of the bomb's fins. These improved the bomb's accuracy by generating a roll rate between 1/3 and 1 revolution per second.

Sandia produced a number of different firing systems and fuzes for the MK 7. The bomb's original MC134 firing-set was a modified MK 6's firing-set. The MC134 was able to accommodate the change to a 92-lens implosion assembly. The bomb was equipped with a pair of radars of the Abee or Albert variety, with their antennas located in the nose of the bomb. It was also equipped with arming and fuzing baro-switches and contact detonators. A safe separation timer prevented detonation of the bomb for a preset time interval. The ground crew could set the

The Republic F-84F Thunderjet (above) was an important tactical nuclear fighter-bomber during the 1950s. It has a T-63 training version of the MK 7 bomb attached under its left wing. Carry of nuclear weapons under the left wing was standard for most aircraft. (Credit: USAF Photo K-9342)

A B-57 Intruder medium bomber. It was a mainstay with Tactical Air Command during the 1960s and the 1970s, replacing the B-45. It had an internal bomb bay. This aircraft is equipped with wing tip fuel tanks. (Credit: USAF)

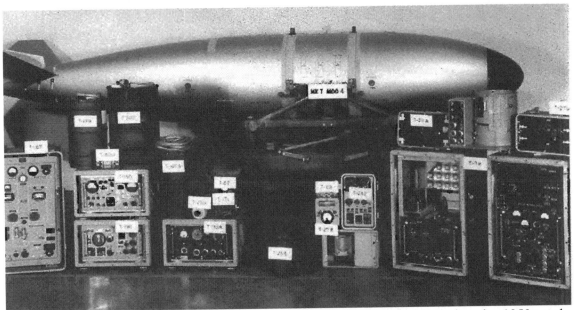

A MK 7 MOD 4 bomb, the primary tactical nuclear weapon deployed during the 1950s and 1960s. This photo shows the testing and maintenance equipment needed to keep it healthy. (Credit: SNL)

A krytron tube, which instantaneously switched the power contained in the charged capacitors into the firing harness of weapons like the Mark 7. (Credit: John Rehwinkel; Color Original)

A Lockheed P-2V "Neptune" maritime patrol aircraft. It carried two MK 90 aerial depth bombs in its weapons bay. (Credit: USN Photo CNIC 042983)

A MK 90 "Alias Betty" aerial depth bomb. It was equipped with a 32-kiloton W7 warhead. Note the flooding ports in the MK 1 MOD 0 afterbody and the Mk 22 MOD 0 parachute pack attached to the tail. (Credit: NAM)

A US Navy Bombardment Aerial Rocket (BOAR) as deployed in 1957. It was developed at the China Lake facility. The forward part of the casing was identical to MK 7 aerial bomb and could accept the bomb's full range of cores to produce a variety of yields. When lofted, the rocket motor provided a range of 6 to 7.5 miles. (Credit: USN)

A McDonnell F-2H Banshee launches a BOAR standoff rocket at the Navy's China Lake test range. BOAR's primary use was to attack groups of enemy shipping. (Credit: USN; Color Original)

A Douglas MGR-1A (M31) Honest John rocket on its M289 launcher vehicle. The Army equipped Honest John with W7 warheads in a variety of yields. (Credit: USA)

A transporter-erector placing a Firestone MGM-5 (M2) Corporate SRBM on its portable launch pad. The Army equipped Corporal with W7 warheads in a variety of yields. (Credit: USA)

An ADM-B type atomic demolition munition on its trailer. Like the ADM-A, it had a waterproof casing and was fired by remote control, radio signal, or a timer. (Credit: LANL)

detonation of the bomb for a preset time interval. The ground crew could set the timer between 1/4 second and 100 seconds. From a cockpit control box, the pilot could select from seven different fusing options that combined height of burst, timer setting, and contact detonation. From the MK 7 MOD 1 onwards, TAC covered the nose of the bomb and its antenna ports with a rubber-deicing cap.

Like the Fat Man family of strategic bombs, pullout switches activated the MK 7's arming circuits at release. For a high-altitude burst, the x-unit arming baro-switch closed at a preset altitude to allow charging of the firing-set's capacitor bank with an electrical current of 1,800 volts. Building a full charge took eight seconds. After a preset delay, a timer activated the bomb's radar receivers and, if a sensor did not detect jamming, it turned on the bomb's radar transmitters – a process that took about 1/4 second. When both of the bomb's radars agreed on a fall to a preset altitude, they sent a signal that detonated the bomb. In the event that the receivers detected jamming, or if one of the radars failed to operate, the bomb detonated when a timer initiated by the arming and fuzing baro-switches closed the firing circuit. This process took less than 3 seconds. For bursts below 1,000 feet, the fuzing baro-switch turned on the bomb's radars and activated charging of the x-unit's capacitor bank. The radars started a timer when the bomb neared a preset detonation altitude that allowed detonation to a height of 300 - 400 feet. The timer fired the bomb between 0 and 5 seconds after activation.

A request from the Armed Forces Special Weapons Project resulted in the development of the MK 7 MOD 1 bomb. It featured battery and radar heaters that accommodated external aircraft power. The requirement to carry the MK 7 on a variety of aircraft resulted in the MOD 2 variant. It had a tail that was rotatable in 15-degree increments to provide the best fit for its fins on different aircraft configurations. The MOD 2 also had an improved in-flight insertion mechanism and came equipped with nickel-cadmium batteries.

Developed in parallel with the MK 7 MOD 0 (TX-7) was the TX7X1. For fuzing, the TX7X1 was equipped with radar, timer, and piezoelectric impact fuzes compatible with the Low-Altitude Bombing System (LABS). Whichever fuze was selected, a safe separation timer provided an interval up to 15 seconds before arming and firing could take place. The TX7X1 was designated the MK 7 MOD 3 in service. Fighter-bombers could deliver it at altitudes from 4,000 to 12,000 feet above its height of burst and at dive angles from 0 to 70 degrees from the horizontal. A small access port added to the forward section of the case permitted arming or safing of the fuzing system during ground handling or loading operations. It also allowed the selection of fuzing options, which the ground crew set before takeoff. The electrical fuzing and firing cartridge of the MOD 3 was mounted forward of the implosion assembly and was physically similar to that used on the MK 5 MOD 0 bomb (Fuze B). The DOD designated training versions of the bomb as T-63.

The MK 7 MOD 4 was a universal device designed for use in both bombs and warheads. The MOD 5 incorporated shock absorption pads for its radars and timers to cushion the forces created by release from the Aero 7A bomb rack. The MOD 6 replaced the MOD 5's mechanical neutron initiator with an electronic external neutron initiator (ENI) and the MOD 7 delayed charging of the x-unit until after the bomb had experienced a normal trajectory as part of a toss-bombing maneuver. Installation of PAL units in the MOD 5 and the MOD 7 resulted in the designations MOD 8 and MOD 9.

Los Alamos conducted the first test of a MK 7 bomb on November 5, 1951, when a B-45 Tornado dropped a TX7E prototype at the Nevada Proving Grounds. Buster Easy exploded at an altitude of 1,314 feet with a yield of 31 kilotons using a composite-levitated pit. This demonstrated that tactical bombs were now more powerful than the strategic weapons used against Japan. Sandia next conducted non-atomic tests to determine whether a MK 7 bomb could reliably transmit impact to an internal piezoelectric fuze. On May 1, 1952, Snapper Dog evaluated the potential of a TX7 pit for deuterium boosting, a concept that was never implemented in an operational design. This was the first test of the MK 7 using its standard ballistic case. Finally, in June 1953, Los Alamos conducted a maximum yield, 61-kiloton test of a MK 7 bomb with a composite Cobra core in a Type D pit as Upshot-Knothole Climax. The AEC used this device as a primary in early thermonuclear weapons.

Like the MK 5 bomb, early MODs of the MK 7 were equipped with internal neutron initiators and in-flight insertion mechanisms for their nuclear cores. MK 7 MODs produced after 1957 were fitted with external neutron initiators, which the Author describes in a later chapter. The ENI simplified bomb maintenance because it did not rely on short-lived isotopic components and could be stored in a weapon for years. On the negative side, the new initiators were at first bulky and added an extra layer of electronic complexity.

As previously noted, Sandia retrofitted the late MODs of the MK 7 with an electromechanical version of the Permissive Action Link known as PAL A. PAL was a means of delaying access to a nuclear weapon if it fell into unauthorized hands. Sandia first made PAL A available in March 1963. It provided positive control over weapons mated to American and NATO aircraft after the Supreme Allied Commander in Europe (SACEUR) instituted the Quick

Reaction Alert Program in 1960. NSAM-160, issued by President Kennedy in June 1962, authorized the first PAL devices. These were widely available in the European Theater by 1964.

The evolution of PAL was as follows:

PAL	A three or five-digit mechanical lock.
PAL A	An electro-mechanical switch that required input of a four-digit unlocking code from a portable electronic device that plugged into the weapon.
PAL B	An advanced version of PAL A developed for the in-flight arming of bombs via AMAC. PAL B permitted re-keying and re-locking. Improved PAL B models were equipped with a limited try feature.
PAL C	A six-digit version of Pal B that saw limited use with the Lance W70-0 warhead.
PAL D	An electro-mechanical switch that required a six-digit unlocking code. Groups of weapons could be unlocked with a single transmission. It also provided for a number of selectable violent or non-violent methods used to disable a bomb. PAL D adjusted yield on weapons equipped with this feature.
PAL E	A special version of PAL D intended for use with weapons in long-term storage. It could remain operable without servicing for a minimum period of eight years. PAL E was never implemented.
PAL F	Similar to PAL D but requires a twelve-digit unlocking code.

PAL A was a coded electromechanical device that interrupted critical electrical circuits necessary for the operation of a weapon. The system consisted of an MC1541 electromechanical encoder, a T1500 decoder, a T1501 re-coder, and a T436 battery power supply. The PAL A system was also compatible with T1520 and T1521 coder-testers. The final production of this support equipment included 578 T1500 decoders, 57 T1501 re-coders, and 274 testers that emulated the performance of an MC1541 coded device.

Installed in pairs in MK 7 bombs, MC1541 encoders occupied a volume of 10 cubic inches and weighed 1.2 pounds. The device was an arrangement of motors and gears that took between 30 and 150 seconds to close a series of electrical switches to render a warhead operable (enable it). Each encoder required seven control and monitor lines, a total of fourteen lines per weapon. The T1500 decoder had four code wheels and plugged into the MC1541 device. This device weighed about 40 pounds when attached to its T436 power supply. The T1500 controlled both encoders simultaneously from two separate channels but indicated when at least one set of MC1541 switches had closed to enable the weapon.

To code or re-code an MC1541 device, it had to be unlocked and a code change solenoid activated with the T1501 re-coder. This device could also act as a decoder. When activated, the MC1541's solenoid allowed the unit's code wheels to rotate to a new position corresponding to the desired code. After a technician entered a code, the solenoid released to store the new code. The technician had to repeat this process for each coded switch in a weapon. Following a re-code, it was necessary to enter an incorrect code with the T1501 to lock in the new code. PAL A was not operable from the cockpit. Ground crew had to enable a weapon before its strike aircraft took off.

A wide variety of Air Force and Navy light bombers and fighter-bombers aircraft carried the MK 7. One of the first of these delivery platforms was the B-57 Intruder, which replaced Tactical Air Command's B-45 Tornado bomber. The B-57 was actually a British Canberra bomber rebuilt to American standards. Powered by two Wright J65-W-5 turbojets, it had a speed of 580 MPH, a ceiling of 55,000 feet but more importantly, a 4,600-mile range. A two-man crew flew it from a tandem cockpit. Although the B-57 was soon retired from Europe, it remained in service in the Far East with the 13th Air Force. Here distances to targets were longer and its twin-engines provided a margin of safety while operating over water.

In addition to the B-57, TAC acquired a substantial inventory of nuclear capable fighter-bombers. It used these to arm forward units serving with the United States Air Force in Europe (USAFE) and the Pacific Air Force (PACAF) in the Far East. In the continental United States, TAC assigned its fighter-bombers to the 9th Air Force, headquartered at Shaw AFB, South Carolina. The 9th Air Force's mission was to protect the continental United States, train aircrews for overseas duty, provide air units for expeditionary missions, and to backfill the USAFE and PACAF as required. After the Korean War, the Air Force developed the Composite Air Strike Force (CASF) concept, with which TAC could rapidly deploy fighter-bombers, troop carriers, tankers, and tactical reconnaissance assets worldwide to support combat units already in place.

The Air Force activated the USAFE on February 1, 1942, and after WW II, headquartered it at Lindsey AS, Wiesbaden, West Germany. During the Berlin Crisis 1n 1947, TAC deployed 210 aircraft to Europe. The United States considered the defense of Europe so important that even during the period of the Korean Conflict; the USAFE received a lion's share of aircraft appropriations. The USAFE attached its nuclear capable fighter-bombers with the 3rd Air Force, (activated on May 1, 1951), which it headquartered at RAF South Ruislip, UK, and with the 17th Air Force, (activated on April 25, 1953), which it headquartered at Ramstein AFB, Germany. The 3rd Air Force was originally responsible for tactical air operations in the UK and Europe. The 17th Air Force was alternately responsible for tactical air operations over North Africa, Portugal, Austria, the Middle East, India, Ceylon, and the Mediterranean Islands. In 1956, USAFE re-established its operational area to include Italy, Greece, and Turkey, and relocated its headquarters to Ramstein AFB, Germany, in 1959. After the move to Ramstein, the Air Force altered the 17th's mission to the defense of central Europe with bases in West Germany, France, the Netherlands, and Italy.

A meeting of NATO's Military Committee in February 1952 established a goal of 28 USAFE Air Wings comprising 3,000 aircraft of all types to support 96 planned NATO Ground Divisions. By 1955, the Air Force had distributed these wings throughout the UK, France, and Germany. An important early component of these wings was Republic's F-84E Thunderjet fighter-bomber. The F-84E was a straight wing fighter-bomber powered by a 5,000-pound-thrust Allison J-35-A-17D turbojet. This engine gave it a maximum speed just over 600 MPH and a tactical range of 2,000 miles.

The improved F-84G followed the F-84E into service in 1951. It had a more powerful 5,600-pound thrust Allison J-35-A-29 engine and could be equipped to carry a single MK 7 bomb. Of the 3,025 F-84G aircraft produced, the United States supplied 1,936 to NATO Allies. These formed the nucleus of 21 fighter-bomber wings. F-84Gs, the most prominent component of these early Air Wings, eventually served with the Air Forces of Belgium, Denmark, France, Italy, Greece, Norway, the Netherlands, Portugal, Turkey, and Taiwan. They United States also exported the F-84 to such "non-aligned" nations such as Iran, Thailand, and Communist Yugoslavia.

Critical to the F-84E and F8-4G's performance was a toss-bombing capability provided through Project Back Breaker. This project resulted from a shift in thinking about atomic bombs. The Air Force had previously viewed atomic bombs as special weapons for use against strategic targets. It now looked to smaller weapons for use against a wide variety of strategic and tactical targets. According to General John Mills, Commander of the Air Force Special Weapons Center, the addition of the MK 5 and MK 7 allowed the "Air Force (to) deliver varied kinds of atomic weapons against varied kinds of targets with varied aircraft types." The main task of USAFE F-84G aircraft was to cripple Russian airpower through the destruction of heavily defended airfields and the disruption of communications facilities.

In terms of tactical weapons, LASL's John Wheeler declared that "we need to increase the scale of production of tactical nuclear weapons to the point where we are aiming at having a million of 'em.'" Before World War I, everybody thought they had enough bombs to kill an inconceivable number of people. A few weeks after the war broke out, all sides had used up their reserves of bombs. They had to build factories to make thousands more bombs to continue the war. Wheeler said that this situation would arise with tactical nuclear weapons. If the Soviets recognized this and the West did not, it would pay the price for its ignorance. (Wheeler was very concerned about Russia dominating the world.)

In April 1966, the USAFE supplemented the 3rd and 17th Air Forces with the 16th Air Force. The 16th Air Force had operated medium bombers in Spain and Morocco before the USAFE moved it to Aviano AFB, Italy. Following TAC's acquisition of the 16th Air Force, the 17th Air Force moved its headquarters to Sembach AFB in 1972. This allowed the transfer of USAFE Headquarters to Ramstein AFB, West Germany, in March 1973. As part of this reorganization, the 3rd Air Force moved its headquarters to RAF Mildenhall, UK, and gained responsibility for US Air Force operations north of the Alps. Its operational areas included the United Kingdom, West Germany, the Netherlands, Norway, Belgium, Denmark, Luxembourg, and portions of France. In turn, the 16th Air Force became responsible for air operations in southern Europe, portions of the Middle East, and northern Africa. Its two main bases were located at Aviano, Italy, and Incirlik, Turkey. It also had four support bases and other sites in Spain, France, Germany, Italy, Greece, Turkey, and Israel.

The Air Force activated PACAF on August 3, 1944, two years after the USAFE. PACAF received its current designation on December 6, 1945. The new command attached its nuclear capable fighter-bombers to the 7th Air Force headquartered at Osan AFB, Korea, and the 13th Air Force headquartered at Clark AFB in the Philippines. The mission of the 7th Air Force was to plan and direct air operations in the Republic of Korea and in the Northwest Pacific. The mission of the 13th Air Force was to conduct missions in Southeast Asia and the South Pacific.

Starting in 1951, TAC carried out the development of a toss-bombing capability to give its fighter-bombers a standoff capability. It did this concurrently with the development of the MK 7 bomb. In fact, the F-84E / MK 7 weapon system was the first joint project between the Air Force (DOD) and the DOE. Before it attempted toss-bombing, TAC perfected level carry and dive-bombing techniques for its aircraft. For these purposes, Air Materiel Command made it possible to attach a MK 7 bomb to either wing tip of an F-84G or to attach it as an under-wing store (on its left wing). However, in an aborted mission, an aircraft had to retain fuel in the wing tank opposite the MK 7 to maintain the aircraft's balance. This limited its radius of action so that the under-wing storage position was preferred. In this position, it was important that neither the pylon nor the bomb shape interfere with airflow about the aircraft. Flight tests conducted by Air Materiel Command soon showed that a fighter-bomber could safely release a bomb carried in level flight and at dive angles up to 70 degrees.

The flight tests also showed that the standard A-1C gun-bomb-rocket sight could not accurately deliver the MK 7. As a temporary measure, AMC developed the A-7 bombsight. As it turned out, the performance of the A-7 was not much better than the A-1C so TAC restricted carry of the MK 7 for release in level bombing by experienced pilots. During carry, a T40 control box monitored the MK 7. The T40 controlled insertion of the bomb's nuclear core, arming of the bomb's electrical circuits and selection of radar range for the bomb's T44 "black box." The T79 later replaced the T40 with improved features.

Following the delivery of the A-7 bombsight, TAC focused attention on aircraft delivery techniques. Of these, it considered toss-bombing the most important because the "three tenths" cloud cover prevalent in Europe throughout most of the year was sufficient to prevent high-altitude level delivery and dive-bombing. Pilots had to deliver their bombs from low-altitude using visual clues and a level, low-altitude release that placed the aircraft very close to the bomb's detonation point. In examining nuclear delivery by fighter aircraft, the Aircraft Laboratory and MIT demonstrated that the maximum overpressure sustainable by fighter aircraft was 2 PSI. This dictated that the strike aircraft maintain an appropriate standoff distance from the detonation point when making a delivery. Other considerations were the danger of radiation to the pilot and the effect of thermal radiation on the aircraft.

Special Weapons Branch engineers outlined a number of possible escape methods for an aircraft after it released a lofted bomb. These included:
- Continuing in a dive straight ahead to the "deck"
- Pulling maximum Gs to a straight and level flight path
- Pulling maximum Gs to a flight path whose angle with the horizontal corresponded to the best angle when the airspeed was dropped to the most effective climb rate
- Pulling maximum Gs straight ahead to gain maximum angle
- Pulling the same maximum maneuvers with a turn toward a radically outbound heading from the point of detonation

So important was toss-bombing (the performance of the 20th Fighter Wing's 49th Air Division was thought to determine the loss or survival of Western Europe in the opening days of WW III) that development of a proper sight was given the highest priority after this technique was chosen as the primary means for delivering tactical nuclear bombs. As the Armament Laboratory's Special Weapons Branch began developing this capability, two technical innovations presented themselves. These were the Navy's MK 3 MOD 3 and MK 3 MOD 5 bomb directors and Sweden's BT-9 dive-toss-bomb computer developed by SAAB. Unfortunately, the Navy's MK 3 MOD 5 fire director used integrating accelerometers instead of the vertical gyro preferred by the Air Force but when the Air Force tried to purchase SAAB's BT-9 in quantity, offshore procurement complications prevented its acquisition. The Air Force was thus on its own to develop a toss-bombing capability.

Captain John Ryan and Captain John Hanlen began loft-bombing research for the Air Force when they proposed the combination of an A-4 rocket-gun-bomb sight with an AN/APG-31 air-to-ground ranging system. Since this system proved rather complicated, they instead decided to wire the A-1C, A-4, and A-7 bombsights to a computer that included a B-10 intervalometer, a Sperry zero reader gyroscope, a quick erection gyro unit, and a relay box. Named the Low-altitude Bombing System (LABS), this latter arrangement could be easily adapted to existing aircraft through rewiring. In support of LABS, the Navy had the Aero Company modify its MK 6A bomb rack into the MK 6B configuration, which it optimized for toss-bombing.

After it acquired an LABS sight, the Air Force developed toss-bombing methodologies as part of Project Red Dog. During a standard attack, an aircraft would approach its target in level fight at a constant speed of about 500 MPH. At a predetermined waypoint, the pilot would press and hold down a "pickle" button to energize the preset intervalometer, which extinguished the sight reticle when it was time to initiate the climb required for the toss-bombing maneuver. The pilot then pulled his aircraft up into a steady 4G climb. Part way into a half loop, the bombing computer exploded a cartridge that propelled the MK 7 away from the aircraft on a trajectory to its target. Release was typically at one of six different standard angles. The release of the bomb activated its safe separation timer. Following that timer's expiration, a second timer detonated the bomb after a duration of 1/4 - 100 seconds. At the top of the half loop, the pilot executed a half roll, kicked in the afterburner, and accelerated away from the target at full speed. Although frequently referred to as an Immelman, this maneuver is properly called a Half-Cuban Eight.

In the event that a suitable waypoint was unavailable, a pilot could execute a backup toss-bombing maneuver in place of the Half-Cuban Eight. In this circumstance, the pilot flew directly over the target at a constant 500 MPH speed and at a specified distance past, commenced his constant 4G climb. The computer released the bomb past the top of the arc so that it flew back from the original direction of travel. The pilot continued through a 360-degree loop and then accelerated away from the target on afterburner on his original heading. With low angle toss-bombing maneuvers, if the pilot missed his climb point, it was possible to continue over the target and transition to an "over the shoulder" bombing maneuver for a second chance at delivery.

The production LABS unit weighed only 10.5 pounds and occupied 420 cubic inches. Its computer was designated the MA-1 LABS computer set and it could release a MK 7 at angles from 0 - 130 degrees. An improved MA-1A LABS computer was supplied with two switches. One switch was for angles between 0 and 130 degrees and the other was for angles between 85 and 130 degrees. As part of On Top Phase 9, Air Materiel Command supplied LABS to the USAFE and then to the Far East Air Force. Recipients included 107 F-84Gs belonging to the 20th Bomb Wing and the remaining 254 F-84Gs awaiting delivery by Republic. TAC also ordered all models of the F-84F follow-on fighter-bomber equipped with LABS.

First delivered in 1954, Republic's F-84F Thunderstreak was a redesign of the F-84E that incorporated sweptback wings and a more powerful 7,220-pound thrust Wright J-65-W-3 turbojet that boosted its speed to 650 miles per hour. However, because of numerous engine problems, TAC soon transferred the Thunderstreak to the National Guard, whereas the F-84G continued in service into the 1960s. The fiasco surrounding the J-65 engine resulted in cancellation of further contracts to Wright Aircraft, which forced the company out of the engine business.

Probably the most important fighter-bomber to carry the MK 7 was the North American F-100 Super Sabre. This aircraft was equipped with a 16,950-pound thrust (afterburner) J-57 turbojet engine manufactured by Pratt & Whitney. The secret of this second-generation jet engine was increased compression provided by an inline combination of low and high-speed compressors. Twin turbines located in the area immediately after the combustion chamber powered the compressors. The J-57 also had a following "afterburner" where it could burn additional fuel to increase power at the expense of range. Activation of the afterburner provided the Super Sabre with a level flight, supersonic dash capability. An interesting adaptation for the F-100D was a zero-length launcher that boosted it directly into the air by means of an Astrodyne XM-34 132,000 pound thrust solid fuel rocket. The booster fell way after a 4-second burn that imparted a 275 MPH speed to the aircraft.

Designed as a single seat interceptor and air superiority fighter, The F-100C was the first model of this aircraft to have an atomic capability. The definitive two-seat F-100D next entered service in the continental United States and at overseas bases in Japan, France, and Morocco in 1956. The F-100D was also nuclear capable and could carry a MK 7 bomb at the port wing intermediate attachment points or the fuselage centerline attachment point. Like the F84G and F84F, the F-100D had an LABS bombing system. In conjunction with an A-4 gyroscopic site, an electronic AN/AJB-1B bombing system calculated the aim point data for the proper release of the MK 7. F-100D production reached 1,274 aircraft, some of which continued in service as late as 1979. Originally, the Air Force acquired 1,137 DCU116/A PAL B cockpit controllers for use with the F-100 and the F-104. It augmented these with 487 DCU/121A cockpit controllers that combined PAL B and AMAC in 1965.

Another important fighter-bomber that carried the MK 7 was the McDonnell F-101 Voodoo. TAC had it equipped with four Colt M39 20mm cannon and provision for a centerline external nuclear store. Republic built its avionics around the MA-7 fire control system, which featured an AN/APS-54 radar and MA-2 LABS. The F-101A was strictly a nuclear weapon delivery system, with no real conventional attack capability except for its cannon. In service, a TACAN radio beacon navigation receiver typically replaced one of its cannons.

In order for the Air Force to use the F-100 and F-101 most efficiently during the critical early period of a nuclear war, it requested volunteers for special missions. It assigned the volunteer pilots targets deep inside Soviet held territory, beyond the normal action radius for their aircraft. After attacking their specified targets, the pilots had to exit the area of radioactive hazard and then bail out. The Air Force trained the pilots in escape and evasion. Air Force planners thought that a nuclear war would be over as soon as Strategic Air Command could deliver a decapitating strike with its heavy bombers. Following the secession of hostilities, the Air Force could repatriate its pilots.

Like the Air Force, the Navy carried the MK 7 on a wide variety of aircraft. The first of these was the Douglas A-1D Skyraider. The last propeller driven fighter-bomber to enter service, the Skyraider was powered by a 2,900 horsepower Wright R-3350-26SWB radial engine, similar to the one used in a B-29 Superfortress. This engine provided the Skyraider with enough power to carry 6,000 pounds of ordnance and contributed to a very long operational lifespan. Although it had a 1,945-mile range with external fuel and an extended loiter time, its maximum speed was an unimpressive 325 MPH. The Skyraider could carry either a MK 7 or a MK 8 nuclear bomb and was equipped with an early version of LABS. The Navy removed the last Skyraider from service in 1971. Other Navy fighter-bombers that carried the MK 7 were the jet-powered McDonnell F2H Banshee, the Grumman F9F-8 Cougar, the McDonnell F3H-2 Demon, and North American's FJ-4B Fury. The last of these was the naval version of the F-86 Sabrejet. Equipped with LABS the F-86J had a maximum speed of 690 MPH and a range of 1,500 miles.

Larger than all these fighter-bombers were the Navy's Douglas A-3 Skywarrior medium bombers. The successor to the AJ-2 Savage, the Skywarrior was capable of takeoffs and landings only on the Navy's larger carriers. Deployed in 1956, a production run of 282 aircraft finally provided the Navy with strategic influence. A pair of Pratt & Whitney's second-generation J57-P-10 turbojet engines powered the A-3. Rated at 10,600 pounds thrust, these engines provided the 35-ton bomber with a 2,300-mile range, a maximum speed of 620 MPH, and a 70,000-foot ceiling. Designed to carry thermonuclear weapons, the A-3 could also fly tactical missions. Like most contemporary aircraft, the Skywarrior was capable of in-flight refueling. The aircrew consisted of a pilot, a bombardier-navigator, and a gunner who operated a pair of rear facing 20mm cannons.

Douglas also produced the A-3 in an Air Force version designated the B-66B Destroyer. The B-66B actually preceded its Navy counterpart into service in March 1965. Conversion from a high level, carrier-based bomber to a low-level Air Force attack aircraft proved more difficult than anticipated. Nevertheless, the performance of the two aircraft was very similar. One major difference was the use of Allison J71-A-11 engines on the B-66B. Seventy-Two B-66B Destroyers replaced the B-57 Intruder in Europe.

In addition to the MK 7 freefall bomb, Los Alamos developed a compact W7 warhead. This warhead measured 30 inches in diameter by 55 inches long and weighed between 900 and 1,100 pounds depending on the yield and MOD. Unlike the W5 warhead, which the DOD used to arm cruise missiles, the DOD incorporated the W7 warhead into a wide variety of tactical weapons for the Army and the Navy. Clearly, the "nuclear genie" had escaped from the bottle and the age of proliferation had begun.

The W7 was among a number of warheads considered for use with missiles (especially antiaircraft missiles) because of its size. The Army or Air Force could explode a warhead of suitable yield at altitudes in excess of 30,000 feet, leaving citizens unscathed in their homes below. Because the design of the W7 was much further along than competing designs, the DOD chose it to fulfill a number of roles. Military characteristics called for a diameter of 29.5-inches, a length of 70 inches (to accommodate a core insertion mechanism), and a weight of 1,500 pounds. The AEC first stockpiled the basic MOD 0 version of the W7 warhead in August 1954. These warheads were converted MOD 1 bombs. Environmental requirements included the ability to withstand longitudinal accelerations of 5 Gs, lateral accelerations of 3 Gs, operational temperatures from −80 °F to +160 °F and altitudes up to 250,000 feet.

A physics package for use in both bombs and warheads entered the stockpile in March 1954 as the W7 MOD 1. The W7 MOD 2, equipped with a dual motor insertion mechanism that was operable under conditions of extreme acceleration or rotational motion, followed the MOD 1 in March 1955. This warhead was not adaptable for use in bombs. In August 1963, the AEC began stockpiling the W7 MOD 3 warhead, which it equipped with PAL A. The W7 MOD 3 had a very fast-acting MC621 IFI that replaced the MC490 IFI in the MOD 2, and a more advanced MC578 x-unit that replaced the MC134. The W7 was compatible with core Types 110, 130, 170, 190, 210, 240, and 260. Not all of these core types were available for use with every weapon system equipped with the W7. Army engineering stockpiled two special low-yield cores, the Type 270 and 280, for use with the W7 based atomic demolition munition.

The buildup of numerically superior Soviet Forces in Eastern Europe meant that NATO Ground Divisions would have to fight a delaying action until SAC's heavy bombers could destroy the Soviet Union and the Communist will to fight. NATO's interim defensive action entailed a slow, fighting withdrawal in the face of a Soviet onslaught. In order to delay the Soviet advance, it was necessary to destroy a variety of bridges, tunnels, mountain passes, rail, and road junctions.

This would hinder the forward movement of Soviet Forces and channel them into choke points. Once concentrated at known locations, NATO could destroy Soviet forces with a combination of conventional and nuclear weapons.

As part of its European withdrawal strategy, NATO had to destroy key facilities such as military bases and airfields as they abandoned them. For this purpose, the Army assigned Special Engineering Platoons the task of accomplishing this objective with Atomic Demolition Munitions, which Los Alamos developed under the auspices of Project "A." The ADM-A was a MK 6 atomic bomb assembly fitted with a firing device. Far more practical than the ADM-A was the ADM-B. This device derived from the W7 warhead and was in service from 1955 to 1963. It was the perfect weapon to get major demolition tasks accomplished. The ADM-B had a variety of yields including a scaled down yield of 90 tons TNT equivalent. A single unit could easily destroy a bridge or runway. This feature released engineering units from the complex and time-consuming task of mining large structures with truckloads of conventional explosives. The detonation of an atomic munition left behind a radioactive zone that advancing troops had to avoid.

ADMs were not pre-positioned for security reasons. Instead, a Combat Engineering Unit had to drive them to their assigned targets from Operational Storage Sites in the event of a Soviet attack. The AEC manufactured approximately 300 ADM-B demolition munitions. In Europe, the Army deployed them with both Divisional and Non-Divisional Combat Engineering Units. The Army delayed deployment to Korea until 1958, at which time the United States began the nuclearization of the peninsula.

Despite the utility of the ADM-B, it was both unwieldy and heavy. It came packed in a shipping container that measured approximately 2.5 feet square by 3.75 feet long and weighed just over 1,500 pounds when loaded. Extracted from its crate, the warhead weighed 1,000 pounds and required a good deal of effort to move to a pre-planned position. An engineering unit detonated it by command or with a timer.

Another application of the W7 warhead was the Navy's MK 90 "Alias Betty" or "Betty" airdropped depth bomb. (An atomic depth bomb was unsuitable for use with the conventional projectors onboard naval vessels because of its power. The depth bomb would have sunk to its operational depth and exploded well before a ship could move to a sufficient distance to escape its effects.) The Betty emerged from a program that evaluated various nuclear devices for use in the antisubmarine role. Also considered were "Alice" and "Alvin" versions of the W12 warhead and "Candy" and "Chris" versions of the W5 warhead.

The Navy demonstrated the ability of an underwater atomic explosion to damage or sink naval vessels of all types during Operation Crossroads in 1946. Crossroads Baker, the explosion of a MK III atomic bomb at a subsea depth of 90 feet, sank an aircraft carrier, two battleships, two submarines, and three auxiliary craft. Crossroads Able, an airburst at an altitude of 523 feet, took a much smaller toll by comparison. This toll consisted of a heavy cruiser, two destroyers, and two attack transports.

The tremendous damage caused by an underwater blast results from a shock wave that separates from the rapidly expanding gas bubble formed around a nuclear fireball. The subaqueous shock wave is similar to a shock wave formed in air, except that its peak overpressures do not fall off as rapidly. Its maximum destructive effects therefore extend over a considerable area. Reflections of the primary shock wave can propagate from both the sea bottom at shallow depths and from the air / water interface. In deep water, these secondary reflections are not significant.

Underwater nuclear explosions produce two main effects. The first is the direct impact of the shock wave on the hull of a ship. This impact cause's distortion to both framing and hull plating and may result in leaks severe enough to cause the vessel to founder. The shock also induces sudden movement, which can damage equipment and violently throw un-stowed articles and crew about. Supporting structures for heavy equipment such as boilers, engines, and armament may collapse with serious consequences. A powerful base surge can also create waves large enough to overturn and sink vessels in proximity to the blast's epicenter.

The MK 90 depth bomb exploited all of these effects. It consisted of a forward body with a parachute pack attached to a perforated afterbody. The forward body contained a 32-kiloton W7 warhead. The afterbody's MK 22 MOD 0 parachute pack was equipped with a drogue chute that pulled out the bomb's 16.5-foot diameter main canopy. The specific version of the W7 incorporated in the MK 90 weighed 983 pounds. Completely assembled, the depth bomb measured 10 feet 2 inches long by 2 feet 7.5 inches in diameter and weighed 1,243 pounds. Two lugs on the upper surface of the warhead body provided suspension. A timer, contact fuzes, and hydrostatic detonators that exploded the bomb at a preset water depth provided fuzing. The AEC produced 225 warheads for this weapon. Its service life spanned the period 1955 to 1960.

It was necessary to activate the MK 90's in-flight core insertion circuits before release. Withdrawal of its pullout cables at release connected power to the bomb's safe separation timer and power inverters. A static line opened the bomb's parachute. Following safe separation time, a switch connected the inverter's output to the x-unit. After several seconds of delay to allow for the charging of the bomb's condensers, a timer circuit closed the firing circuit to the bomb's thyratron trigger circuit. At water entry, the bomb's parachute detached and explosive switches armed the timing fuze, pressure fuze, and contact fuzes. Burning powder generated pressure to operate a piston, which at the end of its travel closed a pair of contacts completing an electrical circuit. Detonation was dependent on whichever fuze activated first.

In order to gather further data on the effects of underwater atomic explosions, the Navy tested a MK 90 bomb as the centerpiece of Operation Wigwam in 1957. It suspended the depth bomb in a waterproof caisson at a 2,000-foot depth from a barge located some 500 miles southwest of San Diego in the Pacific Ocean. The ocean was 16,000 feet deep at the test site and represented the conditions in which a vessel might use the bomb might in the fleet defense roll.

The Navy suspended three instrumented, submarine-scale, pressure hulls along a 6-mile cable that led to the Command Tug, USS Tawasa (ATF 92). Internal cameras recorded the demise of the pressure hulls as the shock wave crushed them. Wigwam showed that 32-kiloton blast at a sea depth of 2,000 feet would collapse the hull of a fleet-type submarine with a 650-foot crush if the sub was caught operating at a depth of 250 feet in the open ocean at a range of 14,000 feet. The base surge from the blast was initially 600 feet high and caused contamination of the USS George Eastman (YAG 39) and USS Granville S. Hall (YAG 40), which were located five miles downwind.

Martin PBM-5 Mariner and Lockheed P2V-5 Neptune aircraft carried nuclear depth bombs in the maritime patrol mission. The bombs were also adapted for use with carrier-based A-1D Skyraiders and Grumman S-2 Trackers in the fleet defense roll. The Grumman Tracker was the premiere carrier-based anti-submarine aircraft for almost 25 years. Introduced in 1952, the last Tracker was not retired until 1976, when the S-3 Viking finally superseded it. The first nuclear capable versions of the Tracker were 77 S-2F2 aircraft introduced in 1954. These were 42 feet long, with a wingspan of almost 70 feet. Fully loaded, they weighed 26,000 pounds. Two Wright 1,525 horsepower, R1820 -82WA Cyclone radial engines provided a top speed of 265 MPH and endurance of 6 hours.

The S-2F2 was equipped with an enlarged bomb bay big enough to accommodate its MK 90 depth bomb internally. The aircraft's detection equipment included a retractable AN/APS-38 search radar that the crew could lower during ASW missions. It also had an extendable nine-foot Magnetic Anomaly Detection boom in its tail. The MAD unit could sense the proximity of a submarine's steel hull. The Tracker also mounted a powerful searchlight under its starboard wing to detect submarines on the surface. It could drop eight SSQ-2 and 2 SSQ-1 sonobuoys from ports at the rear of each engine nacelle.

The Navy first used sonobuoys in the closing months of the Pacific Campaign during World War II. After an aircraft dropped them into the ocean, these floating sonar stations transmitted search data back to the circling anti-submarine aircraft to improve its chances of a successful detection. Grumman also installed an AN/APA-69 electronic countermeasures system in a radome located over the Tracker's cockpit. The later model S-F3 had improved countermeasures mounted in the wings and enlarged engine nacelles that could each hold sixteen sonobuoys.

The ASW PBM-5S conversion of the Martin Mariner "flying boat" performed the first nuclear Maritime Patrol Missions. A pair of Pratt & Whiney R2800-34 radial engines gave it a top speed of 330 MPH and a 20,800-foot ceiling. The PBM-5S had a range approaching 2,150 miles, endurance of 16 hours, and weighed 25 tons in combat configuration. It was retired in 1962.

Far more important than Mariner flying boats were the Navy's P2V Neptune land-based bombers. The Neptune's origins dated back to WW II Navy specifications for a long-range, ocean patrol aircraft. Lockheed received a letter of intent concerning this aircraft in 1943. After the Navy evaluated the resulting design, it awarded Lockheed a production contract in early 1944. Lockheed delivered the first P2V-1 in 1946, too late to see action. A large aircraft, the Neptune had a length of 78 feet and a 100-foot wingspan. It weighed 33,960 pounds empty, 63,080 pounds loaded, and could carry 8,000 pounds of ordnance. Stripped down and provided with auxiliary fuel tanks, a P2V-1 named "Truculent Turtle" made an un-refueled flight of 11,235 miles from Perth, Australia to the Columbus, Ohio, Zoo with a kangaroo for cargo.

Introduced in 1950, the P2V-5 was the first nuclear capable Neptune aircraft. Powered by a pair of 3,750 horsepower Wright R-3350-30W turbo-compound engines, the Neptune had a top speed of 323 MPH, a 23,200-foot ceiling, and a range of 4,750 miles made possible by extra fuel contained in permanently attached wing tanks. Lockheed later fitted a P2V-5F variant with auxiliary 3,400-pound thrust Westinghouse J34-WE-34 turbojets placed outboard of the piston engines. Production of the P2V-5 was significant at 424 aircraft. It was equipped with an AN/APS-20 search radar, a sonobuoy ejector, searchlight, and magnetic anomaly detection gear. It also housed ECM gear in one of its wing tanks. The provision for so much capability required operation by a nine-man crew. The Neptune's bomb bay was large enough to accept two MK 90 bombs.

The primary targets for MK 90 equipped aircraft were Soviet attack and missile submarines. Important Soviet blue water attack submarines in the late 1950s include the diesel-powered Foxtrot and the nuclear-powered November. November class submarines displaced

5,300 tons submerged and their twin pressurized water reactors gave them a maximum speed of 29 knots, with an estimated unrefueled range of 100,000 miles. Standard depth was nearly 300 meters and their armament consisted of eight 533mm torpedo tubes. Fortunately, for the United States, the performance of both Foxtrot and November submarines was impaired because of their conventional hull designs and noisy machinery. They were, however, able to fire a T5 torpedo with a 10-kiloton nuclear warhead. This made it imperative to keep them from closing to a range where the submarine could make this weapon effective.

The Soviet missile submarines hunted by US and NATO maritime patrol aircraft were diesel-powered Whiskey and Zulu Class missile boats. The Soviet Navy introduced the Whiskey in 1952, with total construction of 236 boats. A small submarine, it displaced 1,350 tons submerged. Nevertheless, the Soviet Navy configured some models to carry from 1 - 4 P5 (NATO SS-N-3, Shaddock) cruise missiles. The 600 MPH Shaddock had a 750-kilometer range that was dependent on the specific model and could deliver a 350-kiloton thermonuclear warhead. Like early American cruise missile submarines, the Whiskey had to surface in calm seas to carry out a launch. This made surface radars effective in its detection.

In 1952, the Soviet Navy accepted the first of 26 Zulu Class submarines for service. Larger than the Whiskey submarine, the Zulu displaced 2,350 tons submerged and had a 20,000-mile range. One of these submarines was configured to carry a single R13 (NATO SS-N-4, Sark) ballistic missile based on the R11FM (NATO Scud). The Soviet navy later converted five more vessels to each carry a pair of the missiles. Fueled with a combination of red fuming nitric acid and kerosene, the launch procedure was quite dangerous and required the submarine to surface. The Sark could carry a megaton range warhead over a distance of 600 kilometers. It was not terribly accurate but with its large warhead, it did not have to be. Surface radars were useful in the detection of Zulu submarines while snorkeling or surfaced.

In addition to its use in demolition munitions and depth bombs, the W7 warhead was adapted for carry by a number of rocket-propelled weapons. The Armed Services began the military development of large rocket motors in 1938 when the Army Air Force and the National Academy of Sciences signed a contract with the Guggenheim Aeronautical Laboratory, California Institute of Technology (GALCIT). The Army Air Force directed a group of graduate students led by Frank J. Malina and mentored by Professor Theodore von Karman to look into a solid fuel motor capable of providing 1,000 pounds of thrust for a period of about 30 seconds. The AAF intended to use the motor for aircraft auxiliary propulsion. In order to confirm the theoretical basis for a solid fuel engine of this nature, von Karman assigned the solution of a series of critical equations to Malina. He confirmed that an end burning rocket motor would maintain constant pressure as long as the ratio of the burning area of the propellant charge to the combustion chamber throat area remained constant.

In order to maintain an area of constant burning, it was necessary to develop a stable propellant that bonded to the case walls of a rocket. If burning propagated along the rocket casing's walls or if cracks developed, the burning area of the fuel would rapidly increase and raise pressures to the point that the rocket casing ruptured. Despite limited success in solving this problem, the Army continued its support of the program and the Navy showed its confidence by contracting for rockets of its own. Like the Air Force, the Navy intended its rockets for aircraft auxiliary propulsion and called for 200 pounds of thrust for a period of 8 seconds. The Navy term for the rockets was a Jet Assisted Take Off unit or JATO.

After much experimentation, John Parsons developed a workable propellant from a mixture of asphalt and potassium perchlorate. Because it contains both fuel and an oxidizer, this combination is termed a composite solid propellant. Known as GALCIT 53, Parsons' propellant bonded directly to rocket casings, thus preventing the spread of burning along case walls. GALCIT 53 was gradually improved into GALCIT 61C, which was commercially employed in the manufacture of Navy JATO units. GALCIT 61C was formulated from 76 percent potassium perchlorate and 24 percent fuel. The fuel component was 70 percent Texaco No. 18 asphalt and 30 percent Union Oil Company Pure Penn SAE No. 10 lubricating oil. In 1942, the laboratory spun off a private company named Aerojet to provide critical components for the manufacture of rocket motors when it could not find private suppliers. General Tire & Rubber of Akron, Ohio, later bought out the business and provided funding for expansion under the name of Aerojet General.

Charles Bartley made an important improvement to composite propellants when he replaced asphalt with a castable, elastomeric, polysulfide rubber produced by the Thiokol Chemical Corporation. Thiokol used its LP-2 polysulfide rubber compound in the production of adhesives, gaskets, and as a coating in self-sealing fuel tanks. Bartley's LP-2 based propellant had better temperature tolerances, storage properties, and durability than asphalt based GALCIT 61C. In order to capitalize on Bartley's discovery, Thiokol attempted to sell off its patent for LP-2 to Aerojet and to other rocket manufacturers. When they rebuffed Thiokol's offers due to the product's high sulfur content, the Army encouraged the company to enter the rocket fuel business on its own. Thus, in 1948, Thiokol set up shop at a former ordnance plant in Elkton, Maryland, and at the Redstone Rocket Research and Development Center near Huntsville, Alabama.

Composite propellants were not the only choice for a solid rocket fuel. The Navy run Allegany Ballistics Laboratory experimented with double-base propellants under the direction of NRDC's Section H during WW II. The ABL manufactured double-base propellants from the same combination of nitroglycerine and nitrocellulose used in the production of ammunition. After the war, the Hercules Powder Company took over the management of ABL and made excellent progress in improving double-base propellants. Based on research by John Kincaid and Henry Shuey at NRDC, Hercules developed a process whereby they treated nitrocellulose granules with nitroglycerine solvent, a plasticizer and stabilizers to produce a liquid that they could heat and cure to form a solid mass of unlimited size. The 400-pound, X-201A1 (3-DS-47000) 16-inch diameter motor that provided 50,000 pounds of thrust for 3 seconds was an early product of this research. Formerly, double-base propellants had to be die-extruded and this process was only suitable for small engines.

Hercules designed its X-201A1 motor in conjunction with the Naval Ordnance Test Station (NOTS). The Navy created the Naval Ordnance Test Station in 1943 to provide a proving ground for aviation ordnance and to co-ordinate with CalTech in developing surface-to-air rockets. CalTech was engaged in two rocket development projects at this time – the previously outlined GALCIT Project and Section L. The latter project was the West Coast equivalent of ABL's Section H, named for Charles Lauritsen, CalTech's Vice-chairman for the Division of Armor and Ordnance. Under Lauritsen, a vast complex of laboratories, shops, and munitions bunkers was set up in Eaton Canyon near Pasadena to develop aircraft and bombardment rockets for the Navy. The complex specialized in dry extruded double-based rocket grains. Ultimately, Section L developed a variety of rockets and manufactured approximately 1,000,000 units using a 3,000-man workforce.

The Naval Ordnance Test Station was located at a disused landing strip in a relatively remote area of California near Inyokern and China Lake. With the pressures of war, a small China Lake Pilot Plant (CLPP) implemented for research purposes turned into a manufacturing complex with 100 buildings. On November 18, 1944, the plant began to extrude dual base propellant motors for 5-inch diameter High Velocity Aircraft Rockets. Its facilities included equipment for extruding, machining, testing, and packaging motors, and a complex of shipping docks. The plant had its own administration, machine shop, maintenance, fire-fighting, and safety sections.

In the years that followed WW II, the Navy refocused the Naval Ordnance Test Station on research and development. In addition to propulsion systems for missiles of all sizes, it researched naval weapons in general. Projects that it undertook include guidance systems, aircraft avionics, and weapon integration. Managers for the operation, along with scientists from the Eaton Canyon operation, were transferred from CalTech to NOTS at the end of the war. NOTS' name changed to the Naval Weapons Center in 1967 and the Naval Air Warfare Center Weapons Division in 1992. The AEC established the Salt Wells Pilot Plant (SWPP) at China Lake in the 1950s, to manufacture the explosive lenses used in implosion assemblies for early fission bombs. The merger of CLPP and SWPP created the China Lake Propulsion Laboratories.

When the Navy assigned NOTS the development of a rocket powered standoff missile for the A-1D Skyraider in 1952, the facility produced a compact solid fuel motor rated at 15,000 pounds of thrust for three seconds – very likely with the assistance of the Allegany Ballistics Laboratory that later manufactured it. The rocket motor contained 320 pounds of double-base solid propellant with a specific impulse of 193 seconds. The rocket that used this motor was the 30.5-inch Rocket, MK 1 MOD 0 or "BOAR," a name variously interpreted as "Bureau of Ordnance Atomic Rocket" or "BOmbardment Aircraft Rocket." Originally considered for use with a MK 12 gun-assembled warhead, the NAVY eventually approved the use of the MK 7 implosion warhead with BOAR. At 15-feet-long, BOAR was about the same size as a MK 7 freefall bomb but slightly heavier at 2,070 pounds because of the propellant in its rocket motor.

BOAR comprised a ballistic case, warhead, rocket motor, and tail. Although Douglas Aircraft Corporation designed the ballistic case and the tail assembly under contract to NOTS, the Century Engineering Company of Santa Ana, California, manufactured these components. BOAR's body shape was constrained by the size of its W7 warhead, by its rocket motor diameter, and by aircraft compatibility considerations. It had a number of screw-in suspension lug locations that allowed delivery by a variety of aircraft. A forward cowl equipped with quick opening latches gave access to its 20-kiloton W7 MOD 3 warhead, a fuzing and arming system similar to the MK 7 MOD 3, an in-flight insertion mechanism, and the missile's batteries. In operation, this compartment was pressurized.

NOTS internally fitted BOAR's forward case for attachment of the rocket motor, which it equipped with 50-inch diameter retractable stabilizer fins. The rocket motor ignited a half second after release and propelled BOAR to a velocity of 700 feet per second. A number of small explosive charges assisted in its separation. As previously mentioned, BOAR was primarily intended for LABS delivery by the A-1D Skyraider. It had four fuzing options: contact burst as a freefall bomb; contact burst under rocket power; timed airburst from release under rocket power with contact backup; and timed airburst from the IP in conjunction with LABS and rocket power (contact backup). With a range of 7.5 miles, BOAR increased the Skyraider's standoff distance with a MK 7 bomb by a factor of five.

Prior to takeoff, ground crew set a timer with sufficient interval to insure the safety of the launching aircraft and the accurate delivery of BOAR. A ram air switch prevented accidental ignition of the rocket motor prior to take off. In the air, differential pressure between an intake in the weapon's nose and a second intake located 14 inches further back closed the switch. At a preselected point on approach to a target, the strike aircraft's pilot closed a switch initiating in-flight insertion of the nuclear core and timer activation. When the timer completed its cycle, a red light alerted the pilot to begin a climb during which automatic equipment launched BOAR when the aircraft reached a predetermined angle. Release started three system timers. A drop timer provided a 1/2 second delay prior to rocket ignition. A safe separation timer provided an additional interval before charging the x-unit. Finally, an adjustable interval timer provided a preselected period between release and detonation. A barium titanite fuze provided contact detonation as a primary option or as backup.

NOTS designed and delivered BOAR for testing in just under a year. The Navy released it for production in 1955, and it reached operational status in 1956. Although it was supposed to be an interim weapon with a life span of three years, the Navy did not remove BOAR from service until 1963. At that time, the condition of BOAR's motors had deteriorated to a point where they were no longer reliable and a precision guided nuclear version of the Bullpup B air-to-surface missile was available as a substitute. The AEC supplied Approximately 225 W7 warheads for use with BOAR.

The Army also adopted rocket systems for use with the W7 warhead. These were the Honest John battlefield rocket and the Corporal SRBM. Honest John, the less sophisticated of the two systems, originated with a 1950 requirement by the Army Chief of Ordnance for a large caliber, field artillery rocket capable of carrying an atomic warhead (Project TU-1007D). In establishing requirements for rocket weapons, a technical panel directed by General John R. Hodge guided the Army. The "Hodge Board" concluded that the Army's highest priority should be the acquisition of technically sophisticated weapons; including missiles, atomic munitions, and all necessary support equipment.

The advantages inherent in a field artillery rocket were significant. It would be mobile, reliable, and amenable to inexpensive mass production. In addition, because it did not require a cumbersome guidance system, it would have low training requirements for its crew and a speedy response time. Finally, contractors could build the missile with off the shelf items, thus shortening its deployment time. The DOD considered this last consideration pressing because of concerns over the progress of the Korean War.

The Army began preliminary design of the XM31 "Honest John" rocket at Redstone Arsenal in the fall of 1950. The Army had amalgamated the former Redstone Arsenal with the Huntsville Arsenal under the Redstone name as the Army Ordnance Rocket Center in June 1949. The Army then transferred the Ordnance Research and Development Division Sub-Office (Rocket) to this site from Fort Bliss, Texas, on October 28. By 1951, the facility featured a machine shop, chemical laboratory, mechanical and hydraulic laboratory, and guidance center. Redstone later saw the construction of a static test tower for engine testing, propellant storage facilities, missile assembly buildings, hangers, and a variety of smaller, ancillary buildings. As various missile programs began to take shape, civilian contractors such as Thiokol and Hercules established their presence on the base, creating a military-industrial complex.

The Redstone staff chose a Hercules X-202C6 JATO rocket as the propulsion unit for Honest John. The reason for this choice was the smokeless character and high specific impulse of its double-base propellant. "Specific Impulse" refers to the pounds of thrust per pounds of fuel burned per second in a rocket motor. The greater the specific impulse, the greater the energy of the fuel burned. Contemporary composite propellants had a specific impulse of about 180 seconds, whereas the specific impulse of a double-base propellant was significantly higher at 230 seconds.

Considerably larger than the X-201A1 used in BOAR, the X-202C6 (M6) was a 23-inch diameter, motor that supplied a thrust of 90,325 pounds for a period of 4.39 seconds. In a suitable vehicle, it could deliver a 1,500-pound payload to a range of 20,000 yards. The diameter of Honest John's rocket motor and its W7 atomic warhead largely dictated the weapon's final dimensions. These established a nose diameter of 30.5 inches, a length of 327.5 inches, and a rear wingspan of 104 inches. The missiles weighed 3,780 pounds empty and 5,800 pounds loaded. After establishing the rocket's basic specifications, Redstone Arsenal contacted the Douglas Aircraft Company for a firm quote on five, Type I, fin-stabilized, aluminum missile bodies to be demonstration-fired as a "proof of concept." Redstone chose Douglas as the prime contractor in part because of the company's experience with X-202 rocket motors.

The most serious concern of the Redstone Design Group was achieving a 300-yard accuracy and a lateral dispersion of not more than 200 yards. It considered two methods of stabilization to achieve the specified accuracy. The first method used tailfins canted at 1.5 degrees in conjunction with eight small rockets to produce a spin rate of 300 RPM. The second method incorporated a simple autopilot that established control during the short engine burn at launch. The Redstone Group chose spin stabilization because of its simplicity. Nevertheless, it developed an autopilot in case the spin method could not establish the required accuracy. JPL had on hand the T53 (M7) spin rockets used for the initial tests.

To determine accuracy during the evaluation tests, a spotting charge simulated the detonation of a 1,500-pound dummy warhead. An integrating accelerometer that measured the total impulse of the rocket's acceleration and a 100-second time fuze supplied by Frankford Arsenal provided arming. Conducted during the summer of 1951, the initial field tests made use of a fixed, temporary launcher with a quadrant elevation of 22.5 degrees and a 30-foot launch rail. Development of a mobile launcher was delayed pending final project approval.

While the tests were not flawless, all five missiles demonstrated a level of accuracy within the guidelines provided. The most successful flight occurred with the first test. The Model 1236F test missile flew 22,495 yards and reached a peak altitude of 8,000 feet. The calculated burst of its warhead (which failed because of timer problems) would have occurred only 91 yards short and 5 yards west of the launcher settings! It is remarkable that Army Ordnance, in conjunction with Douglas Aircraft and the subcontractors that supplied the rocket components, could have designed, built, and tested a missile of this complexity in only 9 months!

Two days after the last demonstration test on August 9, the Army Chief of Ordnance put the Honest John program on a "crash" basis. Under an accelerated development program, the Army purchased 40 more rockets from Douglas. It expended 25 of these rockets in the first stage of the accelerated program. It fired 10 rockets with dummy warheads and 15 more with a blast fragmentation warhead. In addition to the blast fragmentation warhead, the Army developed a flash / smoke practice warhead and a cluster CB (chemical-biological) warhead at the Army Chemical Center, Maryland. The M190 chemical cluster warhead could dispense 356 M139 spherical bomblets filled with agent GB (Sarin) at an altitude of 5,000 feet as an alternate to an atomic warhead. Delivery of cluster munitions at supersonic speeds generally produced poor dispersion, but Honest John could provide 50 percent casualties in a mean effect area (MEA) slightly larger than a kilometer in diameter. This effect was comparable to that of a low kiloton atomic warhead, without the destructive effects of heat and blast.

Redstone Arsenal and Sandia Laboratory also proceeded with the adaption of Los Alamos' 20-kiloton W7 atomic warhead to Honest John. In general, the Army applied the aforementioned pattern of blast fragmentation, chemical-biological, and atomic warhead development to the entire family of tactical missiles it produced during the 1950s, and this helps explain the discrepancy between the number of missiles produced and a paucity of atomic warheads. Los Alamos produced only 300 Honest John W7 warheads against 5,023 M31 tactical missiles delivered to the Army and 2,776 tactical missiles provided to NATO under the Military Assistance Plan.

Reliability tests of the W7 warhead established a maximum linear acceleration of 35 Gs and a maximum spin rate of 120 RPM or 7 Gs radial acceleration for dependable operation. The Army altered the spin rockets used in the accelerated development tests to meet this limitation and moved them forward, just behind the nose section. It also reduced the cant of Honest John's fins to 0.5 degrees. The test firing of the upgraded rockets produced encouraging accuracies.

Two events in Honest John's flight required close attention. The first event was the time needed for the in-flight core insertion of its W7 atomic warhead and the second was its detonation. In the first event, acceleration integrator switches initiated in-flight insertion of the warhead's nuclear core 4 seconds after launch. In the second event, a motor timer that operated several seconds after in-flight insertion supplied power to the warhead's inverter to charge the warhead's x-unit. Then, at a preset time, a ground connection applied to the cathodes of the x-unit's thyratrons triggered the detonators.

A series of Research and Development firings followed the accelerated test program. The success of these tests resulted in the release of plans for the Type II tactical missile and a production order for 2,000 rounds. To insure security of supply, the Army had Douglas Aircraft produce half of these rounds and Emerson Electric Manufacturing produce the other half. Principle changes in the Type II round included a pedestal-type warhead mounting and a hinged nose section that facilitated access to the warhead compartment. A second set of R&D firings resulted in a Type III round that received the designation M31 in December 1953. Although the Army accepted this missile, it had a number of shortcomings. The shortcomings resulted in a Product Improvement Program that saw production of the M31A1 variant in 1954. The M31A1 became the Standard Model and the Army reduced the M31 to Substitute Standard.

One of the principle concerns associated with the M31 rocket was the temperature variation allowable for ignition of its M6 rocket motor. Hercules established acceptable temperature limits between 40 °F and 100 °F. For use in weather conditions beyond these limits, the Redstone Arsenal developed the XM1 thermal jacket in conjunction with General Electric. The thermal jacket could insulate or electrically heat a missile that an ammunition depot had preconditioned to an acceptable firing temperature. The XM1 thermal jacket allowed all-weather use of the missile in temperatures ranging from −40 °F to +140 °F. For the M31A1 missile, GE supplied an improved XM2 jacket.

In order to establish wider firing limits, Redstone had Hercules re-grain the M6A1 motor that equipped the M31A1. Quality control efforts succeeded in establishing safe firing limits between 0 °F and 120 °F. "Project Retest" certified M6A1 motors produced before 1956, resulting in an M31A1C designation for those rockets equipped with motors that passed the project's test regimen. These latter rockets were designated Limited Standard, whereas rockets equipped with the new motors were designated Standard A. Redstone equipped the M31A1 and M31A1C with improved Thiokol T53E1 (M7A1) double-base spin motors and a BA605/U thermocell dry battery for ignition. The new battery improved reliability and reduced servicing requirements associated with the Willard NT6 lead acid battery that was previously in use. Other improvements included an improved nose that could mount a variety of fuze types, as well as a forward bulkhead to accommodate the battery required for "special" warheads. Among the fuzes adapted for operational use was a T1400 contact fuze and a T2039 low burst VT or radar fuze.

Continued product improvement resulted in a final round, introduced in 1959. The M31A2 had M7A2B spin motors and a new M6A2 main motor equipped with an igniter that eliminated the need for an external wiring harness. The introduction of the M31A2 resulted in the retirement of the M31 round and the re-designation of the M31A1 and M31A1C rounds as Standard B. Continued improvement of the M31A2 stopped in 1955 because of the XM31E2 program. This latter program resulted in the 1961 deployment of an Improved Honest John M50, which had a superior motor and a W31 fusion-boosted warhead. The M50 was not adapted for use with W7 warheads, which the AEC withdrew from service in late 1960.

The development of a mobile, rail-type launcher by the Rock Island Arsenal proceeded in parallel with missile development. Rock Island began a feasibility study in 1951 that evolved into the final tactical launcher. Initial specifications required that the launcher be capable of transport using standard vehicles, easy to assemble, and air portable. This led to the consideration of four designs: a self-propelled vehicle launcher, a semi-trailer mounted launcher, a ground emplaced trailer mounted launcher, and a ground emplaced air transported launcher. An evaluation board selected the self-propelled launcher as its final choice.

The XM289 mobile launcher developed by Rock Island Arsenal consisted of a 30-foot launching rail supported by an "A" frame on a 5-ton, 6 X 6, XM139 truck chassis. A six-cylinder Continental R6602 gasoline engine with a 602-cubic-inch displacement powered the XM289, which had a 110-gallon fuel capacity. The launcher incorporated mechanisms for traversing, elevating, and leveling, along with an electrical fire control. Redstone proved the ability of the 30-foot rail to provide adequate guidance by means of five demonstration firings. A number of delays incurred during vehicle testing at the White Sands Proving Grounds resulted in a December 1953 delivery date for the first production launchers. The Treadwell Construction Company manufactured 24 launchers under contract to Rock Island.

In 1954, the Army placed follow-on contracts with Rock Island Arsenal for 44 launchers and the Marine Corps placed an order for six launchers. The Army then expanded its contract to 54 units. A second round of contracts to Watertown Arsenal called for six more Marine and nine more Army launchers. Altogether, contractors manufactured 100 M289 launchers, including the final development launcher tested at White Sands.

In addition to its mobile launcher, an M31 launcher unit required a number of supplementary vehicles to support a fire mission. These included an M329 missile trailer towed by an M55 truck, an M62 wrecker with a 9-ton lift capacity, and an AN/MMQ-1 wind set trailer. The M329 missile trailer was a four-wheel, pole-type vehicle used to tow the missile body. The M55 tow vehicle carried the fins and the warhead in its cargo compartment. The tow vehicle also transported the missile's heating blanket. In 1957, the Army standardized an M78 truck equipped with a tie down and heating unit for the transport of missiles and warheads.

After the ground crew assembled a missile, an M62 wrecker hoisted it from a trailer or truck to the launcher. The entire process took six men about 5 minutes, after which the launcher proceeded to a firing site where the driver parked the launcher's sight mount over a pre-established survey mark. The ground crew then lowered the rear and side jacks to steady the vehicle, and elevated and extended the launching beam to the firing position. After the completion of these tasks, a rocket specialist inserted the motor's igniter, sealed the forward end of the rocket with a closure plug, armed the spin rockets, and connected the firing circuits. At the same time, data from the wind set trailer established ambient atmospheric conditions that assisted in calculating the final aiming parameters, which the gunner and cannoneer laid. Pressing the launch button on the remote firing box sent the rocket on its way. Altogether, it took about an hour and a half from arrival at Battery Assembly Area to the firing of Honest John from a dispersed launch site.

The Army declared its first eight Honest John Batteries fully equipped on June 1, 1954, making Honest John the first tactical atomic missile to reach deployment. Army Headquarters transferred six Batteries to the 7th Army in Europe the following year. The Army assigned these at the Divisional level and posted them:

- 1st Field Artillery Battery, 72nd Field Artillery Group to Kitzingen
- 3rd Field Artillery Battery, 18th Field Artillery Group to Ansbach
- 6th Field Artillery Battery, 35th Field Artillery Group to Schwäbisch Gmünd
- 7th Field Artillery Battery, 36th Field Artillery Group to Darmstadt
- 84th Field Artillery Battery, 30th Field Artillery Group to Hanau
- 85th Field Artillery Battery, 35th Field Artillery Group to Leipheim

The Army assigned three launch vehicles and had a complement of eight officers and 127 enlisted men to each battery. Battery organization consisted of a headquarters section, a communications section, a survey section, a radar section and a fire direction center. Rocket assembly and transport platoons supported the launcher sections.

These units were followed over the next 2 years by ten Honest John Missile Battalions that the Army assigned directly to infantry and armored divisions for atomic support. Similar in overall organization to a Missile Battery, a Missile Battalion included an additional launcher, for a total of four. Both types of unit carried out their training as field exercises supplemented by live firing practice. Field exercises required the dispersal of a missile unit and its 50 vehicles to a central area where the rockets were loaded onto their launchers. Drivers then relocated the mobile launchers to isolated firing locations where the missile's crew simulated a launch. A central fire control provided firing data to each launcher by means of radio or a dedicated communications line. The mobile launchers then left their firing sites to re-arm and to avoid possible counter battery fire.

Germany's Grafenwohr Artillery Range, one of the few areas in Europe large enough to accommodate these long-range rockets, conducted live firing exercises about three times a year. A typical Unit might fire a dozen rockets armed with practice or blast fragmentation warheads over the course of the year. The high point of the year for an Honest John Unit was a Corps Artillery Test that culminated in the firing of a series of live conventional rounds.

The Army deployed Honest John Units to Okinawa in 1957. It then transferred these missile Units to South Korea in January 1958, along with a compliment of M65 atomic cannon. As early as 1956, Far East Command had identified locations at Uijongbu and Anyang-Ni as suitable for atomic operations. Since the nuclearization of the Korean peninsula was in violation of the 1953 armistice, the atomic capable units landed at night and rushed forward to prepared positions in a "fait accompli." The US Army exhibited Honest John rockets for the benefit of the South Korean press at Uijongbu on February 3, 1958.

Honest John's ease of use resulted in the transfer of M31A1 equipped missile batteries to NATO allies in 1957, as part of the Mutual Defense Aid Program (MDAP). Belgium, Denmark, France, Greece, Italy, the Netherlands, Turkey, and West Germany all fielded M31 Honest John batteries. Ironically, before its withdrawal from NATO, France operated three Honest John missile Units in the southern part of West Germany.

In 1959, as part of *Operations Plan 100-3*, the 7th Army assigned two Battalions each of Honest John surface-to-surface missiles, Corporal guided missiles, and M65 atomic cannon to its Northern Task Force (NORTAF) for the support of non-NATO allies in Central Europe. The 7th Army based these units in NATO countries. In the event of an alert, the Units were to move forward under the guise of a training exercise. During times of crisis, they were to move more directly to their deployment areas. Each Unit moved with its specified allotment of atomic munitions. They could procure additional atomic munitions from US Advanced Weapons Ammunition Supply Points (AWASP's), which remained under direct US control.

The Army operated Honest John units armed with M31 type rockets and W7 warheads in Europe and Southeast Asia up to 1960. Of the tactical atomic weapons deployed during this period, it was Honest John in conjunction with airdropped MK 7 bombs that made the most credible threat. The large numbers of mobile launchers available to NATO and the US Army would have been able to deploy quickly to meet a threat by the Soviet Union, China, or their allies. M31 units remaining operational after 1960 carried only conventional warheads.

The story of Corporal, the Army's first nuclear tipped, guided ballistic missile, begins in July 1943. Three British reconnaissance photographs of a V-1 flying bomb and intelligence about a large German ballistic missile caused the Army Air Force to request a technical analysis from noted CalTech aerodynamicist, Theodore von Karman. On August 2, von Karman received a follow up request for a paper on long-range rockets from CalTech's Air Corps Materiel Command liaison officer, W. H. Joiner. Frank Malina and Hsue-Shien Tsien, two of von Karman's graduate students, wrote the requested paper, which they delivered in November. It had a forward by von Karman entitled: *Memorandum on the Possibilities of Long-Range Rocket Projectiles*. The paper envisioned an orderly progression in rocket development that the CalTech Aeronautics Engineering Department could provide to the Army.

Colonel Gervais Trichel, Commander of Army Ordnance's newly formed Rocket Research Branch, received a copy of the CalTech proposal from Robert Staver, the institution's Army Ordnance liaison officer. Although the Army Air Force did not respond to the study, Colonel Trichel requested an expanded proposal along with the promise of 3,300,000 dollars in funding. The lack of response from the AAF probably anticipated the McNarney Directive, which assigned the development of ballistic weapons to Army Service Forces.

Malina and von Karman replied to Colonel Trichel on January 22, 1944, with a proposal to develop a solid fuel rocket with a 10-mile range, which they would follow with a liquid fueled rocket that had a 12-mile range. They further recommended a second phase of supporting studies to investigate ramjet propulsion. Phase 3 envisioned a 10,000-pound missile with a range of 75 miles. After much negotiation for deliverables and facilities, the Army issued a contract to Ordnance / California Institute of Technology (ORDCIT) in June 1944, just after the first V-1 attack on London. At this time, Malina leased 7 acres of land at "Arroyo Seco" near Pasadena,

California, to establish a rocket research facility, for which Theodore von Karman chose the name "Jet Propulsion Laboratory." Although the CalTech team provided rocket research, the scientific community considered the term "jet" far more acceptable at this time. Only "crackpots" experimented with rockets.

Development of the 530-pound, Phase I, solid fuel, Private rocket proceeded quickly and JPL began testing it at Fort Irwin in the Mojave Desert in late 1944. At this time, the 14-ton, V-2 ballistic missile designed by Wernher von Braun began a second reign of terror on London, falling out of the sky at supersonic speeds with a 1,650-pound high explosive payload. This development caused Colonel Trichel to issue a second contract for a long-range rocket to General Electric. The Army dubbed the GE project Hermes.

JPL completed the Private program in April 1944, after testing 41 Model A and Model F solid fuel projectiles. The project achieved its goal of providing JPL with basic information about rocket flight. After the Allies announced "Victory in Europe" on May 8, 1944, the Army constructed a 40-mile-wide by 100-mile-long Army Proving Ground at White Sands, New Mexico. JPL tested the missiles for the final phases of ORDCIT at this facility.

With the success of the Private missile in hand, the Army wanted JPL to move to a much larger liquid fuel missile named Corporal. Malina argued that the lab should build a smaller, less ambitious vehicle first, in the form of a "sounding rocket" to explore high-altitudes. The Army agreed, and JPL began development of the WAC Corporal missile. In January 1945, the US Corps of Engineers acquired the JPL facilities, which became a Government Owned Contract-Operated (GOCO) facility supervised by the California Institute of Technology. Under this new arrangement, JPL continued with the development of WAC Corporal. The Army also assigned it to provide support for all of its new missile contracts.

"WAC" meant "Women's Army Corps," the female branch of the service. The name implied that WAC Corporal was Corporal's "little sister." WAC or "baby" Corporal was an unguided, liquid fuel, 0.4 scale version of the full-scale tactical missile. It measured 16.2 feet long and weighed 690 pounds. JPL launched the prototype missile in September 1946. A 1,500-pound thrust liquid fuel motor that burned a red fuming nitric acid oxidizer and an aniline / alcohol fuel powered it. A "Tiny Tim" solid fuel rocket gave it a boost to provide stability during early flight time. WAC Corporal Models A and B provided much needed basic information about the performance and design of liquid fuel motors as well as answering questions about the aerodynamics, structural integrity, and stability of larger missiles.

Corporal E, a full-scale prototype of the 75-mile range tactical missile, first flew on May 22, 1947. The success of this test resulted in cancellation of the ramjet program. By this time, post-war budget cuts had reduced funding slowing the missile program dramatically. The Army therefore downgraded Corporal from a weapons development program to a research project. This change was quite reasonable, considering the number of technical problems that still required solutions.

Douglas Aircraft produced the Corporal E airframes, which were 30 inches in diameter by 39 feet 8 inches long. JPL built the engines using the same propellants pioneered in WAC Corporal B but stored them in separate tanks connected to a multi-bottle pressurization system. Fully fueled, the missile weighed 9,250 pounds. Burst diaphragm valves started the flow of fuel to the motor, which generated 20,000 pounds of thrust for a maximum duration of 60 seconds. The flow of fuel through the hollow walls of the engine bell cooled the motor. Due to burn-through problems in the throat region, JPL had to redesign the motor to improve its cooling characteristics. Remarkably, the replacement motor weighed only 125 pounds. Along with the new motor, JPL produced a redesigned, 52-jet propellant-injection system.

A rudimentary guidance system supplied by Sperry Gyroscope provided attitude control around three axes during vertical ascent and the powered transition to the missile's ballistic trajectory. The autopilot received inputs from its two Gyrosyn gyroscopes to control roll and pitch. Inputs from an A-12 vertical gyro-controlled yaw. An early pneumatic control system proved unsatisfactory and, after considerable delay, JPL replaced it with an electro-pneumatic design. In the new control system, electrical servos in the tail adjusted four moveable fins. Because airflow exerted insufficient pressure for control at low-speed, JPL placed carbon vanes inside the engine bell. Mechanically connected to the fins, the vanes made attitude corrections during early flight time. Telemetry and radar tracking equipment produced by Gilfillan Brothers informed ground control of the missile's trajectory. Following early testing, JPL placed the operational guidance system in development.

Corporal E also saw implementation of a new launching system that JPL carried over to the semi-mobile tactical missile. Four, 10-foot long, spring-loaded steel struts placed equidistant around the small launch pad provided support at a point 1/5 of the way up the missile's body. After the missile had risen approximately four inches on its trajectory, the struts automatically retracted allowing an unobstructed launch. This reduced stress on the lower missile body and allowed the installation of additional inspection hatches that improved the missile's servicing characteristics.

Because of delays in the parallel Hermes program, the Army chose Corporal for a "crash program" to enter service in December 1950. This followed the successful test of a Russian atomic bomb in late 1949. Although the warheads considered for Corporal were conventional, nuclear, chemical, and biological, the Army chose a W7 nuclear warhead based on the missile's accuracy. Without guidance, Corporal could only produce a circular error probability or CEP of 10 miles radius! With terminal guidance, JPL hoped to increase Corporal's accuracy to a highly--theoretical 300 yards.

Corporal E received a lightweight, transistorized guidance unit midway through its test program. Recognizing the rapid pace of electronic advances, JPL had cleverly factored the interchangeability of guidance units into Corporal's design. The configuration of the last round in the E test series provided the basic pattern for the tactical weapon. It had delta shaped fins and a payload section configured to house a W7 tactical warhead. These changes increased the missile's length to 45 feet 4 inches. At this time, the Army reassigned authority for Corporal from JPL to the Ordnance Guided Missile Center (later Army Ballistic Missile Agency) under the command of Colonel Holger N. Toftoy. The Army located the OGMC at Redstone Arsenal near Huntsville Alabama following its formation in April 1950. Although the first 27 operational Type I tactical rounds were manufactured by Douglas, the Army awarded a 200-missile production contract to the Firestone Rubber Co. Testing of these missiles began on August 7, 1952.

The Army launched sixty-four Type I missiles in contractor evaluation and engineering-user test programs prior to deployment. The operational philosophy for Corporal was a vertical launch followed by a tilt of several degrees to one of a series of pre-programmed "zero-lift" or non-maneuvering trajectories after 4 seconds. At the proper time, valves shut off propellant flow to achieve a desired range. JPL quickly realized that to establish accuracy, a high-speed fuel shutoff valve was required, a piece of hardware that was subjected to a continuous series of improvements. Type I Corporal missiles could engage targets at ranges between 30 and 75 miles with a trajectory that reached a maximum altitude of 135,000 feet. The final speed of the missile as it descended on the target was between 1,500 and 2,500 FPS, depending on range.

Accuracy was further improved in the Type I missile by means of a terminal correction maneuver provided by Corporal's new guidance unit. Unlike modern missiles in which the warhead and missile body separate, these components remained joined until impact in Corporal, thus contributing to trajectory variance. A pair of accelerometers added to the existing gyros provided fine control for this arrangement during descent. Exterior control used a modified SC-584 fire control radar designated AN/MPQ-25 to provide trajectory information while two Doppler antennae measured velocity. Between 95 and 130 seconds into the flight, a computer calculated a correction based on telemetry and radar guidance, which it sent to the missile over the pitch control channel of the radar. Guidance control implemented the correction at impact minus 20 seconds. The maximum adjustment possible was 1,200 meters. Despite this corrective system, only a meager 27.1 percent of test rounds fell inside of a 300-meter radius circle. Mechanical and electronic reliability of the Type I missile was a disappointing 47.1 percent.

Initially, a Corporal Battalion consisted of two Batteries or launchers, 250 men, and about 35 vehicles. The combination of JPL and private industry that designed these vehicles, also designed Corporal's shipping containers. Redstone ultimately contracted Firestone for the production of ground equipment. The Firestone contract included a mobile erector-launcher, missile / warhead transporter, propellant, service, compressor, guidance and computer vehicles, and an electronics shop. The first tactical launch made with this equipment took place on July 7, 1953.

Launching Corporal was a complex process that took 8 - 9 hours. After getting within range of a target, the Launch Commander selected a guidance site. Then, in order for the missile to bear on the target, he had his crew position the portable launch pad at a firing site no more than 600 meters distant from the target line and no more than -200 to +2800 meters from the ground radar. After the ground crew positioned the launcher, they removed the 4,400-pound missile body from its storage container at a service site and placed it on a test bed for assembly and the installation of its fins. At this time, they set up the firing station. Following attachment of the missile body to a horizontal rail, a crew wearing bulky protective clothing carried out the extremely hazardous fueling procedure. After fueling, the missile's systems were "peaked" and nuclear specialists mated the W7 atomic warhead to the missile body. The erector-launcher then transported the 11,400-pound, operationally ready missile to the launch site. The erector-launcher lowered Corporal vertically onto its launch pad placing a registration mark on its body into alignment with the target to insure proper guidance. Following pressurization, the attachment of all necessary umbilicals, and a final check, ground crew withdrew Corporal's erector and servicing vehicles and launched the missile.

The W7 warhead mated to Corporal was fitted with a fast firing x-unit with special circuits that prevented arc over at altitude, an in-flight insertion mechanism, and nickel-cadmium batteries. With the exception of the core insertion mechanism, Sandia made all parts of the warhead's electrical system redundant. The warhead also had two sets of piezoelectric contact fuzes. Acceleration switches installed in the warhead closed 4 seconds after launch and applied power to a pair of Abee radars (the Albert was not yet available) and the thyratron circuits. As the missile climbed, a baro-switch opened at an altitude of 35,000 feet (49 seconds flight time) to prevent flashover of the warhead's x-arm relay. Subsequent closure of the relay on descent started an inverter to charge the bomb's condenser bank. A ground station transmitted the arming and core insertion signals together with range corrections. A short time later, the fuzing radars ranged and transmitted a signal to the x-unit to detonate the warhead. The contact fuze acted as backup.

If the missile passed outside of the fixed, 5-degree azimuth beam of the guidance radar, both guidance and command arm capabilities were lost. The delay in transmitting the arming signal until late in the flight prevented accidental detonation of the warhead if the errant missile fell in friendly territory. Even though Corporal lost guidance if it strayed outside of its azimuth beam, the Doppler radars continued to track its flight. If it was determined that the missile would land in enemy territory, a Doppler command backup system could send an arming signal to allow the warhead to detonate.

The warhead's fast firing system (necessary for use with contact fuzes during high-speed impact) included a probe-triggered spark gap to discharge the warhead's x-unit. It replaced a slower acting mechanical arming switch. Operation of this system required the placement of a 2,000-volt potential across the plates of two thyratrons. A ranging signal from the warhead's radars discharged one of these thyratrons to activate a pulse transformer that ionized the spark gap to discharge into the firing harness. The two radars had to fire in series, thus requiring both radars to produce ranging signals. If the radars did not send a firing signal at the proper altitude, a barometric switch reconnected the radars in parallel after a further descent of 1,000 feet. Either radar was then able to command detonation. The contact fuzes worked by applying a voltage to the second thyratron.

In 1955, the AEC approved an updated adaption kit, a dual motor insertion mechanism, and circuit improvements for the MOD 1 version of Corporal's W7 warhead. This program was cancelled for a time but when it was reinstated, the warhead was renamed the W7 MOD 2. The W7 MOD 3 was fitted with a fuze that provided for radar, timer, or contact operation, and the W7 MOD 4 was equipped with a permissive action link and an external electronic neutron initiator. Los Alamos began production of this last MOD in August 1963.

Training and educational material were considered just as important as the development of the missile and its tactical equipment. Operating a sophisticated missile armed with an atomic warhead was far more complex than slapping a shell into a breech and yanking on a lanyard. The first JPL training school began operation in July 1951 with five personnel each from Ordnance and Field Forces. The Army assigned graduates from the first two classes to the Redstone Arsenal, Alabama, and Fort Bliss, Texas, guided missile schools as instructors. A printed maintenance plan for the guided missiles and their ancillary equipment followed. By March 1952, the Army had formed a Direct Support Company and activated three Corporal Field Artillery Battalions.

In February 1955, the Army sent the 259th Missile Battalion and the 96th Direct Support Company to Germany armed with Corporal XM2 (Type I) missiles. The 246th and 247th Corporal Battalions remained behind at Fort Bliss. The 259th was the only Battalion to see overseas service with this missile. A design flaw in the Type I guidance system allowed a 1 KW transmitter operating on the Doppler frequency to jam it and bring down the warhead unarmed. Recognizing the problem, the Army had extensive improvements made to the Doppler system and radio link as well as to the design of new servicing and launcher / erector vehicles. After the Army procured sufficient missiles and ground equipment to equip six Corporal Battalions in late 1954, they were redesignated M2 (Type II). The Army designated Type II missiles equipped with an advanced guidance set produced by Gilfillan Brothers as Type IIa in 1957. A Type IIb (M2A1) missile with an air turbine alternator and quick-disconnect fins went into production in 1958.

Seventy-eight contractor and engineering-user test firings of Type II missiles took place starting on October 29, 1954. These demonstrated a significant increase in accuracy with 46.1 percent of the rounds falling inside a 300-meter radius circle. Performance reliability also had increased to 60.1 percent. The Army reorganized the structure of Corporal Field Artillery Battalions in 1956. Previously, they had a standard organization with a Battalion HHB, two Firing Batteries and a Service Battery. The Battalion now became a single fire unit organization consisting of a Headquarters and Service Battery (HSB) and a Firing Battery. In the spring of 1956, six of the new Corporal Battalions armed with the M2 missile replaced the 259th in Germany. The Army sent two additional units to Italy. The Army now had twelve Corporal Field Artillery Battalions, with eight deployed to Europe and four kept in reserve in the continental United States. It regularly rotated all twelve units to provide them with live firing training at the White Sands Proving Ground.

In 1958, the Army cancelled a Type III Corporal missile with improved guidance. The cancellation related to the planned deployment of Sergeant, a JPL designed solid fuel missile that rectified many of Corporal's shortcomings. Although extensively redesigned during its history, Corporal remained unnecessarily complex because of its transition from a research vehicle to a tactical missile. This led to poor reliability, slow mobilization times, and a low cyclic rate of fire. A single-launcher SETAF Battalion was able to fire only four missiles during its first 24 hours in action and one every 12 hours thereafter. This assumed that the Battalion fired its first missile at zero hour and did not undertake any intermediate moves. Corporal also needed a large number of trained personnel to support a launcher, was susceptible to electronic countermeasures, and did not meet the dispersal distance between guidance and launchers required for security. The Army began demobilizing Corporal in 1963 and the last Battery ceased operation in June 1964. On July 1, the Army declared Corporal obsolete.

Despite Corporal's limited deployment and short service life, the Army still holds this missile in high regard, mainly because it enabled the Army to enter the technological age of warfare. Prior to Corporal, there was no body of established knowledge in the field of rocketry available to either industry or the military. Manufacturers had to learn about the development and fabrication of missiles, which the Army required to function with a high degree of reliability. In turn, the Army had to develop the arts of contract negotiation, execution, and administration. The Army also had to become adept at technical supervision in order to maintain control over its projects. Beyond this was the need to develop educational programs and facilities to train personnel in the maintenance and operation of new weapons. For these reasons, the Army declared Corporal "the embryo of the Army missile program."

CHAPTER 14
GUN-ASSEMBLED FISSION BOMBS AND "ATOMIC ANNIE"

The initial concept for an atomic bomb was a gun-assembled device – the Model 1850 MK I "Little Boy" bomb that the USAAF dropped on Hiroshima in 1945. The weapon measured 10 feet long by 28 inches in diameter and weighed 8,900 pounds. A classic, but incorrect, description of this device indicates that it was based on a 6-foot length of 3-inch caliber antiaircraft barrel, bored out to an inside diameter of 4.25 inches. Considering that such a barrel had a 6.25-inch outer diameter, the smoothbore gun tube would have had a wall thickness just over an inch thick and would have weighed 1,500 pounds. A cylindrical target-assembly, fastened to the front of the barrel, held a set of uranium target rings. At the other end of the barrel, a breechblock held a uranium projectile or "insert." Three MK XV electrically initiated primers ignited 8 pounds of cordite in the breechblock to send the uranium insert down the barrel to fill the hollow space in the target rings and form a critical mass. Little Boy's arming and fuzing system was identical with that of Fat Man and was located in the space between the gun barrel and the forged steel case of the bomb. Eight baro-ports, that connected to the bomb's barometric fuzes, were located equidistant around a tapered section of the bomb casing just ahead of the box tail.

The target adapter that threaded onto the muzzle of Little Boy's gun barrel held the 24-inch diameter, 5,000-pound target-assembly inside a 1-inch-thick steel cylinder. Steel plates, welded to the target adapter and to the inside of the bomb's casing, held and centered the entire gun-assembly. The barrel and the target-assembly's adapter screwed together to insure a secure fit when the device was fired. The target-assembly consisted of two hollow sleeves that fit one inside each other. The outer sleeve consisted of tungsten carbide inserts enclosed at its front end with a circular tungsten plate. About two thirds of the length of the tungsten cylinder comprised a 6-inch circular opening aligned with the center of the gun tube. This held a number of enriched, hollow, uranium rings that formed the "target."

Attached at the other end of the barrel was a breechblock that contained a solid, cylindrical, enriched uranium insert (the projectile). Behind the insert lay a disk of tungsten carbide and a cylindrical steel plate that completed the geometry of the target-assembly after insertion. Los Alamos cast the insert from six separate uranium disks that measured 4 inches in diameter. The insert had a length of 7 inches and contained 56 pounds of uranium. A boron sabot and a 1/16- inch-thick steel jacket held it together, bringing the total diameter of the insert to 4.25 inches – the same diameter as the interior of the smoothbore gun barrel. Four bags of cordite propellant were the last items added to the breech before a screw-in steel plug sealed it from the rear of the weapon.

When ignited, the propellant in the breech accelerated the uranium insert to a velocity of 1,000 feet per second. This velocity was only 1/3 of the requirement for the original plutonium gun. Since the mass of a gun varies with the square of its projectile velocity, the reduction in velocity associated with its projectile allowed a potential decrease in gun mass to 1/9 that needed for the original design. The mass of the final gun-assembly was actually greater than this, allowing a large margin for reliability. Impact with the target rings stripped off the projectile's

The Silverplated Boeing B-29 Bomber – Enola Gay – which dropped the first atomic bomb – Little Boy. The target was Hiroshima, Japan. (Credit: National Archives; Color Original)

The MK I "Little Boy" (L-11) gun-type atomic bomb in a pit on Tinian Island waiting to be loaded onto the B-29 "Enola Gay". (Credit: LANL Photo 114)

Testing the circuits of a Little Boy bomb. Note the connections to the arming plugs on the top of the bomb. Electronics such as batteries, radars, and barometric fuzes were carefully stored around the central gun barrel. (Credit: National Archives)

The 280mm, M65 "Atomic Annie" cannon. Its shell had a yield of 15 kilotons. (Credit: USA)

An M65 cannon fires a 15-kiloton atomic shell at Frenchman Flat on the Nevada Test Range. (Credit: NNSA Photo CIC 0318564; Color Original)

Left, a 280mm (11-inch) diameter nuclear shell. Early shells were equipped with a 15-kiloton W9 warhead, which was later replaced with a W19 warhead of similar yield. (Credit: USA) Right, a 16-inch diameter W23 naval shell, which contained a W9 nuclear mechanism. (Credit NAM)

An aerodynamic test shape for a MK 8 bomb. The ring tail was later replaced with fins in the MK 8 MOD 2. The MK 8 had a single yield in the range of 15 kilotons. (Credit: NAM)

A Mk 8 bomb carried under the wing of a McDonnell F2H Banshee fighter-bomber. Note the nose cap that provided streamlining and the fairing for the fuzing system that ran down the length of the bomb. The F2H was unusual in that it carried its nuclear bomb under its right wing. (Credit: Cropped USAF Photo)

An aerodynamic test shape for the MK 11 bomb. The MK 11 replaced the MK 8 on a unit for unit basis starting in 1956. The MK 11 had a 15-kiloton yield using the internal warhead components of the MK 8. Its Navy designation was MK 91. (Credit: NAM)

A four-man T4 atomic demolition squad advances through the countryside. Credit: USA)

steel and boron sabot and drove the boron steel plug out of the target rings into a recess in the nose of the bomb where it acted as an anvil to absorb the shock of impact. Altogether, the insert had to exert a force of 70,000 pounds to force itself into the uranium target rings, where friction held it fast. To assist in starting the chain reaction, four mechanical neutron initiators known as "Abners" were included in the design. These devices had less active material than Fat Man's Urchin initiator in order to avoid predetonation by a stray neutron.

John Coster-Mullen, a dedicated atomic researcher who had the opportunity to examine a MK I bomb (less its nuclear components) before the DOD had it gutted for security reasons, recently challenged the traditional idea that the gun-assembly fired a solid insert at a stationary stack of enriched uranium target rings. In his correct description of Little Boy, the bomb's gun fired a hollow stack of enriched uranium target rings down its barrel to surround a solid uranium insert, centered in the middle of the target-assembly. A
locknut fastened to a 1-inch diameter steel bolt, which protruded through the end of the target-assembly and out through a 15-inch diameter, 5-inch-thick steel forging, held the insert in place. The steel forging formed the nose of the bomb.

Placing the stack of target rings in the breech allowed the use of more fissile material than would have been possible if the rings had been located in the tamper, as described in the classic scenario. The neutron reflectivity of the tamper would have raised the criticality of the uranium in the target ring stack to unworkable levels in that configuration. The new scenario also makes room to place the four neutron initiators into the annular space between the insert and the tungsten carbide tamper. The projectile assembly weighed 190 pounds and had brass bands to provide a tight gas seal with the barrel.

Coster-Mullen's description further states that Little Boy's smoothbore gun tube was 10.5 inches in diameter with a 6.5-inch (smooth) bore and measured 68 inches long. Los Alamos designed the gun to withstand a maximum pressure of 40,000 psi. The Laboratory likely based the design on a 5-inch naval gun barrel. In fact, the Naval Gun Factory under the supervision of the Bureau of Ordnance manufactured Little Boy's barrel and breechblock. A drawback to the larger diameter barrel would have been its weight. Other drawbacks to this design would have been the need for more propellant to accelerate the larger and heavier target ring stack and the effect on critical mass by having a 1-inch diameter steel bolt run through the center of the fissile assembly.

The rear 33 inches of the gun tube was occupied by a screw-in breech plug requiring 16 turns to seat, a space for cordite propellant, and for the projectile. The 16 1/4-inch long projectile consisted of nine fissile rings backed by a 3.25-inch-thick disk of tungsten carbide, and a 5.5-inch-thick steel disk. The target rings measured 6.5 inches and had a total length of 7 inches. Three brass studs in the barrel prevented the forward acceleration of the projectile in the event of an accident. It took the full force of the cordite charge to sheer off these studs and allow acceleration of the projectile. An inch-thick steel plate, bolted in place over the rear of the barrel, isolated the breech.

The front 6 inches of the gun barrel had threads that accepted the target adapter. Forged from high-strength K46 steel, the 1-inch-thick target adapter had an inner diameter of 13 inches and abutted the steel nose plug. Each end of the target adapter held a 13-inch diameter, 6 1/2-inch-thick, steel plate with a 6.5-inch diameter central boring. The forward section of the target adapter / tamper had four slots that vented air and gases into the bomb casing as the projectile traveled down the barrel.

The target adapter contained a tungsten carbide sleeve made from four inserts with a length of 13 inches each and a total weight of 680 pounds. The tungsten carbide sleeve had the same 6.5-inch diameter central cavity as the steel discs. Into this cavity, at the nose end of the assembly, was inserted a 6.5-inch diameter, 7-inch-long malleable steel disk or "anvil" that helped to absorb the impact of projectile. The bomb's heavy steel nose plate absorbed the projectile's momentum. After the malleable steel cylinder came a 3.25-inch-thick tungsten carbide disk. Both the malleable steel and tungsten carbide disks had 1-inch diameter holes in their center to allow passage of the rod that anchored the enriched uranium insert in the target-assembly's central cavity. The combination of the dense, heavy, tungsten carbide cylinder and the steel target-assembly acted as an inertial tamper and a neutron reflector for the assembled core.

The 141 pounds of enriched uranium used in Little Boy's core represented all of Los Alamos' available stock. Oak Ridge had enriched about 110 pounds to 89 percent U235, and the remainder to 80 percent U235 – an average of 83.5 percent U235. The more highly enriched uranium was included in the insert, which had an average enrichment of 86 percent, whereas the uranium rings had an average enrichment of only 82 percent. In a homogeneous, un-reflected sphere, 141 pounds of 83.5 percent enriched U235 represents about 1.2 critical masses. The neutron reflectivity of the tamper would have increased this value to about 3 critical masses. The tamper also decreased the disassembly time of the core. When used in combat, the MK I produced a yield equivalent of 15 - 18 kilotons, an efficiency of 1.5 percent. This was only 1/10 the efficiency of a MK III plutonium implosion bomb, a factor that resulted in the relegation of gun-assembled designs to special applications after WW II.

This cautionary tale of Little Boy's interpretation demonstrates the spotty reliability of secondary sources in researching atomic weapons, or any subject for that matter. That is why in the production of this book, the Author has tried to rely on actual DOD or DOE documents as primary material, using secondary sources only when no other resource was available. Secondary resources were restricted to documents from researchers (such as Coster-Mullen) whose work has demonstrated a high degree of reliability.

Although the AEC abandoned the development of gun-assembled weapons after WW II, the Navy remained interested in their continued development as a penetrator weapon suitable for the destruction of hardened structures such as submarine pens. It also wanted a bomb it could drop into a group of ships to explode under water. The Baker Shot of Operation Crossroads had shown that underwater explosions were far more effective for maritime attacks than surface bursts. The reason the Navy chose a gun-assembled mechanism for these tasks was its small diameter and its robust nature. Despite the Navy's interest in gun-assembled weapons, SAC discouraged their development because the design was inefficient in its use of fissile material, which SAC thought better employed in strategic weapons for its bombers. As well, Los Alamos showed no inclination to participate in a penetration bomb project because it was otherwise engaged in implosion research.

The Military Liaison Committee, therefore, gave the Navy permission to develop an improved, gun-assembled weapon through the auspices of the Research and Development Division of its Bureau of Ordnance. A key design feature was the survival of the nuclear assembly under extreme impact conditions. The Naval Ordnance Test Station, Inyokern, California, developed the afterbody and nose to meet aerodynamic and water entry requirements and the Naval Powder Factory, Indian Head, Maryland, tested the burning rates of potential propellants. The Naval Proving Ground, Arco, Idaho, conducted reduced and full-scale aerodynamic tests, the Naval Ordnance Plant, Pocatello, Idaho, manufactured the warheads less their nuclear components, and the Naval Proving Ground, Dahlgren, Virginia, conducted a variety of other tests. In addition, the Naval Air Development Center, Johnsonville, Pennsylvania, developed special aircraft installations for AJ type bombers, and the Naval Administration Unit at Sandia Base operated the aircraft used for the bomb's drop tests. The Navy designated the bomb "Elsie" (LC) because it was the next step in evolution after Little Boy (LB). Some authors indicate that LC stood for "Light Case" or "Little Child," when the size of the bomb was compared with the MK I.

In order to insure the bomb's reliability after impact, the Naval Proving Ground set up an elaborate test program in which it dropped several hundred models from half size to full-scale on a variety of surfaces that included concrete, soil, and water. Through this program, the Naval Proving Grounds found that the initial design requirements for the bomb were insufficient. The Bureau of Ordnance had to increase the bomb's impact survivability from 7,700 Gs to 300,000 Gs. The Proving Grounds found the final design to be remarkably robust at its terminal velocity of 1,500 feet per second.

As it became apparent that the Navy was successfully developing a penetrator bomb, Los Alamos stepped in because the AEC had designated it the sole institution for manufacturing and stockpiling nuclear weapons. The Division of Military Application proposed that Sandia Base design the airplane release mechanism, assist in drop tests conducted at the Salton Sea Range, and develop criteria for fuzing. The DMA also requested that Sandia design the bomb's test and handling equipment and produce a crew-training program. Los Alamos and Sandia agreed to this proposal on February 1, 1950, and, to carry out this mandate, Los Alamos formed Ordnance Division C at Sandia Base in April 1950. At this time, the penetration bomb was designated TX8. Sandia then proposed the formation of an inter-laboratory steering committee, similar to a committee that was supervising the TX5 project. Upon acceptance of this proposal, Sandia formed a TXG (gun) Committee. At the same time, it renamed the TX5 Committee to TXN (nuclear), which assumed the broader mandate of developing implosion weapons.

Although the Bureau of Ordnance initially intended to produce a slightly modified Little Boy bomb, it made wholesale changes after it discovered many weaknesses in the original design. The Bureau discarded much of the heavy tamper that Los Alamos had deliberately added to the MK I to insure wartime reliability. It also shortened the gun-assembly's barrel. A proposed W8 warhead for the Regulus missile measured only 63.85 inches long. The design of the TX8's gun-assembly also reversed that of Little Boy, by firing the insert into the target ring stack. This allowed the use of a smaller diameter barrel. The use of highly enriched uranium and the reduced neutron reflectivity of a downsized tamper allowed this change in configuration. In fact, the erroneous classic description for the MK I was likely derived from the TX8 gun mechanism. The reworked gun-assembly reduced the TX8's diameter to 14.5 inches and its weight to 3,260 pounds.

The TX8 took advantage of a progressive burning propellant to achieve the muzzle velocity needed to prevent predetonation in its shortened barrel. Reactivity insertion time for the slug was about 1.0 - 1.5 milliseconds. The warhead achieved criticality even before the slug entered the target. A thin, high-strength barrel that reflected a minimum number of neutrons, a high insertion speed, and a neutron absorbing boron sabot around the slug would all have been important considerations in the bomb's final design. Reportedly, the insert and target ring assembly contained 145 pounds of enriched uranium. A "Cell" internal neutron initiator replaced the four Abners used in Little Boy. Los Alamos designed the initiator and the enriched uranium components to withstand ground and water impact and still function reliably.

The Navy equipped the TX8's firing system with three redundant pyrotechnic fuzes that the ground crew could set to detonate over a 90 - 180 second interval. As fuzing elements became available during the development program, the Naval Proving Ground fired them out of a 14.25-inch caliber railway gun to test their durability. One fuze was located in the nose of the bomb and the other two fuzes were located on its sides. Each fuze attached to a tape that ran through a primer before attaching to the aircraft. Upon release, the tapes pulled the fuzes into contact with their primers, which ignited them. The TX8 had no need for batteries because aircraft power actuated the primers.

The fissile assembly was composed of an enriched uranium insert and a target made of four enriched uranium rings. The MK 8, the TX10 airburst bomb (cancelled), the MK 11 bomb, the W9, W19, W23 projectiles, and the T4 Atomic Demolition Munition all used the same fissile components. In fact, the AEC stored the fissile components without regard for which weapon they were for use in. The yield for all of these weapons was in the range of 15 - 20 kilotons based on the W9 device tested as Upshot-Knothole Grable. A certain amount of yield variation resulted from the nature of the tampers used in these devices – freefall bombs had room for heavier tampers than did artillery shells. Also, some devices fired the rings at the insert and some fired the insert at the rings. Jangle Uncle was a weapons effect test of a sub-surface burst that left a crater measuring 53 feet deep and 260 feet wide from a warhead with a 1.2-kiloton yield. The 17-foot depth of burial acted as a scaled test of a 23-kiloton, ground-penetrating, gun-type weapon like the MK 8. Analysis of the test results indicated that such a weapon would leave a crater measuring 700 feet in diameter and 140 feet deep.

Designed primarily for internal carry, the blunt nosed MK 8 MOD 0 (TX8X1) bomb measured 116 inches long and 14.5 inches in diameter. It broke down into a tail assembly, a nose assembly, and a clamping ring that held the two components together. The nose contained the gun-assembly, tamper, and the projectile (insert). Early versions of the MK 8 had a 25 1/4-inch diameter ringtail to provide stability during freefall. This allowed an aircraft to drop the bomb from altitudes between 500 and 50,000 feet. Its blunt, heavy, nose forging helped it to pierce 40 feet of hard sand, 60 feet of loam or 100 feet of clay at is maximum impact velocity of 1,500 feet per second. The MK 8 MOD 0 began production in November 1951 and it entered the stockpile in January 1952.

For internal carry, the ground crew installed the MK 8 MOD 0 bomb on a T-shaped T-23 "saddle" that remained aboard the delivery aircraft after the bomb was released. The saddle contained the switches and electrical connectors that operated the bomb's primers. In order to arm an internally carried bomb, a blank breech plug was unscrewed from the nose of the bomb to allow the insertion of the enriched uranium projectile and the propellant charge. This indicates that the direction of firing was the reverse of Little Boy, which placed the breech of its gun tube at the rear of the bomb. A member of the aircrew then inserted a breech plug that connected to the delivery aircraft with an arming wire. Unfortunately, available photographs and examples of the MK 8 (such as found at the National Atomic Museum) are aerodynamic "drop shapes" that were not equipped with the breechblock access ports found in the actual bomb.

In 1950, early in the project, the Navy decided to adapt the TX8 for external carry at low speeds and designated this project TX8' (TX8 Prime) or MK II Elsie. The MLC ordered the Bureau of Ordnance to carry out this project with a minimum of design changes. This decision was expedient because the Division of Military Application had placed a new TX11 bomb in development for transport at high speeds. The agency expected it would take some time to develop the TX11 and a modified TX8 could serve in the interim.

In order to streamline the blunt nose of the MK 8 MOD 0 for external carry, Douglas Aircraft designed a frangible nose cap for the TX8' that protected the bomb's fuzing elements and increased its length to 132 inches. In late October 1951, BuOrd divided the TX8' program into the TX8X1 and the TX8X2 programs. The TX8X1 remained the basic TX8' whereas the Navy slated the TX8X2 casing for improved aerodynamics. In order to mount the TX8X1 bomb externally, ground crews installed it in a streamlined T-28 saddle. Like the T23 saddle, the T28 contained the switches and electrical connectors that operated the bomb's primers. BuOrd tested the T-28 with AD4 aircraft but assigned it specifically for use with the F4U5. The changes to the TX8X1 required that an aircraft take off with the bomb in an armed condition. It is not clear whether the MK 8 MOD 0 bomb with nose cap (TX8X1) ever entered the stockpile or whether the T-28 carried the blunt nosed bomb in its stead.

In the TX8X2, Sandia moved the T28 saddle functions to the conical nose of the bomb and hung it on a MK 51 bomb rack with two hooks separated by the standard 30 inches. When contractors failed to produce a suitable arming fuze for the bomb's nose cap, Sandia redesigned the MK 51 bomb rack to accommodate the bomb's arming system. The TX8X2 used MC215 arming control equipment instead of the T28 controls. The MK 51 bomb rack increased the clearance of the MK 8 from the ground. A quick connect system allowed an emergency bomb drop with the tapes in place to prevent ignition of the fuzes. To improve aerodynamics, Sandia installed fairings over the bomb's side fuzes. The fuze tapes entered the bomb through the fairing to protect them from the airstream. The Navy named this new arrangement the T31 fuze. In the event of a crash, it would have taken the fuzes about five minutes to cook off in an ensuing fire, allowing the pilot to escape from the wreckage. Because Sandia decided that these changes related to the fuze and not the bomb, the new weapon was designated MK 8 bomb with MOD 1 fuze instead of MK 8 MOD 2. It entered the stockpile in September 1953.

Because of a Nuclear Safing Program begun in 1950, Sandia later equipped the TX8X2 with an in-flight insertion system in its extended nose. Early drops of the TX8X2 resulted in severe pitching of the bomb due to air loading when released from an aircraft. This may have reflected the weight of the in-flight insertion system in its extended nose. BuOrd resolved the pitching problem by discarding the bomb's original ringtail and replacing it with three tailfins. The TX8X2 with automatic in-flight insertion entered the stockpile in May 1955 as the MK 8 MOD 2.

BuOrd also developed a follow-on TX8X3, which became the MK 8 MOD 3. Intervening drop tests into water at the Salton Sea Test Site had stimulated concerns about the possibility of water entry into the gun barrel of the MK 8. In addition, there were concerns about propellant temperature limitations. Inasmuch as BuOrd had approved the original propellant specifications some time earlier, it decided to defer this problem for solution with the TX11 Bomb Committee. Sandia solved the water leakage concerns by replacing the bomb's cork gaskets with O-rings. The AEC and BuOrd released the updated MK 8 MOD 3 to the stockpile in November 1955. It was equipped with a "Phoebe" internal neutron initiator.

In addition to the MK 8 penetrator, the DMA requested that BuOrd initiate an airburst gun-assembled bomb project on January 21, 1949. The release of the TX10's military characteristics by the Military Liaison Committee on August 17, 1950, was followed by a BuOrd announcement on November 3 that Sandia would develop the bomb. The MLC also requested that Sandia consider the TX7 implosion bomb for the same purpose. On November 18, 1950, Sandia released the lightweight gun-assembled airburst bomb's technical designation as TX10, more popularly known as "Airburst Elsie" (LC) and "Little Freddie" (LF). This begs the question whether there were LD and LE gun-assembled projects.

BuOrd predicated its announcement that Sandia would develop the lightweight gun-assembled bomb on its involvement in the formulation of the MK 8's military characteristics. It had already made a study of lightweight designs. Sandia therefore announced that North American Aviation would develop the case, its supports and pylon whereas the lab would develop the fuzing and firing system, conduct test drops, provide handling equipment and act as project coordinator. The DOD intended the bomb for use with fighters, dive-bombers, and light bombardment aircraft. Navy and Air Force aircraft would employ the bomb in high-altitude attacks that included level, glide, toss, and dive bombing.

The original military characteristics called for a maximum diameter of 14 inches and a weight of 1,200 pounds. Sandia replied that a diameter between 14 and 18 inches and a weight of 1,750 pounds were more realistic. The Bureau of Ordnance nevertheless encouraged Sandia to develop a "quick and dirty" design with a lightened barrel but without nuclear safing. Unfortunately, the final design reached a 17-inch diameter and a weight of 2,000 pounds, which characteristics were little different from the more efficient MK 12 implosion bomb which was nearing the end of its development. For this reason, the Sandia Weapons Development Board cancelled the TX-10 on May 7, 1952.

The MK 11 (Navy MK 91 MOD 0) penetrator replaced the MK 8 MOD 3 bomb on a unit for unit basis, starting in 1956. The MK 11 originated as a concept for a missile warhead during a conference held at Sandia in March 1950. Later, the Department of Military Application showed an interest in a gun-assembled bomb with nuclear safing that a high-speed aircraft could carry externally. Although the MK 8 was already in development, a redesign of its blunt nose would have required a lengthy effort. Thus, on April 17, 1950, the Military Liaison Committee released military characteristics for a bomb that had nuclear safing and that a high-speed aircraft could carry externally at speeds up to Mach 1.2. The DMA then requested that BuOrd develop this weapon with assistance from Sandia, fitting the design to appropriate aircraft. This proposal was accepted and the bomb was designated TX11.

In order to meet impact requirements for the TX11, BuOrd took a two-pronged approach to its development. It assigned Program A personnel the task of providing a weapon that could survive impact with reinforced concrete at a velocity of 1,500 feet per second and it assigned Program B the task of providing a weapon that could survive impact against hard targets at a velocity of 2,700 feet per second. This latter weapon was at first intended for use with missiles before the DMA dropped its requirement for a missile warhead in June 1952. Ultimately, BuOrd produced a single version of the TX11 that could survive impact with a hard target at a velocity between 2,000 and 2,200 feet per second. At this time, the Bureau finalized the bomb's dimensions.

The MK 11 measured 14 inches in diameter, 146 inches long, and had a fin cross-section of 28 inches. It weighed 3,350 pounds, about the same weight as the MK 8. It could pierce 22 feet of reinforced concrete, 90 feet of hard sand, 120 feet of clay, or 4 inches of armor plate before detonating. It was also able to survive a ricochet off hard rock. Its detonating system consisted of two MK 250 time-delay fuzes of a pyrotechnic design similar to those used in the MK 8. It used the same in-flight insertion mechanism for its nuclear cores as the MK 8 MOD 3.

In order to accommodate both low and high-speed delivery aircraft, BuOrd produced two interchangeable fiberglass nose caps for the TX-11, one having low subsonic drag and higher but acceptable supersonic drag for low-speed aircraft, and the other having acceptable subsonic drag and low supersonic drag for high-speed aircraft. The MK 11 bomb was equipped with two suspension lugs to hang it on the same MK 51 bomb rack used to suspend the MK 8 MODs 2 and 3. A nose cap was not necessary for internal carry. The withdrawal of the MK 11 in 1960 ended the use of gun-assembled bombs in the American nuclear stockpile.

In addition to penetration bombs, the Army desired atomic artillery shells. Because the smallest plutonium implosion system in the early 1950s was 30 inches in diameter, it was completely impractical for use in an artillery role. The Button Committee of the Los Alamos Scientific Laboratory, therefore, decided to modify a gun-assembled uranium bomb into a suitable artillery projectile. Although the Army had hoped that its T92 240mm (9.5-inch) self-propelled howitzer might prove suitable for use with atomic shells, LASL indicated that a 280mm bore would be required to meet the Army's yield and range requirements for a 15-kiloton MK 9 shell. A boosted MK 9 shell with a 75-kiloton yield was also proposed; but was found to be impractical for use by short-ranged battlefield artillery.

Watertown Arsenal, Massachusetts undertook the technical supervision for the design and the manufacture of the 280mm M65 atomic gun. Because of the urgency arising from the explosion of the first Soviet atomic bomb, the Army decided in November 1949 to bore out an existing 240mm design in preference to producing a new barrel. The resulting 22-ton T-131 barrel measured 42 feet 8.5 inches (45.5 calibers) long, had 72 lands and grooves, a muzzle velocity of 2,500 feet per second, and an accuracy life rated at 300 rounds. The breech utilized a stop thread, interrupted screw with a T95 dual electrical-percussion firing mechanism. The 280mm shell and a separate 158-pound propellant charge were loaded by means of a powerful hydraulic rammer. Four propellant increments were available to extend the range. A small crane attached to the gun carriage lifted shells into place for loading. The Watervliet Arsenal, New York carried out production of the barrel and its breechblock.

The prototype M30 carriage and barrel mounting assembly were produced by the DRAVCO Corporation for the Pittsburgh Ordnance District and were loosely based on the German "Leopold" K5 280mm railway gun. The Army captured several of these guns, including one of the infamous "Anzio Annies," and removed to the Aberdeen Proving Ground for testing at the end of WW II. Distinctive characteristics of the M30 carriage were a rectangular mount with a large circular firing base and two recoil mechanisms produced by R. Hoe & Company, New York. The first recoil mechanism absorbed the "kick" from the barrel and the second (an innovation) absorbed the forces of primary recoil by travel of the entire weapon (90 percent of carriage weight) along turntable slides. A hydro-pneumatic mechanism attached to the carriage returned it to battery. The carriage was 38 feet 5 inches long, 12 feet 2 inches wide and 10 feet 3 1/2 inches high. It had a road clearance of 36 inches and a weight of 94,000 pounds when emplaced with its barrel.

Based on a ball and socket mechanism, the T131 barrel had an elevation of 55 degrees and a fine traverse (right and left) of 7.5 degrees. Rotating the float provided 360 degrees of coarse traverse. The Frankford Arsenal, Philadelphia, Pennsylvania, undertook adaptation of the fire control system. A single lever powered by an auxiliary generator hydraulically operated elevation of the barrel. In the event of a hydraulic failure, manual elevation was available. Hand wheels and a gear train operated the traversing mechanism. An M1A1 elevation quadrant, M12A7c telescopic sight and M26 or M27 and M28 fuze setters provided on carriage sighting. An M1 aiming circle and M48 B.C. telescopic sight similarly provided off carriage sighting.

The T10 4 X 4 heavy artillery transporter units, M249 front and M250 rear, were designed by the by the Detroit Arsenal, Detroit, Michigan and manufactured by the Kenworth Motor Truck Corporation of Seattle, Washington. The leading transporter was a cab forward design while the trailing transporter had its cab at the rear. A telephone system provided communication between drivers. Both transporters were powered by air-cooled, 375 horsepower, six-cylinder, Ordnance-Continental gasoline engines and together they could transport the carriage at speeds up to 35 miles per hour over an unrefueled range of 20 miles. Specially engineered hydraulic hoists allowed the transporters to pick up and lower the carriage assembly. The entire unit, made ready for transport, measured 84 feet 2 inches long and weighed 166,638 pounds. The design allowed forwards, backwards, or even sideways movement of the gun carriage because the transporters could turn at right angles to the gun and move in parallel. Emplacement of the carriage, ready for firing, took 8 - 12 minutes. Entrainment, ready for travel, took about 15 minutes.

The Army held an official presentation for the M65 at the Aberdeen Proving Grounds, on October 15, 1952. Presiding, were Brigadier General J. L. Holman, Aberdeen's commander and Major General E. L. Ford, Chief of Ordnance, US Army. After an address by General J. Lawton Collins, Chief of Staff, US Army, gunners fired the M65 with conventional shells to compare it with 240mm howitzers and eight-inch guns. The Army's 52nd Field Artillery Group began training soon afterwards at Ft. Sill, Oklahoma. President Dwight D. Eisenhower was sufficiently impressed with the M65 that he had one towed in his inaugural parade on January 20, 1953, like a triumphal Roman commander showing off an elephant from some foreign conquest. (He also had three elephants.)

The high point for the M65 came at 8:30 local time on May 23, 1953 when the Army fired a live atomic round at Frenchman's Flat on the AEC's Nevada Test Site as Upshot-Knothole Grable. The gun crew first fired several conventional rounds to zero the weapon. They then fired the atomic round, which traveled a distance of 11,400 yards before a time fuze exploded it at a height of 524 feet. Placement was only 86 feet west, 137 feet south, and 24 feet above the desired burst point, thus proving the accuracy of this weapon. The yield of the explosion was equivalent to 15 kilotons of TNT, slightly more than the 14 kilotons predicted. After the Grable shot, the press popularized the field piece as "Atomic Annie."

Robert Schwartz of the Picatinny Arsenal, Dover, New Jersey, produced the preliminary design for the first atomic shell in 1949. In a private room at the Pentagon, he laid out the basic specifications for the T124 280mm (11-inch) shell in a period of two weeks. Presumably, the AEC had supplied him with specifications for the shell's W9 atomic payload. The product of his labor was a projectile that measured 54.5 inches long and weighed 805 pounds. He later provided the design for a 600-pound conventional shell. Samuel Feltman, Chief of the Ballistics Section of the Ordnance Department's Research and Development Division championed the atomic projectile until its final approval by the Pentagon.

The W9 nuclear warhead contained in the T124 shell was no less a technical marvel than the M65 gun that fired it. Los Alamos Scientific Laboratory designed it and Sandia Laboratory supplied an evaluation program, certification and developed a training program for gun crews.

The Army Ordnance Corps produced the war reserve, training, and spotter shells, less their nuclear components, field handling equipment, production tooling, and packaging. Picatinny Arsenal developed the fuze.

The Artillery Test Unit at Fort Sill, Oklahoma, carried out preliminary testing of telemetry equipped dummy nuclear and conventional rounds in the spring of 1953. The tests established a range of 31,400 yards for the T124 conventional shell and a range of 26,300 yards for the heavier atomic projectile. In 1958, the T124 atomic shells, designated M354, were remanufactured into T4 Atomic Demolition Munitions and replaced on a round for round basis with an updated T315 projectile that contained a lighter W19 warhead. The new 600-pound shell had a range of 32,700 yards. The Army described the accuracy of projectiles fired from an M65 gun as four times more accurate than pre-war mobile artillery.

The uranium for the projectile inserts and target rings of the MK 8, MK 11, W9 / W19 and W23 warheads was most likely enriched to the modern "weapons grade" standard of 93.5 percent U235. The mass of enriched uranium in the core would have weighed 145 pounds – sufficient for its 15-kiloton yield. At 70,000 dollars / pound in 1950's dollars, the material cost for a core would have rounded out to 10,000,000 dollars per nuclear assembly, against the 800,000-dollar cost of an M65 gun, complete with transporters. Eighty W9 warheads were manufactured over the period April 1952 - November 1953 at a cost exceeding 3/4 billion dollars for material alone. The 40 warheads for the MK 8 aerial bomb would have brought the total cost of gun-assembled nuclear components to 1 billion dollars.

In preparation for firing, the M65 gun crew inserted four fissile target rings into a tungsten carbide tamper at the base of the shell and screwed two circumferentially-threaded, cone-shaped, beryllium-polonium-nickel-barium "Phoebe" initiators into a locking plate behind the target rings, with their tips protruding into the center of the target rings. The then enclosed the target rings to prevent movement and screwed the gun-barrel assembly into the enclosure. The breech of the gun barrel assembly was located in the ogive nose of the shell to optimize the barrel's length. This arrangement maximized projectile velocity and provided room for the tamper that surrounded the target rings at the base of the shell. Next, the gun crew loaded the insert projectile, a powder charge in a perforated can, and a percussive detonator into the gun barrel. Two detents restrained the projectile from forward movement. The detents retracted when the shell began spinning after the gun crew fired it.

The final act of assembly took place when a crewmember inserted an assembly composed of three MT-220 mechanical time fuzes into the shell's tip. The gunner then sat astride the shell on the gun's loading tray and set three dials corresponding to each fuze with a wrench. Initiation of the propellant charge was by means of a percussion detonator triggered by the time fuzes. As a safety measure, a "safety gate" prevented activation of the detonator until centrifugal force caused by the rotation of the shell closed it. The fuze assembly was not equipped for impact detonation.

The Army likely based the assembly procedure for the W9 warhead on the T4 Atomic Demolition Munition, which was in service from 1957 - 1963. Originally conceived as an ADM for offensive operations, the T4 came as a base assembly; a gun-barrel assembly with a preloaded fissile projectile and propellant charge; and four individual fissile target rings. BuOrd may have incorporated a minimum sized tungsten carbide or steel tamper at the target end of the gun barrel. These items were stored with assembly equipment in four, 40-pound man-packs. The idea was that each component was portable and a five-man squad could carry and emplace the device. Team members inserted two neutron initiators into two-inch diameter breech holes in the bottom of a 12-inch diameter base before installing the target rings and the gun-barrel assembly. Various sources describe the gun barrel as made of tungsten and steel, 28 inches long, and 6 inches in diameter. It protruded 23 inches from the base when installed. A timer that had a maximum setting of 7 days detonated the T4.

The Army began replacement of W9 based T124 shells with the lighter W19 based T315 shell in 1952. At this time, the AEC thought it possible to reduce the weight of a nuclear shell to the 600-pound weight of a conventional round. This would eliminate the need for special spotting rounds, firing tables and propellant charges. The new T315 shell required strengthening and redesign of its ignition, fuzing, and safety systems. Settings on the nose and the two internal fuzes of the W19 shell were made with a closed end ratchet wrench in a clockwise direction, facing the nose of the shell, that prevented backlash on firing. This allowed more rapid and accurate positioning. The T315 shell used the same internal nuclear components as the T124 with the exception of its mechanical neutron initiator, which LASL changed from the Phoebe to the Squab. The T315 shell measured 53.62 inches long, about 1 inch shorter than its predecessor. The T315 was retired in 1963, at which time the Army replaced it with the 8-inch T317 nuclear round.

The AEC used the W19 warhead in the construction of 50 MK 23 "Katie" shells for the 16-inch main armament of four Iowa class battleships. The Navy originally built these large battleships to provide naval superiority, but after their eclipse by aircraft carriers, it used them for shore bombardment. Mounted in three triple turrets, their MK 7 guns could throw 1,900-pound, 15-kiloton atomic shells to a distance of 23 miles. Like their land-based counterparts, the MK 23 shells were fuzed for airburst only, because their primary purpose was to support amphibious Marine operations. Each shell contained three identical mechanical time fuzes in its nose. A full broadside would have been quite spectacular! The Navy stockpiled the shells onshore during the period 1956 - 1962, the same period in which they retired their battleships.

The first deployment of an M65 armed unit came on October 12, 1953, when the 868th Field Artillery Battalion arrived in West Germany. The Army followed this deployment by exhibiting several atomic field pieces at Mainz, Germany on October 23. The exhibition was for the benefit of the French, German, and American press. Four additional Field Artillery Battalions, the 867th, 265th, 264th and 59th, followed the 868th, with the last unit arriving on Apr 28, 1954. The 7th Army attached all of these units to its 42nd Field Artillery Group. In early 1955, the *Army, Navy & Air Force Journal* reported that the Army had deployed the 216th Field Artillery Battalion from Fort Bliss to Europe, which meant that the "7th Army now had a force of 36 M65s capable of firing conventional or atomic shells." This is at odds with a frequently quoted figure of only 20 M65 guns manufactured. The second figure implies that some of the 280mm Field Artillery Battalions were under strength with respect to their gun complements during the course of their overseas deployments.

In addition to the M65s sent to Europe, the Army made an additional deployment of M65 cannon to Okinawa in 1955. It assigned three batteries, each composed of two guns, to the 8th Army's 663rd Field Artillery Battalion. The Army later transferred these batteries to Korea as part of the drive to nuclearize the peninsula's defenses. In preparation for the transfer, the Army upgraded the Battalion's equipment during the fall of 1957. It then moved advance parties of the Battalion to Korea on New Year's Day, 1958, to prepare gun positions and to ensure that transportation routes were capable of handling the movement of the monstrous guns.

When all preparations were complete, the guns were loaded onto LST's for their trip to Korea. The Navy timed their arrival for the early evening so that the Army could move the guns in the dark. The Army's timetable called for them to emplaced and operational by the following dawn. The guns could have then dealt with any hostile reaction on the part of the North Koreans in short order. Proud of its accomplishment, the Army displayed a pair of M65 cannons, along with Honest John artillery missiles, at a media presentation near the 1st Corps' Uijongbu Headquarters on February 3, 1958. The six atomic cannons provided a nuclear defense for Korea until the Army deactivated them in 1962.

The 7th Army in Europe based its M65s in the Darmstadt area of southern Germany, near Frankfurt. Later, the Army expanded their deployment toward Nurnberg, closer to the Soviet occupied Czech frontier. NATO termed this area the Central Front, to which it assigned the forces of the Seventh Army as part of the Central Army Group (CENTAG). Opportunities for concealment were excellent in the rugged and wooded terrain of southern Germany. A Soviet advance here would have been easily concentrated along identifiable routes allowing for very effective atomic fire. This area had a lower population density than the northern plains. Surprisingly, the 7th Army never deployed atomic cannons in the northern half of West Germany, on the continental plain. A Soviet thrust in this area would most likely have speedily overrun the guns.

USEUCOM Operations Plan 100-3 made USAREUR responsible for providing ground-delivered atomic support to non-US NATO forces. On the Northern Front, the Army delegated this task to the Northern Task Force (NORTAF), which covered the Northern Army Group (NORTHAG) with two M65 guns, two Honest John rocket Battalions, and two Corporal Missile Battalions. A special agreement covered the movement, operational control, administration, logistical support, and communications of NORTAF. After approval by US CINCEUR / SACEUR, the CINCUSAREUR would have issued orders for the deployment of this force through the Commanding General, US 7th Army.

The aggregate personnel strength of a Field Artillery Battalion totaled 494 personnel, along with six M65 guns deployed in three, two-gun Batteries. Shortly after their arrival in Europe, each Battalion began maneuvers to give their troops practice in simulated atomic actions. They held "Alerts" on a monthly basis, moving their guns to hidden firing positions. The gun crews practiced these drills on a 24-hour basis from the Divisional level to NATO exercises. Troops had to learn to move the guns through the narrow streets of small German towns and along country roads. In a number of instances, the guns ditched rounding turns or overturned due to a tendency to be top heavy. Occasionally the gun carriage ended up so far in the rhubarb that the services of an M60 tank retriever were necessary to extract it. Crews were expected and able to effect repairs on the spot. Modest convoys accompanied the guns during transport. Armored vehicles carried their ammunition and a rifle platoon provided security. Also, in accompaniment, were command and communication vehicles equipped with radar units.

Each unit carried out extensive firing practice, although not as often as desirable. There were only a limited number of firing ranges able to accommodate the big guns. Two areas that received frequent firings were the training areas at Baumholder and Grafenwohr. It was advisable for drivers to park vehicles at an angle to the guns because the concussion of firing could blow out windows. A gunner fired the M65 with a 50-foot long lanyard. In practice, crews fired three - four shells at 5-minute intervals. A 21-man crew served an M65, although there were often not that many personnel in attendance. These crews showed their competence when Battery a 264 Field Artillery Battalion won Best Battery of the Year in 1957. Proud crews ornamented the guns with painted pictures and gave them "pet names."

Following the arrival of the 216th in 1955, the Army reassigned M65 Battalions to the direct command of 7th Army's 2nd Corps. A further reorganization occurred in 1958 under the Combat Arms Regimental System. Reorganized Units included the 2nd Gun Battalion 38th Field Artillery, 3rd Gun Battalion 39th Field Artillery, 3rd Gun Battalion 79th Field Artillery, 3rd Gun Battalion 80th Field Artillery, and the 3rd Gun Battalion 82nd Field Artillery.

The M65 gun soldiered on in Europe until 1963. By this time, the Army was in the process of fielding such advanced weapons as the air-transportable Sergeant missile, which could carry a 200-kiloton W52 warhead over a range of 85 miles. It was also stockpiling 2,000 M454 atomic rounds for its smaller and more maneuverable M110 8-inch self-propelled guns. TAC's all-weather F-100 Super Saber and F-105 Thunderchief fighter-bombers provided backup for the guns and missiles. On November 22, the Army retired "Atomic Annie" or "AWOL" as her crew knew her – the last atomic cannon in Europe. Three generals and a bugler who sounded Taps attended the cannon's last parade ground appearance.

In order to evaluate the impact of the M65, it is necessary to understand its mission and to place it in its historical context. The AEC only manufactured 80 atomic rounds for this gun, so it is clear that the Army never intended to use the M65 to fight a protracted war. The Army deployed the M65 alongside of Honest John rocket Battalions able to carry a 20-kiloton W7 warhead to a range of 15 miles and Corporal guided missile Battalions able to carry the same warhead to a distance of 75 miles. In addition, Tactical Air Command operated 150 Matador mobile cruise missiles and a fleet of fighter-bombers, mostly F-84G Thunderjets and B-45 Tornado light bombers armed with MK 7 tactical bombs.

The purpose of these units was to blunt and slow a Soviet attack. Since TAC's aircraft lacked an all-weather operational capability and had only a limited night fighting ability, the guns and missiles of the 7th Army would have had to stop a thrust by a numerically superior but conventionally armed force of Soviet troops and armor in the event of inclement weather. This would have given Strategic Air Command the time needed to launch its "Sunday Punch." American nuclear planners expected that over a period of several hours, the Soviet Union would be utterly devastated and hostilities would cease. General Collins, the Army's Chief of Staff, stated that the presence of atomic cannon in Europe was a great deterrent to a Soviet offensive.

CHAPTER 15
"BOUNCING THE RUBBLE HIGHER" WITH HYDROGEN BOMBS

President Truman first heard about the hydrogen bomb on October 6, 1949. This information came from Admiral Sydney Souers, his Director of National Security. Souers had acquired this knowledge on the previous day at an informal luncheon with retired Admiral Lewis Strauss, a member of the Board of Directors of the Atomic Energy Commission. Although theoretical work for a hydrogen bomb had been ongoing since 1944, information about this weapon had heretofore been restricted to the scientific community at Los Alamos and a select group of academic consultants.

The main attraction of the hydrogen bomb was its theoretical ability to burn an unlimited quantity of fuel. Critical mass considerations restrict fission bombs to a maximum yield of 500 to-600-kilotons before premature detonation rears its ugly head. If Los Alamos scientists could harness the power of fusion, they realized that they could build weapons of unimaginable power. The most attractive fuel for these weapons was the deuterium isotope of hydrogen, which the scientist planned to convert into helium and energy. Through the process of fusion, one kilogram of deuterium can release eight times more energy than an equal mass of plutonium. It is also much cheaper than fission fuels; deuterium costs only 1/10 of 1 percent as much as plutonium, which at this time was valued at several hundred dollars per gram. Deuterium's only drawback as a fusion fuel was the apparent need to maintain it in a liquid state at cryogenic temperatures.

Fusion reactions are primarily dependent on temperature. After scientists discovered that fission explosions could generate multi-million-degree temperatures, the concept of a thermonuclear bomb emerged naturally from the science of astrophysics. In the United States, the idea first occurred to Enrico Fermi, the same physicist who conceived of the atomic reactor. During the war years, Fermi directed a small team known as F Division at Los Alamos. It was the responsibility of F Division to investigate alternative possibilities to an atomic bomb. Fermi placed Edward Teller, a young and enthusiastic physicist, in charge of the F-1 Group, which was responsible for the "Super," or hydrogen bomb, and related thermonuclear theory. Egon Bretscher, a recruit from the British delegation, supported Teller and his associates with experimental data produced by his F-3 Group.

The basic design pursued by the F-1 Group was a weapon that could convert a cubic meter of liquid deuterium into helium, creating an explosion of 10 megatons yield. The group also assessed the type of damage such a weapon might produce in comparison with an atomic bomb. Exploded at an optimum height, a 10-kiloton atomic weapon could destroy about 1 square mile. In comparison, Los Alamos Scientists expected a 10-megaton Super Bomb to destroy 1,000 square miles, an area comparable to the world's largest metropolitan areas. The group also estimated the effects for a 100-megaton weapon. This was the maximum possible size to make use of blast, which the curvature of the earth's surface limits. Larger weapons can only drive cylinders of the earth's atmosphere upwards and outwards into space at a higher velocity.

The Ivy Mike Sausage device on the left, with the light pipes that measured its effects emerging to the right. The sausage yielded 10.4 megatons and was briefly weaponized as the EC 16. For scale, note the technician seated lower right. (Credit: AEC; Original Color)

The Castle Bravo Shrimp device showing a conical projection at the right end. A series of light pipes are clustered around the projection, with others attached down the length of the bomb. The device produced a yield of 15 megatons due to an unanticipated reaction with the Li7 in its fuel. The device was weaponized as the MK 21 bomb. (Credit: LANL, Photo D92-1)

The MK 14 bomb. A solid fuel device lacking any safety features, it entered the stockpile ahead of the first test of a lithium deuteride fueled device. (Credit: LANL Photo D32-9)

The TX 16 liquid deuterium fueled bomb. It had a very complex delivery procedure. With the success of solid fuel weapons, this device was discontinued after only a few months of service. This bomb was carried by a specially equipped B-36. (Credit: LANL Photo D27-6)

A Boeing B-52D heavy bomber shown with a GAM-72D Quail decoy. The B-52D was the first model of the bomber built strictly for the strategic nuclear bombing mission. (Credit: USAF)

Two MK 15 MOD 0 trainers being hoisted into the weapons bay of a B-52 bomber. The weapons were held in place with slings rather than the later developed clip-in assemblies. (Credit: SNL Photo D 77-162)

A MK 17 thermonuclear bomb on its B2 lift truck. The size and weight of this, the largest nuclear bomb ever manufactured by the United States, relegated it to carry by the B-36. It had a yield of 10 - 15 megatons. The MK 17 and MK 24 used the same casing. (Credit: SNL Photo D 57-787)

MK 36 thermonuclear bomb, which had a "normal" 19-megaton yield. At one time, this weapon represented half the megatonnage in the US stockpile. It used the he same casing as its MK 21 predecessor. Note the partial ogive nose, the double angle fins, and the spoiler bands. (Credit: NAM)

A MK 39 MOD 2 thermonuclear bomb. In a crash over Goldsboro, North Carolina, a B-52 released two operational 3.75-megaton MK 39 bombs. Fortunately, safety characteristics had evolved to the point that neither bomb exploded. (Credit: NAM)

The remains of a MK 39 bomb accidentally dropped near Goldsboro, North Carolina. Its parachute did not deploy and it hit the ground nose-down at a speed of 700 MPH. (Credit: USAF)

Although the earth's curvature limits the effects of blast, the same is not true for thermal radiation or fallout. By exploding a 100-megaton warhead at an altitude of 300 miles, the F-1 group estimated that the effects of its thermal pulse might make themselves felt over an area of 1,000,000 square miles. There was, of course, no means of propelling a bomb to this altitude at this time. In fact, Los Alamos scientists thought that remotely operated submarines or aircraft might be required to transport gigantic Super Bombs to their final destinations. "Brass Ring" was a DOD project aimed at turning a B-47 aircraft into a remotely controlled MB-47 flying bomb. A sister DB-47 bomber would direct the MB-47 in flight. Even more fantastic was a cobalt-jacketed "Doomsday" device that produced enough fallout to exterminate all life on earth. Los Alamos could have made it as large as necessary because it did not require transportation!

Egon Bretscher's F-3 Group carried out the complex experiments needed to determine the reaction cross-sections for deuterium and tritium, the isotopes of hydrogen that had potential as thermonuclear fuels. They made their measurements with a small quantity of tritium that Oppenheimer acquired from Hanford and from a specially constructed Cockcroft-Walton accelerator located at Los Alamos. (Ernest Rutherford, Marcus Oliphant, and Paul Hartack had demonstrated the use of high-energy collisions to fuse deuterium as early as 1932.) The Los Alamos results showed that the current understanding of DD reactions was essentially correct, but that DT reactions were much more reactive than anticipated. This augured well for the inclusion of small amounts of this expensive, substance in thermonuclear fuel.

In order to investigate the possibility of a self-sustaining thermonuclear reaction in a fuel mass, it was first necessary to understand the progress of the fission explosion that would heat it to ignition. Los Alamos did this in a crude way using calculations made with mechanical calculators in a series of projects known as "Baby Hippo" and "Hippo." This program took several years to complete. Expanding the work to incorporate fusion reactions proved much too difficult for simple desktop equipment. Determining whether thermonuclear combustion can propagate and lead to efficient thermonuclear burning is a difficult problem because a large number of physical processes are involved. Los Alamos physicists had to consider questions about energy production rates from competing reactions and energy transport by radiation, electrons, and particles. Fortunately, John von Neumann overcame the mechanical computational barrier after a chance meeting with Herman Goldstine at the Army's Aberdeen Proving Grounds.

A professor at Princeton University, Goldstine was visiting Aberdeen to discuss the use of a radical new "electronic numerical integrator and computer" for determining the ballistic trajectories of artillery shells. ENIAC, the world's first computer, could perform 1,000 computations at the lightning pace of 3 seconds. Von Neumann immediately grasped the significance of this machine and arranged a visit to Philadelphia to further evaluate Goldstine's prodigy. He then sat down and developed a report on how he might use it to perform the calculations necessary to simulate the process of events in a Super Bomb explosion. In doing this, von Neumann created the world's first machine language.

Using ENIAC, Los Alamos scientists carried out a simplified two-dimensional simulation of a Super Bomb explosion during the winter of 1945 - 1946. It made use of information that the F-1 group published in a *Super Handbook*. This was a compendium of thermonuclear data and research available up to October 1945. The simulation proceeded slowly because ENIAC was not fully automatic. The computer required manual rewiring to treat different aspects of the Super Bomb problem. Los Alamos held a secret Super Conference to review the results of the Super Project during the week of April 18 - 20. After the conference, the laboratory published the first major technical report on thermonuclear weapons: *Prima Facie Proof of the Feasibility of the Super.*

The conference focused on what Los Alamos termed the "Runaway" or "Classical Super" – a hydrogen bomb in which a fission explosion would heat and ignite a self-sustaining thermonuclear reaction in an adjacent quantity of deuterium. The conference also considered a "boosted" atomic bomb, in which a fusion reaction initiated by a charge of DT gas or liquid placed inside a hollow plutonium pit provided the neutrons needed to increase its efficiency. To Edward Teller, an unrestrained optimist and H-bomb enthusiast, the results of the conference looked promising and he took pains to note this in the *Prima Facie* technical report that summed up the conference results. To others such as Hans Bethe, the results appeared negative. Although scientists could undoubtedly ignite a fusion reaction, analysis indicated that cooling by several types of radiation emission would quickly quench the reaction.

Despite divided opinion on the feasibility of the Super, Director Norris Bradbury committed Los Alamos to further investigation of fusion. He placed the Theoretical Division under the supervision of physicist J. Carson Mark and ordered him to spend about half of the Division's time on thermonuclear calculations. Mark was to spend the rest of the Division's time on developing improved atomic weapons. This second aspect of the lab's mandate was critical to the success of a hydrogen bomb because the temperatures reached in the fireballs of early atomic weapons were too low to ignite a fusion reaction. In particular, scientists hoped that a levitated or boosted primary would achieve sufficiently high temperatures to make a hydrogen bomb practical. It is interesting to note that the initial concept for a thermonuclear bomb, defined by a meeting of American and European physicists at Berkley, in the summer of 1942, envisioned a gun-assembled enriched uranium primary!

Shortly after Mark's appointment by Bradbury, Teller decided to leave the lab and return to his former teaching position as a professor at the University of Chicago. He wanted to spend more time with his family and was disappointed that Bradbury had not chosen him to lead the Theoretical Division. Bradbury had also overlooked him in favor of John Manley when he incorporated the Research and F Divisions into an expanded Physics Division. However, like many other leading physicists, he found time to consult and returned to Los Alamos on summer sabbaticals to continue his thermonuclear research.

While a number of scientists questioned the technical feasibility of the Super, others wondered whether they should even try to build a weapon capable of such "mass destruction." The Los Alamos scientists had been intimately involved in evaluating the effects of their atomic weapons on Hiroshima and Nagasaki and many saw the destruction and the horrific civilian casualties first hand. While military targets of sufficient size existed to justify the use of an atomic bomb, the only obvious use for a hydrogen bomb was to exterminate large segments of an enemy's population. For this reason, a number of influential scientists, including Oppenheimer, Fermi, and Bethe, openly opposed its development. David Lilienthal, the Chairman of the AEC, also opposed its development. Others, such as AEC board member Lewis Strauss and avowed anti-Communist and Hungarian refugee Edward Teller, argued for intensive H-bomb development to prevent the Russians from obtaining such a weapon first. The explosion of Russia's first atomic bomb was the stimulus to cause a badly frightened Strauss to use Sydney Souers as a back door to reach Truman.

The period leading up to Truman's thermonuclear revelation saw slow technical progress at Los Alamos. Teller worked on a number of Classical Super Bomb variations, one of which was designated TX14. A number of his colleagues provided helpful suggestions. John von Neumann and Klaus Fuchs suggested placing an intermediate charge of deuterium and tritium between the fission bomb and the main mass of deuterium to increase ignition temperatures, whereas Emil Konopinski suggested salting the main deuterium charge with tritium to increase its reactivity. Fuchs also suggested enclosing the intermediate charge of tritium and deuterium inside a radiation proof beryllium oxide container. Fuchs' suggestion recognized the recently discovered problem of radiation emission from a thermonuclear fuel mass. Los Alamos scientists had determined that the deuterium fuel in the Classical Super would quickly ionize, making it transparent to the radiation that was supposed to heat it. Since 19 percent of the energy released by an atomic explosion is associated with the kinetic energy of its bomb debris and 80 percent of the energy is radiation, only the 1 percent of energy associated with the explosion's neutron flux remained for heating.

In order to overcome the problems associated with radiation emission and the ionization of the fuel mass, Teller proposed adding more and more expensive tritium to enhance the reaction and, in one design, he envisioned a 30,000-pound boosted primary at the end of a 30-foot long bomb. He also considered the use of lithium deuteride fuel because the decomposition of lithium generates tritium, which would increase reactivity. Nevertheless, the theoretical problems associated with heating and radiation loss seemed insurmountable. Finally, Teller suggested a pair of atomic tests to provide physical data from the staged implosion of a small quantity of deuterium and from an experimental boosted warhead. In order to prepare for these tests, Teller returned to Los Alamos.

Possibly, as an alternative in the event that the Classical Super proved unworkable, Teller's fertile mind brought forward the concept for a device he called an "Alarm Clock." This was a single stage thermonuclear weapon composed of alternating layers of fissile and fusionable materials that would "wake up the world." Since the developmental problems for this type of weapon were equal in every aspect to a Super Bomb, Los Alamos scientists largely neglected it. Requiring explosive compression for its mechanism (like the core of a fission bomb), it had an upper practical limit of about 1 megaton for its yield.

Scientists at Los Alamos made a number of important mathematical advancements for the simulation of a thermonuclear explosion during the period in which they carried out the initial research on the Classical Super. Stanislaw Ulam, assisted by John von Neumann, devised a new statistical approach known as the "Monte Carlo" method to treat variables as statistical populations. Robert Richtmeyer and von Neumann also introduced a "viscosity treatment" that simplified the mathematics of shock fronts so that they could be analyzed using computer methods. By the time that experimental work derived the equations of state for hydrogen and its isotopes, Nicholas Metropolis had begun building MANIAC, an all-electronic successor to ENIAC. A copy of a sister computer located at Princeton, the Los Alamos based MANIAC crunched the numbers needed for the final evaluation of the Super.

Although thermonuclear theory strongly suggested that the Classical Super was an unworkable design, Teller remained confident that he could somehow overcome its problems. This confidence fueled the debate over whether the United States should build such a weapon. The Joint Chiefs of Staff, who at first had expressed doubts about using a 10-megaton bomb for military purposes, now began to lobby for its development. If it was feasible as Teller suggested, the United States needed such a weapon to regain its strategic advantage over the Russians. The various members of the GAC and JCAE also reached this conclusion. In a discussion with the board members of the AEC on January 19, 1950, this view carried the day and President Truman demanded that work on the weapon proceed in earnest. He also let the cat out of the bag by making a public announcement on that same day. In protest, Lilienthal resigned his position as Director of the AEC. Truman replaced him with Gordon Dean – who was much more sanguine about the development of thermonuclear weapons.

After Truman decided that the United States should pursue the H-bomb, the scientists at Los Alamos wracked their brains for almost a year before Stanislaw Ulam and Edward Teller discovered a solution to the "Super Problem." In attempting to design a more efficient atomic bomb, Ulam considered using the kinetic energy imparted by the neutrons in the shock front of a small atomic explosion to compress a larger fissile core. Conventional explosives can only achieve fuel compression of 2 – 3-fold, but an atomic blast might yield compressions of 10 or even 100-fold. Since critical mass and atomic reactivity (alpha) are proportional to density, Ulam thought that this approach might produce large, efficient explosions from small quantities of fissile material.

In late January 1951, Ulam discussed his idea with Edward Teller who came up with an even better idea. He suggested using the prodigious radiation output of an atomic explosion, which escapes ahead of its neutron flux and explosive debris, to compress a mass of thermonuclear fuel and thus initiate a fusion reaction. Although scientists at Los Alamos had previously considered compression, they had rejected this concept because the 2 or 3-fold increase obtainable with explosives was insignificant to the fusion problem. Teller recognized that a compression of several hundred-fold could significantly alter the equations of state and the physical processes inherent in a mass of thermonuclear fuel. Teller and Ulam published the basic premise of staged, radiation implosion in a still (mostly) classified report entitled *On Heterocatalytic Detonations I: Hydrodynamic Lenses and Radiation Mirrors* (LAMS-1225). Heterocatalytic detonation meant using one explosion to set off a second explosion, hydrodynamic lenses referred to Ulam's shock front, and radiation mirrors referred to the bomb's casing, which absorbed and re-radiated the flow of x-rays from the primary.

In particular, Teller realized that in highly compressed plasma, three-body reactions between photons, electrons, and ionized nuclei would scatter radiation and prevent it from escaping. The fusion process would thus continuously raise the temperature of the fuel mass until it ran away in an uncontrolled reaction. Later in the thermonuclear process, he expected that interactions between deuterium and the charged nuclei of the helium byproduct would transfer about 20 percent of the produced energy to the remaining fuel.

Carson Mark added his own take to the Teller - Ulam thesis. Before deuterium - helium reactions could become important energy contributors, neutronic collisions with the nuclei of light elements predominate. The mean free path for 14.1 MeV neutrons in uncompressed liquid deuterium is 22 centimeters, about 10 inches. In a reservoir measuring 22 centimeters across, as might be found in a thermonuclear secondary, most of the 14.1 MeV neutrons generated by DT reactions would escape without a single collision. If this sphere were compressed 125-fold, its diameter would shrink to 4.5 centimeters and the neutron MFP would reduce to only 0.18 centimeters. In this state, it would be almost impossible for neutrons to escape the fuel mass without depositing their energy.

As surmised above, a proper understanding of the interactions between neutrons, electrons, photons, and ions is just as important as understanding the fusion reactions themselves.

This is because most of the energy released by fusion reactions first transfers from neutrons (n) to ionized nuclei (i):

$$n + i > n' + i'$$

Energy then transfers from ionized nuclei to electrons (e) through elastic (Compton) collisions as the thermonuclear burn proceeds:

$$i + e <> i' + e'$$

Finally, energy transfers from electrons to the electromagnetic field in the form of photons (γ) produced by a process called "bremsstrahlung" radiation emission that involves the interaction of ions with electrons:

$$i + e <> i' + e' + \gamma$$

If the thermonuclear plasma has a low density, the plasma is transparent to electromagnetic radiation and the photons escape. Neutrons also escape.

In dense thermonuclear plasmas, however, neutrons have short mean free paths, thus they cannot escape from the fuel mass. Dense thermonuclear plasmas are also opaque to photons, which cannot readily escape and so they begin to accumulate. The compressed tamper that surrounds the thermonuclear plasma further contributes to the retention of neutrons and photons through a process of back scattering. Because energy density is the sum of the plasma's neutron, electron, ion, and radiation temperatures, the radiation term increasingly dominates the plasma's energy density as photons accumulate. As the plasma's temperature rises, energy begins to transfer from photons back to electrons, and then from the electrons to ions through an inverse bremsstrahlung process and through inverse Compton collisions:

$$e + \gamma <> e' + \gamma'$$
$$i + e <> i' + e'$$

This raises the temperature of the plasma's components such that the electron, ion, and photon temperatures tend towards an "equilibrium". Since ionic temperature determines the fusion reaction rate, the rate of fusion increases rapidly and runs away. Hans Bethe, who had resisted the temptation to contribute to thermonuclear studies, capitulated after learning of the new theory. He realized that the successful development of a hydrogen bomb was now a certainty and, like the Joint Chiefs, wished the United States to obtain such a weapon ahead of the Russians.

Although it was clear that fusion reactions would undergo self-heating in dense plasma, there was concern that radiation induced compression might not generate the 20,000,000 °C temperature needed for ignition of the fusion fuel. In order to overcome this concern, Teller added a refinement to his and Ulam's design in a second classified report *A Thermonuclear Device* (LAMS-1230). Frederick de Hoffmann carried out the calculations for this report, which described a hollow cylinder of plutonium containing a few grams of tritium positioned along the axis of the thermonuclear fuel container (the secondary). The forces that compressed the deuterium fuel in the secondary would also compress the central plutonium "sparkplug" to criticality, causing it to explode. The shock wave from this second fission explosion would collide with the incoming wave of radiation-induced compression, briefly stabilizing in a zone where compressed and heated deuterium fuel would ignite and generate a self-propagating wave of burning. Secondary atomic compression from the fission of the surrounding natural uranium tamper would help maintain the conditions needed to maintain efficient thermonuclear burning.

Operation Greenhouse, carried out at the Pacific Proving Grounds in April 1951, confirmed the Teller-Ulam concept. George, the third test in the series, evaluated the effects of compressing and heating an ounce of liquefied deuterium mixed with an even smaller amount of tritium. The test device, known as the "Cylinder," was a disc approximately 8 feet in diameter and 2 feet thick. It consisted of high explosive lenses that enclosed an enriched uranium core. Implosion of the Cylinder channeled energy through an axial tunnel to a small beryllium oxide flask that held the fusion fuel. This rather large device provided a 220-kiloton fission explosion that was powerful enough to compress and heat the contents of the flask, enabling it to contribute a few kilotons of fusion energy. Diagnostic equipment measured the characteristic energies of fusion neutrons, confirming the test's success.

Item, the fourth Greenhouse test, evaluated the concept of fusion boosting. The device used for this purpose was an implosion warhead that contained a small amount of liquid deuterium and tritium. A special feature of the Item device was a thin copper layer that isolated the initiator chamber and prevented the hydrogen isotopes from reacting with the enriched uranium pit. Plutonium and uranium are highly prone to forming hydrides in the presence of hydrogen and the resulting mix of high and low-Z material would not have reacted efficiently. The test was a success, producing a yield of 45 kilotons, half of which Los Alamos attributed to the effects of boosting.

As soon as Los Alamos made available the results from Greenhouse, Atomic Energy Chairman Gordon Dean called a second Super Conference at Princeton University. The Weapon Subcommittee of the GAC and Dean's fellow AEC board members attended it. Although the staff at Los Alamos planned an orderly presentation, Teller hijacked the session almost immediately. He didn't trust of the motives of the Los Alamos scientific staff and wished to make his personal views known. Despite Teller's behavior, the meeting ended with agreement that the Equilibrium Super was a workable design and that the United States should build it. Misgivings about the morality of this weapon had begun to fade and even Oppenheimer looked forward to the coming technical challenges.

Teller immediately requested that Norris Bradbury place him in charge of the H-bomb project. Although there was consensus that Teller could provide valuable help in accomplishing the project, there was also agreement that his temperament made him unsuitable to act as General Manager. Instead of Teller, Bradbury chose Marshall Holloway. Bradbury recognized Holloway as a steady hand who had contributed significantly to the development of Fat Man's nuclear hardware. In the face of this decision, Teller once again left Los Alamos and began to lobby for a second nuclear design facility where he could carry out his work. He achieved his purpose in 1952, when the University of California at Berkeley formed the Lawrence Radiation Laboratory (LRL) at the abandoned Livermore Naval Station under the direction of Herbert York. Imagine Teller's disappointment when he learned that the laboratory's initial mandate was to provide scientific support for the Los Alamos thermonuclear program.

Holloway chaired the first meeting of the Theoretical Megaton Group or "Panda Committee" on October 5, 1951. The members had to decide what type of thermonuclear fuel they would use, how soon they could conduct a test and whether they were developing a weapon or providing a "proof of principle." They also considered three types of fuel: liquid deuterium, deuterated ammonia (ND_3) and lithium deuteride (LiD). All of these potential fuels had the required concentrations of deuterium necessary for use in a thermonuclear weapon. Although ND_3 and LiD presented far less challenging technical problems than liquid deuterium, the Committee decided to proceed with a cryogenic device because pure deuterium would yield the most straightforward scientific results. The use of LiD, a solid fuel, would in any case have delayed testing since its production required the building of a large-scale manufacturing facility.

Because Richard Garwin had already devised the basic blueprint for a cryogenic bomb on a 3-month sabbatical the previous summer, the Committee was able to agree on the test of an "experimental scientific installation" as early as October 1952.

Los Alamos scheduled the detonation of the thermonuclear device, designated "Mike" for megaton, at the AEC's Pacific Proving Grounds. The device was far too large for testing within the confines of the continental United States, not only because of its yield, but also because of the fallout that it would produce. Technicians assembled Mike inside a large, corrugated aluminum shelter that measured 88 feet long, 46 feet wide, and 61 feet high. Seabees constructed the "shot cab" or assembly building in an area built up with bulldozed sand and gravel on the island of Elugelab at Eniwetok Atoll. They topped the shot cab with a 300-foot high radio / television mast that started the bomb's countdown and transmitted test data. The firing team used three radio links to activate the sequence timer that detonated the device. Base facilities for the test consisted of a deep-water pier and ramp, an airstrip, a 3,000 KW power plant, a cryogenic plant, a machine shop, and accommodations.

To evaluate the test, Los Alamos scientists surrounded the island of Elugelab with some 500 diagnostic instruments and experiments located on 30 separate islands. The largest of these installations was an 8-foot square, 9,000-foot long, aluminum-sheathed plywood shed that ran over a causeway from Elugelab to the adjacent island of Bogon. Technicians packed this shed, known as a Krauss-Ogle box after its inventors, with helium-filled plastic balloons that allowed the un-attenuated passage of gamma rays and neutrons from the exploding thermonuclear device into various scientific instruments. These instruments were sheltered inside a reinforced concrete bunker covered in sand and gravel for added protection. This bunker needed to survive only for a brief instant. The instruments inside retransmitted their data to a survivable bunker that was located further away from ground zero.

The atomic primary chosen for the Mike device was a 40-inch diameter, TX5 implosion system. At the last minute, Mark Rosenbluth developed a concern that its composite-levitated core contained too much plutonium and might pre-detonate with a loss of yield. Los Alamos, therefore, exchanged the core for one that contained a minimum amount of plutonium. Mike's liquid deuterium occupied a cylindrical reservoir about "1 1/2 - 2 feet in diameter and 10 - 12 feet high," a configuration that Teller called "natural" to the thermonuclear design. Its designers suspended it inside a massive, cylindrical radiation case or "hohlraum" nicknamed the "Sausage."

Mark Holloway approved the initial design for the hohlraum in January 1952. Its internal cavity tapered away from the primary in a shape like a bowling pin. It was nicknamed the "Shmoo," after an imaginary creature featured in the "L'il Abner" comic strip. The tapered cavity was supposed to maintain a constant pressure as the radiation flowed down the interior of the casing. It was not long before Carson Mark scrapped the Shmoo in favor of a straight walled design that necessitated a computational rework that he did not complete until March. The new straight-walled design featured a much wider channel to improve the flow of radiation.

American Car and Foundry manufactured the Sausage, which formed the outer casing for the bomb. The 54-ton, unit-welded radiation case supplied the bulk of the 82-ton thermonuclear installation, which Los Alamos constructed onsite at Elugelab. Less the weight of the 12-ton "cryogenic unit," Mike weighed 70 tons. The device stood 244 inches tall and 80 inches in diameter. These dimensions and weights imply the Sausage had 5 to 6-inch-thick steel walls that helped contain the radiation produced by the TX5 primary for as long as possible. Because of its size, American Car and Foundry cast the cylinder in nine sections. Its intermediate length consisted of seven stacked rings; with each of the lowest three rings approximately twice the height of the uppermost four rings. A pair of hemispherical end caps completed the cylinder, with the upper cap ported to allow access for final assembly. Subtracting the 40-inch diameter of the W5 warhead from the 68 - 70 inch inside diameter of the cylinder indicates that a 14 to 15-inch void surrounded the primary.

Occupying the lower inside of the bomb casing was a three-walled stainless-steel thermos flask or "Dewar" that contained the bomb's liquid deuterium fuel. The Dewar maintained a temperature sufficiently low to prevent the liquid deuterium from boiling off. Between its inner and outer walls stood an intermediate copper shield, cooled by a pool of liquid nitrogen placed at the bottom of the cylinder. This intermediate thermal barrier significantly slowed heat loss from the fuel reservoir and reduced the evaporation of its liquid deuterium. Ferdinand Brickwedde, from the Bureau of Standards, supervised the shipment of the deuterium used to fill the bomb's inner Dewar. It arrived as a compressed gas in 2,000-pound tube banks. The deuterium was liquefied onsite using special equipment designed by Samuel Collins and manufactured by Arthur D. Little Inc. of Cambridge, Massachusetts. The liquefier consisted of three transport trailers with a total weight of 75 metric tons. It was capable of producing 100 liters of liquid hydrogen per hour. Technicians charged the bomb with its cryogenic fuel just prior to the test.

Inserted down the axis of the Dewar was a hollow plutonium sparkplug, approximately 1 inch in diameter. Some authors indicate that this sparkplug contained several grams of tritium cooled to a liquid state and Carson Mark refers to the properties of plutonium cooled to liquid hydrogen temperatures. Stainless-steel rods supported the sparkplug at each end. Surrounding the Dewar was a thick pusher or tamper that contained about 5 tons of natural uranium. Assuming the same 40-inch diameter as the TX 5 primary, this mass equates to a cylinder about an inch thick. Stainless-steel cables suspended it and kept it separated from the vertical walls of the hohlraum. Gold leaf on its inner walls helped to minimize thermal emission from the Dewar as did beryllium plates mounted over the Dewar. Between the secondary and the primary, the end of the tamper thickened into a heavy blast shield. The shield improved compression by preventing the primary's radiation flux from preheating the fuel. Boron 10 incorporated in the blast shield prevented heating from the primary's neutron flux.

The thermonuclear age was ushered in when Los Alamos exploded the Ivy Mike device on October 31, 1952. The nuclear fireball produced by the device's 50-kiloton TX5 primary immediately began to radiate soft x-rays. The mean free path of these x-rays was sufficiently long that they were able to shine directly onto the inside of the casing that formed the exterior of the bomb's radiation channel. The blast shield, however, kept the x-rays from affecting the thermonuclear fuel. Anticipating that the x-rays would instantly vaporize the moderate Z steel in the bomb casing, blowing it to bits and prematurely releasing the energy trapped in the radiation channel, Los Alamos weaponeers lined the casing with a high-Z layer of lead. High-Z materials ionize and heat more slowly than moderate Z materials. Lining the steel casing with lead slowed the ablation process to minimize shock.

Thick sheets of polyethylene attached with copper nails covered the lead lining. Polyethylene contains only low-Z atoms of hydrogen and carbon. The x-rays from the primary had to ionize and heat this material into plasma before they could begin to affect the lead lining. This in turn reduced the ablation and shock problems to a point where they occurred too late in the bomb process to affect the implosion of the thermonuclear secondary. The low-Z plasma also prevented jets of ablated opaque material (partially ionized lead and uranium) from obstructing the radiation channel or chopping up the thermonuclear secondary. Furthermore, the ionized hydrogen plasma voraciously absorbed neutrons to reduce fuel heating. Richard Garwin, a physicist with cryogenic experience calculated the parameters necessary to fill the radiation channel with liquid hydrogen as an alternative low-Z material.

Garwin also realized that the casing thickness was twice the thickness it needed to be. It only had to be half as thick as the design thickness because the radiation-pressure-driven shock wave, had to go through the radiation case to blow off the outside of the structural steel casing. But nothing was visible from the inside until a rarefaction wave got back through the casing, doubling the time before the casing blew back into the radiation channel. Because of this design oversight the device was considerably overweight.

As the flow of radiation ionized the polyethylene into plasma, it re-radiated the previously absorbed x-rays, which had by now lost some of their energy from the work accomplished in the heating process. The bomb casing and the tamper then absorbed and re-radiated this energy to bring the environment of the radiation channel into thermodynamic equilibrium at a temperature of about 20,000,000 °C. Inside the casing, material ablated from the surface of the tamper accelerated it across the standoff gap. The ablation process also generated a series of shock waves that changed into kinetic energy when they reached the interface at the far side of the tamper. This process further aided in the tamper's acceleration.

While the external casing for Mike could be overbuilt, the H-bomb's designers had to calculate an optimum thickness for the secondary's tamper. If they made the tamper too thin, progressive ablation would have burned through to the fuel before the fusion reaction got under way and the bomb would have fizzled. If they made the tamper too massive, reduced acceleration would have resulted in a loss of compression and yield. The designers also had to make sure that the tamper applied a sustained and accelerating pressure to the deuterium fuel so that it reached its maximum possible density. If the tamper arrived at too high a velocity, the impact would have converted much of its kinetic energy into unwanted heat. Determining the correct velocity to minimize the initial shock and to provide a sustained pressure required an optimum standoff distance, which, in the case of Mike, was the 8 inches occupied by the Dewar's vacuum space. The sequence of successive shocks provided by multi-layer radiation induced ablation created the sustained pressure increase required for successful adiabatic compression.

As the imploding uranium tamper compressed the fusion fuel, it also compressed the plutonium sparkplug at the center of the fuel mass. Because of the high neutron flux within the bomb environment, it is unlikely that the sparkplug reached more than a 5-fold increase in density before it exploded in a fission chain reaction. This second atomic reaction would have run to completion very quickly, heating and further compressing the thermonuclear fuel. The sparkplug contained 18 kilograms of plutonium, the complete fission of which produced a yield of 300 kilotons, six times the yield of the primary. The explosion of the sparkplug thus created conditions that exceeded the critical temperatures and pressures needed to support fusion burning. The thermal or Marshak wave, initiated by this process, propagated its way through the remaining thermonuclear fuel mass in a matter of a few shakes.

Fusion in the thermal wave was at first dominated by the two key reactions:

$D + D > He3 + n + 3.27$ MeV and
$D + D > T + p + 4.03$ MeV

The tritium produced by the second reaction combined almost instantly with some of the deuterium to produce He4:

$$D + T > He4 + n + 17.59 \text{ MeV}$$

Then, if self-heating reached a temperature of 20 KeV or 200,000,000 °C before explosive disassembly occurred, a fourth reaction, involving He3, took place:

$$He3 + D > He4 + p + 18.35 \text{ MeV}$$

This is the most energetic of the four main fusion reactions and is necessary to provide efficient energy production in a hydrogen bomb.

To ensure that the bomb reached high enough temperatures to support the reaction of He3, Los Alamos made the secondary's tamper out of natural uranium. In a final stage of energy production, a flood of high-energy fusion neutrons caused the fast fission of U238 in the surrounding tamper. Because ablation had tremendously compressed the tamper by this stage, it was almost impossible for the fusion produced neutrons not to interact with the tamper's densely packed nuclei. The energy from this process helped produce and maintain the temperatures needed to sustain the reaction of He3 and to retard the disassembly of the fuel mass.

The total yield from Mike was 10.4 megatons, 77 percent of which came from the fission of its primary, sparkplug, and tamper. This equates to the complete conversion of 400 kilograms of fissile and fissionable material. The 2.4 megatons of yield attributed to fusion equates to the complete reaction of 30 kilograms or 185 liters of liquid deuterium at a density of 162 kilograms per cubic meter. Assuming a thermonuclear efficiency of 25 percent, Mike contained 740 liters (120 kilograms) of cryogenic fuel, enough to fill a 22-inch diameter cylinder that was 10 feet long.

Assuming a 300-fold compression of the secondary's deuterium fuel, the diameter of the secondary would have shrunk to 5 inches, neglecting ablation. Because of their physical properties and their relative positions in the secondary, the tamper, fusion fuel, and sparkplug all achieved different degrees of compression. Theoretically, the plutonium sparkplug would have been compressed 16-fold, and the tamper would have been compressed 8-fold. This being the case, the central plutonium sparkplug would have measured about half an inch in diameter and the surrounding nuclear fuel shell would have measured just over an inch in diameter. Considering the timing of energy release from the sparkplug, these estimates are approximate.

In the years since 1955, data declassifications have supplemented the aforementioned knowledge about Mike's nuclear processes. These declassifications reveal that the new elements einsteinium and fermium (atomic numbers 99 and 100) were produced by multiple neutron captures in U238, the result of the neutron flux released by the fast fission of other U238 nuclei by fusion neutrons. Some authors also state that the neutron concentration in the U238 blanket exceeded 6×10^{23} neutrons / centimeter^3 for 10 nanoseconds. This implies a thermonuclear burn time of 1 shake. The average neutron temperature was estimated at 5 KeV or about 50 million degrees. Had all the deuterium and uranium contained within Mike reacted with complete efficiency, the final yield for the bomb would have approached 90 megatons. There was yet room for tremendous improvement in the hydrogen bomb's efficiency.

On May 22, 1952, Los Alamos and Sandia Laboratories announced the formation of an "Emergency Capability" Committee charged with weaponizing a "wet" EC 16 thermonuclear bomb at the earliest opportunity. Emergency capability (EC) meant that the AEC and DOD did not test the bombs for reliability or safety and that scientific personnel had to assemble and operate them. The Committee also announced plans for a "dry" lithium deuteride based TX14 thermonuclear bomb. The Joint Chiefs endorsed these plans in August. By this time, the Los

Alamos weaponeers were fully aware that high-energy neutron could efficiently decompose Li6 into tritium through the reaction

$$Li6 + n > T + He4 + 4.78 \text{ MeV}$$

Unfortunately, Li6 makes up only 7.4 percent of natural lithium. This forced the Atomic Energy Commission to convert the Y-12 Plant at Oak Ridge Tennessee into an enrichment facility to provide the high-grade fuel needed for thermonuclear assemblies.

In a bomb fueled with lithium deuteride, compression of the secondary decomposes LiD into a plasma of lithium and deuterium ions. At sufficiently high pressures and temperatures, the deuterium ions are free to fuse according to the reactions:

$$D + D > He3 + n + 3.27 \text{ MeV}$$

and

$$D + D > T + p + 4.03 \text{ MeV}$$

Since the first of these reactions produces neutrons, a third reaction:

$$Li6 + n > He4 + T + 4.78 \text{ MeV}$$

breeds tritium from lithium. This tritium then reacts with deuterium according to the reaction:

$$D + T > He4 + n + 17.59 \text{ MeV}$$

Neutrons from the fusion-boosted sparkplug enhance the decomposition of lithium to tritium.

Under conditions of sufficient temperature and pressure, the reactions:

$$Li6 + n > He4 + T + 4.78 \text{ MeV}$$

and

$$n + i <> n' + l'$$

combine into a closed-chain reaction called the "Jetter Cycle" to produce continuous burning:

$$T + D > He4 + n$$
$$\wedge \qquad\qquad \vee$$
$$T + He4 < Li6 + n$$

The relative importance of the Jetter Cycle grows exponentially because of the neutrons produced by thermonuclear reactions. To compensate for neutron losses from the excited plasma, and to accelerate the cycle, a neutron multiplier is required. The bomb's tamper supplies the needed neutrons through fast and slow fission. The downside of producing tritium from lithium is a reduced number of neutrons that might otherwise fission uranium in the secondary's tamper.

The use of lithium deuteride as a nuclear fuel has both advantages and disadvantages. The biggest advantage is that this fuel breeds its own tritium so that tritium's short half-life and high cost do not become an economic factor in the production of the bomb. LiD's main drawbacks are its high ignition temperature and an energy yield only 80 percent that of pure deuterium. Its benign handling characteristics more than compensate for these deficiencies.

In anticipation of the successful test of a thermonuclear bomb, the Air Force assigned Air Materiel Command the task of producing a practical delivery system in 1950. The initial delivery system was based on a remotely controlled unmanned B-47 bomber designated Brass Ring. The Air Force cancelled this project in 1952 when Air Materiel Command demonstrated that it could safely-drop drogued thermonuclear weapons from manned bombers. Air Materiel Command passed this mode of delivery to Air Research and Development Command's Wright Air Development Center for further development. Following an Air Force reorganization, Air Materiel Command passed the manned thermonuclear delivery project to the Air Force Special Weapons Center, which took on the task of producing delivery systems for the prototype thermonuclear weapons (TX14, TX15, TX16, TX17, TX21, and TX 24) scheduled for testing during Operation Castle at the Pacific Proving Ground in the summer of 1954.

As originally envisioned, Air Materiel Command assigned the Weapons Center's Thermonuclear Delivery Project seven tasks:
- Modify four B-36H bombers as prototype carriers for the MK 17 / 24
- Modify two B-47B bombers as prototype carriers for the MK 21 / 26
- Modify and redesign a B-2 bomb lift to MA-1 specifications in order to carry 50,000-pound bombs
- Develop a series of drogue chutes to decelerate the new bombs and to provide time for their delivery aircraft to escape their blast
- Develop practice bombs for training and operational efficiency
- Develop aircraft monitoring and control equipment
- Provide weapon effects support and determine safe delivery techniques

Along with a decision to freeze the size of the MK 17 / 24 bombs at a diameter of 61.5 inches and a weight of 50,000 pounds, AMC added the task of Emergency Operational Support to the project in 1953.

In mid-November 1953, DOD provided the AEC informal guidance as to three general types of thermonuclear weapons it desired:
- A "Class A" weapon weighing not more than 50,000 pounds with a yield of 10 megatons for delivery by the B-36 and B-52
- A "Class B" weapon weighing not more than 20,000 pounds with a yield of 5 megatons for delivery by the B-47
- A "Class C" weapon weighing not more than 8,500 lbs. with a yield of 1 - 3 megatons for delivery by a variety of aircraft and missiles

In February 1954, the MLC added to this list:
- A "Class D" high-yield thermonuclear weapon that weighed no more than 3,000 pounds

These specifications gradually evolved into detailed requirements for weapons with a 1957 deployment date:
- A Class A thermonuclear weapon with a maximum yield exceeding 20 megatons and compatibility with B-36 and B-52 aircraft. Although the weapon could weigh as much as 50,000 pounds, a lighter weight was desirable. The AEC was to determine the bomb's military characteristics when its weight, yield, and cost relationships became available.
- A Class B thermonuclear weapon with a yield of 5 megatons and a maximum weight of 23,000 pounds. The AEC was to give priority to reducing this weight below 15,000 pounds and then increasing the yield to at least 10 megatons. This weapon had to be compatible with the B-47, which had a short bomb bay.
- A Class C thermonuclear weapon with a yield of 2 - 10 megatons, weighing not more than 8,500 lbs. This weapon was to be compatible with B-47, B-66, A3D, and AJ aircraft as a gravity bomb and applicable with missiles as a warhead.
- A Class D weapon with a yield from 0.25 - 3 megatons that weighed no more than 3,000 lbs. The AEC was to adapt this weapon for internal or external

carriage and for release or separation at subsonic and supersonic speeds. It was to be compatible with bombers, fighter-bombers, and applicable missiles.

The Joint Chiefs of Staff desired a 5 - 1 ratio of Class B weapons with respect to Class A weapons. It also desired a 3 - 1 ratio of Class C and Class D (small) weapons with respect to Class A and Class B (large) weapons. Within the small weapon category, the Joint Chiefs desired a 3 -1 ratio of Class D to Class C weapons. These ratios were not to stand in the way of achieving the total number of weapons enumerated by the Secretary of Defense in a December 15, 1953, memorandum: *Desired Composition of the Atomic Stockpile*.

The weaponization of hydrogen bombs to meet this wish list was carried out under the direction of Los Alamos' Theta Subcommittee. Theta (θ) was a Greek letter used to designate thermonuclear activities. In addition to the longer-term projects associated with the MK 17 / 24 and MK 21 / 26 bombs, the DOD wanted B-36H and B-47B bombers adapted for the delivery of "Emergency Capability" EC 14, EC 16, and EC 17 bombs. (The DOD wished to have a small stockpile of weapons on hand, ready for immediate use, if their prototype tests proved successful). This strategy required aircraft modifications that included the addition of sway brace beams to weapons bays, bomb release adapters, pneumatic systems, manual bomb releases, suspension slings, sling retractor and snubbing equipment, arming control support, weapon arming control assemblies, parachute arming control assemblies, weapon and parachute safing handles, and platform junction boxes.

In addition to the aforementioned aircraft modifications, it was also necessary to develop drogue chutes to delay the time between release and detonation of these very powerful weapons. The strategic atomic bombs they replaced had all been freefall weapons. The immediate need was for a parachute capable of retarding a 50,000-pound bomb dropped from an altitude of 35,000 - 40,000 feet at release speeds of 300 - 500 knots. During 1953, the Air Force made over 60 drops from B-36s and B-47s at Kirtland AFB to define suitable bomb shapes and parachute sizes. It also specified the incorporation of four selectable conditions into the parachute designs: release of an armed weapon with parachute deployment, release of an armed weapon without parachute deployment (freefall), release of a safed weapon with parachute deployment and release of a freefall safed weapon. The Air Force did not consider wind drift a problem for delivery because they equipped the bombs with relatively small chutes.

The AEC tested candidate primaries for its proposed hydrogen bombs as part of Operation Upshot-Knothole, in the early summer of 1953. Los Alamos constructed mockups of several solid fuel thermonuclear prototypes, which it equipped with live primaries and dummy secondaries. These included the TX 14, TX16, and TX 17 / 24. The results for the TX14 (Nancy) and the TX16 (Badger) proved unsatisfactory when their deuterium gas-boosted "Racer" primaries failed to consistently reach a design yield of 35 - 40 kilotons. In a subsequent test (Simon) of the TX17 system, an improved Racer IV primary that contained an additional 2 kilograms of enriched uranium supplied satisfactory results. The last test in the series (Climax) used a composite-levitated core in a Cobra primary based on the high-yield MK 7 tactical bomb. It achieved its design yield of 60 kilotons in a Type D pit. Doubts about the reliability of boosted primaries resulted in the final choice of the Cobra TX7 implosion system for use in many of the first deployed thermonuclear weapons.

In January 1954, five EC 16 cryogenic bombs reached the thermonuclear stockpile. Their design yield was 6 - 8 megatons, and they measured 61.4 inches in diameter by 296.7 inches long. In order to achieve a 40,000-pound flyable weight, Sandia significantly reduced the bomb's casing thickness in relation to Mike. This was still one of the heaviest bombs ever produced. Fortunately, Air Materiel Command was able to convert a number of B-2 hydraulic lifts to the MA-1 configuration for loading the EC 16, affectionately known as "Jughead." These lifts were also capable of loading the EC 14 and the MK 17 / 24 bombs. Planned distribution of this lift was to National Storage Sites A, B, C, D, L, - 6 units each, and Site K, - 7units.

Just before technicians loaded the EC 16 into its sling, they would have charged it with its liquid cryogenic deuterium fuel. The venting of highly flammable deuterium gas during standby or in flight required the installation of ducts to clear it from the aircraft. The continuous evaporation of the deuterium fuel limited the time for a strike to 20 hours, when the weapon's fuel supply became seriously depleted.

The cryogenic fuel arrived at the flight line in specially developed trailer-mounted refrigerated transport Dewars. The transport Dewars had a capacity for 2,000 liters of liquid deuterium, which a closed-cycle helium refrigerator cooled. The refrigerator operated on electricity supplied by a diesel generator or by landlines. Provided the refrigeration system was operating, there was no appreciable loss of deuterium from the Dewar. Each transport unit occupied a semi-trailer that weighed 18.1 metric tons. The Air Force also acquired air-transportable Dewars for carrying liquid hydrogen or deuterium at 20 °K to forward sites in the B-36 and B-47. The National Bureau of Standards and H. L. Johnston Inc. developed these Dewars to minimize heat transfer through reduced conduction, convection, and radiation. The Dewars each held 750 liters of liquid deuterium and were so efficient that liquid deuterium boil-off was limited to 1 percent of capacity per day.

Although the Air Force desired to have contact fuzing for the MK 16, Sandia only equipped the EC 16 with an XMC317 barometric fuze. Los Alamos also used this fuze in conjunction with the EC 14 and the EC 17 / 24. Retarded by an 80-foot diameter parachute, the time of fall to the burst altitude of a MK 16 was 200 seconds. This compared to a freefall time of 52 seconds. The EC 16 was never equipped with an afterbody and parachute, leaving its only delivery option as freefall airburst. Because of their similarity, Los Alamos planned to equip the MK 14, MK 16, MK 17, and MK 24 bombs with identical afterbodies, parachutes, primaries, and nose sections.

As it turned out, Air Materiel Command only modified one B-36 as an EC 16 carrier. SAC saw solid fuel thermonuclear weapons as less expensive and easier to maintain than cryogenic weapons. The first solid fuel weapon was the EC 14, which entered service in February 1954, 1 month after the EC 16 and 8 months ahead of the first solid fuel thermonuclear proof test. The EC 14 was somewhat smaller than the EC 16, with a weight of 29,850 pounds, a diameter of 61.4 inches, and a length of 222 inches. A crude design, it had its thermonuclear fuel enriched to 95 percent Li6 and used the same barometric fuze as the EC 16. The TX14 prototype, tested as Castle Union, was equipped with a MK 7 based Cobra primary. The EC 14 derived five megatons of its 6.9-megaton yield from fission, making it a very "dirty" weapon. A 64-foot drogue parachute would have retarded its descent. However, no afterbodies were ever manufactured for the EC 14, so it would have been dropped freefall only. The B-36 and the B-47 were both potential carriers for the EC 14, although fin clearance for a MK 16 in the close confines of the B-47's bomb bay would have been critical. Confidence in the EC 14 design resulted in the demobilization of the EC 16 bomb in April, after only a few months of active service.

A single-stage thermonuclear "Alarm Clock" or "Sloika," which the Soviets successfully tested on August 12, 1953, pushed nuclear work along in the United States. By late 1954, seven prototypes from LASL and one prototype from LRL were ready for assessment. Los Alamos tested Bravo, the first device of the Castle Series at the Pacific Proving Grounds on October 1, 1954. Bravo evaluated a TX21 prototype named Shrimp. Shrimp was a 3/4 scale model of the EC-17 and EC-24 bombs, which were already deployed in an emergency capability mode. The Shrimp test was a validation for weapons containing solid thermonuclear fuel. The Shrimp device was a heavily instrumented, 23,500-pound cylinder that measured 179.5 inches long by 53.5 inches in diameter. Its relatively low weight resulted from the use of a 3 - 4-inch-thick 7075 aluminum alloy case instead of stainless-steel. The weaponized TX21 was suitable for carry by a B-47, which the EC-17 and EC-24 were not. Although the MK 21 could be carried by a B-47, the aircraft required the installation of special "fin" recesses in its bomb bay doors.

The Shrimp device was ignited by a Cobra TX7 primary, located at one end of its casing. This end sprouted a frustum or truncated cone, the base of which was about half the diameter of the casing. A complex series of steel pipes was attached to this projection to carry light from the primary to an instrumented bunker. About halfway down the casing and then about 80 percent of the way down the casing, two pairs of pipes emerged. These pipes carried light from the secondary to the same instrument bunker. Using this technique. scientists could time the ignition of the primary and the secondary and measure temperatures in the radiation channel.

The aluminum ballistic casing was lined with an inch of natural uranium. Natural uranium nails, coated with copper, attached the radiation case to the ballistic case as was done in the Ivy Mike device. The thickness of the radiation channel has not been divulged. Defining the parameters to calculate this value requires substantial test verification and so is a highly-kept secret. The radiation channel was filled with polystyrene plastic foam impregnated with a low-molecular-weight hydrocarbon. Unlike the Ivy Mike device, this material completely filled the radiation channel and acted as a structural component to support the secondary. For this reason, shrimp was detonated horizontally whereas Ivy Mike had to be detonated vertically.

Estimates suggest that the uranium tamper had a 1-inch thickness, similar to that of the uranium that lined the ballistic casing. This meant the force exerted by the vaporized hohlraums filling met with equal resistance on its outer and inner surfaces. The tamper was made of natural, or perhaps low-enriched, uranium. The end of the tamper, lying immediately below the primary, measured about a foot thick and was coated with Boron 10 to prevent predetonation of the sparkplug by stray neutrons emitted from the primary.

The volume of LiD fuel used was approximately 60 percent of the volume of the fusion fuel used in the EC 17 and EC 24 bombs. This was 500 liters (130 U.S. gallons), corresponding to a weight of about 1,125 pounds. The thermonuclear fuel was enriched to 40 percent Li6. The lithium deuteride fusion fuel was placed inside a stainless-steel canister of unspecified thickness. Between the stainless-steel case and the uranium tamper was an air-gap that increased the tamper's momentum during compression. Running down the center of the secondary was a 1/2-inch-thick hollow plutonium sparkplug. The 40-pound sparkplug was boosted with a few grams of tritium, which filled a 1/4-inch -diameter opening that ran down its length.

At least one author states that the sparkplug was timed to detonated when the first generations of neutrons arrived from the primary through a small hole in the end of the pusher. This hole was plugged with Boron10 impregnated paraffin wax to help control the timing of detonation. This made sure that maximum compression and ignition occurred simultaneously. The bottom end of the secondary was locked to the radiation case by a type of mortise and tenon joint. The hohlraum had a projection, which nested in the base of the secondary to lock it in place and provide structural support. Neither of these statements is substantiated by a reference but seem plausible to the author.

In evaluating the solid fuel selected for the Shrimp device, its designers missed a tritium reaction involving Li7:

$$Li7 + n > He4 + T + n - 2.47 \text{ MeV}$$

Because of this oversight, Bravo thus ran away to 15 megatons, the largest nuclear test in US history. The explosion opened up a crater over a mile wide and 250 feet deep, destroying much of the test equipment and trapping some of the observers in their bunkers. These observers noticed a pronounced ground roll before they felt the arrival of the shock wave. Bravo's fireball expanded to 4 miles in diameter before shooting up into the stratosphere. The mushroom cloud then dropped 100,000,000 tons of radioactive debris in a path measuring hundreds of miles long.

The main group of observers, located on ships at a distance of 30 miles from the test site, found they were too close for comfort. Heat from the explosion quickly built to intolerable levels causing some of the scientists to recall a concern that the bomb might ignite atmospheric nitrogen in a global thermonuclear holocaust. As the giant mushroom cloud spread directly overhead, a sudden shift in the wind subjected the ships to a rain of radioactive, white debris. It was vaporized coral condensing from the nuclear fireball as it cooled! This caused the crew and scientific personnel to evacuate to the lower levels of their ships until automatic equipment decontaminated the ships' topsides and hulls. A Japanese fishing trawler, the Lucky Dragon situated 85 miles downwind, was not so lucky. The fallout badly contaminated the ship and resulted in severe radiation sickness for the crew, causing one fatality. To improve safety, the Pacific Proving Grounds expanded the exclusion zone around future tests to a radius of 425 miles, an area equivalent to 1 percent of the earth's land surface.

The success of Bravo caused immediate changes in the order of Castle's lineup. The AEC cancelled Yankee, a proof test of the EC 16 Jughead cryogenic bomb and moved a TX17 device known as Runt into second place. The TX17 exploded with an 11-megaton yield in the Romeo test. If not for the success of Bravo, Los Alamos might not have tested the un-enriched Runt. Yankee, a test of the Li6 enriched TX24 Runt II, followed with a yield of 13.5 megatons.

Operation Castle had only one dud. This was the LRL Morgenstern device, which the AEC evaluated in the Koon test. It was the last weapon designed by Edward Teller. Reportedly, Morgenstern was an attempt to produce a megaton yield from a compact, spherical, solid fuel implosion system. It was equipped a Racer IV rather than a Cobra primary. Morgenstern's thermonuclear sphere central sphere had a number of spikes radiating from its surface giving it the appearance of the medieval weapon with the same name. Teller may have intended that inward propagating implosive jets from for these spikes would compress and ignite a volume of fuel in the sphere's central region. The Morgenstern device may have lacked a sparkplug.

The unit weighed 23,000 pounds and measured 56.4 inches in diameter. With a length of 115.9 inches, it was considerably shorter than the offerings provided by Los Alamos. The reason for its failure was preheating of the secondary that led to poor compression. Teller chose LASL's Racer IV boosted primary for inclusion in the device. It seems the Racer IV generated a much stronger neutron flux than anticipated by Teller. This, and the non-inclusion of boron in Morgenstern's blast shield probably caused the overheating. Pre-initiation may also have been a contributing factor in the bomb's low-yield. LRL staff members stacked water cans around the weapon before its test in a bid to shield it from stray cosmic neutrons. The AEC cancelled the test of UCRL's Ramrod device following this failure. It also canceled the MK 22 production version of the Morgenstern device within a month of its test.

To compensate for the demobilization of Jughead, and based on the success of Castle, the Air Force deployed five EC 17 and ten EC 24 bombs in support of the EC 14 between April and October 1954. The new bombs differed only in the degree of enrichment given to their secondaries. The EC 17 used un-enriched fuel, whereas the EC 24 had lithium deuteride enriched to 95 percent Li6. The bombs were almost 19 feet long and measured just over 61 inches in diameter. Equipped with 3.5-inch-thick steel casings, they weighed in at 21 tons. They used the Cobra TX7 primary and barometric fuzes developed for the EC 14, but lacked such features as in-flight insertion for the primary's core, safe arming and fusing systems, and parachutes. Because they did not have parachutes to delay the time of detonation, the delivery of these bombs was tantamount to a suicide mission.

The overall success of Castle's thermonuclear tests caused both jubilation and paranoia. The notorious atomic spy, Klaus Fuchs, had supplied key design elements of the atomic bomb and plutonium production reactors to the Russians in the late 1940's. This enabled the acceleration of the Soviet atomic program. Fuchs was a German born British physicist who served at Los Alamos in the days of the Manhattan District and was a co-inventor of the mechanical neutron initiator. With the success of the hydrogen bomb program, suspicions were aroused that the physicists who had opposed its development might also supply information about staged radiation implosion to the Russians.

Unbelievably, serious suspicion fell on Robert Oppenheimer, the father of the atomic bomb. In this era of McCarthyism, it did not take much to attract attention and Oppenheimer's flirtation with communism as a liberal professor in the 1930s and his initial opposition to the development of thermonuclear weapons did not sit well with military and government circles. The Air Force had already made an internal decision not to use Oppenheimer's services as a consultant, but personal enmity between Oppenheimer and Lewis Strauss resulted in special hearings to ensure that the AEC revoked his security clearance. The charges were groundless, but the powerful political forces behind H-bomb development had their way. Edward Teller, a former colleague, allowed himself to testify against Oppenheimer. His "thirty pieces of silver" was the assured future of thermonuclear development.

The removal of Oppenheimer's security clearance and the spectacle of the security hearings left the physicist disheartened. To the end of his days, he was on the outside looking in. In an ironic turn of events, the personal situations of his detractors miscarried because of the show trial. Although Teller remained a friend and confidante of generals and admirals, many in the scientific community ostracized him because of his disservice to Oppenheimer. In similar fashion, Lewis Strauss failed in his bid to become Secretary of Commerce, largely because of his handling of the Oppenheimer case.

Thermonuclear development now began to move swiftly. The Air Force cancelled the MK 14 because it used expensive enriched lithium fuel and had a relatively low yield when compared with the MK 17. Sandia then brought the components of the early EC 17 and EC 24 freefall bombs up to certified stockpile quality. The manufacture of 200 MK 17 and 105 MK 24 "Class A" weapons provided the nucleus of a nascent thermonuclear stockpile. SAC carried these weapons on board B-36 aircraft after Air Materiel Command joined their number three and four bomb bays into a continuous unit. The project, named "Bar Room," initially envisioned the carriage of MK 16, MK 14, and 17 bombs, but after the cancellation of the MK 16, it received the designation "Cauterize". Ultimately, 208 bombers were converted to the thermonuclear role. Operationally, SAC tasked the B-36 with attacking primary targets with a Mk 17 / MK 24 bomb and secondary targets with MK 6 and MK 18 atomic bombs carried in its forward bomb bays. Although the B-52 was large enough to carry a MK 17 or a MK 24 bomb, Air Materiel Command considered the shock loading associated with parachute opening unacceptable at the high speeds associated with this aircraft.

Since the bombs did not come equipped with suspension lugs, Air Materiel Command had to develop flexible metal-link slings and sway brace systems for their carriage in a bomb bay. After takeoff, a crewmember installed a safety pin in the bomb's U-2 pneumatic release mechanism that prevented the accidental or deliberate release of the weapon. The crewmember had to remove the pin for an attack. A crewmember also had to remove the pin during takeoffs and landings so the bombardier could jettison the bomb in case of an emergency.

Beginning with the MOD 0 version, MK 17 and MK 24 bombs were fitted with 5-foot diameter guide parachutes, 16-foot diameter deployment chutes, and 64-foot diameter main canopies. Deploying parachutes in multiple stages to slow the fall of the bomb, meant that the Wright Air Development Center could make the main canopy lighter, reducing bulk and weight. The parachute system was equipped with a static line that attached to a mount inside the bomb bay by means of a V ring. Selective in-flight control was available to retain or release the V-ring for freefall or parachute-retarded delivery. Use of the parachutes provided a 2-minute delay as the bomb fell from 36,000 feet to a detonation altitude of about 8,000 feet.

Sandia had to lengthen the MK 17 / 24 MOD 0 casing to accommodate the afterbody for the new parachute package. Its casing then measured 61.4 inches in diameter by 24.8 feet long and had an ogive shaped nose with a flattened end. The afterbody of the bomb that held the parachute was several inches smaller in diameter than the rest of the bomb. The bomb casing had five spoiler bands, three that surrounded the ogive nose, and two on the sloped section where the main bomb casing narrowed in diameter to accommodate the sub-diameter parachute compartment. The bomb had two ports just ahead of the parachute compartment. One port accommodated the bomb's pullout cables and the second accepted a pair of arming wires. On either side of the parachute compartment, hatches gave access to the bomb's batteries.

The MK 17 MOD 0 and MK 24 MOD 0 had manual in-flight core insertion systems. Prior to a flight, the ground crew installed the core in the bomb's receiver through its nose. When the bomb was loaded into the bomb bay, crewmembers had to leave a minimum 5 1/2-inch gap in front to allow for the insertion of a crank. Turning the crank while in flight inserted the core into the primary. To facilitate this operation, Air Materiel Command provided a small catwalk in the bomb bay with lights, oxygen and heated suit outlets, a communications jack, and an emergency alarm bell.

For in-flight monitoring and control, aircrew used a T18 in-flight controller, a T19 in-flight controller, and a T35 flight control box. The T18 monitored the condition of the batteries (load / no load), the baro-switch setting (open / closed), the safety switch setting (open / closed), and the presence of voltage at the inverter (yes / no). It also supplied power to the battery heaters. Aircrew used the T19 to set the MC5 or MC10 baro-switches and to indicate their settings. The function of the T35 was to monitor and control the safety switch. A T23 flight circuit tester checked on the condition of the flight control and monitoring equipment as well as associated aircraft wiring.

Fuzing for the MK 17 / 24 was similar to contemporary fission bombs. Before release, ground crew had to set the weapon for airburst or contact and for freefall or parachute-retarded delivery. If the ground crew selected retarded freefall delivery, pullout wires activated the bomb's batteries and a static line opened the parachute at release. The pullout wires also started the bomb's mechanical timers. After falling from a release altitude of 36,000 feet to a predetermined height, these timers activated the bomb's barometric fuzes and started the charging of the primary's x-unit. If the ground crew selected airburst, the baro-fuzes closed at a pre-selected detonation altitude to release the stored energy of the x-unit into the firing-set. If the ground crew selected contact fuzing, the bomb did not detonate until the impact fuzes contacted the ground.

The MK 17 MOD 1 appeared in March 1955. It had an electro-mechanical, in-flight core insertion mechanism for its TX7 atomic primary, which simplified flight operations. This change required lengthening the bomb's nose by 2 inches. A lightweight 3 1/2-inch-thick aluminum alloy replaced the steel casing used for earlier models, and plastic and steel composites replaced the MOD 0's all steel fins. In order to reduce weight further, Sandia thinned the MOD 1's lead and low-Z plastic liners. To maintain the weapon's case seal, Sandia routed the cables that connected the afterbody to the primary through a plastic conduit with special pressure connections. Other openings that specialists had previously covered with duct tape in the MOD 0 received permanent seals. A positive sealing hatch in the nose facilitated the installation of the nuclear core in the in-flight insertion mechanism prior to a mission.

Sandia fuzed the MOD 1 bomb for barometric airburst only. The DOD then requested Sandia convert all MOD 0 weapons to the new standard by September 1955. The final MOD 2 version of the MK 17 bomb reverted to the MOD 0 fuze with a contact burst capability. The MK 17 MOD 2 had an 18-month stockpile life and a 15-day limit as a ready response weapon. Stored in a separate sealed container, the MK 17's fuze had a 6-month storage capacity. The AEC converted approximately 25 percent of the MK 17 stockpile to the MOD 2 standard during the period June - August 1956.

Whereas a B-36 Peacemaker could comfortably carry a MK 17 or MK 24 bomb, the B-47 Stratojet and the Navy's proposed P6M Seamaster amphibious bomber required a smaller, lighter weapon. Los Alamos created this bomb by streamlining the TX21 (Shrimp) device. To this end, the Theta Subcommittee outlined a plan whereby the external dimensions of the TX21 bomb could be determined by mid-1954; fuzing, firing, and ballistic drops held by early 1955; the final design released by mid-1955; and early production begun in January 1956. The Air Force specified that the bomb be capable of remote controlled automatic nuclear arming and in-flight selectable release in freefall or retarded options and at any time prior to release. It also requested a safe separation timer and the capability for airburst or surface burst, with the latter acting as cleanup in the event of an airburst malfunction. The bomb's power sources were to have a 2 to 5-year shelf life, require no preparation other than preflight installation in the bomb, and no activation requirements other than weapon release.

Los Alamos initially estimated that the TX21's dimensions would be 147 inches long and 47.5 inches in diameter in the parachute-retarded version. It would weigh between 15,000 and 16,000 pounds with a casing made from 1/2-inch-thick aluminum. As different dimensions might be required for different applications, Sandia tested scaled diameters of 48, 51, 52, and 54.44 inches at Mach 0.93 speeds in a wind tunnel. The final version of the Class B MK 21 MOD 0 bomb had a hemispherical nose with a 0.7 caliber flat. It had a 56-inch diameter casing that measured 58.48 inches in diameter over its four spoiler bands and was 148.37 inches long. It weighed 17,600 pounds with a 3 and 1/2-inch-thick aluminum alloy case. The overall design produced good pressure sensing results for the bomb's barometric fuzes.

In freefall, the MK 21 MOD 0 (TX21X1) casing was stabilized by four fins having a 6 / 14-degree double angle wedge, a 39.5-inch chord, and an 80.9-inch span. The forward half of each fin was made of stamped metal and the rear half was made of polyester glass laminate. The plastic portion of each fin incorporated a proximity fuze antenna. The bomb's afterbody, on which the fins were mounted, was an extension of the warhead case. An external bolting flange that acted as a fifth spoiler connected the forward and after bodies into a single unit. The nickel-cadmium batteries, safety switches, and an interconnecting box were all readily accessible through the rear of the bomb after removal of the parachute container.

The AEC planned to produce the TX21 in a "normal" or dirty version, and as the clean TX26. Thus, it referred to this weapon to as the TX21 / 26. As it turned out, the DOD had no interest in a low-yield clean bomb so the AEC restricted production to the MK 21 MOD 0 normal version. The AEC then cancelled advanced versions of the MK 21 (such as the TX21X2 and the TX21X3) in favor of the TX36. Although the AEC never tested a normal TX21 device, it tested a clean TX21C / TX26 device as Redwing Navaho. The test yielded 4.5 megatons with a lead secondary tamper. Its 95 percent fusion yield set a record as the cleanest American thermonuclear test on record and may have been a proof of the TX/WX 26 warhead. In fact, the AEC sometimes mandated the use of a lead tamper to reduce fallout during testing.

The MK 21 had a Cobra primary, thermonuclear fuel enriched to 95 percent Li6, and a partially enriched uranium tamper. Despite its compact size, the normal yield of the MK 21 was likely 15 megatons, an estimate based on Mike test. Like its contemporaries, the MK 21 MOD 0 was barometrically fuzed for parachute-retarded and freefall airburst, with a radar proximity surface burst backup. The parachute had a two-stage deployment system equipped with a 24-foot diameter main ribbon canopy. Reefed, the canopy provided 75 seconds of retardation, and unreefed, it provided 108 seconds of retardation. Aircraft speed at bomb release was limited to 400 knots so as not to exceed an opening shock of 6,000 Gs on the parachute harness.

For freefall or retarded airburst, the ground crew set the MK 21's arming baro-switch to a desired altitude prior to takeoff. During flight, the aircraft supplied power to battery heaters in the bomb. As the bomber approached the target, the bombardier set the firing baro-switch to the selected height of burst, activated the automatic insertion of the nuclear capsule, and armed the safing switch. Withdrawal of the arming wires upon release closed a pullout switch to supply power to the arming fuze. This action also started the inverters and, after a safe separation interval determined by a setting on the arming baro-switch, charged the x-unit and armed the radar proximity fuzes. A timer-controlled 3-second delay prevented firing until the x-unit was fully charged. As the bomb reached its burst altitude, the firing baro-switch actuated the x-units trigger to discharge its capacitor bank into the firing harness and detonate the primary. If this system malfunctioned, the radar proximity fuze detonated the bomb. When SAC desired a near surface burst capability, the firing baro-fuze was set to -3,000 feet leaving the radar proximity fuze to detonate the bomb.

Sandia equipped the MK 21 MOD 1 (TX21X2) with nickel-cadmium batteries and true contact fuzing. For contact fuzing, the laboratory considered barium titanate piezoelectric crystals, fixed or extendable nose probes, and a special insulated double shell that crushed on impact to produce a firing signal. This latter device consisted of a hemispherical nose fairing connected to the bomb's forward spoiler band. It was electrically isolated by means of an insulating rubber gasket. The outer component of the hemispherical shell was composed of a fiberglass laminate. Underneath was an insulated electrode layer, separated from a ground electrode layer by an intermediate layer of insulation. Upon impact, the two electrode layers completed a circuit that triggered the bomb's x-unit to set off a full yield thermonuclear explosion. The MK 21 MOD 2 (TX21X3) was to be equipped with all of the MOD 1 features and a boosted primary. Neither the MOD 1 nor the MOD 2 saw production.

Los Alamos produced and stockpiled 275 MK 21 MOD 0 bombs during the period December 1955 - July 1956. The Air Force began the conversion of B-47s to carry these weapons in January 1955, and by April 1956, more than 1,100 aircraft were available for this purpose. Strike aircraft carried the MK 21 bomb in a sling, in much the same way they carried the MK 17 / 24,

A year after Los Alamos tested the TX15 Zombie device as shot Nectar of Operation Castle, it placed the device into service as the lightweight, Class C, MK 15 bomb. Because its military characteristics originally called for its use in both bombs and missile warheads, the MK 15 MOD 0 was made smaller than contemporary bombs. Depending on its MOD, the blunt nosed MK 15 measured from 34.5 - 35 inches in diameter, from 136 - 140 inches long, and weighed 7,600 pounds. The blunt nose helped to optimize drag. Although classified as a hydrogen bomb, the MK 15 was more of a fusion-boosted, radiation-imploded fission bomb. It originated with Stan Ulam's idea of using the shock wave from one atomic explosion to set off a second larger atomic explosion.

The prototype used for the Castle Nectar test was equipped with a Cobra TX7 primary, but the production MK 15 MOD 0 bomb incorporated a Viper I primary with an all U235 core that required in flight insertion. The explosive sphere for the Viper I weighed 400 pounds and used 92 detonators. The MK 15's secondary incorporated a large amount of tritium and had an enriched uranium pusher that substantially increased the fission percentage of its total yield. The concept behind this weapon started a trend in which Los Alamos used secondary tampers with 40 to 80 percent enriched uranium to boost their bomb's yields. On May 21, 1956, the Redwing Cherokee test produced 3.8 megatons from a Class C TX15X1 (MK 15 MOD 0) airdropped bomb at the Pacific Test Range. This was the first air drop of a thermonuclear weapon.

To be useful as a missile warhead, the MK 15 had to operate reliably in temperatures ranging from −65 °F to 165 °F and at altitudes as high as 60,000 feet. Its in-flight insertion / extraction mechanism had to function within 10 seconds under a 10G acceleration. Configured as a bomb, the MK 15 had to withstand catapulted takeoffs and arrested landings on aircraft carriers. The Navy favored development of this weapon for internal carry with the A-3 Skywarrior from the deck of its aircraft carriers. During its development, the DOD cancelled the TX15's requirement for use as a warhead. Instead, it deployed the MK 15 MOD 0 as an internally carried freefall bomb. In this mode, the bomb was equipped with a barometric fuze that had adjustment for continuous height of burst and a radar proximity fuze that provided a near surface burst capability like the MK 21 MOD 0. Because Sandia installed the bomb's radar antennas in its fins, they gave adequate coverage straight ahead, but triggered the firing system if the bomb fell alongside a building.

The MK 15 MOD 0 bomb had a Type 3 case made from cast iron with a thickness of 2.69 inches. Sandia later reduced this thickness to 1.88 inches of cast iron in the MOD 2. In the MK 39 MOD 0, Sandia further reduced the case thickness to 1.23 inches of steel. The AEC used retired MK 15 casings to drop test a variety of thermonuclear devices because this practice simplified loading and aiming procedures for the Air Force. As these tests proceeded, it became apparent that the thick iron and steel casings were providing a back reflected neutron flux that was absent in devices that had been exploded on the ground. In order to make the drop test results more comparable with previous surface shots, it was necessary to add a boron neutron absorber to the inside of the bomb casings.

A TX15X1 (MK 15 MOD 1) version of the MK 15 was intended to have thermal batteries and improved fuzing. The AEC, however, cancelled the MOD 1 in favor of a proposed TX/XW15X3 (MOD 3). A TX15X2 (MK 15 MOD 2) bomb with a safing device, sealed pit, and external initiation reached deployment in March 1957. It was equipped with a frangible aluminum honeycomb structure for shock absorption where the core insertion mechanism had been situated in earlier models. A 3-foot diameter extractor and 12-foot diameter ribbon chute stored in a sub-diameter afterbody gave the MOD 2 a 110-second time of fall from an altitude of 32,000 feet. The MK 15 was sufficiently versatile that the AEC produced 1,200 units in MODS 0 and 2.

The MK 15 MOD 2 was the first bomb to be equipped with thermal batteries, originally developed by the Germans for use in the V-2. Sandia Laboratories continued their development after WW II. With an indefinite shelf life, thermal batteries ended the repetitive task of NiCad battery maintenance and replacement. Thermal batteries are hermetically sealed, solid-state units that have their individual power cells constructed from electrochemical couples. Their solid electrolytes initially consisted of a combination of calcium and calcium chromate in the MC473 battery used in the MK 15 MOD 2. The MC473 battery had a sealed rectangular case that measured four x four x eight inches and weighed 18 pounds. Its activation time was approximately three seconds, and it provided 30 amperes at 28-volts for a time in excess of a minute. The MC473 was manufactured for the AEC by the Catalyst Research Corporation. Sandia later replaced the carbonate formulation with a lithium / potassium chloride mixture, since superseded by even more efficient lithium silicon / iron disulfide and lithium silicon / cobalt disulfide formulations.

A heat pellet activated by an electric match (Bisch generator) or percussive primer liquefies the solid electrolyte in thermal batteries, which achieve full power in about a second and, depending on their design, can operate for a period of a few minutes to an hour. Internal operating temperatures vary from 500 °C to 700 °C and require an insulating blanket to protect adjacent equipment from damage. Thermal batteries produce power outputs from a few watts to several kilowatts. Their advantages include a high-power density, a low uniform internal impedance, and minimal electronic noise. The use of thermal batteries as a power source in nuclear weapons is now almost universal.

Two advanced TX15 designs were the TX/XW15X3 and TX/XW15X4. The TX/XW15X3 was a modification of the TX15X1 that had reduced weight and incorporated thermal batteries and improved fuzing. The TX15X4 was a TX15X3 device with an externally-initiated, gas-boosted, 40-kiloton Viper II primary with a one-point safe enriched uranium core. The TX/XW15X3 was redesignated as the TX/XW39X1 (which became the MK 39 MOD 0) and the TX/XW15X4 was redesignated the TX/XW39X2 (which became the MK 39 MOD 1).

Continued thermonuclear refinements in the mid-1950s soon made the MK 21 and MK 15 obsolete. The MK 21 was withdrawn after only a year of service, at which time the AEC had its casings recycled into an improved Class B weapon that it designated MK 36 MOD 0 Y1. An important design, the MK 36 had a production run of 920 bombs in two MODs and two yields that at one point represented half the megatonnage in the US arsenal. The normal Y1 yield was 19 megatons and the clean Y2 yield was 6 megatons. Only 250 kilotons of the Y2 yield were the result of fission; it was likely equipped with a lead thermonuclear tamper. The MK 36 Y1 was equipped with a natural uranium thermonuclear tamper. The large production run of MK 36 bombs gave rise to the notion of "bouncing the rubble higher" since they could destroy the Soviet Union several times over.

The MK 36 MOD 0 Y1 had specifications similar to the TX21X2, it measured 56 inches in diameter across its casing, 58 1/2 inches in diameter across its spoiler bands, and 12 feet 6 inches in length. It had a 78-inch finspan. Like the MK 21, the 17,500-pound MK 36 had a 3 and 1/2-inch-thick aluminum case to reduce its weight. The casing alone weighed 12,000 pounds. The bomb was parachute-retarded, equipped with nickel-cadmium batteries, and fuzed for either airburst or proximity delivery. It had a mechanical in-flight insertion mechanism for the composite-levitated core in its primary. Boeing's new B-52 heavy bombers could each carry a pair of MK 36 bombs, a circumstance that resulted in the retirement of the MK 17 and MK 24 bombs, along with the B-36 bomber. Its primary modes of release were freefall ground burst and retarded airburst. The MOD 1 version of the MK 36 replaced proximity fuzing with barium titanate fuzes to give it a true contact burst capability. As well, thermal batteries replaced the NiCad batteries that equipped the MK 36 MOD 0.

The ground crew made arming and firing baro-switch settings using single dial settings. The settings took into account the height of the target above sea level and included a 300-foot offset before firing. Prior to release, the aircraft's radar navigator placed a selector switch in the GND position, initiated in-flight insertion of the nuclear capsule, and armed the bomb. At release, pulse generators activated low voltage batteries, and closed safety switches in the bomb's high voltage circuits. When the arming baro-switch closed, low voltage power activated the 2,500-volt high-power thermal batteries that charged the x-unit and applied plate voltage to the trigger circuit. If the radar-navigator selected airburst, the baro-fuze detonated the bomb at the preset altitude. If not, the contact fuze detonated the bomb on impact. The MK 36 MOD 1 entered the stockpile in October 1956.

Los Alamos and Sandia next began development of an improved MK 36 MOD 2 (TX36X2) bomb that they equipped with a boosted primary and a sealed pit. In addition, they upgraded the bomb's firing system with an external neutron initiator to improve the primary's efficiency. The AEC and DOD intended to convert approximately half of the existing stockpile to this new configuration. However, on January 9, 1959, the AEC cancelled the MK 36 upgrade program and the MK 36 MOD 1 bombs scheduled for renovation remained unchanged until retired from service in 1961. The MK 41 and MK 53 replaced the MK 36 at this time.

In order to keep the smaller MK 15 MOD 0 from reaching obsolescence, the DOD requested that the AEC produce a lightweight version of this weapon. The upgrades for the TX15X3 version of the bomb were to include thermal batteries, true contact fuzing, and a lightened case. Because the MLC thought that the new bomb would have different performance characteristics than the MK 15 MOD 0, it requested a new designation. The Division of Military Application agreed and gave it the designation TX39X1 (MK39 MOD 0). The MK 39 MOD 0 measured 34.5 inches in diameter, 137 inches long, and had four double wedge fins with a box dimension of 39.75 inches. At 6,650 pounds, the MK 39 MOD 0 weighed 850 pounds less than did its MK 15 MOD 0 predecessor.

The MK 39 MOD 0 had the same 3.8-megaton yield as the MK 15 MOD 0 / 2. It was the first thermonuclear weapon carried by means of a clip-in suspension system. The Air Force then adopted this suspension method for the MK 36. Ground crew secured the MK 39 to an MHU-21/C suspension pallet and the MK 36 to an MHU-22/C suspension pallet by tightening aluminum bands around the bombs' casings. The suspension pallets and bombs were then winched into strike aircraft. Explosive bolts released both bomb and pallet after the bomb bay doors opened in flight. One set of bolts separated the suspension pallet from the aircraft and another set separated the bomb from the suspension pallet. Like the MK 15 MOD 2, the MK 39 MOD 0 was equipped with a 3-foot diameter extractor chute and a 12-foot diameter ribbon chute. This gave it freefall airburst, freefall contact, and retarded contact fuzing options.

The MK 39 was controlled and monitored with a DCU-9A in-flight system. At this time, Sandia tailored monitoring and controlling systems to specific weapons and then installed them in the aircraft assigned to carry them. Boeing installed the DCU-9A at the radar navigator's station to serve both of the B-52's bomb bays. The DCU-9A had a rotary selector switch with SAFE, GND, AIR, and OFF positions, a control arm having the positions OS and SGA, a red warning light, a test switch, a dim control, and a holding relay. When placed in the OFF position, the rotary selector switch cut aircraft power to the bomb bays and when placed in either the GND or AIR positions, applied pre-arming power.

When the control arm was in the OS position, the navigator could move the rotary selector from OFF to SAFE and back to OFF but could not rotate it beyond the SAFE position. When the Control Arm was in the SGA position, the navigator could move the rotary selector switch from SAFE to GND or AIR and back to SAFE but could not return it to the OFF position. The warning light served to indicate a malfunction. The test switch verified that the warning light was operational. The rotary selector switch had to be in the SAFE, GND, or AIR position for the test switch circuit to operate. The dim control affected panel illumination. A holding relay-controlled bomb safing and power monitoring even if the navigator returned the rotary selector switch from GND or AIR to OFF faster than the safing cycle could operate.

The DCU-9A panels and the bomb's arming units were interlinked through a DCU-47A readiness panel located at the pilot's (Bomb Commander's) position. The pilot had to first place the DCU-47A switch in the READY position before the radar-navigator could set the DCU-9A panel to AIR or GND in preparation for a strike. This applied aircraft power (28 volts at 3 amps for three seconds) to arm the MC772 ARM / SAFE switch. This partially pre-armed a selected bomb. In order to complete pre-arming, lanyards that ran from the bomb to the crew compartment were pulled to extract safety pins from the bomb's pullout rods before release. In its SAFE mode, the MC772 ARM / SAFE switch allowed the radar navigator to jettison a bomb without concern of accidental detonation.

The ground crew made baro-switch altitude settings for the MK 39 MOD 0 prior to takeoff using a preflight control box that plugged into an electrical receptacle in its casing. The ground crew calculated the baro-setting by adding height of burst to the target's ground level but did not bother to account for meteorological conditions. After takeoff, the radar-navigator turned on the DCU-9A and set it to SAFE to monitor the bomb. Prior to bomb release, when the radar-navigator moved the DCU-9A's rotary selector from SAFE to either AIR or GND, a solenoid in the bomb's ARM / SAFE unit rotated to the armed position and the insertion mechanism moved the nuclear core into the bomb's primary. A second solenoid switch remained closed or opened depending on whether the navigator selected AIR or GND.

Upon release and the pull out of the arming rods, Bisch generators ignited the bomb's low voltage thermal batteries and closed switches in its high voltage circuits. When the arming baro-witches closed, low voltage power ignited a pulse transformer that initiated the high voltage thermal battery pack. The high voltage battery pack provided the 2,500 volts needed to charge the x-unit's capacitors and applied plate voltage to the trigger circuit. If the radar-navigator selected airburst, closure of the firing-baros at a pre-set altitude detonated the bomb. If the radar-navigator selected ground burst, an option switch held the firing circuits open to allow the contact fuze to supply a separate voltage to detonate the bomb.

The MK 39 MOD 1 (TX39X1 / TX15X4), which followed the MK 39 MOD 0, had an externally initiated, boosted, Viper II primary with a core that did not require inflight insertion. The bomb measured 136.25 inches long and was 34.5 inches in diameter. The change to a boosted primary reduced both weight and power requirements, resulting in the use of smaller and lighter batteries. Sandia placed 150 pounds of aluminum "honeycomb" in the bomb's nose to crumple and absorb impact. The honeycomb material occupied the space that formerly contained the MK 39 MOD 0's in-flight insertion mechanism. Sandia also replaced the MOD 1's baro arming system with a timer.

Although the MK 39 MOD 1's Viper II primary was fusion-boosted, it was not one-point safe. Thus, the AEC thought it necessary to add a number of safety mechanisms. Sandia built an integral shear pin into the bomb's pullout rods to prevent accidental removal when the bomb was not in an aircraft. After the ground crew uploaded the bomb to an aircraft, the installation of a non-shearing pin allowed proper rod removal. At least 200 pounds of pull was required to remove the arming rods. As well, Sandia added an MC788 high voltage safety switch to prevent accidental charging of the x-unit and installed thermal fuzes between the pulse igniters and the low and high voltage thermal battery packs to make them inoperable in the presence of abnormally high temperatures. Finally, Sandia incorporated a trajectory-arming switch to prevent accidental detonation of the bomb on the ground or during loading. The AEC upgraded some MOD 0 bombs to the MOD 1 configuration.

Removal of the MK 39 MOD 1's pullout rods upon its release activated its MC685 Bisch (pulse) generator that in turn ignited its MC640 low voltage thermal battery pack. Power from the low voltage battery then activated the bomb's MC543 timers. If the retarded fuzing option was selected a solenoid locked the bomb's static line to the aircraft. An MC788 high voltage safing switch prevented current from reaching the x-unit until the closure of a trajectory-arming switch supplied a continuous signal. After the bomb had fallen for a 42 second period, the timer connected an arm and fire baro-switch to the electrical circuitry. The arm and fire baro-switch was an eight element, fixed offset device that was set manually before a mission. When the high voltage safety switch closed at its preset altitude, power from the low voltage battery closed a second switch that connected the high voltage batteries to the x-unit. At a preselected arming altitude, the arming baro-switch closed to allow power from the low voltage supply to activate the high voltage thermal batteries, which charged the x-unit with 2,500 volts of power. It also initiated the injection of DT gas into the bomb's fusion-boosted primary.

At airburst altitude, the firing baro-switch closed to connect the low voltage 28-volt power supply through a pair of fuzes to saturable transformers in the trigger circuit. When the transformer's main coil had a current applied, the resulting magnetic flux drove the core into a state of saturation. If the coil had a subsequent flux applied in the same direction as the initial pulse, it generated very little current in the transformer's secondary coil. However, if the direction of flux was reversed, there was a large change in flux and the resulting current output was sufficient to trigger a set of thyratrons. After the baro-switch closed, it took only about 100 microseconds for the fuzes to fail, reversing the direction of current in the transformers to detonate the bomb. This device was much simpler and more reliable than available electromechanical systems. In the contact option, the saturable transformers blocked the signal from the firing baro-switches to allow detonation by a signal from its barium titanite contact fuzes.

The MK 39 MOD 1 was initially equipped with a MK 2 MOD 0 parachute system that provided retardation. Drop testing of the MK 39 MOD 1 consisted of eight ballistic and 20 fuzing and firing tests. The Air Force conducted the fuzing and firing tests at a variety of altitudes and selection of speeds, with half of the tests freefall and the other half retarded. This program indicated great reliability and compatibility with the bomb's strike aircraft.

In addition to use as a bomb, the Air Force selected the MK 39 MOD 1 as the basis for a missile warhead. At this time, the Army requested a low-yield to supplement the W39 Y1's 3.8-megaton yield. The Y2 low-yield probably involved replacing the uranium in the bomb's thermonuclear tamper with lead. Sandia produced the warhead by removing the fins, nose cap and rear assembly from the MOD 1 bomb and installing warhead protective covers. The W39Y1 MOD 1 warhead measured 101.5 inches long and weighed 6,230 pounds.

In November 1957, Strategic Air Command urgently requested a MK 39 bomb suitable for low-level release and retarded laydown delivery. This bomb would become the MK 39 MOD 2. Because Sandia could not iron out the problems with a laydown delivery system in a meaningful timeframe, it fuzed the MK 39 MOD 2 for freefall ground burst and retarded ground burst only.

In order to produce the MK 39 MOD 2, the Wright Air Development Center first equipped the MK 39 MOD 1 with a three-parachute system in a 42-inch diameter afterbody produced by Sandia. For attachment purposes, Sandia reduced the center body of the bomb to a cylinder, re-attaching its four fins to the afterbody. In this configuration, the fins had a 55-inch span. This early parachute system did not slow the bomb sufficiently and a 24-foot diameter ribbon chute that extracted a 100-foot diameter solid canopy replaced it. During the deployment of the solid canopy, an ejection mechanism discarded the ribbon chute. Each chute had one reefed and one un-reefed stage. Sandia named this system the MK 39 MOD 1 Big Tail.

Because of the bomb's relatively low impact speed, Sandia could not use barium titanate crystals for its contact fuze. Instead, the laboratory had the nose cap machined to a thickness of 0.08 inches and a metal spike placed behind it with a separation of 1/8 inch. When the nose cap deformed on impact and contacted the spike, it completed an electrical circuit that detonated the bomb. This was superior to a double shell design because it was more resistant to antiaircraft fire. Since the Air Force wanted the MK 39 MOD 1 Big Tail for retarded ground burst only, Sandia installed a cable that bypassed the bomb's airburst fuze. Although the MK 39 MOD 1 Big Tail was not capable of laydown delivery, an aircraft could drop it in retarded delivery from much lower altitudes than the basic MK 39 MOD 1.

In order to improve the MK 39's parachute system, the Wright Air Development Center produced an "Improved Big Tail" parachute system known as the "BAT." The BAT used the same Sandia designed afterbody as the preceding Big Tail bomb. A 6-foot extractor chute, a 28-foot diameter first stage ribbon chute, a 64-inch diameter octagonal deceleration chute, and a 100-foot diameter second stage canopy safely decelerated the bomb to its landing speed. In developing a laydown capability for the MK 39, Sandia considered landing the bomb both vertically and horizontally. To land the bomb horizontally, Sandia designed a thick layer of shock mitigating material that attached to the underside of the bomb. This attachment looked so much like a boat that Sandia called it the "Lone Star," after a local small boat manufacturer. The design weighed 7,500 pounds. Unfortunately, it was too bulky for the B-47 and a number of other aircraft, so the Air Force rejected it and accepted the MK 39 MOD 2 BAT which had fuzing for freefall ground burst and retarded ground burst only.

Prior to release of the MK 39 MOD 2 from a B-52, the radar-navigator rotated the DCU-9A selector to AIR or GND. This required the radar navigator to pull out a knob on his control panel using both hands and shear a copper retaining pin to turn it. This action allowed the operation of the bomb's MC772 ARM / SAFE electrical switch and closed a solenoid that locked the bomb's static line to the aircraft if retardation was selected. The MC772, if left on SAFE, allowed a bomb to be jettisoned without fear of nuclear consequences.

At bomb release, the extraction of pullout rods operated an MC845 Bisch (electrical pulse) generator that ignited the bomb's MC530 low voltage thermal battery pack and started a pair of MC834 timers. If the bomb's MC732 trajectory arming switch experienced a normal trajectory on release, MC834 explosive actuators in the fuze pack fired to apply power to contacts in the MC832 differential pressure switch. The timers could be set so that they would not detonate a bomb that followed a freefall trajectory or it could be set for a time so short that the weapon would detonate, whether or not it had deployed its chute. If retardation was selected, the bomb's static-line initiated parachute deployment.

The bomb's release also opened a pair of 2.5-inch diameter ports in the bomb's afterbody to supply pressure to an MC832 differential pressure switch. After the bomb had fallen sufficiently far to close the MC832 differential pressure switch (a differential pressure equivalent to 1,500 feet at sea level), solenoid coils moved a rotary safety switch in the MC772 ARM / SAFE unit to the armed position. This action disconnected a ground from the x-unit and closed the MC778 high voltage safing switch that completed the path from the high voltage thermal batteries to the x-unit through a set of saturable transformers. It also connected the external neutron initiators and the boost gas injection systems. After the 42 second run of the sequential timer, the low voltage batteries supplied power through the fuze pack to ignite the MC641 high-voltage battery pack. The 2,500-volt output of the high-voltage thermal battery pack charged the MC787 trigger circuit, the capacitors of the x-unit, and the external neutron generator's capacitors through the circuits of the rotary safing switch. It also activated the boost gas injection mechanism. When the bomb hit the ground, the nose cap deformed, supplying a firing signal to the MC787 trigger circuit. This caused a spark gap to break down that discharged the x-unit into the firing harness and activated the neutron generators to set off a thermonuclear explosion.

The release altitude for retarded delivery was between 3,500 and 5,700 feet above the target, with the limits established by aircraft escape time and thermal battery life. For freefall delivery, the bomb discarded its parachute system. The timer setting and the x-unit's charging time dictated a minimum release altitude of 35,000 feet in freefall mode based on capacitor charging and aircraft escape time. A freefall bomb hit the ground at a speed of 700 miles per hour.

The Air Force completed wind tunnel tests of the MK39 MOD 2 in October 1958. The MK 39 MOD 1 could be field converted to the MOD 2 by replacing the nose and afterbody and by realigning the fuzing system. Sandia replaced the arming functions performed by a baro-switch in the MOD 1 with an MC543 timer. To jettison a bomb safely, the radar-navigator set the DCU-9A panel to SAFE, opened the bomb bay doors and released the bomb without removing the safety pins from its pullout rods. This prevented the generation of electrical power in the bomb and left critical switches open so that the bomb fell like an inert log. Production of about 700 MK 39 units in all three MODs ended in March 1959.

After-accident reports of an incident involving a B-52G and two MK 39 MOD 2 bombs over Greensboro, North Carolina provide a glimpse into how the bomb's safety features worked to prevent a nuclear detonation. The aircraft involved in this accident suffered catastrophic structural failure. As the fuselage plunged toward the ground, its two MK 39 MOD 2 bombs flew free. This action served to pull safety pins out of the out of the bombs' arming rods and pull the arming rods out of the bombs. This activated both bombs' low voltage batteries and MC543 timers.

The number one bomb from the rear bomb bay was thrown clear at an altitude of 8,000 feet. The bomb's static line, which was not locked to the aircraft, apparently snagged on something and initiated parachute deployment. The retarded rate of descent allowed the timer to complete its full 42 second sequence, which activated the bomb's MC641 high voltage battery pack. Since the maximum altitude for delivery of a MK 39 bomb in retarded delivery was 5,700 feet above the target, the high voltage batteries and the x-unit were likely dead when it hit the ground. In addition, the bomb's DCU-9A panel was set on SAFE, which meant that bomb's MC772 ARM / SAFE switch remained in the SAFE position. Because of this circumstance, the MC788 high voltage safing switch remained open. Thus, the x-unit remained isolated from the electrical system and its capacitors were not charged. Further, the tritium injection system for the MK 39's boosted primary was not activated. In this configuration, the bomb could not and did not detonate, even though its crush switch was deformed.

The number two bomb, from the forward bomb bay, was flung free at an altitude of 3,500 feet but did not deploy its parachute. Its MC543 timer ran for only 12 - 15 seconds, thus, the MC641 high voltage battery pack did not activate. Its two MC832 differential pressure switches closed before the bomb hit the ground at high-speed, deforming its crush switch. Like the first bomb, its DCU-9A panel was set to SAFE leaving its MC772 ARM /SAFE switch inoperable. Impact physically moved the rotary MC772 ARM /SAFE switch to the ARM position but thereby rendered the unit inoperable. The MC788 high voltage safing switch did not close. (Even if it had closed, there was insufficient time to charge the x-unit's capacitors before the bomb hit the ground.) The lack of high voltage and electrical continuity prevented the x-unit's capacitors from charging and boost gas was not injected into the core. The high explosives in the bomb's primary did not explode and neither the primary nor the secondary released radioactive material into the environment.

As a result of the accident, Sandia performed the ALT 193 and ALT 197 modifications on SAC's stock of MK 39 alert bombs. The ALT 197 modification replaced the MC772 ARM / SAFE switch with the MC1288 ARM / SAFE switch. In addition to the contacts controlled by the MC772, the MC1288 had contacts in the Bisch generator line to ensure that the thermal batteries would not operate if the bomb was dropped in a SAFE condition. This did away with the need for safety pins in the pullout rods. The ALT 193 modification connected the bombs pullout rods to the aircraft's clip in release mechanism because safety pins no longer had to be withdrawn as part of the pre-arming procedure.

The attention to safety paid in the MK 39 weapon system was a watershed in the manufacture of nuclear devices. Previously, design had focused mainly on reductions in size, and improvements to efficiency and operational handling. In the MK 39, the AEC placed as much (or more) attention on safety. This may have been the result of the weapon's adaptation to the W39 warhead, which armed the Redstone Medium Range Ballistic Missile and the Snark Intercontinental Cruise Missile. The reliability of early missiles was such that safety mechanisms were required to prevent an accidental warhead explosion in the event of a launch failure. In particular, a trajectory switch that did not allow final arming until the missile had completed its boost phase was an important feature.

CHAPTER 16
BOOSTED WARHEADS, SEALED PITS, ELECTRONIC NEUTRON INITIATORS, ADVANCED FIRING-SETS, AND DERIVATIVE WEAPON SYSTEMS

Shortly after the weaponeers at Los Alamos started work on thermonuclear weapons, they began work on fusion-boosted fission warheads. Fusion-boosted fission warheads incorporate light elements like deuterium and tritium in hollow fissile cores. Implosion compresses these light elements along with the core. The resulting fission chain reaction then heats and ionizes the light elements. When the temperature in the core reaches 20,000,000 °C, fusion of the light elements takes place, producing high-energy neutrons to enhance the fission process.

High-energy neutrons carry away approximately 80 percent of the energy from fusion reactions. When these high-energy neutrons strike a surrounding fissile nucleus in the core, they induce fission events that release substantially more neutrons than fission evens initiated by low or moderate-energy neutrons. For plutonium, the number of neutrons released by a high-energy event jumps from an average of 2.9 particles to 4.6 particles. This phenomenon increases the multiplication rate (alpha) of the atomic reaction and the total supply of neutrons in the core, phenomena that increase the efficiency of a boosted warhead to 80 or 90 percent. In comparison, the maximum efficiency of un-boosted atomic warheads is about 50 percent, with 35 percent a typical value. After half of the nuclear material in an un-boosted core has undergone fission, the chain reaction has released two daughter nuclei for every fission event, diluting the remaining fissile nuclei on a two for one basis. As stated in Chapter 1, this effect poisons the chain reaction. Although fusion boosting has a significant effect on the efficiency of a fissile core, the energy contribution from fusion typically amounts to only 1 percent of a boosted warhead's total yield.

Because of boosting's increased efficiency, it was possible for weaponeers to reduce the amount of fissile material used to produce a desired yield. It was, nevertheless, necessary to ensure that the initial fission yield was high enough to initiate the fusion process. Yields of a few kilotons can ignite a fusion reaction using pure deuterium as a boosting medium. The more reactive mixture of deuterium and tritium reduces the yield threshold for fusion to 100 - 300 tons. For this reason, it has become the mixture of choice for use in boosted warheads.

During boosting, tritium reacts with deuterium on a three for two basis by weight. That is, for every gram of tritium that undergoes fusion, 2/3 of a gram of deuterium is also required. The reaction of 1 2/3 grams of a DT mixture produces enough neutrons to fission 80 grams of plutonium directly and 440 grams when the subsequently produced neutrons are accounted for. This releases about 8.4 kilotons of energy. In order to completely fission the 3 kilograms of plutonium found in the core of a typical boosted warhead; 12 grams of boost gas mixture is required. This combination releases just under 60 kilotons of energy. More realistically, the warhead would yield about 50 kilotons of energy at 80 percent efficiency.

When a boosted core implodes, the DT gas at its center ends up occupying a sphere less than a centimeter in diameter. Its density is thus many times greater than its solid-phase density at STP. Assuming that the pressure over the DT sphere is equal to the pressure over the sphere of

Testing an XM 113 tactical atomic demolition munition (shown left). It was equipped with a boosted W30 fission warhead and came in 300- and 500-ton yields. (Credit: USA Photo SC 614239)

An improved MGR-31B (M-50) Honest John rocket. It was equipped with W31 boosted fission warheads in a variety of yields. The W31 warhead was almost the same size as the preceding W7 but was much safer to operate. Note the stubby fins on the rocket. (Credit: USA)

A Lockheed P-3 Orion Maritime Patrol aircraft that carried the MK 101 depth bomb. The P-3 Orion was based on the Lockheed Electra airframe. Note the magnetic detector (MAD) boom at the rear of the aircraft, the sonobuoys in the fuselage to the rear of the wing, and the weapons bay at the front of the fuselage. (Credit: USN; Color Original)

A MK 101 "Lulu" aerial depth bomb. It was equipped with an 11-kiloton W34, boosted fission warhead. Note the ring tail and the water inlet ports in the afterbody. (Credit: NAM)

Sailors load a MK 105 "Hotpoint" tactical nuclear / nuclear depth bomb on an aircraft. The Hotpoint was equipped with an 11-kiloton W34 warhead and was the first nuclear weapon with a laydown capability. Note the midbody fins and ejectable nose fairing that uncovered a "cookie-cutter" nose. (Credit: USN)

An MGM-18 (M4) Lacrosse missile, armed with a 10-kiloton, W40, boosted fission warhead, sits on its XM398 launcher vehicle. (Credit: USA; Color Original)

fissile material, the density of the DT gas increases to about 7 kilograms per cubic meter. Calculations show that a compression of 30 times DT's solid density is attainable using a single convergent shock wave and spherical geometry in an imploding warhead. By using more sophisticated implosion technologies, such as shock waves from the collapse of multiple shells, this density can increase to 50 - 100 times the density of solid DT. As with two-stage thermonuclear weapons, a high-density DT fuel is important in producing a fast, thermonuclear burn.

The nature of the implosion process makes possible the performance of boosted fission devices. The time-scale for DT ignition is only a few nanoseconds, whereas the onset of Rayleigh-Taylor instability at the plutonium / DT boundary takes about 100 nanoseconds. The duration of the DT burn is thus significantly less than the plutonium / DT mixing time. This means that the fusion fuel compresses without mixing with the fissile material during the course of implosion. Nevertheless, building a high-efficiency boosted device is much more difficult than building a basic boosted device. This is because the plutonium and the DT in the core do not reach maximum compression at the same time. Thus, weaponeers have to design boosted devices that operate in regions where their performance is sensitive to various internal and external conditions. In this region, extensive testing is required to ascertain the correct timing for ignition.

John von Neumann put forward the concept of fusion boosting in 1944, a process which he and Klaus Fuchs patented in 1946. This application was desirable for reasons other than pure efficiency. An important benefit expected from small, boosted warheads was a compact primary for use in thermonuclear weapons. A significant constraint on hydrogen bombs during their early development was the large size of available primaries. The use of large fission primaries dictated the use of large thermonuclear secondaries to preserve the geometry of the radiation channel. This resulted in very large yields. In a major nuclear exchange, the poor efficiency and high-yields of existing weapons guaranteed that hundreds of tons of unexploded plutonium would rain down on victor and vanquished alike. Large quantities of undesirable radioactive fission byproducts would accompany the plutonium. The DOD could greatly reduce fallout if it could scale the yields of its weapon to the size of their targets and if the AEC could develop high efficiency primaries that functioned with a minimum amount of fissile material.

Although highly efficient boosted warheads were developed, the goal of reduced fallout proved to be an illusion. Strategic developments in the 1960s focused on the production of small thermonuclear warheads, which the Air Force placed in independently targetable reentry vehicles and then crammed onto ballistic missiles in increasing numbers. Whereas thermonuclear fuel yields more energy on a weight for weight basis than does plutonium or uranium, it is bulky. The trick in producing small, powerful thermonuclear warheads is to engineer them with as much enriched uranium in their secondary tampers as possible. The fission of this material offsets any gain from the reduced amount of plutonium in their boosted primaries.

Greenhouse Item, the first test of the boosting principle, experimented with cryogenic deuterium as a boosting medium. Liquefied hydrogen isotopes were, however, no more practical for boosting than they were for use in a hydrogen bomb's secondary. As a result, Los Alamos pursued boosting with lithium deuteride, deuterium gas, and tritium gas. The lab first used deuterium gas in an experimental TX7E warhead exploded as Snapper Dog in 1952. A year later, Upshot Dixie tested a MK 5D bomb that incorporated lithium deuteride in its levitated core. Los Alamos eventually chose gas boosting as the superior technology, because it did away with the requirement for a complex mechanical system that inserted and de-inserted a nuclear core for safety. Until injected as part of the arming process, gas-boosted warheads store their boost gas outside of their cores in high-pressure (5,000 PSI) stainless steel containers. Thus, an accidental explosion relating to this type of warhead is limited to the un-boosted, sub-kiloton yield of its core.

A sub-kiloton yield is, nevertheless, substantially higher than the zero-yield expected from a warhead that stored its explosives and nuclear components separately. This drawback caused the nuclear laboratories to begin the quest for a "one-point safe" design, in which the accidental initiation of the implosion sphere at a single point would not produce a nuclear yield higher than 4 pounds of TNT equivalent. Los Alamos directed one third of the tests it conducted during the period 1956 - 1957 toward proving this concept. The 1958 test series, conducted just prior to the voluntary nuclear test moratorium, was almost entirely devoted to this purpose. The AEC chose a 4-pound nuclear threshold to limit the release of radiation as opposed to controlling the effects of blast and heat.

Although the nuclear labs were originally satisfied with their one-point safety tests, an advanced analysis by Robert Osborne and Arthur Sayer determined that these results were strongly dependent on the location of the initiation point. Based on this information, the AEC called into question the safety of four major weapons introduced in 1958. In order to resolve the identified problems, Norris Bradbury (Director of Los Alamos) and John McCone (Director of the AEC) arranged for a series of hydronuclear tests to sort things out. These tests were able to provide a satisfactory answer for some of the weapons, but not for others. In the W47 thermonuclear warhead developed for Polaris, a mechanical safing device that withdrew a cadmium / boron neutron absorbing wire from the interior of its Robin core was developed. This system proved unreliable and it was not until after the AEC resumed full-scale testing in 1960 that the safety problem was satisfactorily resolved.

The implementation of gas-boosted warheads required the development of sealed pit systems capable of withstanding the high internal pressures needed to produce a useful gas density. Los Alamos weaponeers typically used two metal hemispheres sealed with three-joint welds at their equator to construct the pit's shell. Technicians used electron beam or TIG welding to attach gas injection tubes to steel pressure shells and brazing to attach the tubes to aluminum or beryllium shells.

As previously mentioned, the boost gas mixture is stored outside the pit in a heavy-duty reservoir made from a special stainless-steel alloy. Before loading gas into reservoirs, personnel at the Savannah River site mix deuterium and tritium to an exact ratio. Savannah River fills different types of reservoirs, for different types of warheads, all requiring different gas mixtures. (Remember that higher yield primaries require less tritium.) Mass spectrometers verify that each reservoir receives the required mix of gases. When each reservoir is loaded to the correct pressure, electrodes pinch its fill stem closed before it is resistance-welded. Personnel then inspect the weld using non-destructive methods.

Because of the rapid decay of tritium to He3, the DOD has the DOE replace the boost gas reservoirs in its warheads every few years. The DOD sends recovered reservoirs to Savannah River, which salvages the remaining tritium for reuse. SRS also salvages the deuterium and the He3 contained in the reservoirs. Operators remove gases from the reservoirs using a laser that they direct through a series of containment windows before it strikes the reservoir stem. A series of pinpoint firings cut a hole in the stem, thus allowing the reservoir's gas to expand into a receiving tank. Operators next pass the three-component gas from the reservoir through a hydride bed to separate the He3 from the hydrogen isotopes. They then process the separated tritium / deuterium gas using a thermal cycling absorption process (TCAP) that draws the separated gases from the ends of a column and feeds them into their own hydride storage beds.

Boost gas enters hollow cores through 1/8-inch diameter, ductile, stainless-steel tubes controlled by a gas transfer valve. The gas transfer valve opens a flow path between the reservoir and the core. A gas transfer valve consists of a body, a piston, and an actuator. The actuator burns small amounts of high explosive to generate combustion gases that move a piston. In older valves, Sandia used lead styphnate (an impact-sensitive primary explosive) as fuel. It loads modern valves with 125 - 250 milligrams of HMX (a less sensitive secondary explosive). An exploding bridgewire or slapper acts as the igniter. The explosive burns rapidly or "deflagrates" to create the expanding gases that drive the piston. The piston either cuts the ends off sealed tubes or punctures a diaphragm to release the boost gas so that it can flow around the piston. Piston movement stops when it the strikes the bottom of the valve body.

Technical references often use the term "life" in conjunction with gas boosting systems. "Cycle life" refers to the interval from the filling of a gas reservoir until it no longer meets the warhead's yield certification due to tritium decay. "Extended cycle life" refers to the time that "unacceptable yield degradation" results from tritium decay and the buildup of its He3 by-product. "Limit life" refers to the interval from the time of the reservoir's first fill until corrosion and stress compromise its structural integrity, resulting in the potential for leakage. Limit life includes an appropriate factor for safety. "Stockpile life" refers to the useful exposure life of a reservoir to tritium, thus placing a limit on refills. It was discovered in 1979 that tritium and its He3 decay product can enter the stainless-steel lattice of the reservoir and cause life-cycle reducing embrittlement.

The adoption of a sealed pit design necessitated the development of external neutron initiators to trigger the chain reaction in a compressed pit. Dismantling pressure-welded pits to continually test and replace internal mechanical initiators was simply not practical. Los Alamos weaponeers proposed the external neutron initiator (ENI) in 1949, which they tested in the Snapper Fox (XR1) and Snapper George (XR2) devices in 1952, and the Upshot-Knothole Annie device (XR3) in 1953. (The Snapper Fox test also included an internal Tom initiator.) A betatron accelerated a beam of high-energy electrons through the expanding cloud of explosive gases produced during implosion. As the electrons interacted with the underlying core, they generated high-energy gamma rays, which in turn induced photo-fission. The neutrons supplied by the radiation-induced fission initiated the chain reaction. The betatron was located within the bomb casing, but far enough away from the explosive shell to avoid destruction before high-speed switches activated it.

A significant advantage of external initiators is their ability to supply many more neutrons than internal neutron initiators. These neutrons also have higher energy and are timing independent of events in the core. With an ENI, weaponeers can time initiation so that the all-important final generations of the chain reaction occur at maximum compression, leading to optimum efficiency. The weaponeers increased yield by as much as 50 percent, simply by replacing internal mechanical initiators with external electronic initiators!

Los Alamos successfully tested betatron initiators in a number of subsequent devices, but the device did not catch on. Instead, an "external neutron initiator" known as a "pulsed neutron tube" became the new standard. First suggested by Luis Alvarez; Richard Garwin, Ted Taylor, and Carson Mark designed and patented the device. The descendants of this device are still used in modern weapons, as well as in commercial applications such as oil well tools.

Pulsed neutron tubes are miniature particle accelerators that smash together deuterium and tritium nuclei to generate high-energy neutrons. AEC suppliers manufacture them in the form of vacuum tubes made with a deuterium ion source at one end and a tritium-bearing target at the other. A powerful surge of electricity applied to the ion source generates deuterium plasma, which accelerates toward the target across a steep potential gradient. Upon striking the tritium in the target, some of the deuterium ions undergo fusion and release high-energy neutrons in a radial burst. Like the electrons produced by a betatron, these neutrons are able to penetrate the expanding cloud of bomb debris and initiate a chain reaction in the core.

Advantageously, a pulsed neutron tube can pump out more neutrons than can be generated with a betatron or a mechanical initiator. They also have higher energy (14 MeV vs. 2 MeV). Initiating a chain reaction with a large number of neutrons reduces the time needed to complete a bomb's fission process by 15 percent. Physicists soon recognized they could vary the yield in an externally initiated device by adjusting the timing of neutron injection with respect to the state of compression in the core, by adjusting the amount of tritium and deuterium used in the initiator, or by changing the electrical potential between the initiator's ion source and its target. The latter change varied the velocity of the fusion components to allow better control of the neutron generating fusion reaction.

All of these control methods affect the initial efficiency of the fission chain reaction and its total yield. These control methods are not only important for fission weapons, but also for multi-stage thermonuclear weapons where the yield of each stage depends directly on the yield of the preceding stage. Nuclear weapons are routinely equipped with two or more redundant external initiators, a practice that was not possible with the older internal initiator. Russian physicists' have reported that the switch from internal to timed external initiators doubled the yield for some of their nuclear devices.

GE initially fabricated pulsed neutron targets from titanium with a thin film of tritium hydrided onto their surface. GE has since replaced the targets with scandium, palladium, and erbium tritides deposited onto molybdenum and other backings. Neutron tube activation requires a power source that provides several hundred amps of high voltage. The power source is usually a capacitor. A typical burst from an initiator produces 500,000,000 neutrons per microsecond over a period of 5 - 6 microseconds. This period is comparable to the duration of a chain reaction. Since the source ejects neutrons in all directions, and is located some distance from the core, only a small percentage of the produced neutrons (less than 5 percent) actually enter it.

The challenges faced in developing neutron initiators included timing, miniaturization, ruggedization, and dealing with the He3 byproduct produced by the decay of tritium in their targets. Los Alamos carried out the basic development work but gave the job of producing a practical external initiator to General Electric and tasked Sandia Laboratories with adapting it for use with operational weapons. One of the first units produced in 1953 was 29 3/8-inches long, 15 inches in diameter, and weighed 108 pounds. A year later in 1954, GE had reduced the diameter of its neutron tubes to 2 1/2 inches. Their diameter is currently about 1 1/4 inches.

In addition to neutron tubes, Sandia and GE developed electronic timers that synchronized the neutron burst with a time that produced a desired yield. To achieve this effect, engineers had to consider neutron generator jitter, firing-set jitter, detonator jitter, neutron power generator jitter, system center time shift with temperature, neutron generator time shift with temperature, and the temperature shift in electronic components. Jitter refers to variations in the amplitude or timing of a waveform arising from fluctuations in its voltage supply. For protection, Sandia coated the delicate timing devices with silicone rubber and cast neutron initiators into polymer blocks. Many of the atomic tests carried out during the early 1950s dealt with determining yield versus initiation time for setting neutron timers. The T-2 timing unit in use at this time was adjustable in 0.02 microsecond steps over a 6-microsecond range for service in a variety of weapons. Its warm up period was 15 seconds.

Los Alamos and Sandia successfully tested a number of prototype boosted warheads and pulsed neutron tubes as part Operation Teapot in 1955. LASL used General Electric's S-3 device as the neutron initiator of choice in its boosted warheads during these tests. The warheads tested included the Moth and Hornet versions of the XW30 air defense warhead. The "Boa" device used in the Moth test was 23 inches in diameter and weighed 445 pounds. The explosive envelope consisted of Baratol and Cyclotol. Its nuclear system weighed only 375 pounds – the most compact physics package produced to that date. The Boa delivered an unboosted yield of 2 kilotons, which Los Alamos found sufficient to initiate boosting. The "Hornet" warhead, which contained a 460-pound nuclear system, was slightly heavier than the Moth with a total weight of 500 pounds. It produced a boosted yield of 4 kilotons, which was lower than anticipated. Following further testing during Operation Plumbob, Los Alamos chose the Boa for inclusion in the militarized version of the W30 warhead because of its diminutive size. Also tested were a number of preliminary designs (Apple-1, Apple-2, and Zucchini) that led to the boosted "Python" primary used in the MK 28 Class "D" thermonuclear bomb.

During Operation Redwing, conducted a year later at the Pacific Proving Grounds, the weapons labs used pulsed neutron tubes as initiators in 13 of 17 nuclear tests. Sandia Laboratory made available three variations of an S-5D XR device developed by the General Electric X-ray Department. The common building blocks of this device were a neutron source, a timer control box (Unit A), and a junction box (Unit B). The miniaturized, pulsed neutron generators measured approximately 4 1/2 inches in diameter by 9 inches long and produced neutrons with 14 MeV energies by means of DT reactions. The XR timer control box (Unit A) contained a 115-volt, 400 CPS power supply that operated on 28-volts DC power; an interlocking circuit that confirmed the proper operation of the 115-volt power supply; time delayed connections to Unit B; and cables that distributed power to the rest of the system. Several points in the weapon system could output a signal to initiate the timer. These included the x-unit load plate, the x-unit capacitor plate, an x-unit detonator, or a piezoelectric crystal signal derived from the detonation of the implosion assembly.

The Unit B junction box furnished AC power and a triggering signal to either two or four neutron-generating S-5D units. The control unit supplied the junction box with timer output pulses and 400-cycle, 115-volt power which the junction box in turn fed to the S-5D Units. Each of these units had its own power supply, which charged a high voltage capacitor similar to the models used in the x-unit. In fact, consideration was given to using part of the x-unit firing pulse as a power source for neutron generation. The timer pulses triggered the discharge of the S-5D Units' capacitors. The Unit A timer was set so that it interlaced the triggering pulses with delays of less than a microsecond.

Redwing Lacrosse, detonated in May 1956, was a full-scale test of the gas-boosted, externally initiated Viper primary that weaponeers from LRL used in the MK 27 MOD 0 bomb and that weaponeers from LASL used in the MK 39 MOD 1 bomb. The device weighed 580 pounds and had a yield of 40 kilotons. Immediately after this test, LRL / LASL used the Viper in the Redwing Apache 1.85-megaton proof test of the MK 27 thermonuclear bomb. The MK 27 MOD 0, equipped with a Viper primary and a Zither secondary, entered service in 1958 and the MK 39 MOD 1, equipped with a Viper II primary, entered service in 1959. Operation Redwing established an alternating pattern of usage with the Nevada Test Site. The nuclear laboratories tested nuclear warheads and primaries in the continental United States and then used the primaries for full-scale thermonuclear tests they conducted in the Pacific.

Recognizing the ability of external neutron initiators to simplify maintenance and the stockpile to target sequence, LASL, Sandia, the GE X-Ray Department, the GE Research Laboratory, and the Berkeley lab entered into a joint program to further develop neutron initiators. The output from this program included the S-20, S-40, and the S-41, of which, the S-41 was the most significant initiator. The S-41 met all of the requirements for the boosted weapons proposed in August 1955.

The S-41 comprised three sections: a power supply, an electronic timer, and a neutron source. For protection, the timer and the power supply were encapsulated electronic devices. The power for the S-41 derived from a high voltage thermal battery, which also charged the firing-set. The S-41 was equipped with condensers, a transistorized converter that transformed a 28-volt DC power supply to the 6,000 volts needed to charge the neutron tubes condensers, a time delay circuit consisting of two cold cathode tubes, an inductance-capacitance circuit, the circuits needed to discharge the condensers into the pulse transformer, and the pulse transformer. Production of the S-41 was expected in January 1956. Through the second quarter of 1958, Sandia expected GE to produce 10,385 units. The AEC expected the need for at least 20,000 units through the spring of 1959. For this purpose, GE built a dedicated production plant at Pinellas, Florida, measuring 175,000 square feet in area at a cost of five million dollars.

At about this time, the code designation for neutron initiators changed from XR to "Zipper". The DOD then decided to refer to all nuclear components with the Military Characteristic code "MC" followed by a number. Production of the model MC774 zipper for use with the TX-28 began in January 1957. The TX-28 was the first weapon assigned to use an external neutron initiator. When the MC774 ENI developed high voltage breakdown problems, Sandia replaced it with the MC890. Sandia then had GE place the MC825 in production for the W34 warhead used in the MK 101 LULU nuclear depth charge. By the end of 1960, thirteen different zippers had been put into production. The zippers included nine different MC types representing three design configurations. The AEC expected six more MC types to enter production between 1961 and 1963. By 1964, twenty different warhead types were either equipped, or scheduled to be equipped, with external neutron initiators.

A W25 warhead for the Genie air-to-air missile, stockpiled in 1957, was actually the first operational system to use an external initiated neutron generator. The W25 warhead also ranks as the first warhead to be equipped with a sealed pit. In the W25 warhead, the sealed pit was not part of a boosted system; the AEC adopted a sealed pit for environmental reasons. The AEC retired the last weapon equipped with a mechanical neutron initiator in 1962.

The continued development of electronic neutron initiators included the MC918 introduced in 1960, the MC1418 introduced in 1963, the MC1917 introduced in 1967, and the MC3421 introduced in 1983. Some more recent electronic neutron generators are the MC3540 used in the W81 warhead, the MC3554 used in the MK / B 61 MOD 7 bomb; and the MC4380 used in the W76-1 warhead. More recently, solid state (tubeless) neutron generators have become available for use in life extension programs.

The nuclear laboratories used fusion boosting and external initiation extensively in lightweight warheads for air defense. With this type of warhead, and suitable interceptors, it was possible to disrupt mass raids carrying conventional weapons, or to destroy individual aircraft carrying nuclear weapons. In order to develop this new type of weapon, it was first necessary to investigate the effects of high-altitudes on nuclear arming, fuzing, and firing systems. Los Alamos carried out this work in 1952, in conjunction with the Air Force Special Weapons Center under the auspices of Project Heavenbound. Project Heavenbound's staff initially favored adapting the XW12 warhead for surface to air missiles until it was determined that the warhead would have to undergo a major redesign to operate at high-altitudes. This liability caused the Heavenbound staff to develop the XW30 and XW31 warheads, which had diameters of 22 inches and 30 inches, respectively. Upon the satisfactory completion of the testing of these warheads, the DOD requested their use in a broad variety of weapon systems that are the focus of this chapter.

The AEC proposed the XW31 as a replacement for the XW7 Nike Hercules warhead in 1954. Two years later, in January 1956, a second proposal called for a more powerful warhead to arm the Improved Honest John. This warhead was designated XW37. Since both the XW31 and XW37 used the same 30-inch technology, Sandia proposed new ordnance characteristics in which it named the low-yield, 30-inch diameter warhead the XW31 Y1 and the high-yield 30-inch diameter warhead the XW31 Y2. Sandia further proposed that this nomenclature be adapted for general use. Army Field Command accepted Sandia's proposal on September 17, 1956.

The XW31 / XW37 warhead had a diameter of 29.01 inches, a length between 39.05 and 43.75 inches, and a weight between 900 and 945 pounds – characteristics quite similar to the warhead it replaced. In fact, it used the W7 high explosive implosion system equipped with a pit suitable for boosting and for use with an external initiator. As previously discussed, late model MK 7 bombs were already in use with external initiators.

The W31's operational requirements specified the capability of mission flights up to 2 hours and a total flight time of 50 hours at temperatures between -65 and +165 °F. The firing system had to be capable of operation at altitudes up to 100,000 feet. Weapons were required to remain operational, without servicing, for 30 days. The warhead had a boosted external initiation system and an enriched uranium core because it was envisioned for use in the Nike B antiaircraft missile over friendly territory. Fuzing for Honest John would be by radar, timer, radar-timer and contact. The W31 warhead had a NiCad battery powered x-unit.

Los Alamos carried out its first test of a W31 warhead in 1957. Plumbbob Newton achieved a yield of 15 kilotons with a 28-inch diameter device incorporated in a thermonuclear mockup. After the 5.9-kiloton test of an XW31 Y3 device designated Hardtack Holly, the AEC began serial production of the warhead in October 1958. The Army used the W31 to replace a W7 warhead under development for the Nike Hercules antiaircraft missile (discussed in Chapter 35), to arm the M50 Improved Honest John rocket, and as the basis for the second-generation Atomic Demolition Munition. The AEC produced the warhead in four yields: 2 kilotons (Y1), 12 kilotons (Y4), 20 kilotons (Y3), and 40 kilotons (Y2). The 2-kiloton yield derived from an un-boosted core. Los Alamos achieved the higher yields by varying the amount of boost gas injected into the warhead. It was a standard operating practice of the nuclear labs to reduce the amount of boost gas in underground reliability tests of high-yield warheads to meet treaty obligations during the 1980s and 1990s. In total, the AEC produced 4,500 W31 warheads.

The Army selected the W31 MOD 0 for use with the Honest John M50 because the warhead was designed to withstand the high acceleration associated with the Nike Hercules SAM and was of similar dimensions and weight to the W7 used in the preceding M31 battlefield rocket. The Army briefly considered retrofitting the W31 to the residual stock of Honest John M31 rockets, but due to their short remaining engine life and the need to produce a new adaption kit, it scrapped the idea. The M50 Honest John had a longer range and greater acceleration than the preceding M31, which allowed the Army to arm it with a higher yield warhead than used in the M31. For use with the M50, the Army installed Y1, Y2, and Y3 warheads in M27 (Y2), M47 (Y1), and M48 (Y3) nuclear sections, all ballasted to the same weight. The AEC produced 1,650 warheads for the Improved Honest John Program.

In comparison with the original M31 / W7 weapon system, the M50 / W31 MOD 0 Honest John had improved fuzing. The M50 nuclear kit contained a long-range fuze, contact fuze, control assembly, and warhead. No external power was required for warhead operation. Fuzing was by radar, timer, radar / timer, or contact; the timer setting automatically chose the optimal fuze. Fuzing selections were made through panels in Honest John's nose cover. The selections included Air Burst, Ground Burst, height of burst, and a timer. In order to evaluate Honest John's contact fuzing system, Sandia sled tested three warheads at Holloman AFB and followed this effort with twelve flight tests at White Sands. Nine of these tests were successful, two were partially successful, and one test failed because of crew error. Thus, the Army concluded that the warhead was suitable for employment with the Improved M50 Honest John Rocket. Airburst fuzing allowed for detonation at altitudes up to 6,500 feet.

The drive for safety in nuclear weapons resulted in an April 1959 request by the Division of Military Application to fit the W31 MOD 0 with an environmental safety device. This led to the development of the inertial safety equipped W31 MOD 2. The velocimeter-based safety consisted of a cylinder having a piston with orifices, a pair of normally open contacts, and latching and reset features. By applying an acceleration of 2.5 Gs, a mass held in place by a magnet and a restraining spring moved to release the piston, and to close and latch the contacts to arm the warhead. Since the W31 production run was nearly complete, the AEC decided to refit the Y2 and Y3 warheads with the new device during the regular replacement of their boost gas reservoirs. The AEC subjected the un-boosted Y1 to a refit whenever it required service.

Redstone Arsenal initiated the Honest John XM31E2 (M50) replacement program in 1955. This program was intended to correct a number of deficiencies identified with the original M31. These problems included operating temperatures, accuracy, and maximum and minimum range. The origin of the improvement program was rooted in predecessor projects aimed at producing Honest John Junior for use at ranges less than 15,000 yards and Honest John Senior for use at ranges from 30,000 to 60,000 yards. The Army later developed Honest John Junior into the Little John rocket for use with airborne divisions. It cancelled Honest John Senior with the realization that new propellants could provide the desired range extension in an improved Honest John rocket.

Development of the M50 motor stretched out over a period of 5 years due to a number of setbacks. At least four versions, XM31E1 through XM31E4 were tested. Both the XM31E3 (Industrial Model, Standard B) and XM31E4 (Production Engineered Model, Standard A) were used to power the M50. Primary motor problems were associated with the collapse of the motor's double-base propellant grain and subsequent blockage of the exhaust nozzle.

The first firing of an XM50 rocket equipped with an XM31E3 motor took place in June 1958. Conditional release following eight additional tests that ended in September. An order then followed for 1,200 Limited Production (LP) rockets that the Army quickly increased to 2,000. Ultimately, 6,347 LP models rolled off the production lines. Although slightly smaller than the M6 motor used in the M31, the XM31E3 developed 108,000 pounds of thrust and could propel an M50 rocket to a maximum range of 44,000 yards. The new motor reduced the M50's operational limits from 0 °F to –30 °F at the low end of the temperature scale. Disappointingly, the upper limit for safe operation was only 100 °F, 20 degrees less than the M6A2 motor it replaced.

Redstone carried out the final development of the M50 in parallel with a major overhaul of its management structure. In March 1958, the Army reconstituted Redstone Arsenal as part of Army Ordnance Missile Command under General John B. Medaris. The Army also placed the Army Ballistic Missile Agency, the Army Rocket and Guided Missile Agency, the Jet Propulsion Laboratory, and the White Sands Missile Range under Medaris' direct control. ARGMA took over the former technical functions at Redstone, which then became responsible for housekeeping functions and post support. These changes had the beneficial effects of integrating the functions of research, development, testing, and logistical support under the auspices of a dynamic leader.

It was under the direction of General Medaris that the Army carried out development of the Production Engineered Model XM31-E4. Only after the successful completion of this program did the Army General Staff approve the reclassification of the XM50 to M50. A second change occurred in 1963, with the re-designation of all Honest John rockets. The Army designated the M31 as MGR-1A, the Limited Production M50 as MGR-1B and the Engineered Production Model as MGR-1C. The production of 742 Engineered Models brought the total number of M50 rockets manufactured to 7,089. Of these, DOD supplied 4,503 units to the Army and the Marines, and 2,586 to other customers, including MAP recipients.

The smaller motors resulted in changes to the overall length and weight of the improved rocket. The M50 was 15 inches shorter than the M31 and weighed 1,490 pounds less. As part of an improved spin program, Douglas squared off the M50's fins, decreasing their span to 4 feet 6 inches. After some experimentation, Douglas gave the fins a 1/2-degree cant. The spin program also adopted four M37 (XM37E3) spin rockets, each with a 1,500-pound thrust delivered for 0.2 seconds. The motor assembly, less the warhead, was designated M66 for the Standard A rocket and M66A1 for the Standard B. Douglas Aircraft produced the rocket's metal parts, not all of which were interchangeable with the M31. The M50 had considerably improved accuracy over the M31. At a distance of 25,000 yards, range error for the M50 was 145 yards, altitude error was 140 yards, and deflection error was 200 yards.

Although an M289 launcher could fire the M50 rocket, the Rock Island Arsenal was engaged to develop a pair of improved launch systems. The M386 self-propelled launcher was a redesign of the standard M289 system and the M33 launcher was heli-transportable. Military personnel developed the specifications for the M386 launcher at a conference held at Redstone Arsenal in February 1955. The new specifications called for a shorter guide rail, lighter weight, faster launcher elevation and vehicle emplacement, better ground clearance, and a lower silhouette on a standard M139C truck chassis. The conference also recommended the development of a simple transfer device to move rockets from their transporters to their launchers. The transfer device would eliminate the need for M62 wreckers in the area forward of the Battery Assembly Area.

The Army authorized the Rock Island Arsenal to begin development of the short rail, M386 launcher and an M405 handling trailer to replace the old XM329 pole trailer in late 1955. The development of the M386 proceeded quickly and by mid-1957 it had entered pre-production and been accepted as the new Standard Model. The Army then re-designated the M289 as Limited Standard. The M386 weighed about 10,000 pounds less than the M289, which weighed 43,750 pounds. It was also more compact and, because of its shorter launch rail, deleted the bumper mounted A frame needed for support on the M289. All of these improvements made it more amenable to air transport. An innovative addition was the M56 protective cover with aluminum frame. This nylon waterproof structure gave the launching equipment security from the elements and provided the appearance of a conventional cargo truck. Total production of M386 launchers amounted to 234 units, of which the DOD assigned 102 launchers to American Units and 132 launchers to European MAP Units.

The M405 handling unit did not duplicate Rock Island's success of the M386 launcher. Rock Island began developing the M405 in 1955 as a measure to replace both the M329 pole trailer and the M62 wrecker with a combination trailer / handling unit that mounted a jib crane. The Army intended this unit to transfer rocket bodies to a launcher from itself or accompanying trucks. Unfortunately, both the XM405 and follow-on XM405E1 were slower at this task than the M62 wrecker was. They also had limited lifting capacity and a propensity to tip over. Despite these problems, Rock Island placed both units into production as the M405 and M405A1. At the end of the day, they were unable to eliminate the need for the M62 wrecker and the M405 and M405A1 amounted to nothing more than expensive transport trailers. This misadventure led to a full-scale investigation by the Government Accounting Office and likely contributed to the reorganization at Redstone.

As originally deployed, an Improved Honest John Battery was assigned four M62 5-ton wreckers, eight M55 5-ton trucks, eight M405 handling trailers, 16 M50 rockets, four M386 launchers mounted on a 5-ton, 6 X 6 chassis, two M25-C generator trailers, four AN/MMQ-1B wind measuring trailers, 16 electric blankets, and eight heating and tie down kits. Along with deliveries to Army and Marine units, the United States deployed M50 batteries to Canada, Great Britain, Denmark, France, and West Germany for use in Europe.

The M33 heli-portable "Chopper John" launcher suggests a highly transportable system. Although this system was effective, it was anything but highly mobile. Development of the XM33 launcher system was a result of delays in fielding the Little John system originally intended to accompany airborne troops. When this program fell behind, the General Staff directed the Ordnance Corps to expedite the development of an interim system for Honest John. The wheeled M33 launcher was 29 feet long, 7 feet wide, and stood 4.5 feet high. It weighed 4,375 pounds. The launching beam was 25 feet long and could be elevated between 4 degrees and 62 degrees for use with an M50 rocket. It was able to traverse 10 degrees either right or left. In addition to carry by an H-37 helicopter, it could be parachuted into the field from a cargo plane. A 5- ton, long wheel base truck could tow a loaded M33 launcher cross-country at speeds up to 20 MPH. The truck could tow an empty M33 launcher at speeds up to 35 MPH.

Although the launcher was quite compact, its need for attendant support equipment made the system unwieldy. A Chopper John detachment required four Sikorsky H-37 Mojave helicopters and a smaller Sikorsky H-34 Chocktaw helicopter for a move. Along with the launcher, it was necessary to transport the wind set, electrical generator, test equipment, a jeep, transport dolly, rocket, warhead, and hoisting tripod. Accompanying the equipment was a nine-man launch team. Chopper John's complexity limited production to 33 launcher sets. Of these, four were delivered to the 101st Airborne, four to the 82nd Airborne, four to the Army Artillery and Guided Missile School, eight to the US Army in Europe, four to the US Army in the Pacific, six to the Marines, and three to the Ordnance Corps.

The M50 system gave satisfactory performance up to 1964. At this time, the Army found that 75 percent of the M139 truck frames associated with the M386 launcher had serious cracks. Although the Army considered replacing existing M139F models with improved M139A2F models equipped with multi-fuel engines, it simply decided to repair the existing equipment. By 1965, half of the Chopper John systems had been retired or reissued to National Guard Units. The Army re-equipped its Honest John units with the Lance SRBM in 1973 but continued to operate Honest John with National Guard units until 1982.

In addition to its use in Honest John, Sandia used the W31 warhead to replace the Army's W7 based Atomic Demolition Munitions on a round for round basis. For this purpose, it produced the W31 MOD 1 (XW31X1). Design release for a W31 MOD1 based demolition munition came in June 1960, and by September, it was in production. The new warhead measured 29.01 inches in diameter by 43.75 inches long and weighed 925 pounds. Because of their similar dimensions, the Army stored the new weapon in older W7 ADM shipping containers. The AEC produced the second-generation ADM in Y1, Y2, Y3, and Y4 yields (2, 40, 20, and 12 kilotons). Sandia protected the device with a lock-secured cover and incorporated a thermal plug in its boost gas system that vented tritium in the event of a fire. The design met all of its military characteristics apart from a slightly less than desired reliability figure. The W31 MOD 1 based ADM was in service over the period 1960 - 1965.

In 1962, the AEC launched a Nike Hercules missile from Johnston Island as Operation Dominic Tightrope. It carried a modified W31 MOD 1 Y4 warhead to an altitude of 69,000 feet where it detonated with a yield of 12 kilotons. The warhead for this test measured 29 inches in diameter, 39.3 inches long and weighed 900 pounds. The use of the 12-kiloton ADM warhead for this test was likely to meet the maximum allowable fission content of Operation Dominic.

In a parallel development with the ADM, the Army replaced the T4 ADM with a "Tactical" Atomic Demolition Munition that Sandia based on a non-boosted version of the 22-inch diameter XW30X1. (The Navy used the boosted W30 warhead in its antiaircraft missiles, which the Author describes in Chapter 34). The Army considered the TADM tactical because it weighed considerably less than the T4. Although Sandia originally considered the W25 warhead in the tactical demolition role (Project "C"), it rejected this warhead in favor of the W30. For application with the TADM, the XW30X1's 375-pound nuclear system came in two yields: 300 tons (W30 MOD 1 Y1) and 500 tons (W30 MOD 1 Y2). The 48-inch long nuclear package had an all U235 core and an explosive envelope utilizing PBX9010 and PBX9404. Sandia controlled the warhead's yield by varying the timing, the energy, and the number of neutrons injected into the core.

Sandia mounted the W30 warheads in corrugated XM-113 cases that measured 26 inches in diameter by 70 inches long. The weight of the "sewer pipe" case with warhead and support equipment was 840 pounds. The system's weight and compact size increased the weapon's mobility and allowed its use in a limited offensive role such as originally intended for the T4. The Johnnie Boy shot of the Dominic II test series was a 0.5-kiloton dull yield weapons effect test of the TADM.

The TADM was waterproof to a depth of 50 feet. Engineering units employed a timer that could be set with a delay up to 48 hours or a remote radio / wire system to arm and to command detonate it. Prior to use, the engineering unit leader had to dial in a combination to free a locked switch. After the receipt of the proper signal, the warhead's adaption kit supplied power to close a high-voltage arming relay that ignited the warhead's high voltage thermal battery, which supplied power to the neutron generators and the x-unit. About 11 seconds later, high voltage applied to the trigger transformers exploded the TADM.

In this discussion of early boosted fission warheads, the W34 and W40 nuclear systems also deserve mention. These warheads were adaptations of the Python primary, which Los Alamos developed for the MK 28 thermonuclear bomb. A contemporary of the W31 warhead, the W34 entered into production in August 1958 and was significant in achieving a 2/3 weight reduction over the W31. It was 32 inches long, 17 inches in diameter, and weighed between 311 and 320 pounds in its various MODs. The implosion system was a 40-lens model that made use of melt cast Cyclotol as its primary explosive. The 16-inch diameter nuclear system weighed 144 pounds and produced an 11-kiloton yield in all models. It was proof tested as Plumbob Fizeau on September 14, 1957. The diameter of the physics package most likely reflects the deletion of an aluminum pusher shell and some or all of the uranium tamper. The total weight of the explosives used in the warhead did not exceed 50 pounds – a single pentagonal lens in the outer envelope of the MK IV bomb weighed 95 pounds and the hexagonal blocks weighed even more at 145 pounds each.

Los Alamos and Sandia Laboratories achieved the miniaturization of the W34 and W40 warheads by means of a technical marvel known as the pulsed power generator or explosive to electric transducer (EET). Sandia designed pulsed power generators to provide an instantaneous power charge that could activate a warhead's detonators. The first type of pulsed power generator was the ferro-electric (FE) generator. This type of firing-set consists of an explosive plane wave generator and a poled ferroelectric ceramic element. A pressure pulse produced by the detonation of explosives in the plane wave generator releases a bound charge from the ceramic element. This technology was favored for use as an energy source in neutron generators for warheads.

The slim loop ferroelectric (SFE) generator is a variation of the ferroelectric generator. In this unit, an external power source supplies a poling voltage in the same manner as a CDU storage capacitor. The SFE firing-set produces a high, fast-rising current output used to ignite an exploding bridgewire detonator. It is composed of an explosive lens, a buffer plate, a Mylar layer and electrode interfaces, slim-loop ferroelectric ceramics, an output circuit, and explosively inert regions (everything else). These compose two major subassemblies, the driver, and the transducer. A typical driver consisted of an XTX8003 extrudable explosive composed of 80 percent pentaerythritoltetranitrate (PETN) and 20 percent Sylgard (a polydimethylsiloxane). Sandia used PETN because of its low melting point and thermal stability. Since the explosive driver provides the explosive shock input for the transducer, the explosive subassembly is critical for the reliable function of the overall device.

SFE firing-sets are laminate in construction, having a copper foil electrode bonded to a lead-barium-zirconate-titanate (PBZT) ceramic dielectric. A thin layer of a room temperature cured silicone adhesive (RTVS 11) establishes a contact between the foil electrode and a sputter deposited gold film (with a chrome adhesion layer) on the PBZT. The adhesive is sufficiently conductive to support both rapid charge accumulation and charge release at the Au / PBZT interface. An SFE firing-set is suitable for use with neutron generators and detonators. Sandia used SFE technology in the MK 3 MOD 1, MK 4, and MK 12A bombs. More recently, Sandia released the MC1649 in 1964; the MC1578A in 1968; the MC2043 in 1970; the MC2454 in 1973; the MC2789 in 1977; the MC3040 in 1978; and the MC3388 in 1988. Initially, the the W76 and W78 warheads were equipped with the MC2989, the W80 was equipped with the MC3369, and the W88 was equipped with the MC3818. The MC4255 firing set developed for the W90 warhead, which was to be deployed on the (cancelled) SRAM-T air-to-ground missile, was cancelled.

The ferro-magnetic (FM) firing-set features an explosive charge, a ring of ferromagnetic cores, and excitation and load windings. An explosively driven pressure pulse transduces mechanical energy into electrical energy in the load windings of the firing-set as a shock wave demagnetizes the magnetic cores. This type of firing-set is extremely rapid in operation and is particularly well suited for use in nuclear artillery shells.

Also developed was a compressed ferro-magnetic (CMF) or "LOBO" firing-set. This type of generator is about the size of a flashlight. It could have replaced the entire bank of lead-acid batteries found in the original Fat Man bomb. Known as an "explosively pumped flux compression generator," the device consists of a cylindrical explosive charge sealed in an aluminum armature tube. A helically wound conductive wire surrounds the aluminum tube, separated by an air gap. A current supplied to this winding by a thermal battery generates a magnetic field within the gap. The detonation of the explosive at one end of the device, forces the aluminum armature against the winding as the shock front propagates down its length. This progressively shorts out the winding from the input toward the output end of the device. The motion of the armature through the intervening magnetic field generates an electrical current that acts as a brake to resist the force of the explosion and converts it to a ramped energy pulse in the outer winding. A transformer at the output end taps the energy pulse. LOBO generators can reliably produce thousands of amps, more than enough energy to service the needs of any detonation system. Various agencies have recently developed new geometries to supplement the previously described coaxial arrangement.

CDUs were and still are preferred for use in bombs because it is possible to test their performance on a regular basis. Explosive powered generators are one-time use items. Despite this drawback, they are very compact, so that their use is favored in missile warheads, where bulk and weight are at a premium. A number of factors drive firing-set complexity. These include the need for nuclear safety, ease of radiation hardening, use control, space saving, testability, ease of manufacture, and cost.

The W34 warhead formed the basis for three naval weapons. These were the MK 101 Lulu depth bomb, the MK 105 Hotpoint retarded bomb, and the MK 45 ASTOR torpedo. Los Alamos advertised the W34 warhead as a rugged and more reliable successor to the W7. For this reason, the Navy replaced the MK 90 Betty with the MK 101 Lulu, deploying 2,000 of the depth bombs in five MODs during the period 1968 - 1971. Although the Lulu depth bomb was similar in weight to the Betty, it was more compact. The Lulu had a diameter of 18 inches and a length of 7 feet 6 inches. Two such weapons could easily fit in the bomb bay of a Neptune patrol plane.

Sandia designed the Lulu depth bomb for unrestricted use on a wide variety of aerial platforms. To insure the weapon's reliability, Sandia dropped development models from a 40-foot high tower onto a concrete platform surfaced with a thick steel plate. A lead pad attached to the nose of the bomb helped simulate impact with the sea surface. Sandia found that the bomb's high explosive sphere survived relatively unscathed.

The depth bomb's high explosive case consisted of two hemispherical aluminum end caps and a center seal. A pressure hull capable of withstanding pressures up to 1,000 PSI encased the aluminum sphere. The pressure container also housed a motorized ARM / SAFE switch. The pilot operated the switch from the aircraft cockpit. The bomb's high voltage system contained five thermal batteries with an output of 2,500 volts, two battery squib transformers, two thermal fuzes, five silicon diode nets, and two monitor switches. This system charged an x-unit that was composed of four capacitors, a dual probe gap, and two gap trigger transformers.

Maximum altitude for release of Lulu was 600 feet and maximum speed for external carry was 300 knots. Hydrostatic depth fuzes backed up by a time fuze provided activation. It also had a contact fuze that set the bomb off if it struck the seafloor prior to reaching the depth set for the hydrostatic fuzes. A safety feature prevented the bomb from detonating in less than 30 meters of water. The weapon required servicing after 1,000 hours of operation. Assuming three 4-hour patrols a day, each with a catapulted takeoff and an arrested landing, this equated to a period of three months.

Because of its utility, the Lulu depth bomb was widely dispersed to fleet locations around the world where the Navy flew it on P-2V Neptune and P-3 Orion patrol planes. It was also in demand by NATO partners, including Canada, which also flew it on the P-3 Orion. In the UK, the Royal navy flew it on Avro Shackleton bombers.

Lockheed patterned the P-3 Orion on the four-engined Electra, a turboprop passenger airliner that jet airliners rendered obsolete. In the P-3C Orion configuration, the aircraft measured 117 feet long, had a 100-foot wingspan. Four Allison T56-A-14 turboprop engines, rated at 4,600 horsepower each, provided a maximum 330-knot speed and a 27,000-foot service ceiling. Maximum take-off weight was 139,760 pounds and a fully loaded Orion had an endurance of 17 hours operating on two engines and 12 hours operating on all four engines. Its ferry range was 2,070 nautical miles.

The Navy assigned the P-3C Orion to Maritime Patrol Aviation Squadrons, each having a complement of 9 aircraft, 60 officers, and 250 enlisted personnel. Individual Orion patrol planes carried a five-man flight crew and an operating staff of 11 specialists. Intended as a replacement for the aging Neptune, the Orion was equipped with advanced submarine detection sensors that included directional frequency and ranging (DIFAR) sonobuoys and magnetic anomaly detection (MAD) equipment. A digital computer that integrated the aircraft's avionics with its tactical displays had the ability to provide flight information and automatically launch ordnance. Carried internally and on wing pylons, the P-3C's ordnance included anti-ship missiles, torpedoes, mines and a pair of MK 101 depth bombs. The aircraft could operate alone or in support of other units such as carrier battle groups.

In addition to its airdropped nuclear depth bombs, the Navy desired a general-purpose tactical nuclear bomb suitable for laydown delivery. This program was designated HOTPOINT. Los Alamos had previously investigated the idea of a laydown bomb in the mid-1950s when it considered placing a parachute-retarded W34 warhead inside a MK 28 bomb case. The MK 28 bomb case would have separated to eject a parachute-retarded bomb after release. As it turned out, the AEC was in the midst of its TABLELEG study, which examined possibilities for a family of nuclear weapons with diameters not to exceed 18 inches. The TABLELEG study eventually made recommendations for a three-step development program.

Step one of the TABLELEG program, designated HOTPOINT A, was the development of a laydown weapon slated for operational deployment in 1958. Based on the W34 warhead, aircraft would carry this weapon externally at speeds from Mach 0.6 - Mach 1.4, and release it from altitudes between 50 and 200 feet. It was to be somewhat impact resistant and have a time delay fuze to allow its strike aircraft to escape. The Navy designated this weapon MK 105. Step two of TABLELEG was a more powerful thermonuclear weapon capable of delivery at low-altitude using over the shoulder and toss-bombing techniques. It was to have airburst fuzing with a contact cleanup option, preflight selectability between timer and radar airburst options, and in-flight selectability between radar and contact options. This bomb became the MK / B28. TABLELEG step three was a long-term program intended to combine the best characteristics of Step 1 and Step 2. This nuclear device would be capable of external carry at speeds exceeding Mach 0.9 and delivery at both high and low-altitudes. The Step 3 concept would provide the Navy and Air Force with a single weapon or weapon family having airburst, contact, and delayed laydown capabilities. This bomb became the B43.

The DOD and the AEC approved military characteristics for the MK 105 Hotpoint bomb in October 1957. For use in laydown delivery, a special tail cone with large fins and a 12.5-foot diameter ringslot parachute-assembly replaced the small finned aerodynamic fairing used for airbursts. The parachute slowed the bomb's descent to prevent damage on impact. As it neared the ground, the nose cone blew off to reveal a specially designed, shock absorbent "spike and cookie cutter" nose that also helped to reduce impact and prevent ricochet. Like Lulu, the MK 105 was activated by means of a time fuze that could be configured for airburst or surface burst. Because it was waterproof, the Navy could use it as a depth bomb with hydrostatic fuzes. It measured 7 feet 10 inches long in its internal carriage configuration (without the aerodynamic nose) and 12 feet long for external carriage. The bomb was 18 inches in diameter and weighed 1,500 pounds. For external carriage, Air Force technicians attached stub fins to the weapon's mid-body. Both fighter-bombers and maritime patrol aircraft carried the MK 105. The Navy deployed Hotpoint bombs over the period 1958 - 1964. The AEC produced six hundred units.

The Navy developed the MK 45 Anti-Submarine TORpedo (ASTOR), the third weapon system known to have incorporated a W34 warhead, to meet the threat of high-speed, deep diving submarines. Its story is included in the next chapter, which also describes the development of fast attack, nuclear powered submarines.

The previously mentioned W40 warhead incorporated the same, 144-pound, physics package as the W34, but was slightly larger. The Y1 (10 kiloton) version of the warhead measured 31.64 inches long by 17.9 inches in diameter and weighed 385 pounds. The Y2 un-boosted version had a yield of 1.7 kilotons. The W40 used the same x-unit, removable high voltage battery pack, dual external initiators, dual motor operated arm / safe switches, gas-boosting mechanism, and dual cold-cathode thyratron trigger circuits as did the W28. The adaption kit contained five batteries connected in series, with a pair of silicon diodes shunted across each set of battery terminals to allow normal operation of the other four batteries if one should fail. Thermal fuzes were located along the ground line to isolate the batteries in the case of fire. Dual squib transformers reduced power requirements and provided increased reliability. The high voltage ARM / SAFE switches were motor driven and reversible, affording positive control. The switches were fire safe and could not be accidentally operated by a stray electro-magnetic field. The gap switches of the W40 could be triggered by a low impedance source provided by the adaption kit, and from signals derived from timers, barometric, and impact switches. It could also be triggered by a high impedance source derived from radar fuzes, thyratrons, or impact crystals.

This required a four-probe gap and four pulse-firing transformers, two with low impedance and two with high impedance.

Los Alamos selected the W40 MOD 0 for use in a high-accuracy missile system conceived by the United States Marine Corps. The Corps envisioned the XSSM-N-9 Lacrosse guided missile system as a means to destroy strong point targets such as pillboxes and reinforced bunkers in the immediate post war era. Lacrosse would also supplement aerial bombardment and naval gunfire against troop concentrations, supply columns, and staging areas. Operational characteristics called for the launch of the missile from a rear area, after which a forward observer would provide terminal guidance by means of radio command or a target designator. Armed with a 100-pound conventional warhead, the missile's accuracy requirements called for a circular error probability of not more than 5 yards.

Early in 1947, the Navy's Bureau of Ordnance approved the Marines' application for Lacrosse and the Corps requested feasibility plans from the Applied Physics Laboratory at Johns Hopkins University. The Cornell Aeronautical Laboratory was also engaged to investigate the missile's guidance system and eventually selected as the lead designer. The Corps chose Cornell for the lead role because precision guidance was the dominant and most difficult requirement for the weapon system. The joint APL / CAL feasibility report called for several lines of investigation before the final choice of the guidance system. In their final recommendations, both laboratories called for a high subsonic speed missile capable of carrying a 500-pound shaped-charge warhead. An observer equipped with an optical tracking sight, an optical range finder, and a radio command system would manually guide the missile to its target. The labs would furnish the missile with a missile-to-observer radio range transmitter.

The Navy cancelled the Lacrosse project in late 1949, when the Joint Chiefs of Staff established a policy that placed the Army in charge of ground launched missiles intended to supplant or extend the range of conventional artillery. Although the DOD now tasked the Army with Lacrosse development, limited funding caused it to ignore the project. This policy changed in June 1950, when development funding was forthcoming because of the Korean War. The Army initiated its XSSM-A-12 Army version of Lacrosse with supplemental funding provided by the Navy. The Marine Corps continued to show interest in the project.

The Ordnance Department's first priority was to develop the STEER guidance system, which consisted of a command computer, optical tracker, and radio link. The guidance system was designed to employ optical tracking for fair weather operation and a radar tracking head for all-weather use. The Army also confirmed its desire to continue the project's association with Cornell Aeronautics Laboratory, since CAL had been associated with Lacrosse from its inception. Preliminary guidance tests used a Beechcraft AT-7 airplane to simulate the missile and confirmed the feasibility of the proposed guidance concept. The Army carried out the pilot-monitored tests during the first half of 1951, beginning the formal development of Lacrosse in November.

After the Ordnance Department agreed to power the missile with a solid fuel motor, it set up the Group O Program to complete the development of STEER. The guidance test vehicles used for this purpose were surplus SAM-N-2 and -4 Lark antiaircraft missiles produced by the Navy Bureau of Aeronautics. BuAir had designed Lark with a cruciform wing that was very similar to the final design adopted for Lacrosse. During the period April 1953 through January 1954, the Ordnance Department conducted eight separate tests of forward guidance equipment with unpowered RV-A-22 Lark airframes dropped from a B-26 bomber. At this time, the Bureau of Ordnance transferred the technical control and supervision of the Lacrosse project to Redstone Arsenal.

The 150-pound, AN/ASPS-26 passive tracking radar that emerged from the Group O test program was an X-band unit with an 18-inch diameter rotating-antenna, built by Dalmo Victor under contract to Westinghouse. The airborne beacon used with this radar was a modified AN/DPN-12 X-band, pulse-modulated transmitter produced by Bendix Aviation. Sylvania Electrical Products produced the radio ranging or distance measuring equipment based on a 600-pound Convair system modified by the Cornell engineers. A "ruggedized" Collins AN/ARW-55 transmitter and an AN/ARW-59 airborne receiver comprised the command link. The optical tracker and ranger for the fair-weather system was a pulsed light system developed by the Farland Optical Company. Redstone later simplified the guidance system to an optical tracking unit suitable for all-weather operation. The optical tracker operated in conjunction with a rate prediction computer and a guidance computer.

With the successful completion of the guidance program, Martin began work on the tactical missile as its prime contractor. Although there were some hard feelings on the part of Cornell at this decision, the laboratory continued to play an important role in guidance development. To be fair, Cornell was primarily a research institute that had neither the equipment nor the expertise for production manufacturing.

Despite Cornell's status as a research institute, it was able to produce eighteen missiles and associated ground equipment for the follow-on Group A test program. Cornell powered its missiles with a T52 solid fuel rocket motor produced by Picatinny Arsenal. The motor generated a nominal thrust of 38,000 pounds for 2.8 seconds and accelerated the missiles to a Mach 0.8 speed. The rocket motor defined the dimensions for the fuselage, which was 19 feet 4.2 inches long by 20.5 inches in diameter and of aircraft type construction. It was equipped with four swept wings spaced 90 degrees apart and four tailfins offset at 45 degrees from the wings. Each wing projected 4 feet 8.4 inches from the fuselage. Moveable surfaces located at the rear of each tailfin controlled the missile. Total weight of the Lacrosse when armed with a 500-pound warhead was 2,350 pounds.

The change in lead contractor and a series of increased performance requirements extended the length of time needed to develop Lacrosse. The Army first extended the missile's range from 1,000 yards with a 100-pound warhead to 7,000 yards with a 500-pound payload. The Army then extended the maximum desired range to 30,000 meters. Redstone Arsenal established system accuracy at 5 meters for a range of 8,000 meters and 30 - 40 meters at a 20,000-meter range. By 1955, fusion-boosted warheads with a weight less than 500 pounds were on the horizon. This allowed Lacrosse planners to include a nuclear warhead in a selection of munitions that included a variety of shaped-charge warheads. The 400 W40 warheads eventually produced for Lacrosse had yields of either 1.7 or 10 kilotons.

Martin flew the Group A Lacrosse program at the White Sands Missile Range in the latter half of 1954 and throughout 1955. It incorporated pre-production XSSM-A-12 missiles with complete guidance systems that evaluated their performance and accuracy. Since Redstone intended the program to test the entire system, it required Cornel to produce a prototype launcher early in the program. The Army terminated the Group A program after 15 firings successfully demonstrated the Lacrosse employment concept.

To follow up the Group A program, Redstone gave Martin Aircraft a contract to produce four sets of ground equipment and 50 Lacrosse I missiles. Redstone had to delay this contract because of the lead-time required for Martin to set up tooling. Martin next flew an interim Group B program in the first half of 1956 using pre-production ground equipment to bridge the gap between the development program and the tactical evaluation. The Group B program established the ability of pre-production guidance equipment to strike an intended target with acceptable accuracy. Redstone placed its emphasis in developing the Group B forward guidance system and ground equipment on ruggedness, ease of maintenance, and weight.

The Forward Guidance Unit comprised a guidance jeep, a radio jeep, and a cargo jeep and trailer. It was crewed by an officer and a twelve-man team. The guidance jeep carried two main instrument groupings. The first consisted of an AN/MSW-5 missile guidance center, a target survey indicator, a secondary power supply, and battery eliminators secured in a special shock mounted container. The second was a target ranging set located next to the driver in place of a passenger seat. The cargo jeep transported an MX-3029 angular tracker set. Four men were required to set up the guidance system but it required only one power supply operator to run it. Test equipment was self-checking and built in. For use away from the forward guidance Jeep, Redstone packaged the guidance equipment in nine man-packs. The system could operate in three observational modes: direct observed fire, indirect observed fire, and indirect unobserved fire.

A Lacrosse Unit situated its remaining ground equipment at a Battery Assembly Area located behind the Forward Guidance Unit post but within range of potential targets. The central piece of equipment at this site was Cornell's missile launcher. The prototype was an XM387 square rail mobile launcher based on an M44, 6 X 6, 2 1/2-ton truck chassis. Cornel used this launcher extensively for testing but replaced it prior to tactical production. The successor M389 launcher incorporated a novel 6.5-foot long, helical launch rail mounted on the bed of an M45 6 X 6, 2 1/2-ton truck. The rail had a 6-degree twist that rotated the missile through 42 degrees and imparted a 480 degree per second spin as the missile left the launcher. This stabilized the missile without the need for spin rockets. The launch rail could be elevated to a maximum angle of 790 mils and was able to traverse 290 mils both left and right. In order to withstand the stress of a helical launch, the guide rail had to be extremely robust. This resulted in a number of manufacturing difficulties. A time and motion study of missile loading indicated that a four-launcher battery could sustain a rate of fire averaging 12 missiles per hour.

The M389 launcher was accompanied to the Battery Assembly Area by an M108 crane and a number of M54, 6 X 6, 5-ton, cargo trucks. The trucks were each capable of transporting a pair of M1 missiles in their M374 shipping containers and M139 warhead sections in their M471 shipping containers. At the BAA, the missile crew transferred the Lacrosse missile bodies from their truck transports to their launchers by means of the M108 crane. As soon they had accomplished this task, they attached the warhead section and fins to the missile body, which they subjected to a pre-launch checkout using an AN/DMS-59 Test Set. Following the check out, the forward observer could remotely launch the Lacrosse missile. After the forward tracking unit detected the incoming missile, the unit's operator could place it on a ballistic path that ended with the target. Accuracy was, therefore, dependent on the skill of the operator. Although capable of reasonable accuracy in experienced hands, Lacrosse could not hit a moving target.

The AEC worked on three MODs of the W40 for Lacrosse. It produced 400 10-kiloton W40 MOD 0 units during the period September 1959 - May 1962. The AEC stockpiled a few MOD 1 warheads with an environmental sensing device in 1964 but cancelled a MOD 2 warhead with a PAL device because of the system's early demise. Fuzing for all warheads, including conventional models, was by radio command operators in an M242 trailer stationed near the

M389 missile launcher at the BAA. The missile's electrical system provided power for the warhead. Arming took place in less than 1/2 seconds but was not available during the first 5 seconds of flight. An acceleration switch used in conjunction with a command signal prevented accidental arming from a transient electrical impulse when the missile was on its launcher or during early flight time. An M33C antiaircraft fire control system located in the M242 trailer tracked Lacrosse during its early flight.

Martin divided its development program into a sequence of "Tasks" that were intended to produce the final tactical design. Task I was a feasibility study that reviewed the materials, methods, and processes it would use in the manufacturing process. In Task II, Martin flew and evaluated eight pre-production missiles to confirm that a minimum 60 percent of the ultimate missile design was satisfactory. The pre-production missiles were equipped with Thiokol M10 rocket motor that did not prove entirely satisfactory. Because of this problem, Picatinny Arsenal began development of an improved T52E1 motor, whereas Thiokol began an M10E1 improvement program. Ultimately, Martin powered all of the Lacrosse missiles with either M10 or M10E1 motors. Four of the missiles tested satisfactorily, while a fifth exhibited a minor range error. A sixth missile failed due to hydraulic problems and the final pair exhibited significant accuracy problems arising from component malfunctions that produced errors of pitch and yaw.

By the summer of 1957, sufficient Type I Martin missiles were on hand to conduct the Task III Program. Task III consisted of three unguided firings, two special firings, and sixteen R&D firings. Cornell, which had begun to develop a lightweight, helicopter portable launcher, conducted the three unguided firings. It had also begun preliminary research into an airborne control system to provide guidance against targets that were not visible to front line observers. The Army eventually cancelled both of these programs. The R&D firings had mixed results. Only five of the sixteen R&D test flights achieved acceptable accuracy. Guidance and control malfunctions plagued the remaining flights.

Completion of the Task III Program ended participation by Cornell, leaving Martin to conduct the Task IV program. Martin ran the program in conjunction with a "composite test program" intended to supply an approved weapon system in the shortest time possible. At the beginning of each quarter, ARGMA scheduled a meeting of the firing test subcommittee to evaluate results and to develop priorities. The Task IV Program was essentially a continuation of the firings executed in Tasks II and III.

The first Task IV launch took place in March 1958 and by 1960, Martin had conducted 181 scheduled tests. These were a mix of R&D firings, Artillery Board evaluations, WSMR Ordnance Mission System Test Division evaluations, and Army acceptance tests. The fact that Martin had not been able to establish functional reliability for Type I Lacrosse became evident very early in the program. The Army graded 55 percent of the early firings as unacceptable. It then implemented Operation Pickle Barrel to clean up the missile for its acceptance trials by the Artillery Board. The Army held these in early 1959, obtaining approval for the release of Lacrosse in July. Nevertheless, disappointment with the system resulted in funding cutbacks that reduced its planned deployment from sixteen to eight Army Battalions, and 20 sets of ground equipment. Final missile procurement ended at 1,194 rounds.

In parallel with development of the Type I tactical system, the Federal Telecommunications Laboratory began work on a MOD 1 guidance system that incorporated electronic countermeasures to prevent the jamming of the missile's guidance signals. The Marines anticipated the need for electronic countermeasures in their initial outline for the weapon system. This requirement had, however, been neglected in the rush to develop the missile's operational guidance system. The MOD 1 program immediately experienced slippage due to technical problems and by 1959, the projected cost of the project had ballooned to 100,000,000 dollars over the comparable cost of the Type I system. In an environment of cutbacks following the end of the Korean War, the Army decided to reduce Lacrosse funding and to deploy the Type I system only. The impaired reliability resulting from the cancellation of the MOD 1 guidance upgrade caused the Marines to withdraw from the Lacrosse project.

Just before the final release of Lacrosse, the Army kicked its training and documentation programs into high gear. Contracts to Martin called for the delivery of 150 staff operation manuals, 100 guidance manuals, 75 mechanical manuals and 225 operator's manuals. The Army also contracted Martin to provide 4,072 hours of classroom instruction. Martin initiated the training program with 10 instructors at the Redstone New Equipment Training Center in late 1956. As soon as Redstone Arsenal had acquired a cadre of trained personnel, it began the training of individuals to support the Lacrosse firing units. The Army carried out resident training at the Redstone Ordnance Guided Missile School.

The Army declared the 5th Battalion, 41st Field Artillery operational with Type I tactical equipment in December 1959. At this time, Lacrosse was designated Guided Missile, Field Artillery, M4. It was re-designated MGM-18A in 1963. The 5th Battalion comprised two M389 Lacrosse launchers, five generator trailers, five AN/DSM-59 guided missile test sets loaded on M37 3/4-ton trucks, two trucks towing M242 trailers equipped with M33C antiaircraft fire control systems, four MX-3029 automatic angular missile trackers loaded on M37 3/4-ton trucks, two primary power units, four M38 Willys Jeeps, two target survey units, and two AN/MSW-5 forward guidance stations. During 1960, seven more Lacrosse Battalions reached operational deployment. Of these, the 5th Battalion, 42nd Field Artillery; 4th Battalion, 28th Field Artillery; 5th Battalion, 39th Field Artillery; 5th Battalion, 33rd Field Artillery and 2nd Battalion, 22nd Field Artillery were posted to Europe, while the 6th Battalion 8th Field Artillery was posted to Korea. The Army assigned the 5th Battalion, 40th Field Artillery to Strategic Command and distributed the remaining 18 sets of ground equipment across its new units. In addition, it assigned each unit its own Ordnance Support Unit equipped with an M109 shop van that carried the parts and equipment to perform routine field maintenance on missiles and launch equipment.

The ability of Army Ordnance to support Lacrosse units in the field never developed. The number of technical deficiencies that still required attention was overwhelming and the documentation and equipment needed to address the problems was generally lacking. Reduced funding also hampered corrective actions. The Army, therefore, decided to fund only the currently deployed units in 1961, and to deactivate them as soon as possible. The Lacrosse inventory was reduced to 95 missiles per Battalion (as opposed to a full strength of 120), with other stores reduced in proportion. The Army stood the last Lacrosse Battalion down in 1963, after a service life of only 4 years.

While a sound idea, the technology did not exist in the late 1950s to produce a lightweight tactical system as originally envisioned by the Marines. Although it is true that Lacrosse suffered from both accuracy and dependability issues, the Army might have overcome these problems if it had allocated further funding to produce a more reliable system. A much more likely reason for the demise of Lacrosse was the inordinate number of men and machines needed for its support. In 1961, the same year that the Army began the retirement of Lacrosse, a jeep-mounted recoilless launcher that could toss a 20-ton yield, mini-nuclear warhead to a distance of 2.5 miles entered service. It required only a four-man crew for operation. It is no wonder that the Army cancelled Lacrosse!

CHAPTER 17
HUNTER-KILLERS WITH NUCLEAR CLAWS

In the aftermath of World War II, Dictator Joseph Stalin ordered the Soviet Navy to improve and expand its fleet of submarines. With this program, he planned to counter the power of the fast carrier groups that gave the United States Navy de facto control of the world's oceans. The Soviet Union did not have the financial or industrial means to build its own aircraft carriers and support vessels at this time and had to rely on mass produced submarines to provide a credible deterrent.

The Soviet Navy based its new submarines on German technology, which it captured at the end of WW II. The prize booty included very advanced, ocean-going submarines known as Elektrobootes. Bearing the designation Type XXI, the German vessels measured just over 251 feet long, displaced 2,100 tons submerged and had a range of 15,500 nautical miles. They Kriegsmarine equipped them with six 533mm bow torpedo tubes that had a rapid (hydraulic) reloading capacity for 23 torpedoes carried. An advanced sonar capability allowed these submarines to aim their torpedoes when fully submerged. Streamlined and provided with an enlarged battery capacity, the Type XXI had a 15 1/2 knot speed on the surface and 16 knots submerged. This compared with a submerged speed of 7 - 7 1/2 knots for earlier Type VII or Type IX submarines. An Elektroboote could travel submerged for 3 days at the leisurely speed of 5 knots. It then had to rise to snorkel depth where its diesel engines recharged the submarine's batteries in about 5 hours. The crew had much better amenities than earlier designs with improved food storage and a shower.

The first Soviet blue water adaptation of the Type XXI Elektroboote was the Project 611 submarine, better known to western navies as Zulu. It had a length of 295 feet, a displacement of 2,387 tons submerged, and six bow and four stern 533mm torpedo tubes – with 22 torpedoes carried. The vessel had a submerged speed of 16 knots, a range of 20,000 nautical miles, and a nominal endurance of 58 days at sea. The design was, however, not a big success and resulted in a production run of only 26 vessels – unusually small by Soviet standards. Its three propellers and associated machinery made Zulu extremely noisy and its hull construction limited its operational depth to 200 meters. The Soviet Navy later converted six of these vessels into ballistic missile submarines that could hot launch R-11FM (Scud) missiles through their sail.

The Soviet Navy followed Zulu with Project 633 Romeo and Project 641 Foxtrot submarines. The Romeo was a small, two-screw design that served with the Soviet and Chinese Navies, the latter producing it under license. The very successful Foxtrot was an improved Zulu Class submarine that displaced 2,475 tons submerged. The Soviet Navy produced 75 Foxtrots over the period 1957 to 1983. They served with the Soviet Navy and such aligned nations as India, Libya, and Cuba. The USSR supplied a number of used Foxtrots to Poland, North Korea, and the Ukraine. Constructed with a QT28 nickel steel alloy pressure hull that allowed it to dive to a maximum depth of 300 meters, a Foxtrot could run submerged at speeds up to 17 knots. It had an unrefueled range of 20,000 nautical miles and came equipped with T5 nuclear torpedoes, including four Foxtrots that appeared off the coast of the United States in the midst of the Cuban Missile Crisis.

The USS Nautilus (SSN 571) – The first nuclear powered submarine. (Credit: USN)

Nautilus nuclear reactor prototype undergoing testing in a brine-filled pool to determinre its neutron flux when the submarine was submerged. (Credit: AEC)

SS Permit (SSN 594) – the first class of nuclear submarine (SSN) to have the full characteristics of a modern fast attack submarine. (Credit: USN Photo NY9-51928-7-61)

A MK 45 "Astor" torpedo (armed with a 10-kiloton W34 warhead) aboard a test platform. A cowling covers the twin contra-rotating propellers. (Credit: USN)

A Navy technician checks out a UUM-44A SUBROC missile. The weapon was launched from a torpedo tube and equipped with a 250-kiloton W55 warhead. (Credit: USN)

Aware of the significant role its own submarines played against Japanese shipping during WW II, the United States Navy quickly detected Soviet intentions. As a counter to the Soviet submarine menace, it acquired nuclear depth bombs for its maritime patrol aircraft and escort vessels. It also developed the "hunter-killer" submarine. The hunter-killer was a submarine optimized for anti-submarine warfare. It operated in the same medium as its prey and could track an adversary down through the various layers of the ocean depths. However, to develop a submarine effective in this role, the Navy had to surmount a number of technical challenges. These included a propulsion system that could operate for long periods while submerged, an improved hull to take advantage of the new propulsion system, more sensitive detection equipment, and new weapons.

Novel means of propulsion have always interested the United States Navy. When the First Lord of the Admiralty, Sir Winston Churchill, switched the British Navy from coal to oil fired turbines in 1912, the USN was already building the oil-fired battleships, USS Nevada (BB 36) and USS Oklahoma (BB 37). Congress authorized these vessels on March 4, 1911. Beginning in 1939, the United States Navy made available the first small allocations of capital to study the potential of nuclear power reactors as a means of shipboard propulsion.

Nuclear power reactors are very different from plutonium production reactors. The production reactors built by the Manhattan District at Hanford in the early 1940s were low temperature, open systems that drew and returned their coolant water to the Columbia River. In a power reactor, heat transferred from the reactor's core to its coolant produces steam, which a turbine uses to generate electricity. In a ship, steam can also drive turbines connected to propeller shafts. In a submarine, this type of propulsion system is particularly important because it does not consume air. Using nuclear propulsion, the length of time a submarine can operate under water is limited only to the endurance of the crew and their food stocks.

Following World War II, the Navy sent a number of officers to the Oak Ridge atomic facility to study nuclear engineering. Notable among these men was Captain Hyman G. Rickover, formerly in charge of the Bureau of Ships' Electrical Division. In that position, Rickover developed standards for Navy equipment and worked with civilian manufacturers to ensure that the equipment they produced met his specifications. On his watch, the Electrical Division grew from 23 individuals with a poorly defined set of responsibilities to a staff of 343 that included highly trained engineers. During his stay at Oak Ridge, Rickover grasped the idea that power reactors were the wave of the future and he decided to get in on the ground floor.

Upon returning to Washington, the Navy assigned Rickover to the staff of Admiral Earle Mills, Chief of BuShips. He received the title: Special Assistant for Nuclear Matters. In moving forward with shipboard nuclear propulsion, Rickover faced two serious problems. The Atomic Energy Commission had taken charge of reactor development and the Navy was prepared to accede to the AEC's program. Unfortunately, the AEC was intent on building bulky civilian power plants that were not serviceable for use in submarines.

Rickover tackled these problems by first taking on the Navy. He wrote letters outlining the benefits of nuclear propulsion to senior naval officials such as Admiral Chester Nimitz and Navy Secretary, John L. Sullivan. He also able persuaded Admiral Mills to give a speech on the subject of nuclear powered submarines at the 1948 Undersea Warfare Symposium. After much lobbying, he established a requirement for a nuclear-powered submarine with unlimited endurance at high-submerged speeds. He also established the Bureau of Ships as the Navy's agency for carrying out the development of the proposed submarine.

Rickover now turned to face the AEC, which had established Chicago's Argonne National Laboratory as the nation's hub for reactor development. Walter Zinn, the Lab's director, made it abundantly clear that he would apportion responsibility for naval reactor development unequally; the AEC would take charge of any such program. Because of Zinn's rebuff, Rickover looked elsewhere for help. Two companies drew his attention. These were General Electric, which the Navy had contracted to develop a liquid metal heat transfer system in 1946, and Westinghouse, which was interested in a pressurized water reactor. To meet his requirements, Rickover contracted GE for further work at their Knolls Atomic Power Laboratory at Schenectady, New York, and arranged for Westinghouse to build a new atomic laboratory located at the disused Bettis Airfield near Pittsburgh, Pennsylvania.

With their plans moving forward, Mills and Rickover were ready for a second round of negotiations with the AEC. During these sessions, Donald F. Carpenter, the Chairman of the Military Liaison Committee, championed the Navy men. After much discussion, the AEC and the Navy decided that they would share equally the responsibility for developing a naval power reactor. In recognition of Rickover's outstanding performance in this matter, Admiral Mills placed him in charge of the Nuclear Power Branch of BuShip's Reactor Division. Rickover finally had the authority to create his naval propulsion reactor.

He immediately set about acquiring the staff necessary to accomplish this task, signing on many of the same officers who had worked with him at Oak Ridge. One of the first decisions made by Rickover and his staff was the choice of power reactor technology. Their choices included a liquid metal fast breeder, a high temperature gas cooled reactor and a pressurized light water reactor. The high temperature gas-cooled reactor (HTGR) had a large footprint and

Rickover quickly rejected it. The main disadvantage of the liquid metal fast breeder reactor (LMFBR) was its sodium metal coolant. Sodium turns into a liquid at a temperature of 100 °C and is an efficient coolant, but it burns in air and explodes in water! This was hardly an ideal material for use in a submarine.

The pressurized water reactor (PWR) thus became the primary design for a submarine thermal reactor (STR); so-called because it relies on moderated neutrons for its operation. The reactor Division chose Westinghouse as the industrial supplier for this piece of equipment in 1950 and the Argonne Laboratory went to work to provide some initial design concepts. To reduce the impact of a nuclear accident, the Navy had Westinghouse built its MARK I STR prototype at the National Reactor Test Station (NRTS), located on the desolate lava plains of Idaho. Rickover required that the reactor's development take place inside a hull mockup set in a tank of seawater so that when finished, the he could install the design without modification inside a working submarine. He also had the Bettis staff reject and re-work many of the Argonne concepts, because the civilian AEC staff did not have an appreciation of the stringent requirements needed for military hardware. As a backup in case the STR proved unworkable, Rickover contracted GE to develop a liquid-metal cooled submarine intermediate reactor (SIR) at Knolls in New York.

A Submarine Thermal Reactor is contained inside a large, welded steel vessel capped with a heavy lid to maintain the necessary pressure to keep its light water coolant from boiling. Water boils at higher temperatures if kept under pressure. A typical civilian pressurized water reactor operates with 5 percent enriched uranium at coolant outlet temperatures of 300 - 400 °C and at pressures of 150 atmospheres. The higher the coolant temperature, the greater is the efficiency of a reactor. For this reason, compact military designs that fit inside a submarine operate with uranium enriched to 50 percent $U235$ or higher. They also operate at higher temperatures and pressures than their civilian cousins. In terms of mechanical safety, Rickover had a choice of bolting or welding the pressure lid to reactor vessels. It is a tribute to the Admiral that he did both. Safety was an overriding concern during the entire reactor development process and paid off handsomely. The hastily developed HEN military reactor used in early Soviet submarines omitted many safety features and caused a number of nasty, fatal accidents.

The Navy pioneered the concept of uranium enrichment when it built a thermal diffusion facility during WW II. Nevertheless, Oak Ridge produced 50 percent of the enriched uranium fuel used in Navy power reactors after the war. Power reactors contain jacketed, enriched uranium fuel rods fabricated from metal or metal oxide pellets. Oxides are more stable at high temperatures than pure metal. The aluminum jackets used for plutonium production reactors and even stainless-steel were not amenable to the high temperatures generated in a power reactor. This caused Sam Untermeyer, a nuclear engineer at Oak Ridge, to suggest using the rare element, zirconium, as an alternative fuel cladding.

Zirconium is highly resistant to corrosion, has a melting temperature of 1,855 °C, and is almost completely transparent to thermal neutrons – an ideal set of properties. The Navy almost overlooked it as a cladding material because the few pounds of metal produced up to this time contained 2 percent of the element hafnium. Chemically and physically similar to zirconium, hafnium is a notorious absorber of neutrons. Upon detecting this anomaly, a pair of entirely new industries developed. One industry produced purified zirconium to clad fuel rods and as a structural element for reactor cores. The other industry focused on the extraction of hafnium for use in the manufacture of reactor control rods.

The core for a light water reactor is contained entirely within its pressure vessel and in the case of the MK I, had its fuel rods oriented vertically. The control rods also ran in the same orientation, penetrating the reactor lid for remote insertion or withdrawal. The light water that surrounded the core acted as a heat absorber, a moderator, and a neutron reflector. The withdrawal of the control rods increased reactivity to heat the surrounding water. The hot water then circulated through a heat exchanger in a pressurized circuit known as a "hot leg" and returned to the core through a circuit known as a "cold leg." Within the exchanger, the heat from the cooling circuit transferred across adjacent pipes into a low-pressure circuit. The heat converted the water in the low-pressure circuit to steam, which passed through a turbine to drive the submarine's propellers, and through a turbo-generator to produce electricity. Both loops required pumps and pressurizers to accommodate the effects of condensation and thermal expansion.

The overall design for the pressurized water reactor was clever because it had a built-in safety feature. In the event of a coolant loss, the core also lost its moderator causing the atomic reaction to shut down. Nevertheless, in fault conditions, the high power-density within the core could result in fuel melting. Fortunately, the radioactive products in such a situation could not escape the containment area through the cooling system because it was a closed circuit, located entirely within the area of shielding. Only the low-pressure steam circuit traversed the reactor shielding on its way to the turbines.

Although the Navy never reported a serious accident with an American submarine reactor during normal shipboard operation, such a reactor demonstrated its power during an accident at NRTS on January 3, 1961. Three Navy technicians had shut down the SL-1, a 3 MW prototype military reactor for servicing. They had fully inserted its control rods and disconnected their drives. In attempting to free a stuck rod, one or more of the technicians accidentally heaved the rod up a distance of 20 inches. The resulting increase in reactivity instantly fried the core and flashed some of the reactor's coolant into steam. This drove a solid slug of water up against the cap of the pressure vessel ejecting some of the control rods, which pinned one of the men to the roof of the containment building. The sudden surge of radioactivity within the containment area killed the other two men. The Navy buried the three technicians in lead lined coffins.

Congress authorized construction of the first nuclear powered submarine in July 1951. This allowed Rickover to choose a submarine builder. Initially, he contacted the naval shipyard at Portsmouth, Maine, and arranged a visit by his staff and Charles Weaver, the head of the Bettis Atomic Laboratory. In a meeting with Admiral Ralph McShane, the Admiral told him that he had neither the resources nor the inclination to build a nuclear-powered submarine. Rickover then asked if he could use the admiral's telephone, which he used to call the Electric Boat Company, the only other submarine builder on the East Coast. O. P. Robinson, the general manager informed Rickover that he would be pleased to build the new submarine. Bidding farewell to the naval shipyard, Rickover availed himself of the Westinghouse limousine to drive himself and his staff over to Electric Boat, where they consummated a deal.

The size and upright design of the nuclear power reactor required a three-deck submarine; one more deck than the post war Tang Class of conventional fast attack boats. The three-deck design turned into an advantage because of the storage requirements for a submarine that would remain submerged for months at a time. The reactor shielding also added a lot of weight. When its thickness and placement proved difficult to calculate, reactor engineers measured shielding effectiveness empirically, using various setups with the NTRS prototype. A particular engineering concern was backscattering of radiation into the submarine's living space from the seawater surrounding the hull.

The lower hull plates (representing a keel) for SSN 571 were laid by President Truman on June 14, 1952. The submarine was named USS Nautilus after the famous submarine in the classic Jules Verne novel *20,00 Leagues Under the Sea*. Electric Boat did not install her reactor until after the MARK I had simulated a full transatlantic trip at NRTS. In Navy jargon, the MARK 1 was designated S2W. The "S" stood for submarine, the "W" for Westinghouse and the "2" indicated this was the second design produced by the specified manufacturer. The first was the MARK I prototype. The reactor connected to a pair of steam turbines that developed 15,000 shaft horsepower and turned two propellers. During the submarine's building, Rickover's group trained her captain and crew and produced the necessary technical documentation for the propulsion system. After 18 months of construction, Electric Boat launched Nautilus on January 21, 1954. First Lady Mamie Eisenhower, wife of the new president, christened her.

Nautilus measured 323.8 feet long and displaced 4,092 tons submerged. She was capable of 25 knots surfaced, 28 knots submerged, and had a test or maximum operational depth of 700 feet. Although primarily a research vessel, the Navy equipped her with six bow MK 59 torpedo tubes. The Navy commissioned the vessel on September 30, 1954, and on January 17, 1955, Captain Eugene P. Wilkinson announced that his ship was "underway on nuclear power." Nautilus had a complement of 12 officers and 92 enlisted personnel. The first shakedown cruise took Nautilus from New London, Connecticut, to San Juan, Puerto Rico. The boat traveled submerged for a distance of 1,381 miles in a time of 84 hours.

Upon her return, Rickover learned that not everything had gone well. Commander John Ebersole, the Navy doctor assigned to Nautilus, told him about a problem with the ship's atmospheric conditioning equipment. The carbon dioxide scrubbers did not work and the carbon monoxide burners tended to explode in flames. The burners also decomposed Freon leaks into hydrochloric and hydrofluoric acid fumes, which corroded the ships fittings and irritated the lungs and throats of the crew. The Navy Bureau of Medicine had difficulty in grasping the problem because they could not visualize the extended period of time that nuclear submarines could stay submerged. Although the problem was not strictly under his jurisdiction, Rickover had C. B. "Doc" Jackson from the Mine Safety Appliance Company flown out to the ship while it was on maneuvers. MSA was the national leader in air purification technology and eventually solved the atmosphere problem.

After steaming 60,000 miles, Nautilus received a second core for her reactor. Navy Headquarters then ordered her to reconnoiter under the Arctic ice pack in August 1957. As she approached within 180 miles of the North Pole, both her magnetic compass and gyrocompass failed and Nautilus had to return by means of dead reckoning. The following year, Nautilus made a second successful attempt at navigating under the ice. She made a submerged passage from Honolulu to England across the top of the world, firmly establishing the pressurized water reactor as a safe and reliable means of submarine propulsion. During this journey, the crew navigated by means of a MK 2 Submarine Inertial Navigation System (SINS), which North American Aviation that had previously developed as the N6A guidance unit for the Navaho supersonic cruise missile. The SINS measured and recorded the movements of the vessel while a computer plotted a record of the vessel's progress.

The incredible success of Nautilus and her pressurized water propulsion system convinced Rickover that the Navy no longer needed a liquid metal cooled reactor. Nevertheless, political momentum saw the General Electric S2G intermediate reactor installed in USS Seawolf (SSN 575), a sister ship to Nautilus. The reactor was termed intermediate because it operated with fast neutrons. Liquid metal reactors have coolant outlet temperatures that are 300 degrees hotter than the outlet temperatures of a comparable PWR. This allowed Seawolf to operate with very efficient superheated steam; as opposed to the low temperature saturated steam used by Nautilus. The sodium leg of the coolant cycle also operated at atmospheric pressure, reducing the need for high-pressure piping within the reactor's core.

In an intermediate configuration, designers can make a reactor core extremely compact because there is no need to disperse a moderator between the fuel elements. The basic challenge in designing this type of reactor is to achieve a very high neutron density. It takes 400 times as many fast neutrons as it does thermal neutrons to produce each fission event. Most of the fast neutrons in an intermediate reactor simply reflect off U235 nuclei. Others are absorbed to produce a stable nucleus of U236 by emitting a gamma ray. In order to increase the probability of fission events, the reactor has to burn fuel with U235 enrichment between 25 and 30 percent.

During operation, neutron capture transmutes some of the sodium used in an intermediate reactor into radioactive Na24. Na24 is a particularly active gamma ray emitter and its presence necessitated increasing the weight of the shielding around Seawolf's reactor core. In a preliminary concept, GE proposed placing a blanket of depleted uranium around the core. The U238 in the blanket would have absorbed neutrons to breed plutonium. Properly configured, the reactor would have bred as much or more plutonium fuel as the U235 it consumed. In the end, Rickover rejected this concept as too complex and substituted an intermediate non-breeding design.

The additional shielding needed to protect a submarine's living space from radiation counteracted the space saving benefits of intermediate reactors. In addition, the short-term radioactivity induced in the sodium coolant prevented maintenance for at least a week after the reactor shut down. By contrast, a pressurized water reactor transforms its light water coolant into non-radioactive deuterium. Sodium coolant also required electrical "superheaters" to keep it liquid in the event of a reactor shut down. Although this was not a serious problem onshore where auxiliary power was available, at sea, the tremendous amount of power required to run such heaters would have depleted the submarine's batteries in an emergency.

An important issue associated with liquid metal coolant relates to the consequences of a leak in the heat exchange system. The combination of sodium and water could produce corrosive lye or possibly result in an explosion. For this reason, the heat exchange tubes in the prototype reactor were double-walled and filled with mercury. Sensitive mercury detection devices were located within the system to detect any possible leaks. In Seawolf's S2G reactor, a sodium-potassium alloy that remained liquid at room temperature replaced mercury. The presence of toxic mercury within the enclosed environment of a submarine was not desirable.

Although Electric Boat laid her keel on September 7, 1953, Seawolf did not reach operational capability until March 30, 1957. The delay was a result of dealing with leaks detected in the submarine's superheaters. These were eventually isolated because they were not essential to the ship's operation. The ship then put to sea under restricted power and maneuvering limits in case leaks should develop in the heat exchange system. As it turned out, Seawolf operated successfully for 2 years without the need for its superheaters and her heat exchange system never developed any leaks. Nevertheless, the superheater problems and an accident with a liquid metal prototype power reactor at NRTS contributed to Rickover's decision to phase out intermediate reactors from the submarine program.

On November 29, 1955, technicians carried out an experiment that entailed interrupting the flow of the sodium-potassium coolant in the SBR-1 MARK II breeder reactor. It had shown a positive temperature coefficient of reactivity in which a rise in temperature was associated with increased reactivity. Since this positive feedback situation had the potential to get out of control, the Reactor Division was investigating this effect. Unfortunately, in shutting down the reactor, an operator used the slow operating control rods instead of the high-speed scram rods. The 20-second difference between these systems caused the temperature in the core to soar to over 1,100 °C and resulted in partial melting of the fuel, which slumped to the bottom of the containment vessel. Since the uranium fuel was highly enriched, this situation might have resulted in a critical configuration in which a reacting mass of fuel burned its way through the bottom of the reactor – on its way to who knows where. IPC Films made this much-dreaded scenario into a movie called *The China Syndrome.*

Even before Seawolf finished her trials, Rickover had her supply of supplementary fuel rods destroyed, thus leaving pressurized water reactors for sole approved use in submarines. He had already made the decision to build USS Triton (SSRN 586) as a twin, pressurized water reactor design for use as a radar picket. He also had USS Halibut (USGN 587), constructed with a pressurized water reactor. Halibut served as America's first nuclear powered guided missile submarine. In addition, Rickover placed the Skate Class of nuclear attack submarines in development as a simpler, cheaper, and smaller design based on the knowledge gained from operating Nautilus. The ships in this class were USS Skate (SSN 578), built by Electric Boat (General Dynamics), USS Sargo (SSN 583) and USS Swordfish (SSN 579) built by the Mare Island Naval Shipyard and USS Seadragon (SSN 584), built by the Naval Shipyard at Portsmouth.

S3W reactors powered Skate and Seadragon, whereas S4W reactors powered Sargo and Swordfish. Both types of thermal reactor provided steam for a single 13,200 HP turbine. The S4W was a straightforward improvement on the Nautilus S2W power plant and for this reason, Rickover awarded the building of Sargo and Swordfish to the Mare Island Shipyard. The S3W had an improved layout, so Rickover awarded the building of Skate and Seadragon to the more experienced Electric Boat and Portsmouth shipyards. A significant problem identified in Nautilus was the arrangement of reactor shielding. The S2W layout involved a horizontal heat exchanger and a large number of pipes penetrating the reactor shield that reduced access to the boiler. The S3W incorporated a vertical heat exchanger with a shielded access tunnel in which personnel could walk safely around the compartment. The success of this design led to the S5W layout that became a standard fixture in following classes of ballistic missile and attack submarines.

The Skate Class was essentially a nuclear-powered version of the Tang fast attack submarine, from which Nautilus and Seawolf drew their inspiration. The Skate Class measured 267.4 feet long and displaced 2,861 tons submerged. These were the first nuclear submarines equipped with a single shaft and propeller. In this regard, Admiral Rickover took a lot of convincing because of his concerns for safety – especially during under ice operations. Eventually, he realized that the need for silent operation in combat was more important than a redundant propulsion capability. Skate had the same 700-foot test depth as Nautilus but could make only 22 knots submerged – a consequence of the single propeller. Like USS Tang, the Navy equipped each Skate Class submarine with six bow and two aft MK 59 torpedo tubes, with 48 torpedoes carried. The crew complement included 8 officers and 75 enlisted personnel.

At about the same time that the Navy declared nuclear propulsion safe and reliable for use in submarines; new hull designs made their appearance. The redesign of submarine hulls was an outgrowth from the Greater Underwater Propulsion program that reverse-engineered the captured Type XXI U-boats: U2513 and U3008. The GUPPY Program led to four goals: increased battery capacity, hull streamlining, air breathing snorkels, and improved fire control systems. The Navy ordered USS Albacore (AGSS 569) as a hydrodynamic test vessel in late 1950 to experiment with a circular, tear drop shaped hull. Admiral Charles "Swede" Momsen sponsored Albacore.

Prior to Albacore, submarines ran on the surface and submerged only during an attack or when attacked. Thus, submarines exhibited serious drag when fully submerged. Admiral Momsen expected Albacore's hull design to improve submerged performance at the expense of surface stability. He considered this an acceptable tradeoff because increased battery storage and an air-breathing snorkel meant Albacore could spend most of her time submerged. Although the British Royal Navy experimented with cylindrical hulls as early as WW I with their Type R submarines, both the Royal Navy and other navies eventually abandoned the design in favor of hulls optimized to run on the surface.

In addition to her hydrodynamically shaped hull, Albacore had a streamlined sail that replaced the traditional conning tower. This complemented a lack of external fittings, such as deck guns, which the submarine did not need for its underwater mission. The Navy had Albacore's hull fabricated from low carbon, HY80 steel, which has a high tensile strength. This gave the vessel a maximum test depth of 600 feet and a collapse depth of 1,170 feet. A single propeller centered at the rear of the boat provided propulsion. Four fins set at right angles around the screw and a dorsal rudder built into the rear of the sail provided control. Powered by two diesel engines on the surface and a pair of battery powered, 7,500 HP electric motors submerged, Albacore could make more than 30 knots while fully submerged. Sound isolation mountings connected most of Albacore's internal fitting to her hull in order to suppress the submarine's noise signature. Albacore also introduced a fiberglass sonar dome in its bow, which is now standard equipment on submarines.

The success of Albacore resulted in the construction of the Barbel Class of attack submarines. Three vessels comprising this Class were in operation by 1956. Apart from their conventional propulsion, they were modern in almost all aspects. In addition to the stern control system developed for Albacore, they sported a pair of dive planes mounted on the sail. Like Albacore, the Navy had their hulls fabricated from HY80 steel to give them a test depth of 712 feet and a crush depth of 1,050 feet. USS Barbel (SS 580) was notable in having an "attack center," which combined the ship's control and combat operations into a single entity. The Barbel Class of submarines mounted six MK 59 torpedo tubes in the bow.

USS Skipjack (SSN 585) and her four sisters were the first submarines to combine both nuclear propulsion and a teardrop shaped hull. They measured 252 feet long and had a submerged displacement of 3,153 tons. Their control surfaces were in the same arrangement found on Barbel and they used the same HY80 steel in the construction of their hulls. Each vessel had an S5W pressurized water reactor connected to a pair of 7,500 HP steam turbines that gave them a surfaced speed of 15 knots and 29 knots submerged. The reactor cores could drive the ships 100,000 miles before they required refueling. Their high speed allowed the Skipjacks to act for the first time as submarine escorts to carrier groups. Previous Classes of submarine lacked either the speed or the endurance for this role. The Navy armed each Skipjack with six MK 59 21-inch bow torpedo tubes.

The only military hardware that Skipjack lacked was a thoroughly modern acoustic detection system. Although a number of scientists had experimented with the properties of sound in water during the 1800s, it took almost a century before Lewis Nixon developed the first underwater detection system in 1906. It provided advanced warning of icebergs in a ship's path. Specialized equipment for use in detecting submarines first appeared in 1915. This system passively listened for underwater sounds. Active systems that sent out a pulse of sound or "ping" and then listened for an echo only became available near the end of WW I. During WW II, the United States developed much more sophisticated systems, which the Navy dubbed SONAR, for SOund NAvigation and Ranging.

Sonar operates by means of transducers that convert one form of energy into another. In passive systems, a hydrophone or hydrophone array listens for underwater sounds, which electronics convert into electrical energy for analysis. Active systems use a transducer to convert an electrical signal into sound waves, which a submarine can bounce off potential targets. The disadvantage of an active system is that it immediately gives away a submarine's location. The frequency of sound in an active system can be either constant or variable. Lower frequencies have the longest range, whereas higher frequencies exhibit narrower beam widths. Varying power output across the face of a transducer enables directional output.

Sonar made a great leap forward when the Navy began construction of USS Tullibee (SSN 597) as a dedicated anti-submarine vessel in 1956. Tullibee was the development concept for a deep diving, ultra-quiet submarine with a long-range detection capability. Launched in 1960, Tullibee had the first integrated sonar system, the AN/BQQ-1. The sonar consisted of a bow mounted BQS-6 active array that spread about 1,000 transducers over the face of a large sphere. A BQR-7 low frequency passive array wrapped around the BQS-6 for listening. The AN/BQQ-1 interfaced with a MK 113 fire-control system that aimed the torpedoes in Tullibee's four 21-inch torpedo tubes. The Navy had to move these tubes to an amidships position and angle them to fire out through the sides of the hull because the spherical BQS-6 sonar occupied the entire bow. The AN/BQQ-1 had five operating modes: passive detection, passive tracking, passive localization, active localization, and active detection and tracking. Each of these modes had a specified probability of success associated with a given mission and together they provided tremendous operational flexibility.

Spherical sonar systems were desirable because they provided a wide area of detection around the submarine. The only place they could not see was directly behind the submarine, in an area known as the baffles. For this reason, Tullibee also experimented with towed arrays of hydrophones, payed out behind the submarine from a special reel. In order to make use of a towed array, the submarine had to maintain its forward motion. Passive towed arrays had equipment similar in design to the sensors used for SOSUS. TB-16, the first such operational towed sonar, consisted of a 1,400 pound, 240-foot-long detector array attached to the end of a 2,400-foot long cable.

All of the lessons learned from Skipjack and Tullibee came together in USS Permit (SSN 594). Ordered from the Mare Island Shipyard in 1958 and commissioned in 1962, Permit was the second vessel of the Thresher Class. The SSN-593, which gave its name to this Class, was lost in a diving accident on April 10, 1963. Rickover equipped Permit with an SW5 reactor and a single

15,000 HP steam turbine. Slightly longer than Skipjack at 278.5 feet and displacing 4,300 tons submerged, she could make 30 knots. The vessel's increased size, improved quieting, and operational efficiency started a trend that continued in the following Sturgeon and Los Angeles Class attack submarines.

Permit's hull design gave the vessel a maximum operating depth of 1,300 feet and a crush depth of 1,900 feet. In her bow, she carried an AN/BQQ-2 spherical sonar, the operational version of the BQQ-1 tested in Tullibee. Compared to any submarine in service at this time, her detection and tracking capability were truly phenomenal. Like SSN-590, the Navy equipped her with a MK 113 fire control system and four amidships MK 59 21-inch torpedo tubes. Permit was the first submarine to receive nuclear tipped torpedoes.

The United States Navy entered WW II with only one heavyweight submarine torpedo: the EC 14. It measured 20 feet 6 inches long, weighed 3,280 pounds and carried 643 pounds of Torpex in its warhead. Torpex was a powerful explosive manufactured from a combination of TNT, RDX and aluminum powder. Designed with dual speeds, the EC 14 had a range of 4,500 yards at a speed of 46 knots and a 9,000-yard range at a speed of 31 knots. A water-injected alcohol and air fueled turbine engine drove the torpedo. In consequence, it was designated "steam" powered.

Although the Navy deployed the EC 14 for 10 years prior to the Japanese attack on Pearl Harbor, it never tested the torpedo. This was the result of pre-war financial restrictions. Consequently, three very serious flaws went undetected. The torpedo's contact exploder was not sufficiently robust to survive a 90-degree collision with a target at high-speed; its magnetic influence exploder, designed to go off as the torpedo passed under the keel of a target and break its back, exhibited erratic performance; and it had poor depth keeping, which could cause it to dive harmlessly under a target. Incredibly, during the early days of WW II, the Bureau of Ordnance ignored these problems as poor marksmanship until Vice Admiral Charles A. Lockwood intervened to rectify the situation.

Despite rehabilitation of the EC 14, the demands of wartime combat called for better torpedoes. A combined effort from the Navy, research institutions such as the NRDC, and civilian contractors met this problem with new designs. These designs diverged along two different lines: gyro controlled, set depth, conventional torpedoes, and homing torpedoes. Conventional torpedo improvements were primarily associated with their propulsion systems. One of the most successful was a hydrogen peroxide / alcohol or NAVOL powered system used in the 21-inch diameter, MK 16 torpedo developed near the end of WW II. Approximately the same size as the EC 14, it had a range of 11,000 yards at a speed of 46 knots! Following the capture of a German battery powered G7e torpedo, the Navy began development of their own electric torpedo. This design did not leave telltale exhaust bubbles in its wake. In order to produce a lighter version of this weapon, the MK 26 used seawater as an electrolyte in a battery designed by Bell Telephone and built by General Electric. The torpedo only powered up after the bow cap of its tube opened to allow flooding with seawater.

The Navy began designing homing torpedoes in December 1941. Long-range aiming could be difficult in rough weather and a torpedo that auto-maneuvered to attack its target was highly desirable. Early efforts were passive designs that homed in on the cavitation noise of their target's propellers. In order to increase their effectiveness, they were battery powered. Electric motors were so quiet they did not mask target sounds. Pairs of hydrophones located on the left and right side and later the top and bottom hull, controlled the torpedoes. These compared the strength of incoming sounds and generated signals that steered the rudder toward the stronger. The Westinghouse / Bell Telephone MK 28 torpedo that first saw service in 1944 used this system. Later developments concentrated on active homing, in which a sound generator in the torpedo bounced a signal off the target.

None of the torpedoes developed during WW II were especially effective against a submerged and maneuvering target. When the US Submarine Force chose to concentrate its efforts in the anti-submarine role at the end of the war, it required new and improved weapons. Research by the Harvard Underwater Sound laboratory (HUSL) and the Penn State Ordnance Research Laboratory (ORL) eventually culminated in the Westinghouse-ORL MK 37 MOD 0 torpedo. This was a 19-inch diameter, 135-inch-long, 1,430-pound weapon with a 330-pound warhead. A two-speed electric motor gave it a range of 10,000 yards at a speed of 26 knots and a 23,000-yard range at 17 knots. Guidance consisted of a straight, preset gyro-controlled run on a calculated intercept course, followed by a passive acoustic search pattern. After target acquisition, a Doppler-enabled active search system took over for the final attack. The Doppler component prevented the torpedo from homing on false, stationary targets. Concerns about target motion during the lengthy enabling run led to the wire guided MOD 1. This 161-inch long upgrade had a spool of fine wire, which reeled out the back of the torpedo and remained connected to the submarine. A guidance computer sent out mid-course corrections if the submarine's sonar detected evasive movement of the target.

The development of high-speed, deep diving submarines by the Soviet Union countered the effectiveness of the MK 37 and resulted in a high-speed successor known as the MK 45 Anti-Submarine TORpedo (ASTOR). Early in the development of this weapon, the Navy decided to arm it with a 10-kiloton W34 nuclear warhead. Other warheads considered were the W5 and W12. Of these, neither was as compact or as efficient as the W34. The AEC released the W34 MOD 1 for the MK 45 MOD 0 torpedo in October 1960, with MODs 2, 3, and 4 for the MOD 1 torpedo following later. The W34 MOD 1 warhead had thermal batteries, stored end over end in their battery clamps until required for use. Eventually the Navy restricted ASTOR for use with MODs 3 and 4 of the W34 warhead. The MOD 3 had a 15-ohm bleeder resistor installed in its neutron generators and a shielded pullout cable and filter pack that attenuated induced currents. The MOD 4 increased the resistance of the bleeder resistors to 30 ohms. Concerns relating to the positive control of a nuclear warhead resulted in the deletion of a homing capability for ASTOR. Instead, the DOD had it equipped with a command guidance wire, with target tracking accomplished by means of the submarine's sonar. The torpedo did not have contact or influence exploders.

The torpedo consisted of a tail cone, afterbody, battery compartment, and nose / warhead compartment. A seawater-activated battery that was connected to a 160 HP electric motor provided power. The 250-volt, 550-amp battery provided sufficient power for 9 minutes of operation. The rate at which water flowed through the battery controlled its voltage. A pair of contra-rotating propellers connected to the electric motor produced speeds up to 40 knots over a range of 11,000 - 15,000 yards. The MK 45 MOD 0 was 225 inches long, 19 inches in diameter and weighed 2,330 pounds with a dry battery. It had 21-inch guide rails that allowed it to fit inside a standard MK 59 torpedo tube. The guide rails gave the 21-inch torpedo a silent, swim out capability. The 2,213-pound MOD 1 version of the MK 45 was 2 inches longer than the MOD 0. Sandia supplied the MOD 1 with a more reliable integrator than was available for the MOD 0. The Navy designated the W34 warhead for the MK 45 MOD 0 torpedo as the MK 34 MOD 0, whereas it designated the warhead provided for the MOD 1 torpedo, as the MK 102 MOD 0. Command detonation was only possible after a run of 2,050 yards, although only a desperate skipper would have activated the weapon at this range.

The MK 45 torpedo received guidance commands from a MK 113 Fire Control System, which in turn received inputs from Western Electric's BQR-15 passive towed array, a BQS-4 active sonar, and a Raytheon's AN/BQQ-2 bow mounted sonar system. Upon launch, the MK 45 torpedo immediately dove or climbed to a depth of 100 feet for quiet running on a preset course. When it reached a speed of 27 knots, a velocity switch closed and the integrator started measuring distance. Feedback from the torpedo to the submarine provided information to determine if the torpedo was running properly. A malfunctioning torpedo could be jettisoned. At 2,050 yards, the minimum safe detonation range, ASTOR dove to a pre-set depth, no greater than 1,000 feet. At this time the torpedo was at its maximum speed of 40 knots and the integrator armed the torpedo. A depth monitor function kept it from diving too deep. Using wire guidance commands, the MK 113 Fire Control System could change the range and course of the torpedo during the run to the target. An anti-circular run function prevented it from turning back itself. At a distance of 200 yards from the target, the integrator locked out further course changes and at the calculated burst point, it command detonated the warhead.

The Navy never carried out a live test of the nuclear-armed MK 45 ASTOR. However, the Soviet Foxtrot Submarine B-130 fired at least two live T5 10-kiloton torpedoes near Novaya Zemlaya Island in the Arctic Ocean. Captain Second Rank Nikolai Shumkov reported that at a distance of 10 miles, the effects on his submarine were so violent that the crew almost lost control of the vessel. American submariners declared that the MK 45 torpedo had a dual kill capability: its target and the launching submarine!

The DOD gave production authorization for the nuclear-armed ASTOR in March 1959. Later that year, the Navy made four successful fuzing / firing runs of the complete system from a Permit Class SSN. It approved ASTOR for service in 1961 and took its first deliveries in 1963. Westinghouse manufactured approximately 1,000 torpedoes and the AEC built 600 W34 warheads. Under normal circumstances, an attack submarine carried a pair of these weapons, along with a load of EC 14 / MK 16 anti-ship and MK 37 ASW conventional torpedoes. Two MK 45s were lost along with the submarine USS Scorpion (SSN 589), when it sank in the mid-Atlantic under mysterious conditions while returning from a patrol in the Mediterranean.

The Navy withdrew the MK 45 from service during the period 1972 to 1976. This was a result of the care needed to employ the weapon in close proximity to a Task Group and because of the introduction of the conventional Gould / Honeywell MK 48 torpedo. This successor to the MK 37 and MK 45 is a fast, deep diving, wire-guided torpedo with both active and passive homing. This torpedo measures a full 21 inches in diameter, is 230 inches long, and weighs 3,440 pounds with a warhead containing 650 pounds of PBXN-103. In order to achieve the speeds necessary to close with high-speed submarines and surface targets, Gould / Honeywell powered it with an OTTO fuel external combustion axial-piston engine. This gives it a range of 35,000 yards at a speed of 55 knots. The torpedo is capable of attacking targets at depths of 2,500 feet. An "advanced capability" or ADCAP version has even higher speeds with improved guidance and logic.

After its retirement, Westinghouse reconfigured the nuclear MK 45 as the conventional "Freedom" torpedo and made it available for foreign military sales. Although Westinghouse built a few demonstration models, it never sold any because improved versions of the MK 37 homing torpedo were superior in performance.

SUBROC, a thermonuclear tipped, standoff, antisubmarine missile joined the ASTOR torpedo in mid-1965. The Navy conceived SUBROC in 1953 as part of an antisubmarine project focused on developing a long-range weapon for use by a submarine outside the effective range of an enemy submarine's torpedoes. The original concept for this weapon (Project Marlin) was a solid fuel nuclear bombardment missile that a submarine could launch from a torpedo tube against either maritime or land-based targets. This project considered weapons with ranges up to 100 nautical miles and warheads with yields up to 500 kilotons. The Navy then scaled back this concept to a dedicated antisubmarine weapon with a secondary airburst capability to speed up the project. It also considered a Wet ASROC Missile (WAM) for submarine launch but this weapon did not have the range required.

The Navy began the feasibility study for SUBROC in September 1957, and initiated development of the MK 28 "SUBmarine ROCket" at the US Naval Ordnance Laboratory in White Oak, Maryland, in June 1958. At the same time, the Navy began development of the Permit Class submarine with its AN/BQQ-2 sonar as SUBROC's primary weapon platform. It chose Goodyear Aerospace of Akron, Ohio, for systems integration and production, Singer Kearfott of Little Falls, New Jersey, for the inertial guidance system, AiResearch of Los Angeles, California, for auxiliary power, and Thiokol of Elkton, Maryland, for the composite solid fuel rocket motor. The AEC assigned LRL the task of producing SUBROC's W55 thermonuclear warhead.

After hydrodynamic testing in the Thomas Garfield 48-inch diameter water tunnel at Penn State University, SUBROC's first test launch took place in August 1959. The ensuing development program did not go smoothly, requiring solutions to a large number of technical problems. These included poor solid fuel performance because of the missile's residence time in a torpedo tube flooded with ice-cold seawater. In addition, the air pulse that accompanied ejection of SUBROC from a torpedo tube created bubbles that interfered with motor ignition and its control surfaces. Problems also developed with vibrations generated during SUBROC's transition from water-to-air and its subsequent reentry.

Because of these mechanical issues, the Naval Ordnance Test Station conducted an inertial guidance sled test program at the Supersonic Naval Ordnance Research Track (SNORT). It designed the program to obtain engineering performance data on the guidance and control system in an environment of sustained acceleration similar to those expected during the launch and boost phases of the flight trajectory. Three instrumentation checkout runs and eight guidance tests demonstrated the successful operation of SUBROC's Singer-Kearfott SD-510 inertial guidance system under simulated flight conditions.

The Navy gave Singer-Librasope of Glendale, California, responsibility for integrating the missile's control system with Permit's MK 113 fire control system. The MK 113 received inputs from Western Electric's BQR-15 passive towed array, BQS-4 active sonar, and Raytheon's AN/BQQ-2 bow mounted sonar system. The AN/BQQ-2 was the first sonar with sufficient detection range to provide targeting data for SUBROC. The submarine had to sprint to several locations after detecting a target to provide a target motion analysis or TMA. An operator then fed this data into the MK 113 Fire Control, which had a combination analog / mechanical computer that estimated a firing solution based on target bearing, range, and speed. Target limits were 72,000 yards and a 50-knot speed.

Located in the Combat Control Center to the right of the periscopes, the MK 113 consisted of two large "position-keepers" that displayed both the ship's and the target's relative positions. Behind the position-keepers stood the actual computer that fed data to a weapon console on either side. To the inside of these lay the main torpedo and SUBROC consoles that programmed firing parameters into those weapons.

Beginning in 1964, the Navy carried out operational testing of SUBROC aboard the USS Plunger (SSN 595), based out of Pearl Harbor, Hawaii. By this time, the UUM-44 designation had replaced MK 28. Unfortunately, Plunger could not detect a snorkeling submarine at half of SUBROC's maximum range. This resulted in a DOD threat to cancel the Permit Class submarine if its sonar problems were not rectified. Following a brief scramble, the Navy corrected Permit's problem and declared SUBROC operational in 1965. The Navy acquired 431 SUBROC missiles, 89 of which it expended in development.

As deployed, SUBROC was a two-piece missile / nuclear depth charge combination that measured 21 feet long, 21 inches in diameter and weighed 4,000 pounds. It had a reentry bomb weight of 700 pounds. A submarine could launch SUBROC from a standard 21-inch torpedo tube immediately after the MK 113 Fire Control System had uploaded targeting instructions to its Kearfott SD-510 guidance system. Several seconds after the weapon left the launch tube, a 36,500-pound-thrust Thiokol TE-260G solid fuel motor ignited, propelling SUBROC out of the water and onto a supersonic trajectory to its target. SUBROC's minimum range was 5 nautical miles and its maximum range was 35 nautical miles. Four jet deflectors in the rocket exhaust provided directional control. At a predetermined distance along the weapon's trajectory, explosive bolts activated to separate the nuclear depth charge from the missile body. The Singer-Kearfott SD-510 inertial guidance system, located in the forward section, continued to control the missile using small aerodynamic surfaces until it hit the water. At this point an aerodynamic plastic fairing shattered to reveal a flat nose that improved underwater stability. Under water, the depth charge sank to a specified depth, where its warhead detonated.

The AEC began development of SUBROC's W55 MOD 0 thermonuclear warhead in the mid 1950's, conducting a concept / feasibility test in June 1958. Hardtack Olive afforded a yield of 202 kilotons with a Gnat primary and a Tuba secondary. The device measured 32 inches long by 12.7 inches in diameter and weighed 218.7 pounds. Its yield to weight ratio was 2 kilotons per kilogram. The finalized W55 warhead measured 13 inches in diameter by 39.4 inches long and weighed 465 pounds. Much of the increased weight came from a high tensile steel case with a yield point of 160,000 pounds per square inch. Fifteen pounds of its weight was ballast to provide the proper center of gravity for use with SUBROC. LRL weaponized the W55 with a Kinglet primary, which it tested in an XW-55 mockup as Nougat Black at the NTS on April 27, 1962. The AEC never tested a complete W-55 warhead.

The Kinglet primary ascribed to the W55 was a two-point initiated, beryllium clad, boosted design featuring and MC1324 pit. It had a chopper / converter, capacitor-type x-unit, and acceleration sensing safety devices. The AEC tested the Kinglet approximately ten times during Operation Dominic. The full thermonuclear yield of the W55 is stated as 250 kilotons, using a Tuba secondary. The W55 full warhead set a new standard in yield to weight ratio for lightweight weapons. The AEC produced 285 SUBROC warheads in two runs: January 1964 - March 1968 and March 1970 - April 1974. It was the first time that the AEC produced a warhead in two separate runs.

From the above description, it is apparent that SUBROC used the same fission package as the W58. In addition to the same primary, both warheads were equipped with thorium radiation cases. Th232 is radioactive and decays to lead 208 through a series of alpha and beta particle emissions. Near the end of the decay process, thallium 208 decays to lead 208 with the emission of a 2.63 MeV gamma ray. This created a problem for sailors who had to share the same environment as SUBROC. The Navy therefore put a plan in place to store SUBROC missiles under a submarine's berthing compartment. This required placing six inches of composite lead and polyurethane shielding around the SUBROC storage compartment.

SUBROC had a number of advantages and one disadvantage when compared with the MK 45 torpedo. The compressed air pulse used to expel the torpedo from its launch tube created a noticeable transient that its target might detect. Fortunately, this did not give the enemy much time to react since SUBROC's Mach 1.5 supersonic speed allowed it to reach its target in a very short time. Accuracy was not a great concern because the quarter megaton warhead could sink most vessels within a 5-mile radius of its burst point. During SUBROC's deployment, it served on board Permit Class, Sturgeon Class, and Los Angeles Class attack submarines. A typical load out varied from four to six missiles. Fleet ballistic missile submarines could carry one or two SUBROCs for self-defense. The Navy carried out crew training at a facility in Akron, Ohio.

The missile was so successful that the Navy proposed to extend its life by re-graining its deteriorating solid fuel motors and replacing its analog guidance unit with a digital system in 1972. Congress, however, vetoed this idea and called for its replacement with Sea Lance, an entirely new standoff missile. Authorized in 1982, Congress intended Sea Lance to replace both the SUBROC and ASROC (discussed in the next chapter) systems. However, the Sea Lance program suffered so many delays that the Navy cancelled it in 1990, when it forswore the use of tactical atomic weapons. In order to maintain its anti-submarine capability while waiting for Sea Lance, the Navy replaced 224 SUBROC motors during the period 1977 - 1981, extending the weapon's lifespan by 15 years. The Navy removed its last SUBROC missile from service in 1992, along with its other tactical nuclear weapons.

CHAPTER 18
BERYLLIUM TAMPERS, LINEAR IMPLOSION, PLASTIC BONDED EXPLOSIVES, MINI-NUKES, AND DERIVATIVE WEAPON SYSTEMS

The mid 1950s were a busy time for the nuclear laboratories. In addition to developing fusion technology and external initiators, they were investigating the use of beryllium as a neutron tamper / reflector in fission warheads. The usefulness of any material as a neutron reflector is inversely proportional to the amount of space between its nuclei, a measurement known as its "atomic density." Low-Z materials tend to have the highest atomic densities because they have only a few electrons to take up space. Of the light materials, the most efficient neutron reflector is diamond, which forms a close packed, face centered, cubic structure. Unfortunately, diamond is both expensive and difficult to work. Next in utility is beryllium, which weapon designers favor because it has many desirable physical characteristics. The metal is lighter than aluminum, stronger than steel, highly compressible, has a high melting point, and exhibits excellent dimensional stability. A serious drawback to beryllium's industrial use is its toxicity, a property that manufacturers have to account for during its handling. Other important neutron reflecting materials include beryllium oxide, beryllium carbide, tungsten carbide, vanadium, and graphite.

The term "reflector" is a bit misleading. After encountering a nucleus of beryllium, a neutron does not just reflect back into the core; it scatters in a random direction. Several scattering events are typically required before the neutron reenters the core. The amount of time taken by this process is dependent on beryllium's mean free path for scattering and the velocity of the scattered neutrons. Because beryllium is highly compressible, a well-designed warhead insures that implosion simultaneously compresses both the neutron reflector and the core to reduce their mean free paths for neutron scattering. Measurements have shown that a 5-centimeter-thick beryllium shell is sufficient to reduce critical mass by 50 percent when compressed in a warhead. Larger thicknesses are of little practical use and only serve to enlarge the dimensions and weight of the warhead.

There are a number of advantages to using beryllium as a neutron reflector. Beryllium has only 1/10 the density of an inertial tamper material like uranium, allowing for a significant weight reduction in comparable systems. It further allows designers to make size and weight reductions by eliminating the aluminum pusher shell. Because beryllium has a density similar to aluminum, it preserves the density contrast geometry that manages Rayleigh-Taylor instabilities and intensifies the shock wave during implosion. A single layer of beryllium is thus able to replace both the uranium tamper and the aluminum pusher that were a characteristic of Fat Man type weapons. With much less mass to compress in a beryllium reflected pit, the nuclear laboratories could also reduce the amount of explosive used in implosion assemblies. If a plutonium pit substitutes for an equivalent uranium pit (which weighs 5 times as much), it is further possible to reduce the amount of explosive in the implosion sphere – for which reason plutonium and beryllium have become the materials of choice for use in thermonuclear primaries and compact warheads. Finally, a beryllium Finally, a beryllium reflector provides shielding from external neutrons, reducing a core's susceptibility to predetonation.

A MK 12 nuclear bomb. This was the first weapon to incorporate a beryllium tamper in its pit. Note the removable nose cap and folded fins. The bomb had an in-flight insertion device for its pit and could accept a variety of different cores. (Credit: NAM)

A Navy FJ-4B Fury carries a MK 12 bomb shape on its left wing at the China Lake Test Range. (Credit: USN)

An RUR-5A ASROC missile leaves its box launcher on the foredeck of a destroyer. It was equipped with either a MK 17 depth bomb armed with a 10-kiloton W44 warhead or (as shown) with a MK 44 torpedo with a conventional warhead. (Credit: USN; Color Original)

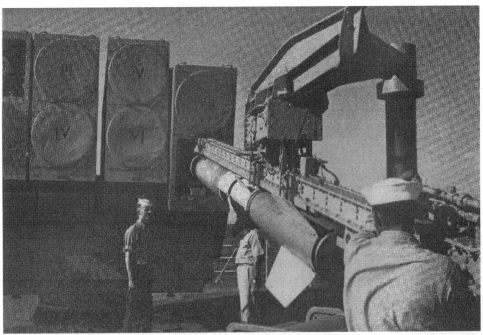
Loading an ASROC missile into an M16 box launcher. The launcher held eight missiles in four double cells. As shown, each double cell could be elevated independently. (Credit: USN)

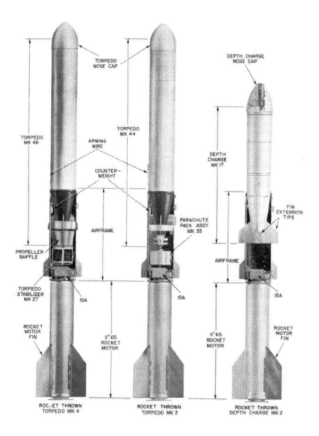

ASROC Missiles. Shown on the right is the nuclear depth bomb equipped version. (Credit: USN)

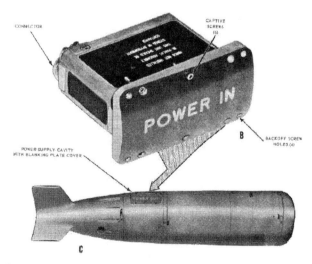

At the top of the illustration is an ASROC nuclear Depth Bomb power supply. Bottom shows where it fits in the Depth Bomb. The power supply was kept isolated from the depth bomb while in storage. (Credit: USN)

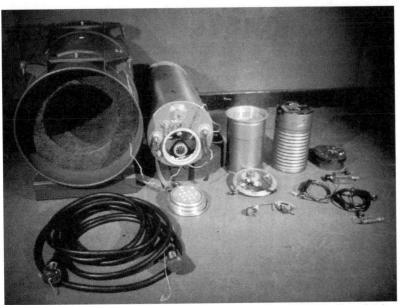

A MOD 3 medium atomic demolition munition (MADM), which was equipped with a W45-3 warhead in a variety of yields. From the left are the H815 packing container, M17X warhead, firing unit, and M5 coder / decoder. The cover has been removed from the warhead showing the receptacle for the firing cable, which lies in front. (Credit: DOD; Color Original)

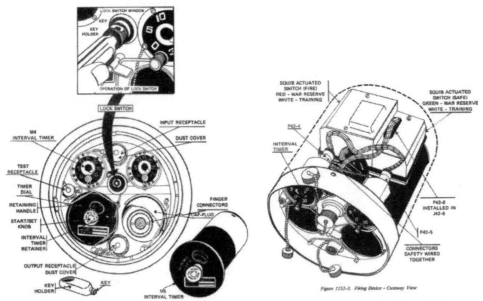

Control Panel (left) and cutaway (right) for a MOD 3 medium atomic demolition munition's M96 Firing Device (Credit: USA FM9-55G1/2)

424

H-911 SADM shipping container. (Credit: USA)

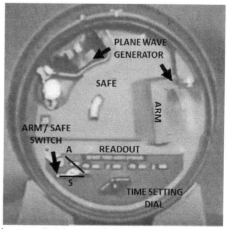

A man-portable, H-912 transport pack for transporting a B54, special atomic demolition munition, (SADM). The B54 had a variety of yields to a maximum of 1 kiloton. The MOD 1 SADM's control panel is shown on the right. (Left Credit: SNL, Right Credit: USA, Annotation: Author)

An M29 Davy Crockett recoilless launcher. Its W54 warhead is attached at right. An M113 transport vehicle can be seen in the rear. (Credit: USA)

Equipment storage inside an M29 equipped M113 tracked vehicle. (Credit USA)

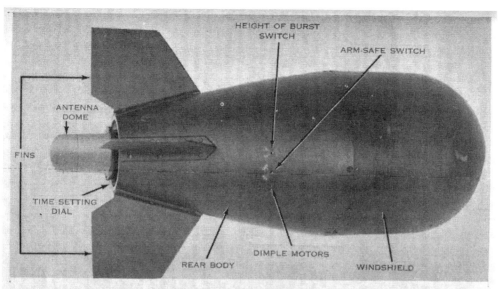

Figure 28. Projectile, atomic, supercaliber, 279-mm, dummy, XM421.

An XM421 practice projectile for the M388 nuclear projectile, which was used in both Davy Crockett recoilless launchers. Its W54 warhead had a nominal 20-kiloton yield. (Credit: USA FM23-20)

An MGR-3 (M51) "Little John" rocket on its M34 towed launcher. It was equipped with a W45 nuclear warhead. (Credit: USA)

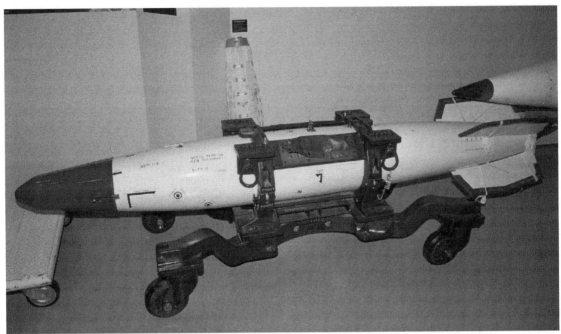

A B57 multi-purpose nuclear bomb and its laydown / retardation chute. It was based on the boosted W44 (Tsetse) warhead. Note the protective nose cap. (Credit: Clifford Bossie with Permission; Color Original)

Part of a convoy carrying double stacked B57 bombs. The bombs have just been removed from the on-base nuclear vault (shown behind) and are on their way to a carrier or the flight line. (Credit: Navy Nuclear Weapons Association with Permission; Color Original, Cropped)

Shown is a remotely operated Navy DSN QH-50 DASH Gyrodyne antisubmarine helicopter carrying a pair of Mk 43 torpedoes. It could also carry ASROC's MK 17 nuclear depth charge or a B57 nuclear depth bomb. (Credit USN)

A disadvantage of beryllium is the moderation of neutron velocities produced with each scattering event in the reflection process. After 3.5 collisions with a beryllium nucleus, a neutron's energy reduces to one-half its initial value. Since a neutron's energy is largely dependent on its velocity, this serves to increase the time it takes the neutron to reenter the core. Because the chain reaction grows exponentially with time, it is important that neutrons reenter the system as soon as possible. Fortunately, the reaction:

$$Be9 + n > Be8 + 2n - 2.4 \text{ MeV}$$

multiplies neutrons in the reflector shell. This phenomenon, known as "neutron amplification," helps to compensate for beryllium's moderating effects.

The first weapon to incorporate a beryllium reflector (at least in a prototype) was the MK 12 bomb. Los Alamos exploded two prototypes of this weapon as part of Operation Tumbler-Snapper in 1952. The TX12 Brok-1 device, tested as Snapper Easy, had the standard depleted uranium tamper found in contemporary weapons. The nuclear system weighed 550 pounds in a device that weighed 625 pounds. It produced a yield of 12 kilotons. One month later, Los Alamos tested a beryllium reflected TX12 Scorpion prototype as Snapper How. The Scorpion device shaved 80 pounds off the weight of Brok-1 and produced a yield of 14 kilotons. The closely spaced tests likely compared the relative efficiencies of beryllium and uranium tampers. Some authors indicate that both the Brok-1 and the Scorpion devices had beryllium tampers and that their weight and yield differences resulted from the use of two different pits.

It is interesting to evaluate the implosion process of a plutonium core surrounded by beryllium. Since plutonium decays primarily by alpha particle production, a plutonium core surrounded by beryllium is essentially an oversized neutron initiator. The nickel plating of the plutonium core prevents alpha particles from interacting with beryllium prior to implosion. It would be important to prevent Rayleigh-Taylor instabilities from mixing core and tamper materials during implosion. If not properly designed, a neutron flux from beryllium - alpha particle interactions early in the implosion process could lead to predetonation and a reduced yield.

The Air Force requested development of the MK 12 bomb because its fighter-bombers could not carry the MK 7 bomb externally at high speed. It wanted a weapon optimized for high-speed external carriage and low-level attack including horizontal, dive, toss, and glide bombing. Military characteristics called for carry at speeds up to Mach 1.4, release at altitudes up to 50,000 feet, and internal carriage if that characteristic did not interfere with the bomb's primary mission. (The final maximum release altitude was 30,000 feet). The weight of the bomb was not to exceed 1,200 pounds and its physics package could be no more than 22 inches in diameter.

The deployed MK 12 MOD 0 bomb, derived from the TX12 and TX12X1, measured 155 inches long by 22 inches in diameter, and weighed 1,050 pounds. Its tail fins folded horizontally to provide clearance for wing flaps, and the entire tail mechanism rotated in 22.5-degree increments to solve aircraft installation issues. Fin tabs spun the bomb to provide stability in freefall. Sandia changed the design of the bomb case shortly before production by adding a strengthened center to accommodate sway bracing. This changed the program designation from TX12 to TX12X1. The strengthened (hard) case had two different suspension systems, one system had two lugs separated by 30 inches, and the other system had three lugs spaced at 20-inch intervals.

A conventional implosion device, the MK 12 bomb was equipped with an MC220A linear in-flight insertion mechanism for its core, a 92-lens implosion assembly, and a TOM internal neutron initiator. The 22-inch diameter implosion system had a volume that was only 40 percent of the MK 7's. The bomb had a single radar fuze, two timers, and a contact fuse. Service personnel could access the in-flight insertion mechanism behind a hinged nose that housed the bomb's radar antenna. The radome nose was equipped with buried heating wires. Behind the nuclear explosive package, a removable shelf held a miniaturized x-unit along with additional components of the firing system. The ground crew set the bomb's safe separation timers through a hatch in the bomb casing before takeoff.

For power, Sandia equipped the MK 12 with MC271 silver peroxide - zinc alkaline batteries, which first became available in April 1955. These were stored in an MC272 battery box. Silver peroxide - zinc alkaline batteries are rechargeable and have very high energy densities – they can withstand one of the largest loads of secondary power sources. The MC271 battery system included batteries, electrolyte, and an electrolyte-filling syringe sufficient for use with one MK 12 weapon.

Like its MK 7 predecessor, Sandia supplied the MK 12 with seven different air and surface burst fuzing options that the pilot could select in flight. These included:
- Radar airburst altitude setting 1 with safe separation time setting 1
- Radar airburst altitude setting 2 with safe separation time setting 2

- Radar airburst altitude setting 1 with safe separation time setting 2
- Radar airburst altitude setting 2 with safe separation time setting 1
- Timer airburst setting 1 with safe separation time setting 3
- Timer airburst setting 2 with safe separation time setting 4
- Contact burst with safe separation time setting 5

The Air Force stockpiled Type 110, 130, 170, 240, and 260 cores for use with the MK 12. Thus, the MK 12 would have been available in a variety of yields to a maximum of some 20 kilotons. Intended as a high-speed external replacement for the MK 7, the Air Force retired the MK 12 before it removed the last MK 7 from service. The MK 12 MOD 0 reached operational capability in early 1955, with limited production of 250 units. A number of authors attribute its early retirement and low production numbers to nuclear inefficiency, since its 5-inch-thick implosion system was much less powerful than that of the MK 7. Its low production numbers may also have resulted from a scarcity of beryllium at the time of its manufacture.

During the operational life of the MK 12, air units began to report fin damage during takeoffs and landings. As a result, Sandia instituted a TX12X2 program that developed stronger fins, a more powerful fin deployment motor, and a more robust gear chain. Sandia supplied the modification kits needed to upgrade the MOD 0 bombs to the MOD 1 configuration in February 1957. The Navy used these kits to modify the MK 12 stockpile through November 1957. This upgrade increased the maximum fin-operating aircraft speed from 180 knots to 220 knots.

The MK 12 bomb equipped both Air Force and Navy fighter-bombers. In fact, the Navy acquired the North American F-86F-35-NA variant of the Sabre fighter equipped with LABS especially for use with this weapon. At the time, the Air Force was retiring the F-86 from front line service in the air superiority role, making it available for conversion to a fighter-bomber. The F-86F carried the MK 12 under its port wing and a drop tank under its starboard wing. The carriage of a nuclear bomb under the port wing was standard operating procedure for many aircraft. Navy aircraft that carried the MK 12 include the F9F, F2H-3 & 4, F3H-2, FJ-4B, and A-4. Although the MK 12 was capable of supersonic carry, only one of these aircraft, the F3H-2 Banshee, possessed the ability to fly at transonic speeds.

Prior to takeoff, it was necessary to set the MC220A in-flight insertion mechanism, set radar ranges S1 and S2 for airburst, set the MC348 safe separation timer, service and install the bomb's batteries, and load the bomb's nuclear core on an externally hung bomb. When the aircraft was in flight, the pilot prepared the bomb for use with the help of a T-79 or T-145 controller. The pilot extended the bomb's fins, turned on the bomb's radar, selected one of seven fuzing options using a rotary selector switch, and set the ARM / SAFE Switch to ARM. This latter action activated the in-flight insertion mechanism. After release, the bomb's arming wires pulled out to start the MC73 timing sequencer. This included the operation of the MC348 safe separation timer at the start of the timing sequence. When the safe separation timer expired, the bomb's inverters began charging the x-unit, and other switches closed depending on the detonation sequence selected.

Following Air Force acceptance of the MK 12, the AEC began planning for the wide scale use of beryllium in nuclear weapons. In 1956, it awarded contracts to a pair of domestic companies to produce 100 tons of beryllium per year each for a period of 5 years. The use of beryllium as a weapon material actually diverged along two separate paths. In the first, Los Alamos used beryllium in tampers to produce compact warheads and, in the second, it used beryllium as a structural material and heat absorber in warheads and reentry vehicles. Like the preceding programs that created the first generation of boosted warheads, the beryllium weaponization program produced both boosted fission warheads and primaries for use in compact thermonuclear weapons.

One of the first beryllium reflected devices produced was the Swan boosted primary. It was the result of LRL research into linear implosion warheads for use in air-to-air missiles and artillery shells. LASL conducted research into cylindrical implosion during WW II but concluded that detonation synchronization, the lack of homogeneity in available high explosives, complex geometries, and other intrinsic factors militated against the success of this design. After the war, LASL ignored research on linear implosion systems, in part because the mathematics were not very tractable and in part because linear implosion was inherently inefficient. LRL, however, saw linear implosion as a way to differentiate itself from LASL. In May 1954, it began conducting experiments using U238 cylinders at LASL'S Sawmill Site and at China Lake, California.

Linear implosion uses an explosive assembly to reshape a sub-critically configured cylindrical or elliptically-shaped plutonium core into a super-critical sphere – without the aid of compression. In order to convert an elliptical core into a sphere, LRL had to create a shock wave that reached the ends of the core slightly before it converged on the sides. The initial contact with the shock wave shaped the core into a sphere and the formation of the complete shock wave maintained the sphere's shape long enough for a fission reaction to proceed to completion.

LRL accomplished this feat with a watermelon-shaped explosive envelope. It equipped both ends of the explosive envelope's long axis with detonators – a concept called two-point detonation. LRL then achieved super-criticality through a combination of transformation from the δ phase to the α-phase in the core's plutonium metal and a reduction in the surface area of the core by deforming it into a sphere. Methods for producing a shock wave with the preceding characteristics include the addition of inert wave shapers at each end of the explosive envelope and varying the composition (velocity characteristics) of the explosives used to form the implosion envelope.

The primary means of shaping the shock wave in American weapons seems to have been the addition of mechanical wave shapers. The most useful materials for wave shapers are void-forming, low-Z plastic foams that soft x-rays can completely ionize. Wave shapers have velocity characteristics that differ from those of the explosives that surround them. The watermelon shape of the explosive envelope thus results from the placement of wave shapers and (possibly) a greater thickness of explosives at the end of each hemisphere of the implosion assembly. The wave shapers and explosive thickness thus shape the detonation wave to its desired characteristics. LRL at one time worked on a system that placed a hollow form with inert wave shapers around a nuclear core. A mechanism squeezed the warhead's explosive, which had the consistency of toothpaste, into the empty mold during the arming procedure. In this way, the bomb was incapable of developing any nuclear yield prior to arming.

Designers have also used explosives to form wave shapes, such as the lenses of the Fat Man implosion system. Explosive wave shapers require two different types of explosive or a single explosive with modified velocity characteristics. Explosive manufacturers can "whip" air bubbles into explosive slurries or generate them in situ using a chemical like sodium borohydride. This varies the explosive's density and thus its velocity profile. Alternatively, if a manufacturer plans to press an explosive after casting, glass or plastic microspheres added to the casting slurry can preserve a desired density gradient. Linear implosion techniques, which involve the bonding and forming of hemispherical shells from alternating layers of explosive and inert material to create an anisotropic detonation wave, have been discussed by the French and the Chinese.

LRL exploded Cleo I, its first linear implosion device, as Teapot Tesla on March 1, 1955. Cleo I measured 10 inches in diameter by 39.5 inches long, weighed 785 pounds, and produced a yield of 7 kilotons with a solid cylindrical core. It was equipped with a "zipper" or external neutron initiator. A considerably reduced, 322-pound Cleo II yielded 2 kilotons as Teapot Post on April 9. It took only a year before LRL attempted to boost a linear implosion device with the intent of increasing its yield and efficiency. This involved the use of a hollow, asymmetrical core. On May 27, 1956, LRL tested the Swift device as Redwing Yuma. The device proved a failure when it failed to boost at a yield of 0.19 kilotons. Undaunted, LRL tried again with the aforementioned Swan device as Redwing Inca on June 21. Inca was a complete success with a yield of 15.2 kilotons. LRL later used the Swan device as a control for evaluating new types of thermonuclear assemblies. A typical Swan device measured 11.6 inches in diameter, 22.8 inches long, and weighed 105 pounds. Advantageously, it had a simplified detonating system, reduced power requirements, and insensitivity to predetonation. The asymmetric shape of the core also contributed to nuclear safety since accidental detonation was unlikely to produce a critical configuration.

In addition to its use as a primary, LRL weaponized the Swan as the lightweight multi-purpose W45 warhead. Preliminary tests leading to the W45 warhead include the 10-kiloton Wilson and the 8-kiloton Morgan XW45X1 shots conducted during Operation Plumbbob in 1957. These were composite, boosted, two-point initiated devices with hollow, beryllium reflected, asymmetric cores. The W45 was extensively tested a year later during Operation Hardtack I; tests that led to two designs: the XW45 and the XW45X1. Because of the 1958 test moratorium, the AEC at first decided to weaponize only the XW45 as the W45 MOD 0. LRL used the W45 MOD 0 in a range of weapons that included the Terrier SAM missile, Bullpup ASM missile, the Little John SSM, and the Medium Atomic Demolition Munition. The production version of the W45 MOD 0 measured 27 inches long by 11.5 inches in diameter and weighed 150 pounds. It had an all plutonium core and six yields: Y1 (1 kiloton), Y2 (un-boosted 0.5 kilotons), Y3 (5 kilotons), Y4 (8 kilotons) Y5 (10 kilotons), and Y6 (15 kilotons).

Despite its utility, the W45 warhead suffered from a number of problems while in the stockpile. The first was corrosion of its fissile core, a problem that also dogged the Robin primary used in the W47 thermonuclear warhead. This led to the test of a deliberately corroded warhead, which established how much deterioration the DOD could allow before it had to remove a warhead from the stockpile. Redesigned W45 warheads did not exhibit the corrosion problem. Another technical issue associated with the W45 was shrinkage of the warhead's PBX9404 explosives due to the loss of volatile components. Since there was no easy fix for this problem, the AEC had to rebuild every warhead with a new explosive formulation.

Engineering Units transported the W45 warhead as the Medium Atomic Demolition Munition (MADM) in an H815 shipping container that was 42.5 inches long, 24.5 inches wide, and 28 inches deep. It weighed 350 pounds. In order to remove the lid of the shipping container, it was necessary to enter a five-digit code, split into a pair of four-digit inputs, into an MC1885 mechanical combination lock as part of the two-man security concept. Crews quickly added up the two sets of digits so they could enter a single five-digit combination to speed up the arming process. The shipping container contained all the necessary electronics for use with the warhead, including XM3 and M4 code transmitters, an M5 decoder-receiver and an M96 firing device. A six-man engineering team emplaced and fired the "ditch digger" by means of a time fuze or remote control. The weapon had four nylon straps attached so it could be carried to its target.

The AEC produced the MADM's W45 warhead in two MODS (1 and 3) and three yields from 1 to 15 kilotons. The Army designated these M167, M172, and M175. The M172 and MC175 warheads used limited life deuterium-tritium gas containers designated MC1188A and MC1189A. The AEC released the MOD 1 for production in September 1960. The MOD 3 entered service in July 1965 with a digit combination lock and had a case that was watertight to a depth of 30 meters. The AEC manufactured 350 medium demolition units during the period 1962 to 1966. They were gradually retired during the period 1967 to 1988.

The ancillary equipment described above can be considered the MADM's adaption kit. The munition could be fired by a timer or on command. The M96 firing device was connected to the W45-3 warhead and the M5 decoder receiver was attached to the M96 firing device. The code transmitters XM3 or M4 could send commands to the M5 which decoded them, verified them, and retransmitted them to the M96 which acted on them. The code transmitter and coder / decoder all required the installation of batteries and operational checks. One of the code transmitters operated wirelessly.

The control panel for the M96 firing device came packed in an H1186 container. It had an input receptacle at its 12-o'clock position for the cable from the M5 coder decoder and an output receptacle at the 6-o'clock position for a cable that ran to the W45-3 warhead. It was also equipped with redundant, removable T5 timers with thermal batteries at the 4-o'clock and the 8-o'clock positions, and a pair of built-in T4 interval timers at the 10-o'clock and 2'o-clock positions. The T5 timer units could be removed to service their thermal batteries and then replaced to be held in place with retaining mechanisms. The T5 timers could be set in 10-minute intervals up to 12 hours. When the timers ran down, they set off a squib that closed a switch to send a firing pulse to the slim loop ferroelectric firing device in the W54-3 warhead. This activated its two detonators, its neutron generators and the tritium injection system. The coder senders could also send timer override commands to fire or to disable the firing device – rendering the warhead inoperable.

The T4 timers could be set for a time span up to 30 minutes. Presumably they acted as safe separation devices in the event a T5 timer ran down prematurely. These timers would have closed switches in the firing circuit to arm the firing device. Finally, an opening below the input receptacle could accept a special key. Inserting and turning the key was the final act necessary to activate the M96 firing device. Turning it reciprocally would interrupted the countdown allowing a specialist to reset the timers or to cancel the operation.

Together with the smaller Special Atomic Demolition Munition (SADM), the United States stockpiled 608 ADMs at the height of the Cold War. It stored 372 of these weapons in Europe (West Germany and Italy), 21 in the Far East (Korea and Guam) and 215 in the Continental United States. Seven hundred and fifty Army ADM specialists and 250 Navy and Marine specialists operated these weapons. The United States made ADMs available to Belgian, British, Greek, Italian, Dutch, Turkish, and West German Commando and Engineering Units. The Army retired the last MADM in 1987. The primary reason for its retirement was the realization that military doctrine called for its use early in any confrontation, an action that might have led to unwanted nuclear escalation.

In 1953, Redstone Arsenal and the Douglas Aircraft Company conceived of Honest John Junior, a 17.5-inch diameter, short-range version of the M31 Honest John rocket. The proposed

Honest John Junior artillery rocket was to carry a 1,200-pound W12 warhead to ranges from 5,000 to 15,000 yards. In late 1954, the Office, Chief of Ordnance, redefined the program under the name of Little John. The redefined program called for a choice between three alternative weapon systems. These were a 17.35-inch diameter rocket with an XW25 warhead; a 12.5-inch diameter rocket with an XW 33 warhead; or a 12.5-inch diameter rocket with an XW45 warhead. On October 20, 1954, the Army approved the small diameter XW45 implosion warhead for use with a 12.5-inch diameter Little John rocket and approved its military characteristics in 1955. The AEC authorized UCRL to begin engineering development of the XW45 on January 27, 1956, and on June 22, 1956, tested a Swan device as Redwing Inca. This prototype of the XW 45 measured 11.6 inches in diameter by 22.8 inches long, weighed 105 pounds, and had a 16-kiloton yield. Succeeding tests of the XW45 include Hardtack II Ganymede and Nougat Antler, Codsaw, Hoosic, Hudson, and Arikaree.

A month after the Redwing Inca test, Redstone Arsenal awarded a contract to Emerson Electric to provide 20 airframes for a 12.5-foot long XM47 missile. The Army asked the company to provide sufficient tooling for production of 24 airframes per month. Redstone completed testing of the XM47 in December 1957 but found the accuracy of the vehicle unsatisfactory. As a result of this determination, the Army reserved the XM47 for training. Following the failure of the XM47, Redstone began development of the XM51. With a length of 14 feet 5.7 inches and a diameter of 12.5 inches, the XM51 was slightly larger than the XM47. Four stubby tail fins gave it a span of 1 foot 11.75 inches. A 53,999-pound thrust Hercules XM-26 solid fuel motor powered the rocket to a maximum range of 10 nautical miles. Armed with its W45 warhead and M77 adaption kit, the XM51 (designated MGR 3 by the Army) weighed 779 pounds.

Redstone contracted the General Electric Company to produce the M77 adaption kit for which GE manufactured a number of metal parts and provided final assembly. The Philco Corporation, Philadelphia, Pennsylvania, manufactured the T-2074 fuze, the Frankford Arsenal manufactured the M8 sequential timer, the Minneapolis-Honeywell Regulator Company built the option selector, and the Magnavox Company, Urbana, Illinois, provided the acceleration switch. The Catalyst Research Corporation, Baltimore, Maryland, and the Wurlitzer Company, Tonawanda, New York, supplied the kit's thermal batteries. After fabrication, General Electric shipped the M77 kits to the Seneca Ordnance Depot, Romulus, New York, where technicians mated them to W45 warheads to produce M50 and M78 Little John warhead section. The M50 warhead section featured a W45Y1 warhead and the M78 warhead section had a W45Y2 warhead. The AEC produced 500 W45 warheads in 8-kiloton (Y4), 10-kiloton (Y5), and 15-kiloton (Y6) yields.

Although the Army originally intended Little John for use with a self-propelled, multi-launch unit, it only deployed it on the single-launch XM34 heli-portable / air droppable launcher with XM449 trailer. The XM449 trailer measured 6 feet tall, 4 feet 10 inches wide, 4 feet 8 inches long, and weighed 1,350 pounds with its XM34 launcher mounted. The XM34's launch rail could be elevated from 0 to 65 degrees and traversed 15 degrees right and left. Unlike the Honest John rocket, which incorporated spin rockets that ignited after launch, the Little John rocket had its stabilizing spin imparted before launch by a method known as "spin on straight rail" or "saucer" (SOSR). Prior to firing, Little John's crew wound a flat coil spring mounted under the rear of the launching rail 8 1/2 turns. A lanyard released the spring, which turned a drive shaft that rotated a pinion gear meshed to a ring gear attached to the rocket nozzle. When the rocket reached a spin rate of 210 RPM, inertial switches attached to a pair of thermocell batteries closed a relay to activate the rocket's igniter. As the spinning rocket cleared the launch rail, it ejected a front attachment shoe and retracted a rear attachment shoe into the fin barrel at the base of the rocket. Little John had canted fins to maintain its rate of spin.

Because of significant delays in developing the W45 warhead and concerns arising from political developments in Lebanon, the DOD requested that the AEC develop an emergency capability warhead for Little John in 1956. The AEC in turn passed this task to Los Alamos, which assigned the modification of the W33 warhead used in the M422 artillery shell to Leon F. Scanlin. A later chapter describes the W33 warhead. Instead of an artillery shell fuze, Scanlin produced a missile-type adaption kit for the W33, which Frankford Arsenal produced. The new arming system used accelerometers that opened arming valves after launch. Activation of a gas generator that passed gas through these valves completed the warhead's arming cycle. After the Army successfully flew the W33 and its adaption kit on a Little John at White Sands, a number of emergency capability adaption kits entered the nuclear inventory. The W33 re-development work took place over an 18-month period.

By October 1959, development of the M51 Little John missile, W45 warhead, M77 adaption kit, launcher, and handling equipment were complete. The Army then designated the M51 as Limited Production Model Standard A. Redstone carried out final testing of the M51 at the Yuma Test Station. Between July 19, 1961, and December 7, 1961, it conducted sixteen launches. Seven rockets were equipped with pre-production, war reserve M77 adaption kits and inert XW45 warheads modified to include telemetry equipment, and nine rockets were equipped with pre-production, war reserve, M77 adaption kits and war-reserve XW45 warheads with high explosives and non-nuclear components. To certify the weapon for cold-weather operations, Redstone launched one rocket from Fort Greely, Alaska. Douglas Aircraft, the prime contractor for Little John, built 2,343 production missiles.

The Army intended to place Little John into general usage but engineering - test programs conducted with the M47 rocket and elements of the 101st Airborne Division demonstrated a need for higher levels of training than anticipated. Because of this finding, Little John saw limited deployment with seven Battalions, each of which the Army organized with four launchers and a complement of 123 officers and enlisted personnel. One Battalion served at Fort Sill, Oklahoma, in a training capacity. The 1st Battalion, 57th Artillery and the 25th Infantry Division's 2nd Battalion, 21st Artillery served in the Pacific on Okinawa and Guam; one Battery each served with the 82nd Airborne Division and the 101st Airborne Division; another Battalion served at Fort Bragg, North Carolina; and the last Battalion served with the 4th Infantry Division at Fort Lewis, Washington.

As deployed, the Little John system was very mobile. A jeep could tow its launcher accompanied by a truck with a missile-handling crane and two fully assembled rockets covered in XM85E3 conditioning kits (thermal blankets). In addition to using the missile-handling crane, the ground crew could set up a portable tripod to facilitate loading and, if it became necessary, eight men could pick up the rocket and place it on the launcher. Once loaded, launchers could be towed or heli-lifted from a Battery Assembly Area to a Firing Area. Despite Little John's ease of use and reliability, the Army withdrew it from service in 1969, displaced by equally portable artillery pieces armed with nuclear rounds having superior accuracy.

A contemporary of UCRL's W45 warhead was LASL's boosted, two-point initiated, beryllium reflected, one point safe "Tsetse." LASL used the Tsetse as the basis for its W44 warhead, as the primary in the B43 thermonuclear bomb, and as the primary in its W50 and W59 thermonuclear warheads, which the Author describes separately in several following chapters. The original purpose for the Tsetse appears to have been as a primary in the B43 bomb and the W50 warhead. Hardtack Fir was an 18-kiloton test of a TX43 primary at the Pacific Proving Grounds conducted in 1958. Los Alamos followed Fir with Linden, which may have been the test of an XW50 primary. Hardtack I also saw a full-scale evaluation of the B43 freefall bomb (Elder) and two XW50 prototypes (Tobacco and Pisonia). Immediately following Operation Hardtack, President Eisenhower declared a unilateral American moratorium on nuclear testing. Because of this moratorium and their relatively brief experience with beryllium reflectors, the nuclear laboratories had to design tactical warheads around the boosted devices that they had originally intended for use as primaries.

The Tsetse device was revolutionary in using a new class of "plastic (or polymer) bonded explosives" (PBX) developed by Los Alamos in the late 1940s and early 1950s. Plastic bonded explosives consist of a powdered explosive held together by a few percent of synthetic polymer binder. PBX binders are most effective if they have intrinsic explosive properties. One of the first LASL explosives was a polystyrene - RDX formulation known as PBX 9205. This formulation developed instabilities, ultimately requiring correction for safety reasons.

Important early plastic bonded explosives used in nuclear warheads were PBX 9010, PBX 9011, PBX 9404, and PBX 9501. Los Alamos developed all of these explosives. Although PBX-9010 incorporated RDX as its primary explosive, Los Alamos based all of its following formulations on HMX, a more powerful explosive. Los Alamos formulated PBX 9404, as incorporated in the Tsetse primary, from 94 percent HMX, 3 percent nitrocellulose, and 3 percent chloroethyl phosphate (CEF). It had a detonation velocity of 8,800 meters per second. Redwing Blackfoot tested the first device to use plastic bonded explosives on June 11, 1956, – a combination of PBX 9501 and PBX 9404. PBX 9404 later developed sensitivity problems that required correction.

The Lawrence Radiation Laboratory also manufactured plastic bonded explosives. These incorporated the prefix LX for Livermore eXplosive. Some important contributions from LRL were LX-04, LX-07, LX-09, LX-10, and LX-14. LRL chose HMX as the primary explosive in all its formulations. LX-09 was an explosive similar to PBX-9404 with improved thermal stability. Nevertheless, LX-09 suffered from ageing problems that showed up in the primaries of W68 Poseidon missile warheads. LX-10, an explosive in the same energy class as LX-09, and PBX-9404, used thermally stable Vitron A as a binder. LX-14 had energy characteristics similar to LX-10 and used Estane as a binder, which had excellent thermal characteristics.

The most common method used in the production of plastic bonded explosives is a water / slurry technique. A manufacturer mixes powdered explosive (RDX or HMX) with water and a water immiscible "lacquer" composed of a thermoplastic resin dissolved in solvent. The solvent wets the surface of the powdered explosive, after which a distillation process removes it. This process deposits a coating of thermoplastic resin on the explosive particles. Further stirring causes the plastic-explosive mixture to form "beads." A filtering and drying process then removes the water from the mixture. The final product of this process is an explosive molding powder.

Various agencies at first formed PBX molding powders into predetermined shapes using steel molds or hydrostatic presses. In the latter method, the hydrostatic presses applied high fluid confining pressures to rubber sacks loaded with explosive powder. Both forming processes required a vacuum. Once formed, plastic bonded explosives have a number of advantages over conventional explosives. If the polymer matrix has elastic properties, the resulting product has reduced sensitivity. Other binders can produce a hard PBX that is suitable for machining into complex shapes. PBX pressed in molds at room temperature can achieve densities close to that of the explosive base. Disadvantages associated with PBX production include toxic ingredients, long curing times, complex formulas, and the need for special mixing equipment.

Since 1960, the Holston Army Ammunition Plant, Kingsport, Tennessee has controlled the final production of the high explosives used in American nuclear weapons. After production, Holston ships its plasticized explosive powders to the Pantex Final Assembly Plant in Amarillo, Texas. Pantex presses the powder into billets, which it then x-rays to detect internal cavities, bubbles or foreign matter. Pantex then machines acceptable billets into requested shapes and inspects them. Samples from a given production lot of explosives are test fired for certification. After certification, Pantex assembles components into implosion spheres, a procedure that sometimes requires sub-assembly, or the mating of attachments used in the assembly of the PBX high explosives and the pit. It then installs fuzing and firing components such as detonators and wiring.

When the United States government lifted the voluntary nuclear moratorium in 1961, one of the first weapons that the AEC tested was a W44 MOD 0 warhead based on the Tsetse primary. Considering the novelty of this warhead, the Los Alamos scientists should not have been surprised when Nougat Chena had disappointing results. The rapid decay of tritium in the warhead's gas reservoir produced He3, a substance that absorbed neutrons to defeat any attempt at boosting. It thus became periodically necessary to remove boost gas reservoirs of warheads in order to reprocess their contents to filter out unwanted He3.

On May 11, 1962, the AEC successfully tested a redesigned W44 warhead as Dominic Swordfish. Swordfish was a live fire exercise of the Navy's new rocket propelled anti-submarine ASROC weapon for surface ships. The warhead used for this exercise measured 13.75 inches in diameter by 25.3 inches long and weighed 170 pounds. Its primary contained both PBX 9011 and PBX 9404 and had a nominal yield of 10 kilotons.

The SAROC anti-submarine weapon was the product of a post WW II Navy program to provide escort vessels with standoff weapons, which were eventually equipped with nuclear warheads. The first such weapons were a pair of rocket-assisted conventional depth charges that the Navy assigned for development to its China Lake Naval Ordnance Test Station. "Weapon A" carried 250 pounds of explosives whereas "Weapon B" carried 50 pounds. It quickly became apparent that the latter weapon lacked sufficient power to attack emerging classes of Soviet submarines, so the Navy cancelled it.

Weapon A, later designated RUR-4A, used a solid propellant rocket to boost its warhead to a maximum range of 800 yards. Actual range varied slightly between the MK 1 and the MK 2; the latter had a slightly more powerful motor. The projectile measured 12.75 inches in diameter by 8 feet 6.5 inches long and weighed 525 pounds. It had a plastic casing that did not interfere with its magnetic fuze. Flight time was a little over 10 seconds and after entering the water, the depth charge sank at a rate of 38 feet per second. A single barrel, MK 108 turret that resembled a 5-inch gun mount fired the RUR-4A. It did not take long to develop the MK 108 / RUR-4A weapon system, and in 1951, the rocket-propelled depth charge saw deployment on Navy destroyers. The MK 108's maximum rate of fire was 12 rounds per minute.

Not satisfied with just a standoff depth charge, the Navy also began development of a "Rocket Assisted Torpedo" known as RAT. NOTS tested the first such weapon, RAT A, with a

MK 24 aerial homing torpedo in 1952. RAT B followed in 1954 and used the larger MK 34 torpedo. Although the torpedo could be fired to a range between 1,500 and 3,000 yards by means of a solid fuel booster, the accuracy of the system was sufficiently abysmal that the initial RAT configurations were cancelled and a RAT C project initiated at NOTS in 1955.

At the same time that NOTS was developing RAT C, the Navy's Bureau of Ordnance released a study of four different anti-submarine systems, which it forwarded to the Division for Military Application. The systems included a ballistic rocket (ASROC), a gun type weapon (ASGUN), a torpedo (ASTOR) and a wire-guided drone (ASDRONE). BuOrd subsequently chose ASROC for immediate development and requested that the AEC form a joint ASROC Coordinating Committee with the DOD to produce the RUR-5A Anti-Submarine ROCket. Development Characteristic AS07101-1 called for a nuclear-armed depth charge and a conventionally armed homing torpedo as armaments.

The Navy selected the AN/SQS-23 escort sonar system, built by Sangamo Electric, for use with ASROC. The sonar had a 20-foot diameter, azimuth scanning transducer array mounted in a bulbous dome at the base of an anti-submarine vessel's bow. The AN/SQS-23 had both search and attack modes and was effective to a range of more than 10,000 yards. The size of the sonar mounting confined the system's deployment to destroyers, destroyer escorts, and cruisers.

The Honeywell Company developed ASROC as its prime contractor. The Navy contracted the Librascope Division of General Precision to provide fire control and the Armament Division of the Universal Match Corporation to supply the MK 16 Launcher Group. NOTS developed the MOD 0 booster engine, which the Navy initially mated to a MK 44 aircraft type torpedo (MOD 3). Later, it added the capability to deliver a MK 17 MOD 1 depth charge armed with a W44 warhead (MOD 5). The Navy then replaced the original homing torpedo with the MK 46 torpedo (MOD 4). The MOD 1 was an inert training aid, whereas the MOD 2 was a practice weapon equipped with a simulated warhead, weighted with lead shot and sea dye. The ASROC motor was able to propel depth bombs, torpedoes, and simulated warheads to a range of 10,000 yards. The AN/SQS 23 sonar provided target range and bearing to ASROC's fire control group, which could also accept radar target position information. Wind speed and direction, ship's speed and course were fed directly to the fire control group from the ships anemometers, compass and log. The fire control group then computed elevation and bearing, which it fed to the launcher.

A Mk 44 torpedo equipped ASROC missile measured 13.25 inches in diameter by 180 inches long and weighed 957 pounds. A MK 46 torpedo equipped ASROC missile had a diameter of 13.25 inches, a length of 14 feet 9 inches, and weighed 1,072 pounds. Four stubby wings gave the torpedoes a span of 2 feet 9.25 inches. ASROC's MK 46 torpedo measured 8 feet 6 inches long and weighed 508 pounds. The ASROC booster that propelled the torpedoes measured 6 feet 4 inches long and weighed about 500 pounds. It had a thrust of 11,000 pounds that propelled an ASROC missile to a maximum Mach 0.8 velocity. The torpedoes and the MK 17 depth charge are said to have had similar shapes, weights, and flight characteristics – although the depth charge equipped ASROC missile was somewhat shorter with a length of 12 feet 11 inches and a weight of 825 pounds. The depth charge measured 13.75 inches in diameter by 6 feet 7 inches long.

ASROC's W44 MOD 0 (XW44 then XW44X1) nuclear warhead measured 25.3 inches long by 13.75 inches in diameter and weighed 170 pounds. It was installed in a 30-inch-long desiccated, unpressurized, waterproof case. The warhead had a boosted, plutonium, Type 74 pit. (LANL equipped the Tsetse primary with Type 74 and Type 100 pits.) Sandia had to move the low voltage power supply and the x-unit to the rear of the warhead in order to produce the proper center of gravity for the depth charge. The warhead's firing system was composed of high and low voltage thermal batteries, a rotary chopper / converter, a condenser equipped x-unit, and external neutron initiators. The condition of the external neutron initiators could be checked at any time. The DOE considered pulsed power generators for the W44, but at this time, they were too bulky. For safety, the W44's gas boosting system was equipped with a fusible plug. In the event of an abnormal thermal environment, the fusible plug failed, venting the warhead's boost gas into the environment. This feature limited the accidental explosion of a warhead to its maximum un-boosted yield – less than a kiloton. All of the warhead's electrical components were cushioned by foam mountings and featured a continuity loop to ensure at assembly that all internal electrical connections were correct. The warhead required no external power other than short bursts of 28 volts DC through the launcher. Hydrostats, contact crystals, and a timer provided fuzing.

The Navy used the USS Sarsfield (DD 837), a Gearing Class destroyer, to carry out initial testing of the ASROC system. After the first ASROC production order in 1959, the USS Norfolk (DL 1) carried out a 2-month evaluation of the weapon. After the success of this evaluation, the Navy installed ASROC on 253 American vessels including 27 cruisers (CG, CGN), 78 destroyers (DD, DDG), and 65 frigates (FF, FFG). ASROC also saw service on 69 allied warships.

The octuple "matchbox" MK 112 launcher for the MK 16 MOD 7 launcher group was the primary launcher used with ASROC. Leahy CG, California DLGN / CGN, Spruance DD, Charles F. Adams DDG, Decatur DDG, Mitscher DDG, Bronstein DE / FF, Garcia DE / FF, Knox DE / FF, and Brooke class DEG / FFG vessels all mounted it. The Spruance class was notable for its MK 16 ASROC launcher system which feature a MK 112 octuple launcher mounted on a recycled 3"/50 deck mount behind the main forward gun. Beneath it, a MK 4 ASROC Weapons Handling System (AWHS) held 16 assembled missiles (two 8-round reloads), giving the ship a 24-round capacity. A number of the smaller vessels mounted a MK 112 launcher without a reload capability.

The launcher crew could load eight rockets into the MK 112 "pepperbox" launcher that, together with ancillary equipment, weighed 22.3 tons. The launcher incorporated four adjacent guides, each of which carried a pair of "over and under" launch assemblies. A launcher crew could elevate the guides (one at a time) to an angle between -3 and 85 degrees at a rate of 25 degrees per second. The training limits were plus or minus 179 degrees, with a rate of traverse at 42 degrees per second. The unit operated electrically with a pneumatic backup. Each launcher had a temperature control that insured optimum performance of ASROC's solid fuel booster, and gyrostabilizers that maintained a constant launch angle and azimuth on a pitching boat. Prior to firing, a launch hatch opened and a guide extended to direct the missile onto its selected path. A soft closure, expended by rocket back-blast, covered the rear of each launcher. The launcher's firing rate was three missiles per minute.

The Navy produced the MK 16 Launcher Group in eight MODs, some of which it equipped with remotely operated magazines. The magazines were located in the superstructure behind the MK 112 launcher and had forward facing hatches. Reloading machinery automatically lifted missiles onto a roof-mounted crane and rail system that aligned and carried them forward into the launcher. Otherwise, crewmembers moved the missiles manually with a MK 42 hand truck. The MK 16's launch control station was operated by a two-man crew and included a MK

199 Launcher Captain's control panel, a Mark 198 power control panel, a MK 153 power drive amplifier, a MK 36 interface control panel, and a MK 6 launcher and missile simulator.

ASROC was also carried on vessels equipped with the MK 9 and MK 10 Terrier launcher systems. In the case of these vessels, ASROC was fitted with an adapter to provide for attachment to the launcher rails. The MK 9 launcher was mounted on Providence CLG vessels where the missile magazine was located behind the launcher on the main deck. MK 10 launchers were carried on Farragut DDG, Leahy CG, and Belknap CG class ships with their magazines situated below decks. ASROC rounds could be loaded into the MK 10's two 20-round upper magazines provided they were separated by Terrier rounds. Thus, each vessel could accommodate a maximum twenty rounds. Each ASROC round loaded was one less Terrier missile that could be carried by these air defense vessels.

The Mk 26 Guided Missile Launching System (GMLS) was a United States Navy automated system for stowing, handling, and launching a variety of missiles, including ASROC. The Mk 26 had a short reaction time and extremely fast firing. Different MODS 0 – 5 of this system had magazine capacities of 24 to 64 missiles. In the late 1970s, Virginia CGN, Kidd DDG, and Ticonderoga DDG class vessels were adapted for use with the ASROC missile.

Except for the attachment of its fins, ASROC units arrived onboard completely assembled in a container. It was, therefore, not necessary to disassemble ASROC in the checkout area. Before loading a depth charge equipped missile into a launcher, four tapes that covered inlet ports had to be removed. The lead foil tapes were placed over the hydrostatic ports for the depth charge fuzes to prevent entry of dirt or contaminants. The tapes were not a safety device. The depth charge was armed by removing a thrust neutralizer installed on its booster, which prevented it from firing. The thrust neutralizer was stored for replacement in the booster if the missile was placed back into the storage magazine. In addition, the warhead's high-power thermal battery had to installed. This could be done before or after loading. In storage, the power supply was removed from the depth charge and replaced with the blanking plate and seal. These were placed in storage when th battery was installed.

In preparing for an ASROC Depth Charge launch, a Safety Plug had to be inserted in the Launcher Captain's control panel. The plug was kept in the custody of the Captain, the ASW officer or locked into a dummy receptacle on the panel. Prior to launch, a crewmember rotated an ARM / SAFE switch on the warhead to the ARM position. This allowed the warhead's low voltage supply to become active at launch. After launch, the missile's booster propelled a torpedo or depth charge to a predetermined subsonic velocity (Mach 0.8) and an explosive charge separated the booster from the payload. A second explosive charge deployed a parachute near the top of its trajectory to reduce the velocity of torpedoes prior to water entry. An explosive charge released a depth charge from a clamshell type airframe to allow it to fall free. In both payloads, a frangible plastic nose cap shattered on impact with the sea surface, activating the weapon's sensors.

The MK 17 nuclear depth charge had a sensor that determined if it had reached a minimum safe range and prevented arming if the warhead did not achieve safe separation. In order to determine the missile's ballistic trajectory, an attitude sensor became active for a period of 1 second after warhead separation, providing that the missile had attained its desired velocity. If the attitude sensor determined that the weapon had not reached a total attitude angle in excess of 6 degrees during the 1 second interval, it safed the fuzing system to prevent the detonation of the warhead. At the end of the missile's flight, impact with the sea-surface initiated arming and fuzing of the depth charge. A 2-second delay during the rapid deceleration that followed water entry provided enough time to complete arming. Shallow water hydrostats confirmed that the depth charge had entered the water. When the warhead reached its preset depth, or made bottom contact, it exploded. A timer detonated the warhead if the hydrostats or contact fuzes failed for any reason. An unloading cycle is necessary after every firing of an ASROC missile from a Terrier system because the adapter must be returned to the magazine tray.

Dominic Swordfish was the only test of this system with a nuclear payload. USS Agerholm (DD 826) launched a depth bomb at a raft positioned 4,348 yards distant. The Navy conducted this test 450 miles southeast of San Diego in an area of the Pacific Ocean with a water depth of 17,000 feet. A MK 111 fire control system provided the interface between the launcher and the targeting sonar. (Depending on the vessel, the Navy later upgraded the MK 111 to the MK 114 and the MK 116 configurations). The MK 17 depth charge landed only 20 yards from the raft. It then took 40 seconds to sink to a depth of 650 feet where it exploded. The Navy placed observation vessels between 2,200 yards and 4,600 yards from ground zero. These included a surfaced submarine and four destroyers. The resulting blast threw up a spray dome that reached an altitude of 2,100 feet. The test was a complete success and provided useful data for the MK 45 torpedo, which had a similar yield.

Total production was 35,020 RUR-5A rockets, of which the Navy expended 20 in development. The AEC produced 575 W44 warheads for use in MK 17 depth bombs. This implies two nuclear depth bombs carried on each ASROC equipped vessel. The Navy withdrew the nuclear depth bombs from service in 1989 as part of an initiative to reduce the deployment of tactical warheads. The MK 16 launcher was phased out when rocket assisted torpedoes were adapted for use in the MK 41 Vertical Launch System.

In addition to ship-mounted launchers, the Navy acquired the Gyrodyne DSN/QH-50 Drone Anti-Submarine Helicopter to carry ASROC's MK 17 depth charge. The Navy initiated the DASH program in the late 1950s to exploit the AN/SQS-26 ASW sonar, which had a much greater range than ASROC. DASH was an unmanned, B43 torpedo-carrying helicopter that a destroyer could deploy to a mission radius of 26 nautical miles. DASH had a pair of contra-rotating propellers and when fully loaded weighed about 1,000 pounds. The Gyrodyne DSN-1 (QH-50A) made its first successful unmanned flight in August 1960.

A UHF line of sight data-link controlled the DASH drone. During takeoffs and landings, a "pilot" flew the helicopter visually from the deck, whereas during the mission, the destroyer's Combat Information Center tracked the drone with radar and directed it to the sonar target. Control of the drone was transferable from one ship or ground station to another during a mission. When the drone was over the target, remotely transmitted commands armed and released the torpedo. Initial versions of the drone had piston powered engines, which Boeing replaced with a T50-BO-8A turbo-shaft engine in the QH-50C. The Navy preferred the turbo-shaft engine because its kerosene fuel did not have the volatility of the gasoline required for piston engines. Performance of the QH-50C was marred by the urgent requirement that it be ready for deployment under the under the Fleet Rehabilitation and modernization (FRAM) program.

An improved version of the -50C, the QH-50D first flew in 1965. Gyrodyne equipped it with a higher performance T50-BO-12 engine, lighter fiberglass rotor blades, increased fuel capacity for a mission radius of 40 nautical miles and provisions for easier weapon loading and aircraft maintenance. The first tailless drone in the QH series, the -50D had a loaded weight of 2,330 pounds. These changes gave the QH-50D better performance, improved reliability, and all-weather capability. Altogether, Gyrodyne produced 360 QH-50D drones, with first deployment in 1966. The Navy cancelled the DASH program in 1970 for a variety of reasons that included an inability to provide training, lackluster performance, limited deployment of the AN/SQS-26 sonar system, and the difficulty in managing nuclear weapons aboard small vessels.

The Navy's multi-purpose MK / B57 (TX57) bomb was another Tsetse-based weapon. Informally called "Dr. Pepper", the B57 was the lightest freefall nuclear bomb ever produced by the United States, with a weight between 490 and 520 pounds depending on MOD. The use of 0.35-inch-thick 6060-T6 aluminum alloy for the case of its center subassembly made this possible. This very common alloy is frequently used in aircraft construction because of its good mechanical properties and weldability. Stabilized by four fins, the B-57's streamlined casing measured 9 feet 10 inches long by 1 foot 2.75 inches in diameter and its fin span was 25.2 inches. Three lug points allowed for 14 and 30-inch suspension spacing on its upper surface.

Like its B28 predecessor, the B57 consisted of interchangeable basic subassemblies, nose subassemblies, and rear subassemblies. It also had a preflight selection subassembly. N57-0 (ballistic) and N57-2 (high friction) nose subassemblies were used for depth bomb and laydown delivery respectively, although they were interchangeable. Sandia designed the latter nose to remove a low-level delivery prohibition associated with N57-0 nose. The N57-1 nose section contained dual-channel fuzing radars, antennas, and two MC1589 thermal batteries for power (early versions instead had a pair of solid fuel turbo-generators). The nose was 10.7 inches in diameter at its base and 19.4 inches long. Its tip was made of epoxy resin impregnated fiberglass coated with neoprene. It was primarily associated with airburst delivery but could also be used for laydown or depth bomb delivery. All three noses were field interchangeable with two water entry flooding ports for depth bomb delivery.

The preflight selection subassembly consisted of a pullout switch, CF/HRU cable assembly connection, an access door for weapon setting, a strike mode selector switch panel, and a safe-separation timer with setting dials. These controls allowed the bomb to be fuzed for immediate or delayed parachute deployment, for parachute-retarded depth bomb delivery, or for free-fall airburst. The preflight selection subassembly also contained a 28-volt depth bomb / laydown timer to delay its burst. This gave rise to four fuzing options: free-fall airburst, parachute-retarded airburst, retarded laydown surface burst and retarded depth bomb. For antisubmarine applications, aircraft released the B57 at altitudes between 50 and 1,000 feet and at speeds from 0 to 400 knots. This release envelope prevented ricochet of the bomb. A zero-knot release option was available for helicopter delivery. For laydown delivery, aircraft released the bomb at altitudes between 50 and 2,000 feet and at speeds from 130 knots to Mach 0.95. Sandia designed the bomb not to penetrate soil or concrete because this would have degraded it blast effect, which was its main means of offense. Delivery methods included level flight, dive, toss, over the shoulder, glide, and roll ahead. Selection of yield and firing mode was done before takeoff.

The two basic subassemblies B-57-1 and B57-2 were identical except for a PAL mechanism associated with the B57-2. Sandia equipped both basic assemblies with boosted plutonium Type 104 nuclear pits and two internal power sources: a fast rise, 12-volt MC1192 thermal battery and a pair of 28-volt MC1262 thermal batteries. The MC1192 battery was initiated by aircraft power through a pulse plug that provided electricity for all of the bomb's initial arming functions. A strike enable plug, ready / safe switch, pullout switch, breakaway pulse connector assembly that set time delay for parachute deployment, inertial switch, velocity sensing switch, trajectory arming switch, and safe-separation timer provided weapon and aircraft safety.

In the B57's rear subassembly or after section, Sandia could install an optional, 50-pound, 12 1/2-foot diameter Kevlar parachute package that provided retardation for delayed airburst or laydown delivery. An MC1362 gas generator deployed the parachute via a telescoping axial deployment tube.

For laydown operations, the ground crew installed the B57's strike enable plug before takeoff. They also set an option switch to RETARDED. After takeoff, the pilot switched the Aircraft Monitoring and Control (AMAC) switch to ARM and set the selector switch to GROUND. Upon release of the bomb, plugs in the breakaway connecter assembly initiated the 12-volt thermal battery and the safe separation timer. A pullout switch enabled the trajectory arming sensor and parachute deployment. At release plus 0.08 seconds (fast) and 0.8 seconds (slow) the parachute deployed. Trajectory arming occurred when the shock of parachute opening closed an inertial switch. The inertial switch had to remain closed for at least 0.25 seconds for arming to take place. (The inertial switch remained closed above an applied force of 6.2 Gs). Three seconds after release, the 28-volt component of the thermal battery reached its operating voltage, which disabled the bomb's 12-volt circuitry, started the chopper / converter motors, and isolated the rear subassembly from power input. About 25 seconds after release, the sequential timer sent signals to fire the x-unit and the neutron generators to detonate the bomb as it lay on the ground.

Retarded airburst operations of the B57 were very similar to the actions associated with its laydown operation except that the ground crew set ae burst height switch to HIGH or LOW and then set the desired safe separation time. In flight, after activating the aircraft monitor, the Pilot set the selector switch to AIR. The safe separation timer was initiated at release and if a long safe separation time was selected, a pulse pin actuated a transfer switch that opened (deactivated) the short time circuit. If a short separation time was selected, the circuit remained closed. After the parachute opened and the inertial timer armed the bomb, the safe separation timer activated x-unit charging and connected the radar fuze. When the radar fuze detected the selected height of burst, it sent a firing signal to the warhead.

For freefall operation, the ground crew did not install the optional parachute. They set the bomb for FREEFALL operation, the burst height switch to LOW or HIGH, and the safe separation time. In flight, after activating the aircraft monitor, the Pilot set the selector switch to AIR. After release, the safe separation timer activated x-unit charging and connected the radar fuze. When the radar fuze detected the selected height of burst, it sent a firing signal to the warhead.

Depth bomb operation of the B-57 was very similar to laydown delivery right up to water entry except that the parachute was deployed at 1 second (fast) and 1.4 seconds (delayed.) This delay was necessary to prevent accidental parachute jettisoning while the bomb was in the air. At entry, a pressure switch closed as water entered through the bomb's nose ports. This activated a detonator that jettisoned the parachute, enabled the 28-volt depth bomb circuits, started a sequential timer, and disabled 12-volt circuitry. Seven seconds provided by the interval timer prevented premature arming from accidental hydrostat closure upon water entry or from bottom impact in shallow water. Detonation occurred at a preset water depth or by timer if the hydrostats failed to close. The typical detonation depth of the depth bomb was about150 feet below surface.

The AEC manufactured 3,100 B57 bombs in seven MODs between January 1963 and May 1967. The MOD 0 had a yield of 5 kilotons; MODs 1 and 2 were 10 kilotons; MODs 3 and 4 were 15 kilotons; and MODs 5 and 6 were 20 kilotons. The Navy deployed the MOD 0 bomb only in the retarded airburst and freefall options. The MOD 1 added the radar airburst option and a radio frequency filter pack. Otherwise identical to the MOD 1, the MOD 2 incorporated Category B PAL. In November, 1965, the AEC released the MK 57 MOD 3 and MK 57 MOD 4 bombs with PAL B and a new pit. The MK 57 MOD 5 was an upgrade of the MOD 1, designed for compatibility with the B-52 bomber, and the MOD 6 was an upgrade of the MOD 3 to provide the same capability. These MODs required an increase in the time delay from release to parachute deployment.

The B57 was so small that any nuclear designated attack aircraft could carry it. These aircraft included the A-4 Skyhawk, the A-6 Intruder, the A-7 Corsair II, and the F/A-18 Hornet. The Navy's P-3 Orion, S-3 Viking, S-2 Tracker and S-3 Viking aircraft, and the SH-3D/H Sea King and SH-60F Seahawk helicopters also carried it in the maritime patrol / fleet defense role. The B57's versatility and popularity maintained it in the stockpile until the DOD withdrew the last of the Navy's tactical nuclear weapons in 1992.

Probably the most important attack aircraft to carry the B-57 was the Douglas A-4D/E/M Skyhawk. Douglas' Chief Designer, Douglas Heinemann, conceived this aircraft in 1952, as an independent project to provide the Navy with an inexpensive, lightweight means of delivering the short-lived MK 12 tactical bomb. The United States has always deployed high tech equipment operated by highly trained personnel, so the quantity and quality implications of Skyhawk were positively "Soviet" in their conception. Nevertheless, the Navy fully embraced this aircraft for itself and for the Marine Corps, accepting its first deliveries in October 1956. Ultimately, the Navy accepted more than 3,500 Skyhawks for service, most of which were nuclear capable.

The Skyhawk measured 40 feet 3 inches long with a 27-foot 6-inch deltaform wingspan and had a maximum loaded weight of 22,500 pounds. A single Wright J65-W-4 turbojet, rated at 7,700 pounds thrust powered early models whereas a Pratt & Whitney J52-P-408 turbojet rated at 11,187 pounds thrust, powered later models. These provided a cruising speed of 500 MPH, a service ceiling of 45,000 feet and a combat radius of 400 miles. The Skyhawk had a 20mm cannon in each of its wing roots and could carry up to 8,000 pounds of ordnance on one centerline and four under wing pylons. A single, streamlined B57 was the perfect complement to this aircraft. The pilot hardly knew he had one attached.

Its large number of Skyhawks and its vast arsenal of B57 bombs gave the Navy a tremendous tactical nuclear capability at a time when the Kennedy Administration was cancelling its strategic mission. The employment of such a massive nuclear capability was comparable with the Army's overzealous plans for the expenditure of 400 nuclear rounds per day in a European offensive. A hypothetical example based on a nuclear-armed task force investing the Japanese held Island of Iwo Jima during WW II might help elucidate the Navy's thinking.

Iwo Jima is an 8-square-mile island in the South Pacific. The 550-foot height of Mount Suribachi, an extinct volcano, overlooks the island at its south end. In early 1944, it possessed two Japanese airfields that the United States Army Air Force wanted as a safe haven for damaged B-29s returning to the Marianas from attacks on Japan. The Japanese dispersed a garrison of 22,000 men, 22 tanks, and 500 major artillery pieces in concrete fortifications and tunnels to defend the island. In our hypothetical scenario, the battleships USS Missouri (BB 63) and USS Iowa (BB 61) position themselves about 10 miles off the southwestern and northeastern ends of the island. Along with a landing force and a protective screen of cruisers and destroyers, the aircraft carriers USS Lexington (CV 16), USS Hornet (CV 12) and USS Yorktown (CV 10) position themselves upwind to the east of Iwo Jima. At 8:00 AM, 1 hour before H-Hour on the morning of February 19, the Skyhawk Groups launch from their carriers and begin to form up in attack formations.

At 8:00 AM, a single Skyhawk detaches itself from the airborne force circling Iwo Jima to begin the attack. It carries a MK 8 penetration bomb, which it drops with pinpoint accuracy on the buried headquarters of General Tadamichi Kurabayashi, the Japanese commander. The bomb easily crashes through the Command Post's cover of volcanic rocks to explode with a yield of 15 kilotons. In an instant, the blast annihilates General Kurabayashi and his command staff, utterly disrupting communications throughout the island.

This is the signal for the remaining Skyhawks to attack. The first contingent drops 20-kiloton B57s fuzed for airburst over the island's airfields. The airbursts prevent serious physical damage to the runways, which are Iwo Jima's major prize. The heat from the nuclear blasts vaporizes surface installations, hangars, personnel, and aircraft. The remaining Skyhawks next drop 5, 10, and 15-kiloton B57s fuzed for contact and laydown delivery in preplanned attacks on command installations, communications centers, strong points, troop accumulations, supply depots, and targets of opportunity. By 9:00 AM, the aircraft begin recovery.

Finally, at H-hour, as the Marines in their lead-lined landing craft begin their approach to the beaches, the 16-inch guns of the battleships open up to provide covering fire. The battleships' aim their guns at the commanding heights of Mount Suribachi and any other locations thought to imperil the landing. They are firing 1,900 pound 15-kiloton MK 23 nuclear shells fuzed for a near surface burst. Within minutes, the guns have fired one round each – a cumulative yield of 1/4 megaton. The low-altitude at which the shells explode transfers their tremendous power to buried Japanese fortifications and tunnels.

As the air begins to clear following the battleship's bombardment, the Marines offloading onto the beaches stare blankly at the devastation. The Navy has plastered Iwo Jima with half a dozen nuclear blasts per square mile. Nothing remains alive, except a few deeply buried, radiation resistant microbes. The former Japanese command posts are now gaping craters and the upper half of Mount Suribachi has buried the fortifications on its lower slopes. It becomes clear to the men present that there is no longer room for close quarter battle or personal valor in the age of nuclear combat. It also dawns on these men that if the Navy had extended its bombardment for another hour, they would have required Scuba gear to plant a flag on the remains of Suribachi!

Though the preceding story may seem a trifle fantastic, consider in a few short years, the Navy supplemented the A-4 Skyhawk with A-6 Intruders, each able to carry three 70-kiloton MK 28 bombs on a single mission. The A-7 Corsair II, which replaced the Skyhawk, could carry four megaton-range B43 bombs on a single mission – approximately 3 times the yield delivered by the capital ships and supporting task force in the preceding scenario. Given a couple of aerial refuelings, a Navy Corsair flying from the Subic Bay Naval Base in the Philippines could have wiped out the Japanese defenders on Iwo Jima without any help from a naval task force!

Although weapons that were more capable supplanted the B57 for land attack, it remained the Navy's only nuclear depth bomb until deleted from this role in 1991. Not only could Navy attack and patrol aircraft carry the B57 depth bomb, it provided a nuclear capability for anti-submarine helicopters. The first such nuclear weapon system was Sikorsky's SH-3D/H Sea King. A large aircraft, it measured 72 feet 8 inches long by 16 feet 10 inches high and had a 61-

foot diameter main rotor. A pair of General Electric T-58-GE-10 turbo wave engines powered the 3D Model and a pair of General Electric T-58-GE-402 turbo shaft engines powered the 3H. A pilot, copilot, and a pair of sensor operators crewed the Sea King. It came equipped with a variety of detection equipment that included an AN/AQS-13 dipping sonar that could penetrate below the ocean's surface acoustic layers for better resolution, and its armament combined depth charges, homing torpedoes, and a B57 bomb. The aircraft had to react very quickly after it dropped a B57 depth charge to avoid destruction by the massive spray dome from its blast. Sikorsky produced 104 aircraft.

The Sikorsky SH-60F Seahawk, which replaced the Sea King in 1988, had a very short life carrying nuclear ordnance because the Navy withdrew the B57 depth bomb from service in 1991. With a crew of four, the SH-60F had a 50-nautical-mile range and 3 hours of endurance on station. A pair of General Electric T700-GE-401 turbo-shafts powered the Seahawk to a top speed of 155 MPH. Apart from its improved AN/AQS-18 dipping sonar, the Seahawk's detection equipment and armaments were similar to the Sea King. Sikorsky produced only 23 Seahawk helicopters.

Even smaller than the B57 were ultra-compact warheads on which the nuclear laboratories began development in the mid to late 1950's. The AEC tested an LASL "Gnat" prototype primary as Redwing Blackfoot in June 1956. The physics package for the Blackfoot device was 11.5 inches in diameter and weighed 62.7 pounds. LRL repeatedly tested its "Quail" (XW51) device during Operations Hardtack I and Hardtack II (Quince, Fig, Hamilton, and Humboldt) in 1958. The Quail device measured 10 - 11 inches in diameter by 11.5 inches long. It weighed about 35 pounds.

LASL carried out a number of one-point safety tests of its Gnat device (weight 64.6 pounds, dimensions 11.75 x 15 inches) as a follow on to Blackfoot. LASL's first test was at the bottom of an un-stemmed, 485-foot-deep shaft as Pascal-A. One-point safety tests verified a weapon's ability to generate a nuclear yield equivalent to less than 4 pounds of TNT in case of an accident. The actual yield was 55 tons, considerably higher than expected. Because the AEC conducted the test at night, a glowing blue radioactive jet ascended high into the air from the vertical test shaft. Fearing for their safety, the scientists and engineers present at the test site hopped into their cars and drove away at breakneck speed.

A second Gnat test, Pascal-B, had much more spectacular results, demonstrating the difficulty encountered in producing one-point safe warheads. Like Pascal-A, the AEC carried out the Pascal-B retest at the bottom of a 500-foot-deep shaft. This time, workers poured a concrete plug with a mass of several tons immediately above the test device and capped the shaft with a 4-inch-thick steel plate. Instead of a few pounds yield, Pascal B exploded with the surprising force of 300 tons of TNT equivalent. The blast vaporized the concrete plug and sent a mass of superheated debris roaring up the shaft. This blew the steel pate off the shaft with such force that no one ever saw it again. A scenario exists where the force of the blast propelled it into orbit, to beat Sputnik by three months! The Wheeler and Pascal tests demonstrate the difficulties encountered in developing reliable, compact, low-yield weapons.

On February 19, 1957, the Assistant Secretary of Defense requested that the AEC develop a warhead that would be suitable for use on the Falcon air-to-air missile and on a surface-to-surface missile for the Army. In a program designated NUTCRACKER, the Army was evaluating concepts for recoilless weapons able to launch mini nuclear warheads with yields in a range from 10 tons - 1,000 tons to distances between 2,000 and 11,000 yards.

The Army developed its surface-to-surface launcher to fill a short-range niche in an array of nuclear weapons intended to blunt a Soviet attack in Europe. For this purpose, it desired a simple, low cost, lightweight, delivery system for use by frontline combat troops. The initial military characteristics for this weapon called for a man portable weapon having a range from 500 to 4,000 yards, and a sub-kiloton nuclear warhead as armament. With these characteristics in mind, the Army assigned Picatinny Arsenal the task of exploring delivery systems capable of propelling a 12-inch diameter, 35-pound nuclear warhead. It also narrowed the nuclear characteristics for the Army's surface-to-surface warhead to a yield between 10 and 100 tons.

The Army carried out detailed evaluations of more than 20 possible launchers. The most important of these designs comprised a 155mm (6.1 inch), spigot type, portable, recoilless rifle with a maximum range of 2,000 yards; a full caliber (11 - 12 inch), portable, recoilless rifle with a maximum range of 2,000 yards; and a full caliber, recoilless rifle with a maximum range of 4,000 yards. The Army also considered disposable single launch fiberglass tubes. Eventually it chose a spigot type system that it named Davy Crockett for final development. Ordnance Special Weapons Ammunition Command at Picatinny Arsenal, Ordnance Weapons Command (made up of the Rock Island and Watertown Arsenals) and eleven other installations participated in the development program. Directly responsible to the Chief of Ordnance, the Diamond Ordnance Fuze Laboratory, Washington, DC, performed research and development activities for all of the ordnance installations.

The AEC assigned warhead development for Falcon and Davy Crockett to LRL as the XW51 on September 24, 1958, possibly because of its previous work on the Wee Gnat. Four months later, on January 19, 1959, it re-assigned the warhead program to the Los Alamos Scientific Laboratory in a redistribution of the labs' responsibilities. LASL re-designated the warhead as XW54 and began the weaponization of the Gnat as the Scarab. Beginning with the Storax Wolverine shot in 1962, the AEC conducted at least a dozen tests of the Scarab.

The Air Force quickly established the yield for the Falcon air-to-air missile warhead at 250 tons and the Army finalized the yield for the Davy Crockett warhead at 10 - 20 tons. The Army referred to this yield as 25 tons in its Davy Crockett instructions manuals. At the behest of the Army, LASL later developed a B54 warhead with a range of yields between 10 and 1,000 tons for the special atomic demolition munition, the widest (relative) range of yields ever developed for an atomic warhead.

LASL likely based the 250-ton version of the W54 warhead for Falcon on an unboosted version of a boosted primary. The nuclear laboratories designed primaries for a yield of 300 - 500 tons prior to boosting, fairly close to the desired yield for Falcon. The final dimensions of the W54 warhead were a diameter of 10.862 inches and length of 15.716 inches. It weighed 50.9 pounds. This was marginally heavier than specified. One Author asserts that the W54 warhead contained a 19-pound plutonium core surrounded by a 20-pound, 36-point implosion system. Weapon disassembly data from Pantex does not indicate whether the warhead had a beryllium tamper.

Envisioning the design of the 20-ton yield W54 warhead developed for Davy Crocket requires a little more consideration. Undoubtedly, LASL substituted its core into the same multi-point implosion system developed for the Falcon version of the W54 warhead. Adjusting the timing of neutron initiation may have assisted in lowering the warhead's yield. LASL tested a number of one-point safe, 10-inch diameter primaries during Operation Nougat in 1961 - 1962. These included Stoat, Agouti, Ermine, Chinchilla I / II, and Armadillo.

Providing the W54 warhead with an inertial safety device suitable for use with the 40G acceleration of Falcon and the 2,000G acceleration associated with the Davy Crockett recoilless rifle was a primary area of concern for Sandia Laboratory. The laboratory therefore decided to produce two versions of the W54, identical but for their sensing devices. Sandia assigned the XW54 designation to the Falcon warhead in 1959 and the XW54X1 designation to Davy Crockett in 1960. Sandia also began development of an XW54X2 for Davy Crockett, which had no safety-sensing device. At the Army's request, this latter device insured early delivery of the weapon system. Sandia design-released the XW54 warhead for Falcon as the W54 MOD 0. It design-released the XW54X2 for Davy Crockett as the W54 MOD 1. LASL and Sandia later released the XW54X1 as the W54 MOD 2. The MOD 0 and the MOD 2 were interchangeable through the substitution of their safety devices. Slippage of parts production then delayed the W54, resulting in the cancellation of the W54 MOD 1. The AEC began production of W54 MOD 2 warheads for use with the Davy Crockett launcher and W54 MOD 0 warheads for the Falcon air-to-air missile in April 1961.

The Development and Proof Services Division of the Aberdeen Proving Ground carried out the primary testing of the Davy Crockett system. The Erie Ordnance Depot, Ohio; Forts Wainwright and Greely, Alaska; and the Yuma Test Station, Arizona, provided additional facilities for the program. The title "Battle Group Atomic Delivery" came into general use when development work started in August 1958. The first prototype tube for the lightweight system arrived at Picatinny Arsenal in November 1958. At this time, the Arsenal outlined the following development plan: a feasibility study completed by March 1959, engineering design completed by January 1961, engineering tests completed by April 1962, service tests completed by May 1962, and Type Classification in June 1962.

The Davy Crockett system met its development timetable and the Army deployed it in two different models. A light, 120mm M28 launcher with a length of 56.5 inches and a weight of 103.5 pounds and a heavy, 155mm M29 launcher with a length of 97.5 inches and a weight of 311 pounds. The M28 had a 459 FPS muzzle velocity and the M29 had a 656 foot per second muzzle velocity. Both weapons used a hand-operated wheel to elevate their barrels.

At first, the launchers fired warheads loaded with conventional explosives for spotting. Later, in 1963, the Army fitted a 37mm cannon that fired conventional rounds below the heavy weapon and a 20mm cannon that fired depleted uranium M101 spotting rounds beneath the barrel of the light weapon. Both launchers fired the same 76-pound, 31-inch-long (25 inches without fins) by 11-inch diameter M388 round incorporating a W54-2 fission warhead. The ballistic shape of the round resembled a miniature MK 6 bomb.

Davy Crockett's M388 round comprised a W54-2 warhead, rear body, fin assembly, and windshield. The windshield and the fins were made of plastic whereas the rear body was made of aluminum. A manually operated ARM / SAFE switch interrupted power to the warhead when in the SAFE position and a HI / LO switch provided for a medium height burst or near surface detonation (about 20 feet above ground level). The Diamond Ordnance fuze incorporated a mechanical timer at the rear of the projectile to set the time of flight and to provide safe separation distance. The timer dial had graduations in 1/2 second increments from S (Safe) to 50 seconds and numbers printed at 5 second intervals. If the crew left the timer on Safe, the projectile did not arm under any condition. For proper function, the gunner had to set the timer to a minimum of 1 second.

The projectile had a rear antenna dome and an XM1117 fuze of the radio proximity type that detonated the projectile at either of two preset altitudes after the mechanical timer completed operation. If the timer was improperly set and the projectile impacted before the timer ran out, the warhead would dud. The projectile contained thermal batteries that provided operating power for the warhead through its adaption kit. The Diamond Ordnance Fuze Company began testing the XM 388 in late 1959. Its personnel fired projectiles from 155mm mortars at accelerations of 1,800 - 2,000 Gs to distances of 1,500 yards with no mechanical problems.

When a Davy Crockett Atomic Battle Unit received a firing order, a five-man squad drove the weapon to a designated firing site, if it was not already located there. The squad consisted of a squad leader, a communications specialist, and three infantrymen who carried the weapon and its projectile in three port-a-packs. Pack A (XM1) held the projectile, Pack B (XM2) held the barrel assembly and Pack C (XM3) held the tripod. The three infantrymen set up the small M28 launcher, starting with assembly of the tripod. They then installed the 120mm M65 smoothbore barrel. A squad could fire the larger M29 launcher directly from a jeep that was especially equipped for this purpose or dismounted after removal from a specially equipped M113 APC. Although the setup for the M29 was similar to that of the M28, it took five men to lift its 155mm M64 smoothbore barrel into position. Davy Crockett squads were equipped with shaped charges to destroy their atomic projectiles and thermite grenades to destroy their launchers if they were in danger of capture.

After a Davy Crockett crew set up its launcher, they loaded it. The M29 used either an M76 or an M77 charge, which came in a perforated case, complete with its own firing mechanism. The design for the propellant casing was already in use with conventional recoilless rifles. The charge was loaded from the muzzle. The M76 charge propelled the M388 warhead to a range between 540 and 1,900 meters (Zone I) and the M77 charge propelled the warhead to a range between 1,700 and 4,000 meters (Zone II). After the crew had inserted the firing charge, they inserted the 6-inch diameter, 32-pound, hollow, aluminum M2 launching piston and positioned its index tab. A dome shaped plate with a hole in its center at the piston's lower end provided a seal. The open upper end extended about 2 inches beyond the barrel's muzzle. The M28 used a 4-inch diameter launching piston and a charge that launched the M388 projectile to a distance between 540 and 2,000 meters.

After inserting the launching piston, the crew loaded the spotting rifle and prepared the M388 projectile for firing. The M388 round came packed in a 55-gallon drum along with some related electronics. (In 1964, the AEC supplied the Army with PAL locks to secure the projectile's storage drums). The crew removed the projectile's rear cover to uncover the timing dial inside the tail well. They then checked the projectile's "dimple motors," and set and verified the projectile's timer setting. The dimple motors verified that the projectile's batteries had not previously discharged and the timer provided for a minimum safe range of 500 meters. The crew then attached the M388 projectile to a pair of studs at the front end of the launching piston. This procedure required aligning the projectile's bayonet pins with bayonet slots on the launch piston's adapter. The projectile was turned to the right and down for a positive lock. The crew then set the height of burst switch and armed the warhead.

After the warhead was armed, the crew laid the weapon for firing by rotating and elevating its barrel with instructions supplied by an observer. They then removed the protective covering from the rear of the propellant charge, extracted the firing spool and ran out the firing cord to the firing position. The crew placed the firing position behind a hill or dug it in to protect themselves from the direct burst of radiation emitted by the exploding warhead. As soon as all was ready, the crew fired the spotting charge. The round had to land within 100 feet of the target. If it did not, the crew re-laid the gun and fired a second spotting round to verify the setting. The crew then super elevated the barrel to a second position associated with the firing of the nuclear projectile, insured that the back-blast area to the rear of the weapon was clear, pulled the safety pin from the front of the M388 projectile, and took cover.

The weapon fired when a crewmember pulled a pin at the end of the firing cord. This released a stab firing pin into the fuze lighter. This action ignited the mild detonating cord, which in turn ignited the propellant charge in the barrel's open breech. The gases from the burning propellant expelled the launching piston and exited the rear of the barrel with equal force to stabilize the weapon. Like all recoilless weapons, the back blast raised a large cloud of highly visible dust. As the launching piston left the barrel, gases entering the hole in its base raised its internal pressure to 800 PSI. This caused two pins to shear, separating the M388 round from the launching piston about 10 - 13 feet from the weapon's muzzle.

Launch acceleration activated the XM388's fuze timer mechanism, inertial switches, and 28-volt thermal batteries. This in turn activated the x-unit's chopper / converter motors, and upon reaching a preset time, closed four sets of switches in the timer. Two of the switches armed the warhead and activated more thermal batteries that provided power to the chopper / converter to charge the x-unit. The other two switches connected the thermal batteries to the warhead's timer fuze. Upon the radio proximity fuze sensing the selected burst height, a set of squib switches triggered a firing signal. The timer insured that the warhead exploded on the way down and not on the way up.

Los Alamos at first equipped the W54 warhead with an MC1319 chopper / converter firing-set to permit early fabrication of war reserve projectiles prior to the availability of an MC1258 electrostatic device. Except for replacing the MC1258 ESD with a dummy weight, the MC1319 was physically and electrically equivalent to the later MC1291. Inclusion of a premolded cavity filled with a dummy weight provided interchangeability with the MC1258 ESD, when it later became available. This permitted a common design for the MC1319 and the MC1291, since an MC1107 ESD fit into the MC1319 to make an MC1291.

The XM388 round came in two MODs with a single yield: 20 tons. This statement is subject to the Office of Science and Technology's pronouncement that "note should be taken of the great importance of safeguarding the yield of the Davy Crockett." There have been reports that the XM388 round was produced in "dial-a-yield" versions with upper yields of 250 tons or even 1 kiloton. Operating instructions for the warhead do not mention yield setting. The confusion over the yield for the Davy Crockett round may be associated with the subsequently produced Special Atomic Demolition Munition.

The AEC conducted two XM388 tests at the NTS during Operation Sunbeam in 1962. A static firing of Little Feller II produced a 22-ton yield. In Little Feller I, witnessed by President John F. Kennedy and General Maxwell Taylor, an M29 crew fired a nuclear round to a distance of 1.7 miles, where it detonated 40 feet above the surface with a yield of 18 tons. The nominal, 20-ton blast of an XM388 round, while destructive to structures within a radius of 150 meters (7-PSI overpressure), was not its deadliest feature. Instead, it subjected exposed individuals to a 10,300 REM dose of radiation at 150 meters and a 500 REM dose of radiation at 400 meters. The first radiation dose is immediately incapacitating whereas the second is fatal within days for about 50 percent of those exposed. For this reason, many authors consider Davy Crockett a forerunner to enhanced radiation weapons (neutron bombs). The AEC manufactured 400 XM388 nuclear rounds over the period April 1961 to February 1965, with the non-nuclear components produced by the Rock Island Arsenal.

The Army never deployed Davy Crockett as extensively as it originally envisioned. Although it established funding for 6,247 guns, it only produced 2,100. The limited production of nuclear rounds further limited the deployment of this weapon. The Army shipped only 171 examples to Europe during 1962 - 1963 for use by the 5th and 7th Corps of the 7th Army. Instead of hundreds of Davy Crockett armed Squads roaming the nuclear battlefield, the Army assigned a limited number of Atomic Battle Groups to the Heavy Mortar Platoons of Battalion Headquarters Companies. The Army withdrew Davy Crockett from West Germany in 1967. Other deployments include Guam from January 1965 - June 1969, Hawaii from April 1964 - June 1969, Okinawa from April 1964 - December 1968, and South Korea from July 1962 - June 1968.

The Battalion Davy Crockett Section was originally composed of a Section Leader (normally a First Sergeant), two Squad Leaders, two Fire Direction Computers, two Gunners, two Assistant Gunners and support personnel that included truck drivers. A commissioned officer later commanded them. For equipment, each section received two 120mm M28 launchers, two 155mm M29 launchers and a pair of M38 (later M151) Jeeps for weapons transport. The section also had a third Jeep utility vehicle and two more 3/4-ton cargo trucks for general transport. Eventually the Davy Crockett Units standardized on four M29 launchers.

With the four launchers of an Atomic Battle Group under his control, a Battalion Commander could rapidly initiate a nuclear fire mission. Even so, the effective use of Davy Crockett depended on decentralized control, which European Field Commanders and the Kennedy Administration viewed quite negatively. These concerns, plus test firings described as unable to deliver acceptable accuracy, even with a low-yield nuclear warhead, led to a short deployment for Davy Crockett. Retirement of the weapons system took place over the period 1967 to 1971. The 55th and 56th Infantry Platoons, attached to the US 82nd Airborne Division, were the last units equipped with the M-29 Davy Crockett weapons system. In an interesting bit of historical trivia, the Army intended to defend a planned lunar military outpost known as Project Horizon with Davy Crockett launchers. In the low-gravity lunar environment, the launchers would have been able to project their warheads to a considerable distance.

The final weapon to make use of the W54 warhead was the B54 Special Atomic Demolition Munition (SADM) M129, M129E1, M129E2 / M159, M159E1, M1595E2. The primary purpose of an ADM was to breach enemy defenses (enhance mobility) or to deny various facilities and routes to an advancing enemy (counter mobility). The Army deployed the B54 concurrently with the larger W45 based Medium Atomic Demolition Munition (MADM). The B54 had yields in a range from 10 tons to 1 kiloton and came complete in a hard shipping case that measured 35 inches by 26.6 inches by 26.2 inches and weighed 163 pounds when fully outfitted.

The B54 device was a cylinder, 12 inches in diameter by 18 inches long with a weight of 58.5 pounds. The warhead was equipped with an MC1984 cover and MC1827 mechanical lock for security. The cover prevented access to the warhead's arming and fuzing mechanism. The lock used the same type of five-digit split code input as MADM and had self-illuminated numbers for use at night.

Although the B54 had multiple yields, it was "not adjustable in the field." Data from the dismantlement of Special Atomic Demolition Munitions at Pantex indicates LANL used a Type

81 pit in the B54 warhead and a Type 96 pit in the B54-1 warhead. Both of these pits required cleaning as part of their dismantlement procedure. The term "cleaning" generally denotes contamination with tritium. Thus, for their higher yields, the B54-0 and B54-1 likely were boosted. This should not have been difficult to accomplish because the LASL W54 / B54 warhead descended from LRL's boosted W51 warhead. In order to provide proper yield for a mission, technicians would have installed an appropriate pit and boost gas reservoir in a B54 unit before they sent it to the field.

SADM's B54 warhead was mounted in a sealed housing and equipped with a mechanical timer that required auxiliary illumination if it had to be set in total darkness. A contemporary version of US Army Field Manual 5-26 states that the mechanical timer had delays from 5 minutes to 12 hours on a MOD 1 warhead and from 5 minutes to 24 hours on a MOD 2 warhead. The warhead had a high explosive based nuclear system, which was detonated by a slim loop ferroelectric firing-set and a pair of external neutron initiators. Sandia considered renaming this warhead the WX58 because it was a self-powered warhead, as opposed to the earlier W54 versions that received external power through an adaption kit. However, its family resemblance to the preceding W54 warheads, resulted in a continuation of the W54 designation.

The weapon's case consisted of a front section manufactured from a forged aluminum and molded fiberglass laminate; a thin, forged aluminum end cap; and a die forged, thin wall, rear section. Foam pads inserted between the case walls and the nuclear explosive assembly provided shock protection and absorbed differential thermal expansion. The AEC assembled the center case parts with a pressed fit and riveted the front and rear sections, which were sealed with O-rings. The case was desiccated but not pressurized and had a negative 5-pound buoyancy in water. The control panel permitted underwater arming After removing the cover and exposing the control panel, an operator set the timer using knurled dials, which were located at the six-o'clock position. In order to do this, the ARM / SAFE lever had to be set to SAFE. The operator next extracted an explosive plane wave generator from a SAFE well located in the 10-o'clock position and placed it in an ARM well located in the 2-o'clock position. The operator then moved an ARM / SAFE switch located at the 7-o'clock position to the ARM position. With the control switch in the armed position, the operator could not change the detonation time. If the operator returned the ARM / SAFE switch to SAFE, all of the warhead's interlocks and safety provisions reengaged. After completion of manual arming and expiry of the preset time interval, the timer provided a firing signal that activated the warhead's detonators and neutron generators.

The Army equipped the B54 nuclear device with an H-912 backpack, with an inner H911 bag, for offensive operations. Special Forces troops practiced training missions in which a SADM armed team parachuted into enemy terrain where they used the weapon for harassment and interdiction. The Army practiced the two-man rule for nuclear weapons even here! It trained its Special Forces to swim the device into an enemy harbor for attacks on shipping and infrastructure. In Europe, the placement of the munition to blunt a Soviet thrust was preplanned, although the units were not pre-positioned, being kept under centralized control. The AEC manufactured SADM during the period August 1964 to June 1966, producing 300 units. Starting in June 1965, the AEC placed a 70-pound MOD 2 version of SADM in production.

Teapot Ess exploded a Ranger Able U235 core in a MK 6 HE assembly in March 1955. Ess (Effects Sub-Surface) had a 1-kiloton yield, similar to the upper yield of the B54 device. The test evaluated the cratering potential of an atomic demolition munition. Test personnel placed the 8,000-pound bomb in a 70-foot deep shaft lined with corrugated steel. NTS personnel then plugged the shaft with sandbags and soil prior to firing. The explosion produced a 300-foot wide crater that was 128 feet deep. Ten minutes after the blast, monitoring of the crater indicated a radiation level of 8,000 roentgens per hour. The success of this test insured the further development of the ADM concept that culminated in SADM.

Although the Army retired the last SADM from service in 1989, the legacy of this weapon still lives on. Its technology is highly sought after by terrorists desiring to prosecute a nuclear attack. Many failed one-point safety tests make it clear that it is difficult not to obtain a nuclear yield in the range of tens of tons from an explosive encased, solid, near-critical plutonium core. Although multi-kiloton yields are a much more difficult venture, as evidenced by the need to explode some 20,000 explosive lenses over a period of several years' year in the Fat Man program, a yield in the Davy Crockett range would be attainable for a motivated individual or group able to obtain an adequate supply of plutonium. Fortunately, most modern cores are hollow boosted designs that would require remanufacture to a simpler, solid design for use in a terrorist bomb. Properly placed, a terrorist weapon with an explosive yield in the 20-ton range could bring down many of the world's largest structures and its radiation could kill thousands or even tens of thousands of people. It would seem that the final price for the XW51 / W54 is eternal vigilance.

SELECTED BIBLIOGRAPHY

The Author researched much of this book on the World Wide Web. Unfortunately, websites are ephemeral – they come and go with the whims and the resources of their authors. Some of the material used to write this book came from websites that have long since disappeared. Other websites appeared after the initial research was completed, making this book's completion a moving target that required continual rewrites. Despite the transient nature of the World Wide Web, many of the reports contained in this bibliography are still accessible on the Web or are available from libraries. The Author has organized the Bibliography into a variety of topics to make it as useful as possible to Readers. The government has heavily redacted many of the documents listed here.

NUCLEAR WEAPONS

NUCLEAR PHYSICS

Agnew, Harold M. 1977. A Primer on Enhanced Radiation Weapons. Bulletin of the Atomic Scientists, December Issue.

AJW. 1997. Inertial Confinement Fusion. Plasma Physics, August 16 Issue.

Al'tshuler, L.V., Ya. B. Zel'dovich, and Yu. M. Styazhkin. 1997. Investigation of Isentropic Compression and Equations of State of Fissionable Materials. Physics-Uspekhi, January Issue.

American Institute of Physics. Oral History Interviews.
https://www.aip.org/history-programs/niels-bohr-library/oral-histories/browse

Badalov A.F. and V. I. Kopejkin. 1988. Uranium and Plutonium Energy Release per Fission Event in a Nuclear Reactor (TR). International Atomic Energy Agency, Nuclear Constants 2.

Barnes, Cris et al. 1996. Inertial Confinement Fusion at Los Alamos: The Pursuit of Ignition and Science-Based Stockpile Stewardship. Paper for the American Nuclear Society Embedded Topical Meeting on Fusion Technology, June 17 - 20. Los Alamos LA-UR-96-2131.

Bethe, Hans A. and R. F. Christy. 1944. Los Alamos Handbook of Nuclear Physics. Los Alamos Report LA 11.

Bernstein, Jeremy. 2008. Nuclear Weapons: What You Need to Know. Cambridge University Press.

Cartwright, David D. 1985. Summary of Classified Research for the Inertial Confinement Fusion Program at Los Alamos National Laboratory. Los Alamos LACP-95-0992.

Center for X-ray Optics and Advanced Light Sources. 2001. The X-ray Data Booklet. Lawrence Livermore National Laboratory.

Chadwick, James. 1932. Possible Existence of a Neutron. Nature, February 27.

De Geer, Erik-Lars. 1999. The Radioactive Signature of the Hydrogen Bomb. Science and Global Security, February Issue.

Department of Energy. 1993. Nuclear Physics and Reactor Theory: Volumes 1 - 4. DOE-HDBK-1019.

Diamond, H., P. R. Fields, C. S. Stevens, M. H. Studier, S. M. Fried, M. G. Inghram, D. C. Hess, G. L. Pyle, J. F. Mech, W. M. Manning, A. Ghiorso, S. G. Thompson, G. H. Higgins, G. T. Seaborg, C. I. Browne, H. L. Smith, and R. W. Spence. 1960. Heavy Isotope Abundances in Mike Thermonuclear Device. Physics Review, September 15 Issue.

Diven, Ben C., John H. Manley, and Richard F. Taschek. 1983. Nuclear Data: The Numbers Needed to Design the Bombs. Los Alamos Science, Number 7.

Fitzpatrick, Anne. 1999. Igniting the Light Elements: The Los Alamos Thermonuclear Weapon Project 1942 - 1952. Los Alamos LA-13577-T.

Frisch, O. R. 1939. Physical Evidence for the Division of Heavy Nuclei under Neutron Bombardment. Nature.

Frolov, Alexei M. 2006. Atomic Compression of the Light Element Plasma to Very High Densities. Canadian Journal of Physics / Review of Canadian Physics, September Issue.

Frolov, Alexei M. 2000. Influence of Tritium B-Decay on Low-temperature Thermonuclear Burn-up in Deuterium-Tritium Mixtures. Physics Review, September Issue.

Frolov, Alexei M. 1998. The Thermonuclear Burn-up in Deuterium - Tritium Mixtures and Hydrides of Light Elements. Plasma Physics and Controlled Fusion, August Issue.

Frolov, Alexei M., Vedene H. Smith Jr., and Gary T. Smith. 2002. Deuterides of Light Elements: Low Temperature Thermonuclear Burn-up and Applications to Thermonuclear Fusion Problems. Canadian Journal of Physics, January Issue.

Glasstone, Samuel and Ralph H. Lovberg. 1960. Controlled Thermonuclear Reactions: An Introduction to Theory and Experiment. Van Nostrand.

Goldhaber, Maurice. 1993. Reminiscences from the Cavendish Laboratory in the 1930s. Annual Review of Nuclear and Particle Science, December Issue.

Goncharov, G. A. 1996. American and Soviet H-bomb Development Programmes: Historical Background. Physics-Uspekhi, October Issue.

Gsponer, André and Jean Pierre Hurni. 1998. The Physical Principles of Thermonuclear Explosives, Inertial Confinement Fusion, and the Quest for Fourth Generation Nuclear Weapons. Geneva: International Network of Engineers and Scientists against Proliferation. Technical Report No. 1. Third Printing.

Hafemeister, David W. 2007. The Physics of Societal Issues: Calculations on National Security, Environment, and Energy. Springer.

Haight, Robert C., Mark B. Chadwick, and David J. Vieira. 2006. Fundamental Data for Pinning Down the Performance of Nuclear Weapons. Los Alamos Science, Number 7.

Jetter, U. 1950. Die Sogenannte Superbombe. Physics Blaetter, June Issue. (In German)

Kaler, James B. 1998. Stars. Scientific American Library.

Krappe, Hans J. and Krzysztof Pomorski. 2012. Theory of Nuclear Fission. Springer.

Lederman, Leon M. and David N. Schramm. 1995. From Quarks to the Cosmos. Scientific American Library.

Los Alamos National Laboratory. T2 Nuclear Information Service. http://t2.lanl.gov/

Los Alamos National Laboratory Technical Reports. http://www.fas.org/sgp/othergov/doe/lanl/index1.html

Marshak, R. E. 1958. Effect of Radiation on Shock Wave Behaviour. Physics of Fluids, January Issue.

McPhee, John. 1994. The Curve of Binding Energy: A Journey into the Awesome and Alarming World of Theodore B. Taylor. Farrar, Strauss, and Giroux.

Meitner, Lise, and Otto R. Frisch. 1939. Disintegration of Uranium by Neutrons: A New Type of Nuclear Reaction. Nature, February 11.

Mihalas, Dmitri and Barbara Weibel Mihalas. 1984. Foundations of Radiation Hydrodynamics. Oxford University Press.

Paxton, H. C. and N. L. Pruvost. 1986 Revision. Critical Dimensions of Systems Containing 235U, 239Pu, and 233U. Los Alamos LA-10860-MS.

Poss, H. L., E. O. Salant, G. A. Snow, and L. C. L. Yuan. 1952. Total Cross-section for 14-MeV Neutrons. Physical Review Letter, July Issue.

Pritzker, A. and W. Halg. 1981. Radiation Dynamics of a Nuclear Explosion. Journal of Applied Mathematics and Physics, January Issue.

Properties of the Elements. http://www.webelements.com/webelements/

Sandmeier, H.A., M.E. Battat, and G. E. Hansen. 1982. How to Calculate the Effects of Low-Yield Enhanced-Radiation and Fission Warheads. Los Alamos LA-9434.

Serber, Robert. 1992. The Los Alamos Primer. University of California Press.

Schoenberg, Kurt F. and Paul W. Lisowski. 2006. LANSCE: A Key Facility for National Science and Defense. Los Alamos Science, Number 30.

Sublette, Carey. 2001. Nuclear Weapons Frequently Asked Questions. http://nuclearweaponarchive.org/Nwfaq/Nfaq0.html

Sagie, D. and I. I. Glass. 1982. Explosive Driven Hemispherical Implosions for Generating Fusion Plasmas. UTIAS Technical Note 233, March.

Tuck, James L. 1954. Thermonuclear Reaction Rates. Los Alamos, LAMS-1640.

Watt, Bob, et al. 2009. Energy Balance in Fusion Hohlraums. Nuclear Weapons Journal, Issue 2.

Weast, Robert C. and Samuel M. Selby. Ed. The Handbook of Chemistry and Physics. The Chemical Rubber Co.

Winterberg, Friedwardt. 2010. The Release of Thermonuclear Energy by Inertial Confinement: Ways Toward Ignition. World Scientific Publishing.

Winterberg, Friedwardt. 1981. The Physical Principles of Thermonuclear Explosive Devices. Fusion Energy Foundation.

Winterberg, Friedwardt. 1975. Production of Dense Thermonuclear Plasmas by Intense Ion Beams. Journal of Plasma Physics, January Issue.

Younger, Stephen M. 1993. The High-density Regime of Weapon Physics. Los Alamos Science, Number 21.

Zeldovich, Y. B. and Y. P. Raizer. 2002. Physics of Shock Waves and High-Temperature Hydrodynamic Phenomena. Dover.

NUCLEAR LABORATORIES / WEAPON DESIGN / ENGINEERING

Alexander, F.C. Jr. 1969. Early Thermonuclear Weapons Development: The Origins of the Hydrogen Bomb. Sandia Corporation, SC-WD-68-334.

Alldred, David D., Mathew B. Squires, R. Steven Turley, Webster Cash, and Ann Shipley. 2002. Highly Reflective Uranium Mirrors for Astrophysics Applications. Society of Photographic Instrumentation Engineers, July Issue.

Allison, Samuel K. 1946. A General Survey of the Work of the Los Alamos Laboratory. Los Alamos Technical Series.

Anderson, J. W. 1958. Welding of Plutonium. Los Alamos LA-2220.

Anderson, R. C. 1952. "Vade Mecum" of Implosion Bombs. Los Alamos Laboratory.

Aquino, Michael A. 1980, 1982. The Neutron Bomb.
http://www.xeper.org/maquino/nm/NeutronBomb.pdf

Atomic Energy Commission. 1953. Part III. Weapons Progress Report to the Joint Committee; June through November 1953.

Badash, Lawrence, Joseph O. Hirschfelder, and Herbert P. Broida, ed. 1980. Reminiscences of Los Alamos, 1943 - 1945. Studies in the History of Modern Science 5. D. Reidel Publishing Company.

Ban the Bomb. 2008. Fogbank.
http://209.85.173.132/serch?q=cache:RqxHf-js_fUJ:www.banthebomb.org/newbombs/fogbank%2520material.doc+fogbank+nuclear&cd=2&hl=en&ct=clnk&gl=ca

Barsamian, Ara. No Date. Compressibility of Uranium and the Minimum Quantity for a Fission Weapon. Nuclear Non-Proliferation Institute.

Barsamian, Ara. 2008. Proliferation and Weapon Design – How Important? Nuclear Non-Proliferation Institute.

Bickes, R. W. Jr., M. C. Grubelich, S. M. Harris, J. A. Merson, and W. W. Tarbell. 1997. Semiconductor Bridge, SCB, Ignition of Energetic Materials. The Combustion Institute Spring Meeting.

Bostick, W. H., H. L. Freeman, J. C. King, and A. R. Martinez. 1986. Design of Pulse Transformers for Isolation, Protection, and Higher Speeds in the Remote Firing of Bridge Wires. Los Alamos, LA-10586-MS.

Brundige, E. L., J. M. Taub, G. S. Hanks, and D. T. Doll. 1958. Welding Thin Walled Uranium Cylinders. Technical Information Service Atomic Energy Commission. TID 8019

Bryant, Theodore C. and Gary R. Mowrer. 1979. Neutron Generator Instrumentation at the Department 2350 Neutron Generator Test Facility. Sandia SAND78-1188.

Caird, R. S., C. M. Fowler, D. J. Erickson, B. L. Freeman, and W. B. Garn. 1979. A Survey of Recent Work on Explosive-Driven Magnetic Flux Compression Generators. Los Alamos LA-UR-79-1202.

Cameron, Bruce. 2014. The History and Science of the Manhattan Project. Springer.

Carr, B. J. 1957. A Proposed Zipper System for Small Weapons. Sandia Corporation.

Carr, B. J. 1956. XR Systems for Operation Redwing. Sandia Corporation.

Christman, Al. 1998. Target Hiroshima: Deak Parsons and the Creation of the Atomic Bomb. US Naval Institute Press.

Comyn, Raymond H. 1962. Pyrotechnic Research at DOFL. Part II: Pyrotechnic Delays. Diamond Ordnance Fuze Laboratories.

Cook, Robert. 1995. Creating Microsphere Targets for Inertial Confinement Fusion Experiments. Energy Technology Review, April Issue.

Cotter, D. R. 1959. Addendum to SCDR 185-59, Characteristics and Development Report for the MC-1191 and MC-1291 Firing-sets. December 30.

Crompton, Joseph. 1956. Parachute Retardation Studies for New Class B Weapon. Part I – Conventional Parachute System. SC-TM-180-56-51.

Crompton, Joseph. 1956. Parachute Retardation Studies for New Class B Weapon. Part II – Laydown Parachute System. SC-TM-180-56-51.

Crumb, C.B. 1945. Effect of Voids on Detonation Waves. Los Alamos, LA-214.

Davis, Willim C. 1979. Introduction to Detonation Phenomena. Los Alamos, LA-UR-79-2880.

Debel, Charles A. 1950. A Historical Account of the Contractual, Organizational, and Technical Development (of the Abee Fuze). Motorola Incorporated.

Demmie, Paul M. No Date. Modeling and Simulation of Explosive Driven Electromechanical Devices. Sandia.

Department of the Army. 1984. Military Explosives. TM 9-1300-214.

DeVolpi, Alexander. 1989. Born Secret: The H-bomb, the Progressive Case, and National Security. Pergamon Books Ltd.

Drell, S. and R. Jeanloz. 1999. Remanufacture. JASON Report JSR-99-300.

E2V Technologies. No Date. Hydrogen Thyratrons.

EaglePicher Technologies. No Date. Thermal Battery Design.

Elsener F. 1998. Thickness distribution for Gold and Copper Electroformed Hohlraums. Fusion Technology.

Fakley, Dennis C. 1983. The British Mission. Los Alamos Science, Number 7.

Federation of American Scientists. No Date. Nuclear Weapon Design.

Federation of American Scientists. No Date. Nuclear Weapons Technology.

Fehner, Terrence R. and Jack M. Holl. 1994. The United States Department of Energy: A Summary History, 1977 - 1994. History Division, Executive Secretariat, Human Resources and Administration, Department of Energy.

Fetter, Steve, Valery A. Frolov, Oleg F. Prilutsky, and Roald Z. Sagdeev. 1990. Appendix A: Fissile Materials and Weapon Design. Science and Global Security, Volume 1.

Fowler, C. M. 1977. Pulsed Electrical Power from Explosives. Los Alamos, LASL 77-12.

Fowler, C. M., R. S. Caird and W. B. Garn. 1975. An Introduction to Explosive Flux Compression Generators. Los Alamos, LA-5890-MS.

Francis, Sybil. 1995. Warhead Politics: Livermore and the Competitive System for Nuclear Weapon Design. MIT Department of Political Science Ph.D.

Fritz, J. N. 1990. A Simple Plane Wave Generator. Los Alamos, LA-11956-MS.

Germain, Lawrence S. 1991. The Evolution of US Nuclear Weapon Design Trinity to King. Los Alamos, LA-11403.

Glasstone, Samuel. 1954. Weapons Activities of Los Alamos Scientific Laboratory Part I, Chapters 1 - 7. Los Alamos, LA-1633.

Glasstone, Samuel. 1954. Weapons Activities of Los Alamos Scientific Laboratory Part II, Chapter 8: Thermonuclear Devices. Los Alamos, LA-1633.

Glasstone, S. and L. M. Redman. 1972. An Introduction to Nuclear Weapons. WASH-1038.

Goncharov, German A. 1996. Milestones in the History of Hydrogen Bomb Construction in the Soviet Union and the United States. Physics Today, November Issue.

Gosling, F. G. 1999. The Manhattan Project: Making the Atomic Bomb. History Division, Department of Energy.

Grenko, J. D., B. J. Carr, Ellis A. Turner, and T. S. Church. 1953. Development Specification for XMC-64 (Modulated Neuron Initiator). Sandia, SC-5098.

Grimm, Gordon D. 1993. Characteristics and Development of the W89 High-Voltage, Low Inductance, Interconnect. Sandia, SAND92-2674.

Groves, Leslie R. Ed. 1948. Manhattan District History, Manhattan Project. AEC. https://ia902303.us.archive.org

Hafemeister, David. 2007. How Much Warhead Reliability Is Enough for a Comprehensive Nuclear Test Ban Treaty? Physics and Society, April Issue.

Hansen, Chuck. 2001. Beware the Old Story. Bulletin of Atomic Scientists, March Issue.

Hansen, Chuck. 2000. The Oops List. Bulletin of Atomic Scientists, November Issue.

Hargitai, Istvan and Magdolna Hargitai. 2006. Chapter 23. Richard L. Garwin in Candid Science VI: More Candid Conversations with Famous Scientists. Imperial College Press.

Harlow, Francis H. and N. Metropolis. 1983. Computing and Computers: Weapons Simulation Leads to the Computer Era. Los Alamos Science, Number 7.

Harms, J. L. and P. Z. Bulkeley. No Date. Energy Storage Devices – Electroexplosive Devices. Stanford University Department of Mechanical Engineering Design Division.

Hawkins, David. 1961. Manhattan District History Project Y. Vol. I. Inception until August 1945. Los Alamos, LAMS-2532.

Hoddeson, Lillian, Paul W. Henriksen, Roger A. Meade, and Catherine Westfall. 1993. Critical Assembly: A Technical History of Los Alamos during the Oppenheimer Years, 1943 - 1945. Cambridge University Press.

Holt W. H., T. L. Holt, and W. Mock Jr. 1980. Explosive shock de-poling of PZT Ferroelectric Ceramics for the Pulse Charging of Capacitors to High Voltage. Naval Surface weapons Center, NWSC TR3891.

Howard, Joseph S., Edward J. Palaneck, John L. Richter, Richard R. Sandoval, Frank L. Smith, Richard H. Stolpe, Larkin E. Garcia and L. Warren. 1983. Modular Weapons Systems and Insertable Nuclear Components: A Compendium of Requirements, Technology, Applications, and Utility. Los Alamos, LA-9866-MS.

Hennings, George N, Dieter K. Teschke, and, Richard K. Reynolds. 2005. Energetic Material Initiating Device Using Exploding Foil Initiated Ignition System with Secondary Explosive Material. US Patent Number: 6,923,122 B2.

Huizenga, David G. 1999. Management of Rocky Flats Environmental Technology Metal Parts at the Savannah River Site. Nuclear Material and Facility Stabilization Office of Environmental Management.

Jacobsen, Albin K. 1982. MC644 Detonator Development Status Report for the Period Ending October 30, 1981. Sandia, Sand81-2527.

Jones, Suzanne L. and Frank N. von Hipple. 1998. The Question of Pure Fusion Explosions under the CTBT. Science and Global Security.

Karr, Hugh J., John H. McQueen, and Barney J. Carr. 1953. Status of the External Initiator Development Program. XR Steering Committee, November 3.

Karr, Hugh J., John J. Wagner, Barney J. Carr, and Norman J. Ellison. 1954. Status of the External Initiator Development Program. XR Steering Committee, July 15.

Kristensen, Hans M. 2010. Congress Receives Nuclear Warhead Plan. FAS Strategic Security Blog, December 6.

Kristensen, Hans M. 2017. How US Nuclear Force Modernization is Undermining Strategic Stability: The Burst-Height Compensating Super-Fuze. Bulletin of the Atomic Scientists, March Issue.

Johnson, Leland. 1997. A History of Exceptional Service in the National Interest. Sandia, SAND 97-1029.

Libby, Stephen B. 1994. NIF and National Security. Energy & Technology Review, December Issue.

Lilmatainen, T. M. 1959. Progress in Miniaturization and Microminiaturization. Diamond Fuze Laboratories, PR 60-1.

Lillard, Jennifer. 2009. Fogbank: Lost Knowledge Regained. Nuclear Weapons Journal, Issue 2.

Loyola, Vincent M. and Sandra E. Klassen. 1998. Materials and Performance Evaluation of Accelerated Aged XTX8003. Sandia, SAND 97-2721C.

Magraw, Katherine. 1998. Teller and the "Clean Bomb" Incident. Bulletin of the Atomic Scientists, May Issue.

Maienschein, Jon L. and Jeffrey F. Wardell. 2003. Deflagration Behavior of HMX-Based Explosives at High Temperatures and Pressures. UCRL-CONF-201132

Makhijani, Arjun and Hisham Zerritti. 1998. Dangerous Thermonuclear Quest: The Potential of Explosive Fusion Research for the Development of Pure Fusion Weapons. Institute for Energy and Environmental Research.

Mark, J. Carson. 1974. A Short Account of Los Alamos Theoretical Work on Thermonuclear Weapons, 1946 - 1950. Los Alamos, LA-5647-MS.

Mark, C., Hunter, R. and J. Wechsler. 1983. Weapons Design: We've Done a Lot but We Can't Say Much. Los Alamos Science, Number 7.

Martin, L. Peter, Jeffrey H. Nguyen, Jeremy R. Patterson, Daniel Orlikowski, Palakkal P. Asoka-Kumar, and Neil C. Holmes. 2006. Tape Casting Technique for Fabrication of Graded Density Impactors for Tailored Dynamic Compression. Materials Research Society, Fall Meeting.

Mateev, S., E. Mateeva and E. Kirchuk. 2009. Plasma Assisted Combustion Technologies. Proceedings of the European Combustion Meeting.

Mathieu, Jorg and Hans Stucki. 2004. Military High Explosives. Chimia, June Issue.

MAUD Committee. 1941. Report by MAUD Committee on the Use of Uranium for a Bomb.

McNamara, Laura Agnes. 2001. Ways of Knowing about Weapons: The Cold War's End at the Los Alamos National Laboratory. University of New Mexico Ph.D. Dissertation.

Milani, K. No Date. The Scientific History of the Atomic Bomb.
http://www.hibbing.tec.mn.us/programs/dept/chem/abomb/index.html

Mello, Greg. 1990. That Old Designing Fever. Bulletin of the Atomic Scientist, January / February Issue.

Moniak, Dan. 2001. Plutonium: The Last Five Years - Part III: Plutonium in Pits. Blue Ridge Environmental Defense League.

De Montmollin, J. M. and W. R. Hoagland. 1961. Analysis of the Safety Aspects of the MK 39 MOD 2 Bombs Involved in B-52G Crash Near Greensboro, North Carolina. Sandia, SCDR 81-61.

Morland, Howard. 1995. Born Secret. Cardozo Law Review, March.

Morland, Howard. 1999. The Holocaust Bomb: A Matter of Time.
http://www.fas.org/sgp/eprint/morland.html

Morland, Howard. 2005. The Progressive Case. http://www.fas.org/sgp/eprint/cardozo.html

Morland, Howard. 2000. What's Left to Protect? Bulletin of the Atomic Scientist, November / December Issue.

Morgan, Michael. 2013.Tritium Effects on Reservoir Materials. Savannah River Tritium Focus Group.

Murphy, Charles H. 1977. Angular Motion of Projectiles with a Moving Internal Part. Ballistic Research Laboratory, Memorandum Report No. 2731, February.

NASA Space Vehicle Design Criteria. 1971. Solid Rocket Motor Igniters. NASASP-8051.

Neal, Timothy N. 1993. The Explosive Regime of Weapons Physics. Los Alamos Science, No. 27.

Office of the Secretary of Defense. 1978. History of the Deployment of Nuclear Weapons July 1945 through September 1977.

Oppenheimer, Andy. 2004. Mininukes: Boom or Bust. Bulletin of the Atomic Scientists, September / October Issue.

Palsey, John. 1995. Pulse Power Switching Devices – An Overview. http://nuclearweaponarchive.org/Library/Pasley1.html

Parker, Ann. 2002. Fifty Years of Innovation through Nuclear Weapon Design. LLNL Science and Technology Review, January / February Issue.

Pollin, Irvin. 1959. Barometric Devices and Fuze Design. Diamond Ordnance Fuze Laboratory, TR-763.

Pollock, Raymond. 1991. A Short History of the US Nuclear Stockpile 1945 - 1985. Center for National Security Studies, Los Alamos National Laboratory. LA-11401.

Pyrochemie. Historic Survey of the Chemical-Technical Development of Detonators and Primers. http://www.pyrochemie.at/e_historie.htm

Rhodes, Richard. 1986. The Making of the Atomic Bomb. Simon and Schuster.

Rhodes, Richard. 1995. Dark Sun: The Making of the Hydrogen Bomb. Simon and Schuster.

Salzbrunner, Dick, Mark Perra, and Wendy Cieslak. 1998. Materials Aging in the Enduring Nuclear Weapon Stockpile. Sandia National Laboratories.

Sandia Corporation Proving Group. 1951. Operation Greenhouse – Science Directors Report Part II, Mechanical Assembly. Sandia, WT-SAN-73.

Sandia National Laboratories. 1999. Nuclear Weapons. Sandia Lab News, February 12.

Sandia National Laboratory. Power Sources Technology Group. http://www.sandia.gov/pstg/

Sandia National Laboratory. Lab Accomplishments Archive. http://www.sandia.gov/LabNews/labs-accomplish/archive.html

Sandia National Laboratory. Nuclear Weapons Journal Archive. http://www.lanl.gov/science/weapons_journal/index.shtml

Sandoval, John. 2008. Enhanced Surveillance of Gas Transfer Valves for Stockpile Stewardship. Nuclear Weapons Journal, Issue 1.

Science and Global Security. 2007. Trying to lift the Veil on the Reliable Replacement Warhead (It's not Easy). Science and Global Security, June 6.

Scudder, David W., Stephanie A. Archuleta, Evan O. Ballard, Gerald W. Barr, J. C. Bucky Cochrane, Harold A. Davis, Jeffrey R. Griego, E. Staley Hadden, William B. Hinckley, Kieth W. Hosack, John E. Martinez, Diann Mills, Jennifer N. Padilla, Jerald V. Parker, W. Mark Parsons, Robert E. Reinovsky, John L. Stokes, and M. Clark. 2001. Atlas - A New Pulsed Power Tool at Los Alamos National Laboratory. LANL LA-UR-01-3136

Thompson, C. Y. Tom, Fred J. Wysocki, Billy N. Vigil

Selden, R. W. 1969. An Introduction to Fission Explosives. Lawrence Livermore National Laboratory, UCID–15554.

Shelton, Frank H. 1988. Reflections of a Nuclear Weaponeer. Shelton Enterprises, Inc.

Simpson, Randy, Larry Fried, Francis Ree, and Jack Reaugh. 1999. Unraveling the Mystery of Detonation. Science & Technology Review, June Issue.

Smith, P. D. 2007. Doomsday Men: The Real Dr. Strangelove and the Dream of the Superweapon. Allen Lane. (NWFAQ).

Smyth, Henry de Wolf. 1945. Atomic Energy for Military Purposes.

Stephens, Douglas R. 1992. Fire-resistant Pits: Reducing the Probability of Accidental Plutonium Dispersal from Fuel Fires. UCRL-ID-11056.

Sublette, Carey. 2001. Nuclear Weapons Frequently Asked Questions. (NWFAQ).
http://nuclearweaponarchive.org/Nwfaq/Nfaq0.html

Taylor, Theodore B. 1961. Third Generation Nuclear Weapons. Scientific American, April Issue.

Teller, E. and S. Ulam. 1946. On Heterocatalytic Detonations I: Hydrodynamic Lenses and Radiation Mirrors. LAMS-1225.

Truslow, Edith C. and Ralph Carlisle Smith. 1961. Manhattan District History Project Y. Vol. II. August 1945 through December 1946. Los Alamos, LAMS-2532.

Tucker, T. J. 1972. Explosive Initiators. Proceedings of the 12th Annual Symposium of the New Mexico Section of the ASME.

Ulrich, Rebecca A. 2003. Cold War Context Statement: Sandia National Laboratories California Site. Sandia, SAND2003-0112.

Union of Concerned Scientists. 2013. Kansas City Plant. Fact Sheet.

Union of Concerned Scientists. 2013. Lawrence Livermore National Laboratories Fact Sheet.

Union of Concerned Scientists. 2013. Los Alamos National Laboratories. Fact Sheet.

Union of Concerned Scientists. 2013. Nevada National Security Site. Fact Sheet.

Union of Concerned Scientists. 2013. Pantex Plant. Fact Sheet.

Union of Concerned Scientists. 2013. Sandia National Laboratory. Fact Sheet.

Union of Concerned Scientists. 2013. Savannah River Site. Fact Sheet.

Union of Concerned Scientists. 2013. Y-12 National Security Complex. Fact Sheet.

Van Dorn, W. G. 2008. Ivy-Mike: The First Hydrogen Bomb. Xlibris.

Wellerstein, Alex. 2012. Soviet Drawings of an American Bomb.
http://nuclearsecrecy.com/blog/2012/11/30/soviet-drawings-of-an-american-bomb/

Yong, Kristina. 2005. Focusing Optics for X-Ray Applications. MSc. Thesis, Graduate College of the Illinois Institute of Technology.

Younger, Stephen, Irvin Lindemuth, Robert Reinovsky, C. Maxwell Fowler, James Goforth, and Carl Ekdahl. Scientific Collaborations between Los Alamos and Arzamas-16 Using Explosive-Driven Flux Compression Generators. Los Alamos Science Number 24.

Zavadil, Kevin R. 1997. Evaluation of the Metal / Adhesive Interface in the MC2370 Firing-set. 21st Aging, Compatibility and Stockpile Stewardship Conference.

NUCLEAR BOMBS / WARHEADS

Ackland, Len and Steven McGuire, eds. 1986. Assessing the Nuclear Age. Chicago: Educational Foundation for Nuclear Science.

Asselin, S. V. 1966. B-52 / KC135 Collision near Palomares, Spain. Sandia Laboratory SC-DR-66-397.

Coster-Mullen, John. 2006. Atom Bombs: The Top Secret Inside Story of Little Boy and Fat Man. Self-Published.

Fields, K. E. 1955. Status of the XW-30 Development Program. Atomic Energy Commission, August 2.

Flaherty, Ted. 1997. Nuclear Weapon Database: United States Arsenal. CDI.
http://www.cdi.org/nuclear/database/usnukes.html

Glasstone, Samuel. 1954. Weapons Activities of Los Alamos Scientific Laboratory: Part I. Los Alamos LA-1632.

Godwin, J. 1997, 2003. The Mark 6 Atomic Bomb in the US Stockpile.
http://members.aol.com/nucinfo/

Gsponer, Andre. 2008. The B61 Based "Robust Nuclear Earth Penetrator": Clever Retrofit or Headway Towards Fourth-Generation Nuclear Weapons? arXiv, February 2.

Hansen, Chuck. 1998. A Hindenburg in the Bomb Bay: The Transition from Liquid to Solid Fueled Thermonuclear Weapons, 1951 - 1954. Second Los Alamos International History Conference.

Hansen Chuck. 2008. The Swords of Armageddon. Chucklea Publications.

Hansen, Chuck. 1988. US Nuclear Weapons: The Secret History. Crown Publishing.

Johnson, William Robert. 2009. Multimegaton Weapons.
http://www.johnstonsarchive.net/nuclear/multimeg.html

Joint Strategic Plans Committee. 1954. Mark 15 Program.

Kristensen, Hans M. 2012. B61 LEP: Increasing NATO Nuclear Capability and Precision Low-Yield Strikes. FAS Strategic Security Blog, June 15.
http://www.fas.org/blog/ssp/2011/06/b61-12.php

Kristensen, Hans M. 2012. B61-12: The NNSA's Gold-Plated Nuclear Bomb Project. FAS Strategic Security Blog, July 26.
http://www.fas.org/blog/ssp/2012/07/b61-12gold.php

Kristensen, Hans M. 2005. The Birth of a Nuclear Bomb: B61-11. Nuclear Information Project.
http://www.nukestrat.com/us/afn/B61-11.htm

Los Alamos and Sandia National Laboratories. 1992. B57 Special Study Report. RS5115/92/00016.

Medalia, Jonathan. Nuclear Weapons: 2007. The Reliable Replacement Warhead Program and the Life Extension Program. CRS Report for Congress, December 3.

Millikin, F. W., R. D. Lindsey, and L. A. Suber. 1965. Test Plan for TX-61 Weapon Test 100-16. Sandia Laboratory.

National Resources Defense Council. Archive of Nuclear Data from NRDC's Nuclear Program.
http://www.nrdc.org/nuclear//nudb/datainx.asp

Norris, Robert S. 2003. The B61 Family of Bombs. Bulletin of the Atomic Scientists, January / February Issue.

Norris, Robert S., Thomas B. Cochran, and William M. Arkin. 1985. History of the Nuclear Stockpile. Bulletin of the Atomic Scientists, August Issue.

Nuclear Weapons Archive. Complete List of All US Nuclear Weapons.
http://nuclearweaponarchive.org/Usa/Weapons/Allbombs.html

Padilla, Michael. 2006. Sandia Engineers Test Cruise Missile to Qualify W80-3 in Electromagnetic Environments. Sandia LabNews, March 14.

Parsch, Andreas. Designations of US Nuclear Weapons.
http://www.designation-systems.net/usmilav/nuke.html

Pretzel, C. W. 1995. E S & H Development Activities for the W89 Warhead. Sandia Laboratories, DAND95-8232.

Ramsey, Norman F. and Raymond L. Brin. 1946. Nuclear Weapons and Engineering Delivery. Los Alamos, LA-1161.

Ramsey, Norman F. 1945. The History of Project A.

Sandia Corporation. 1952. Final Design Status of the TX-7-X1 Weapon. SC-2166 (TR).

Sandia Corporation. 1962. Interim Development Report for the TX-53 Basic Assembly and the TX-53 Bomb. SC-4621 (WD).

Sandia Laboratories Information Research Division. 1967. History of the Mark 4 Bomb. AEC Atomic Weapons Data, SC-M-67-544.

Sandia Laboratories Information Research Division. 1967. History of the Mark 5 Bomb. AEC Atomic Weapons Data, SC-M-67-545.

Sandia Laboratories Information Research Division. 1967. History of the Mark 5 Warhead. AEC Atomic Weapons Data, SC-M-67-546.

Sandia Laboratories Information Research Division. 1967. History of the Mark 6 Bomb: Including the TX/XW 13, MK 18, and TX 20. AEC Atomic Weapons Data, SC-M-67-726.

Sandia Laboratories Information Research Division. 1967. History of the Mark 7 Bomb. AEC Atomic Weapons Data, SC-M-67-547.

Sandia Laboratories Information Research Division. 1967. History of the Mark 7 Warhead. AEC Atomic Weapons Data, SC-M-67-548.

Sandia Laboratories Information Research Division. 1967. History of Gun Type Bombs and Warheads: The Mark 8, 10, and 11. AEC Atomic Weapons Data, SC-M-67-658.

Sandia Laboratories Information Research Division. 1967. History of Gun Type Artillery Fired Atomic Projectiles: The Mark 9, 19, 23, 32, and 33 Shells. AEC Atomic Weapons Data, SC-M-67-659.

Sandia Laboratories Information Research Division. 1967. History of the Mark 12 Weapon. AEC Atomic Weapons Data, SC-M-67-660.

Sandia Laboratories Information Research Division. 1967. History of Early Thermonuclear Bombs: MKs 14, 15, 16, 17, 24, and 29. AEC Atomic Weapons Data, SC-M-67-661.

Sandia Laboratories Information Research Division. 1967. History of Early Thermonuclear Bombs: MKs 21, 22, 26, and 36. AEC Atomic Weapons Data, SC-M-67-662.

Sandia Laboratories Information Research Division. 1967. History of the Mark 25 Warhead. AEC Atomic Weapons Data, SC-M-67-663.

Sandia Laboratories Information Research Division. 1967. History of the Mark 27 Weapon. AEC Atomic Weapons Data, SC-M-67-664.

Sandia Laboratories Information Research Division. 1968. History of the Mark 28 Weapon. AEC Atomic Weapons Data, SC-M-67-665.

Sandia Laboratories Information Research Division. 1968. History of the Mark 30 Warhead. AEC Atomic Weapons Data, SC-M-67-666.

Sandia Laboratories Information Research Division. 1968. History of the Mark 31/37 Warhead. AEC Atomic Weapons Data, SC-M-67-667.

Sandia Laboratories Information Research Division. 1968. History of the Mark 34 Warhead. AEC Atomic Weapons Data, SC-M-67-668.

Sandia Laboratories Information Research Division. 1968. History of the Mark 35 Warhead. AEC Atomic Weapons Data, SC-M-67-669.
Sandia Laboratories Information Research Division. 1968. History of the Mark 38 Warhead. AEC Atomic Weapons Data, SC-M-67-670.
Sandia Laboratories Information Research Division. 1968. History of the Mark 39 Weapon. AEC Atomic Weapons Data, SC-M-67-671.
Sandia Laboratories Information Research Division. 1968. History of the Mark 40 Warhead. AEC Atomic Weapons Data, SC-M-67-672.
Sandia Laboratories Information Research Division. 1968. History of the Mark 41 Warhead. AEC Atomic Weapons Data, SC-M-67-673.
Sandia Laboratories Information Research Division. 1968. History of the Mark 42 Warhead. AEC Atomic Weapons Data, SC-M-67-674.
Sandia Laboratories Information Research Division. 1968. History of the Mark 43 Bomb. AEC Atomic Weapons Data, SC-M-67-675.
Sandia Laboratories Information Research Division. 1968. History of the Mark 44 Warhead. AEC Atomic Weapons Data, SC-M-67-676.
Sandia Laboratories Information Research Division. 1968. History of the Mark 45 Warhead. AEC Atomic Weapons Data, SC-M-67-677.
Sandia Laboratories Information Research Division. 1968. History of the Mark 46 Warhead. AEC Atomic Weapons Data, SC-M-67-678.
Sandia Laboratories Information Research Division. 1968. History of the Mark 47 Warhead. AEC Atomic Weapons Data, SC-M-67-679.
Sandia Laboratories Information Research Division. 1968. History of the Mark 48 Shell. AEC Atomic Weapons Data, SC-M-67-680.
Sandia Laboratories Information Research Division. 1968. History of the Mark 49 Warhead. AEC Atomic Weapons Data, SC-M-67-681.
Sandia Laboratories Information Research Division. 1968. History of the Mark 50 Warhead. AEC Atomic Weapons Data, SC-M-67-682.
Sandia Laboratories Information Research Division. 1968. History of the Mark 52 Warhead. AEC Atomic Weapons Data, SC-M-67-684.
Sandia Laboratories Information Research Division. 1968. History of the Mark 53 Weapon. AEC Atomic Weapons Data, SC-M-67-685.
Sandia Laboratories Information Research Division. 1968. History of the Mark 54 Weapon. AEC Atomic Weapons Data, SC-M-67-686.
Sandia Laboratories Information Research Division. 1968. History of the Mark 55 Warhead. AEC Atomic Weapons Data, SC-M-68-47.
Sandia Laboratories Information Research Division. 1968. History of the Mark 56 Warhead. AEC Atomic Weapons Data, SC-M-68-49.
Sandia Laboratories Information Research Division. 1968. History of the Mark 57 Bomb. AEC Atomic Weapons Data, SC-M-68-48.
Sandia Laboratories Information Research Division. 1968. History of the Mark 58 Warhead. AEC Atomic Weapons Data, SC-M-68-50.
Sandia Laboratories Information Research Division. 1968. History of the Mark 59 Warhead. AEC Atomic Weapons Data, SC-M-68-51.

Sandia Laboratories Information Research Division. 1971. History of the TX 61 Bomb. AEC Atomic Weapons Data, SC-M-71-0339.
Sandia Laboratories Information Research Division. 1968. History of the XW 35 Warhead. AEC Atomic Weapons Data, SC-M-67-669.
Sandia Laboratories Information Research Division. 1968. History of the XW 51 Warhead. AEC Atomic Weapons Data, SC-M-67-683.
Sandia Laboratory. 1949. Final Evaluation Report MK IV MOD 0 FM Bomb. SL-82.
Sandia National Laboratories. 1993. B53-1 Special Study Report. Sandia SAND93-0943.
Sandia National Laboratories. 1993. Defense Programs: B83 MOD 1 Retrofits are Under Way. A Sandia Weapon Review Bulletin, Summer Issue.
Sandia National Laboratories. 1986. Final Development Report for the B61-7 Bomb. Sandia SAND85-0474.
Sandia National Laboratories. 1989. Interim Development Report for the B61-6, -8 Bombs. Sandia SAND88-2986.
Sandia National Laboratories. 1966. Post-Mortem Examination of MK 28 Components from the Palomares, Spain Accident. RS2131/86.
Soden, Jerry M. and Richard E. Anderson. 1998. W88 Integrated Circuit Shelf Life Program. Sandia National Laboratories, SAND98-0029.
US Air Force Nuclear Weapons Accident / Incident Report. 1961. Ammunitions Letter No. 136-11-56G, Goldsboro, North Carolina.
US Atomic Energy Commission, Albuquerque Operations Information Division. Developing and Producing the B61. (Film)
US Department of Energy Nuclear Explosive Safety Study Group. 1993. Nuclear Explosive Safety Study of B53 Mechanical Disassembly Operations at the USDOE Pantex Plant.
Walker, Stephen. 2005. Shockwave: The Countdown to Hiroshima. John Murray.

NUCLEAR TESTS / NUCLEAR TEST SITES / NUCLEAR WEAPON EFFECTS

NATO Handbook on the Medical Aspects of Defensive NBC Operations. 1966. FM8-9.
Advisory Committee on Human Radiation Experiments Part II: Case Studies. 1995. US Government Printing Office.
Bainbridge, K.T. 1976. Trinity. Los Alamos, LA-6300H.
Barasch, Guy E. 1979. Light Flash Produced by an atmospheric Nuclear Explosion. LASL-79-84.
Berkhouse, L., S.E. Davis, F.R. Gladeck, J.H. Hallowell, C.B. Jones, E.J. Martin, R.A. Miller, F.W. McMullan and M.J. Osborne. 1962. Operation Dominic I. Defense Nuclear Agency.
Born, D, and E. Woodward. 1962. Instant Fireball Yield. UCRL-ID-124700.
Brode, Harold L. 1964. Fireball Phenomenology. RAND.
Brode, Harold L. 1954. Height of Burst for Atomic Bombs (After Upshot - Knothole). Rand Corporation, June 1.
Campbell, Bob, Ben Diven, John McDonald, Ben Ogle, and Tom Scolman. 1983. Field Testing: The Physical Proof of Design Principles. Los Alamos Science, Spring Issue.
Cochran, Thomas B. and Christopher E. Paine. 1994. Hydronuclear Testing and the Comprehensive Test Ban: Memorandum to Participants, JASON Summer Study.
Cochran, Thomas B. and Christopher E. Paine. 1995. The Role of Hydronuclear Tests and Other Low Yield Explosions and their Status under a Nuclear Test Ban. Natural Resources Defense Council, Inc.

Cohen, S. T., J. O. Hirschfelder, M. Hull, and J. L. Magee. 1946. Crossroads Handbook of Explosion Phenomena. Los Alamos, LA-4550.
Dahlman, Ola and Hans Israelson. 1997. Monitoring Underground Nuclear Explosions. Elsevier Science Ltd.
Daugherty, William, Barbara Levi, PH.D., and Frank Von Hippel, PH.D. 1986. Casualties Due to the Blast, Heat, and Radioactive Fallout from Various Hypothetical Nuclear Attacks on the United States. Report #PU/CEES 198, Princeton University's Center for Energy and Environmental Studies.
Defense Special Weapons Agency 1947 - 1957: The First Fifty Years of National Service. No Date.
Defense Threat Reduction Agency. 2007. Hiroshima and Nagasaki Occupation Forces.
Defense Threat Reduction Agency. 2006. Operation Argus Fact Sheet.
Defense Threat Reduction Agency. 2006. Operation Buster-Jangle Fact Sheet.
Defense Threat Reduction Agency. 2007. Operation Castle Fact Sheet.
Defense Threat Reduction Agency. 2007. Operation Crossroads Fact Sheet.
Defense Threat Reduction Agency. 2007. Operation Dominic I Fact Sheet.
Defense Threat Reduction Agency. 2007. Operation Dominic II Fact Sheet.
Defense Threat Reduction Agency. 2007. Operation Greenhouse Fact Sheet.
Defense Threat Reduction Agency. 2007. Operation Hardtack I Fact Sheet.
Defense Threat Reduction Agency. 2007. Operation Hardtack II Fact Sheet.
Defense Threat Reduction Agency. 2007. Operation Ivy Fact Sheet.
Defense Threat Reduction Agency. 2007. Operation Plowshare Fact Sheet.
Defense Threat Reduction Agency. 2007. Operation Plumbbob Fact Sheet.
Defense Threat Reduction Agency. 2007. Operation Ranger Fact Sheet.
Defense Threat Reduction Agency. 2009. Operation Redwing Fact Sheet.
Defense Threat Reduction Agency. 2007. Operation Sandstone Fact Sheet.
Defense Threat Reduction Agency. 2007. Operation Trinity Fact Sheet.
Defense Threat Reduction Agency. 2007. Operation Tumbler-Snapper Fact Sheet.
Defense Threat Reduction Agency. 2007. Operation Teapot Fact Sheet.
Defense Threat Reduction Agency. 2007. Operation Upshot-Knothole Fact Sheet.
Defense Threat Reduction Agency. 2007. Operation Wigwam Fact Sheet.
Demos, D. 1998. (film) Castle Bravo Shot. US Department of Energy.
Department of Defense. 1983. Major Range and Test Facilities Base: Summary of Capabilities. DOD 3208.11-D.
Department of Defense and Nuclear Analysis Center. 1972. Handbook of Underwater Nuclear Explosions: Volumes 1 & 2. DNA-1240.
DeWitt, Hugh et al. 1955. A Compilation of Spectroscopic Observations of Air Around Atomic Bomb Explosions. Los Alamos, LAMS-1935.
Dieke, G. H. 1955. Spectroscopy of Bomb Explosions. Los Alamos, LA-132.
DODD 5105.31 Defense Special Weapons Agency.
    http://www.fas.org/nuke/guide/usa/doctrine/dod/dodd-5105_31.htm
DOE/NV-209. 2000. United States Nuclear Tests, July 1945 through September 1992.
DuPont, Daniel, G. 2004. Nuclear Attacks in Orbit. Scientific American.
Eden, Lynn, 2006. Whole World on Fire: Organizations, Knowledge, and Nuclear Weapons Devastation. Cornell Studies in Security Affairs.

Fehner, Terrence R. and F. G. Gosling. 2006. Battlefield of the Cold War, The Nevada Test Site Volume I, Atmospheric Nuclear Weapons Testing. MA-0003.

Fehner, Terrence R. and F. G. Gosling. 2000. Origins of the Nevada Test Site. MA-0518.

FEMA. 1990. Nuclear Attack Planning Base – 1990. NAPB-90.

NATO Handbook on the Medical Aspects of Defensive NBC Operations. 1966. FM8-9.

Gallery of US Nuclear Tests. http://nuclearweaponarchive.org/Usa/Tests/index.html

Gellert, E. 1963. Fireball Yield from Fractional Intensity Diameters. UCRL-ID-124699.

General Principles of Nuclear Explosions. 1977.
http://www.cddc.vt.edu/host/atomic/nukeffct/enw77a.html

Glasstone, Samuel and Philip J. Dolan. 1977. The Effects of Nuclear Weapons, Third Edition. US Government Printing Office.

Headquarters, Department of the Army. 2000. Treatment of Nuclear Warfare Casualties and Low-level Radiation Injuries. Field Manual 8-283.

Hoerlin, Herman. 1976. United States High-altitude Test Experiences. LASL Monograph.

Joint Defense Science Board / Threat Reduction Advisory Committee Task Force. 2000. The Nuclear Weapons Effects National Enterprise.

JTF-8 Ad Hoc Group for Nuclear Safety. 1962. Technical Nuclear Safety Study of Project Dominic B-52 Airdrops. Safety Analysis and Development Division Development Directorate Air Force Special Weapons Center Kirtland Air Force Base New Mexico.

Kunkle, Thomas and Byron Ristvet. 2013. Castle Bravo: Fifty Years of Legend and Lore. DTRA Special Report.

Langford, R. Everett. 2004. Introduction to Weapons of Mass Destruction: Radiological, Chemical and Biological. John Wiley and Sons.

Lawrence Livermore National Lab. No Date. Handbook for United Nations Observers, Pinon Test, Eniwetok. University of California Radiation Laboratory UCRL-5367.

Kerr, George D., Robert W. Young, Harry M. Cullings, and Robert F. Christy. 2005. Chapter 1: Parameters in Reassessment of the Atomic Bomb Radiation Dosimetry for Hiroshima and Nagasaki. Radiation Effects Research Foundation.

Kistiakowsky, George B. 1980. Trinity – A Reminiscence. Bulletin of the Atomic Scientists, June Issue.

Light, Michael. 2003. 100 Suns 1945 - 1962. Knopf Books.

Malik, John. 1985. The Yields of the Hiroshima and Nagasaki Nuclear Explosions. Los Alamos, LA-8819.

May, Michael and Zachary Haldeman. 2004. The Effectiveness of Nuclear Weapons against Buried Biological Agents. Science and Global Security, January Issue.

McDonnell Douglas Astronautics. 1982. Summary of Trapped Electron Data: Final Report. AD-B069354 (AFWL·TR·81·223).

Miller, George H., Paul S. Brown and Carol T. Alonso. 1987. Report to Congress on Stockpile Reliability, Weapon Remanufacture, and the Role of Nuclear Testing. LLNL.

National Resources 1997. Defense Council. End Run: Simulating Nuclear Explosions under the Test Ban Treaty.
http://www.nrdc.org/nuclear/endrun/erinx.asp

Naval Historical Center. 1976. Operations Crossroads Fact Sheet.

Norris, Robert S. and Thomas B. Cochran. 1994. United States Nuclear Tests: July 1945 to 31 December 1992. NWD94-1.

Nuclear Testing and Weapons Design. 1995. Science for Democratic Action, Spring Issue. Office of Technology Assessment. 1979. The Effects of Nuclear War.

Ogle, William E. 1985. An Account of the Return to Nuclear Weapon Testing by the United States after the Test Moratorium 1958 - 1961. NVO-291.

Porzel, Francis B. 1953. Preliminary Hydrodynamic Yields of Nuclear Weapons. Los Alamos Scientific Laboratory, WT-9001.

Radiation Effects Research Foundation. 2005. Reassessment of the Atomic Bomb Radiation Dosimetry for Hiroshima and Nagasaki.

Sandia National Laboratory, Official List of Underground Nuclear Explosions in Nevada. http://nuclearweaponarchive.org/Usa/Tests/Nevada.html

Semkow, Thomas M., Pravin P. Parekh, and Douglas K. Haines. 2006. Modeling the Effects of the Trinity Test. Chapter 11 in Effects of the Trinity Testing Applied Modeling and Computations in Nuclear Science. ACS Symposium Series. Vol. 945.

Solomon, Frederic and Robert Q. Marston. 1986. The Medical Implications of Nuclear War. Institute of Medicine, National Academy of Sciences.

Thorn, Robert N. and Donald R. Westervelt. 1987. Hydronuclear Experiments. Los Alamos, LA10902-MS.

Turco, R. P., 0. B. Toon, T. P. Ackerman, J. B. Pollack, and Carl Sagan. 1983. Global Atmospheric Consequences of Nuclear War. NASA-TM-101281.

Undersecretary of Defense for Research and Engineering. 1983. Major Range and Test Facility Base Summary of Capabilities. DOD3200.11-D.

United States Department of Energy. No Date. Plowshare Program.

US Congress, Office of Technology Assessment. 1989. The Containment of Underground Nuclear Explosions. OTA-ISC.

NUCLEAR WARHEAD MATERIALS / PRODUCTION

AEC / DOD. 1953. An Agreement between the AEC and the DOD for the Development, Production, and Standardization of Atomic Weapons.

Arkin, William M., Thomas B. Cochran, and Milton M. Hoenig. 1982. The US Nuclear Stockpile: Materials Production and New Weapons Requirements. Arms Control Today, April Issue.

Atomic Energy Commission 1946 - 1977. http://www.u-s-history.com/pages/h1813.html

Brown, Donald W., Robert E. Hackenberg, David F. Teter, Mark A. Bourke, and Dan Thoma. 2006. Aging and Deformation of Uranium-Niobium Alloys. Los Alamos Science, Number 30.

Carlisle, Rodney P. and Joan M. Zenzen. 1994. Supplying the Nuclear Arsenal: Production Reactor Technology, Management, and Policy 1942 - 1992. US Department of Energy History Division.

Chronology of Important FOIA Documents: Hanford's Semi-Secret Thorium to U-233 Production Campaign. No Date. http://www.hanfordchallenge.org/cmsAdmin/uploads/Chronology_of_thorium_to_U-233_FOIA_Docs.pdf

Cochran, Thomas B. 1996. US Inventories of Nuclear Weapons and Weapons-Usable Fissile Materials. Washington, DC: Natural Resources Defense Council.

Cochran, T., W. Arkin, and M. Hoenig. 1984. Nuclear Weapons Databook Volume II: US Nuclear Warhead Production. Natural Resources Defense Council.

Cochran, T., W. Arkin, and M. Hoenig. 1984. Nuclear Weapons Databook Volume III: US Nuclear Warhead Facility Profiles. Natural Resources Defense Council.

Cochran, Thomas B., Christopher E. Paine, and Robert S. Norris. 1991. Awash in Tritium: Maintaining the Nuclear Weapon Stockpile under START. NWD91-2.

Cohen, William S. and Bill Richardson. 2000. Nuclear Skills Retention Measures within the Department of Defense and the Department of Energy. DOE Response to Chiles Commission.

Department of Energy Office of Environmental Management. 1997. The Legacy Story: A History of the US Nuclear Weapons Complex.

Department of Energy. 2001. Striking a Balance: A Historical Report on the United States Highly Enriched Uranium Production, Acquisition and Utilization Activities from 1945 through September 30, 1966. National Nuclear Security Administration.

Dobratz, B. M. and P.C. Crawford. 1985. LLNL Explosives Handbook: Properties of Chemical Explosives and Simulants. UCRL-52997.

Erickson, W. C. and G. E. Jaynes, D. J. Sandstrom, R. Seegmiller, and J. M. Taub. 1972. Evaluation of Uranium Alloys. Los Alamos, LA-5002.

Federation of American Scientists. Plutonium: The First 50 Years. http://www.fas.org/sgp/othergov/doe/pu50y.html

Federation of American Scientists. Uranium Production. http://www.fas.org/nuke/intro/nuke/uranium.htm

Formation of Uranium Ore Deposits. 1974. International Atomic Energy Agency.

Forsberg, C. W., C. M. Hopper, J. L. Richter, and H. C. Vantine. 1998. Definition of Weapons Usable Uranium 233. Oak Ridge National Laboratory, ORNL/TM-13517.

Gerber, M. S. 1966. The Plutonium Production Story at the Hanford Site: Processes and Facilities History. Westinghouse Hanford.

Hanford Cultural and Historic Resources Program. 2002. History of the Plutonium Production Facilities at the Hanford Site Historic District, 1943 - 1990. DOE/RL-97-1047.

Hartley, Richard S. 1988. Neutron Multiplication in Beryllium. University of Austin, Texas. M.Sc. Thesis.

Harvey, David. No Date. History of the Hanford Site 1943 - 1990. Pacific Northwest National Laboratory.

Hecker, Siegfried S. 2000. Plutonium and its Alloys: From Atoms to Microstructure. Los Alamos Science, Number 26.

Hecker, Siegfried S. and Joseph C. Martz. 2000. Ageing of Plutonium and its Alloys. Los Alamos Science, Number 26.

Hecker, Siegfried S. and Michael F. Stevens. 2000. Mechanical Behavior of Plutonium and its Alloys. Los Alamos Science, Number 26.

Institute for Energy and Environmental Research. Physical, Nuclear and Chemical Properties of Plutonium. http://www.ieer.org/fctsheet/pu-props.html

Institute for Energy and Environmental Research. Uranium: Its Uses and Its Hazards. http://www.ieer.org/fctsheet/uranium.html

Kang, Jungmin and Frank N. von Hippel. 2001. U-232 and the Proliferation-Resistance of U-233 in Spent Fuel. Science & Global Security.

Kimberley, M. M. 1978. Short Course in Uranium Deposits: Their Mineralogy and Origin. Mineralogical Association of Canada.

Loeber, Charles R. 2004. Building the Bombs: A History of the Nuclear Weapons Complex. Diane Publishing Company.

Lonadier, Frank D. and Joseph F. Griffo. 1963. The Preparation of Plutonium 238 Metal. Monsanto Research Corporation.

Lovins, A. B. 1980. Nuclear Weapons and Power Reactor Plutonium. Nature, Volume 283.

Makhijani, Arjun, Lois Chalmers, and Brice Smith. 2004. Uranium Enrichment: Just Plain Facts to Fuel an Informed Debate on Nuclear Proliferation and Nuclear Power. IEER.

Mello, Greg. 2010. US Plutonium Pit Production: Additional Facilities Production are Unnecessary, Costly, and Provocative. Los Alamos Study Group, March 2.

Messer, Charles E. 1960. A Survey Report on Lithium Hydride. Atomic Energy Commission, NYO-9470.

Michaudon, André F. and Ileana G. Buican. 2000. A Factor of Millions: Why we Made Plutonium. Los Alamos Science, Number 26.

National Academy of Sciences. 1965. The Radiochemistry of Plutonium. NAS - NS 3058.

Quist, Arvin. 2000. A History of the Classified Activities at Oak Ridge National Laboratory. ORCA-7.

Reavis, J. G. 1984. Plutonium Metal and Alloy Production by Molten Chloride Reduction. Los Alamos, LA-UR-84-3810.

Report of the Defense Science Board Task Force. 2005. Nuclear Weapon Effects, Test, Evaluation, and Simulation.

Report of the Secretary of Energy Advisory Board, Nuclear Weapons Complex Infrastructure Task Force. 2005. Recommendations for the Nuclear Weapons Complex of the Future.

Rosenthal, Murray W. 2010. An Account of Oak Ridge National Laboratory's Thirteen Nuclear Reactors - Revised. Oak Ridge National Laboratory.

Rothe, Robert E. 2005. A Technically Useful History of the Critical Mass Laboratory at Rocky Flats. Los Alamos, LA-UR-05-3247.

Savannah River. 2011. About the Savannah River Site.

Smith, Roger L. and James W. Miser. 1963. Compilation of the Properties of Lithium Hydride. NASA Technical Memorandum X-483.

Ulrich, Rebecca A. 1998. Technical Area II: A History. Sandia, SAND98-617.

US Department of Energy. 1992. Production Reactors: An Outline Overview, 1944 - 1988. DOE/NP/00092T-H1.

US Department of Energy. Savannah River Site. Tritium Extraction Facility.

US Department of Energy National Nuclear Security Administration Office of the Deputy Administrator for Defense Programs. 2001. Highly Enriched Uranium: Striking a Balance. Historical Report on the United States Highly Enriched Uranium Production, Acquisition, and Utilization Activities from 1945 through September 30, 1966.

Wahlen, R. K. 1989. History of 100-B Area. Westinghouse Hanford Company, WHC-EP-0273.

West, George T. 1992. United States Nuclear Warhead Assembly Facilities 1945 - 1990. Pantex.

NUCLEAR STORAGE / TRANSPORT

Alling, Frederick A. 1956. History of Atomic Logistics. Historian, Historical Division, Office of Information Services, Air Materiel Command, Wright-Patterson AFB.

Arkin, William, Robert Norris and Joshua Handler. 1998. Taking Stock: Worldwide Nuclear Deployments. Natural Resources Defense Council

Brown, L. A. and M. C. Higuera. 1977. Weapon Container Catalog Volumes 1 and 2. SAND 97-017.

Burr, William. ed. 2006. How Many and Where Were the Nukes? The National Security Archive Electronic Briefing Book.

Cary, Lyle W. 2000. Evaluating US Air Force Nuclear Weapons Storage Area Security in the Post-Cold War Environment. Air University.

Van Citters, Karen and Kristen Bisson. 2003. National register of Historic Places Historic Context and Evaluation for Kirtland Air Force Base, Albuquerque, New Mexico. US Air Force.

DASA. 1968. Custody of Atomic Weapons: Historical Summary of Principle Actions.

Department of Defense. 2008. The Defense Science Board (DSB) Permanent Task Force on Nuclear Weapons Surety Report on Unauthorized Movement of Nuclear Weapons.

DODD 4540.5. Movement of Nuclear Weapons by Non-Combat Delivery Vehicles.

DODD 4540.5-M. DOD Nuclear Weapons Transportation Manual.

Kristensen, Hans M. 2005. A History of US Nuclear Weapons in South Korea. Nuclear Information Project.

Kristensen, Hans M. 2005. US Nuclear Weapons in Europe. Natural Resources Defense Council.

Naval Training Command. Gunners Mate M1&C Rate Training Manual. 1973. NAVTRA 10200-B, United States Government Printing Office.

Naval Training Command. 1972. Gunners Mate 2&3 Rate Training Manual. NAVTRA 10199-B, United States Government Printing Office.

Norris, Robert S., William M. Arkin, and William Burr. 1999. Where They Were. Bulletin of the Atomic Scientists, November / December Issue.

Norris, Robert S. and Hans M. Kristensen. 2006. Where the Bombs Are, Nukestrat. http://www.nukestrat.com/us/where.htm

Office to the Assistant of the Secretary of Defense. 1981. History of the Custody and Deployment of Nuclear Weapons July 1945 through September 1977.

Paxton, H. C. 1975. Capsule Storage and Density Analog Techniques. Los Alamos, LA-5930-M.

The Safeguard System: Transporting and Safeguarding Special Nuclear Material. No Date. Milnet Brief.

Schwarz, Stephen I. 2002. Bombs in the Back Yard: Bases and Facilities with Significant or Historical US Nuclear Weapons or Naval Nuclear Propulsion Missions. The Brookings Institution.

Stallings, Patricia and Edward G. Salo. 2010. Clarksville Base Historic Context. US Army

Wolf, Ron. 1985. On the Road with Plutonium. The APF Reporter.

MAINTENANCE / SURVEILLANCE

Abrahamson, James L and Paul H. Carew. 2002. Vanguard of American Deterrence: The Sandia Pioneers 1946 - 1949. Praeger.

Air Force Enlisted Job Descriptions: 2W2X1 – Nuclear Weapons Specialist. http://usmilitary.about.com/od/airforceenlistedjobs/a/afjob2w2x1.htm

Arms Control Association. 2012. Fact Sheet: US Nuclear Modernization Programs. http://www.armscontrol.org/factsheets/USNuclearMODernization

Bierbaum, R. L., J. J. Cashen, T. J. Kerschen, J. M. Sjulin, and D. L. Wright. 1999. DOE Nuclear Weapon Reliability Definition: History, Description, and Implementation. Sandia Report 99-8240.

Fetter, Steve. 1987 / 88. Stockpile Confidence under a Nuclear Test Ban. International Security. Winter.

Fishbine, Brian. 2007. Shelf Life Guaranteed: Extending the Life of Nuclear Weapons. Los Alamos Research Quarterly, August Issue.

Government Accounting Office. 1996. Nuclear Weapons: Improvements Needed for DOE's Nuclear Weapons Stockpile Surveillance Program. GAO/RCED-96-216.

Government Accounting Office. 1997. Nuclear Weapons: Capabilities of DOE's Limited Life Component Program to Meet Operational Needs. GAO/RCED-97-52.

Hemley, R. J., D. Meiron, L. Bildsten, J. Cornwall, F. Dyson, S. Drell, D. Eardley, D. Hammer, R. Jeanloz, J. Katz, M. Ruderman, R. Schwitters and J. Sullivan. 2007. Pit Lifetime. The MITRE Corporation, JSR-06-335.

Johnson, Kent, Joseph Keller, Carl Ekdahl, Richard Krajcik, Luis Salazar, Earl Kelley, and Robert Paulsen. 1996. Stockpile Surveillance: Past and Future. Sandia, SAND95-2751.

Ketter, Clarke. No Date. 3084th Aviation Depot Squadron. Memories of Experiences on Stony Brook Air Force Station: An Operational Storage Site of the United States Air Force 1954 - 1971. http://www.usafnukes.com/Documents/Memories%20including%20Kepler.pdf

Lemay, James. 1999. A Better Picture of Aging Materials. Science and Technology Review.

Los Alamos National Laboratory. 2009. W88 Pit Certification without Testing. Los Alamos Science and Technology Magazine, August Issue.

Lundberg, Anders W. 1966. High Explosives in Stockpile Surveillance Indicate Consistency. Lawrence Livermore National Laboratory.

Maggelet, Michael H. No Date. USAF Nuclear Weapons Specialists Homepage: Career Field History. http://www.geocities.com/usaf463/course.html

Martz, Joseph C. and Adam J. Schwarz. 2003. Plutonium: Aging Mechanisms and Weapon Pit Lifetime Assessment. Journal of Materials, September Issue.

Morris, Ted A. 2001. Munitions Maintenance Squadrons in the Strategic Air Command during the Cold War. http://www.zianet.com/tmorris/mms.html

National Nuclear Security Administration. 2010. FY 2011 Stockpile Stewardship and Management Plan Summary. DOE.

Oskins, Jim. No Date. Nuclear Weapons Maintenance – the Early Years. 35th MMMS at Biggs AFB, TX, March 1956 - July 1959. http://www.geocities.com/usaf463/biggs.html

Rosengren, J. W. 1986. Stockpile Reliability and Nuclear Test Bans: A Reply to a Critic's Comments. Arlington, Virginia. R&D Associates Report RDA-TR-138522-001.

Russ, Harlow W. 1990. Project Alberta: The Preparation of Atomic Bombs for use in WW II. Exceptional Books Ltd.

Secretary of the Air Force. 2009. Nuclear Weapons Maintenance Procedures. Air Force Instruction 21-204, November 30.

Tyler, James. 2001. Annual Certification takes a Snapshot of Stockpile's Health. Science & Technology Review, July / August Issue.

US Department of Energy / National Nuclear Security Administration. 2010. Annex A, FY 2011 Stockpile Stewardship Program.

Walter, Katie. 2004. A Better Method for Certifying the Nuclear Stockpile. LLNL Science and Technology Review.

Wines, Glenn. 2003. Nuclear Weapons Maintenance in the Early Years. The Mark 5 Bomb.

Zimmerman, Jonathan. 2005. The Trouble with Bubbles: Modeling the Material Degradation Caused by Helium Bubble Growth. Sandia Research Highlight.

COSTS AND CONSEQUENCES

Department of Energy. 2007. Five Year Plan: FY 2008 - FY 2012. Office of the Chief Financial Officer.

Dagget, Stephen. 2010. Costs of Major US Wars. Congressional Research Service, January 29.

Department of Energy. 1997. Linking Legacies: Connecting the Cold War Nuclear Weapons Production Processes to Their Environmental Consequences. The Environmental Management Information Center, DOE/EM-03 I9.

Krieger, David. 2011. The Costs of Nuclear Weapons. Nuclear Age Peace Foundation.

Rumbaugh, Russell and Nathan Cohn. 2012. Resolving Ambiguity: Costing Nuclear Weapons. The Henry L. Stimson Center.

Schwarz, Stephen I. 1998. Atomic Audit: The Costs and Consequences of US Nuclear Weapons since 1940. Brookings Institution Press.

Shapiro, Charles S., Ted F. Harvey, and Kendall Peterson. 1985. Radioactive Fallout. UCRL-93835.

Simon, Steven L. 2006. André Bouville and Charles E. Land. Fallout Risks from Nuclear Weapons Tests and Cancer Risks. American Scientist.

U.S. Congress, Office of Technology Assessment. 1993. Dismantling the Bomb and Managing the Nuclear Materials, OTA-O-572. U.S. Government Printing Office.

MISCELLANEOUS

Ainslie, John. The Future of the British Bomb.
http://www.banthebomb.org/future.doc

Bethe, Hans. 1982. Comments on the History of the H-bomb. Los Alamos Science, Fall Issue.

Bostick, W. D. 2010. Chemical and Radiological Properties Affecting the Control of Tc-99 Contamination during K-25 and K-27 D & D Activities. Material and Chemistry Laboratory. K25-10-050.

Bradbury, Norris. 1983. The Bradbury Years: 1945 - 1970. Los Alamos Science, Winter / Spring Issue.

Carter, Ashton B., John D. Steinbruner, and Charles A. Zraket eds. 1987. Managing Nuclear Operations. The Brookings Institution.

Center for Defense Information. 2000. Nuclear Weapons Database.

Clearwater, John M. 2008. Broken Arrow #1: The World's first Lost Atomic Bomb. Hancock House Publishers Ltd. e-book.

Defense Threat Reduction Agency. 2002. Defense's Nuclear Agency 1947 -1997. US Department of Defense.

Department of Defense Briefing. 1967. History of the Phase Out of Large Yield Bombs.

Dyson, Freeman. 1979. Disturbing the Universe. Basic Books.

Federation of American Scientists. Los Alamos Technical Reports and Publications Web Site. http://fas.org/sgp/othergov/doe/lanl/

Federation of American Scientists. Militarily Critical Technologies List (MCTL) Part II: Weapons of Mass Destruction Technologies. http://www.fas.org/irp/threat/mctl98-2/

Fermi, R. and E. Samra. 1997. Picturing the Bomb. Harry N. Adams Inc.

Gibson, James N. 1986. Nuclear Weapons of the United States: An Illustrated History. Schiffer.

von Hipple, Frank. 1989. The 1969 ACDA Study on Warhead Dismantlement. Princeton University.

IAEA. 2005. International Workshop on Environmental Contamination from Uranium Production Facilities and their Remediation. Proceedings of an international Workshop on Environmental Contamination from Uranium Production Facilities and their Remediation held in Lisbon, 11 - 13 February 2004.

Joint Reporting Structure, Nuclear Weapon Reports. 1995. CJCSM 3150.04.

Joint Strategic Plans Committee. 1953. Military Requirements for Atomic Bombs. JCS 1823/136, May 4.

Jones, Vincent C. 1985. Manhattan: The Army and the Atomic Bomb. Center of Military History United States Army, Special Study.

Large, John. 1995. Dual Capable Nuclear Technology. Large and Associates.

Lathrop, Judith M. 1983. The Oppenheimer Years: 1943 - 1945. Los Alamos Science, Winter / Spring Issue.

Mahaffey, James. 2009. Atomic Awakening: A New Look at the History and Future of Nuclear Power. Pegasus Books.

Manhattan Project Heritage Preservation Association Web Site. http://www.childrenofthemanhattan Project.org/index.htm

National Atomic Museum. No Date. Weapon History Exhibits Factsheet.

National Nuclear Security Agency. Freedom of Information Act Virtual Reading Room. http://www.doeal.gov/opa/FOIAReadRmLinks.aspx

Norris, Robert S. 2002. Racing for the Bomb: General Leslie R. Groves, The Manhattan Project's Indispensable Man. Steerforth Press.

Noshkin, Victor E. and William L. Robison. 1997. Assessment of a Radioactive Disposal Site at Enewetak Atoll. Health Physics, July.

Nuclear Weapon Archive. http://nuclearweaponarchive.org

Nuclear Energy Agency and the International Atomic Energy Agency. 2002. Environmental Remediation of Uranium Production Facilities.

Office of the Deputy Assistant to the Secretary of Defense for Nuclear Matters. 2008. Nuclear Matters: A Practical Guide.

Office of the Deputy Assistant to the Secretary of Defense for Nuclear Matters. 2011. The Nuclear Matters Handbook: Expanded Edition.

Office of Radiation & Indoor Air Radiation Protection Division. 2005. Uranium Location Database Compilation. EPA 402-R-05-009.

Patterson, Walter C. 1976. Nuclear Power. Penguin.

Ragheb, M. 2008. Nuclear World. University of Illinois at Urbana-Champagne.

Ragheb, M. Nuclear Marine Propulsion. 2009. University of Illinois at Urbana-Champagne.

Report of the Secretary General, Department for Disarmament Affairs. 1991. Nuclear Weapons: A Comprehensive Study. United Nations.

Sloop, John L. 1978. Liquid hydrogen as a Propulsion Fuel, 1945 - 1959 (Chapter 4. Hydrogen Technology from thermonuclear Research.) The NASA History Series.

Survey of Weapon Development and Technology (WR708). 1988. Corporate Training and Development. http://www1.cs.columbia.edu/~smb/wr708/wr708.pdf

US Department of Energy, Office of Declassification. Restricted Data Declassifications 1946 to Present.

US Nuclear Regulatory Commission: Our History. No Date. http://www.nrc.gov/who-we-are/history.html

Vaughan, Edgar M. 2007. Recapitalizing Nuclear Weapons. Walker Paper No. 8. Air University Press.

Widner, Thomas. 2010. Final Report of the Los Alamos Historical Document Retrieval and Assessment Project. Center for Disease Control and Prevention.

NUCLEAR DELIVERY SYSTEMS / TECHNOLOGY

ARMY WEAPON SYSTEMS

Aberdeen Archives. 1952. Program for the Presentation of the Army's 280mm Gun, October 15.

Aberdeen Proving Grounds Bulletin 232. No Date. Gun, Heavy Motorized, 280mm, M65, Carriage, 280mm, M30.

Arkin, William H. 1985. Nuclear Backpack. Bulletin of the Atomic Scientists, April Issue.

The Artillery School. No Date. MAT 324.1. Description of the 280mm Gun.

Bell Laboratories. 1975. ABM Research and Development at Bell Laboratories: Project History.

Bell Laboratories. 1975. Safeguard Data Processing System. Bell System Technical Journal Special Supplement.

Bragg, James W. 1961. Development of the Corporal: Embryo of the Army Missile Program. Historical Monograph No. 4, Vol. I.

Bragg, James W. 1961. Development of the Corporal: Embryo of the Army Missile Program. Historical Monograph No. 4, Vol. II.

Brookings Institute. No Date. Atomic Demolition Munitions. http://www.brook.edu/FP/projects/nucwcost/madm.htm

Brookings Institute. No Date. The Davy Crockett. http://www.brook.edu/FP/projects/nucwcost/DAVYC.HTM

Bullard, John W. 1965. History of the Redstone Missile System. US Army Missile Command Historical Monograph AMC 23M.

Cagle, Mary T. 1959. Development, Production and Deployment of the Nike Ajax Guided Missile System 1945 - 1959. US Army Missile Command Historical Monograph.

Cagle, Mary T. 1964. History of the Basic (M31) Honest John Rocket (U) 1950 - 1964. US Army Missile Command Historical Monograph, AMC 7 M, Part I.

Cagle, Mary T. 1962. History of the Lacrosse Guided Missile System (U) 1947 - 1962. US Army Missile Command Historical Monograph.

Cagle, Mary T. 1967. History of the Little John Rocket System, 1953 - 1966. US Army Missile Command Historical Monograph.

Cagle, Mary T. 1973. History of the Nike Hercules Weapon System. US Army Missile Command Historical Monograph.

Cagle, Mary T. 1971. History of the Sergeant Weapon System. US Army Missile Command Historical Monograph AMC 55M.

Cagle, Mary T. and Elva W. McLin. 1965. History of the Improved (M50) Honest John Rocket (U) 1954 - 1965. US Army Missile Command Historical Monograph, AMC 7 M, Part II.

Chrysler Corporation, Missile Division. No Date. This is Redstone.

Connors, Chris. The Armored Fighting Vehicle Database.
http://afvdb.50megs.com/index.html

The Corporal Program, White Sands Missile Range. No Date.
http://www.wsmr.army.mil/pao/FactSheets/corppr.htm

Corporal Type I, White Sands Missile Range. No Date.
http://www.wsmr.army.mil/pao/FactSheets/ctype1.htm

Corporal Type II, White Sands Missile Range. No Date.
http://www.wsmr.army.mil/pao/FactSheets/ctype2.htm

Department of the Army. 1963. Operation and Employment of the Davy Crockett Weapon Battlefield Missile, XM28/XM29. FM 9-11.

Department of the Army. 1961. Davy Crockett Weapon System in Infantry and Armor Units. FM 23-20.

Department of the Army. 1969. 155mm Howitzer, M109, Self-Propelled. FM 6-88.

Department of the Army. 1962. 155mm Howitzer, M114, Towed. FM 6-81.

Department of the Army. 1968. 175mm Gun, M107, Self-Propelled, and 8-Inch Howitzer, M110, Self-Propelled. FM 6-94.

Department of the Army. 1979. Air Transport Procedures. Transport of Atomic Projectile, M422, By US Army Helicopters. Transport of Atomic Projectile, M422, Complete mission Load by US Army CH47 Helicopters. F 55-218.

Department of the Army. 1979. Air Transport Procedures. Transport of M454 Atomic Projectile by US Army Aircraft. Transport of M454 Nuclear Projectile Complete Mission Loads by US Army CH47 Helicopter. FM 55-204.

Department of the Army. 1983. Air Transport Procedures. Transport of XM753 Nuclear Projectile by US Army Helicopter. Transport of XM75 Nuclear Projectile Complete Mission Loads by US Army CH47 Helicopter. FM 55-220

Department of the Army. 1984. Employment of Atomic Demolition Munitions. FM 5-106.

Department of the Army. 1978. Field Artillery Battalion, Lance. FM 6-42.

Department of the Army. 1968. Field Artillery Honest John / Little John Rocket Gunnery. FM 6 40-1.

Department of the Army. 1962. Field Artillery Missile, Redstone. FM 6-35.

Department of the Army. 1960. Field Artillery Missile, Redstone Firing Procedures. FM 6-36.

Department of the Army. 1962. Field Artillery Missile, Sergeant. FM 6-38.

Department of the Army. 1960. Field Artillery Missile Group, Redstone. FM 6-25.

Department of the Army. 1990. Nuclear Weapons Specialist, Maintenance Specialist MOS55G, Skill Levels 1 and 2. FM9-55G1/2.

Department of the Army. 1981.Nuclear Weapons Electronics Specialist, Skill Levels 1 and 2. MOS35F. FM9-35F

Department of the Army. 1961. Operator and Organizational Maintenance Manual: Weapon System, Atomic, Battle Group, M28 (XM28) (Lightweight) (Portable and Vehicle-

Mounted) and Weapons System, Atomic, Battle Group, M29 (XM29) (Heavy) (Vehicle-Mounted). TM 9-1000-209-12

Department of the Army. 1986. Pershing II Firing Battery. FM 6-11.

Department of the Army. 1985. Pershing II Weapon System: Operator's Manual. TM 9-1425-386-10-1.

Department of the Army. 1959. The Redstone Missile System.

Fagen, M. D. ed. 1978. A History of Engineering and Science in the Bell System. Vol. 2: National Service in War and Peace 1925 - 1975. Bell Laboratories
M65 280mm "ATOMIC CANNON" V Corps Weapon in Support of the 3AD during 1957 - 1963. http://3ad.com/

Field Artillery in the European Theater. http://www.usarmygermany.com/Sont.htm

Forrest, R. M. ed. M29 "Davy Crockett" Nuclear Delivery System. http://3ad.com/history/cold.war/nuclear.index.htm

Gawell, Mike. 2003. Lance Tactics and Organization. http://www.scaleworkshop.com/workshop/m667mm_1.htm

Grimwood, James M. and Frances Strowd. 1962. History of the Jupiter Missile System. Historical Monograph.

Hemphill, Donald F. 1972. Communications for Safeguard. Air University Report 4606.

Historical Branch, Rock Island Arsenal. 1964. Project Management of the Davy Crockett Weapon Systems 1958 - 1962.

Hopkins, Norman B. 1960. Artillery's "Topkick" Missile – Sergeant. Field Artillery Magazine, November Issue.

Jolliff, Elizabeth C. 1974. History of the Pershing Weapon System. Historical Monograph AMC 76M, US Army Missile Command, Redstone Arsenal, Alabama.

Keller, Morris J. 1960. Little John: The Mighty Mite. Field Artillery Magazine, July Issue.

Kimball, Robert H. 1979. Lance Tactical Concepts: Positioning and Movement. Field Artillery Magazine, July - August Issue.

Kinard, Jeff. 2007. Artillery: A History of its Impact. ABC-CLIO Inc.

Lockwood, Myron D. 1964. Sergeant Electrodynamics. Flight International, April 23 Issue.

Nike Missile System. The Nike Historical Society Website. http://www.nikemissile.org/

Martin Marietta Aerospace. 1974. Pershing IA System Description.

McKenney, Janice E. 2007. The Organizational History of Field Artillery 1775 - 2003. US Army Center for Military History.

McMullen, Richard F. History of Air Defense Weapons, 1946 - 1962. ADC Historical Study No. 14. USAF Air Defense Command.

McMullen, Richard F. 1963. Interceptor Missiles, 1962 - 1963. ADC Historical Study 18, USAF Air Defense Command.

Artillery Trends. 1964. A Picture of Sergeant. December Issue.

Mentzer, William R. 1998. Test and Evaluation of Land Mobile Missile Systems. Johns Hopkins APL Technical Digest. Winter Issue.

Missile Defense Agency. 2006. Nike Zeus: America's first Anti-Ballistic Missile.

Morgan, Mark L. and Mark A. Berhow. 2002. Rings of Supersonic Steel: Air Defenses of the United States Army: 1950 - 1979. Hole in the Head Press.

National Infantry Museum, Davy Crockett Weapon System. https://www.infantry.army.mil/museum/inside_tour/photo_tour/18_davy_crockett.htm

Ney, Virgil. 1969. Evolution of the U. S. Army Division, 1939 - 1968. Fort Belvoir, Virginia: Headquarters, U. S. Army Combat Development Command.

Novak, David. The Stanley R. Mickelsen Safeguard Complex Site. http://srmsc.org/

Nuclear ABMs of the USA.
http://www.paineless.id.au/missiles/

O'Connell, John P. 1959. Organization and Tactics for Field Artillery Missile Units. Field Artillery Magazine, February Issue.

Redstone Arsenal. No Date. Historical Summary of the Lance Missile System.

Rocketdyne Division Rockwell International. No Date. The Lance Rocket Engine.

Rocketdyne Division Rockwell International. No Date. Santa Susana Test Facility.

Rosenkranz, Robert B. 1979. The "Nuclear" ARTEP: An Idea Whose Time has Come. Field Artillery Magazine, August Issue.

Ryan, Jim. My Army Redstone Missile Days. No Date. http://www.myarmyredstonedays.com/

Satterfield, Paul H. and David S. Akens. 1958. The Army Ordnance Satellite Program. Historical Monograph.

Schwarz, Stephen I. 1997. Nuclear Davy Crockett. The Washington Times, January 26.

Shearer, Robert L. 1980. Development of Pershing II. Field Artillery Magazine, June Issue.

Somerville, Paul F. 1963. The Section Leader of the Battalion Davy Crockett Section.
https://www.benning.army.mil/monographs/content/Papers%201900%20Forward/STUP6/SomervillePaulF%20%20CPT.pdf

Strobridge, Truman R. and Bernard C. Nalty. 1980. The Roar of the 8-Incher. Field Artillery Magazine, March Issue.

Thelen, Ed. Nike Missile Web Site.
http://ed-thelen.org/index.html

Thesing, John W. 663rd Field Artillery BN (280mm) US 8th Army.
http://www.koreaatourofduty.org/8thArmy/663rdFABN280mm.html

US 3rd Armored Division History Website.
http://www.3ad.com/

US Army. 1975. History of Strategic Air and Ballistic Missile Defense, 1945 - 1955: Volume I.

US Army. 1975. History of Strategic Air and Ballistic Missile Defense, 1956 - 1972: Volume II.

US Army Artillery and Missile School. 1959. Lacrosse – From Bunker Busting to General Support. Artillery Trends, April Issue.

US Army Field Artillery School. 1967. History of the Field Artillery School Fort Sill, Oklahoma: Volume IV 1958 - 1967. USAFAS/MSL/FS400005.

US Army Field Artillery School. 1970. The Sergeant Guided Missile System M-2700.

US Army in Germany Website. http://www.usarmygermany.com/Sont.htm

US Army Ordnance Corps and General Electric Company. 1959. Hermes Guided Missile Research and Development Project, 1944 - 1954.

AIR FORCE WEAPON SYSTEMS

Geodetic Survey Squadron (Missile).
http://www.geocities.com/harlyd13/gss/history/gsshistory.html

321st Missile Wing at Grand Forks: Cold War Legacy.
http://srmsc.org/pdf/005201p0.pdf

38th Tactical Missile Wing.

http://www.mace-b.com/38TMW/

Alling, Frederick A. 1955. History of modification of USAF Aircraft for Atomic Weapon Delivery, 1948 - 1954. Historian, Historical Division, Office of Information Services, Air Materiel Command, Wright-Patterson Air Force Base.

AMMS Alumni. AGM-28 Hound Dog Missile History / Data. http://www.ammsalumni.org/html/agm-28_history_data.html

Baker, Jim. 2009. MMRBM – WS325A. AAFM Newsletter, March Issue.

Bell Aircraft Corporation. 1950. Rascal MX-776B Design Proposal.

The Boeing Corporation. 1973. Technical Order 21M-LGM30G-1-1: Minuteman Weapon System Description. Boeing Aerospace.

The Boeing Corporation. 1973. Technical Order 21M-LGM30G-1-22: Minuteman Weapon System Operations. Boeing Aerospace.

The Boeing Corporation. 1994. Technical Order 21M-LGM30G-2-1-7: Organizational Maintenance Control, Minuteman Weapon System. Boeing Aerospace.

The Boeing Corporation. 2007. Technical Order 21M-LGM30G-2-1-19: Launch Facility Personnel Access Systems (After ICBM Security Modernization Program). Boeing Aerospace.

The Boeing Corporation. 2001. Technical Order 21M-LGM30G-2-28-1: Ancillary Mechanical Systems VAFB, Wing 1, Squadron 4. Boeing Aerospace.

The Boeing Corporation. 2006. Technical Order 1B-B52H-1 Flight Manual.

The Boeing Corporation. 2006. Technical Order 1B-B52H-1-12: Radar Navigator's / Navigator's Manual.

The Boeing Corporation. 2006. Technical Order 1B-B52H-1-13 Electronic Warfare Officer's Manual.

Boelling, Donald. Titan II ICBM History Web Site. http://www.titan2icbm.org/

Bowman, Martin. 2006. Stratofortress: The Story of the B-52. Pen and Sword.

Boyne, Walter J. 2000. The Man Who Built the Missiles. Air Force Magazine Online, October Issue.

Buchonnet, Daniel. 1976. MIRV: A Brief History of Minuteman and Multiple Reentry Vehicles. Lawrence Livermore Laboratory.

Charles, H. ed. 1971. Proceedings of the Second Meeting of the Minuteman Computer Users Group. Systems Laboratory Report TSL-3-71.

Chun, Clayton K. S. 2000. Shooting Down a Star: Program 437, the US Nuclear ASAT System and Present-Day Copycat Killers. Air University Press.

Coffey, Thomas M. 1988. Iron Eagle: The Turbulent Life of General Curtis E. LeMay. Avon Books.

Congressional Budget Office. 1988. The B-1B Bomber and Options for Enhancements. Special Study.

Davies, Steve and Doug Dildy. 2007. F-15 Eagle Engaged: The World's most Successful Jet Fighter. Osprey Publishing.

Del Papaya, Michael E. 1975. Strategic Air Command Missile Chronology 1939 - 1973. Office of the Historian, Headquarters Strategic Air Command.

Department of Defense. 1983. Strategic Forces Technical Assessment Review.

Department of the Air Force. 1999. US Air Force White Paper on Long-range Bombers Forecast International. 1991. MGM-118A Peacekeeper.

Gandy, C. L. and I. B. Hanson. 1963. Mercury Atlas Launch Vehicle Development and Performance. NASA SP-45, Chapter 5.

Gardner, Trevor. 1957. US Tries Hard to Catch Up, But We Are Still Lagging: What We Can Do About It. Life Magazine, November 4 Issue.

Gertler, Jeremiah. 2011. F-35 Joint Strike Fighter (JSF) Program: Background and Issues for Congress. Congressional Research Service, April 26.

Gibson, James N. 2000. The Navaho Missile Project: The Story of the "Know-How" Missile of American Rocketry. Schiffer Publishing.

Goodchild, Geoff. 2017. Thor: Anatomy of a Weapon System. Fonthill Media.

Greene, Warren E. 1962. The Development of the SM-68 Titan. Historical Office Air Force Systems Command.

Greenwood, John T. 1979. Space and Missile Systems Organization, a Chronology: 1954 - 1979. Air Force History Support Office.

Hall, Al. 2006. The AIR-2A Genie. USAF Nuclear Weapons Specialists Nuclear Weapons Historical Series.

Hartt, Julian Norris. 1961. The Mighty Thor; Missile in Readiness. Duell, Sloan and Pearce.

Headquarters Air Force Safety Center; Weapons, Space and Nuclear Safety Division. 1997. Operational Safety Review of the F-15E and F-16C/D Weapon System. USAF NWSSG 97-1.

Headquarters Air Force Space Command. 2002. Final Mission Needs Statement (MNS) Land-Based Strategic Nuclear Deterrent. AGSPC 001-00.

Headquarters Strategic Air Command. 1980. Strategic Air Command Weapons Acquisition 1964 - 1979. Volumes I - IV. Office of the Historian Headquarters Strategic Air Command.

Headquarters, United States Air Force. 2009. United States Air Force Unmanned Aircraft Systems Flight Plan 2009 - 2047.

Historical Division, Office of Information, Aeronautical Systems Division, Air Force Systems Command. 1961. Development of Airborne Armament, 1910 - 1961, Volume I, Bombing Systems.

Historical Division, Office of Information, Aeronautical Systems Division, Air Force Systems Command. 1961. Development of Airborne Armament, 1910 - 1961, Volume III, Fighter Fire Control.

ICBM Prime Team. 2001. Minuteman Weapon System History and Description. TRW Systems.
    Jackson, Robert. No Date. Thor Missile Deployment in the UK. Strike Force - The USAF in Britain since 1948.
    http://harringtonmuseum.org.uk/ThorUK.htm

Jenkins, Dennis R. 2008. Magnesium Overcast: The Story of the Convair B-36. Specialty PR Pub & Wholesalers.

Jenkins Dennis R. and Don Pyeatt. 2010. Cold War Peacemaker: The Story of Cowtown and Convair's B-36. Specialty Publishing.

Knaack, Marcel Size. 1978. Encyclopedia of US Air Force Aircraft and Missile Systems. Volume 1. Post-World War II Fighters 1945 - 1973. Office of Air History.

Knaack, Marcel Size. 1978. Encyclopedia of US Air Force Aircraft and Missile Systems. Volume 1. Post-World War II Bombers 1945 - 1973. Office of Air History.

Kohut, F. A. and J. Crompton. 1954. Aircraft modifications for the MK 17 and the MK 24 Atomic Bombs. Sandia Corporation, SC-3417 (TR)

Lanning, Randall L. 1992. United States Air Force Ground Launched Cruise Missiles: A Study in Technology, Concepts, and Deterrence. Air War College.

Leitenberg, Milton. No Date. Case Study 1: The History of US Anti-Satellite Weapons.

Lewis, Donald E., Bruce W. Don, Robert M. Paulson, and Willis H. Ware. 1986. A Perspective on the USAFE Collocated Operating Base System. The RAND Corporation.

Little, R. D. A. 1959. History of the Air Force Atomic Energy Program 1943 - 1953. Introduction and Chapter I. Project Silverplate 1943 - 1946. Air University Historical Liaison Office.

Little, R. D. A. 1959. History of the Air Force Atomic Energy Program 1943 - 1953. Volume II. Foundations of an Atomic Air Force and Operation Sandstone 1946 - 1948. Air University Historical Liaison Office.

Little, R. D. A. 1959. History of the Air Force Atomic Energy Program 1943 - 1953. Volume III. Building an Atomic Air Force. Air University Historical Liaison Office.

Little, R. D. A. 1959. History of the Air Force Atomic Energy Program 1943 - 1953. Volume IV. The Development of Weapons. Air University Historical Liaison Office,

Little, R. D. A. 1959. History of the Air Force Atomic Energy Program 1943 - 1953. Volume V. Atomic Delivery Systems. Air University Historical Liaison Office.

Lonnquest, John C. and David F. Winkler. 1997. To Defend and Deter: The Legacy of the United States Cold War Missile Program. USACERL Special Report 97/01.

Lloyd, Alwyn T. 2000. A Cold War Legacy: A Tribute to Strategic Air Command – 1946 - 1992. Pictorial Histories Publishing Company.

Lloyd, Alwyn T. 2006. Boeing's B-47 Stratojet. Specialty Press.

Logan, Don. 1998. General Dynamics F-111 Aardvark. Schiffer Publishing.

Mann, Robert A. 2009. The B-29 Superfortress Chronology: 1934 - 1960. McFarland.

The Martin Company. 1960. Titan I and Titan II.

Medalia, Jonathan E. 1977. MX Intercontinental Missile Program. Foreign Affairs and National Defense Division. Issue Brief IB77080.

Miller, J. 1997. Convair B-58 Hustler: The World's First Supersonic Bomber. Midland Publishing.

Mindling, George and Robert Bolton. 2009. US Air Force Tactical Missiles, 1949 - 1969: The Pioneers. Lulu.com.

The Minuteman ICBM System. No Date. National Park Service. http://www.nps.gov/archive/mimi/history/srs/history.htm

Missile Base Exploration. No Date. http://www.siloworld.net/

Mitchell, Douglas D. 1982. Bomber Options for Replacing B-52s. Foreign Affairs and National Defense Division Issue Brief Number IB81107.

Morris, Ted A. 2000. Flying the Aluminum and Magnesium Overcast. http://www.zianet.com/tmorris/b36.html

Moore James T. 1981. Low-altitude Defense: An Analysis of its Effect on MX Survivability. Air Force Institute of Technology, Wright-Patterson School of Engineering. Master's Thesis.

Nalty, Bernard C. 1966. The Quest for an Advanced Manned Strategic Bomber: UASF Plans and Policies, 1961 - 1966. USAF Historical Division Liaison Office.

Nalty, Bernard C. 1965. USAF Ballistic Missile Programs 1962 - 1964. USAF Historical Division Liaison Office.

Nalty, Bernard C. 1967. USAF Ballistic Missile Programs 1964 - 1966. USAF Historical Division Liaison Office.

Nalty, Bernard C. 1969. USAF Ballistic Missile Programs 1967 - 1968. Office of Air Force History.

Narducci, Henry M. 1988. Strategic Air Command and the Alert Program: A Brief History. Office of the Historian, Headquarters, Strategic Air Command.

Narducci, Henry M. ed. SAC Missile Chronology 1949 - 1988. Office of the Historian, Headquarters, Strategic Air Command, 1989.

Neal, Roy. 1962. Ace in the Hole: The Story of the Minuteman Missile. Doubleday & Company.

Neufeld, Jacob. 1990. The Development of Ballistic Missiles in the United States Air Force 1945 - 1960. Office of the Air Force History.

Neufeld, Jacob. 1971. USAF Ballistic Missile Programs 1969 - 1970. Office of Air Force History.

Noonan, John. (2011) In Nuclear Silos, Death Wears a Snuggie.
https://www.wired.com/2011/01/death-wears-a-snuggie/

Office of the Historian, Headquarters, Strategic Air Command. 1991. Alert Operations and the Strategic Command: 1957 - 1991.

Office of the Historian, Headquarters, Strategic Air Command. 1990. From Snark to Peacekeeper: A Pictorial History of Strategic Air Command Missiles. University of Michigan Library.

Office of the Historian, Headquarters, Strategic Air Command. 1990. SAC Missile Chronology, 1939 - 1988.

Office of the Armament Product Directorate. 2004. 2003 - 2004 Weapons File.

Office of Technology Assessment. 1981. MX Missile Basing. NTIS order #PB82-108077

Oskins, Jim. 2002. History of the Snark Missile.
http://www.geocities.com/usaf463/SNARK.html

Park, Wally Lee. 2006. Standing Watch: Deer Park's Atlas ICBM. Clayton Historical Society.

Parsch, Andreas. 2002 – 2009. Designations of US Air Force Projects.
http://www.designation-systems.net/usmilav/projects.html#_MX

Patillo, Donald M. 2000. Pushing the Envelope: The American Aircraft Industry. The University of Michigan Press.

Petersen, Nikolaj (2008). The Iceman That Never Came: Project Iceworm, the Search for a NATO Deterrent, and Denmark, 1960 - 1962. Scandinavian Journal of History, March Issue.

Penson, Chuck. 2008. The Titan II Handbook: A Civilian's Guide to the Most Powerful ICBM America Ever Built. Self-Published.

Piper, Robert F. 1962. The Development of the SM-80 Minuteman. DCAS Historical Office, Deputy Commander for Aerospace Systems, Air Force Systems Command.

Pomeroy, Steven 2006. Echoes That Never Were: American Mobile Intercontinental Ballistic, Missiles, 1956 - 1983. Ph.D. Auburn University.

Ray, Thomas W. 1963. BOMARC and Nuclear Armament, 1951 - 1963. Air Defense Command Historical Study No. 21.

Ray, Thomas W. 1963. Nuclear Armament: Its Acquisition, Control, and Application to Manned Interceptors 1951 - 1963. ADC Historical Study No. 20.

Rendall, Ivan. 2000. Rolling Thunder: Jet Combat from World War II to the Gulf War. Dell Publishing.

Reynolds, Bobby J. 1968. On Alert with the 308th. SAC Combat Crew Magazine.

Robinson, John C. No Date. AFSWC's Part in Atomic Warhead – Guided Missile Marriages, 1952 - 1955. AFWSC.

Rosenberg, Max. 1960. Plans and Policies for the Ballistic Missile Initial Operational Capability Program. USAF Historical Division Liaison Office.

Rosenberg, Max. 1960. USAF Ballistic Missiles 1958 - 1959. USAF Historical Division Liaison Office.

Saunders, Dudley F. No Date. The Development of Thermonuclear Weapon Delivery Techniques. Air Force Special Weapons Command.

Schaff, Jeff, Lambeth Blalock, Matt Bille, and Stan Bailey. 2000. Future Ballistic Missile Requirements: A First look. American Institution of Aeronautics and Astronautics.

Schwiebert, Ernest G. 1965. A History of the US Air Force Ballistic Missiles. Praeger.

Senior, Tim. 2002. The Air Force Book of the F-16 Fighting Falcon. Osprey Publishing.

Shattuck, J. Wayne. 1992. Technology and the Peacekeeper. American Institution of Aeronautics and Astronautics Space Programs and Technologies Conference, March 24 - 27, Huntsville, Alabama.

Sherman, Tom. 1961. Hound Dog Streaks to Target Guided by Stars. Popular Science, June Issue.

Simpson, Charlie. 2012. Launch Pads, Gantries, Shelters, Coffins, Silos, TELs, and Shelters. AAFM Newsletter, June Issue.

Simpson, Charlie. 2005. The Early Missiles: Snark, Navaho, BOMARC and Goose. AAFM Newsletter, June Issue.

Simpson, Charlie. 2004. GLCM - From Concept to Mission Complete. AAFM Newsletter, December Issue.

Simpson, Charles G. 1998. History of Air-launched Missiles. AAFM Newsletter, September Issue.

Slattery, Christina, Mary Ebeling, Erin Pogany, and Amy R. Squitieri. 2005. Minuteman Missile Historic Resource Study.

Slattery, Christina, Mary Ebeling, Erin Pogany, and Amy R. Squitieri. 2003. The Missile Plains: Frontline of America's Cold War. United States Department of the Interior Historical Resource Study.

Smith, Richard K. 1998. Seventy-Five Years of Air Force In-flight Refueling. Air Force History and Museums Program.

Spick, Mike. 2006. Brassey's Modern Fighters: The Ultimate Guide to In-Flight Tactics, Technology, Weapons, and Equipment. Potomac Books Inc.

Stanbery, Charles E. et al 1975. AGM 69A SRAM Explosive Components Surveillance Program. Summary Report and FY74 Service Life Estimate. Aeronautical Systems Division Wright-Patterson Air Force Base, Ohio.

Stine, George Harry. 1991. ICBM: The Making of the Weapon that Changed the World. Orion.

Strat-X Volume 4. 1997. Design - Land Mobile Missile Systems. Report R-122, Institute for Defense Analysis Research and Engineering Support Division.

Strat-X Volume 12. 1967. Reaction - Land Mobile Missile Systems. Report R-122, Institute for Defense Analysis Research and Engineering Support Division.

Stumpf, David K. 1996. Titan II: The History of a Cold War Missile Program. Turner Publishing.

Stumpf, David K. 1993. Titan II ICBM Missile Site 8 (571-7). National Historic Landmark Nomination.

Headquarters, US Tactical Air Command. 1952. History of the Tactical Air Command, 1 July through 31 December 1951. Volume VII, Special Weapons Activities.

Tagg, Lori S. 2004. The Development of the B-52: The Wright Field Story. History Office Aeronautical Systems Center Air Force Materiel Command.

Taube, L. J. 1972. B-70 Aircraft Study Final Report Volume I. North American Rockwell.

Thornborough, Anthony. 1995. Modern Fighter Aircraft Technology and Tactics: Into Combat with Today's Fighter Pilots. Patrick Stephens.

Trester, Delmer J. 195. Atomic Weapon Delivery Developments for Fighter Aircraft. Historical Division, Office of Information Services, Wright Air Development Center, Air Research and Development Command.

Twigge, Stephen and Len Scott. 2005. The Other Missiles of October: The Thor IRBMs and the Cuban Missile Crisis. Electronic Journal of International History.

United States Air Force. 1960. Military Specification Clip-in Subassembly MHU-20/C. MIL-C-25830.

United States Air Force. 1961. Military Specification Clip-in Subassembly MHU-29/C. MIL-C-27027.

United States Air Force. 1986. AGM-69 Short Range Attack Missile. Fact Sheet 86-61.

United States Air Force. 2008. Report in ICBM Industrial Base Capabilities to Maintain, Modernize, and Sustain Minuteman III Through 2030 and Provide a Replacement Land-based Strategic Deterrent System after 2030.

United States Air Force. 1963. TM76A Guided Missile: Launch Area Operations. T.O. 21-TM76A-1-2.

United States Air Force. 1962. TM76B Guided Missile: Launch Area Operating Instructions. T.O. 21-TM76B-1-4

United States General Accounting Office. 1982. The Costs and Benefits of a Common Strategic Rotary Launcher should be Reassessed before Further Funds Are Obligated. GAO-MASAD-83-3.

United States General Accounting Office. 1997. B-2 Bomber: Costs and Operational Issues. GAO/NSIAD-97-181.

US Army Engineer Research and Development Center and National Park Service. 2002. Historic American Engineering Record of Space Launch Complex 10, Vandenberg Air Force Base, California. Historic American Engineering Record, Pacific Great Basin Support Office, National Park Service.

US National Committee on Tunneling. 1982. Design and Construction of Deep Underground Basing Facilities for Strategic Missiles. Report of a Workshop Conducted by the U.S. National Committee on Tunneling Technology, Commission on Engineering and Technical Systems, National Research Council, Volume 2.

USAF. 1964. Technical Manual CGM-16E Missile Weapon System Operational Manual. T.O.21M-CGM16E-1.

USAF. 1964. Technical Manual HGM-16F Missile Weapon System Operational Manual. T.O.21M-HGM16F-1.

USAF. 1964. Technical Manual HGF-25A Missile Weapon System Operational Manual. T.O.21M-HGF-25A-1.

USAF. 1964. Technical Manual LGM-25C Missile Weapon System Operational Manual. T.O.21M-LGM-25C-1.

USAF 2012 ALMANAC. Guide to Air Force Installations Worldwide.

USAF 2012 ALMANAC. Major Commands.

Van Staaveren, Jacob. 1964. USAF Intercontinental Ballistic Missiles, Fiscal Years 1960 - 1961. USAF Historical Division Liaison Office.

Walker, Chuck, with Joel Powell. 2005. Atlas - The Ultimate Weapon. Apogee Books.

Webster B. D., L. C. Yang, and Charles Pyles. 1995. Evolution of Ordnance Subsystems and Components Design in Air Force Strategic Missile Systems. American Institute of Aeronautics and Astronautics.

Weitze, Karen J. 1999. Cold War Infrastructure for Strategic Air Command: The Bomber Mission. Air Combat Command.

Weitze, Karen J. 1999. Cold War Infrastructure for Air Defense: The Fighter and Command Missions. Air Combat Command.

Withington, Thomas and Mark Styling. 2006. B-1B Lancer Units in Combat. Osprey Publishing.

Withington, Thomas and Mark Styling. 2006. B-2 Spirit Units in Combat. Osprey Publishing.

Wohlstetter, A. J., F. S. Hoffman, R. J. Lutz, and H. S. Rowen. 1954. Selection and Use of Strategic Airbases. The RAND Corporation R-266.

Worman, Charles G. 1967. History of the GAM-87 Skybolt Air-to-Surface Ballistic Missile. Historical Division, Information Office, Aeronautical Systems Command.

NAVY WEAPON SYSTEMS

Alldridge, Bob. 2002. US Trident Submarine and Missile System: The Ultimate First Strike Weapon. PLRC-01111-7D.

Barlow, Jeffrey G. 1994. Revolt of the Admirals: The Fight for Naval Aviation, 1945 - 1950. Washington: Naval Historical Center.

Barnes, James, P. 1999. A Dive's Eye View. Undersea Warfare, Fall Issue.

Batchelor, J., Preston, A., and L. S. Casey. 1979. Sea Power: A Modern Illustrated History. Phoenix Publishing.

Bureau of Naval Personnel Information Bulletins. 1957. Guided Missile Navy. All Hands Magazine, March Special issue.

Congressional Budget Office. 1993. Rethinking the Trident Force. Staff Working Paper.

Congressional Budget Office. 1998. Trident II Missile Test Program. Staff Working Paper.

Congressional Budget Office. 1987. Trident II Missile Test Program: Implications for Arms Control. Staff Working Paper.

Cote, Owen R. Jr. 2000. The Third Battle: Innovation in the US Navy's Silent Cold War Struggle with Soviet Submarines. MIT Press.

Ford, Dan. Douglas A1 Skyraider as Nuclear Bomber. http://www.warbirdforum.com/toss.htm

Gyrodyne. ASROC and QH-50 DASH. http://www.gyrodynehelicopters.com/

Spinardi, Graham. 2003. From Polaris to Trident: The Development of Fleet Ballistic Missile Technology. Cambridge University Press.

Friends of Albacore. USS Albacore: Forerunner of the Future. http://www.ussalbacore.org/

Friedman, Norman. 1982. US Naval Weapons. Naval Institute Press.

Friedman, Norman. 1994. US Submarines since 1945. Naval Institute Press.

Handler, Joshua and William N. Arkin. 1990. Nuclear Warships and Naval Nuclear Weapons 1990: A Complete Inventory. Greenpeace.

Hayes, Phillip R. 2011. History of the Talos Missile. USS Oklahoma City Website. http://www.okieboat.org/Talos%20history.html

Hughey, Charles M. POMFLANT Remembered. http://www.multiwebs.net/pr/history.html

Johns Hopkins University. 1950. Semiannual Report of Bumblebee Project, July - December 1949. Bumblebee Series Report 123.
Lewis, Jeffrey. 2009. A Problem with Nuclear Tomahawk. New America Foundation.
Lockheed Missiles and Space Company Inc. 1989. A History of the FBM System. LMSC-F255548.
MacKenzie, Donald and Graham Spinardi. 1998. The Shaping of Nuclear Weapon System Technology: US Fleet Ballistic Missile Guidance and Navigation. Part I. From Polaris to Poseidon. Social Studies of Science, March Issue.
MacKenzie, Donald and Graham Spinardi. 1998. The Shaping of Nuclear Weapon System Technology: US Fleet Ballistic Missile Guidance and Navigation: Part II Going for Broke – The Path to Trident II. Social Studies of Science. April Issue.
Milford, Frederick J. 1966. US Navy Torpedoes. The Submarine Review.
Miller, Jerry. 2001. Nuclear Weapons and Aircraft Carriers: How the Bomb Saved Naval Aviation. Smithsonian Institution Press.
Miralgia, Sebastien. 1998. Nuclear Development and Military Technology: The Case of the Fleet Ballistic Missile Programme. Norwegian Institute for Defense Studies.
Nau, Evan D. 1998. The Bumblebee Project.
http://www-personal.umich.edu/~buzznau/bmblbee.html
Naval Historical Center. History of Submarine Technology Website.
https://wrc.navair-rdte.navy.mil/warfighter_enc/History/Subs/histtech/tech3.htm
Naval Training Command. 1972. Gunners Mate 2&3 Rate Training Manual. NAVTRA 10199-B, United States Government Printing Office.
Naval Training Command. 1972. Gunners Mate 1&C Rate Training Manual. NAVTRA 10199-B, United States Government Printing Office.
Nelson, Warren E. 1950. Wind Tunnel Investigations of 1/6-Scale Model of the Bumblebee XPM Missile at High Subsonic Speeds. Ames Aeronautical Laboratory.
O'Rourke, Ronald. 2010. Navy SSBN(X) Ballistic Missile Submarine Program: Background and Issues for Congress. Congressional Research Service, July 27.
Pakistan Military Consortium. RUR-5A ASROC/RUM-139 VLA.
http://www.pakdef.info/pakmilitary/navy/asroc.html
Polmar, Norman. 2003. Cold War Submarines: The Design and Construction of US and Soviet Submarines, 1945 - 2001. Potomac Books Inc.
Polmar, Norman. 2003. The Polaris: A Revolutionary Missile System and Concept.
http://www.history.navy.mil/colloquia/cch9d.html
Polmar, Norman. 2005. Naval Institute Guide to Ships and Aircraft of the US Fleet, 18th Edition. Naval Institute Press.
Polmar, Norman and Donald M. Kerr. 1986. Nuclear Torpedoes. Naval Institute Proceedings.
Ponton, David A., James V. Tyler, Donald D. Tipton, Douglas R. Henson, Karl B. Rueb and Barry W. Hannah. 1994. Joint DOD/DOE Trident MK4 / MK5 Reentry Body Alternate Warhead Phase 2 Feasibility Study Report.
Program Executive Officer for Unmanned Aviation and Strike Weapons. 2009. Tomahawk Cruise Missile RGM/UGM-209 System Description, Revision 15. Technical Manual SW820-AP-MMI-010, March 27.
Raytheon. 2008. Deep Siren Communications System.

Refuto, George J. 2011. Evolution of the US Sea-Based Nuclear Missile Deterrent. Xlibris Corporation.
Rockwell, Theodore. 1992. The Rickover Effect: The Inside Story of How Adm. Hyman Rickover Built the Nuclear Navy. John Wiley and Sons.
Sapolsky, Harvey M. 1972. The Polaris System Development. Harvard University Press.
Smith, Henry B. 2007. Keepers of the Dragon. Navy Nuclear Weapons Association. http://www.navynucweps.com/History/kodhistory.htm
Sokolsky, Joel Jeffrey. 1984. The US Navy and Nuclear ASW Weapons. US Naval Institute Proceedings.
Stumpf, David K. 1996. Regulus: America's First Nuclear Submarine Missile. Turner Books.
Sumrall, Robert F. 2000. The Anti-Submarine Rocket. (ASROC). Tin Can Sailors.
System Planning Corporation. 1980. An Assessment of Small Submarines and Encapsulization of Ballistic Missiles – Phase II Survey. SPC 648.
Sword of Damocles. 2009. Sharpening Trident.
Talos Issue. 1982. Johns Hopkins APL Technical Digest, April - June Issue.
Thomas, Valerie. 1989. Verification of Limits on Long-Range Nuclear SLCMs. Science and Global Security, January - February Issue.
Uboataces. 2005. Rocket U-boat Program. http://www.uboataces.com/articles-rocket-uboat.shtml
United States Navy. 1998. Introduction to Naval Weapons Engineering: U/W Acoustics and SONAR. ES310 Course Syllabus.
US Navy Active Ship Force Levels, 1917 - Present. http://www.history.navy.mil/branches/org9-4.htm
US Navy. 1963. ASROC Missile - Description and Instructions for Assembly, Inspection, and Stowage. NAVWEPS OP 2963.
US Navy Training Publications Center. 1959. Principles of Guided Missiles and Nuclear Weapons. NAVPERS 10784.
Waller, Douglas C. 2001. Practicing for Doomsday. Time Magazine, March 4 Issue.
Waterman, Mark D. and B. J. Richter. No Date. Development of the Trident I Aerospike Mechanism. Lockheed Missile and Space Company.
Watson, John M. 1998. The Origin of the APL Strategic Systems Department. Johns Hopkins APL Technical Digest.
Wong D. G. and C. A. Vollerson. 1976. Advanced Material Applications on the Trident I Missile. in. S. W. Tsai ed. Composite Materials: Testing and Design. American Society for Testing and Materials.
Weir, Gary R. 2000. Deep Ocean, Cold War. Undersea Warfare.
Young, Charles. 2003. Statement before the Strategic Subcommittee Senate Armed Services Committee, April 8.

GUIDANCE / CONTROL TECHNOLOGY

Barbour, Neil M., John M. Elwell and Roy H. Setterlund. 1992. Inertial Instruments: Where to Now. AIAA Guidance, Navigation and Control Conference, Hilton Head Island, SC, August 10 - 12.
Bellamy P. 2012. Thor Missile Guidance: What, Where and How. Airfield Information Exchange.

http://www.airfieldinformationexchange.org/community/showthread.php?7436-Thor-Missile-Guidance-What-Where-and-How

General Accounting Office. 2004. Trident II (D5) MK 6 Guidance System. File B-292895.2.

General Electric. No Date. GERTS: General Electric Radio Tracking System. Radar Systems Department.

Grewal, Mohinder S., Lawrence R. Weill, and Angus P. Andrews. 2007. Global Positioning Systems, Inertial Navigation, and Integration. Wiley-Interscience.

Gibson, John P., and Stephen P. Yaneck. 2011. The Fleet Ballistic Missile Weapon System: APL's Efforts for the US Navy's Strategic Deterrent System and the Relevance of Engineering Systems. Johns Hopkins APL Technical Digest, April Issue.

Gulick, J. F. and J. S. Miller. 1982. Missile Guidance: Interferometer Homing using Fixed Body Antennas. Johns Hopkins APL Technical Digest, August Issue.

Hoselton, Gary A. 1989. The Titan I Guidance System. AAFM Newsletter, March Issue.

ION Editorial Review Committee assisted by Jim Noll, Neil Wood and Charles White. No Date. The N5G Inertial Navigation System in the B-52 Hound Dog Missile: A System Overview. (ION).

The Institute of Navigation Virtual Museum (ION).
http://www.ion.org/museum/

Juang, Jeng-Nan and R. Radharamanan. No Date. Evaluation of Ring Laser and Fiber Optic Gyroscope Technology. American Society for Engineering Education.

King, A. D. 1998. Inertial Navigation: Forty Years of Evolution. GR and EC Review.

Lawrence, Anthony. 1998. Modern Inertial Technology: Navigation, Guidance, and Control. Springer Verlag.

Lefebvre, Michael, and Robert Stewart. 2006. History of Altimetry, 1960 - 1992. Ocean Surface Topography Science Team Meeting, Venice.

Mackenzie, Donald. 1992. Inventing Accuracy: A Historical Sociology of Missile Guidance. MIT Press.

McMurran, Marshall William. 2008. Achieving Accuracy: A Legacy of Missiles and Computers. Xlibris Corporation.

McRuer, Duane and Dunstan Graham. 2003. A Flight Control Century: Triumphs of the Systems Approach. Systems Technology Inc.

Pace, Scott, Gerald Frost, Irving Lachow, David Frelinger, Donna Fossum, Donald K. Wassem, and Monica Pinto. 1995. The Global Positioning System: Assessing National Policies. RAND Critical Technologies Institute.

Plotnick, Daniel. 2003. Atlas Guidance. AAFM Newsletter, September, Issue.

Ringlein, Mark J. Neal J. Barnett, and Marvin B. May. 2000. Next Generation Strategic Submarine Navigator. AIAA-2000-XXXX.

Scott, Jeff. 2004. Missile Guidance. Aerospace Web.

Shi-Xue, Tsai. 1996. Introduction to the Scene Matching Missile Guidance Technologies. National Air Intelligence Center.

SINS MK 2 MOD 6. No Date. ION.

Siouris, George M. 2004. Missile Guidance and Control Systems. Springer Verlag.

Silverstone, Paul. 2009. The Navy of the Nuclear Age, 1947 - 2007. Routledge.

Sorenson, H. W. 1990. Range and Guidance Accuracy Capability of the Atlas Missile System. General Dynamics.

Wuerth, J.M. 1976. The Evolution of Minuteman Guidance and Control. Navigation, Spring Issue.

Ekutekin, Vedat. 2007. Navigation and Control Studies on Cruise Missiles. Ph.D. Thesis in Mechanical Engineering, Middle East Technical University.

Yionoulis, Steve M. 1990. The Transit Satellite Geodesy Program. Johns Hopkins APL Technical Digest, January Issue.

PROPULSION

Abbot Aerospace. 1971. Solid Propellant Characterization and Selection. NASA, SP-8064.

Abbot Aerospace. 1971. Solid Propellant Processing Factors in Rocket Motor Design. NASA, SP-8075.

American Institute of Aeronautics and Astronautics. No Date. Thiokol Chemical Division, Elkton Division: Historic Aerospace Site.

Andrepont, W. C. and R. M. Felix. 1994. The History of Large Solid Rocket Motor Development in the United States. AIAA-94-3057, 30th AIAA / ASME / SAE / ASEE Joint Propulsion Conference, June.

Bedard, Andre. No Date. Double-Base Solid Propellants.
http://www.astronautix.com/articles/doulants.htm

Bedard, Andre. No Date. Composite Solid Propellants.
http://www.astronautix.com/articles/comlants.htm

Blomshield, Fred S. 2006. Lessons Learned in Solid Rocket Combustion Instability. Naval Air Warfare Center Weapons Division China Lake, American Institute of Aeronautics and Astronautics, Missile Sciences Conference, Monterey, California, November.

Carroll, P. Thomas. 1974. Historical Origins of the Sergeant Missile Power Plant. Eighth History of Astronautics Symposium of the International Academy of Astronautics, Amsterdam.

Davenas, Alain. 2003. Development of Modern Solid Propellants. Journal of Propulsion and Power, November - December Issue.

Douglass, Howard W. et.al. 1972. Solid Propellant Grain Design and Internal Ballistics, NASA Space Vehicle Design Criteria (Chemical Propulsion). Monograph SP-8076.

Ehresman, Charles M. No Date. Aerojet Engineering Corporation First Manufacturing Plant, Pasadena, California. American Institute of Aeronautics and Astronautics.

Fry, Ronald S. A 2004. Century of Ramjet Propulsion Technology Evolution. Journal of Propulsion and Power, January - February Issue.

Geisler, Robert. 2008. Solid Rocket Motor Overview. University of Alabama, Huntsville.

Guibet, J. C. 1999. Fuels and Engines. Institut Francais du Petrole Publications.

Hunley, J.D. 2007. The Development of Propulsion Technology for US Space Launch Vehicles 1926 - 1991. Texas A&M University Press.

Hunley, J.D. 1999. The History of Solid Propellant Rocketry: What We Do and Do Not Know. American Institute of Aeronautics and Astronautics.

Huzel, Dieter K. and David H. Huang. 1992. Modern Engineering for Design of Liquid-Propellant Engines. American Institute of Aeronautics & Aeronautics.

Kraemer, Robert S. Rocketdyne: 2005. Powering Humans into Space. American Institute of Aeronautics and Astronautics.

Launius, Roger D. and Dennis R. Jenkins eds. 2002. To Reach the High Frontier: A History of US Launch Vehicles. University of Kentucky Press.

Lethbridge, Cliff. 2000. History of Rocketry. Spaceline. http://spaceline.org/rockethistory.html

Leyes, Richard A. and William A. Fleming. 1999. The History of North American Small Gas Turbine Aircraft Engines. Smithsonian Institution Press.

Moore, Thomas L. Allegany Ballistics Laboratory: No Date. Historic Aerospace Site. American Institute of Aeronautics and Astronautics Morton - Thiokol. Rocket Basics. http://www.aeroconsystems.com/thiokol_rocket_basics.htm

North, B. F. and In-Kun Kim. 1981. Thrust Augmentation for Tomahawk Cruise Missile. Proceedings of the Navy Symposium on Aeroballistics. (12th) Held at the David Taylor Naval Ship Research and Development Center, Bethesda, Maryland on 12 - 14 May. Volume I.

Sutton, George P. 2005. History of Liquid Propellant Rocket Engines. American Institute of Aeronautics and Astronautics.

Sutton, George P. and Oscar Biblaz. 2010. Rocket Propulsion Elements, 8th Edition. John Wiley & Sons.

Umholtz, William D. No Date. The History of Solid Rocket Propulsion and Aerojet. Air Force Research Laboratory.

Von Doehrn, Paul J. 1966. Propellant Handbook. Air Force Propulsion Laboratory, Edwards, California.

Waltrup, Paul J., Michael E. White, Frederick Zarlingo, and Edward S. Gravlin. 1997. History of Ramjet and Scramjet Propulsion Development for US Navy Missiles. Johns Hopkins APL Technical Digest, February Issue.

REENTRY VEHICLES / MRV / MIRV / MARV

Allen, Julian H. and A. J. Eggers Jr. 1958. A Study of the Motion and Aerodynamic Heating of Ballistic Missiles Entering the Earth's Atmosphere at High Supersonic Speeds. NACA Report 1831.

American Institute of Aeronautics and Astronautics. No Date. GE Reentry Systems, Philadelphia, Pennsylvania: Historic Aerospace Site. AIAA.

Caston Lauren, Robert S. Leonard, Christopher A. Mouton, Chad J. R. Ohlandt, S. Craig Moore, Raymond E. Conley, and Glenn Buchan. 2014. The Future of the US Intercontinental Ballistic Missile Force. RAND Corporation.

Day, Dwayne No Date. Early Reentry Vehicles: Blunt Bodies and Ablatives. US Centennial of Flight Commission.

Day, Dwayne A. No Date. Advanced Reentry Vehicles. US Centennial of Flight Commission.

Federation of American Scientists. No Date. Targets and Decoys. http://www.fas.org/spp/starwars/program/targets.htm

Field, A. L., Jr. 1994. Analytical Studies of Beryllium Ablation and Dispersion during Reentry. Journal of Spacecraft and Rockets, January Issue.

Greenwood, Ted. 1973. Qualitative Improvements in Offensive Strategic Arms: The Case of MIRV. C/73-7, Center for International Studies, Massachusetts Institute of Technology.

Handler, Francis A. 1989. Maneuverable Reentry Vehicle Trajectory Footprints: Calculation and Properties. Lawrence Livermore Laboratory.

Hartunian, Richard A. 2003. Ballistic Missiles and Reentry Systems: The Critical Years. Crosslink, Spring Issue.

Johannessen, Karl R. 1964. Least Dispersion of the MK 12 Reentry Vehicle. United States Air Force Air Weather Service.

Kendall, David M., Kaz Niemiec, and Richard A. Harrison. 2003. A Modeling and Simulation Approach for Reentry Vehicle Aeroshell Simulation. TRW Systems, Missile Defense Division.

Leitenberg, Milton. 2010. Case Study 3: The Origin of MIRV.

Scala, Pete E. 1996. A Brief History of Composites in the US – The Dream and the Success. Journal of Materials, February 1 Issue.

Sessler, Andrew M. (Chair of the Study Group), John M. Cornwall, Bob Dietz, Steve Fetter, Sherman Frankel, Richard L. Garwin, Kurt Gottfried, Lisbeth Gronlund, George N. Lewis, Theodore A. Postol, and David C. Wright. 2000. Countermeasures: A Technical Evaluation of the Operational Effectiveness of the Planned US National Missile Defense System. Union of Concerned Scientists.

Tammen. R. L. 1973. MIRV and the Arms Race: An Interpretation of Defense Strategy. Praeger.

Yengst, William. 2010. Lightning Bolts: First Maneuvering Reentry Vehicles. Tate Publishing.

York, Herbert F. 1975. The Origins of MIRV. Report Number 9, Stockholm: PRIO.

TEST UNITS / TEST SITES / TEST FACILITIES

6555th Aerospace Test Wing.
https://www.patrick.af.mil/heritage/6555th/6555toc.htm

American Institute of Aeronautics and Astronautics. 2008. Cape Canaveral, Florida. Historic Aerospace Site.

Cleary, Mark C. The 6555th: Missile and Space Launches through 1970.
http://www.fas.org/spp/military/program/6555th/6555toc.htm#TOC

Encyclopedia Astronautica. http://www.friends-partners.org/partners/mwade/index.htm

Johnson, Leland. 1996. Tonopah Test Range: Outpost of Sandia National Laboratories. Sandia, SAND96-0375.

Lethbridge, Cliff. 2000. The History of Cape Canaveral. Spaceline.
http://www.spaceline.org/capehistory/1a.html

Parker, Loyd C., Jerry D. Watson, and James F. Stephenson. 1989. Final Baseline Assessment Western Space and Missile Center (WSMC). Research Triangle Institute Center for Systems Engineering, Florida Office.

US Navy. 1993. From the Desert to the Sea: A Brief Overview of the History of China Lake. The Rocketeer.

White Sands Missile Range History. No Date. http://www.wsmr-history.org/History.htm

MISCELLANEOUS – BALLISTIC MISSILES

Baker, David, 1978. Rocket: The History and Development of Rocket and Missile Technology. Crown Publishing.

Berhow, Mark A. and Chris Taylor. 2005. US Strategic and Defensive Missile Systems 1950 - 2004. Osprey Publishing Limited.

Convair.1957. Characteristics of Tactical, Strategic and Research Missiles.

Department of Defense. 2004. Model Designation of Military Aerospace Vehicles. DOD 420.15-L.

Gunston, Bill. 1992. The Illustrated Encyclopedia of the World's Rockets and Missiles: A Comprehensive Technical Directory and History of the Military Guided Missile Systems of the 20th Century. Salamander Books.

Heppenheimer, T. A. 2007. Facing the Heat Barrier: A History of Hypersonics. NASA History Series, NASA SP-2007-4232.

Kennedy, Gregory P. 2009. The Rockets and Missiles of White Sands Proving Ground 1945 - 1958. Schiffer Publishing.

Kopp, Carlo. No Date. Genesis of the Surface to Air Missile. Australian Airpower.
http://www.ausairpower.net/DT-MS-1006.pdf

McCullough, Roy. 2001. Missiles at the Cape: Missiles on Display at the Air Force Space and Missile Museum Cape Canaveral Air Force Station, Florida. ERDC/ECRL SR-01-22.

Parsch, Andreas. Directory of US Military Rockets and Missiles.
http://www.designation-systems.net/

Reed, Chris. 2002. US Nuclear Missiles – CD found in United States Air Force: A Chronological History and Guide to Resources. Dataview.

Savage, Melvyn. 1961. Launch Vehicle Handbook. NASA-TM-74948.

Schneider, Bruce et al. 1999. ISST Structure with SSIFCON – HfC-2 Test. Engineering Research Institute, University of New Mexico, Albuquerque, New Mexico.

United States Army Field Artillery School, Gunnery Department. 1998. Introduction to MLRS. GS01AA Course Notes.

MISCELLANEOUS – CRUISE MISSILES

Conrow, E. H., G. K. Smith and A. A. Barbour. 1982. The Joint Cruise Missile Project: An Acquisition History. The RAND Corporation.

Goebel, Greg. Cruise Missiles. 1994.
http://www.vectorsite.net/twcruz.html

Kopp, Carlo. Cruise Missiles. 2005. Australian Airpower.

Werrell, Kenneth P. 1985. The Evolution of Cruise Missiles. Air University Press.

MISCELLANEOUS – AIRCRAFT

Baugher, J. American Military Aircraft.
http://home.att.net/~jbaugher/uscombataircraft.html

Global Aircraft Website.
http://www.globalaircraft.org/

Goebel, Greg. 2009. Air Vectors.
http://www.vectorsite.net/indexav.html

Younossi, Obaid, Mark V. Arena, Richard M. Moore, Mark A. Lorell, Joanna Mason, and John C. Graser. 2002. Military Jet Engine Acquisition: Technology, Basics and Cost Estimating Methods. The RAND Corporation.

MISCELLANEOUS – OTHER

Aspin, Les. 1980. Judge Not by Numbers Alone. Bulletin of the Atomic Scientist, June Issue.

Bright, Christopher J. 2010. Defense in the Eisenhower Era: Nuclear Antiaircraft Arms and the Cold War. Palgrave MacMillan.

Burrows, William E. 1998. This New Ocean. Random House.

Chant, Christopher. 1987. A Compendium of Armaments and Military Hardware. Routledge.
Cochran, T., W. Arkin and M. Hoenig. 1984. Nuclear Weapons Databook Volume I: US Nuclear Forces and Capabilities. Natural Resources Defense Council.
Constant, James N. 1981. Fundamentals of Strategic Weapons Offense and Defense Systems. Martinus Nijoff Publishers.
Dranidis, Dimitris V. 2003. Shipboard Phased Array Radars: Requirements, Technology and Operational Systems. Waypoint Magazine, February Issue.
Federation of American Scientists. United States Weapon Systems.
http://www.fas.org/man/dod-101/sys/index.html
Gibson, James N. 1986. Nuclear Weapons of the United States: An Illustrated History. Schiffer.
Goebel, Greg. 2007. Race to the Moon: 1957 - 1975.
http://www.vectorsite.net/tamrc.html
JED. The Military Equipment Directory.
http://www.jedsite.info/
Missile Technology Control Regime. 1993.
http://www.mtcr.info/english/index.html
United States Air Force Chief Scientist. 2010. A Vision for Air Force Science and Technology During 2010 - 2030. AF/ST-TR-10-01-PR, May 15.
Van Atta, Richard H., Sidney Reed, and Seymour J. Deitchman. 1991. DARPA Technical Accomplishments Volume II: An Historical Review of Selected DARPA Projects. IDA Paper P-2429.
Von Karman, Theodore, and the Army Air Force Scientific Advisory Group. 1946. Towards New Horizons. Thirteen Volumes. Headquarters, Air Materiel Command.
Wade, Mark. Encyclopedia Astronautica.
http://www.astronautix.com/

NUCLEAR POLICY / ORGANIZATION

SAFETY / SECURITY
Bellovin, Steven M. 2006. Permissive Action Links.
http://www.cs.columbia.edu/~smb/nsam-160/pal.html
Bleck, Mark E. and Paul R. Souder. 1982. PAL Control of Theater Nuclear Weapons. Sandia, SAND82-2436.
Center for Defense Information. 1981. US Nuclear Weapons Accidents: Danger in Our Midst. I.S.S.N. #0195-6450.
Cochran, Thomas B. 1989. Technological Issues Related to the Proliferation of Nuclear Weapons. The Non-Proliferation Policy Center.
Chambers, W. H., H. F. Atwater, J. T. Caldwell, W. E. Mauldin, N. Nicholson, T. E. Sampson, G. M. Worth, and T. H. Whittlesey. 1970. MRV Verification by On-site Inspection. Los Alamos, LA-4577.
Crandall, David. 2002. The Essential Role of Credible Correct Simulation in Assuring the Safety of America's Nuclear Stockpile. National Nuclear Safety Administration.
DODD 3150.2 DOD Nuclear Weapon System Safety Program. December 23, 1996.
DODD 3150.3 Nuclear Force Security and Survivability. August 16, 1994.

DODD 5210.41 Security Policy for Protecting Nuclear Weapons. Jul 13, 2009.

Elliot, Grant. 2005. US Nuclear Weapon Safety and Control. MIT.

Harvey, John R. and Stefan Michalowski. 1994. Nuclear Weapons Safety: The Case of Trident. Science and Global Security.

JTF-8 Ad Hoc Group for Nuclear Safety. 1962. Technical Nuclear Safety Study of Project Dominic B-52 Airdrops.

Kidder, Ray. 1991. Nuclear Warhead Safety. UCRL-LR-107454.

Maggelet, Michael H. and James C. Oskins. 2008. Broken Arrow: The Declassified History of US Nuclear Weapons Accidents. Lulu.com.

McHugh, Michael L. No Date. The SSBN Security Program. Federation of American Scientists Website.
http://www.fas.org/nuke/guide/usa/slbm/ssbn-secure.htm

Mueller, Curt, Stan Spray, and Jay Grear. 1992. The Unique Signal Generator for Detonation Safety in Nuclear Weapons. Sandia, SAND91-1269.

Norris, Robert S. and William Arkin. 1999. US Nuclear Weapons Safety and Control Features. Bulletin of the Atomic Scientist, May / June Issue.

Plummer, David W. and William H. Greenwood. 1998. The History of Nuclear Weapon Safety Devices. Sandia, SAND-98-1184C.

Sagan, Scott D. 1993. The Limits of Safety: Organizations, Accidents, and Nuclear Weapons. Princeton University Press.

United States Air Force. 1996. Safety Rules for the Intercontinental Ballistic Missile Weapon System. Air Force Instruction 91-114.

United States Air Force. 1997. Safety Rules for US Strategic Bombers. Air Force Instruction 91-111.

United States Air Force. 1996. Safety Rules for US Strike Aircraft. Air Force Instruction 91-112.

COMMAND / CONTROL / COMMUNICATIONS

Air Force Doctrine Document 2-1.5. Nuclear Operations. 1998.

Alberts, David S. and Richard E. Hayes. 2006. Understanding Command and Control. DOD Command and Control Research Program.

Blair, Bruce G. 2005. Strategic Command and Control: Redefining the Nuclear Threat. Brookings Institution Press.

Burr, William ed. 2001. Launch on Warning: The Development of US Capabilities, 1959 - 1979. The National Security Archive Electronic Briefing.

Cimbala, Stephen J. 1984. US Strategic C3I: A Conceptual Framework. Air University Review, November - December Issue.

Critchlow, Robert F. 2006. Nuclear Command and Control: Current Programs and Issues. CRS Report for Congress.

Gregory Shaun R. and Shaun Gregory. 1996. Nuclear Command and Control in NATO: Nuclear Weapons Operations and the Strategy of Flexible Response. Palgrave Macmillan.

Hamre, John J., Richard H. Davison, and Peter T. Tarpgaard. 1981. Strategic Command, Control, and Communications: Alternative approaches for Modernization. Congressional Budget Office, October.

Jablonsky, David. 2000. Eisenhower and the Origins of the Unified Command. Joint Forces Quarterly. Autumn - Winter Issue.

Johnson, Spencer. 2002. New Challenges for the Unified Command Plan. Joint Forces Quarterly, Summer Issue.
Joint Chiefs of Staff. 2005. Doctrine for Joint Nuclear Operations. JP3-12.
Pearson, David E. 2000. The World Wide Military Command and Control System: Evolution and Effectiveness. Air University Press.
Sturm, Thomas A. 1967. The Air Force Command and Control System 1950 - 1966. USAF Historical Division Liaison Office.
Sturm, Thomas A. 1966. The Air Force and the Worldwide Military Command and Control System 1961 - 1965. USAF Historical Division Liaison Office.
USSTRATCOM/CSH HQ. 2004. History of the United States Strategic Command, June 1, 1992 to October 1, 2002.
Wainstein, L., C. D. Cremeans, J. K. Moriarity, and J. Ponturo. 1975. The Evolution of US Strategic Command Control and Warning: 1945 - 1972. Institute for Defense Analysis.

INTELLIGENCE / TARGETING / SIOP
Air Force Targeting. 1993. Air Force Instruction 14-307.
Alldridge, Robert C. 1983. First Strike: The Pentagon's Strategy for Nuclear War. South End Press.
Andronov, A. Tr. Allen Thompson. 1993. American Geosynchronous SIGINT Satellites. Zarubezhnoye Voyennoye Obozreniye, December Issue.
Arkin, William M. and Hans Kristensen. 1999. The Post-Cold War SIOP and Nuclear Planning. NRDC Nuclear Program.
Bamford, James. 2002. Body of Secrets: Anatomy of the Ultra-Secret National Security Agency. Anchor Books.
Binninger, Gilbert C., Paul J. Castleberry Jr., and Patsy M. McGrady. 1974. Mathematical Background and Programming Aids for the Physical Vulnerability System for Nuclear Weapons. Defense Intelligence Agency AP-550-1-2-69-INT.
Burr, William ed. 2004. The Creation of SIOP-62: More Evidence on the Origins of Overkill. National Security Archive Electronic Briefing.
Burr, William ed. 2005. The Nixon Administration, the SIOP and the Search for Limited Nuclear Options, 1969 - 1974. National Security Archive Electronic Briefing.
Burr, William ed. 2005. SAC Nuclear Planning for 1959. National Security Archive Electronic Briefing.
Burrows, William E. 2002. By Any Means Necessary: America's Heroes Flying Secret Missions in a Hostile World. Plume.
Burrows, William E. 1986. Deep Black: The Startling Truth Behind America's Top Secret Spy Satellites. Random House.
Cimbala, Stephen J. 1988. The SIOP: What Kind of War Plan? Aerospace Power Journal, Summer Issue.
Cloud, John. 2002. American Cartographic Transformations During the Cold War. Cartography and Geographic Information Science, March Issue.
Farquhar, John Thomas. 2004. Need to Know: The Role of Air Force Reconnaissance in War Planning, 1945 - 1953. Air University Press.
Feiveson Harold A. ed. 1999. The Nuclear Turning Point: A Blueprint for Deep Cuts and De-alerting of Nuclear Weapons. The Brooking Institution.

Haines, Gerald K. and Robert E. Legget Eds. 2001. Watching the Bear: Essays on the CIA's Analysis of the Soviet Union. Center for the Study of Intelligence.

Hayes, Peter. 1990. Pacific Powderkeg: American Nuclear Dilemmas in Korea. Lexington Books History and Research Division, Headquarters Strategic Air Command. No Date. History of the Joint Strategic Planning Staff: Background and Preparation of SIOP 62.

Kohler, Robert J. 2007. The Decline of the National Reconnaissance Office. Center for the Study of Intelligence.

Kovich, Andrew S. 2007. ICBM Strike Planning. AAFM Newsletter, June Issue.

Kristensen, Hans M. 1997. Targets of Opportunity: How Nuclear Planners found New Targets for Old Weapons. Bulletin of Atomic Scientists, September / October Issue.

Kristensen, Hans M. 2000. The Awakening Asian Tiger, China in US Nuclear War Planning. The Nautilus Institute.

Kristensen, Hans M., Robert S. Norris, and, Matthew G. McKinzie. 2006. Chinese Nuclear Forces and US Nuclear War Planning. The Federation of American Scientists and the Natural Resources Defense Council.

McKinley, Cynthia A. S. 1996. When the Enemy has Our Eyes. School of Advanced Airpower Studies.

Pringle, Peter and William Arkin. 1983. SIOP: The Secret US Plan for Nuclear War. Norton.

Reporting Manual for Joint Resources Assessment Database System (JRDS), Joint Chiefs of Staff, March 15, 1999.

Richelson, Jeffrey T. 1999. The US Intelligence Community, 4th ed. Westview Press.

Richelson, Jeffrey T. 1999. US Satellite Imagery 1960 - 1999. The National Security Archive Electronic Briefing Book.

Ross, Steven T. 1996. American War Plans 1945 - 1950. Frank Cass & Co. Ltd.

Sontag, Sherry and Christopher Drew. 1999. Blind Man's Buff: The Untold Story of American Submarine Espionage. Harper.

Thomson, Allen. No Date. US Naval Space Command Surveillance System.
http://www.fas.org/spp/military/program/track/spasur_at.htm

United States Air Force. 1998. USAF Intelligence Targeting Guide. Air Force Pamphlet 14-210.

United States Air Force. 1990. Target Intelligence Handbook Unclassified: Targeting Principles. AFP 200-18, Volume 1, U.S. Department of Defense, October 1.

Whitman, Edward C. 2005. SOSUS: The Secret Weapon of Underwater Surveillance. Undersea Warfare, Winter Issue.

EARLY WARNING / NUCLEAR AEROSPACE DEFENSE

20th Space Control Squadron.
http://www.peterson.af.mil/library/factsheets/factsheet.asp?id=4730

Air / Aerospace Defense Command.
http://www.zianet.com/jpage/airforce/history/majcoms/adc.html

Air Defense Radar Museum. http://www.radomes.org/museum/

Bellany, Ian and Coit D. Blacker, ed. Antiballistic Missile defense in the 1980s. Psychology Press.

Boyne, Walter J. 1999. The Rise of Air Defense. Air Force Magazine Online, December Issue.

Buderi, Robert. 1966. The Invention that Changed the World. Simon and Schuster.

Burr, William. 2000. Missile Defense Thirty Years Ago: Déjà vu all over Again? National Security Archive.

Carter, Ashton B. and David N. Schwartz. eds. 2005. Ballistic Missile Defense. Brookings Institution Press.

Cheyenne Mountain Operations Center: Defending North America.
https://www.cheyennemountain.af.mil/

Cimbala, Stephen J. 1985. The Strategic Defense Initiative: Political Risks. Air University Review November - December Issue.

Cornett, Lloyd H. Jr. and Mildred W. Johnson. 1980. A Handbook of Aerospace Defense Organization 1946 - 1980. Office of History, Aerospace Defense Center.

Department of the Army. 1965. US Army Air Defense Employment. FM44-1.

Fenn, Alan J., Donald H. Temme, William P. Delaney, and William E. Courtney. 2000. The Development of Phased Array Radar Technology. Lincoln Laboratory Journal.

Gallery of Old Iron. SAGE.
http://www.thegalleryofoldiron.com/SAGE.HTM

Goebel, Greg. 1994. Missile Defense. http://www.vectorsite.net/twabm.html

The Heritage of Space Command. No Date.
http://www.afspc.af.mil/heritage/index.asp

Hubbard, Bryan. 2009. Missile Early Warning: Peeking over the Curtain.
http://www.military.com/Content/MoreContent1/?file=cw_midas

IBM Military Products Division. 1959 -1965. Introduction to the AN/FSQ-7 Combat Direction Central and AN/FSQ-8 Combat Control Central.

Kristensen, Hans M., Matthew G. Mckinzie, and Robert S. Norris. 2004. The Protection Paradox. Bulletin of the Atomic Scientists, March / April Issue.

McMullen, Richard F. 1960. Aircraft in Air Defense, 1946-1960. ADC Historical Study No. 12, Historical Division, Directorate of Information, Air Defense Command.

Moeller, Stephen P. 1965. Vigilant and Invincible. Air Defense Artillery Magazine, May - June Issue.

Nikunen, Heikki. 1997. Air Defense in Northern Europe. Finnish Defense Studies 10.

NORAD: Selected Chronology. http://www.fas.org/nuke/guide/usa/airdef/norad-chron.htm

Papp, Daniel R. 1977 – 1978. From Project Thumper to SDI. The Role of Ballistic Missile Defense in US Security Policy. Airpower Journal, Winter Issue.

Schaffel, Kenneth. 1990. The Emerging Shield: The Air Force and the Evolution of Air Defense 1945 - 1960. Office of Air Force History.

US Congress, Office of Technology Assessment. 1985. Ballistic Missile Defense Technologies, OTA-ISC-254. US Government Printing Office, September.

Walker, James, Lewis Bernstein, and Sharon. 2003. Seize the High Ground – The Army in Space and Missile Defense. (Lang Historical Office, US Army Space and Missile Defense Command.

Werrell, Kenneth P. 2005, Archie to SAM: A Short Operational History of Ground-Based Air Defense. Air University Press.

Werrell, Kenneth P. 2002. Hitting a Bullet with a Bullet: A History of Ballistic Missile Defense. Airpower Research Institute, Research Paper 2000-02.

Wilson, L. S. The DEW line Sites in Canada, Alaska and Greenland.
http://www.lswilson.ca/dewline.htm

Winkler David F. Searching the Skies: 1997. The Legacy of the United States Cold War Defense Radar Program. Air Combat Command.

STRATEGIC NUCLEAR POLICY / ANALYSIS
Anders, Roger M. ed. 1987. Forging the Atomic Shield: Excerpts from the Office Diary of Gordon E. Dean. University of North Carolina Press.
Bennett, W. S., R. R. Sandoval, and R.G Shreffler. 1974. United States National Security Policy and Nuclear Weapons. Los Alamos, LA-5785-MS.
Brodie, Bernard. ed. 1946. The Absolute Weapon: Atomic Power and World Order. Harcourt.
Brodie, Bernard. 1959. Strategy in the Missile Age. Princeton University Press.
Buchan, Glen C. 1994. US Nuclear Strategy for the Post-Cold War Era. The RAND Corporation.
Chairman of the Joint Chiefs of Staff. 2006. Joint Operation Planning and Execution System (JOPES). Joint Publication 5-0.
Chairman of the Joint Chiefs of Staff. 2006. Overview of Joint Operation Planning Briefing Script.
Cimbala, Stephen J. 1987. US Strategic Nuclear Deterrence: Technical and Policy Changes. Air University Review, January-March Issue.
Cimbala, Stephen J. 2005. Nuclear Weapons and Strategy: The Evolution of American Policy. Routledge.
Cole, Ronald H., Walter S. Poole, James F. Schnabel, Robert J. Watson, and Willard J. Webb. 1995. The History of the Unified Command Plan 1946 - 1993. Office of the Chairman of the Joint Chiefs of Staff Joint History Office.
Collins, John M. 1981. US Strategic Nuclear Force Options. Issue Brief Number IB77046.
Dunn, Lewis A. 2007. Deterrence Today: Roles Challenges and Responses IFRI Proliferation Paper 19.
Ernst, Stephen P., Kang Hosug, Frank J. Rossi Jr., and Keith A. 1996. Thompson. Nuclear Strategy and Arms Control: A Comparison. Air Command and Staff College.
Finn, Christopher and Paul D. Berg. 2004. Anglo-American Strategic Air Power Co-operation in the Cold War and Beyond. Air & Space Power Journal, Winter Issue.
Friedman, Norman. 2007. The Fifty-Year War: Conflict and Strategy in the Cold War. Naval Institute Press.
Grotto, Andrew and Joe Cirincione. 2008. Orienting the 2009 Nuclear Posture Review: A Roadmap. Center for American Progress.
Freedman, Lawrence. 2003. Evolution of Nuclear Strategy. Palgrave Macmillan.
Kahn, Herman. 1978. On Thermonuclear War. Greenwood Press.
Kahn, Herman. 1960. The Nature and Feasibility of War and Deterrence. RAND, P-1888-RC.
Kaminski, Paul G. 1995. Sustaining the US Nuclear Deterrent in the 21st Century. US Strategic Command Strategic Systems Industrial Symposium, Offutt Air Force Base, Nebraska, Aug. 30.
Kotch, John B. 1967. NATO Nuclear Arrangements in the Aftermath of MLF. Air University Review, March - April Issue.
Kristensen, Hans M. 2006. Globalstrike: A Chronology of the Pentagon's New Offensive Strike Plan. American Federation of Scientists.
Kristensen, Hans M. 2005. The Airborne Alert Program over Greenland. Nuclear Information Project.

Kristensen, Hans M. 2005. Nuclear Mission Creep: The Impact of Weapons of Mass Destruction Proliferation on US Nuclear Policy and Planning. The Program on Science and National Security, Princeton.

Kristensen, Hans M. 2004. US Nuclear Planning after the 2001 Nuclear Posture Review. Presentation to the Center for International and Security Studies at University of Maryland.

Kristensen, Hans M. 1998. Nuclear Futures: Proliferation of Weapons of Mass Destruction and US Nuclear Strategy. British American Security Information Council.

Kristensen, Hans M., Robert S. Norris, and Ivan Oelrich. 2009. From Counterforce to Minimal Deterrence: A New Nuclear Policy on the Path Toward Eliminating Nuclear Weapons. Federation of American Scientists & The Natural Resources Defense Council, Occasional Paper No. 7.

Kunsman, David M. and Douglas B. Lawson. 2001. A Primer on US Strategic Nuclear Policy. Sandia, Sand 2001-0053.

Lieber, Keir A. and Daryl G. Press. 2009. The Nukes We Need: Preserving the American Deterrent. Foreign Affairs, November / December Issue.

Lieber, Keir A. and Daryl G. Press. 2006. The End of MAD: The Nuclear Dimensions of US Primacy. International Security, Spring Issue.

Lemmer, George F. 1967. The Air Force and Strategic Deterrence 1951 - 1960. USAF Historical Division Liaison Office.

Lemmer, George F. 1963. The Air Force and the Concept of Deterrence 1951 - 1960. USAF Historical Division Liaison Office.

May, Ernest R., John D. Steinbruner, and Thomas W. Wolfe. 1981. History of the Strategic Arms Competition Part I. Office of the Secretary of Defense Historical Office.

May, Ernest R., John D. Steinbruner, and Thomas W. Wolfe. 1981. History of the Strategic Arms Competition Part II. Office of the Secretary of Defense Historical Office.

McDonough, David S. 2009. Tailored Deterrence: The 'New Triad' and the Tailoring of Nuclear Superiority. Canadian International Council, Strategic Data-link No. 9.

McKinzie, Mathew G., Thomas B. Cochran, Robert S. Norris, and William M. Arkin. 2001. The US Nuclear War Plan: A Time for Change. National Resources Defense Council.

McNamara, Robert. 1967. Mutual Deterrence. Speech by Secretary of Defense Francisco, September 18.

Medalia, Jonathan. 1981. Assessing the Options for Preserving ICBM Survivability. Congressional Research Service, Report 81-222F.

Oelrich, Ivan. 2005. Missions for Nuclear Weapons after the Cold War. Federation of American Scientists. Occasional Paper No. 3.

Office of the Secretary of Defense, History Office. No Date. History of the Strategic Arms Competition 1945 - 1972.

Office of the Under Secretary of Defense for Acquisition, Technology, and Logistics. 2006. Report of the Defense Science Board Task Force on Nuclear Capabilities: Report Summary.

Perry, William J. and James R. Schlesinger. 2009. The Final Report of the Congressional Commission on the Strategic Posture of the United States. United States Institute of Peace Press.

Piotrowski, John L. 2002. Strategic Synchronization: The Relationship between Strategic Offense and Defense. The Heritage Foundation.

The Posture of the US Strategic Command (USSTRATCOM). Hearing before the Strategic Forces Subcommittee of the Committee on Armed Services House of Representatives, March 8, 2007.

Potter, William C. 2005. Trends in US Nuclear Policy. IFRI Security Studies Department.

Rosenberg, David A. 1981 / 1982. "A Smoking, Radiating Ruin at the End of Two Hours": Documents on American Plans for War with the Soviet Union 1954 - 1955. International Security, Winter Issue.

Rosenberg, David A. No Date. Constraining Overkill: Contending Approaches to Nuclear Strategy, 1955 - 1965. Naval Historical Center Colloquium on Contemporary History Project.

Sapolsky, Harvey M. No Date. The US Navy's Fleet Ballistic Missile Program and Finite Deterrence.
http://web.mit.edu/ssp/faculty/sapolsky/TARGETING%20POLARIS.pdf

Sokolsky, Henry D. 2004. Getting MAD: Nuclear Mutual Assured Destruction, Its Origins and Practice. DIANE Publishing.

Taylor, Maxwell. 1974. The Uncertain Trumpet. Greenwood Press.

United States Air Force Nuclear Task Force. 2008. Reinvigorating the Air Force Nuclear Enterprise. Headquarters United States Air Force.

United States Navy. 2002. Naval Power 21… A Naval Vision.

Vandevanter, E. Jr. 1963. Nuclear Forces and the Future of NATO. The RAND Corporation.

Walsh, David M. 2007. The Military Balance in the Cold War: US Perceptions and Policy 1976 - 1985. Routledge.

Wolk, Herman S. 2003. The "New Look." Air Force Magazine Online, August Issue.

Woolf, Amy. F. 2009. US Strategic Nuclear Forces: Background, Developments, and Issues. Report for Congress RL 33640.

TACTICAL NUCLEAR POLICY / ANALYSIS

Alexander, Brian and Alistair Miller eds. 2003. Tactical Nuclear Weapons: Emergent Threats in an Evolving Security Environment. Potomac Books.

Bennett, W. S., R. P. Gard, and G. C. Reinhardt. 1974. Tactical Nuclear Weapons: Objectives and Constraints. MS 562.

Dyson, F. J., R. Gomer, S. Weinberg, and S. C. Wright. 1967. Tactical Nuclear Weapons in SE Asia. Institute for Defense Analysis.

Kleckner, Richard. 2006. GLCM and its Role in the INF Treaty. AAFM Newsletter.

Kristensen, Hans M. 1995. The 520 Forgotten Bombs: How US and British Nuclear Weapons in Europe Undermine the Non-Proliferation Treaty. Greenpeace International.

Headquarters, Department of the Army. 1996. Nuclear Operations. FM 100-30

Headquarters, Department of the Army. 1986. Nuclear Weapons Employment, Doctrine, and Procedures. FM 100-31-3

Kristensen, Hans M. 2012. Non-Strategic Nuclear Weapons. Federation of American Scientists.

Leitenberg, Milton. 1982. The Genesis of the Long-range Theatre Nuclear Forces Issue in NATO. In: The Military Balance in Europe, Conference Papers No. 2. The Swedish Institute of International Affairs.

Midgely, John J. Jr. 1986. Deadly Illusions: Army Policy for the Atomic Battlefield. Westview Special Studies in National Security and Defense Policy, Westview Press.

Shreffler, R. G. and W. S. Bennet. 1970. Tactical Nuclear Warfare. Los Alamos, LA-4467-MS.

Snow, Robert. 1979. The Strategic Implications of Enhanced Radiation Weapons. Air University Review.

US Congress, Office of Technology Assessment. 1987. New Technology for NATO: Implementing Follow-On Force Attack. OTA-ISC-309 US Printing Office.

Wightman, Richard O. Jr. 1989. Soviet Reactions to Follow-on to Lance (FOTL). US Army War College.

Woolf, Amy F. 2006. Nonstrategic Nuclear Weapons. Report for Congress, RL 32572.

US NUCLEAR FORCES COMPOSITION / ORGANIZATION / HISTORY

Air Force Association. 2005. The Air Force and the Cold War. Air Force Magazine, September Issue.

Baumgardner, Neil. US Armed Forces Order of Battle. http://www.geocities.com/Pentagon/9059/usaob.html

Cartwright, James E. 2006. USSTRATCOM: A Command for the 21st Century. JFQ, 3rd Quarter

Chelimsky, Eleanor. 1993. The US Nuclear Triad: GAO's Evaluation of the Strategic Modernization Program. GAO/T-PEMD-93-5.

Congressional Budget Office. 2015. Projected Costs of U.S. Nuclear Forces, 2015 to 2024.

Crouch, J. D. 2005. Challenges of a New Capability-Based Defense Strategy: Transforming US Strategic Forces. Department of Defense.

Futrell, Frank Robert. 1971. Ideas, Concepts, Doctrine: Basic Thinking in the United States Air Force 1907 - 1960. Volume 1. Air University Press.

Futrell, Frank Robert. 1989. Ideas, Concepts, Doctrine: Basic Thinking in the United States Air Force 1961 - 1984. Volume 2. Air University Press.

Government Accounting Office. 2008. Military Transformation: DOD Needs to Strengthen Implementation of its Global Strike Concept and Provide a Comprehensive Investment Approach for Acquiring Needed Capabilities.

Gunziger, Mark A. 2011. Sustaining America's Strategic Advantage in Long Range Strike Center for Strategic and Budgetary Assessments.

Headquarters Strategic Air Command. 1991. (Phoenix) Force Structure Study.

Hogler, J. L. 1996. USSTRATCOM Whitepaper on Post START II Arms Control.

Jussel, Paul C. 2004. Intimidating the World: The United States Atomic Army 1956 - 1960. Ohio State University.

Kedzior, Richard W. 2000. Evolution and Endurance: The US Infantry Division in the Twentieth Century. Rand Monograph MR-1211-A.

Kristensen, Hans M. 2000. The Matrix of Deterrence: US Strategic Command Force Structure Studies. Nautilus Institute.

Lord, Carnes ed. Reposturing the Force: 2006. US Overseas Presence in the Twenty-First Century. Naval War College Press: Newport Paper 26.

McFarland, Stephen L. 1996. The Air Force in the Cold War 1945 - 1960: The Birth of a New Paradigm. Conference on Interservice Rivalry and the Armed Forces.

Memorandum for the President. 1961. Recommended Long-range Nuclear Delivery Forces 1963 - 1967.

Nagy, Paul. 1998. One Insider's look at the Quadrennial Defense Review. DFI International.

Nalty, Bernard C. 1997. Winged Shield, Winged Sword: A History of the USAF. Air Force History and Museums Program.

National Resources Defense Council. Archive of Nuclear Data from NRDC's Nuclear Program. http://www.nrdc.org/nuclear//nudb/datainx.asp

Norris, Robert S. and Hans M. Kristensen. 2000. US Nuclear Forces, 2000. Bulletin of the Atomic Scientists, May / June Issue

Norris, Robert S. and Hans M. Kristensen. 2001. US Nuclear Forces, 2001. Bulletin of the Atomic Scientists, March / April Issue.

Norris, Robert S. and Hans M. Kristensen. 2002. US Nuclear Forces, 2002. Bulletin of the Atomic Scientists, May / June Issue.

Norris, Robert S. and Hans M. Kristensen. 2003. US Nuclear Forces, 2003. Bulletin of the Atomic Scientists, May / June Issue.

Norris, Robert S. and Hans M. Kristensen. 2004. US Nuclear Forces, 2004. Bulletin of the Atomic Scientists, May / June Issue.

Norris, Robert S. and Hans M. Kristensen. 2005. US Nuclear Forces, 2005. Bulletin of the Atomic Scientists, January / February Issue.

Norris, Robert S. and Hans M. Kristensen. 2006. US Nuclear Forces, 2006. Bulletin of the Atomic Scientists, January / February Issue.

Norris, Robert S. and Hans M. Kristensen. 2007. US Nuclear Forces, 2007. Bulletin of the Atomic Scientists, January / February Issue.

Norris, Robert S. and Hans M. Kristensen. 2008. US Nuclear Forces, 20087. Bulletin of the Atomic Scientists, January / February Issue.

Norris, Robert S. and Hans M. Kristensen. 2009. US Nuclear Forces, 2009. Bulletin of the Atomic Scientists, March / April Issue.

Norris, Robert S. and Hans M. Kristensen. 2010. US Nuclear Forces, 2010. Bulletin of the Atomic Scientists, May / June Issue.

Norris, Robert S. and Hans M. Kristensen. 2011. US Nuclear Forces, 2011. Bulletin of the Atomic Scientists, March / April September Issue.

Norris, Robert S. and Hans M. Kristensen. 2012. US Nuclear Forces, 2012. Bulletin of the Atomic Scientists, May /June Issue.

Norris, Robert S. and Hans M. Kristensen. 2013. US Nuclear Forces, 2013. Bulletin of the Atomic Scientists, February Issue.

Norris, Robert S. and Hans M. Kristensen. 2014. US Nuclear Forces, 2014. Bulletin of the Atomic Scientists, January Issue.

Norris, Robert S. and Hans M. Kristensen. 2015. US Nuclear Forces, 2015. Bulletin of the Atomic Scientists, February Issue.

Norris, Robert S. and Hans M. Kristensen. 2016. US Nuclear Forces, 2016. Bulletin of the Atomic Scientists, February Issue.

Norris, Robert S. and Hans M. Kristensen. 2017. US Nuclear Forces, 2017. Bulletin of the Atomic Scientists, February Issue.

Norris, Robert S. and Hans M. Kristensen. 2011. US Tactical Nuclear Weapons in Europe, 2011. Bulletin of the Atomic Scientists, March / April Issue.

USSTRATCOM Web Page. http://www.stratcom.mil/

USSTRATCOM. 1992. Strategic Nuclear Forces: STRATCOM's View. Briefing Notes to Secretary of Defense.
USSTRATCOM. 1993. Sun City Study into Future US Force Structures.
USSTRATCOM. 1994. Sun City Extended Study into Future US Force Structures.
USSTRATCOM. 1996. The Warfighter's Assessment – Post START II Arms Reductions.

ARMS CONTROL AGREEMENTS / TREATIES
Federation of American Scientists. Arms Control Agreements. http://www.fas.org/nuke/control/
Handler, Joshua. 2001. The September 1991 PNIs and the Elimination, Storing, and Security Aspects of TNWs. Princeton University.
Woolf, Amy, Mary Beth Nikitin, and Paul K. Kerr. 2013. Arms Control and Non-Proliferation: A Catalog of treaties and Agreements. Congressional Research Service Report for Congress, February 20.

NUCLEAR DOCUMENTS / COMMISSIONS / COMMENTARY
Brown, Harold. 1980. Report of the Secretary of Defense Harold Brown to the Congress on the FY 1981 Budget, the FY 1982 Authorization Request, and FY 1981 - 1985 Defense Programs, January 19.
Chairman of the Joint Chiefs of Staff. 1993. Roles, Missions, and Functions of the Armed Forces of the United States.
Carter, Jimmy. 1980. Nuclear Weapons Employment Policy. PD-59
Clark, William. 1985. Report of the President's Blue-Ribbon Task Group on Nuclear Weapons Program Management.
Crever, R. H., M. K. Drake, J. T. Mcgahan, J. F. Schneider, and E. Swick. 1979. The Feasibility of Targeting Population. Defense Nuclear Agency 001-78-C-0061.
Department of Defense. 1971. World-Wide Military Command and Control System (WWMCCS). Department of Defense Directive S-5100.30
Department of Defense. 2002. 2002 Nuclear Posture Review (Excerpts).
Department of Defense. 2010. 2010 Nuclear Posture Review.
Department of Defense. 2008. Nuclear Weapons Inspections for the Strategic Nuclear Forces. Report of the Defense Science Board Permanent Task Force on Nuclear Weapons Surety
Department of Defense. 2001. 2001 Quadrennial Defense Review Report. Office of the Secretary of Defense.
Department of Defense. 2006. 2006 Quadrennial Defense Review Report. Office of the Secretary of Defense.
Department of Defense. 2010. 2010 Quadrennial Defense Review Report. Office of the Secretary of Defense.
Department of Defense. 2014. 2014 Quadrennial Defense Review Report. Office of the Secretary of Defense.
Department of Defense.1983, Strategic Forces Technical Assessment Review, March 31.
Department of Defense and Department of Energy. 2008. National Security and Nuclear Weapons in the 21st Century. Joint Working Group of AAAS, the American Physical Society, and the Center for Strategic and International Studies.
Dropshot – The American Plan for War with the Soviet Union, 1957.

http://www.allworldwars.com/Dropshot%20-%20American%20Plan%20for%20War%20with%20the%20Soviet%20Union%201957.html

Federation of American Scientists. Presidential Directives and Executive Orders.
http://www.fas.org/irp/offdocs/direct.htm

Energy Reorganization Act of 1974. Public Law 93-438.

Foelber, Robert. 1983. One Cheer for the Scowcroft Commission. The Heritage Foundation.

Joint Chiefs of Staff. 1985. Emergency Action Procedures of the Joint Chiefs of Staff: Nuclear Control Orders.

National Defense Strategy of the United States of America (NDS). 2008.

National Military Strategy of the United States of America (NMS). 2004.

National Security Action Memoranda. John F. Kennedy Presidential Library and Museum.
http://www.jfklibrary.org/Historical+Resources/Archives/Reference+Desk/NSAMs.htm

National Security Archive. First Documented Evidence that US Presidents Pre-Delegated Nuclear Weapons Release Authority to the Military.
http://www.gwu.edu/~nsarchiv/news/19980319.htm

National Security Archive. Presidential Directives on National Security from Truman to Clinton.
http://nsarchive.chadwyck.com/pdessayx.htm

National Security Council. 1955. Discussion at the 258th Meeting of the National Security Council, "Report to the President by the Technological Capabilities Panel of the of the Science Advisory Committee" September 8.

National Security Council. 1950. NSC 68: United States Objectives and Programs for National Security, April 14.

National Security Council. 1953. NSC-162/2 Basic National Security Policy.

The National Security Strategy of the United States of America. 2006.

Office of Defense Mobilization, Science Advisory Committee, Security Resources Panel. 1957. Deterrence and Survival in the Nuclear Age (Gaither Report).

Perry, William J. and James R. Schlesinger. 2009. Congressional Commission on the Strategic Posture of the United States.

Rumsfeld, Donald H. 2001. Guidance and Terms of Reference for the 2001 Quadrennial Defense Review. Department of Defense.

Schlesinger, James R. 2008. Secretary of Defense Task Force Report on Department of Defense Nuclear Weapons Management. Phase I: The Air Force's Nuclear Mission.

Schlesinger, James R. 2008. Secretary of Defense Task Force Report on Department of Defense Nuclear Weapons Management. Phase II: Review of the DOD Nuclear Mission.

Schlesinger, James. 1974. Policy Guidance for the Employment of Nuclear Weapons. NUWEP-1.

Technical Capabilities Panel. 1955. "Meeting the Threat of Surprise Attack." Scientific Advisory Committee, Washington D.C., February 14.

US National Security Council, Net Evaluation Subcommittee. 1963. The Management and Termination of War with the Soviet Union.

MISCELLANEOUS
Air University. 2003. Space Primer.

Anthony, Ian, Camille Grand, Lukasz Kulesa, Christian Molling, Mark Smith. ed. Jean Pascal Zanders. 2010. Nuclear Weapons after the 2010 NPT Review Conference. Chaillot Papers.

The Avalon Project at Yale Law School: The Atomic Bombing of Hiroshima and Nagasaki. http://www.yale.edu/lawweb/avalon/abomb/mp06.htm

Bennett, Bruce W. 1980. Assessing the Capabilities of Nuclear Forces: The Limits of Current Methods. The RAND Corporation, Note N-1441-A.

Cochran, T., W. Arkin, and M. Hoenig. 1984. The Bomb Book: The Nuclear Arms Race in Facts and Figures. Natural Resources Defense Council.

DEFCON: DEFense CONdition. http://www.fas.org/nuke/guide/usa/c3i/defcon.htm

Duke, Simon. 1989. United States Military Forces and Installations in Europe. Oxford University Press.

Gulley, Bill. 1980. Breaking Cover. Simon & Schuster.

Hewlett, Richard G. and Jack M. Holl. 1989. Atoms for Peace and War: 1953 - 1961, Eisenhower and the Atomic Energy Commission. University of California Press.

Jordan, Robert S. 2000. Norstad: Cold War NATO Supreme Commander – Airman, Strategist Diplomatist. Palgrave McMillan.

Khrushchev, Sergei. Khrushchev on Khrushchev: My Father the Reformer. CNN Cold War Special. Episode 7: After Stalin.

Kugler, Richard L. 1991. The Great Debate: NATO's Evolution in the 1960s. A RAND Corporation Note.

Kitts, Kenneth. 2005. Presidential Commissions and National Security: The Politics of Damage Control. Lynne Rienner Publishers.

Los Alamos National Laboratory. Nuclear Weapons Journal. (Various)

McNamara, Robert. Select Papers on Theater and Strategic Nuclear Weapons 1967 - 1969. http://www.dod.mil/pubs/foi/reading_room/426.pdf

Muolo, Michael J. 1993. Space Handbook: A War Fighter's Guide to Space. Air University Press.

National Resources Defense Council. Archive of Nuclear Data from NRDC's Nuclear Program. http://www.nrdc.org/nuclear//nudb/datainx.asp

Norris, Robert S. and Hans M. Kristensen. 2009. US and Soviet / Russian Intercontinental Ballistic Missiles 1959 - 2008. Bulletin of the Atomic Scientist, January / February Issue.

Office of the Under Secretary of Defense for Acquisition, Technology, and Logistics. 2006. Report of the Defense Science Board Task Force on Future Strategic Strike Skills.

Secretary of Defense Histories. http://www.defenselink.mil/specials/secdef_histories/

Taylor, A. J. P. ed. 1974. History of World War II. Octopus Books.

Williams, J. L. 2008. Oil Price History and Analysis. WTRG Economics. http://www.wtrg.com/prices.htm

Wolk, Herman S. 1998. Revolt of the Admirals. Air Force Magazine Online.

Worden, Mike. 1998. Rise of the Fighter Generals: The Problem of Air Force Leadership 1945 - 1982. Air University Press.

MUSEUMS / DISPLAYS

Air Force Space and Missile Museum. Cape Canaveral, Florida.
    http://afspacemuseum.org/
American Museum of Science and Energy. Oak Ridge, Tennessee
    http://amse.org/
Atomic Heritage Foundation Museum. Hanford, Washington.
    http://www.atomicheritage.org
The Atomic Testing Museum. Las Vegas, Nevada.
    http://www.atomictestingmuseum.org/
Castle Air Museum. Atwater, California.
    http://www.castleairmuseum.org
Bradbury Science Museum. Los Alamos, New Mexico.
    http://www.lanl.gov/museum/exhibits/defense.shtml
Civil Defense Museum. Greenbrier, West Virginia.
    http://www.civildefensemuseum.com/greenbrier/index.html
Experimental Breeder Reactor National Historic Site. Arco, Idaho.
    http://www.atomictourist.com/ebr.htm
F.E. Warren AFB Intercontinental Ballistic Missile and Heritage Museum. Cheyenne, Wyoming.
    http://www.warrenmuseum.com/
Hill Aerospace Museum. Ogden, Utah.
    http://www.hill.af.mil/library/museum/
Historic Ship Nautilus and Submarine Force Museum. Groton, Connecticut.
    http://www.ussnautilus.org/
Minuteman Missile National Historic Site. Wall, South Dakota.
    http://www.nps.gov/mimi/index.htm
Mound Museum. Miamisburg, Ohio.
    http://moundmuseum.com
The National Atomic Museum of Science and Technology. Albuquerque, New Mexico.
    http://www.nuclearmuseum.org/
Missiles and More Museum, Historical Society of Topsail Island. Topsail Island, North Carolina.
    http://www.topsailhistoricalsociety.org/
National Museum of the Air Force. Dayton, Ohio.
    http://www.nationalmuseum.af.mil/
Naval Museum of Armament and Technology. China Lake, California.
    http://www.chinalakemuseum.org/
Naval Undersea Museum. Keyport, Washington.
    http://www.history.navy.mil/museums/keyport/index1.htm
Nike Missile Site Museum. Sausalito, California.
    http://www.nps.gov/goga/nike-missile-site.htm
Patriot's Point Naval and Maritime Museum. Charleston, South Carolina.
    http://www.patriotspoint.org/
Pima Air and Space Museum. Tucson, Arizona.
    http://www.pimaair.org/

Smithsonian National Air and Space Museum. Washington, D.C.
http://www.nasm.si.edu/

Strategic Air and Space Museum. Omaha, Nebraska.
http://www.sasmuseum.com/visit/

Titan II Missile Museum. Sahuarita, Arizona.
http://www.titanmissilemuseum.org/

US Army Ordnance Museum. Aberdeen Proving Grounds, Maryland.
http://www.ordmusfound.org/

USS Bowfin Submarine Museum. Honolulu, Hawaii.
http://www.bowfin.org/

USS Hornet Museum. San Francisco, California.
http://www.uss-hornet.org/

USS Midway Museum. San Diego, California.
http://www.midway.org/

White Sands Missile Range Museum. White Sands, New Mexico.
http://www.nps.gov/goga/nike-missile-site.htm

Wings Over the Rockies Air and Space Museum. Denver, Colorado.
https://wingsmuseum.org/

X-10 Graphite Reactor National Historic Landmark. Oak Ridge, Tennessee.
http://www.ornl.gov/info/news/cco/graphite.htm

ACRONYMS

A/D	Analog to Digital
A&T	Assembly and Test (Platoon)
AAD	Army Air Defense
AAM	Antiaircraft Missile
ABL	Allegheny Ballistics Laboratory
ABM	Anti-Ballistic Missile
ABMA	Army Ballistic Missile Agency
ABRES	Advanced Ballistic Reentry System
ACM	Advanced Cruise Missile
ACN	Acetonitrile
AC&W	Aircraft Control and Warning (System)
ADAPT	Advanced Design and Production Technologies
ADCAP	Advanced Capability (Torpedo)
ACIU	Advanced Central Interface Unit
ACU	Avionics Control Unit
AD	Air Defense
ADC	Air Defense Command
ADCAP	Advanced Capability (Torpedo)
ADIZ	Air Defense Identification Zone
ADM	Atomic Demolition Munition
ADMS	Air Defense Missile Squadron
ADTAC	Air Defense Tactical Air Command
AEAO	Airborne Emergency Action Officer
AEC	Atomic Energy Commission
AEW	Airborne Early Warning
AFA	Analog Filter Assembly
AFB	Air Force Base
AFCP	Air Force Command Post
AFGSC	Air Force Global Strike Command
AFOAT	Air Force Office of Atomic Testing
AFOSR	Air Force Office of Scientific Research
AFRD	Air Force Research Division
AFS	Air Force Station
AFSATCOM	Air Force Satellite Communications System
AFSC	Air Force Systems Command
	Air Force Safety Command
AFSPC	Air Force Space Command
AFSTRAT-GS	Air Force Strategic - Global Strike
AFSTRAT-S	Air Force Strategic – Space
AFSWP	Armed Forces Special Weapons Project
AFV	Armored Fighting Vehicle

AGM	Air-to-ground Missile
AHFM	Alternate High Frequency Material
AHTF	Advanced Hydro Test Facility
AIAA	American Institute of Aeronautics and Astronautics
AICBM	Anti-Intercontinental Ballistic Missile
AIIB	Asian Infrastructure Investment Bank
AIRCOMNET	Air Force Communications Network
AIRS	Advanced Inertial Reference Sphere
AITP	Advanced Individual Training Program
ALA	Army Launch Area
ALCM	Air-launched Cruise Missile
ALCS	Airborne Launch Control System
ALT	Alteration (of nuclear weapon)
AMC	Air Materiel Command
	Air Mobility Command
AMSA	Advance Manned Strategic Aircraft
ANMCC	Alternate National Military Command Center
AO	Action Officer
AOG	Action Officer Group
AOGMA	Army Ordnance and Guided Missile Agency
AOU	Azimuth Orientation Unit
APL	Applied Physics Laboratory (Johns Hopkins)
APS	Active Protection System
ARAACOM	Army Antiaircraft Command Acoustic
A-RCI	Rapid Commercial off the Shelf Insertion (Program
ARADCOM	Army Air Defense Command
ARDC	Air Research and Development Command
ARGMA	Army Rocket and Guided Missile Agency
ARPA	Advance Research Projects Agency
ARS	Automatic Reference System
ARU	Automatic Reference Unit
AS	Active Stockpile
	Air Station
ASA	Automated SIOP Allocation
ASAT	Anti-Satellite (System)
ASC	Advanced Simulation and Computing (Program)
ASO	Air Support Operations
ASOC	Air Support Operations Center
ASPT	Annual Service Practice Test
ASROC	Anti-Submarine Rocket
ASTM	American Society for Testing and Materials
ASW	Anti-Submarine Warfare
ATB	Advanced Technology Bomber

ATBM	Anti-Tactical Ballistic Missile
ATK	Alliant Techno-Systems (Company)
ATRAN	Automatic Terrain Matching and Navigation
AVE	Aerospace Vehicle Equipment
AVLIS	Atomic Vapor Laser Isotope Separation
AWACS	Aircraft Warning and Control System
AWPS	Allocated Windows Planning System
AWASP	US Advanced Weapons Ammunition Supply Point
B	Bomb Designation
BA	Basic Assembly
BAA	Battery Assembly Area
BALPARS	Ballistic Parameters
BAS	Bomb Alarm System
BALTAP	Baltic Approaches (Army Group)
BCA	Budget Control Act
BCC	Battery Control Central
BCE	Battlefield Control Element
BDA	Bomb Damage Assessment
BDHSA	Bomb Director for High-speed Aircraft
BE	Basic Encyclopedia
BEQ	Bachelor Enlisted Quarters
BIT	Built In Test
BLT	Boundary Layer Transition
BMAT	Ballistic Missile Analyst Technician
BMD	Ballistic Missile Defense
BMDC	Ballistic Missile Defense Center
BMDOR	Ballistic Missile Defense Operations Room
BMEWS	Ballistic Missile Early Warning System
BNS	Bomb Navigation System
BOB	Bombs on Base (Program)
	Broad Ocean Bathymetry
BOQ	Bachelor Officer Quarters
BPU	Battery Power Unit
BRML	Bomb Release Mechanical Lock
BRS	Bomb Release System
BT	Beam-riding Tail-controlled
BTL	Bell Telephone Laboratories
BTU	British Thermal Unit
BTV	Burner Test Vehicle
BUCS	Backup Computer System
BUIC	Backup Interceptor Control System
BuOrd	(Navy) Bureau of Ordnance
BVL	Butterfly Valve Lock

BVLC	Butterfly Valve Lock Control
BW	Beam-riding, Wing-controlled
BWCC	Bomb to Warhead Conversion Components
BWP	Basic War Plan
C	Celsius
C^3	Command, Control, and Communications
C^3I	Command, Control, Communications, and Intelligence
C^4ISR	Command and Control, Communications, Computers, Intelligence, Surveillance, and Reconnaissance
C/RD	Confidential / Restricted Data
CAC	Compartmented Advisory Committee
CAL	Cornel Aeronautical Laboratory
CAP	Command Authorities Planning
CAS	Combat Alert Status
	Control Augmentation System
CASF	Composite Air Strike Force
CCCS	Central Command and Control System
CCD	Charge Coupled Device
CCIP	Common Configuration Implementation Program
CD	Command Disable
CDS-M	Command Signal Decoder-Missile (Device)
CDU	Capacitor Discharge Unit
CER	Complete Engineering Release
CERCLA	Comprehensive Environmental Response, Compensation, and Liability Act
CERCLIS	Comprehensive Environmental Response, Compensation, and Liability Information System
CENTAG	Central (European) Army Group
CENTO	Central Treaty Organization
CEOI	Communications-Electronics Operation Instructions
CEP	Circular Error Probability
CER	Complete Engineering Release
CERCLA	Comprehensive Environmental Response, Compensation, and Liability Act
CFIS	Critical Function Interrupt Switch
CFE	Conventional Forces in Europe (Treaty)
CFM	Compressed Ferro-magnetic
CHE	Chemical High Explosive (sensitive)
CHISOP	Chinese Integrated Strategic Operational Plan
CIP	Captain's Indicator Panel
CINCEUR	Commander in Chief European Command
CINCLANT	Commander in Chief Atlantic Command
CINCNORAD	Commander in Chief North American Air Defense

CINCONAD	Commander in Chief Continental Air Defense Command
CINCSAC	Strategic Air Command
CINCSTRATCOM	Commander in Chief Strategic Command
CLAW	Clustered Atomic Warhead
CLC	Central Logic and Control
CLIP	Cancel Launch In Progress
CLPP	China Lake Pilot Plant
CMDB	Composite Modified Double-Base (propellant)
CMGS	Cruise Missile Guidance Set
CMRR	Chemistry and Metallurgy Research Replacement (Project)
CMSA	Cruise Missile Support Activities
CNA	Computer Network Attack
CNC	Central Navigation Computer
CNPG	Computational Nuclear Physics Group
CNO	Chief of Naval Operations
CNPG	Computational Nuclear Physics Group
CNWDI	Critical Nuclear Weapon Design Information
COB	Co-located Operating Base
COC	Combat Operation Center
COG	Continuity of Government
COLEX	Column Exchange (Process)
COMINT	Communications Intelligence
CONAD	Continental Air Defense
CONPLAN	Contingency Plan
CONUS	Continental United States
CORM	Congressional Commission on Roles and Missions
CPFL	Contingency Planning Facilities List
CPIF	Cost Plus Incentive Fee
CRC	Control and Reporting Center
CRP	Carbon Reinforced Plastic
CSDL	Charles Stark Draper Laboratory
CSD-M	Command Signal Decoder-Missile
CSES	Canister Safe / Enable Switch
CSOC	Consolidated Space Operations Center
CSRL	Common Strategic Rotary Launcher
CTB	Comprehensive Test Ban (Treaty)
CTPB	Carboxyl Terminated Poly-Butadiene
CTV	Control Test Vehicle
CUCV	Commercial Utility Cargo Vehicle
DARHT	Dual Axis Radiographic Hydro-Test (Facility)
DAB	Defense Acquisition Board

DASO	Demonstration and Shakedown Operation
DBDB	Digital Bathymetric Data Base
DCAM	Directional Control Automatic Meteorological
DCDPS	Defense Center Data Processing System
DCSS	Display and Control Sub System
DCU	Digital Computer Unit
DDRE	Director of Defense Research and Engineering
DDR&E	Deputy Directorate of Defense Research and Engineering
DEFCON	Defense Condition
DE	Damage Expectancy
DESS	Data Entry Sub System
DEW	Distant Early Warning
DGZ	Desired Ground Zero
DIA	Defense Intelligence Agency
DIAMEX	Diamide Extraction
DIFAR	Directional Finding and Ranging
DISCOS	Disturbance Compensation System
DMA	Division of Military Application
DMCCC	Deputy Missile Combat Crew Commander
DMS	Defensive Management System
DNA	Defense Nuclear Agency
DOD	Department of Defense
DOE	Department of Energy
DPS	Digital Processing System
DR	Detection Radar
	Discrimination Radar
DRAAG	Design Review and Acceptance Group
DSARC	Defense Systems Acquisition Review Council
DSP	Defense Satellite Program
	Defense Support Program
	Digital Signal Processing / Processor
DSTP	Deep Siren Tactical Paging System
DTRA	Defense Threat Reduction Agency
E2P	Enhanced Effectiveness Program
EA	Electronics Assembly
EAM	Emergency Action Message
EBW	Exploding Bridge Wire (Detonator)
EC	Emergency Capability
ECA	Electronic Component Assembly
ECM	Electronic Countermeasures
ECS	Environmental Control System
EDB	Enterprise Database

EEI	Enhanced Electrical Isolation
EFI	Exploding Foil Initiator
	Enhanced Fidelity Instrumentation
EET	Explosive to Electric Transducer
EGISS	Enhanced Guidance Interface Sub System
EL	Erector Launcher
ELF	Extremely Low Frequency
ELEX	Electrochemical Exchange
ELINT	Electronic Intelligence
EM	Engineering Model
	Environmental Management (Office)
EMERGCON	Emergency Condition
EMP	Electromagnetic Pulse
ENDS	Enhanced Nuclear Detonation Safety
ENEC	Extendable Nozzle Exit Cones
ENI	External Neutron Initiator
EPA	Environmental Protection Agency
EPW	Earth Penetrator Weapon
ER	Enhanced Radiation
ERCS	Emergency Rocket Communication System
ERDA	Energy Research and Development Administration
ESG	Electrostatically Supported Gyroscope
ESGM	Electrostatically Supported Gyroscopic Monitor
ESTS	Electronic Systems Test Set
ESU	Electronic Sequencing Unit
EVS	Electro-Optical Viewing System
EW	Electronic Warfare
EWG	Environments Working Group
EWO	Emergency War Order
EWP	Emergency War Plan
EXPO	Extended-Range Poseidon
F	Fahrenheit
FA	Firing Azimuth
FAMAG	Field Artillery Magazine
FAR	Forward Acquisition Radar
FAS	Federation of American Scientists
FASV	Field Alert Status Verification
FB	Firing Battery
FBIA	Force Balance Integrating Accelerometer
FBM	Fleet Ballistic Missile
FBMS	Fleet Ballistic Missile Submarine
FBW	Fly-by-wire
FCET	Follow-on Commander Evaluation Test

FCS	Fire Control System
FDS	Fire Direction System
FE	Ferro-Electric
FEBA	Forward Edge of Battle Area
FEMA	Federal Emergency Management Agency
FGT	Functional Ground Test
FLTSATCOM	Fleet Satellite Communications System
FLIR	Forward-looking Infrared
FFA	Freefall Airburst
FFAR	Folding Fin Aircraft Rocket
FFG	Freefall Ground Burst
FGT	Functional Ground Test
FM	Ferro-Magnetic
FMMS	Field Missile Maintenance Squadron
FMTS	Field Maintenance Test Station
FMS	Foreign Military Sales
FNMOC	Fleet Numerical Meteorological and Oceanography Center
FOBS	Fractional Orbital Bombardment System
FOF	Follow-on Forces
FOT	Follow-on Operational Test
FPU	First Production Unit
FRAM	Fleet Rehabilitation and Modernization (Program)
FRD	Formerly Restricted Data
FRP	Fire-resistant Pit
FSL	Field Storage Location
FWDR	Final Weapon Development Report
FYDP	Future Years Defense Program
GAC	General Advisory Committee (to the Atomic Energy Commission)
GALCIT	Guggenheim Aeronautical Laboratory, California Institute of Technology
GAMA	GLCM Alert and Maintenance Areas
GAPA	Ground to Air Pilotless Aircraft (program)
GAO	Government Accounting Office
GARS	Garrison Alert Reaction Status
G&CC	Guidance and Control Computer
GBSD	Ground based Strategic Deterrent
GCA	Gyro-Compass Assembly
GEBO	General Bomber (Study)
GEDS	Ground-based Electro-Optical Deep Space (Surveillance System
GEMS	General Energy Management Steering

GIEU	Ground Integrated Electronics Unit
GIIPS	Geographic Installation Intelligence Production Specifications
GIS	Geographic Information System
GLAS	Gust Load Alleviation System
GLCM	Ground Launched Cruise Missile
GMLS	Guided Missile Launching System
GMTI	Ground Moving Target Indicator
GNO	Global Network Operations
GOC	Government Owned Contract Operated
GOR	General Operational Requirement
GP	Group
GRP	Guidance Replacement Program
GRU	Soviet Main Intelligence Directorate
GSP	Guidance System Platform
GTS	Gas Transfer System
GUPPY	Greater Underwater Propulsion Program
GZ	Ground Zero
HAC / RMPE	Higher Authority Communications / Rapid Message Processing Element
HARDS	High-altitude Radiation Detection System
H / C	Halite / Centurion (Program)
HEMTT	Heavy Expanded Mobility Tactical Truck
HF	High Frequency
	Hydrofluoric Acid
HHB	Headquarters and Headquarters Battery
HICS	Hardened Inter-Site Cable System
HIPAR	High Power Acquisition Radar
HLOS	Horizontal Line of Sight (Pipe)
HOB	Height of Burst
HP	Horse Power
HRF	High Readiness Forces
HSB	Headquarters and Service Battery
HSBD	High-speed Bomb Director
HTGR	High Temperature Gas-cooled Reactor
HTPB	Hydroxyl-Terminated Poly-Butadiene
HUD	Heads Up Display
HUSL	Harvard Underwater Sound laboratory
HVAR	High Velocity Antiaircraft Rocket
IAEA	International Atomic Energy Agency
IAP	Improved Accuracy Program
IBDL	Interim Battery Data Link
ICF	Internal Confinement Fusion

ICU	Integrated Control Unit
ICBM	Intercontinental Ballistic Missile
IEER	Institute for Energy and Environmental Research
IFC	Integrated Fire Control (Area)
IFF	Identification Friend or Foe
IHE	Insensitive High Explosive
ILCS	Improved Launch Control System
IMINT	Image Intelligence
IMPSS	Improved Minuteman Physical Security System
IMU	Inertial Measurement Unit
INF	Intermediate Nuclear Forces (Treaty)
INI	Internal Neutron Initiator
IOC	Initial Operating Capability
IONDS	Integrated Operational Nuclear Detonation (NUDET) System
IRBM	Intermediate Range Ballistic Missile
IRFNA	Inhibited Red Fuming Nitric Acid
IRIG	Inertial Rate-Integrating Gyroscope
IS	Inactive Stockpile
ISMP	ICBM Security Modernization Program
ISPAN	Integrated Strategic Planning and Analysis Network
ISST	ICBM Super-high-frequency Satellite Terminal
IBMSST	Intercontinental Ballistic Missile Silo Super-Hardening Technology
IUQS	Intent Unique Signal (Generator)
IUSS	Integrated Undersea Surveillance System
JATO	Jet Assisted Take Off
JCAE	Joint Committee on Atomic Energy
JCMPO	Joint Cruise Missile Project Office
JFCC-GS	Joint Functional Component Command - Global Strike
JFCC-GSI	Joint Functional Component Command - Global Strike and Integration
JFCC-IMD	Joint Functional Component Command - Integrated Missile Defense
JFCC-ISR	Joint Functional Component Command - Intelligence, Surveillance, and Reconnaissance
JFCC-NW	Joint Functional Component Command - Network Warfare
JFCC-SPACE	Joint Functional Component Command – Space
JCP	Joint Change Proposal
JCS	Joint Chiefs of Staff
JHU	Johns Hopkins University
JIOWC	Joint Information Operations Warfare Center
JIPP	Joint Integrated Product Plan

JOWP	Joint Outline War Plan
JPL	Jet Propulsion Laboratory
JSCP	Joint Strategic Capabilities Plan
JSCP-N	Joint Strategic Capabilities Plan - Nuclear Supplement
JSP	Joint Staff Planners
JSR	Joint Surety Report
JSS	Joint Surveillance System
JSTPS	Joint Strategic Target Planning Staff
JTA	Joint Test Assembly
JTF	Joint Task Force
JWPC	Joint War Plans Committee
KCP	Kansas City Plant
KGB	Soviet Committee for State Security
KT	Kilotons Equivalent TNT
LAC	Lightning Arrestor Connection
LACEF	Los Alamos Critical Experiment Facility
LANL	Los Alamos National Laboratory
LANSCE	LANL Neutron Scattering Center
LANTFLT	(US Navy) Atlantic Fleet
LANTIRN	Low-altitude Navigation and Targeting Infrared for Night (System)
LAO	Limited Attack Options
LASL	Los Alamos Scientific Laboratory
LAUT	Launch Area Utility Tunnel
LC	Launch Complex
LCC	Launch Control Center
LCCFC	Launch Control Complex Facilities Console
LCF	Launch Control Facility
LECS	Launch Encoded Control System
LEGG	Launch Ejection Gas Generator
LEP	Life Extension Program
LES	Launch Enable System
LF	Launch Facility
LFSB	Launch Facility Support Building
LLC	Limited Life Component
LLM	Launcher Loader Module
LLNL	Lawrence Livermore National Laboratory
LLTV	Low-light Level Television Camera
LMFBR	Liquid Metal Fast Breeder Reactor
LOFAR	Low Frequency Analysis and Recording
LOPAR	Low Power Acquisition Radar
LORAN	Long Range Navigation
LOX	Liquid Oxygen

LP	Limited Production
LPE	Launch Preparation Equipment
LPEC	Launch Preparation Equipment Chamber
LPEV	Launch Preparation Equipment Vault
LPI	Low Probability of Intercept
LPT	Loaded Pylon Test
LRCA	Long-Range Combat Aircraft
LRL	Lawrence Radiation Laboratory
LRU	Line Replaceable Unit
LS	Launch Station
LST/SCAM	Laser Spot Tracker / Strike Camera
LWF	Light Weight Fighter
LWL	Light Weight Launcher
MAC	Military Airlift Command
MAD	Mutually Assured Destruction
	Magnetic Anomaly Detection
MADM	Medium Atomic Demolition Munition
MAO	Major Attack Option
MAP	Military Assistance Program
MAR	Major Assembly Release
	Missile Acquisition Radar
MARC	Missile Armed Response Team
	Michigan Aerospace Research Center
MART	Missile Armed Response Team
MC	Military Committee
MCC	Missile Combat Crew
MCCC	Missile Combat Crew Commander
	Mobile Consolidated Command Center
MCCS	Multi-Code Coded Switch
MCIS	Multiple Corridor Identification System
MCM	Mission Control Module
MDAP	Mutual Defense Aid Program
MEA	Mean Effect Area
MEECN	Minimum Essential Emergency Communications Network
MED	Manhattan Engineering District
MEEV	Mechanical and Electrical Equipment Vault
MET	MCCS Encryption Translator
MFD	Multi-Function Display
MFP	Mean Free Path
MFT	Missile Facilities Technician
MGPS	Missile Graphic Planning System
MGS	Missile Guidance System

MGTT	Motor Guidance Transport Trailer
MICOM	(Army) Missile Command
MIDAS	Missile Detection and Alarm System
MIDB	Modernized Integrated Database
MILDEC	Military Deception
MILSTAR	Military Strategic Tactical and Relay (UHF Network}
MIR	Major Impact Report
MIRV	Multiple Independently-targetable Reentry Vehicle
MIT	Massachusetts Institute of Technology
MK	Mark
MLC	Military Liaison Committee
MLF	Multi-Lateral Force
MLU	Mid-Life Update
MMH	Mono-Methyl Hydrazine
MMRBM	Mobile Mid-Range Ballistic Missile
MMTI	Maritime Moving Target Indicator
MOB	Major Operating Base
MOD	Modification (of Nuclear Weapon)
MON	Mixed Oxides of Nitrogen
MPH	Miles Per Hour
MPMS	Missile Position Measurement System
MPSS	Minuteman Physical Security System
MPT	Missile Procedures Trainer
MRA	Missile Radar Altimeter
MRBM	Medium Range Ballistic Missile
MRV	Multiple Reentry Vehicles
MSA	Mine Safety Appliance (Company)
MSD	Minimum Safe Distance
MSAD	Mechanical Safing and Arming Device
MSCB	Missile Site Control Building
MSDP	Missile Site Digital Processor
MSR	Missile Site Radar
	Maritime Silk Road
MSS	Multi-Spectral Scanner
MT	Megatons Equivalent TNT
MTC	Magnetic Tape Cartridge
MTR	Missile Tracking Radar
MTRE	Missile Test and Readiness Equipment
MUNMS	Munitions Maintenance Squadrons
MVLS	Molecular Vapor Laser Separation
MX	Missile (Experimental)
NAA	North American Aviation
NACA	National Advisory Committee for Aeronautics

NACC	NORAD Automated Control Center
NAM	National Atomic Museum
NAOC	National Airborne Operations Center
NAS	National Academy of Sciences
NASA	National Aeronautics and Space Administration
NASM	National Air and Space Museum
NATO	North Atlantic Treaty Organization
NAMFI	NATO Missile Firing Installation (Crete)
NAVDAC	Navigation Data Assimilation Computer
NAVFACS	Naval Facilities
NAVSPSUR	Naval Space Surveillance and Tracking System
NAVSUBFOR	Naval Submarine Forces
NAVSUBLANT	Naval Submarine Forces Atlantic Fleet
NAVWPNSTA	Naval Weapon Station
NB	Naval Base
NCA	National Command Authorities (The President and the Secretary of Defense)
NCC	NORAD Control Center
NCS	National Communications System
NDP	National Defense Panel
NDRC	National Defense Research Council
NEA	Nuclear Employment Authority
NEACP	National Emergency Airborne Command Post
NECAP	National Emergency Command Post
NECPA	National Emergency Command Post Afloat
NIF	National Ignition Facility
NMCC	National Military Command Center
NME	National Military Establishment
NMISS	Navigation and Missile Interface Sub-System
NNSS	Nevada National Security Site
NSA	National Security Archive
NSS	Navigation Sonar System
NNSA	National Nuclear Security Agency
NORAD	North American Aerospace Defense
NORTAF	Northern Task Force
NORTHAG	Northern (European) Army Group
NORTHCOM	Northern Command
NPG	(NATO) Nuclear Planning Group
NPR	Nuclear Posture Review
NPS	National Park Service
NRAS	Nuclear Release Authentication System
NRDC	National Research Development Council
NRO	National Reconnaissance Office

NRTS	National Reactor Testing Station
NSAM	National Security Action Memorandum
NSA	National Security Agency
NSC	National Security Communication
	National Science Council
NSNF	Non-Strategic Nuclear Forces
NSWCDD	Naval Surface Warfare Center Dahlgren Division
NSDD	National Security Decision Directive
NSDM	National Security Decision Memorandum
NWSM	Nuclear Weapons Stockpile Memorandum
NSNF	Non-Strategic Nuclear Forces
NSS	National Storage Site
NSWC	Naval Surface Weapons Center
NSWU	National Special Weapons Unit
NTS	Nevada Test Site (originally Nevada Proving Grounds)
NUDET	Nuclear Detonation
NWC	Nuclear Weapons Complex
	Nuclear Weapons Council
NWCSC	Nuclear Weapons Council Standing Committee
NWCSSC	Nuclear Weapons Council Standing and Safety Committee
NWCWSC	Nuclear Weapons Council Weapons Safety Committee
NWRWG	Nuclear Weapons Requirements Working Group
NWS	North Warning System (formerly DEW Line}
NWSM / RPD	Nuclear Weapons Stockpile Memorandum / Requirements and Planning Document
NWSS	Nuclear Weapons Support Section
OAD	Operational Availability Date
OAS	Offensive Avionics System
OBOR	One Belt One Road Initiative
OEM	Office of Environmental Management
OGMC	Ordnance Guided Missile Center
OLD	Operations and Logistics Directorate
OMMS	Organizational Missile Maintenance Squadron
OMTS	Organizational Maintenance Test Station
OPEVAL	Operations Evaluation
OPLAN	Operational Plan
OPNAV	Office of the Chief of Naval Operations
OPORD	Operational Order
OPSEC	Operations Security
ORALLLOY	Oak Ridge Alloy (Weapon Grade Uranium)
ORD	Operational Requirements Document
ORDCIT	Ordnance / California Institute of Technology

OREX	Organic Exchange
ORL	(Penn State) Ordnance Research Laboratory
ORT	Operational Readiness Test
OSD	Office of the Secretary of Defense
OSDP	Operational System Development and Production
OSP	Ocean Survey Program
OSR	(Air Force) Office of Scientific Research
OSRD	Office of Scientific Research and Development
OSS	Operational Storage Site
OST	Office of Secure Transport
OSTF	Operational Suitability Test Facility
OT&E	Operational Test and Evaluation
OTH-N	Over-The-Horizon Network
OT	Operational Test
OTC	Ordnance Technical Committee
OTS	Operational Test System
PA	Probability of Arrival
	Pre-Arm
PACAF	Pacific Air Force
PACBAR	Pacific Barrier System
PACC	Post Attack Command and Control
PACCS	Post Attack Command and Control System
PACFLT	(US Navy) Pacific Fleet
PACS	Programmable Armament Control Set
PAG	Polyalkylene Glycol
PAL	Permissive Action Link
PAM	Plume Avoidance Maneuver
PAR	Phased Array Radar
PARBRO	Peacetime Airborne Reconnaissance Program
PARCS	Perimeter Acquisition Radar System
PAS	Protective Aircraft Shelter
PAWS	Phased Array Warning System
PBAA	Poly-Butadiene Acrylic Acid
PBAN	Poly-Butadiene acrylic acid AcryloNitrile
PBCS	Post-Boost Control System
PBPS	Post-Boost Propulsion System
PBV	Post-Boost Vehicle
PBX	Plastic Bonded Explosive
PCC	Platoon Control Center
PCL	Positive Control Launch
PCTAP	Positive Control Turn Around Point
PD	Presidential Directive
	Probability of Damage

PDD	Presidential Decision Directive
PDM	Program Decision Memorandum
PEG	Polyethylene Glycol
PEM	Product Evaluation Missile
PERT	Program Evaluation and Review Technique
PFP	(Hanford) Plutonium Finishing Plant
PHOTINT	Photographic Intelligence
PIGA	Pendulous Integrating Gyro Accelerometer
PIP	Precision Instrumentation Package
PIPA	Pulsed Integrating Pendulous Accelerometer
PLS	Pre-Launch Survival
PM	Production Model
PMI	Primary Mission Inventory
PMO	Project Management Office
PNI	Presidential Nuclear Initiative
POA	Polaris Analysis and Performance (Requirements)
POE	Polaris Evaluation and Testing
POG	Program Officers Group
POM	Program Objectives Memorandum
POMFLANT	Polaris Missile Facility, Atlantic
POMFPAC	Polaris Missile Facility, Pacific
POS	Polaris Satellite Navigation Development
PPG	Pacific Proving Grounds
PPI	Plan Position Indicator
PRC	People's Republic of China
PRF	Pulse-Doppler Fire-Control Radar
PRGB	Point Reference Guide Book
PRP	Propellant Replacement Program
PSE	Power Station Equivalent
PSRE	Propulsion System Rocket Engine
PSYOP	Psychological Operations
PTP	Probability to Penetrate
PTS	Programmer-Test Station
	Propellant Transfer Station
PTV	Propulsion Test Vehicle
PUREX	Plutonium Uranium Recovery by Extraction
PV	Physical Vulnerability (to Nuclear Explosion)
PWDR	Preliminary Weapon Development Report
PWR	Pressurized Water Reactor
QA	Quality Assurance
QART	Quality Assurance and Reliability Testing
QDR	Quadrennial Defense Review
QM	Quantified Milestones

	QMU Quantification of Margins and Uncertainties
QOR	Qualitative Operational Requirement
RAD	Radiation Absorbed Dose
RADAR	Radio Detection and Ranging
RADINT	Radar Intelligence
RAM	Regulus Assault Missile
RASCAL	RAdar SCAnning Link
RB	Reentry Body
RD	Restricted Data
RDCT	Remote Data Change Target (Command)
RDT&E	Research, Development, Tests, and Evaluation
REA	Retarded Airburst
REACT	Rapid Execution and Combat Targeting (System)
REDOX	Reduction and Oxidation
REG	Retarded Ground Burst
REM	Radiation Equivalent Man
RFHCO	Rocket Fuel Handler's Clothing Outfit
RFI	Radio Frequency Interference
RFML	Rapid Fire Multiple Launch (Site)
RFNA	Red Fuming Nitric Acid
RFP	Request for Proposal
RGAP	Rate Gyro / Accelerometer Package
RH-TRUWa	Remote Handled - TRansUranic Waste
RIM	Rocket Intercept Missile
	Reception, Inspection and Maintenance (Building)
RISOP	Russian (Red) Integrated Strategic Operational Plan
	RLE Remote Launch Equipment
RLOB	Remote Launch Operations Building
RMUC	Reference measuring Unit and Computer
RNEP	Robust Nuclear Earth Penetrator
RNO	Regional Nuclear Option
RNWMC	Regionalized Nuclear Weapons Maintenance Concept
ROAD	Reorganization Objectives, Army Divisions
ROCC	Regional Operations Control Centers
ROSA	Report on Stockpile Assessments
RPD	Requirements and Planning Document
RPM	Revolutions Per Minute
RRW	Reliable Replacement (Nuclear) Warhead
RSGF	Reference Scene Generation Facilities
RSL	Remote Sprint Launch (Site)
RSOP	Reconnaissance, Selection, and Occupation of Position
RSS	Ready Storage Shelter
RTB	Reentry Thermal Battery

RTV	Recoverable Test Vehicle
RUP	Radar Updated Path Length (Fuze)
RV	Reentry Vehicle
RVA	Remote Visual Assessment (Camera)
S/RD	Secret / Restricted Data
SAC	Strategic Air Command
SAGE	Semi-Automatic Ground and Environment
SACCS	Strategic Automated Command and Control System
SADM	Special Atomic Demolition Munition
SAINT	SATellite INTerceptor
SALT	Strategic Arms Limitation Treaty
SAMOS	Satellite and Missile Observation System
SANEX	Selective Actinide Extraction
SAO	Selected Attack Option
SAR	Synthetic Aperture Radar
SARH	Semi-Active Radar Homing
SAS	Sealed Authenticator System
SASS	Special Aircraft Service Stores
SATS	SLBM Adaptive Targeting System
SBIRS	Space Based Infrared System
SBSS	Science Based Stockpile Surveillance (Program)
SCAMP	Sergeant Contractor Assisted Modification Program
SCC	Super Combat Center
SCC-WMD	USSTRATCOM Center for Combating Weapons of Mass Destruction
SCSFCH	Standard Coding System Functional Classification Handbook
SCTS	System Components Test Station
SDMH	Symmetrical Di-Methyl Hydrazine
SEAD	Suppression of Enemy Air Defense
SEATO	Southeast Asia Treaty Organization
SECOM	Secure Communications (Network)
SEP	Stockpile Evaluation Program
	Sustaining Engineering Program
SEPW	Strategic Earth Penetrating (nuclear) Weapon
SERV	Safety Enhanced Reentry Vehicle
SFE	Slim-Loop Ferro-Electric
SFIR	Specific Force Integrating Receiver
SFT	Stockpile Flight Test
SI	Smithsonian Institution
SICBM	Small Intercontinental Ballistic Missile
SIGINT	Signals Intelligence
SILEX	Separation of Isotopes by Laser EXcitation

SILVER	Strategic Installation List of Vulnerability Effects and Results
SINS	Ship's Inertial Navigation System
	Submarine Inertial Navigation System
SIOP	Single Integrated Operational Plan
SIP	Stockpile Improvement Program
	Support Improvement Program
SIPS	SLBM Integrated Planning System
SIR	Submarine Intermediate Reactor
SLBM	Submarine Launched Ballistic Missile
SLCM	Submarine Launched Cruise Missile
SLFCS	Survivable Low Frequency Communication System
SLMS	Surface Launched Missile System
SLT	Stockpile Laboratory Test
SLTF	Silo Launch Test Facility
SMAP	Sergeant Modification Assistance Program
SMW	Strategic Missile Wing
SNDV	Strategic Nuclear Deliver Vehicle
SNL	Sandia National Laboratory
SNORT	Supersonic Naval Ordnance Research Track
SOB	Super Oralloy Bomb
SOC	Sector Operating Center
SOCS	Strategic and Operational Control System
SONAR	Sound Navigation and Ranging
SORT	Strategic Offensive Reduction Treaty
SOSUS	Sound Surveillance System
SP	Self-Propelled
SPCC	Ships Parts Control Center
SPO	Strategic Projects Office
SREB	Silk Road Economic Belt
SRF	(Russian) Strategic Rocket Forces
SRS	Savannah River Site
	SLBM Retargeting System
SSPO	Strategic Systems Project Office
SPADATS	Space Detection and Tracking System
SRAM	Short Range Attack Missile
SRBM	Short Range Ballistic Missile
SREMP	Source Region Electromagnetic Pulse
SSB	Single Side Band (Radio)
SSBN	Nuclear Powered Ballistic Missile Submarine
SSM	Surface-to-Surface Missile
SSN	Nuclear Powered Attack Submarine
SSNT	Shape Stable Nose Tip

SSRS	STRATCOM Secure Re-Code System
SST	Safe Separation Time
	Safe Secure Trailer / Safe Secure Train
STAR	Steam Assisted Regulus (Dolly)
START	Strategic Arms Reduction Treaty
STAS	Safe to Arm Signal
STAWS	Submarine Tactical Anti-Ship Weapons System
STO	Special Technical Operations
STR	Submarine Thermal Reactor
STRIKWARN	Nuclear Strike Warning
STRATCOM	Strategic Command
STRATJIC	Strategic Joint Intelligence Center
STS	Stockpile to Target Sequence
STV	Supersonic Test Vehicle
SUBGRU	Submarine Group
SUBLANT	Submarine Forces Atlantic Fleet
SUBPAC	Submarine Force Pacific Fleet
SUBROC	Submarine (Launched) Rocket
SUBRON	Submarine Squadron
SWC	Special Weapons Command
SWCF	SLBM Weapons Control Facility
SWD	Special Weapons Directorate
SWESS	Special Weapons Emergency Separation System
SWFLANT	Strategic Weapons Facility Atlantic
SWFLPAC	Strategic Weapons Facility Pacific
SWPP	Salt Wells Pilot Plant
SWPS	Strategic War Planning System
SWPS-EDB	Strategic War Planning Systems Enterprise Database
TA	Terrain Avoidance
	Technical Area
TAC	Tactical Air Command
TACAMO	Take Charge and Move Out
TACAN	Tactical Air Navigation
TACC	Tactical Air Control Center
TACMAR	Tactical Missile Acquisition Radar
TADM	Tactical Atomic Demolition Munition
TAINS TERCOM	TERCOM Aided Navigation System
TAPS	Tomahawk Afloat Planning System
TASM	Tactical Air-to-Surface (nuclear) Missile
TASS	Telegraph Agency of the Soviet Union
TATTE	Talos Tactical Test Equipment
TCAP	Thermal Cycling Absorption Process
TCC	Transformation Coordinating Committee

TCM	Targeting Control Message
TCP	Two-Component Pod
TDD	Target Detection Device
TDI	Target Data Inventory
TEL	Transporter Erector Launcher
TELINT	Telemetry Intelligence
TERCOM	Terrain Contour Matching
TEWS	Tactical Electronic Warfare System
TF	Terrain Following
TIHB	Target Intelligence Hand Book
TLAM-C	Tomahawk Land Attack Missile - Conventional
TLAM-D	Tomahawk Land Attack Missile - Dispenser
TLAM-N	Tomahawk Land Attack Missile - Nuclear
TMA	Target Motion Analysis
TN	Thematic Mapper
TNDV	Tactical Nuclear Delivery Vehicle
TNS	Tactical Missile Squadron
TNCP	Trident Navigation Commonality Program
TOR	Tracking Only Radar
TP	Tactical Prototype
TPBAR	Tritium Producing Burnable Absorber Rods
TPFDD	Time Phased Force and Deployment Data
TPI	Technical Proficiency Inspection
TPP	Trans-Pacific Partnership
TRAM	Target Recognition Attack Multi-sensor
TRR	Target Ranging Radar
TRUEX	Transuranic Extraction
TSD	Transportation Safeguards Division
TS/RD	Top Secret / Restricted Data
TSSG	Trajectory Sensing Signal Generator
TTR	Tonopah Test Range
	Target Tracking Radar
TUQS	Trajectory Unique Signal
TVS	Target Value System (Manual)
TVCS	Thrust-Vector Control System
TX	Experimental Bomb
UCRL	University of California Radiation Laboratory
UDMH	Unsymmetrical Di-Methyl Hydrazine
UHF	Ultra High Frequency
ULF	Ultra Low Frequency
ULMS	Undersea Long-Range Missile System
UMB	Universal Missile Building
UNEX	Universal Extraction

UPF	Uranium Processing
UREX	Uranium Extraction
USA	United States of America
	United States Army
USACC	US Air Combat Command
USACE	US Army Corps of Engineers
USAFE	US Air Forces in Europe
USJFCOM	US Joint Forces Command
USEUCOM	US European Command
USG	Unique Signal Generator
USGS	Universal Space Guidance System
USGS	United States Geological Survey
USNORTCOM	US Northern Command
USSPACEAF	US Space Air Force
USSPAECOM	US Space Command
USSTRATCOM	US Strategic Command
VHF	Very High Frequency
VLF	Very Low Frequency
VLR	Very Long-Range
VMPB	Vertical Missile Packaging Building
VN	Vulnerability Number
VNIIEF	Russian Scientific Research Institute for Experimental Physics
VPM	Virginia Payload Module
VPRS	Velocity and Position Reference System
VSA	Vibrating String Accelerometer
W	Warhead Designation
WAC	Woman's Army Corps
WADC	Wright Air Development Center
WDCR	Weapon Design and Cost Report
WDD	(Air Force) Western Development Division
WEO	Weapons Engineering Officer
WEU	Western European Union
WFNA	White Fuming Nitric Acid
WHB	Warhead Handling Building
WIPP	Waste Isolation Pilot Plant
WIU	Warhead Interface Unit
WLTR	Western Launch and Test Range
WMD	Weapons of Mass Destruction
WMT	Weapon Maintenance Truck
WPS	War Planning System
WR	War Reserve
WS	Weapon System

WSCE	Weapon System Control Element
WSEG	Weapon System Evaluation Group
WSMC	Western Space and Missile Center
WSMR	White Sands Missile Range
WSR	Weapon System Reliability
WSV	Weapon Storage Vault
X	Experimental Designation
XPM	Experimental Propulsion Missile
XR	External (neutron) Initiator
XRL	Extended Range Lance
XW	Experimental Warhead
Y	Yield
ZAR	Zeus Acquisition Radar
ZEL	Zero Length
ZI	Zone of the Interior (Continental United States)

APPENDICES

APPENDIX 1
ORDER OF BATTLE – UNITED STATES NUCLEAR FORCES

Department of Defense (DOD), Pentagon, Washington, D.C.
US Strategic Command (USSTRATCOM) Offutt AFB, Nebraska
    20th Air Force / Task Force 214 F.E. Warren AFB, WY
        90th Space Wing (LGM 30G) Warren AFB, WY
            319th Missile Squadron
            320th Missile Squadron
            321st Missile Squadron
        91st Space Wing (LGM 30G) Minot AFB, ND
            740th Missile Squadron
            741st Missile Squadron
            742nd Missile Squadron
        341st Space Wing (LGM 30G) Malmstrom AFB, MT
            10th Missile Squadron
            12th Missile Squadron
            490th Missile Squadron
    Joint Functional Component Command for Global Strike
        (JFCC-GS) / 8th Air Force / Task Force 204 Barksdale AFB, LA
    Joint Functional Component Command for Space (JFCC-SPACE) / 14th Air Force / Air Force Strategic - Space Vandenberg AFB, CA
        21st Space Wing (Missile Warning) Peterson AFB, CO
    Naval Submarine Forces (NAVSUBFOR) Norfolk NB, VA
        Submarine Forces Atlantic Fleet (SUBLANT)
            Submarine Group 10 (SUBGRU 10) (SSBN) King's Bay, GA
        Submarine Forces Pacific Fleet (SUBPAC)
            Submarine Group 9 (SUBGRU 9) (SSBN) Bangor, WA
US Air Force Air Combat Command (USACC) Langley AFB, Virginia
    8th Air Force Barksdale AFB, LA
        2nd Bomb Wing (B-52H) Barksdale AFB, LA
            11th Bomb Squadron
            20th Bomb Squadron
            96th Bomb Squadron
        5th Bomb Wing (B-52H) Minot AFB, ND
            23rd Bomb Squadron
            69th Bomb Squadron
        509th Bomb Wing (B-2A) Whiteman AFB, MO
            13th Bomb Squadron
            393rd Bomb Squadron

9th Air Force Shaw AFB, South Carolina
    4th Tactical Fighter Wing (F-15E) Seymour Johnson AFB, NC
        333rd Fighter Squadron
        334th Fighter Squadron
        335th Fighter Squadron
        336th Fighter Squadron

US Northern Command (USNORTHCOM) Peterson AFB, Colorado
    North American Aerospace Defense Command (NORAD) Peterson AFB, CO
    Cheyenne Mountain Directorate Peterson AFB, CO

US European Command (USEUCOM) Stuttgart-Vaihingen, Germany
    US Air Forces in Europe (USAFE), Ramstein AFB, Germany
        *52nd Fighter Wing (F-16C/D) Spangdhalem AFB, German
            480th Fighter Squadron
        31st Fighter Wing (F-16C/D) Aviano AFB, Italy
            510th Fighter Squadron
            555th Fighter Squadron
        *48th Fighter Wing (F-15E) RAF Lakenheath, England
            492nd Fighter Squadron
            494th Fighter Squadron
        *39th Air Base Wing (F-16C/D) Incirlik, Turkey

* Require the forward deployment of nuclear weapons.

APPENDIX 2
US STRATEGIC WAR PLANS

Year	War Plan	National Guidance	OSD Guidance	Strategic Policy
1945	Totality			
1946	Pincher, Makefast			
1947	Broiler, Charioteer			
1948	Halfmoon, Fleetwood	NSC 30		
1949	Trojan, Offtackle	NSC 20/4		
1950	Shakedown, EWP 1-50	NSC 68		
1951	EWP 1-51			
1952				
1953	EWP 1-53	NSC 162/2		Massive Retaliation
1954	50-54			
1955	EWP 1-55			
1956				Counterforce
1957	BWP 1-58, Dropshot			
1958				
1959				
1960	BWP 1-60			
1961	SIOP 62			Flexible Response
1962	SIOP 63			
1963				
1964	SIOP 64			Assured Destruction
1965				
1966	SIOP 4			
1967	SIOP 4			
1969	SIOP 4C			
1969	SIOP 4E/F			
1970	SIOP 4G/H			Strategic Sufficiency
1971	SIOP 4I/J			
1972	SIOP 4K/L			Mutual Destruction
1973	SIOP 4M/N			
1974	SIOP 4O/Ox	NSDM 242	NUWEP-74	
1975	SIOP 4P			
1976	SIOP 5/5A			
1977	SIOP 5B	PD-18		Countervailing
1978	SIOP 5C			
1979	SIOP 5D			
1980	SIOP 5E	PD-59	NUWEP-80	

1981	SIOP 5F	NSDD-13		Prevailing
1982	SIOP 5G		NUWEP-82	
1983	SIOP 6			
1984	SIOP 6A		NUWEP-84	
1985	SIOP 6B			
1986	SIOP 6C			
1987	SIOP 6D		NUWEP-87	
1988	SIOP 6E			
1989	SIOP 6F			
1990	SIOP 6G			
1991	SIOP 6H			
1992	SIOP 93		NUWEP-92	
1993	SIOP 94			Cooperative Engagement
1994	SIOP 95	PDD-37		
1995	SIOP 96			
1996	SIOP 97			
1997	SIOP 98			
1998	SIOP 99			
1999	SIOP 00	PDD-60	NUWEP-99	
2000	SIOP 01			
2001	SIOP 02			
2002	SIOP 03			
2003	OPLAN 8044 Rev 3			
2004	OPLAN 8044 Rev 4		NUWEP-04	
2005	OPLAN 8044 Rev 5			
2006	OPLAN 8044 Rev 6			Tailored Deterrence
2007	OPLAN 8044 Rev 7			
2008	OPLAN 8010 Rev 8	GEF-08		
2009	OPLAN 8010 Rev 9			
2010	OPLAN 8010 Rev 10			Minimum Deterrence
2011	OPLAN 8010 Rev 11			
2012	OPLAN 8010 Rev 12			
2013	OPLAN 8010 Rev 13			
2014	OPLAN 8010 Rev 14			
2015	OPLAN 8010 Rev 15			
2016	OPLAN 8010 Rev 16			
2017	OPLAN 8010 Rev 17			

APPENDIX 3
CURRENT AND HISTORIC DELIVERY SYSTEMS OF THE STRATEGIC TRIAD AIR FORCE STRATEGIC BOMBERS

B-29 SUPERFORTRESS MODELS A & B
    Manufacturer: Boeing
    Number Manufactured: 3,960 – 100 converted to "Silverplate" nuclear-capable model
    In Service: 1944 - 1954

SPECIFICATIONS MODEL B
    Length: 99 feet
    Span: 141 feet 3 inches
    Height: 27 feet 9 inches
    Weight: Max 137,500 pounds
    Powerplant: four 2,200 HP radial engines
    Speed: 363 MPH
    Ceiling: 31,850 feet
    Range: 4,200 miles
    Crew: Eleven
    Defensive Armament: Twelve 0.5 caliber machine guns in remotely controlled turrets
    Offensive Armament: One strategic nuclear weapon

B-50 SUPERFORTRESS MODELS A, B, C, & D
    Manufacturer: Boeing
    Number Manufactured: 370 – 100 converted to "Silverplate" nuclear-capable model
    Maximum Operational: 224
    In Service: 1945 - 1955

SPECIFICATIONS
    Span: 141 feet 3 inches
    Height: 34 feet 7 inches
    Weight: Max 137,500 pounds
    Powerplant: four 3,500 HP radial engines
    Speed: 395 MPH
    Ceiling: 35,000 feet
    Range: 4,800 miles with 10,000-pound load
    Crew: Eleven
    Defensive Armament: Thirteen 0.5 caliber machine guns in remotely controlled turrets
    Offensive Armament: One strategic nuclear weapon

B-36 PEACEMAKER MODELS A, B, C, D, E, F, G, H, & J
    Manufacturer: Convair
    Number Manufactured: 385
    Maximum Operational: 209

In Service: 1948 - 1959

SPECIFICATIONS MODEL H
    Length: 162 feet 1 inch
    Span: 230 feet
    Height: 47 feet
    Weight: Max 357,000 pounds
    Powerplants: six 3,800 HP radial engines and four 5,200-pound thrust turbojets
    Speed: 439 MPH
    Ceiling: 44,000 feet
    Range: 6,225 miles with a 10,000-pound load
    Crew: Sixteen
    Defensive Armament: Sixteen 20mm cannon in remotely controlled turrets
    Offensive Armament: Two strategic nuclear bombs at short range or one at long-range

B-47 STRATOJET MODELS B & E
    Manufacturer: Boeing
    Number Manufactured: 2,041
    Maximum Operational: 1,367
    In Service: 1951 - 1967

SPECIFICATIONS MODEL E
    Length: 109 feet 10 inches
    Span: 116 feet
    Height: 27 feet 11 inches
    Weight: Max 206,700 pounds
    Powerplant: six 6,000-pound thrust turbojets
    Speed: 606 MPH
    Ceiling: 40,500 feet
    Range: 4,000 miles
    Crew: Three
    Defensive Armament: Two 20mm cannon in tail
    Offensive Armament: One strategic nuclear or thermonuclear weapon

B-52 STRATOFORTRESS MODELS B, C, D, E, F, G & H
    Manufacturer: Boeing
    Number Manufactured: 744
    Maximum Operational: 264
    In Service: 1955 - Present

SPECIFICATIONS MODEL D
    Length: 156 feet 7 inches
    Span: 185 feet
    Height: 48 feet 4 inches
    Weight: Max 450,000 pounds

Powerplant: eight 12,100-pound thrust turbojets
Speed: 630 MPH
Ceiling: 47,300 feet
Range: 7,180 miles with a 10,000-pound bomb load
Crew: Six
Defensive Armament: Four 0.5 caliber machine guns or two 20mm cannons in tail
Offensive Armament: One or two strategic thermonuclear Bombs

SPECIFICATIONS MODEL G
Length: 160 feet 11 inches
Span: 185 feet
Height: 40 feet 9 inches
Weight: Max 488,000 pounds
Powerplant: eight 13,750-pound thrust turbojets
Speed: 636 MPH
Ceiling: 47,000 feet
Range: 8,200 miles with 10,000 a pound bomb load
Crew: Six
Defensive Armament: Four 0.5 caliber machine guns in tail, Quail decoy
Offensive Armament: Twenty (max.) SRAMs & four thermonuclear bombs

SPECIFICATIONS MODEL H
Length: 156 feet
Span: 185 feet
Height: 40 feet 8 inches
Weight: Max 488,000 pounds
Powerplant: eight 17,000-pound thrust turbofans
Speed: 632 MPH
Ceiling: 47,700 feet
Range: 9,650 miles
Crew: Six
Defensive Armament: One 20mm Vulcan cannon in tail
Offensive Armament: Twenty ACMs

B-58 HUSTLER
Manufacturer: Convair (General Dynamics)
Number Manufactured: 116
Maximum Operational: 94
In Service: 1960 – 1970

SPECIFICATIONS
Length: 96 feet 9 inches
Span: 56 feet 10 inches
Height: 31 feet 5 inches
Weight: Max 163,000 pounds

Powerplant: four 15,600-pound thrust turbojets
Speed: 1,345 MPH
Ceiling: 63,400 feet
Range: 4,450 miles
Crew: Three
Defensive Armament: One 20mm Vulcan Cannon in tail
Offensive Armament: One thermonuclear pod, and four thermonuclear bombs

FB-111A AARDVARK
Manufacturer: General Dynamics
Number Manufactured: 79
Maximum Operational: 79
In Service: 1969 - 1991

SPECIFICATIONS
Length: 73 feet 6 inches
Span: 33 feet 11inches swept; 70 feet full
Height: 17 feet 4 inches
Weight: Max 100,000 pounds
Powerplant: two 20,350-pound thrust turbojets
Speed: 1,450 MPH
Ceiling: 60,000+ feet
Range: 4,000 miles
Crew: Two
Defensive Armament: Electronics
Offensive Armament: Six thermonuclear bombs or six SRAMs

B-1B LANCER
Manufacturer: Rockwell International
Number Manufactured: 100
Maximum Operational: 95
In Service: 1985 - present

SPECIFICATIONS
Length: 147 feet
Span: 78 feet 2.5 inches, Swept, 136 feet 8.5 inches fully extended
Height: 27 feet 11 inches
Weight: Max 477,000 pounds
Powerplant: four 30,000-pound thrust turbofans
Speed: 825 MPH at altitude, 600 MPH at low-level
Ceiling: 60,000 feet
Range: 7,455 miles
Crew: Four
Defensive Armament: Electronics
Offensive Armament: Twenty-four internal thermonuclear bombs

Aircraft now relegated to the conventional row

B-2A SPIRIT
- Manufacturer: Northrop
- Number Manufactured: 21
- Maximum Operational: 20
- In Service: 1993 - Present

SPECIFICATIONS
- Length: 69 feet
- Span: 172 feet
- Height: 17 feet
- Weight: Max 400,000 pounds
- Powerplant: Four 17,300-pound thrust turbofans
- Speed: High subsonic
- Ceiling: 50,000 feet
- Range: 11,515 miles with one refueling
- Crew: Two
- Defensive Armament: Electronics
- Offensive Armament: Sixteen internal thermonuclear bombs

NAVY STRATEGIC BOMBERS

AJ-1 & AJ-SAVAGE
- Manufacturer: North American
- Number Manufactured: 148
- In Service: 1951 - 1960

SPECIFICATIONS AJ-1
- Length: 63 feet 1 inch
- Span: 71 feet 5 inches
- Height: 20 feet 5 inches
- Weight: Max 50,900 pounds
- Powerplant: two 2,300 HP radials engines and one 4,000-pound thrust turbojet
- Speed: 449 MPH
- Ceiling: 40,800 feet
- Range: 1,730 miles
- Crew: Three
- Defensive Armament: None
- Offensive Armament: One nuclear or thermonuclear

A-3 SKYWARRIOR
- Manufacturer: Douglas
- Number Manufactured: 282

In Service: 1956 - 1965

SPECIFICATIONS
- Length: 76 feet 4 inch
- Span: 72 feet 6 inches
- Height: 22 feet 9 inches
- Weight: Max 82,000 pounds
- Powerplant: Two 10,500-pound thrust turbojets
- Speed: 610 MPH
- Ceiling: 41,000 feet
- Range: 2,100 miles
- Crew: Three
- Defensive Armament: Electronics, two 20mm cannon in tail
- Offensive Armament: One or two thermonuclear bombs

A-5 VIGILANTE
- Manufacturer: North American
- Number Manufactured: 79
- In Service: 1961 - 1964

SPECIFICATIONS
- Length: 76 feet 6 inch
- Span: 53 feet
- Height: 19 feet 5 inches
- Weight: Max 62,950 pounds
- Powerplant: Two 16,500-pound thrust turbojets
- Speed: 1,320 MPH
- Ceiling: 52,100 feet
- Range: 2,580 miles with drop tanks
- Crew: Two
- Defensive Armament: Electronics
- Offensive Armament: One thermonuclear bomb in the central bomb bay and one thermonuclear bomb under each wing

INTERMEDIATE RANGE BALLISTIC MISSILES

THOR (SM-68)
- Manufacturer: McDonnell Douglas
- Missiles Deployed: 60
- In Service: 1959 - 1963 (1975)

SPECIFICATIONS
- Length: 65 feet
- Diameter: 8 feet
- Weight: 110,000 pounds

Range: 1,500 miles
Single stage, liquid fuel, 160,000-pound thrust
Guidance: Inertial
CEP: About 1 mile
Warhead: W49 in MK 2 RV

JUPITER (SM-68)
Manufacturer: Glen L. Martin
Missiles Deployed: 45
In Service: 1961 - 1963

SPECIFICATIONS
Length: 60 feet
Diameter: 8 feet 9 inches
Weight: 110,000 pounds
Range: 1,850 miles
Single Stage, Liquid Fuel, 150,000-pound thrust Rocketdyne S-3D engine
Guidance: Inertial
CEP: about 1 mile
Warhead: W49 in MK 1 RV

INTERCONTINENTAL BALLISTIC MISSILES

ATLAS SM-65 Model D
Manufacturer: Convair
In Service: 1959 - 1964
Missiles Deployed: 30 Model D

SPECIFICATIONS
Length: 75 feet
Diameter: 10 feet
Weight: 260,000
Range: 5,500 - 9,000 miles
One and a half stage liquid fuel missile
Two 150,000-pound thrust boosters & one 57,000-pound thrust sustainer
Guidance: radio command
CEP: 1 mile
Warhead: Squadron 1 - W49 in MK 2 RV, Squadrons 2, 3 & 4 - W49 in MK 3 RV

ATLAS SM-65 Models E & F
Manufacturer: Convair
In Service: 1961 - 1965
Missiles Deployed: 27 Model E & 72 Model F

SPECIFICATIONS

Length: 82 feet 6 inches
Diameter: 10 feet
Weight: 268,000 pounds
Range: 5,500 - 9,000 miles
CEP: 2 miles
One and a half stage liquid fuel missile
Two 165,000-pound thrust boosters & one 57,000-pound thrust sustainer
Guidance: Inertial
Warhead: W38 in MK 4 RV

TITAN (SM-68)
Manufacturer: Glen L. Martin
Missiles Deployed: 54
In Service: 1962 - 1964

SPECIFICATIONS
Length: 98 feet
Diameter: 10 feet
Weight: 220,000 pounds
Range: 5,500 - 6,300 miles
CEP: 1.25 miles
Two-stage liquid fuel
First Stage: Two 150,000-pound thrust boosters
Second Stage: One 80,000-pound thrust sustainer
Guidance: Radio Command
Warhead: W38 or W49 in MK 4 RV

TITAN II (LGM-25C)
Manufacturer: Martin Marietta
Missiles Deployed: 54
In Service: 1963 - 1987

SPECIFICATIONS
Length: 103 feet
Diameter: 10 feet
Weight: 330,000 pounds
Range: 13,000 miles
CEP: 0.65 miles
Two-stage liquid fuel missile
First Stage: Two 215,000-pound thrust boosters
Second Stage: One 100,000-pound thrust sustainer
Guidance: Inertial
Warhead: W53 in MK 6 RV

MINUTEMAN IA (LGM-30A)

Manufacturer: Boeing Aerospace
Missiles Deployed: 150
In Service: 1962 - 1967

SPECIFICATIONS
Length: 53 feet 7 inches
Diameter: 5 feet 5 inches
Weight: 65,000 pounds
Range: 5,000 miles
CEP: 6,000 feet
Three-stage solid fuel missile
First Stage: One 210,000-pound thrust booster
Second Stage: One 60,000-pound thrust sustainer
Third Stage: One 35,000-pound thrust sustainer
Guidance: Inertial
Warhead: W56 in MK 11 RV

MINUTEMAN IB (LGM-30B)
Manufacturer: Boeing Aerospace
Missiles Deployed: 650
In Service: 1963 - 1974

SPECIFICATIONS
Length: 55 feet 9 inches
Diameter: 5 feet 5 inches
Weight: 65,000 pounds
Range: 5,500 miles
CEP: 2,600 feet
Three-stage solid fuel missile
First Stage: One 210,000-pound thrust booster
Second Stage: One 60,000-pound thrust sustainer
Third Stage: One 35,000-pound thrust sustainer
Guidance: Inertial
Warhead: W56 in MK 11 RV

MINUTEMAN II (LGM-30F)
Manufacturer: Boeing Aerospace
Missiles Deployed: 500
In Service: 1967 - 1995

SPECIFICATIONS
Length: 57 feet 6 inches
Diameter: 5 feet 5 inches
Weight: 72,780 pounds

Range: 5,500 miles
CEP: 2,600 feet
Three-stage solid fuel missile
First Stage: One 210,000-pound thrust booster
Second Stage: One 60,300-pound thrust sustainer
Third Stage: One 35,000-pound thrust sustainer
Guidance: Inertial
Warhead: W56 in MK 11 RV

MINUTEMAN III (LGM-30G)
Manufacturer: Boeing Aerospace
Missiles Deployed:
In Service: 1967 - 1995

SPECIFICATIONS
Length: 59 feet 9.5 inches
Diameter: 5 feet 5 inches
Weight: 79,000 pounds
Range: 6,000+ miles
CEP: 700 feet
Three-stage solid fuel missile
First Stage: One 210,000-pound thrust booster
Second Stage: One 60,300-pound thrust sustainer
Third Stage: One 34,400-pound thrust sustainer
Post-boost: One 315-pound thrust re-startable liquid fuel engine
Guidance: Inertial
Three W62 warheads in MK 12 RVs
Three W78 warheads in MK 12a RVs

PEACEKEEPER (LGM-118A)
Manufacturer: Marin Marietta
Missiles Deployed: 50
In Service: 1986 - Present

SPECIFICATIONS
Length: 70 feet
Diameter: 7 feet 8 inches
Weight: 192,300 pounds
Range: 6,870 miles
CEP: 400 feet
Three-stage Solid Fuel
First Stage: One 500,000-pound thrust booster
Second Stage: One 275,000-pound thrust sustainer
Third Stage: One 65,000-pound thrust sustainer
Post-boost: One 2,000-pound thrust re-startable liquid fuel engine

Guidance: Inertial
Warhead: Ten W87 warheads in MK 21 RVs

SUBMARINE LAUNCHED BALLISTIC MISSILES

POLARIS A-1 (UGM-27A)
Manufacturer: Lockheed Missile and Space
Missiles Deployed: 80   on board George Washington Class FBMs
In Service: 1961 - 1965

SPECIFICATIONS
Length: 28 feet 6 inches
Diameter: 4 feet 6 inches
Weight: 28,800 pounds
Range: 1,380 miles
CEP: 3 - 4 nautical miles
Two-stage Solid Fuel
First Stage: 63,000-pound thrust booster
Second Stage: 31,000-pound thrust sustainer
Guidance: Inertial
Warhead: One W47Y1 warhead in MK 1 RV

POLARIS A-2 (UGM-27B)
Manufacturer: Lockheed Missile and Space
Missiles Deployed: 192 on Ethan Allen Class FBMs
In Service: 1962 - 1974

SPECIFICATIONS
Length: 31 feet
Diameter: 4 feet 6 inches
Weight: 32,500 pounds
Range: 1,700 miles
CEP: 2 nautical miles
Two stage solid fuel missile
First Stage: 63,000-pound thrust booster
Second Stage: 31,000-pound thrust sustainer
Guidance: Inertial
Warhead: One W47Y2 warhead in MK 1 RV

POLARIS A-3 (UGM-27C)
Manufacturer: Lockheed Missile and Space
Missiles Deployed: 644 on George Washington, Ethan Allen & Lafayette Class FBMs
In Service: 1964 - 1981

SPECIFICATIONS

Length: 32 feet 4 inches
Diameter: 4 feet 6 inches
Weight: 35,700 pounds
Range: 2,880 miles
CEP: 2,000 feet
Two-stage solid fuel missile
First Stage: 63,000-pound thrust booster
Second Stage: 31,000-pound thrust sustainer
Guidance: Inertial
Warhead: Three W58 warheads in MK 2 RVs

POSEIDON C-3 (UGM-73A)
Manufacturer: Lockheed Missile and Space
Missiles Deployed: 304 on Lafayette Class FBMs
In Service: 1971 - 1992

SPECIFICATIONS
Length: 34 feet 1.2 inches
Diameter: 6 feet 2 inches
Weight: 64,400 pounds
Range: 3,280 miles
CEP: 2,000 feet
Two-stage solid fuel missile
First Stage: ? pounds thrust
Second Stage: ? pounds thrust
Guidance: Inertial
Warhead: Ten W68 warheads in MK 3 RVs

TRIDENT C-4 (UGM-96A)
Manufacturer: Lockheed Missile and Space
Missiles Deployed: 384 on Ohio Class FBMs
In Service: 1979 - 2005

SPECIFICATIONS
Length: 34 feet 1.2 inches
Diameter: 6 feet 2 inches
Weight: 73,000 pounds
Range: 4,600 miles
CEP: 1,250 feet
Three-stage solid fuel missile
First Stage: 175,000-pound thrust booster
Second Stage: 82,500-pound thrust sustainer
Third Stage: 35,000-pound thrust sustainer
Guidance: Inertial
Warhead: Six W76 warheads in MK 4 RVs

TRIDENT II D-5 (UGM-133)
    Manufacturer: Lockheed Missile and Space
    Missiles Deployed: 336 on Ohio Class FBMs
    In Service: 1989 - Present

SPECIFICATIONS
    Length: 70 feet
    Diameter: 7 feet 8 inches
    Weight: 130,000 pounds
    Range: 6,870 miles
    CEP: 400 feet
    Three-stage Solid Fuel
    First: 350,000-pound thrust booster
    Second: 100,000-pound thrust sustainer
    Third: 27,500-pound thrust sustainer
    Guidance: Inertial
    Warhead: Eight to fourteen W88 warheads in MK 5 RVs

BALLISTIC MISSILE SUBMARINES

GEORGE WASHINGTON CLASS
    No. Deployed: 5
    In Service: 1958 - 1990

SPECIFICATIONS (598) USS GEORGE WASHINGTON
    Length: 382 feet
    Width: 33 feet
    Height 29 feet
    Displacement: 6,700 tons submerged
    Speed: 30 knots submerged
    Powerplant: S5W pressurized light water boiling reactor
    Crew Complement: 12 officers, 127 enlisted
    Armament: 16 steam ejection tubes for Polaris missiles
    Six 21-inch torpedo tubes

ETHAN ALLEN CLASS
    No. Deployed: 5
    In Service: 1961 - 1991

SPECIFICATIONS (608) USS ETHAN ALLEN
    Length: 410 feet
    Width: 33 feet
    Height 32 feet

Displacement: 7,880 tons submerged
Speed: 30 knots submerged
Powerplant: S5W pressurized light water boiling reactor
Crew Complement: 15 officers, 130 enlisted
Armament: 16 Air ejection tubes for Polaris missiles
Four 21-inch torpedo tubes

LAFAYETTTE (& BEN FRANKLIN) CLASS
No. Deployed: 31
In Service: 1962 - 1995

SPECIFICATIONS (616) USS LAFAYETTE
Length: 425 feet
Width: 33 feet
Height 31.5 feet
Displacement: 8,260 tons submerged
Speed: 30 knots submerged
Powerplant: S5W pressurized light water boiling reactor
Crew Complement: 14 officers, 130 enlisted
Armament: 16 steam ejection tubes for Polaris or Poseidon missiles
Four 21-inch torpedo tubes

OHIO CLASS
No. Deployed: 18
In Service: 1981 – Present

SPECIFICATIONS (726) USS OHIO
Length: 560 feet
Width: 42 feet
Height 35.5 feet
Displacement: 18,740 tons submerged
Speed: N/A
Powerplant: S8G pressurized light water boiling reactor
Crew Complement: 14 officers, 146 enlisted
Armament: 24 steam ejection tubes for Trident & Trident II missiles
Four 21-inch torpedo tubes

APPENDIX 4
HISTORICALLY DEPLOYED NUCLEAR BOMBS AND WARHEADS

MK I: Los Alamos Designed Strategic Atomic Bomb
Uranium Gun Assembled Mechanism
In Service 1945 - 1950
Five Produced
Principal Delivery Vehicles: B-29, T1 ADM
Length: 120 inches
Diameter: 20 inches
Weight: 8,900 pounds
Yield: 15 Kilotons
Fuzing Options: Radar Altimeter Triggered Airburst with Barometric Backup

MK III: Los Alamos Designed Strategic Atomic Bomb
Plutonium Implosion Mechanism
In Service 1945 - 1950
120 Produced in three MODs
Principal Delivery Vehicles: B-29 & B-50
Length: 128 inches
Diameter: 60.25 inches
Weight: 10,300 pounds
Four Yields: 18, 21, 37, & 49 Kilotons
Fuzing Options: Radar Altimeter Triggered Airburst with Barometric Backup

MK IV: Los Alamos Designed Strategic Atomic Bomb
Composite Implosion Mechanism
In Service 1949 - 1953
550 Produced in four MODs
Principal Delivery Vehicles: B-50, B-36, & B-47
Length: 128 inches
Diameter: 60 inches
Weight: 10,300 pounds
Seven Yields: 1, 3.5, 8, 14, 21, 22 & 31 Kilotons
Fuzing Options: Radar Altimeter Triggered Airburst with Barometric Backup

MK 5: Los Alamos Designed Lightweight Strategic Atomic Bomb
Composite Implosion Mechanism
In Service 1952 - 1963
140 Produced in four MODs
Principal Delivery Vehicles: Light and Medium Bombers
Length: 129 - 132 inches
Diameter: 43.75 inches

Weight: 3,025 - 3,175 pounds
Six Yields: 6, 16, 55, 60, 100, & 120 Kilotons
Fuzing Options: Radar Altimeter Triggered Airburst with Barometric and Contact Backup

W5: Los Alamos Designed Atomic Warhead
Composite Implosion Design
In Service 1954 - 1963
100 Produced
Principal Delivery Vehicles: Snark & Regulus Cruise Missiles
Length: 76 inches
Diameter: 39 inches
Weight: 2,405 - 2,650 pounds
Yields: Same as Mark 5
Fuzing Options: Radar Altimeter Triggered Airburst with Barometric and Contact Backup

MK 6: Los Alamos Designed Strategic Atomic Bomb
Composite Implosion Mechanism
In Service 1951 - 1962
1,100 Produced in seven MODs
Principal Delivery Vehicles: B-50, B-36, B-47
Length: 128 inches
Diameter: 61 inches
Weight: 7,600 - 8,500 pounds
Seven Yields: 8, 22, 26, 31, 80, 154, & 160 Kilotons
Fuzing Options: Radar Altimeter Triggered Airburst with Barometric and Contact Backup

MK 7: Los Alamos Designed Tactical Atomic Bomb
Composite Implosion Design
In Service 1952 - 1967
1,800 Produced in ten MODs
Principal Delivery Vehicles: Fighter-Bombers & Light Bombers
Length: 183 inches
Diameter: 30.5 inches
Weight: 1,645 - 1,700 pounds
Six Yields: 8, 19, 22, 30, 31, & 61 Kilotons
Fuzing Options: Radar Altimeter Triggered Airburst with Barometric and Contact Backup

W7: Los Alamos Designed Atomic Warhead
Composite Implosion Mechanism
In Service 1952 - 1967
1,350 Produced in four MODs
Principal Delivery Vehicles: ADM-B, Betty Depth Bomb, BOAR ASM, Corporal SRBM,
        Honest John SSM
Length: 54.8 - 56 inches
Diameter: 30.5 inches

Weight: 900 - 1,100 pounds
Seven Yields: 90 Tons, 2, 5, 10, 20, 32, & 40 Kilotons
Fuzing Options: Airburst, Contact, Hydrostatic

MK 8: Los Alamos Designed Penetration Bomb
Uranium Gun Assembled Mechanism
In Service 1953 - 1957
40 Produced in two MODs
Principal Delivery Vehicles: Navy Fighter-Bombers
Length: 116 - 132 inches
Diameter: 14.5 inches
Weight: 3,230 - 3,280 pounds
Yield 15 - 20 Kilotons
Fuzing Options: Pyrotechnic Delay

W9: Los Alamos Designed Atomic Artillery Shell
Uranium Gun Mechanism
In Service 1952 - 1957
80 Produced
Principal Delivery Vehicle: T-124 Shell Fired from M65 Cannon, T4 ADM
Length: 54.8 inches
Diameter: 280mm
Weight: 803 pounds
Yield: 15 Kilotons
Fuzing Options: Mechanical Time Delay Airburst

MK 11: Los Alamos Improved Mark 8
Uranium Gun Assembled Mechanism
In Service 1956 - 1960
40 Produced in four MODs
Principal Delivery Vehicles: Navy Fighter-Bombers
Length: 147 inches
Diameter: 14 inches
Weight: 3,210 - 3,250 pounds
Yield 15 - 20 Kilotons
Fuzing Options: Pyrotechnic Delay

MK 12: Los Alamos Designed Tactical Atomic Bomb for Supersonic Carry
Plutonium Implosion Mechanism
In Service 1954 - 1969
50 Produced in two MODs
Principal Delivery Vehicles: Fighter-Bombers
Length: 155 inches
Diameter: 22 inches
Weight: 1,100 - 1,200 pounds

Several Yields: Low Kilotons
Fuzing Options: Contact or Timer

EC 14: Los Alamos Designed Strategic Thermonuclear Bomb
Cylindrical Thermonuclear Implosion Mechanism
In Service 1954
5 Produced
Principal Deliver Vehicle: B-36
Length: ca. 222 inches
Diameter: 61.4 inches
Weight: ca. 30,000 pounds
Yield: 6.9 Megatons
Fuzing Options: Parachute-retarded Airburst

MK 15: Los Alamos Designed Strategic Thermonuclear Bomb
Cylindrical Thermonuclear Implosion Mechanism
In Service 1955 - 1965
1,200 Produced in four MODs
Principal Delivery Vehicles: B-47, B-52, A1-J, A-3
Length: ca. 136-140 inches
Diameter: ca. 35 inches
Weight: 7,600 pounds
Yield: 3.8 Megatons
Fuzing Options: Parachute-retarded Airburst & Contact

EC 16: Los Alamos Designed Liquid Deuterium Fueled Thermonuclear Bomb
Cylindrical Thermonuclear Implosion Mechanism
In Service 1954
5 Produced
Principal Delivery Vehicle: B-36
Length: 296.7 inches
Diameter: 61.4 inches
Weight: ca. 40,000 pounds
Yields: 7 Megatons
Fuzing Options: Freefall Airburst

Mk 17: Los Alamos Designed Strategic Thermonuclear Bomb
Cylindrical Thermonuclear Implosion Mechanism
In Service 1954 - 1957
200 Produced in three MODs
Principal Delivery Vehicle: B-36
Length: 296.7 inches
Diameter: 61.4 inches
Weight: 41,400 - 42,000 pounds
Yield: 15 - 20 Megatons

Fuzing Options: Parachute-Retarded Airburst & Contact (MOD 2)

Mk 18: Los Alamos Improved Mark 6 Atomic Bomb
Uranium Implosion Mechanism
In Service 1953 - 1956
90 Produced
Principal Delivery Vehicles: B-47, B-52
Length: 128 inches
Diameter: 60 inches
Weight: 7,600 pounds
Yield: 500 Kilotons
Fuzing Options: Freefall Radar Altimeter Triggered Airburst with Barometric & Contact Backup

W19: Los Alamos Improved W9
Uranium Gun Assembled Mechanism
In Service 1955 - 1963
80 Produced
Principal Delivery Vehicle: T-315 Shell Fired from M65 Cannon
Length: 54 inches
Diameter: 280mm
Weight: 600 pounds
Yield: 15 Kilotons
Fuzing Options: Mechanical Time Delay Airburst

MK 21: Los Alamos Designed Strategic Thermonuclear Bomb
Cylindrical Thermonuclear Implosion Mechanism
In Service 1955 - 1957
275 Produced in one MOD
Principal Delivery Vehicle: B-36
Length: 148.4 inches   Diameter: 56 inches around casing, 58.5 inches around spoiler bands
Weight: 17,600 pounds
Yield: Y1 15 Megatons, Y2 4.5 Megatons (TX26)
Fuzing Options: Parachute-Retarded Airburst & Contact

W23: Los Alamos Designed Artillery Shell
Uranium Gun Assembled Mechanism
In Service 1956 - 1962
50 Produced
Principal Delivery Vehicle: 16 Inch Naval Shell
Length: 64 inches
Diameter: 16 inches
Weight: 1500; 1900 pounds
Yield: 15 Kilotons
Fuzing Options: Mechanical Time Delay Airburst

MK 24: Los Alamos Designed Strategic Thermonuclear Bomb
Cylindrical Thermonuclear Implosion Mechanism
In Service 1954 - 1956
105 Produced in two MODs
Principal Delivery Vehicle: B-36
Length: 296.7 inches
Diameter: 61.4 inches
Weight: 41,400 - 42,000 pounds
Yield: 15 - 20 Megatons
Fuzing Options: Parachute-Retarded Airburst

W25: Los Alamos Designed Warhead
Sealed Composite Implosion Mechanism
3,150 Produced in two MODs
In Service 1957 - 1984
Principal Delivery Vehicle: Genie AAM
Length: 25.6 - 26.6 inches
Diameter: 17.4 inches
Weight: 218 - 221 pounds
Yield: 1.7 Kilotons
Fuzing Options: Time Delay

MK 27: Livermore Designed Strategic Thermonuclear Bomb
Cylindrical Thermonuclear Implosion Mechanism
In Service 1958 - 1965
700 Produced in three MODs
Principal Delivery Vehicles: A-3, A-5
Length: 75 inches
Diameter: 30.25 - 31 inches
Weight: 2,800 pounds
Yield: 2 Megatons
Fuzing Options: Airburst & Contact

W27: Radiation Laboratory Designed Thermonuclear Warhead
Cylindrical Thermonuclear Implosion Mechanism
In Service 1955 - 1957
20 Produced
Principal Delivery Vehicle: Regulus Cruise Missile
Length: 136 - 140 inches
Diameter: 56.2; 58.5 inches
Weight: 15,000 - 17,700 pounds
Yield: 2 Megatons
Fuzing Options: Airburst & Contact

MK 28: Los Alamos Designed Strategic Thermonuclear Bomb

Cylindrical Implosion Mechanism
In Service 1958 - 1991
4,500 produced in six MODs
Principal Delivery Vehicles: B-47, B-52, Fighter-Bombers
Length: 96 - 170 inches
Diameter: 20; 22 inches
Weight: 1,700 - 2,320 pounds
Five Yields: 15, 70, & 350 Kilotons, 1 & 1.5 Megatons
Fuzing Options: Full Fuzing Options Available Depending on Model

W28: Los Alamos Designed Thermonuclear Warhead
Cylindrical Thermonuclear Implosion Mechanism
In Service 1958 - 1976
1,000 Produced in five MODs
Principal Delivery Vehicles: Hound Dog ASM, Mace Cruise Missile
Length: 60 inches
Diameter: 20 inches
Weight: 1,500 - 1,725 pounds
Five Yields: 15, 70, & 350 Kilotons, 1 & 1.5 Megatons
Fuzing Options: Airburst & Contact

W30: Los Alamos Designed Atomic Warhead
Boosted Implosion Mechanism
In Service 1959 - 1979
600 Produced in four MODs
Principal Delivery Vehicles: TADM, Talos
Length: 48 inches
Diameter: 22 inches
Weight: 438, 450, & 490 pounds
Three Yields: 300 & 500 Tons, 4.7 Kilotons
Fuzing Options: Airburst, Contact, & Time Delay

W31: Los Alamos Designed Tactical Atomic Warhead
Boosted Implosion Mechanism
In Service 1952 - 1967
4,500 Produced in three MODs
Principal Delivery Vehicles: Honest John, Nike Hercules, ADM
Length: 39 - 43.75 inches
Diameter: 29.01 inches
Weight: 900 - 945 pounds
Five Yields: 2, 12, 20, & 40 Kilotons
Fuzing Options: Airburst, Timer, & Contact

W33: Los Alamos Designed Artillery Shell
Uranium Gun Assembled Mechanism

In Service 1956 - 1962
2,000 Produced in two MODs
Principal Deliver Vehicle: T-317 shell from 8 Inch Howitzer
Length: 31 inches
Diameter: 8 inches
Weight: 240 - 243 pounds
Four Yields: 0.5, 5, 10, & 40 Kilotons
Fuzing Options: Time Delay Airburst

W34: Los Alamos Designed Tactical Atomic Warhead
Boosted Implosion Mechanism
In Service 1958 - 1972
3,200 Produced in four MODs
Principal Delivery Vehicles: Lulu Depth Bomb, Astor Torpedo, Hotpoint Bomb
Length: 32 inches
Diameter: 17 inches
Weight: 311, 312, & 320 pounds
Yield: 11 Kilotons
Fuzing Options: Airburst, Timer, Contact, & Command Detonation

MK 36: Los Alamos Improved MK 21 Strategic Thermonuclear Bomb
Cylindrical Thermonuclear Implosion Mechanism
In Service 1956 - 1962
920 Produced in three MODs
Principal Delivery Vehicles: B-47, B-52
Length: 150 inches
Diameter: 56 inches around casing, 58.5 inches around spoiler bands
Weight: 17,500 - 17,700 pounds
Two Yields: 6 & 19 Megatons
Fuzing Options: Parachute-Retarded Airburst & Contact

W38: Radiation Laboratory Designed Strategic Thermonuclear Warhead
Cylindrical Thermonuclear Implosion Mechanism
In Service 1961 - 1965
180 Produced
Principal Delivery Vehicles: Atlas, Titan
Length: 82.5 inches
Diameter: 32 inches
Weight: 3,080 pounds
Yield: 3.75 Megatons
Fuzing Options: Airburst & Contact

MK 39: Los Alamos Designed Strategic Thermonuclear Bomb
Cylindrical Thermonuclear Implosion Mechanism
In Service 1957 - 1966

700 Bombs and Basic Assemblies Produced in three MODS
Principal Delivery Vehicles: B-47, B-52, B-58
Length: 136 - 140 inches
Diameter: 35 inches
Weight: 6,650 - 6,750 pounds
Yields: 3.8 Megatons
Fuzing Options: Parachute-Retarded Airburst, Contact, & Laydown

W39: Radiation Laboratory Designed Strategic Thermonuclear Warhead
Cylindrical Thermonuclear Implosion Mechanism
In Service 1958 - 1965
90 Produced
Principal Delivery Vehicles: Redstone MRBM and Snark ICCM
Length: 105.7 inches
Diameter: 34.5 - 35 inches
Weight: 6,230 - 6,400 pounds
Two Yields: 1 & 3.8 Megatons
Fuzing Options: Airburst & Contact

W40: Los Alamos Designed Atomic Warhead
Boosted Implosion Mechanism
In Service 1956 - 1964
750 Produced in three MODs
Principal Delivery Vehicles: BOMARC SAM, Lacrosse SSM
Length: 31.64 inches
Diameter: 17.9 inches
Weight: 350; 385 pounds
Yield: 1.7 & 10 Kilotons
Fuzing Options: Airburst & Contact

MK 41: Radiation Laboratory Designed Strategic Thermonuclear Bomb
3-Stage Cylindrical Thermonuclear Implosion Mechanism
In Service 1960 - 1976
500 Produced
Principal Delivery Vehicles: B-47, B-52
Length: 148 inches
Diameter: 52 inches
Weight: 10,500 - 10,670 pounds
Two Yields: 10 & 25 Megatons
Fuzing Options: Parachute-Retarded Airburst & Contact

B43: Los Alamos Designed Strategic Thermonuclear Bomb
Cylindrical Thermonuclear Implosion Mechanism
In Service 1961 - 1991
1,000 produced in two MODs

Principal Deliver Vehicles: B-52, FB11A, Fighter-bombers
Length: 150 - 164 inches
Diameter: 18 inches
Weight 2,060 - 2,125 pounds
Five Yields: 70, 100, 200, 500 Kilotons & 1 Megaton
Fuzing Options: FUFO

W44: Radiation Laboratory Designed Tactical Atomic Warhead
Boosted Plutonium Implosion Mechanism
In Service 1961 - 1989
575 produced in three MODs
Principal Delivery Vehicle: ASROC
Length: 25.3 inches
Diameter: 13.75 inches
Weight: 350; 385 pounds
Yield: 10 Kilotons
Fuzing Options: Hydrostatic

W45: Radiation Laboratory Designed Tactical Atomic Warhead
Boosted Linear Implosion Mechanism
In Service 1956 - 1964
1,700 produced in three MODs
Principal Delivery Vehicles: Terrier SAM, Bullpup ASM, Little John SSM, MADM
Length: 27 inches
Diameter: 11.5 inches
Weight: 150 pounds
Six Yields: 0.5 -15 Kilotons
Fuzing Options: Airburst, Contact Timer, & Command

W47: Radiation Laboratory Designed Strategic Thermonuclear Warhead
Cylindrical Thermonuclear Implosion Mechanism
In Service 1960 - 1974
1,360 Produced in four MODs
Principal Delivery Vehicles: Polaris A1 & A2 SLBM
Length: 46.6 inches
Diameter: 18 inches
Weight: 717 - 720; 733 pounds
Two Yields: 600 Kilotons & 1.2 Megatons
Fuzing Options: Airburst & Contact

W48: Radiation Laboratory Designed Artillery Shell
Linear Plutonium Implosion Mechanism
In Service 1963 - 1992
1,060 Produced
Principal Delivery Vehicle: 155mm Howitzer

Length: 33 inches
Diameter: 6 inches
Weight: 118 - 128 pounds
Yield: ~100 Tons
Fuzing Options: Timer and Preset Time Delay (High and Low Burst)

W49: Los Alamos Designed Strategic Thermonuclear Warhead
Cylindrical Thermonuclear Implosion Mechanism
In Service 1958 - 1975
95 Produced in Various MODs
Principal Delivery Vehicles: Thor & Jupiter IRBM's, Atlas D ICBM's
Length: 54.3 - 67.9 inches
Diameter: 20 inches
Weight: 1,640 - 1,680 pounds
Two Yields: 1.1 & 1.5 Megatons
Fuzing Options: Airburst & Contact

W50: Los Alamos Designed Tactical Thermonuclear Warhead
Cylindrical Thermonuclear Implosion Mechanism
In Service 1963 - 1991
280 Produced in two MODs
Principal Delivery Vehicles: Pershing MRBM, Nike Zeus
Length: 44 inches
Diameter: 15.4 inches
Weight: 409 - 410 pounds
Three Yields: 50, 200, & 400 Kilotons
Fuzing Options: Airburst & Contact

W52: Los Alamos Designed Tactical Thermonuclear Warhead
Cylindrical Thermonuclear Implosion
In Service 1962 - 1978
300 Produced in three MODs
Principal Delivery Vehicle: Sergeant SRBM
Length: 56.7 inches
Diameter: 24 inches
Weight: 925 pounds
Yield: 200 Kilotons
Fuzing Options: Airburst & Contact

W53: Los Alamos Designed Strategic Thermonuclear Bomb
Cylindrical Thermonuclear Implosion Mechanism
In Service 1962 - 1997
300 Bombs and 40 Basic Assemblies Produced in 1 MOD
Principal Delivery Vehicles: B-52, B-58
Length: 144 - 150 inches

Diameter: 50 inches
Weight 8,850 - 8,900 pounds
Two Yields: 3 & 9 Megatons
Fuzing Options: FUFO

W53: Los Alamos Designed Strategic Thermonuclear Warhead
Cylindrical Thermonuclear Implosion Mechanism
In Service 1962 - 1997
65 Produced
Principal Delivery Vehicle: Titan II ICBM
Length: 103 inches
Diameter: 37 inches
Weight 6,200 pounds
Yield: 9 Megatons
Fuzing Options: Airburst & Contact

W54: Los Alamos Adaptation of Livermore W51 Tactical Warhead
Boosted Plutonium Implosion Mechanism
In Service 1961 - 1972
1,700 Produced in three MODs
Principal Delivery Vehicles: Davy Crockett, Falcon AAM & SADM
Length: 15.7 - 17.6 inches
Diameter: 10.75 inches
Weight: 50 - 59 pounds
Yields: 10 & 20 Tons, 250 Tons, Variable: 10 Tons - 1 Kiloton
Fuzing Options: Time Delay, Proximity & Contact

W55: Radiation Laboratory Designed Tactical Thermonuclear Warhead
Spherical Implosion Mechanism
In Service 1964 - 1990
200 Produced in three MODs
Principal Delivery Vehicles: SUBROC
Length: 39.4 inches
Diameter: 13 inches
Weight: 470 pounds
Two Yields: 5 and 250 Kilotons
Fuzing Options: Hydrostatic

W56: Radiation Laboratory Designed Strategic Thermonuclear Warhead
Cylindrical Thermonuclear Implosion Mechanism
In Service 1963 - 1993
1,000 produced in four MODs
Principal Delivery Vehicles: Minuteman IB & II ICBMs
Length: 47.3 inches
Diameter: 17.4 inches

Weight: 600; 680 pounds
Yield: 1.2 Megatons
Fuzing Options: Airburst & Contact

MK 57: Los Alamos Designed Tactical Atomic Bomb
Boosted Implosion Mechanism
In Service 1963 - 1993
3,100 Produced in six MODs
Principal Delivery Vehicles: Fighter-bombers
Length: 118 inches
Diameter: 14.75 inches
Weight 490 -510 pounds
Three Yields: 10, 15, & 20 Kilotons
Fuzing Options: FUFO & Hydrostatic

W58: Radiation Laboratory Designed Strategic Thermonuclear Warhead
Spherical Thermonuclear Implosion Mechanism
In Service 1964 - 1982
1,400 Produced in three MODs
Principal Delivery Vehicle: Polaris A3 SLBM
Length: 40.3 inches
Diameter: 15.6 inches
Weight: 257 pounds
Yield: 200 Kilotons
Fuzing Options: Airburst & Contact

W59: Los Alamos Designed Strategic Thermonuclear Warhead
Cylindrical Thermonuclear Implosion Mechanism
In Service 1962 - 1969
150 Produced
Principal Delivery Vehicle: Minuteman IA ICBM
Length: 16.3 inches
Diameter: 47.8 inches
Weight: 550 - 553 pounds
Yield: 1 Megaton
Fuzing Options: Airburst & Contact

B61: Los Alamos Designed Tactical / Strategic Thermonuclear Bomb
Cylindrical Thermonuclear Implosion Mechanism
In Service 1966 - Present
3,150 Produced in twelve MODs
Principal Delivery Vehicles: B2, B52, Fighter-bombers
Length: 141.64 inches
Diameter: 13.3 inches
Weight: 695 - 716 pounds

Four Yield Ranges: Variable 0.3 - 340, 0.3 - 170, 0.3 - 45, & 0.3 - 80 Kilotons
Fuzing Options: FUFO

W62: Livermore Designed Strategic Thermonuclear Warhead
Spherical Thermonuclear Implosion Mechanism
In Service 1970 - 2010
1,725 Produced in three MODs
Principal Delivery Vehicles: Minuteman III ICBM
Length: 39.3 inches
Diameter: 19.7 inches
Weight: 253 pounds
Yield: 170 Kilotons
Fuzing Options: Airburst & Contact

W66: Los Alamos Designed Thermonuclear Warhead
Spherical Thermonuclear (ER) Implosion Mechanism
In Service 1975
70 Produced
Principal Delivery Vehicles: Sprint ABM
Length: 35 inches
Diameter: 18 inches
Weight: 150 pounds
Yield: 10 Kiloton Enhanced Radiation
Fuzing Options: Airburst & Contact

W68: Livermore Designed Strategic Thermonuclear Warhead
Spherical Thermonuclear Implosion Mechanism
In Service 1970 - 1991
5,250 Produced in three MODs
Principal Delivery Vehicles: Poseidon SLBM
Length: NA
Diameter: NA
Weight: 367 pounds (RV & Warhead)
Yield: 40- 50 Kilotons
Fuzing Options: Airburst & Contact

W69: Los Alamos Designed Tactical Thermonuclear Warhead
Spherical Thermonuclear Implosion Mechanism
In Service 1971 - 1994
1,500 Produced
Principal Delivery Vehicles: SRAM
Length: 30 inches
Diameter: 15 inches
Weight: 275 pounds
Yield: 170 Kilotons

Fuzing Options: Airburst & Contact

W70: Livermore Designed Tactical Thermonuclear Warhead
Linear Fission & Spherical Thermonuclear Implosion Mechanisms
In Service 1973 - 1992
900 Fission & 380 Enhanced Radiation Produced; MODS 0, 1, & 2 fission, & MOD 3 Enhanced Radiation
Principal Delivery Vehicle: Lance SRBM
Length: 41 inches
Diameter: 18 inches
Weight: 270 pounds
Three Yields: Variable 1 - 100 Kilotons, 0.75 & 1.25 Kiloton Enhanced Radiation
Fuzing Options: Airburst & Contact

W71: Livermore Designed Tactical Thermonuclear Warhead
Cylindrical Thermonuclear Implosion Mechanism
In Service 1974 - 1975
30 Produced
Principal Delivery Vehicle: Spartan ABM
Length: 101 inches
Diameter: 42 inches
Weight: 2,850 pounds
Yield: 5 Megatons
Fuzing Options: Airburst, Command, Delay Timer, & Contact

W72: Los Alamos Improved W54 Tactical Warhead
Plutonium Implosion Mechanism
In Service 1970 - 1979
300 Produced
Principal Delivery Vehicle: Walleye Glide Bomb
Length: 79 inches
Diameter: 15 inches
Weight: 825 pounds
Yield: 625 Tons
Fuzing Options: Contact

W76: Los Alamos Designed Strategic Thermonuclear Warhead
Cylindrical Thermonuclear Implosion Mechanism
In Service 1978 - Present
3,200 Produced in three MODs
Principal Delivery Vehicles: Trident I & Trident II SLBM
Length: N/A
Diameter: N/A
Weight: 363 pounds
Yield: 100 Kilotons

Fuzing Options: Airburst & Contact

W78: Los Alamos Designed Strategic Thermonuclear Warhead
Cylindrical Thermonuclear Implosion Mechanism
In Service 1979 - Present
1,083 Produced
Principal Delivery Vehicle: Minuteman III ICBM
Length: 41.7 inches
Diameter: 12.4 inches
Combined Weight: 400 pounds
Yield: 335 Kilotons
Fuzing Options: Airburst & Contact

W79: Livermore Designed Artillery Shell
Linear Fission & Cylindrical Implosion Thermonuclear Mechanisms
In Service 1981 - 1990
225 Fission & 325 Enhanced Radiation produced in two MODs
Principal Delivery Vehicle: XM753 shell for 8" Howitzer
Length: 44 inches
Diameter: 8 inches
Weight: 200 pounds
Yields: 0.8 Kilotons Fission & Variable 100 Tons - 1.1 Kilotons Enhanced Radiation
Fuzing Options: Airburst & Contact

W80: Los Alamos Designed Tactical Thermonuclear Warhead
Cylindrical Thermonuclear Implosion Mechanism
In Service 1981 – Present
367 MOD 0 & 1,750 MOD 1 Produced
Principal Delivery Vehicles: SLCM MOD 0, ALCM, and ACM MOD 1
Length: 31.4 inches
Diameter: 11.8 inch
Weight: 290 pounds
Yield: Variable 5 - 170 Kilotons,
Fuzing Options: Airburst & Contact

B83: Livermore Designed Strategic Thermonuclear Bomb
Spherical Thermonuclear Implosion Mechanism
In Service 1983 - Present
650 Produced in two MODS
Principal Delivery Vehicle: B2
Length: 145 inches
Diameter: 18 inches
Weight: 2,400 pounds
Yield: Variable Low Kiloton - 1.2 Megatons
Fuzing Options: FUFO

W84: Livermore Designed Tactical Thermonuclear Warhead
Spherical Thermonuclear Implosion Mechanism
In Service 1983 - 1988
400 Produced
Principal Delivery Vehicle: GLCM
Length: 34 inches
Diameter: 13 inches
Weight: 388 pounds
Yield: Variable 0.2 - 150 Kilotons
Fuzing Options: Airburst & Contact

W85: Los Alamos Designed Tactical Thermonuclear Warhead
Cylindrical Thermonuclear Implosion Mechanism
In Service 1983 - 1991
750 Produced
Principal Delivery Vehicle: Pershing II MRBM
Length: 41.7 inches
Diameter: 12.4 inches
Weight: 400 pounds
Yield: Variable 5 - 80 Kilotons
Fuzing Options: Airburst & Contact

W87: Livermore Designed Strategic Thermonuclear Warhead
Spherical Thermonuclear Implosion Mechanism
In Service 1986 - Present
525 Produced
Principal Delivery Vehicle: Peacekeeper, Minuteman III ICBM's
Length: 68.9 inches
Diameter: 21.8 inches
Weight: between 440 and 600 pounds
Yield: 300 Kilotons upgradeable to 475 Kilotons
Fuzing Options: Timer, Proximity Airburst, & Contact

W88: Los Alamos Designed Strategic Thermonuclear Warhead
Spherical Thermonuclear Implosion Mechanism
In Service 1988 - Present
400 Produced, Currently in Production at low rate
Principal Delivery Vehicle: Trident II SLBM
Length: 68.9 inches
Diameter: 21.8 inches
Weight: < 800 pounds
Yield: 475 Kilotons
Fuzing Options: Timer with Path Length Correction, Airburst, & Contact

APPENDIX 5
US NUCLEAR TEST PROGRAM

OPERATION	YEAR	DETONATIONS	LOCATION(S)
Trinity	1945	1	New Mexico
Crossroads	1946	2	Bikini Atoll
Sandstone	1948	3	Enewetak Atoll
Ranger	1951	5	Nevada
Greenhouse	1951	4	Enewetak Atoll
Buster-Jangle	1952	7	Nevada
Tumbler-Snapper	1952	7	Nevada
Ivy	1952	2	Enewetak Atoll
Upshot-Knothole	1953	11	Nevada
Castle	1954	6	Bikini Atoll
Teapot	1955	14	Nevada
Wigwam	1956	1	Eastern Pacific
Redwing	1956	17	Bikini, Enewetak Atoll
Plumbbob	1957	30	Nevada
Hardtack I	1958	35	Bikini, Enewetak Atoll, Johnston Island
Hardtack II	1958	37	Nevada
Argus	1958	3	South Atlantic

Voluntary Testing Moratorium 1959 – 1960

Nougat	1961 - 62	32	Nevada, New Mexico
Dominic	1962	36	Christmas & Johnston Islands, Central Pacific
Storax	1962 - 63	56	Nevada

Limited Test Ban Treaty (1963) – Prohibits Atmospheric and Space Based Testing

Niblick	1963	42	Nevada
Whetstone	1964 - 65	48	Nevada, Mississippi
Flintlock	1965 - 66	48	Nevada, Alaska
Latchkey	1966 - 67	37	Nevada, Mississippi, New Mexico
Crosstie	1967 - 68	56	Nevada
Bowline	1968 - 69	57	Nevada, Colorado
Mandrel	1969 - 70	77	Nevada, Alaska
Emery	1970	23	Nevada
Grommet	1971 - 72	38	Nevada, Alaska
Toggle	1972 - 73	38	Nevada

Arbor	1973 - 74	19	Nevada
Bedrock	1974 - 75	28	Nevada
Anvil	1975 - 76	19	Nevada

Threshold Test Ban Treaty (1976) – Limits Test Size to 150 Kilotons

Fulcrum	1976 - 77	23	Nevada
Cresset	1977 - 78	22	Nevada
Quicksilver	1978 - 79	15	Nevada
Tinderbox	1979 - 80	13	Nevada
Guardian	1980 - 81	13	Nevada
Praetorian	1981 - 82	19	Nevada
Phalanx	1982 - 83	18	Nevada
Fusileer	1983 - 84	15	Nevada
Grenadier	1984 - 85	15	Nevada
Charioteer	1985 - 86	15	Nevada
Musketeer	1986 - 87	15	Nevada
Touchstone	1987 - 88	15	Nevada
Cornerstone	1988 - 89	16	Nevada
Aqueduct	1989 - 90	12	Nevada
Sculpin	1990 - 91	8	Nevada
Julin	1991 - 92	8	Nevada

Comprehensive Test Ban Treaty – Prohibits Nuclear Tests

APPENDIX 6
BASIC PROPERTIES OF NUCLEAR MATERIALS

URANIUM (Element 92)

USES

    Uranium is a silvery white metal used as a reactor fuel, a breeding material for plutonium, a nuclear explosive, a tamper, and a neutron reflector. Depleted uranium is also used for nuclear shielding because of its density.

PROPERTIES
    Molar Volume 12.56 Centimeters^3 / Mole
    Density 19.1 (α-phase stable to 667 °C)
    Melting Point 1,132 °C

ISOTOPES
  U233
    Atomic Mass: 233.03962 (AMU)
    Binding Energy: 1,771,728 KeV
    Atomic Abundance: N/A
    Half Life: 159,200 Years
    Specific Activity: 9.636 Curies / Kilogram
    Decay Heat: 0.2804 Watts / Kilogram
    Fission Cross-section: 1.946 Barns
    Spontaneous Neutron Multiplicity: 2
    Induced Neutron Multiplicity: 2.65
    Primary Decay Mode: Alpha Emission to Th229
        Decay Energy: 4.909 MeV
    Secondary Decay Mode: Spontaneous Fission
        SR Rate: 0.47 (Fissions / Second / Kilogram)
        Neutron Emission Rate: 0.94 (Neutrons / Second / Kilogram)
    Critical Mass: 16 Kilograms
  U235
    Atomic Mass: 235.04392 (AMU)
    Binding Energy: 1,783,870 KeV
    Atomic Abundance: 0.720 percent
    Half Life: 0.704 Billion Years
    Specific Activity: 2.161 x 10^{-3} Curies / Kilogram
    Decay Heat: 5.994 x 10^{-5} Watts / Kilogram
    Fission Cross-section: 1.235 Barns
    Spontaneous Neutron Multiplicity: 2.0
    Induced Neutron Multiplicity: 2.61
    Primary Decay Mod: Alpha Emission to Th231
        Decay Energy: 4.679 MeV

Secondary Decay Mode: Spontaneous Fission
SR Rate: 5.6 x 10^{-3} (Fissions / Second / Kilogram)
Neutron Emission Rate: 1.1 x 10^{-2} (Neutrons / Second / Kilogram)
Critical Mass: 48 Kilograms

U236
Atomic Mass: 236.04556 (AMU)
Binding Energy: 1,790,415 KeV
Atomic Abundance: N/A
Half Life: 23.42 Million Years
Specific Activity: 6.476 x 10^{-2} Curies / Kilogram
Decay Heat: 1,753 Watts / Kilogram
Fission Cross-section: 0.594 Barns
Spontaneous Neutron Multiplicity: 1.8
Induced Neutron Multiplicity: 2.53
Primary Decay Mode: Alpha Emission to Th232
Decay Energy: 4.572 MeV
Secondary Decay Mode: Spontaneous Fission
SR Rate: 2.3 (Fissions / Second / Kilogram)
Critical Mass: >167 Kilograms

U238
Atomic Mass: 238.05078 (AMU)
Binding Energy: 1,801,694 KeV
Atomic Abundance: 99.2745 percent
Half Life: 4.468 Billion Years
Specific Activity: 0.336 Curies / Kilogram
Decay Heat: 8.508 x 10^{-6} Watts / Kilogram
Fission Cross-section: 1.89 Barns
Spontaneous Neutron Multiplicity: 1.97
Induced Neutron Multiplicity: 2.60
Primary Decay Mode: Alpha Emission to Th234
Decay Energy: 4.270 MeV
Secondary Decay Mode: Spontaneous Fission
SR Rate: 5.51 (Fissions / Second / Kilogram)
Neutron Emission Rate: 10.8 (Neutrons / Second / Kilogram)
Critical Mass: None

PLUTONIUM (Element 94)

USES

Plutonium is a dense, silvery metal used as a reactor fuel and as a nuclear explosive. It is an environmental hazard because of its extreme radio-toxicity.

PROPERTIES
Molar Volume 12.06 Centimeters^3/ Mole

Density 19.8 (α-phase stable below 122 °C)
Density 15.9 (Δ-phase stable in the range 319 - 476 °C)
Melting Point 639 °C

ISOTOPES
Pu239
Atomic Mass: 239.05216 (AMU)
Binding Energy: 1,783,870 KeV
Atomic Abundance: N/A
Half Life: 24,110 Years
Specific Activity: 62.03 Curies / Kilogram
Decay Heat: 1.929 Watts / Kilogram
Fission Cross-section: 1.89 Barns
Spontaneous Neutron Multiplicity: 2.9
Induced Neutron Multiplicity: 3.12
Primary Decay Mode: Alpha Emission to U235
    Decay Energy: 5.245 MeV
Secondary Decay Mode: Spontaneous Fission
    SR Rate: 10.1 (Fissions / Second / Kilogram)
    Neutron Emission Rate: 29.3 (Neutrons / Second / Kilogram)
Critical Mass: 10.5 Kilograms

Pu240
Atomic Mass: 240.05381 (AMU)
Binding Energy: 1,813,454 KeV
Atomic Abundance: NA
Half Life: 6,564 Years
Specific Activity: 227 Curies / Kilogram
Decay Heat: 7.07 Watts / Kilogram
Fission Cross-section: 1.357 Barns
Spontaneous Neutron Multiplicity: 2.19
Induced Neutron Multiplicity: 3.06
Primary Decay Mode: Alpha Emission to U236
    Decay Energy: 5.256 MeV
Secondary Decay Mode: Spontaneous Fission
    SR Rate: 478,000 (Fissions / Second / Kilogram)
    Neutron Emission Rate: 1,047,000 (Neutrons / Second / Kilogram)
Critical Mass: 40 Kilograms

Pu241
Atomic Mass: 241.05685 (AMU)
Binding Energy: 1,818,696 KeV
Atomic Abundance: NA
Half Life: 14.35 Years
Specific Activity: 1.033×10^5 Curies / Kilogram
Decay Heat: 129.4 Watts / Kilogram
Fission Cross-section: 1.648 Barns

Spontaneous Neutron Multiplicity: N/A
Induced Neutron Multiplicity: 3.14
Primary Decay Mode: Beta Emission to Am241
    Decay Energy: 021 MeV
Secondary Decay Mode: Alpha Emission to U237
    Decay Energy: 0.021 MeV
Tertiary Decay Mode: Spontaneous Fission
    SR Rate: <0.8 (Fissions / Second / Kilogram)
Critical Mass: 12 Kilograms

NEPTUNIUM (Element 93)

USES

Neptunium is a dense, silvery metal that could be used as a nuclear explosive. It is an environmental hazard because of its extreme radio-toxicity.

PROPERTIES

Molar Volume 11.59 Centimeters^3/ Mole
Density 20.25
Melting Point 644 °C

ISOTOPES

Np237
Atomic Mass: 237 (AMU)
Binding Energy: 7,574,982 KeV
Atomic Abundance: NA
Half Life: 2.144×10^6 Years
Specific Activity: 0.7034 Curies / Kilogram
Decay Heat: 0.02068 Watts / Kilogram
Fission Cross-section: 1.335 Barns
Spontaneous Neutron Multiplicity: 2
Induced Neutron Multiplicity: 2.889
Primary Decay Mode: Alpha Emission to U237
    Decay Energy: 4.959 MeV
Secondary Mode of Decay: Spontaneous fission
    SF rate (Fissions / Second / Kilogram) < 5.E-2
    Neutron emission rate (Neutrons/ Second / Kilogram) < 1.E-4
Critical Mass: 60 Kg

THORIUM (Element 90)

USES

Thorium is a silvery-white metal used as a reactor fuel and as a target to breed U233, which is a nuclear explosive. It can also substitute for plutonium in hydronuclear experiments

PROPERTIES
>   Molar Volume 19.8 Centimeters³/ Mole
>   Density 11.72
>   Melting Point 1,842 °C

ISOTOPES
>   **Th232**
>   Atomic Mass: 232.03805 (AMU)
>   Binding Energy: 1,766,691 KeV
>   Atomic Abundance: 100 percent
>   Half Life: 1.4 Billion Years
>   Specific Activity: 1.097E-4 Curies / Kilogram
>   Decay Heat: 2.654E-6 Watts / Kilogram
>   Primary Decay Mode: Alpha Emission to Ra222
>   Decay Energy: 4.083 MeV
>   Secondary Decay Mode: Spontaneous Fission
>   SR Rate: <3E-3 (Fissions / Second / Kilogram)
>   Neutron Emission Rate: <6E-3 (Neutrons / Second / Kilogram)
>   Critical Mass: None

POLONIUM (Element 90}

USES
>   Polonium is a vigorous alpha particle emitter that the DOEs predecessors used in the fabrication of mechanical neutron initiators. The alpha particles knocked neutrons loose from a small hollow sphere of beryllium that was mixed with the polonium. Modern weapons use external electronic neutron initiators. The element is so radioactive that it actually evaporates through the loss of daughter products produce by fission.

PROPERTIES
>   Molar Volume 12.56 Centimeters / Mole
>   Density 9.2
>   Melting Point 254 °C

ISOTOPES
>   **Po210**
>   Atomic Mass: 209 (AMU)
>   Binding Energy: 1,783,870 KeV
>   Atomic Abundance: N/A
>   Half Life: 138.4 Days
>   Specific Activity: 4,490,000 Curies / Kilogram
>   Decay Heat: 140,000 Watts / Kilogram
>   Primary Decay Mode: Alpha Emission to Pb206

Decay Energy: 5.407 MeV

HYDROGEN (Element 1)

USES

Hydrogen isotopes are important boosting components used in thermonuclear fuel and in neutron generators. When bonded with oxygen in heavy water, deuterium is an excellent neutron moderator and has a low cross-section for neutron absorption.

PROPERTIES
Molar Volume: 13.1 Centimeters^3/ Mole (solid 13 °K)
Density: 0.0763 (solid to 13 °K)
Melting Point: 13.96 °K
Boiling Point: 20.39 °K

ISOTOPES
H2 (Deuterium)
Atomic Mass: 2.01410 (AMU)
Binding Energy: 2,224.573 KeV
Atomic Abundance: 0.015 Percent
Stable Isotope
Melting Point: 20.4 °K
Boiling Point: 23.67 °K
H3 (Tritium)
Atomic Mass: 3.01605 (AMU)
Binding Energy: 8,481.821 KeV
Atomic Abundance: NA
Half Life: 12.33 Years
Specific Activity: 9,613,000 Curies / Kilogram
Decay Heat: 1,059 Watts / Kilogram
Primary Decay Mode: Beta Emission to H3
	Decay Energy: 18.591 KeV

LITHIUM (Element 3)

USES

The DOE uses lithium as a solid thermonuclear fuel when chemically combined with deuterium as lithium deuteride. Lithium deuteride contains more deuterium per unit volume than liquid deuterium. When struck by a neutron, the isotope Li6 forms tritium and He3. Li7 forms tritium, He3 and a neutron. The DOE has also used lithium to shield warheads against external neutron fluxes in the form of lithium hydride. Although the weapons laboratories used Li7 in early weapons, they discontinued its use in favor of Li6, which has better energetics. Lithium is also useful as a neutron absorber or as an x-ray reflector.

PROPERTIES
- Molar Volume 13.0 Centimeters^3/ Mole
- Density 0.534
- Melting Point 180.54 °C

ISOTOPES

Li6
Atomic Mass: 6.01512 AMU)
Binding Energy: 31,994.0475 KeV
Atomic Abundance: 7.42 percent
Stable Isotope

Li7
Atomic Mass: 7.01600 (AMU)
Binding Energy: 39,244.526 KeV
Atomic Abundance: 92.58 percent
Stable Isotope

BERYLLIUM (Element 4)

USES

The DOE uses beryllium as a neutron reflector and as a moderator in nuclear warheads and primaries. Be9 converts to Be8 plus a neutron and gamma radiation when struck by a high-energy alpha particle and for this reason Be9 was an essential component in early mechanical neutron initiators. Beryllium is light, strong, has a high melting temperature, is highly compressible, and exhibits excellent dimensional stability. For these reasons, the DOE uses it as a structural material and as a heat sink in missile reentry vehicles. It is the most toxic of the naturally occurring elements and requires care in its handling.

PROPERTIES
- Molar Volume 4.8775 Centimers^3/ Mole
- Density 1.8477
- Melting Point 1,287 °C

ISOTOPES

Be9
Atomic Mass: 9.01218 (AMU)
Binding Energy: 58,164.907 KeV
Atomic Abundance: 100 percent
Stable Isotope

APPENDIX 7
CHEMICAL AND NUCLEAR REACTIONS

CHEMICAL REACTIONS
The explosion of a TNT molecule
$$C_6H_2(NO_2)_3CH_3 \rightarrow 3.5\ CO + 3.5\ C + 2.5\ H2O + 1.5\ N2 + 5\ eV$$

NEUTRON PRODUCING REACTIONS
The following reaction generates the neutrons in mechanical initiators
$$Be9 + He4 \rightarrow Be8 + He4 + n$$

The following reaction generates neutrons in a neutron reflecting tamper
$$Be9 + n \rightarrow Be8 + 2n - 2.4\ MeV$$

URANIUM FISSION REACTIONS
More than 370 daughter nuclides with atomic masses between 72 and 161 are formed by the induced fission of U235 or Pu239, including the examples below.
$$U235 + n \rightarrow U236 \rightarrow Sr90 + Xe144 + 2n + 202\ MeV$$
$$U235 + n \rightarrow U236 \rightarrow Ba139 + Kr94 + 3n + 202\ MeV$$
$$U235 + n \rightarrow U236 \rightarrow Cs142 + Rb90 + 4n + 202\ MeV$$

FUSION REACTIONS
Deuterium Reactions
$$D + D \rightarrow He3 + n + 3.27\ MeV$$
$$\underline{He3 + D \rightarrow He4 + p + 18.35\ MeV}$$
$$D + D + D \rightarrow He4 + n + p + 21.62\ MeV$$

$$D + D \rightarrow T + p + 4.03\ MeV$$
$$\underline{D + T \rightarrow He4 + n + 17.59\ MeV}$$
$$D + D + D \rightarrow He4 + n + p + 21.62\ MeV$$

Tritium Reactions
$$T + T \rightarrow He4 + 2n + 11.27\ MeV$$

Lithium Reactions
$$Li6 + n \rightarrow He4 + T + 4.78\ MeV$$
$$Li7 + n \rightarrow He4 + T + n - 2.47\ MeV$$

APPENDIX 8
YIELDS OF NUCLEAR MATERIALS

FISSION

      Fission of 1-kilogram U233 = 17.8 kilotonnes TNT equivalent
      Fission of 1-kilogram U235 = 17.6 kilotonnes TNT equivalent
      Fission of 1-kilogram Pu239 = 17.2 kilotonnes TNT equivalent

FUSION

      Fusion of 1-kilogram D = 82.2 kilotonnes TNT equivalent
      Fusion of 1-kilogram DT/T (50/50) = 80.4 kilotonnes TNT equivalent
      Fusion of 1-kilogram Li6D = 64.0 kilotonnes TNT equivalent

3.75 grams of DT gas produces enough neutrons to fission 1 kilogram of plutonium if subsequently produced neutrons from fission are included.

NUCLEAR CONVERSION

      Conversion of 1 kilogram of matter to energy = 21.4 megatonnes TNT equivalent

DEFINITIONS

      One eV = One Electron Volt (a very small unit of energy) = 1.6×10^{-19} Joules
      One KeV = One Thousand Electron Volts
      One MeV = One Million Electron Volts

The fission of a single uranium atom releases about 202 MeV of energy, which is just sufficient to make movement in a grain of sand visible to the eye.

1 short ton = 2,000 pounds
1 long ton = 2,200 pounds
1 (metric) tonne = 1,000 kilograms = 2,204.6 pounds

APPENDIX 9
TABLES OF NUCLEAR EFFECTS

KILOTON YIELDS

Yield (KT)	20	50	100	250	500
Fireball Radius (mi)					
Surface	0.18	0.26	0.35	0.50	0.66
Airburst	0.14	0.20	0.26	0.38	0.50
Thermal Injury Radius (mi)					
1st Degree Burn	2.2	3.2	4.1	5.9	7.6
2nd Degree Burn	1.7	2.5	3.3	4.8	6.3
3rd Degree Burn	1.4	2.0	2.7	3.9	5.1
Radiation Injury Radius (mi)					
600 REM Dose	0.8	1.0	1.1	1.3	1.4
Blast Damage Radius (mi)					
1 PSI Overpressure	3.5	4.8	6.0	8.2	10.3
3 PSI Overpressure	1.6	2.2	2.7	3.7	4.7
5 PSI Overpressure	1.1	1.5	1.9	2.6	3.3
10 PSI Overpressure	0.7	1.0	1.2	1.7	2.1
20 PSI Overpressure	0.5	0.6	0.8	1.0	1.3

MEGATON YIELDS

Yield (MT)	1	5	10	25	50
Fireball Radius (mi)					
Surface	0.87	1.66	2.19	3.16	4.16
Airburst	0.66	1.26	1.66	2.40	3.16
Thermal Injury Radius (mi)					
1st Degree Burn	9.9	18.3	23.8	33.8	43.9
2nd Degree Burn	8.3	15.7	20.8	30.0	39.6
3rd Degree Burn	6.8	13.2	17.5	25.5	33.9
Radiation Injury Radius (mi)					
600 REM Dose	1.7	2.2	2.5	3.0	3.4
Blast Damage Radius (mi)					
1 PSI Overpressure	12.9	21.9	27.6	37.3	46.9
3 PSI Overpressure	5.9	10.0	12.5	17.0	21.3
5 PSI Overpressure	4.2	7.1	8.9	12.0	15.1
10 PSI Overpressure	2.6	4.5	5.6	7.6	9.6
20 PSI Overpressure	1.6	2.8	3.5	4.7	6.0

ACKNOWLEDGEMENTS

To my (mostly) patient wife…who thought this was a good project to keep me out of her hair during the winter. However, from time to time she indicated that she was waiting for the Men in Black to come and take me away. Or possibly it was the Men in the Little White Coats.

My thanks to the staff of Canadian Natural Resources' Structural Geology Group (especially Roland Dechesne and Ralph White), who encouraged me in this endeavor. They were hopeful that the Men in Black might come and take me away, thus bringing peace to the office. Keith Bottriel didn't want any recognition for his contribution because he was sure the Men in Black would get him – go Men in Black go.

Further thanks to Ron Visser and David Hoover who encouraged me and made some very helpful suggestions.

Made in the USA
Middletown, DE
26 July 2019